ELECTROMAGNETIC FIELDS AND RELATIVISTIC PARTICLES

International Series in Pure and Applied Physics

Adler, Bazin, and Schiffer: *Introduction to General Relativity*
Azároff (Editor): *X-Ray Spectroscopy*
Azároff, Kaplow, Kato, Weiss, Wilson, and Young: *X-Ray Diffraction*
Becker: *Introduction to Theoretical Mechanics*
Bjorken and Drell: *Relativistic Quantum Fields*
Bjorken and Drell: *Relativistic Quantum Mechanics*
Carmeli: *Group Theory and General Relativity*
Collin: *Field Theory of Guided Waves*
Evans: *The Atomic Nucleus*
Fetter and Walecka: *Quantum Theory of Many-Particle Systems*
Fetter and Walecka: *Theoretical Mechanics of Particles and Continua*
Feynman and Hibbs: *Quantum Mechanics and Path Integrals*
Hall: *Introduction to Electron Microscopy*
Harrison: *Solid State Theory*
Itzykson and Zuber: *Quantum Field Theory*
Konopinski: *Electromagnetic Fields and Relativistic Particles*
Krall and Trivelpiece: *Principles of Plasma Physics*
Leighton: *Principles of Modern Physics*
Morse and Feshbach: *Methods of Theoretical Physics*
Morse and Ingard: *Theoretical Acoustics*
Newton: *Scattering Theory of Waves and Particles*
Park: *Introduction to the Quantum Theory*
Richtmyer, Kennard, and Cooper: *Introduction to Modern Physics*
Schiff: *Quantum Mechanics*
Schwartz: *Principles of Electrodynamics*
Slater: *Insulators, Semiconductors, and Metals: Quantum Theory of Molecules and Solids, Vol. 3*
Slater: *The Self-consistent Field for Molecules and Solids: Quantum Theory of Molecules and Solids, Vol. 4*
Smythe: *Static and Dynamic Electricity*
Stratton: *Electromagnetic Theory*
Tinkham: *Group Theory and Quantum Mechanics*
Tinkham: *Introduction to Superconductivity*
Wang: *Solid-State Electronics*

The late F. K. Richtmyer was Consulting Editor of the series from its inception in 1929 to his death in 1939. Lee A. DuBridge was Consulting Editor from 1939 to 1946; and G. P. Harnwell from 1947 to 1954. Leonard I. Schiff served as consultant from 1954 until his death in 1971.

ELECTROMAGNETIC FIELDS AND RELATIVISTIC PARTICLES

Emil J. Konopinski

Professor of Physics
Indiana University

McGraw-Hill Book Company

New York St. Louis San Francisco Auckland Bogotá Hamburg
Johannesburg London Madrid Mexico Montreal New Delhi
Panama Paris São Paulo Singapore Sydney Tokyo Toronto

This book was set in Times Roman. The editors were John J. Corrigan and James W. Bradley; the production supervisor was John Mancia. The drawings were done by J & R Services, Inc.
Kingsport Press, Inc. was printer and binder.

ELECTROMAGNETIC FIELDS AND RELATIVISTIC PARTICLES

2 3 4 5 6 7 8 9 0 K P K P 8 9 8 7 6 5 4 3 2 1

Library of Congress Cataloging in Publication Data

Konopinski, Emil Jan, date
Electromagnetic fields and relativistic particles.

(International series in pure and applied physics)
Bibliography: p.
Includes index.
1. Electromagnetic fields. 2. Particles, Relativistic. I. Title.
QC665.E4K66 530.1'41 80-16405
ISBN 0-07-035264-X

CONTENTS

PREFACE

It is impossible to lecture repeatedly on even a long-established, classical subject without coming upon a variety of ways in which the most fundamental conceptions about the field are continually modified by newer developments, including ones beyond the classical realm. Some cases in point are given a preliminary review in the Introduction to this volume. Many have been given too little or no attention in most extant texts. An example is the fact that the classical theory of free electromagnetic fields constitutes the correct quantum wave mechanics of photons. Another example is the fact that the ubiquitous vector-potential field is amenable to direct measurement and physical interpretation despite the arbitrariness with which it may be gauged. It is such considerations, added to demands by students for more clarification of the physical meaning of each piece of mathematics, that have impelled the writing of this text.

My colleague and friend, Professor L. M. Langer, has provided immeasurably valuable help in eliminating occasional misstatements, overstatements, and even some understatements from this book.

Emil J. Konopinski

ELECTROMAGNETIC FIELDS AND RELATIVISTIC PARTICLES

INTRODUCTION

Classical electromagnetic theory, the primary subject of this book, uses representation by *fields*, and much of what is known about the description of fields was first learned in developing the theory. Fields are conceived to be continuous distributions over space. As such, they contrast with the point *particles* already adopted as basic objects of description in the mechanics of matter. The success of field theory in electromagnetism has led to those two concepts, field and particle, becoming the ultimate objects of description throughout all physical theory.

Particles have also come to play a leading role in electromagnetic theory, especially after the discovery of electrons and ions. Thus, treating the interactions of electromagnetic fields with discrete point charges must be made a central concern. On the other hand, much about electromagnetism was first learned from observing interactions of the fields with matter in bulk and treating the matter as having *continuous* distributions of charge. Paying attention to this will also be necessary.

Electromagnetism has influenced descriptions by particles not only through the fact that they can be charged (and/or possess electric or magnetic moments) but also through its part in transforming notions about relativity at high velocities. Extended attention will therefore be paid to high-speed *relativistic* particles, as the secondary part of the book's title implies. The treatment of relativity and electromagnetism together is quite customary and adds much to the understanding of both.

FREE ELECTROMAGNETIC FIELDS

Electric and magnetic fields were initially regarded as little more than convenient ways of representing the forces that would be felt at the various points in the vicinities of their sources by any charged matter that might be placed at the points. However, after Maxwell completed a unification of electricity and magnetism as different aspects of resultant electromagnetic fields, these began to take on increasingly substantial attributes, properties that can be assigned to each field as an entity, a continuously extended one that can even exist by itself, in isolation from matter and the sources it provides. Presenting the properties of fields must be a principal objective but because they are products of a gradually accumulated sophistication about fields, their presentation must be spread over many chapters. It will therefore be helpful to undertake a preliminary review of the ultimate results, to enhance appreciation of the significance of each development as it comes along. Properties of what are called "free" (electromagnetic radiation) fields will be outlined first, because such fields can exist uncomplicated by interactions with matter.

1. An immediate finding from Maxwell's unifying equations was a possibility that oscillating electromagnetic wave fields can exist out in free space, forming spread-out entities that can sustain themselves independently of any matter. Moreover, it was found that parts of the fields generated by nonstatic charges residing in matter can become divorced from the matter, forming electromagnetic radiation. These radiation fields can continue to exist without further interaction with their sources, becoming self-sustaining free fields.
2. Electromagnetic fields possess *energy* that can be detected whenever it is transferred to charged matter (as is evident to anyone who has been warmed by sunlight or has used a microwave oven). It is through energy exchanges with matter that quantitative electromagnetic field energies prove identifiable. They form continuous distributions, making field energy available at every volume element of space in which a field exists. No free field has a static distribution, yet—as long as it remains isolated from matter—it conserves a time-independent *total* energy to itself.
3. Every nonstatic field will also be found to possess a continuous distribution of *momentum* that can be transferred to and from charged matter (as manifested in radiation pressure, for example). The field momentum carried by a *plane* electromagnetic wave is particularly simple to describe, since it is uniformly distributed and has the same direction everywhere in the wave. However, an ideal plane wave has an infinite extent and can never exist by itself as a properly physical field of finite extent. On the other hand, an arbitrary finite field can always be resolved into a superposition (sum) of mutually interfering plane-wave constituents (it is destructive interference that cancels out the infinitely extended parts of every plane-wave component). Then the total field momentum becomes an integrated resultant of constituent plane-wave momenta and is appropriately called the center-of-mass momentum of the entire field (discussed in Chap. 13).
4. The momentum-containing fields naturally can harbor *angular* momentum as well, since it only takes multiplication of linear momentum elements by their lever arms relative to any chosen momental center to form an angular momentum. Whatever the center chosen, the total field angular momentum about it will be found to be conserved, in each free field, isolated from interactions with charged matter. An important finding about the field angular momenta is that they can be resolved into orbital and intrinsic-spin components. It is the existence of the intrinsic spin that makes it possible for beams of radiation to carry angular momentum indefinitely far from their sources, available for transfer to matter through torques on it.

The treatment of such phenomena requires the mathematical formalism of vector spherical waves, which are relatively unfamiliar and sometimes complex. For that reason, the formalism is entirely restricted to one chapter (Chap. 9) that can be skipped in a first reading. It is nevertheless essential to some of the problems in physics, chemistry, and engineering (of spherical cavity resonators, hohlraums, and aerosol detection, for example).

5. The electromagnetic fields gained added substance after Einstein's discovery that every energy is equivalent to a mass—his famous relation $E = Mc^2$. Actually, the attribution of mass to electromagnetic fields was anticipated in findings by Lorentz and Abraham, as will be seen in the last paragraphs of Chap. 4. The mass equivalent of purely electromagnetic field energy has even been found subject to gravitation. A terrestrial confirmation will be reviewed in Sec. 11.4.

The properties thus enumerated indicate that electromagnetic fields have all the essential attributes of ordinary matter; i.e., an entire field retains to itself a definite, conserved mass, as well as energy, momentum, and angular momentum so long as it remains isolated from interaction with other systems. Interactions can change the mass of a field, but this will also be found true of the ordinary material particles, whenever their (mass-) energies are depleted or augmented by interactions.

That fields can be treated on the same footing as the ordinary mechanical systems was known long before the last development. Their motions (field redistributions) are governed by equations of motion (Maxwell's field equations or equivalent equations for electromagnetic potentials) that follow from Hamilton's principle, like the equations of motion for the particles of ordinary matter (or, perhaps more pertinently, the field equations for motions in continuous fluids). All this will be shown in Chap. 13, together with a demonstration that the existence of field energy-momenta and of field angular momenta is required by the symmetries that must be expected of any description in space and time.

ATTACHED FIELDS

The earliest formulations were concerned with electro*static* and magneto*static* fields, the primary examples being the Coulomb fields outside spherically symmetric charges and the fields around elementary magnetic dipoles. Electromagnetic sources are detectable as such only through the fields attending them, which can be said to be permanently *attached* to their sources. All static fields are so attached, being steadily maintained by their sources and in equilibrium with them. (The free fields, mentioned above as being *de*tached from their sources in the course of their radiation, must have oscillating, nonstatic components to be self-sustaining.)

Energy must also be attributed to the static field distributions, equal to the work that must be done to establish the fields. Particularly familiar is the directly measurable potential energy accumulated by bringing two charges into each other's presence. This proves to be identifiable with a field energy, one that is determined by an interference of the overlapping fields attached to the individual charges. The interaction depends on the distance separating the charges and is measured by a scalar *potential* field from which the force fields are derivable.

One of the charges may be put into a magnetic field instead, and then a field *momentum* is also generated. This represents the momentum of the moving source charges that are fundamentally responsible for magnetism. (Remember that source properties manifest themselves through fields.) The field momentum serves as a potential momentum that can be transferred to motions of the charge and is measured by a suitable *vector* potential field from which the magnetic force fields are derivable. As a consequence, an arbitrary electromagnetic field is as well characterized by the directly measurable energy and momentum it makes available to a charge, representable by an appropriate potential field, as it is by the forces on the charge. (These matters will be discussed at the end of Chap. 2 and in Sec. 6.2.)

SELF-FIELDS

The field attached to a source *particle* is called its self-field, the energy the self-field contains being known as the particle's electromagnetic self-energy. The self-field of a particle must move with the particle, and the moving Coulomb field attached to a simple charged particle is discussed in Chap. 10.

Since it contains energy, the field attached to a particle also has mass, continuously distributed over the field. Such an electromagnetic mass is a particularly important part of the observed mass of a charged particle as light as the electron. It must contribute to the inertia of the particle, in its response to accelerations, and also to its gravitation. The fact that the mass of a charge is not confined to the site of the charge but is spread out wherever its self-field reaches gives rise to difficulties in treating the electron as a point particle when constructing its equation of motion. How those difficulties are circumvented by classical renormalization procedures that lead to the Lorentz-Dirac equation is recounted in Chap. 14.

RELATIVITY AND QUANTUM MECHANICS IN ELECTROMAGNETISM

It is a remarkable fact that classical electromagnetic theory—as represented by the Maxwell equations or their equivalents, the equations for the potentials—was developed in consistency with Einstein's relativity before this was known and, at least as applied to the free radiation fields, it also formed the correct quantum wave mechanics of photons before photons made themselves evident.

Einstein's relativity is introduced in Chap. 11, and the consistency of the electromagnetic theory with it is manifested through Lorentz-covariant reformulations of it in Chap. 12. Here also are to be found necessary modifications of the newtonian equations of motion for applicability to high speeds, i.e., to relativistic particles.

Einstein's modification of the connection between energy and momentum made it possible to contemplate an existence of entities that have mass only while moving with the velocity of light. How this helped identify the rest-mass-less photons as quanta of electromagnetic field energy will be reviewed in Sec. 11.4.

The last step led to the quantum theory of all matter, but taking into account the quantum-mechanical strictures on the detailed behavior of all matter cannot be undertaken within the limited space of this volume. What distinguishes the theory to be reviewed here as "classical" is that it leaves "unquantized" the charged matter with which the electromagnetic field interacts. That still allows for a vast range of situations in which the theory retains validity. Problems of forming expectations comparable to what is measured in only classical detail still predominate. This refers to observations in which effects of the motions of *individual* atoms of matter, within the volume elements that are discriminated, remain undetected. (Quite comparably, it is not necessary to know the motions of every rock on the moon's surface in order to predict the moon's orbit as well as it can be measured.)

The fact is that quantum mechanics is more comprehensive than classical mechanics and *contains* it, classical mechanics remaining valid in situations at what is called the correspondence limit. Results at this limit, where the classical and quantum-mechanical treatments merge, are discussed in the last two sections of Chap. 14.

Finally, the classical electromagnetic theory of free fields is completely consistent with the quantum wave mechanics of photons. This means that many of the classical wave field results have characteristics considered typically quantum-mechanical. Chapter 7 discusses classical complementarity relations that correspond precisely to the famous uncertainty relations basic to the quantum theory. An exercise following Chap. F exhibits a tunneling considered typical of quantum-mechanical behavior. Another, following Chap. 9, finds discrete frequency levels inside an electromagnetic cavity resonator that correspond to the discrete quantum-mechanical energy levels of rest-mass-less, unit-spin particles in a deep potential well, even with regard to distinctions by quantum numbers that measure discretely valued angular momenta. Also in Chap. 9 it is shown that the linear and angular momenta in a classical electromagnetic wave field can be calculated from matrix elements of displacement and rotation operators, exactly as in quantum mechanics.

APPLICATIONS OF THE ELECTROMAGNETIC THEORY

The evidence for the theoretical developments outlined above comes from observations on the real world. It was gathered incidentally to applications of the theory in deriving expectations that correspond to measurements in specific physical situations. These deal with phenomena that have interest in themselves, and much of the book will be devoted to them.

For applications leading to the most basic findings about the electromagnetic field itself, including all those outlined above, the electromagnetic properties of ordinary matter need be represented only by charges and their currents. This is essentially what is done throughout the main text (Chaps. 1 to 14), which is based on equations (sometimes called the Maxwell-Lorentz equations) having charge and current as the only sources or modifiers of the fields. However, the introduction of otherwise specified electromagnetic properties of matter in bulk has helped account for some of the best-known and interesting phenomena. The corresponding applications have served to reveal more about various materials than about the fields, but their striking successes have been important to establishing the great confidence that exists in predictions from the theory.

Maxwell himself added into his original equations special provisions for representing bulk materials, with results that are only phenomenological. Maxwell treated matter as forming a continuous *medium* for the residence and transmission of fields. Media representing different materials are to be distinguished by characteristic conductivities and polarizabilities which since the discovery of electrons and ions have become *derivable*, to various degrees of approximation, through averaging over the more basic, atomic constituents of the matter, treating them as the entities furnishing the charges and currents in the Maxwell-Lorentz equations, in place of Maxwell's original ones. For such reasons, but more to keep clear the basis being used in the main body of the text, bulk-material treatments have been confined to Supplementary Chapters B to G. They are based on the phenomenological Maxwell equations, which have long proved more than adequate for most of what can be observed about the electromagnetic interactions with bulk materials. Most of the supplementary chapters are quite elementary and have been marked to indicate the early junctures in the main text at which they are designed to be read. The order in which all the chapters can be read is A, 1 to 3, B to D, 4, E, 5 to 7, F, G, and 8 to 14.

Some of the more purely mathematical developments likely to be already familiar to many readers have been put into Supplementary Chapters A and C, marked to show when they can be read with profit. They may be essential to some students and will provide convenient bodies of reference for others.

The reader should also be aware that many important, though relatively specialized, results are stated in the Exercises after each chapter. The corresponding exercises consist of carefully framed instructions for obtaining those results by applying the fundamentals presented in the text. This is perhaps the most potent way to elicit understanding of such results, as well as an economical way of presenting them.

CHAPTER
ONE

THE MAXWELL EQUATIONS†

Their observational basis as embodied in the Coulomb, Gauss, Ampère, Ohm, and Faraday laws · Charge conservation and Maxwell induction

All classical electromagnetic phenomena are encompassed by the *Maxwell equations*, in the sense that these equations provide an adequate starting basis for deducing all the verifiable expectations about such phenomena.

This chapter† will review observations which suggest the equations, as promising ones to try. It will be made as concise as seems consistent with some awareness of just why each step in the development is taken. The equations have really been established, and a high degree of confidence exists in predictions from them, by the multitude of their successes in applications, the most important of which will be reviewed in succeeding chapters. It should also be acknowledged that some of the confidence placed in the equations has always been due to their beautiful reasonableness.

1.1 ELECTRIC CHARGE

Bodies can be subjected to various processes (rubbing against cats, for example) by which they are electrified into manifesting forces of attraction for, or repulsion to, each other. The state of electrification is found to be transferable from one body to another by contact, and this suggests conceiving something conservable that is being transferred and calling it *electric charge*.

Experimentation shows that similar bodies similarly electrified always repel each other, whereas dissimilarly electrified bodies sometimes attract. This led to the classification of charge into two varieties, called positive and negative, the like charges of either type repelling each other while the unlike charges attract. This classification also permits understanding the electrifying processes as a means of separating positive and negative charges, which together account for the normally neutral states of matter.

† Supplementary Chapter A is designed to be read first, but only by those not already familiar with the language of gradients, divergences, and curls.

The Coulomb Law

The quantitative treatment of electricity starts from measurements on the forces through which it manifests itself.† Conclusions from such measurements are embodied in Coulomb's law

$$\mathbf{F} = \frac{\hat{\mathbf{r}} q q'}{r^2} \tag{1.1}$$

for the mutual force between two charged bodies as a function of their distance r apart. The unit vector $\hat{\mathbf{r}} \equiv \mathbf{r}/r$ points from one charge to the other and, like r, is least ambiguous when each charge is confined within a dimension which is small in comparison to r. The expression is accurate only insofar as the charged bodies can be approximated as points. (It will later develop that it is also accurate between centers of *extended* charge distributions if each is spherically symmetric; see Exercise 1.2.)

The symbols q and q' stand for *quantities* of the respective charges. Each is defined to be proportional to the force manifested when the other is fixed, together with all other circumstances to which there appears to be sensitivity. It is expressed in units which can be established, for example, by working with two like bodies charged by precisely similar processes and taking $|q| = (r^2 F)^{1/2}$ as the measure of each charge.

Once each quantity of charge has been measured by any such means, the result (1.1) is straightforward to establish. Whenever the force found between dissimilarly charged bodies turns out to be a mutual attraction, q and q' are given opposite signs. Which of the several processes of charging are classified as producing charge to be called positive and which negative is purely a matter of convention. What is significant is that once assignments have been made, all pairwise combinations exhibit attractions or repulsions in conformity with the sign of the *product* qq'.

The Electric Field

Coulomb's law can be expressed as a finding about a given point charge q by defining the force per unit of positive *test charge* q'

$$\mathbf{E}_q(\mathbf{r}) = \hat{\mathbf{r}} \frac{q}{r^2} \tag{1.2}$$

where $\mathbf{r} = \hat{\mathbf{r}} r$ is the position vector, based on q, of any point at which the test charge might be placed. This actually defines an electric *vector field*, since it assigns a vector $\mathbf{E}_q$ to every point $\mathbf{r}$ in the space about the point charge at $\mathbf{r} = 0$.

† A prior establishment of classical mechanical theory is presumed, so that coming to conclusions about forces is no problem.

Note well that the test charge is *not* part of the system that $\mathbf{E}_q(\mathbf{r})$ is to help describe.†

Entities classifiable as forces must be vectorially superposable into vector resultants. Thus, the total electric field arising from several point charges q_1, q_2, $\ldots, q_s, \ldots$ static at positions $\mathbf{r}_1, \mathbf{r}_2, \ldots, \mathbf{r}_s, \ldots$ is

$$\mathbf{E}(\mathbf{r}_f) = \sum_s \hat{\mathbf{R}} \frac{q_s}{R^2} \tag{1.3}$$

with $\mathbf{R} = R\hat{\mathbf{R}} \equiv \mathbf{r}_f - \mathbf{r}_s$ different in each term. Implicit here is a *principle of superposition* that characterizes electric fields (Exercises 2.2 and 2.3).

As far as measurements of classical detail can discriminate, charge is better treated as being continuously distributed, in some generally variable density $\rho(\mathbf{r}_s)$, over a continuum of points $\mathbf{r}_s$. Any one element of it $dq = \rho(\mathbf{r}_s)\, dV(\mathbf{r}_s)$ acts as a point charge contributing a vector element of electric field

$$d\mathbf{E}(\mathbf{r}_f) = \frac{\hat{\mathbf{R}}\rho(\mathbf{r}_s)\, dV(\mathbf{r}_s)}{R^2} \tag{1.4}$$

and the vector resultant of such elements

$$\mathbf{E}(\mathbf{r}_f) = \oint \frac{\hat{\mathbf{R}}\rho(\mathbf{r}_s)\, dV(\mathbf{r}_s)}{R^2} \tag{1.5}$$

is then the field arising from the *entire* charge

$$q = \oint \rho(\mathbf{r}_s)\, dV(\mathbf{r}_s) \tag{1.6}$$

that has been integrated over. In these formulas $\mathbf{R} \equiv \mathbf{r}_f - \mathbf{r}_s$ is a continuously variable vector, not only because it can be evaluated for any of the continuum of *field* points $\mathbf{r}_f$ but also because it is varied in the course of the integration over the charge. The unit vector $\hat{\mathbf{R}} \equiv (\mathbf{r}_f - \mathbf{r}_s)/|\mathbf{r}_f - \mathbf{r}_s|$ changes in direction during the course of the integration, since it must in turn point from each of the various source elements toward the field point. In (1.5) and (1.6), the notation $\oint dV$ is introduced to stand for volume integration extensible to all space because it encloses all points at which the integrand does not vanish.

The Gauss Law

When a field-line representation (see the discussion leading to Fig. A.2) is adopted for the field $\mathbf{E}_q(\mathbf{r})$ arising from a single point charge q, the directed lines must emanate radially from the point of the charge when it is positive and

† The fields due to *two* point charges (as detected by a third, *test*, charge) are discussed in connection with Fig. 6.3. Experimentation can make sure that the test procedure does not alter the sources of the field being measured.

converge radially upon it when it is negative. If the lateral densities of the field lines are to indicate the strength of the field in the various regions, i.e., the magnitude of the force per unit test charge there, they must be uniformly distributed in direction. Moreover, if a number of lines per unit of cross-sectional area is used to represent that number of field-strength units, each of the succession of spheres centered on the source point, with radii r, will have $4\pi r^2 \mathsf{E}_q(\mathbf{r}) = 4\pi q$ lines penetrating it.

The last result implies a restriction on fields which can have their magnitudes indicated by lateral densities of continuous field lines. Continuous radial lines must be the same in total number through every sphere centered at $r = 0$ and when distributed uniformly in direction have a "natural" density proportional to $1/r^2$ at the successive spheres. Thus, only fields with magnitudes inversely proportional to the square of the distance from isotropic point sources of them, like the electric (or gravitational!) fields, can be given the measure in question. Conversely, an inverse-square law of force corresponds to a uniform flux of impulse into all directions.

The quantity $4\pi r^2 \mathsf{E}_q$ is, of course, an evaluation of the *flux* $\oint d\mathbf{S} \cdot \mathsf{E}_q(\mathbf{r})$ through the sphere of radius r. The result

$$\oint d\mathbf{S} \cdot \mathsf{E}(\mathbf{r}) = 4\pi q \tag{1.7}$$

is an instance of what is called the *Gauss law*. [This is a physical assertion, to be distinguished from the more purely mathematical Gauss theorem (A.14).]

The result (1.7) is also valid when the flux is evaluated for all the area elements on an *arbitrary* surface enclosing the point charge, and not only for the spheres. One way to see this is to notice that

$$\frac{d\mathbf{S} \cdot \hat{\mathbf{r}}}{r^2} \equiv d\Omega \tag{1.8}$$

is just the element of solid angle subtended at $r = 0$ by *any* area element $d\mathbf{S}$ at the point $\mathbf{r} = r\hat{\mathbf{r}}$. For instance, $d\mathbf{S} \cdot \hat{\mathbf{r}} \equiv dS_r = r^2 \sin \vartheta \, d\vartheta|_0^\pi \, d\varphi|_0^{2\pi}$ in spherical coordinates, and so

$$d\Omega \equiv d(\cos \vartheta)\Big|_{-1}^{1} d\varphi\Big|_0^{2\pi} \tag{1.9}$$

with $\oint d\Omega = 4\pi$. Since the field lines about a point charge are uniformly distributed in direction, the same number, equal to q on the scale adopted above, must be directed into each unit of solid angle. Thus $\oint q \, d\Omega = \oint d\mathbf{S} \cdot \mathsf{E}_q = 4\pi q$ regardless of the enclosing surface referred to in expression (1.8) of the solid-angle elements.

The arbitrary surface may even be convoluted in such a way that a given field line leaves the enclosed volume at one point and reenters it again at another, as indicated by Fig. 1.1. The projections of the area-element vectors

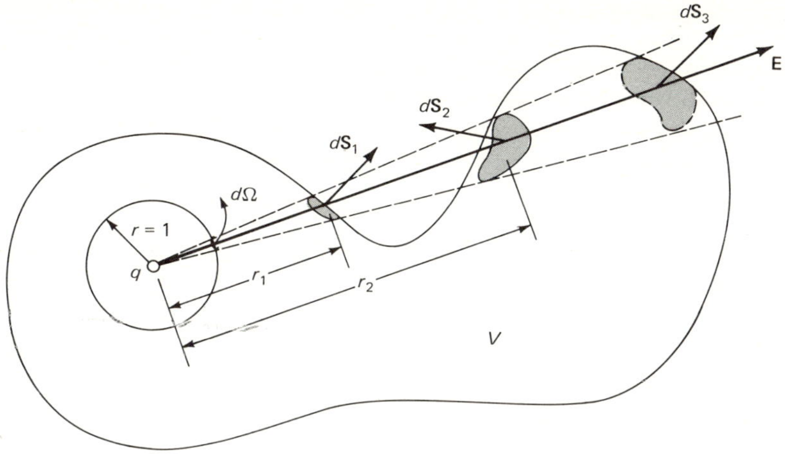

Figure 1.1

indicated as $d\mathbf{S}_1$ and $d\mathbf{S}_2$ on the radial direction $\hat{\mathbf{r}}$ plainly have opposite signs, and their contributions to the flux cancel each other

$$\frac{q(d\mathbf{S}_1 \cdot \hat{\mathbf{r}})}{r_1^2} = -\frac{q(d\mathbf{S}_2 \cdot \hat{\mathbf{r}})}{r_2^2} = \frac{q(d\mathbf{S}_3 \cdot \hat{\mathbf{r}})}{r_3^2} = q\, d\Omega$$

The figure also has noted in it the fact that $d\Omega$ itself is just an area element subtended on a sphere of *unit* radius (just as a plane element of angle $d\varphi$ is an arc length on a *circle* of unit radius; whereas $\oint d\Omega = 4\pi$, $\oint d\varphi = 2\pi$).

The Gauss law (1.7) is also valid when the charge q is distributed arbitrarily within the surface of integration in some density distribution $\rho(\mathbf{r})$ instead of being concentrated at a point. Any one element $\Delta q = \rho\, \Delta V$ of the distributed charge can be treated as a point charge, and so

$$\oint d\mathbf{S} \cdot \Delta\mathbf{E} = 4\pi\, \Delta q$$

is as valid as it is for a point charge if $\Delta\mathbf{E}$ is a field element like (1.4) arising from Δq alone. Then summing the contributions of all such elements to both sides of the relation produces an expression exactly like (1.7), with q now the *total* charge contained (1.6).

Only the charges *enclosed* by the surface of integration should be counted as part of q, even if part of the field $\mathbf{E}$ used in evaluating the total flux arises from sources outside the enclosure. Any field lines contributed, as elements $\Delta\mathbf{E}$, by any exterior source point, penetrate through the enclosed volume, in and out again, and their *net* flux vanishes on the same grounds as those discussed in connection with Fig. 1.1.

The First Maxwell Equation

The flux integral in the expression of the Gauss law (1.7) can be transformed into a volume integral with the help of the Gauss theorem (A.14), so that

$$\int_V dV(\mathbf{r})\,\nabla\cdot\mathbf{E}(\mathbf{r}) = 4\pi q = 4\pi\int_V \rho(\mathbf{r})\,dV(\mathbf{r}) \tag{1.10}$$

if V is a volume enclosing q. Since V can be chosen arbitrarily, it can be reduced to a point, and then

$$\nabla\cdot\mathbf{E} = 4\pi\rho \tag{1.11}$$

follows. This is the first of Maxwell's set of equations. It amounts to an expression of the Gauss law at a point, for a divergence is just a net flux per unit volume at a point (A.13).

The result (1.11) expresses the fact that charges are to constitute the sources and sinks of what is to be called an *electric* force field. It is charge-density distributions $\rho(\mathbf{r})$ multiplied by the complete solid angle 4π that serve as the scalar source distributions like $s(\mathbf{r})$ of (A.30) for the vector field here. The factor 4π is a result of having used $q = \pm(r^2F)^{1/2}$, where F is the mutual force between two equal charges q, as the measure of quantity of charge; so-called gaussian units are then being employed. Some theorists prefer to use units that are introduced by replacing r^2 with the spherical surface area $4\pi r^2$ in the expression (1.1) of the Coulomb law.† Then what they regard as the more basic equation, (1.11), becomes $\nabla\cdot\mathbf{E} = \rho$. Other physicists prefer to retain the factor 4π in the latter equation as a reminder, at its level, of the essential part played by geometry in distributing the field. This will be the practice followed here, simply because it came first in the course of the abstractions from experience.

The introduction of electric field divergences here invites drawing upon what has been learned about divergences of vector fields in general (Sec. A.3). Thus, it is known that points of nonvanishing divergence are points of discontinuity in field lines and hence that electric field lines are continuous everywhere outside their sources—the charges. Moreover, the net flux out of volumes containing no points of divergence must vanish. This furnishes another way to come to the same conclusions as those obtained in the preceding subsection with the help of solid-angle elements. Thus, the flux through an arbitrary surface enclosing a point charge is the same as the flux found through an enclosing sphere, since the volume wedged in between surface and sphere contains no charge and hence the fluxes through the two surfaces must balance. Likewise, a charge external to any closed surface produces no net flux through it since any net flux is accounted

† Then elementary texts replace the factor $1/4\pi$ by some symbol like ϵ_o, as a step toward introducing practical units. Such practices tend to obscure fundamentals and have never had any place in further (particularly relativistic) reaches of the theory.

for by just the interior divergences, if any, according to the Gauss theorem. Lines from exterior sources must accordingly penetrate in and out of any closed surface, and their net flux must balance to zero.

The differential equation (1.11) helps provide a better general starting basis for the problems of electromagnetic theory than the integral expression (1.5) for the field, despite the fact that the latter is already an explicit "solution," which gives the field as soon as the distribution of all its sources is given. One reason is that this distribution may not always be given, yet sufficient other information may be available for a *co*determination of field and source distribution to be possible with the help of (1.11). The differential equation has a greater variety of solution forms than just (1.5), as will be reviewed in succeeding chapters.

Perhaps the most important reason for putting the Maxwell equation (1.11) at the basis of the theory is that it is more safely generalizable to *nonstatic* situations than such integral expressions as (1.5) or the Gauss law (1.7). The latter have been abstracted from experience with static charge distributions, and there is ambiguity in generalizing them to situations in which the distribution changes with time, as represented by some $\rho(\mathbf{r}, t)$. It is then not clear as just which times t_s the various charge elements $\rho(\mathbf{r}_s, t_s)\, dV(\mathbf{r}_s)$ should be evaluated when they are contributing to the field $\mathbf{E}(\mathbf{r}_f, t_f)$ at some definite moment t_f. If the evaluations were all made for some one moment, with all $t_s = t_f$, there would be a presumption of instantaneous response to source changes at various $\mathbf{r}_s$, no matter how remote they might be from the field point $\mathbf{r}_f$.

On the other hand, the relation

$$\nabla \cdot \mathbf{E}(\mathbf{r}, t) = 4\pi\rho(\mathbf{r}, t) \tag{1.12}$$

entails no such presumption, since both sides of it are to be evaluated at the same point of space, and hence no questions about propagation from one point to another need be answered in it. This leaves no further immediately discernible questions about the tenability of the generalization because (1.12) can now be taken to *define* electric charge. It should be remembered that it is just through the detectable forces they produce that the existence of charges manifested itself in the first place.

Considering the Maxwell equation to be a definition may seem to nullify it as a basis for prediction. However, using the same relation as a definition and also as a basis from which fresh consequences are to be deduced is commonplace in physical description. Even Newton's basic equation of motion $m\ddot{\mathbf{r}}(t) = \mathbf{F}$, from which the motions that will take place under a given force are to be derived, is also used to define force. The point is that such relations are first used for *investigating* what appears in what circumstances, in which phase they are being used as definitions. The information so gained then provides a basis for *postulating* characterizations of new circumstances, in which predictions are to be made. As a theory of force, electromagnetic theory substitutes for the characterization by some force, as required by newtonian theory proper, a description of the circumstances in terms of *sources* of force (in an important class of circumstances).

Charge Conservation and Electric Current Density

Electric charge owes its conception to the surmise that it might be something conservable in the transfers of electrification which had been observed. The concept has survived because an exceptionless adherence to a *principle of charge conservation* has been maintainable ever since, in all the experience with the multiplicity of applications in which the charge has been involved.

The introduction of charge distributions $\rho(\mathbf{r}, t)$ which vary in time makes it essential to formulate explicitly their part in charge conservation. Flows of charge must be described, to deal with its transfers from point to point. One way to formulate a density of the flow [see (A.16)] requires ascribing a unique velocity $\mathbf{u}(\mathbf{r}, t)$ to the charge at the point $\mathbf{r}$ at time t, and then there is a *current density*

$$\mathbf{j}(\mathbf{r}, t) = \rho\mathbf{u} \tag{1.13}$$

at the point, measuring the quantity of charge passing the point per unit of lateral area (normal to the direction of $\mathbf{u}$). The conservation of the charge is now to be represented by a continuity equation (A.19)

$$\frac{\partial\rho}{\partial t} + \nabla \cdot \mathbf{j} = 0 \tag{1.14}$$

expressing the balance that must exist between accumulations or diminutions of the charge density at a point and the net influx or efflux, per unit volume, at the point.

The formulation $\mathbf{j} = \rho\mathbf{u}$ of the electric current density may be adequate when dealing with charge of one sign alone, but in more general situations less straightforward interpretations of the conservation equation (1.14) may be required. It might even be maintained that, as far as macroscopic observations can discriminate, positive and negative charges are not individually conserved, for electrostatically neutral states are obtainable by superposing equal amounts of positive and negative charge which might be said to destroy each other in producing the neutrality. However, this notion is contradicted† by a finding that it is possible to have spaces which are completely neutral, as far as can be detected through any appearance of electrostatic force, yet a presence of electric current in the space may be manifested. One way in which currents prove to be detectable in electrostatically neutral matter is through a heating up of the material. A usually more sensitive way was provided after the connection of magnetism to electricity was discovered, as reviewed in the following section.

† Processes in which positive and negative charges do annihilate each other, in the sense that neither charge nor current is afterwards detectable, were eventually found. They occur in circumstances which can be properly described only at a subatomic level and have no effects as detected in only classical detail, the concern here. Even the annihilation processes are consistent with the conservation of *total* charge ($q = |q_+| - |q_-|$), as required by (1.14).

The existence of electric current without a coexistence of net charge can be understood as arising when there are exactly equal densities $\rho_+ = \rho_-$ of positive and negative charge at each "point" (of volume $\Delta V \approx 0$) that can be discriminated in a space but the velocities of the two types of charge are not the same: $\mathbf{u}_+ \neq \mathbf{u}_-$. Then $\mathbf{j} = \rho_+\mathbf{u}_+ - \rho_-\mathbf{u}_- \neq 0$ despite $\rho_+ = \rho_-$. This is indeed what is found to prevail in many situations, and particularly in so-called *metallic conductors*, when investigation is pressed beyond ordinary classical detail to an atomic level. It is found that in the conductors, the positive charges are more or less fixed ($\mathbf{u}_+ = 0$) in the crystalline lattice of the metal, whereas the negative charge flows past the positive charge with a net *drift velocity* $\mathbf{u}_-$. Thus, through the accidents by which the names positive and negative were chosen, it turns out that a positively directed current results, in conductors, from a reverse flow of negative charge: $\mathbf{j} = -\rho_-\mathbf{u}_-$!

1.2 MAGNETISM

Like electricity, magnetism manifests itself through observable attractions and repulsions between bodies. Nevertheless, the two phenomena initially appeared to be quite unrelated, since magnetized bodies do not in general attract or repel electrified bodies.

A rather superficial distinction was implicit in the fact that magnetism was found among naturally occurring ores and so can be at least quasi-permanent, whereas electrification was an artificially induced temporary state, readily discharged.† Magnets are only quasi-permanent because extreme measures, such as melting down and boiling the body, *can* destroy a state of magnetization. (More usually, demagnetization is achieved by imposing external magnetic fields, as discussed in connection with Fig. E.2.)

Magnetic Poles

More significant distinctions from electricity are found when the forces between magnetic bodies are explored in detail. The characteristic result is that whenever some point on a magnetic body is attracted to some part of a second magnetic body, another part of the second body can always be found which repels the same point of the first body. A magnetic body can be so cut that practically all the detectable responses in the product are confined to highly localized *poles* but never so that there are not at least two, oppositely responding, poles in each body. If an attempt is made to isolate a pole of either type, to obtain something

† Quasi-permanent *electrets*, comparable to magnets but having electric rather than magnetic polarizations "fixed" in them, have finally been produced in modern laboratories. Their existence will be ignored in this book since they played no part in the development of the general theory yet fit into it quite comfortably.

analogous to an electric point charge of a given sign, by cracking off a part of a body containing one pole, the fragment immediately acquires a second pole, of the opposite type. The oppositely responding centers are respectively called north and south poles for familiar reasons.

Poles can still be quasi-isolated by forming long, thin magnetic needles having a north and south pole far from each other at the endpoints of each. Then the mutual force between one pole of one needle and a pole of a second needle can be studied in much the same way as the forces between point charges. Quite analogous results are obtained

$$\mathbf{F}_M = \frac{\hat{\mathbf{r}} q_M q'_M}{r^2} \tag{1.15}$$

where q_M and q'_M are magnetic pole strengths which can be measured in the same way as suggested for the electric charges; for example, $|q_M| = (r^2 F_M)^{1/2}$, in a suitable situation. However, Eq. (1.15) cannot play as basic a role for magnetism as Coulomb's law did for electricity. It ignores the characteristic property of the magnetism, the finding that is represented in magnetic needles by the necessary coexistence with any pole strength q_M of a second pole of strength $-q_M$ at the other end of the needle.

The importance of the quasi-isolated poles in the development of magnetic theory lies primarily in their usefulness as test poles in yielding a quantitative measure of *magnetic field*

$$\mathbf{B}(\mathbf{r}) = \frac{\mathbf{F}_M}{q'_M} \tag{1.16}$$

in terms of directly measurable forces $\mathbf{F}_M$ on a test pole q'_M. A more basic measure is found only after further development of the theory, to be left to Chap. 4.

It should still be pointed out here that a development of a Gauss law for a magnetic pole, analogous to (1.7) for a point charge, would require using an enclosure crossing magnetic material to which the quasi-isolated pole is attached. Thus, an equation like $\nabla \cdot \mathbf{B} = 4\pi\rho_M$ could at best be given only a narrowly provisional meaning.†

The Magnetic Field

With test poles available, the vector field $\mathbf{B}(\mathbf{r})$ in the free space around an isolated magnetic body can be mapped. The directions attributed to $\mathbf{B}$ are, by convention, the directions of the forces on *north* test poles. The characteristic finding is that every line of the magnetic field emanating from any part of the

† It will be seen in Chap. E that in treating magnetized materials it turns out to be convenient to separate off, quite artificially, a magnetic field *component*, symbolized by $\mathbf{H}(\mathbf{r})$, that has the property $\nabla \cdot \mathbf{H} = 4\pi\rho_M$ *by definition*.

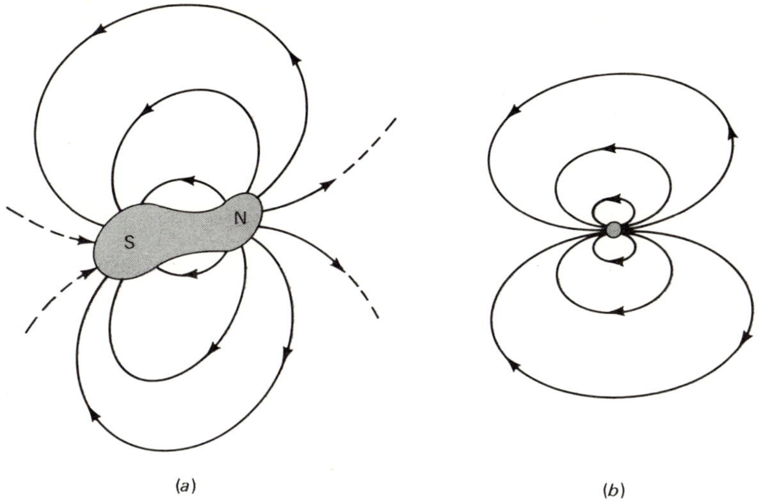

Figure 1.2

isolated magnetic body can be traced out continuously until it reenters the surface of the body in some other part of it (Fig. 1.2*a*). Centers of emanation are the north poles and centers of termination the south poles of the body.

More and more complete circuits of the magnetic field lines are traceable when material in their paths is removed until the body is reduced to what is practically a point (Fig. 1.2*b*). It can then be concluded that a magnetic field line necessarily leaves every point that a field line enters and hence (compare Fig. A.2) that

$$\nabla \cdot \mathbf{B} = 0 \tag{1.17}$$

everywhere. However, there is a jump to the conclusion here, for to have $\mathbf{B} \neq 0$ there must still remain some magnetic material which interrupts the tracing of the field line.

A final conclusion that magnetic field lines must be divergenceless and occur only in uninterrupted loops was reached after the first connection between magnetism and electricity was uncovered. Oersted found that currents of electric charge are surrounded by magnetic fields to which test magnets respond exactly as they do to magnetic bodies. About the most definitive results are obtained around currents which are conducted along metallic filaments (wires), and it is found that the field lines occur in uninterrupted loops encircling the wires, in conformity to the divergencelessness (1.17). The currents themselves can be constructed in loops, and then there are complete magnetic field loops which link the current loops. Ampère formed the hypothesis that the magnetism of materials also originates from current loops, tiny circulations of charge within atomic domains, so that it can be held that magnetic field lines really have no divergences anywhere.

The conclusion in (1.17) forms a second Maxwell equation. It can be said to express the fact that no true magnetic charge comparable to the electric charge responsible for the electric field divergences (1.12) has ever been discovered.†

Ampère's Law

The Maxwell equation $\nabla \cdot \mathbf{B}(\mathbf{r}, t) = 0$, expressing a characteristic of any magnetic field, gives no help toward determining the overall magnitude it will have in any specific circumstances; it is plainly invariant to any scale factor in $\mathbf{B}(\mathbf{r}, t)$. A basic determination of the magnitude should relate it to the sizes of the currents which may be generating the field.

What is usually referred to as *electric current* I, as against current density $\mathbf{j}$, is the flux of $\mathbf{j}$ through some definite area S

$$I = \int_S d\mathbf{S} \cdot \mathbf{j} \tag{1.18}$$

Such a quantity, equal to the charge passing through a definite area S per unit time, is particularly useful in characterizing experimental circumstances which have been found to yield the most definitive conclusions. They have already been mentioned in the preceding subsection as having current conducted along a *wire*. This is a filament-shaped conductor, with a cross-sectional area S small enough to require no analysis into the current-density distribution within it, for a sufficient precision of observation, and $S \to 0$ can be assumed even when $I \neq 0$ (in an idealized wire, serving as an adequate model of a material wire).

No mention has been made so far of how definite currents can be produced, except for an allusion to the finding that charge is transferable from one body to another by contact. The most easily controllable currents are produced when the contact is stretched into a uniform conducting wire. To get steady currents, which yield the most simply interpretable effects, steady reservoirs of charge must plainly be maintained. One way of achieving this was made possible by discoveries of Volta leading to the development of electric batteries, which supply separations of charge through chemical reactions. Closer inquiry into such matters is best left to treatises on experimental electricity. Here, it will suffice to know that when a uniform, steady current has somehow been produced in a wire, this can be unambiguously recognized through the uniformity and steadiness of all effects of it that can be detected.

† Efforts to find magnetic charges, or isolated magnetic poles of one sign, continue to the present day. Recent attempts have largely been inspired by new theoretical speculations about quantized magnetic monopoles. (Ref. 8; see also Exercise 6.10.) (Numbered references appear at the end of the book.) The efforts have only made it clear that if the magnetic monopoles do turn out to exist, it will be only in extremely special circumstances, difficult to arrange. They thereby become quite as foreign to the normal concerns of electromagnetic theory as quarks, for example. The equation $\nabla \cdot \mathbf{B} = 0$ will here continue to be read as a "nonexistence of magnetic charges."

That currents produce magnetic fields which encircle the current in uninterrupted loops has already been reported (and formulated as $\nabla \cdot \mathbf{B} = 0$). This automatically leads [see the discussion of (A.23)] to nonvanishing circulation integrals $\oint d\mathbf{r} \cdot \mathbf{B}$ around geometrical loops enclosing the current. In connection with the magnetic fields, each such integral is called the *magnetomotive force*, (mmf) in the loop used for the evaluation

$$\oint d\mathbf{r} \cdot \mathbf{B}(\mathbf{r}) \equiv \text{mmf} \tag{1.19}$$

It furnishes a measure of the magnetic field strength generated by a steady current that turns out to be independent of how the current and its field are distributed, each in its own space.

Quantitative investigation is simplest to carry out for the magnetostatic field around a long, straight filament carrying a definite steady current I. As to be expected from symmetry, the field lines are found to form circles, with the filament as a common axis. The field *strengths* are found to be proportional to the current used and inversely proportional to the distance s from the filament, as is representable by

$$\mathbf{B} = \mathbf{B}_\varphi = \frac{2I}{cs} \tag{1.20}$$

in cylindrical coordinates (s, φ, z) with the filament as (z) axis. The notation $2/c$ for the proportionality constant here becomes convenient in later developments. The result yields mmf's, $2\pi s\mathbf{B}_\varphi = 4\pi I/c$, that are the same on every circle about the filament! Moreover, a field like (1.20) is curlless outside the filament, since its curl derivative has the magnitude $|\nabla \times \mathbf{B}| = |(\nabla \times \mathbf{B})_z| = |s^{-1}\, \partial(s\mathbf{B})/\partial s| = 0$, according to (A.5). This means that line integrals over arbitrarily shaped loops around the filament, like the one marked A in Fig. 1.3, differ from line integrals over the circles (like C) only by line integrals (as over a) that vanish according to the Stokes theorem (A.21). The consequent invariance of an mmf to variations in

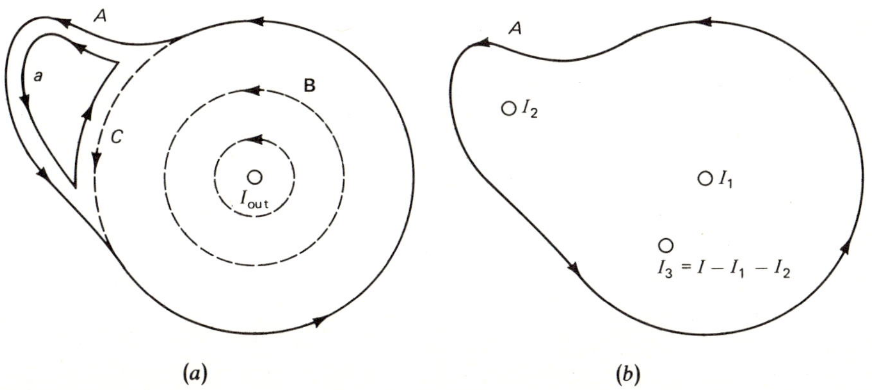

Figure 1.3 The mmf in circuit A is unchanged when the current I is redistributed in any such way as indicated in (b).

the distances of its elements from the current implies that it cannot matter how the same current I is distributed over the area as long as it remains enclosed by the loop on which that mmf is evaluated. The general result is the simple proportionality

$$\oint d\mathbf{r} \cdot \mathbf{B}(\mathbf{r}) = \frac{4\pi I}{c} \tag{1.21}$$

between the mmf on any loop and the resultant flux of charge $I = \int d\mathbf{S} \cdot \mathbf{j}$ through the entire areas enclosed by the loop. This is called Ampère's (integral) law.

Expression (1.21) correctly describes the *direction* found for the magnetic field, relative to the sign of the flux I, if the sense of the line integration is like that of a right-handed screw advancing in the direction the flux area is taken to be facing. This conforms to what is called the right-hand rule in elementary physics.

The measure of the magnetic field $\mathbf{B}$ was introduced in (1.16) as a force per unit test pole. However, it can also be measured by a force it exerts on a test current (actually used by Ampère in his investigations). After such forces on currents are introduced in Chap 4, it will be found that the constant c can be evaluated independently of the units used for the field $\mathbf{B}$, with a result expressable as $c = 3 \times 10^{10}$ cm/s. The effective dimensions of $\mathbf{B}$ corresponding to this are the same as for the electric field, i.e., charge divided by a squared length. That can be seen from the Ampère relation (1.21) when it is recognized that I has the dimensions of charge per unit time and c the dimensions of a velocity.

Ampère's law was abstracted from experience with the *static* magnetic fields $\mathbf{B}(\mathbf{r})$ generated by *steady* currents I, and its validity is restricted to magneto*statics*, just as the Gauss law (1.7) is restricted to electro*statics*. Just as the Gauss law could be reduced to a point form (1.11) with the help of the Gauss theorem (A.14), so Ampère's law can be reduced to a point form with the help of the Stokes theorem (A.21). Thus, the integral form (1.21) can be rewritten as

$$\int_S d\mathbf{S} \cdot (\nabla \times \mathbf{B}) = \frac{4\pi I}{c} = \frac{4\pi}{c} \int_S d\mathbf{S} \cdot \mathbf{j}$$

where S is any area enclosed by the integration loop in (1.21). This is supposed to be valid for any loop and any S enclosed by it, and hence

$$\nabla \times \mathbf{B}(\mathbf{r}) = \frac{4\pi}{c} \mathbf{j}(\mathbf{r}) \tag{1.22}$$

is the point form of Ampère's law.

It is evident that Ampère's law plays a role in magnetostatics comparable to that of the Gauss law in electrostatics. However, the point form (1.12) of the Gauss law can also be held valid in nonstatic situations, whereas Ampère's law cannot be. This is readily seen to follow from the fact that any curl is divergenceless (A.25), and hence (1.22) can apply only to current sources with the property $\nabla \cdot \mathbf{j} = 0$. Yet, in a general situation there may be time-dependent charge distributions $\rho(\mathbf{r}, t)$, as in (1.12), and then the charge conservation (1.14) requires that $\nabla \cdot \mathbf{j} = -\partial\rho/\partial t \neq 0$. Plainly, the magnetostatic Ampère law (1.22) will need modification before it can become a Maxwell equation, applicable to time-dependent situations as well.

Before going on with the development, it should be noted that Ampère's law makes no *explicit* provision for calculating fields arising from magnetic bodies, as characterized by pole strengths q_M, for example. Taking the law to be a general one for static fields therefore presumes that naturally occurring currents will eventually prove identifiable in magnetic bodies, accounting for their magnetism and sometimes permitting a *derivation* of equivalent pole strengths. Further inquiry into this will be left to Chap. E.

1.3 INDUCTION PROCESSES

If the equations $\nabla \cdot \mathbf{E} = 4\pi\rho$ and $\nabla \cdot \mathbf{B} = 0$ for the divergences of electromagnetic fields are to be used as the basis of the theory, the basis should be completed by statements about the curls of the vector fields (as suggested by the considerations in Sec. A.5). About the curl of magnetic field, only Ampère's law (1.22) has been developed so far, and this is restricted to magneto*static* situations. No assertions about the curls of electric field have yet been made.

Electro*static* fields are to be described as curlless

$$\nabla \times \mathbf{E}(\mathbf{r}) = 0 \tag{1.23}$$

One way in which this becomes evident is through the fact that any electrostatic field is derivable from a scalar potential (A.28) via

$$\mathbf{E}(\mathbf{r}) = -\nabla\phi(\mathbf{r}) \tag{1.24}$$

and any gradient field is automatically curlless (A.27). The fact is plainest for the point-charge field $\mathbf{E}_q$ of (1.2), which is obviously the negative radial gradient of

$$\phi_q = \frac{q}{r} \tag{1.25}$$

It must then follow that the vector integral (1.5), which represents the field of an *arbitrary* distribution of charges, is derivable from

$$\phi(\mathbf{r}_f) = \oint \frac{dV(\mathbf{r}_s)\rho(\mathbf{r}_s)}{R} \tag{1.26}$$

This can be checked directly with the help of the operation in

$$\nabla_f g(R) \equiv \nabla_f g(|\mathbf{r}_f - \mathbf{r}_s|) = \frac{dg}{dR}\frac{\mathbf{r}_f - \mathbf{r}_s}{|\mathbf{r}_f - \mathbf{r}_s|} = \hat{\mathbf{R}}\,\frac{dg}{dR} \tag{1.27}$$

valid for any function $g(R)$ of $R = |\mathbf{r}_f - \mathbf{r}_s|$. The curllessness can also be demonstrated without the explicit intervention of a potential by showing that any loop integral of the field (1.5) vanishes, simply because

$$\oint \frac{d\mathbf{r}_f \cdot \hat{\mathbf{R}}}{R^2} = -\oint d\frac{1}{R} = 0$$

That can be so only for curlless fields, as follows from the Stokes theorem and also as recorded in (A.12), for example.

The fact is then that no electro*static* field is formed in divergenceless loops. However, Faraday discovered *nonstatic* situations in which electric field *can* arise in such loops through a process of *induction.*

Faraday Induction

Faraday quite deliberately set out to complement Oersted's discovery that magnetic fields are generated from electrical currents by somehow producing a current from magnetism. He looked for such currents in loops of wire placed near magnets but found none in any static configuration of those elements. However, he did notice that transient currents flowed in the loops while they, or the magnets, were being moved into position and only while such relative motions were going on. He thus discovered the generation of electrical effects by *changing* magnetic fields.

Quantitative investigation of the phenomenon requires finding the relation of the current in a loop to some measure of the magnetic effects on it. As might be expected, it is the *flux* of magnetic field lines through the loop

$$\Phi \equiv \int d\mathbf{S} \cdot \mathbf{B} \tag{1.28}$$

that turns out to be pertinent. Measurements then lead to the simple finding that the instantaneous size $I(t)$ of the current in the wire circuit is directly proportional to the rate at which the magnetic flux threading it is being changed: $I \sim d\Phi/dt$. Unfortunately for the generality of this relationship, the ratio of these proportional quantities is found to depend on the particular material used for the wire.

Eliminating the dependence on the particular material requires knowledge about the conditions under which currents flow in any material wire. The principal experimental findings about such circumstances are embodied in Ohm's law. According to it, to have a current I flow between two points $\mathbf{r}_1$, $\mathbf{r}_2$ of a wire a potential difference

$$\phi_1 - \phi_2 \equiv \int_{\mathbf{r}2}^{\mathbf{r}1} \nabla\phi \cdot d\mathbf{r} = \int_{\mathbf{r}1}^{\mathbf{r}2} \mathbf{E} \cdot d\mathbf{r} \tag{1.29}$$

must be maintained between the two points, and then the magnitude of the current will be proportional to it, as represented by

$$\phi_1 - \phi_2 = RI \tag{1.30}$$

The proportionality constant R is found to be characteristic of the particular piece of material wire and is called its *ohmic resistance*. The name stems from the interpretation of Ohm's discovery (1.30). This implies that not only must electric force, like that arising from the imposed potential gradient (1.29), be applied in order to *start* the motion of charges constituting a current but work by it must be *continued*, to maintain the transport of charge along a wire. Continued expenditure of energy just to maintain a steady motion is understood as being needed to offset frictional losses; it is therefore presumed that there are resistive forces of the wire material to the passage of charges through it to be overcome. The expended energy should then reappear as a thermal energy, and a heating up of the wire is to be expected. This *Joule heating* has actually been observed and, moreover, an equality of the thermal energy that appears to the work expended by the electric force was eventually checked quantitatively.

A usual way to generate current is to provide such a potential difference as (1.29) by inserting a chemical battery into a wire circuit. The battery is then said to be supplying an *electromotive force*† equal to the potential difference it maintains

$$\phi_1 - \phi_2 = \int_{\mathbf{r}1}^{\mathbf{r}2} \mathbf{E} \cdot d\mathbf{r} \equiv (\text{emf})_{12} \tag{1.31}$$

The analogy of this definition to the mmf of (1.19) is apparent.

Faraday's discoveries in effect provided a new way to introduce current-generating emf into a circuit. The new source does not require interrupting the circuit and can be described with the help of an

$$\text{emf} \equiv \oint \mathbf{E} \cdot d\mathbf{r} \tag{1.32}$$

defined on a complete closed loop. This is related to the current I, as observed in any given wire loop, by the Ohm law, emf $= IR$, where R is the resistance of the particular loop. Thus, the above finding of the proportionality $I \sim d\Phi/dt$ can in every case be reexpressed as a proportionality emf $\sim d\Phi/dt$. The advantage is that the proportionality constant found for the new expression turns out to be independent of the particular wire loop used, e.g., its resistance R. The outcome is a general way of writing *Faraday's law of induction*

$$\oint \mathbf{E} \cdot d\mathbf{r} = -\frac{1}{c}\frac{d}{dt}\int \mathbf{B} \cdot d\mathbf{S} \tag{1.33}$$

for the emf induced in any loop by a given rate of change of the magnetic flux through it. As indicated, the universal proportionality constant found for this relation turns out identical with the $1/c$ which has already occurred in the expression (1.21) of the Ampère law when the fields are measured in the same

† Clearly emf's and mmf's do not have the dimensions of force, and so purists prefer using names like *electromotance* and *magnetomotance* instead.

units for both relations. A point of view from which this is no accident, but only to be expected, will be discussed in connection with Fig. 4.3.

The negative sign in expression (1.33) gives correctly the *direction* which the induced electric field $\mathbf{E}$ is found to have relative to the conventions already adopted about the signs of the integrals (the sense of the line integration being that of a right-handed screw advancing into the direction which the enclosed area is taken to be facing). The outcome, called the *Lenz law*, is often described by saying that the current induced by a given flux change is such as to tend to produce new magnetic flux that compensates for, or opposes, the applied flux change. The significance of this will be discussed further in connection with the magnetic field energy, in Chap. 4.

Electromotive force arises whether the flux change is produced by rotating a wire coil between fixed magnetic poles or by shuttling a bar magnet in and out of a fixed coil. Only a *relative* motion of circuit and magnet is needed. The problems implicit here, about electromagnetic phenomena as viewed from relatively moving reference frames, are left for later consideration. For the present, it will be simplest to consider a *fixed loop* and area enclosed by it so that

$$\frac{d}{dt}\int \mathbf{B}(\mathbf{r}, t) \cdot d\mathbf{S}(\mathbf{r}) \equiv \int \frac{\partial \mathbf{B}}{\partial t} \cdot d\mathbf{S}(\mathbf{r}) \tag{1.34}$$

as follows when every element $d\mathbf{S}$ is fixed in space. Then using the Stokes theorem to transform the loop integral in (1.33) leads to

$$\int d\mathbf{S} \cdot (\nabla \times \mathbf{E}) = -\int d\mathbf{S} \cdot \frac{\partial \mathbf{B}}{c\,\partial t} \tag{1.35}$$

If Faraday's law (1.33) is to be valid for any loop and any enclosed area, it can be given the point form

$$\nabla \times \mathbf{E}(\mathbf{r}, t) = -\frac{\partial \mathbf{B}}{c\,\partial t} \tag{1.36}$$

This is the third of the Maxwell equations, additional to the divergence conditions (1.12) and (1.17). Notice that it is properly consistent with the curllessness (1.23) electric fields are supposed to have in static situations. The equation is also taken to imply that when no material circuits are present to carry such induced currents as those found by Faraday, the same electric field lines will exist out in free space. The Faraday currents have only served as detectors of an induced electric field!

Maxwell Induction

The equations developed so far amount to descriptions of results found by Coulomb, Gauss, Oersted, Ampère, Faraday, and others, but it is Maxwell's name that is attached to the entire set. Maxwell *completed* the set, with a bold hypothesis that enormously expanded the range of phenomena to which the theory applies.

To complete the set of equations, a generally valid one for the curl $\nabla \times \mathbf{B}(\mathbf{r}, t)$ of the magnetic field is still needed. The Ampère law (1.22) was found to hold for static situations but must be modified for nonstatic ones because it is restricted to divergenceless currents, with $\nabla \cdot \mathbf{j} = 0$, like the steady ones forming complete loops. As already discussed, this is a consequence of the fact that anything being equated to a curl must be divergenceless. Maxwell sought a quantity to equate to $\nabla \times \mathbf{B}(\mathbf{r}, t)$ which would properly reduce to Ampère's $4\pi\mathbf{j}/c$ in static situations but which would continue to be divergenceless, like any curl, in the nonstatic circumstances when

$$\nabla \cdot \mathbf{j}(\mathbf{r}, t) = -\frac{\partial}{\partial t}\rho(\mathbf{r}, t) \neq 0$$

is required for charge conservation. This relation itself suggests the proper divergenceless quantity when it is recognized that the charges are sources of electric field divergence (1.12), so that the relation can be reexpressed as

$$\nabla \cdot \mathbf{j}(\mathbf{r}, t) = -\frac{\partial}{\partial t}\frac{\nabla \cdot \mathbf{E}}{4\pi} = \nabla \cdot \left(-\frac{\partial \mathbf{E}}{4\pi\, \partial t}\right)$$

In accordance with what this suggests, Maxwell advanced the hypothesis that the magnetic curl should be equated to the divergenceless quantity on the right side of

$$\nabla \times \mathbf{B}(\mathbf{r}, t) = \frac{4\pi}{c}\left(\mathbf{j} + \frac{\partial \mathbf{E}}{4\pi\, \partial t}\right) \tag{1.37}$$

Thus, the "real," material current $\mathbf{j}(\mathbf{r}, t)$ is in general to be supplemented by what Maxwell called a *displacement current* $\partial\mathbf{E}/(4\pi\, \partial t)$. The name stems from a certain mechanical picture of free space which Maxwell had in mind but which has since been found to be unnecessary and misleading. What has perhaps become a more apt name is *field current*, since $\partial\mathbf{E}/(4\pi\, \partial t) \neq 0$ requires only the presence of a changing electric field; it can even exist in vacuo, with no material charges present.

An aesthetically appealing consequence of Maxwell's hypothesis is the symmetry it introduces between the electric and magnetic fields in free space, where $\rho = 0$ and $\mathbf{j} = 0$. There *both* fields are divergenceless, and (1.37) becomes

$$\nabla \times \mathbf{B} = \frac{\partial \mathbf{E}}{c\, \partial t} \qquad \text{where } \mathbf{j} = 0 \tag{1.38}$$

This differs from the point form (1.36) of the Faraday law only in sign when $\mathbf{E}$ and $\mathbf{B}$ are interchanged. The integral form corresponding to (1.33), as obtained by integrating (1.38) over some area and employing the Stokes theorem, can be seen to be

$$\frac{d}{c\, dt}\int \mathbf{E} \cdot d\mathbf{S} = \oint \mathbf{B} \cdot d\mathbf{r} \tag{1.39}$$

Thus, a changing electric flux through any circuit induces mmf in it, just as magnetic-flux changes induce emf. A new source of mmf is provided, besides the flux of material charge represented in the Ampère law (1.21).

A host of phenomena deducible from the existence of Maxwell induction has by now been observed, but perhaps the most spectacular consequence is an expectation that there must exist electromagnetic waves which can propagate themselves across space devoid of matter, with just the velocity of light c and identifiable with light in general. The formal demonstration that this is an expected consequence is left to Chap. 7, but it is already possible to visualize how such a propagation can launch itself across free space quite independently of how it might originate in charged matter. The propagation consists of alternating inductions of electric field lines by changes in magnetic flux and of magnetic field lines by changing electric flux, all out in free space and in some such way as suggested† by Fig. 1.4. It can be appreciated that both laws of induction, Faraday's (1.33) and Maxwell's (1.39), are equally essential to producing a periodically repeating chain which can extend indefinitely far in space and time, as needed for propagation.

The results of this chapter have been embodied in four Maxwell equations, which, in a notation indicated in $\partial\rho/\partial t \equiv \mathring{\rho}(\mathbf{r}, t)$, for partial time derivatives, are summarized by

$$\nabla \cdot \mathbf{E} = 4\pi\rho \qquad \nabla \times \mathbf{E} = -\frac{\mathring{\mathbf{B}}}{c}$$
$$\nabla \cdot \mathbf{B} = 0 \qquad \nabla \times \mathbf{B} = \frac{4\pi\mathbf{j}}{c} + \frac{\mathring{\mathbf{E}}}{c} \tag{1.40}$$

Much of the remainder of this book will consist of deductions from these equations.

Actually, Maxwell produced what he considered to be a more generally valid set of equations, different from (1.40) in that more explicit provisions are made for applicability to other media than free space. Various media were to be distinguished as having different polarizabilities (see Chaps. D and E). However, after electrons and ions were discovered, Lorentz began demonstrating that all

† Propagations closely resembling such linked chains can be produced in wave guides (see Fig. 7.5).

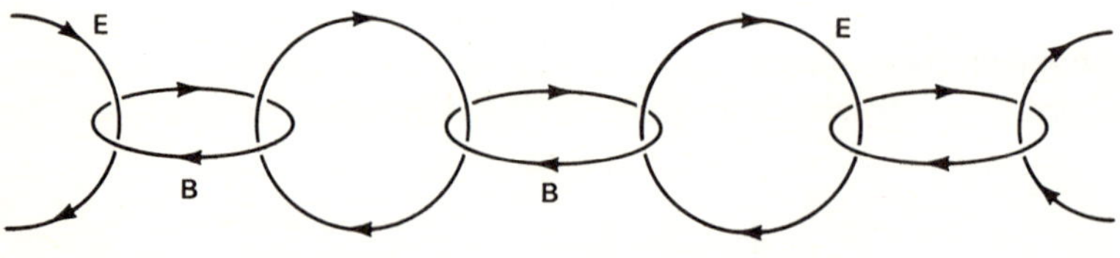

Figure 1.4

effects of material media should be deducible from Eqs. (1.40) as they stand, with charges and currents arising from the atomic constituents of the materials. The polarizabilities become *derivable*, through averaging procedures that make them only approximate descriptions. Relations (1.40), sometimes distinguished as *Maxwell-Lorentz equations*, become the more basic ones. The macroscopic Maxwell equations continue to be highly useful and will be relegated to the supplementary chapters simply to keep the more basic lines of development clearer, as higher-order abstractions from macroscopic observations.

The Maxwell equations plainly couple the electric and magnetic fields to each other, in general, since they have the curl of each field depending on time variations in the other. However, in time-*independent* situations, when $\mathring{\mathbf{E}} = 0$ and $\mathring{\mathbf{B}} = 0$, the equations separate into a pair for a curlless electrostatic field alone and another pair for a (divergenceless) magnetostatic field alone. This makes the subjects of electro*statics* and magneto*statics* quite independent.

EXERCISES

1.1 Two like point charges Q are *fixed* at a distance $2l$ apart. The midpoint between them is obviously an equilibrium point for any particle of charge q that may be placed there. Compare the *stabilities* of that equilibrium to displacement of the particle along the line of the charges and to displacement perpendicular to this line for both the cases $qQ \gtrless 0$. (That every position in an electrostatic field is unstable to *some* direction of displacement is known as *Earnshaw's theorem*.)

1.2 Verify the parenthetical statement about (1.1) in the following steps:

(*a*) Let a charge q_1 be distributed inside a sphere of radius a_1 in a density $\rho_1(s)$ that varies with distance s from the sphere center; then show how the vector integral (1.5) yields $\mathbf{E} = \mathsf{E}_r = q_1/r^2$ at all $r > a_1$.

(*b*) Let a second such charge distribution, this one with q_2, $\rho_2(s)$, and a_2, be placed with its center at a distance $r > a_1 + a_2$ from the center of q_1; then calculate the resultant force on each charge.

1.3 (*a*) A toroidal (doughnut-shaped) surface has a ring of circular cross sections of uniform radius a in all planes through the axis of the ring and all centered at a distance $R > a$ from that axis. Find the solid angle subtended by the part of the surface visible from the ring center. *Ans.* $4\pi a/R$.

(*b*) Suppose a point charge q (>0) is placed within the surface at $R - a < r < R + a$ in the plane normal to the axis and through the ring center. What is the net electric flux out of the entire surface?

1.4 The Gauss *integral* (1.7) over a whole surface of points can be used to find a field magnitude at any one point whenever the charge is distributed with some symmetry, making it possible to identify simple surfaces normal to the field lines on which the magnitude can be expected to have a constant value. Use this familiar trick of elementary physics to deduce the electric field:

(*a*) Arising from an infinitely long line of uniformly distributed charge λ per unit length.

(*b*) Inside and outside a spherical shell on which a charge q is uniformly distributed.

(*c*) *Outside* a sphere within which charge q is distributed in a spherically symmetric density $\rho(r_s)$.

(*d*) *Inside* the sphere of (*c*) for $\rho = \text{const}$ and $\rho \sim r_s$.

1.5 (*a*) Use the Gauss law to find the electrostatic field inside and outside an indefinitely long cylinder of radius a that has charge λ per unit length distributed homogeneously throughout its interior.

(*b*) Make a comparable use of the Ampère integral to find the magnetostatic field inside and outside such a cylinder when it carries a total current $I = \lambda c$ distributed uniformly throughout the interior and flowing parallel to the cylinder axis.

1.6 Now suppose that the cylinder of Exercise 1.5 has a cylindrical *hollow* of radius $b < a$ made in it. The axis of the hollow is parallel to the axis of the outer cylinder, but the two axes are separated a distance $d(<a - b)$, so that the hollow is off center by d.

(*a*) Find the magnitude and direction of $\mathbf{E}$ everywhere *inside the hollow* if the charge λ per unit length is homogeneously distributed within the remainder of the outer cylinder.

(*b*) Find $\mathbf{B}$ inside the hollow if the current $I = \lambda c$ is carried by the cylinder remainder.

(*c*) What happens in cases (*a*) and (*b*) when the hollow is *not* off center ($d = 0$)?

1.7 A total charge q is distributed rigidly in some azimuthally symmetric density $\rho(r, \vartheta)$. It is set spinning bodily about its axis of symmetry with an angular velocity ω_o. Find the total flux of current I about the axis in terms of q and ω_o.

1.8 A magnetic field is produced inside a cylindrical, solenoidal, space by steady currents encircling it [each element of current in the φ direction about the (z) axis of the cylinder]. The current windings are distributed to produce a z component of field that strengthens linearly with distance along the axis: $\mathbf{B}_z = \alpha z$, where α is a constant field gradient.

(*a*) Show that another component of $\mathbf{B}$ must exist in the region and find it.

(*b*) Find an equation $s(z)$ that describes the field lines geometrically and make a sketch of them. (The result is a simple type of magnetic bottle; see Sec. 5.4.)

1.9 Suppose that a circular wire loop of radius a is placed inside the magnetic bottle of Exercise 1.8. The loop is carefully centered on the z axis of the field and oriented so that it faces in the z direction. It is then pushed along the z direction without changing its orientation and with a uniform speed u ($\ll c$). Find the current induced in the loop if its resistance is R. How will this current vary, and what sense will it have relative to the sense of the currents producing the field? What would the current be if the field were uniformly distributed instead?

1.10 A square wire loop of side a is placed in a uniform field $\mathbf{B}_o$ facing in the direction of the field. A uniform rotation of the loop about one of its sides as axis is then started and maintained. What is the current induced in the loop if its resistance is R and its angular velocity is ω_o?

1.11 Two students report their conclusions about the electrostatic field inside a certain sphere. Student A reports the rectangular components

$$\mathbf{E}_x = \alpha xz \qquad \mathbf{E}_y = \alpha yz \qquad \mathbf{E}_z = 2\alpha z^2$$

relative to a frame based on the sphere center. Student B uses cylindrical coordinates, with the same z axis, and reports

$$\mathbf{E}_s = \alpha sz \qquad \mathbf{E}_\varphi = 0 \qquad \mathbf{E}_z = \tfrac{1}{2}\alpha(3z^2 + s^2)$$

with the same value for α. Which student must be wrong and why? What must the charge distribution within the sphere have been? Express α in terms of the total *positive* charge q and the sphere radius a.

1.12 A point charge is moving with a uniform speed u ($\ll c$). Show how the Maxwell induction integral (1.39) can be used to find the magnitude of the magnetic field in terms of the instantaneous distance r from the charge and the angle ϑ that the radius vector $\mathbf{r}$ makes with the direction of the motion.

1.13 A central source is ejecting charge isotropically, so that through every sphere centered on it there is an outward flux of current having density $j(r)$ at the radius r.

(*a*) Form the relation between $j(r)$ at the moment and the rate at which the charge $q(r)$ still inside r is changing.

(*b*) Show how result (*a*) can help in coming to conclusions about the magnetic field in the space.

CHAPTER

TWO

THE ELECTROSTATIC FIELD

The Maxwell and potential descriptions of electrostatic fields from prototype sources · Monopole, dipole, and quadrupole effects · Electrostatic field energy and the potential energies arising from the interference of fields · Forces on monopoles, dipoles, and quadrupoles and interactions between them · The interpretation of scalar potentials

The basis provided by the Maxwell equations (1.40) for deducing expectations about electrostatic fields is

$$\nabla \cdot \mathbf{E}(\mathbf{r}) = 4\pi\rho(\mathbf{r}) \qquad \nabla \times \mathbf{E} = 0 \tag{2.1}$$

For the first deductions, it will be supposed that some static charge distribution $\rho(\mathbf{r})$ has been *given.*

2.1 SUPERPOSITIONS OF COULOMB FORCES

When the interest is only in whatever fields arise from some given charges, what amounts to a superposition of point-charge effects, the vector integral (1.5), can be used to calculate them. The integral can be regarded as the appropriate solution† of the Maxwell equations (2.1) to use and is actually the more immediately grounded on experience. After all, it was part of the basis from which the generalizations to Maxwell's equations began, with a chain of inferences interposed.

The results for various simple, idealized charge distributions will be valuable to review. They help indicate the kinds of fields to be expected in more realistic situations, in that they may approximate the ideal ones or superpositions of them. The results will only be quoted and remarked on, since the integrations needed are quite straightforward (Exercises 2.1 and 2.2).

Next to superpositions of discrete point charges, as in (1.3), the simplest charge distribution is perhaps one forming a uniform linear continuum, with some λ units of charge per unit length, so that the element of charge $dq = \rho(\mathbf{r}_s)\, dV(\mathbf{r}_s)$ can be reexpressed as

$$dq = \lambda\, dz, \qquad \text{with } -\tfrac{1}{2}L < z < \tfrac{1}{2}L$$

† The formal demonstration will be taken up again in Sec. 3.1 but is already evident from the discussion in Sec. A.5 and in connection with (1.5).

for a straight *line* of length L. Opposite the middle of the line, at a perpendicular distance s from it, the field is normal to the line and has the magnitude

$$\mathsf{E} = \mathsf{E}_s = \frac{2\lambda}{s}\left(1 + \frac{4s^2}{L^2}\right)^{-1/2} = \frac{\lambda L}{s^2}\left(1 + \frac{L^2}{4s^2}\right)^{-1/2} \tag{2.2a}$$

The last of these forms makes it obvious that at very large distances ($s \gg \frac{1}{2}L$) the field becomes characteristic of a point charge ($q = \lambda L$, here), as any finite charge distribution should at distances sufficiently remote. At the opposite extreme, very close to the line ($s \ll \frac{1}{2}L$) or for a very long line ($L \to \infty$)

$$\mathsf{E} \to \frac{2\lambda}{s} \tag{2.2b}$$

With L effectively infinite in the latter result, any point can be considered opposite the middle, and so it is valid in all regions in which end effects can be considered negligible. (Advantage can be taken of this "symmetry" to obtain the field from the Gauss integral, as in Exercise 1.4.)

When a circular *ring* of radius a has charge λ per unit length, it presents the finite total charge $q = 2\pi a\lambda$. On the axis of such a ring the field is directed along the axis and can be represented as

$$\mathsf{E}_z = \frac{qz}{(z^2 + a^2)^{3/2}} \gtrless 0 \qquad \text{for } qz \gtrless 0 \tag{2.3}$$

if $|z|$ is the distance along the axis from the ring center. The *magnitude* is symmetrical about the center and properly reduces to $\mathsf{E} = |q|/z^2$ for $|z| \gg a$.

A uniform circular *disk* of charge σ per unit *area* can be treated as a superposition of the rings, with the charge $dq = 2\pi s\sigma\, ds\,|_0^a$ on the ring element at radius s. Integration then yields the magnitude

$$\mathsf{E} = 2\pi\sigma\left[1 - \frac{|z|}{(z^2 + a^2)^{1/2}}\right] \tag{2.4}$$

for the field on the axis. Its directions can be indicated by $\mathsf{E}_z(z \gtrless 0) = \pm\mathsf{E}$. For $|z| \gg a$, $\mathsf{E} \to \pi a^2\sigma/z^2$ properly.

Results for an infinite plane of charge, or in regions near any partial uniform plane where edge effects can be considered negligible, can be obtained by letting $a \to \infty$ for the disk

$$\mathsf{E} = 2\pi\sigma \tag{2.5a}$$

with $\mathsf{E}_z(z \gtrless 0) = \pm 2\pi\sigma$. The same results follow from the Gauss law (2.31) or by superposing infinite *lines* of charge (equivalent to using cartesian, rather than polar, coordinates for an integration). The uniform fields here have opposite directions on the two sides of the surface and so have a discontinuity

$$\Delta\mathsf{E}_z = 4\pi\sigma \tag{2.5b}$$

at the surface. Such a discontinuity is really to be expected at any surface on which a smoothly varying charge distribution lies, for then a charge element $\sigma\, dS$ is equivalent to an infinite plane of charge for points close enough to it.

Results for a uniform spherical charged surface are most elementary to obtain with the help of the Gauss law (Exercise 1.4), but an integration can demonstrate consistency with the ring results, since the sphere can be treated as a superposition of rings. The field is radial and everywhere *outside* the sphere has a magnitude q/r^2, exactly as for a point charge at the sphere center. *Inside* the sphere, $\mathbf{E} = 0$. There is thus a discontinuity at the surface ($r = a$) by $\Delta \mathsf{E}_r = \mathsf{E}_r(r = a+) = q/a^2 = 4\pi\sigma$ when expressed in terms of $\sigma = q/4\pi a^2$.

A homogeneous sphere of charge containing the uniform density $\rho = q/\frac{4}{3}\pi a^3$ again has the radial field q/r^2 outside the charge, as if arising from a point charge at the center. The field inside the sphere is proportional to the distance from the center

$$\mathsf{E} = \mathsf{E}_r(r \le a) = \frac{qr}{a^3} \tag{2.6}$$

This is continuous with the outside field at the surface $r = a$, since here a vanishingly thin element of surface has a vanishing fraction of the charge. Integrations (1.5) can check consistency of the sphere results with superpositions of disk or spherical-shell results.

2.2 THE ELECTROSTATIC POTENTIAL

When deductions about an electrostatic field are initiated by solving the Maxwell equations (2.1), it is best to begin with the equation expressing the curllessness, for this is independent of what the specific sources $\rho(\mathbf{r})$ may be. The general solution of $\nabla \times \mathbf{E} = 0$ is $\mathbf{E} = -\nabla\phi(\mathbf{r})$, with any $\phi(\mathbf{r})$ [see (A.28)]. There remains only a problem of solving the Poisson equation following from $\nabla \cdot \mathbf{E} = 4\pi\rho$

$$\nabla^2\phi = -4\pi\rho(\mathbf{r}) \tag{2.7}$$

The result will be a description of the electrostatic field by a scalar potential $\phi(\mathbf{r})$, in place of $\mathbf{E}(\mathbf{r}) = -\nabla\phi$.

The solution of the Poisson equation that will characterize the potential field arising from a given static charge distribution $\rho(\mathbf{r})$ has already been recorded in (1.26) as $\phi = \oint dV \rho/R$. This amounts to a superposition of point-charge potentials like (1.25), gauged by the work needed to bring a positive unit of test charge from infinity to the field point, as for

$$\phi_q = -\int_\infty^r \frac{q}{r^2}\, dr = \frac{q}{r} \tag{2.8}$$

Such a gauge will be found suitable for any charge distribution of finite extent.

The integration $\phi = \oint dq/R$ is then quite straightforward, and results for some of the situations treated in the preceding section will be discussed.

For the finite line of charge,

$$\phi(r, \vartheta) = \lambda \ln \frac{(r^2 - Lr \cos \vartheta + \frac{1}{4}L^2)^{1/2} + \frac{1}{2}L - r \cos \vartheta}{(r^2 + Lr \cos \vartheta + \frac{1}{4}L^2)^{1/2} - \frac{1}{2}L - r \cos \vartheta} \tag{2.9}$$

in terms of spherical coordinates centered on the middle of the line, with the line also serving as the polar axis from which ϑ is measured. This result is not confined to the median plane ($\vartheta = \frac{1}{2}\pi$), as (2.2*a*) was, simply because the more general result was here obtainable at very little extra cost. This is a characteristic advantage of the potential description. The electric vector field everywhere around the line can now be obtained as $\mathbf{E}_r = -\partial\phi/\partial r$ and $\mathbf{E}_\vartheta = -\partial\phi/(r\,\partial\vartheta)$.

For a very short line or at sufficiently large distances from the line $r \gg \frac{1}{2}L$ and $\phi \to \lambda \ln (1 + L/r) \to \lambda L/r$. Only the radial component of the electric field is then appreciable: $\mathbf{E} \to -d\phi/dr \approx \lambda L/r^2 = q/r^2$.

The exact result in the median plane is

$$\phi(r, \tfrac{1}{2}\pi) = \lambda \ln \frac{1+f}{1-f} \qquad \text{if } f \equiv \frac{L}{2(r^2 + \frac{1}{4}L^2)^{1/2}} \tag{2.10}$$

The radial gradient of this yields the result (2.2*a*).

For the ring of charge, in terms of spherical coordinates with origin at the ring center and the ring axis as polar axis,

$$\phi(r, \vartheta) = \lambda a \int_0^{2\pi} \frac{d\varphi}{[r^2 + a^2 - 2ar \sin \vartheta \cos \varphi]^{1/2}} \tag{2.11a}$$

The integral here cannot be expressed in terms of any of the most elementary functions, and so-called *elliptic functions* have therefore been defined, just for such purposes. Since the elliptic functions are themselves best defined as certain integrals, it is scarcely less satisfactory to consider (2.11*a*) to be the final answer which can be evaluated for any particular case by numerical integration. Its gradients can also be presented as quadratures, e.g.,

$$\mathbf{E}_\vartheta = -\frac{\partial\phi}{r\,\partial\vartheta} = -\lambda a^2 \cos \vartheta \int_0^{2\pi} d\varphi \cos \varphi/[\cdots]^{3/2}$$

with the contents of the square bracket here the same as in the square bracket of (2.11*a*). Such an expression is sufficient to make it evident, for instance, that $\mathbf{E}_\vartheta$ vanishes in the plane of the ring ($\vartheta = \frac{1}{2}\pi$) and also on the axis ($\vartheta = 0$) since $\int_0^{2\pi} d\varphi \cos \varphi = 0$.

An analytic integration of (2.11*a*) becomes trivial on the ring axis, yielding

$$\phi(r, 0) = \frac{2\pi a\lambda}{(r^2 + a^2)^{1/2}} \tag{2.11b}$$

there. This has the gradient along the axis needed for agreement with (2.3).

The field around a disk of charge likewise generally requires expression in terms of elliptic integrals (any circular shape looks like an elliptical one when viewed from a point off axis). On the axis of the disk

$$\phi = 2\pi\sigma[(a^2 + z^2)^{1/2} - |z|] \tag{2.12}$$

in terms of the coordinate z ($\gtrless 0$) along the axis. This is symmetrical in $z \leftrightarrow -z$ (imagine a plot of ϕ vs. $-\infty < z < \infty$), and that property makes it easy to make correct choices of signs when representing the gradient $\mathsf{E}_z(z \gtrless 0) = -d\phi/dz = \pm\mathsf{E}$, as in (2.4). Whereas the gradient derivative $-\mathsf{E}_z$ is discontinuous by $4\pi\sigma$ in a passage through the disk (at $z = 0$), the potential (2.12) in a given gauge is *continuous*, with the value $\phi = 2\pi\sigma a = 2q/a$ at $z = 0$ [these properties can also be clarified by the plot of $\phi(z)$ vs. z].

For the spherical shell of radius a

$$\phi(r \geq a) = \frac{q}{r} \qquad \text{and} \qquad \phi(r \leq a) = \frac{q}{a} \tag{2.13}$$

This is continuous at the surface $r = a$, whereas the force per unit charge has the discontinuity $4\pi\sigma = q/a^2$.

For the solid sphere of charge

$$\phi(r \geq a) = \frac{q}{r} \qquad \text{and} \qquad \phi(r \leq a) = \frac{q}{2a^3}(3a^2 - r^2) \tag{2.14}$$

continuous through the surface (as is $\mathsf{E} = \mathsf{E}_r$). These outer and inner potentials are comparable to the *gravitational* potentials outside and inside the earth.

Evaluation of the integral (1.26) leads to difficulties for charge distributions which are idealized (sometimes overidealized) by being given an infinite extent. The difficulties are such as are easily circumvented simply by changing the way the potential is gauged. They arise only because not even a point at infinity is sufficiently remote from a charge of infinite extent for the force field to be negligible there; it then takes an infinite amount of work to bring a test charge all the way from infinity.

When the integral $\phi = \oint dq/R$ is evaluated for an infinitely long line of charge, it yields

$$\phi = -2\lambda \ln s + 2\lambda \ln (L \to \infty) \tag{2.15}$$

if the s in this result is the radial distance from the line as axis. The infinity of the expression does not interfere with its having a finite gradient

$$\mathsf{E} = \mathsf{E}_s = -\frac{d\phi}{ds} = -\frac{d}{ds}(-2\lambda \ln s) = \frac{2\lambda}{s}$$

in agreement with (2.2*b*). A definite potential is nevertheless desirable and can be obtained from

$$\phi = -\int_{s_o}^{s} \frac{2\lambda}{s}\, ds = 2\lambda \ln \frac{s_o}{s} \tag{2.16}$$

in which it is gauged by the work of carrying a test-charge from *any finite* radius s_o.

A potential which is finite at any finite distance from the infinite plane of charge in the preceding section is

$$\phi = -2\pi\sigma|z| = -\mathsf{E}|z| \tag{2.17}$$

It obviously yields the uniform field $\mathsf{E}_z(z \gtrless 0) = -d\phi/dz = \pm 2\pi\sigma$ properly. Here the potential is gauged by work done in starting from the source itself. Compare the result in Exercise A.8, a potential of any uniform field.

2.3 STATIC MULTIPOLES

The field arising from any static source $q = \oint \rho(\mathbf{r}_s)\, dV(\mathbf{r}_s)$ which is finite in extent (*localized*) but may be arbitrarily distributed within the source volume can be explored as follows. Start with the obvious fact that far enough away, any finite source will look like a point charge; that is, $\phi(r \to \infty) = q/r$ if $q \neq 0$ and $r \equiv r_f$ is the distance to the field point from some origin chosen to lie within the finite limits of the source. Then consider the corrections to $\phi \approx q/r$ which must be made to represent the field closer in. They can be found by treating $1/r$ as only a first approximation to the *propagation factor* $1/|\mathbf{r}_f - \mathbf{r}_s|$ in the general expression (1.26) of the potential. Better approximations follow from the power-series expansion†

$$|\mathbf{r}_f - \mathbf{r}_s|^{-1} = (r_f^2 - 2\mathbf{r}_f \cdot \mathbf{r}_s + r_s^2)^{-1/2} = \frac{1}{r_f}\left(1 - 2\frac{\mathbf{r}_f \cdot \mathbf{r}_s}{r_f^2} + \frac{r_s^2}{r_f^2}\right)^{-1/2}$$

$$\approx \frac{1}{r_f}\left[1 + \frac{\hat{\mathbf{r}}_f \cdot \mathbf{r}_s}{r_f} + \frac{3(\hat{\mathbf{r}}_f \cdot \mathbf{r}_s)^2 - r_s^2}{2r_f^2} + \cdots\right] \tag{2.18}$$

The unit vector $\hat{\mathbf{r}}_f \equiv \mathbf{r}_f/r_f$, the direction to the field point, has been introduced here in order to make it plainer that this is a series in successive powers of the ratio of magnitudes r_s/r_f and hence valid for any $r_f > |\mathbf{r}_s|_{\max}$ (anywhere outside the source).

† The famous binomial expansion

$$(1 + \epsilon)^n = 1 + n\epsilon + \tfrac{1}{2}n(n-1)\epsilon^2 + \cdots + \frac{n!}{s!\,(n-s)!}\epsilon^s + \cdots$$

can be used here despite the noninteger n, even though it results in an infinite series (convergent if only $\epsilon < 1$). It is then equivalent to a *Taylor expansion*

$$f(\epsilon) = f(0) + \epsilon f'(0) + \tfrac{1}{2}\epsilon^2 f''(0) + \cdots + \frac{\epsilon^s}{s!}f^{(s)}(0) + \cdots$$

since $n!/(n-s)! \equiv n(n-1)\cdots(n-s+2)(n-s+1)$ also for noninteger n.

When (2.18) is substituted into the integral for the potential, the result is a series in inverse powers of the distance $r \equiv r_f$ from source to field point

$$\phi = \frac{q}{r} + \frac{\hat{\mathbf{r}} \cdot \mathbf{D}}{r^2} + \frac{\hat{\mathbf{r}} \cdot \mathbf{Q} \cdot \hat{\mathbf{r}}}{2r^3} + \cdots \tag{2.19}$$

For this expression, various quantities q, $\mathbf{D}$, $\mathbf{Q}$, ..., called the *multipole moments* of the source, are defined. The vector $\mathbf{D}$ is called the *dipole moment* and is defined by

$$\mathbf{D} \equiv \oint \mathbf{r}_s \rho(\mathbf{r}_s)\, dV(\mathbf{r}_s) \equiv \oint dq\mathbf{r}_s \tag{2.20}$$

The symbol $\mathbf{Q}$ stands for a *second-rank tensor*, called the *quadrupole moment tensor*, of the source. It is defined to have nine components, any one of which can be written in a cartesian representation as

$$Q_{ij} = \oint dq(3x_s^i x_s^j - \delta_{ij} r_s^2) \qquad (= Q_{ji}) \tag{2.21a}$$

with $x_s^{1,2,3} \equiv x_s, y_s, z_s$, respectively, so that

$$Q_{11} \equiv Q_{xx} = \oint dq(3x_s^2 - r_s^2) \qquad \text{and} \qquad Q_{12} \equiv Q_{xy} = 3\oint dq x_s y_s$$

for example. Sometimes a *dyadic* notation is used:

$$\mathbf{Q} = \oint dq(3\mathbf{r}_s\mathbf{r}_s - \mathbf{1}r_s^2) \tag{2.21b}$$

where $\mathbf{1}$ is the *unit tensor* with components $\mathbf{1}_{ij} \equiv \delta_{ij}$ (δ_{ij}, the Kronecker delta, is 1 for $i = j$ and 0 otherwise). Products like $\mathbf{a} \cdot \mathbf{Q} \cdot \mathbf{b}$ of a tensor with *two* vectors are *defined* as having the scalar result

$$\mathbf{a} \cdot \mathbf{Q} \cdot \mathbf{b} \equiv \Sigma_{i,j} a_i Q_{ij} b_j \equiv \oint dq[3(\mathbf{a} \cdot \mathbf{r}_s)(\mathbf{r}_s \cdot \mathbf{b}) - (\mathbf{a} \cdot \mathbf{b}) r_s^2] \tag{2.22}$$

With these definitions, it can readily be checked that the substitution of the third term of (2.18) into the integral for the potential yields just the term $\mathbf{r} \cdot \mathbf{Q} \cdot \mathbf{r}/2r^5$ of expression (2.19). Of course, an infinite sequence of multipole moments should be defined to carry on with the series (2.19). That will actually be done at the end of Chap. 3, after a concise way of representing the general moment has been introduced. The most important property of all the multipole moments is that they are characteristic of the source itself, being independent of the field point, with regard to both the distance $r \equiv r_f$ to it and the direction $\hat{\mathbf{r}} \equiv \hat{\mathbf{r}}_f$ from which an observer at it views the source.

In the context of the multipole expansion (2.19), the term $\phi_q = q/r$, the largest one to survive as $r \to \infty$, is called the *monopole effect* of the source. The next two multipole effects will be discussed in greater detail.

The Static Electric Dipole

The word "dipole" was first coined for a source idealized as having two equal and opposite point charges at some distance l apart. Let $\mathbf{l}$ be a vector originating at a negative point charge $-q$ and extending to the positive point charge $+q$. Then, at a position $\mathbf{r}$ relative to the midpoint between the charges (so that they have the positions $\mathbf{r}_s = \pm\frac{1}{2}\mathbf{l}$),

$$\phi(\mathbf{r}) = \frac{q}{|\mathbf{r} - \frac{1}{2}\mathbf{l}|} - \frac{q}{|\mathbf{r} + \frac{1}{2}\mathbf{l}|} \tag{2.23a}$$

A multipole expansion of this can be formed by treating the inverse lengths here as suggested by (2.18):

$$\phi \approx \frac{q\mathbf{l} \cdot \hat{\mathbf{r}}}{r^2} + q(\mathbf{l} \cdot \hat{\mathbf{r}}) \frac{5(\mathbf{l} \cdot \hat{\mathbf{r}})^2 - 3l^2}{8r^4} + \cdots \tag{2.23b}$$

The source being considered here has no monopole effect, since it is neutral in total charge. Its largest effect at distances $r > l$ arises from the dipole moment

$$\mathbf{D} = q(\tfrac{1}{2}\mathbf{l}) - q(-\tfrac{1}{2}\mathbf{l}) = q\mathbf{l} \tag{2.24}$$

as calculated from the definition (2.20). There is no quadrupole effect proportional to $1/r^3$, in expression (2.23b), and this also follows from an evaluation of Q (2.21) for the source here, but octupole ($\sim 1/r^4$) and still higher-order effects do exist. All effects except

$$\phi_D \equiv \mathbf{D} \cdot \frac{\hat{\mathbf{r}}}{r^2} \tag{2.25}$$

disappear only for an ideal dipole, a source which it is adequate to treat as a *point* ($l \to 0$) but with a finite dipole moment nevertheless, i.e., a source behaving as if $l \to 0$ and $q = D/l \to \infty$ in such a way that $ql \to D < \infty$. Clearly, the octupole and higher-order terms of (2.23b) will vanish as $l \to 0$ even though $q = D/l \to \infty$ if D remains finite, since $ql^n = Dl^{n-1} \to 0$. This is the basis on which the term of the form (2.25), in the general expression (2.19), is called the dipole effect of the arbitrary source.

It is also useful to be acquainted with the *electric vector* field arising from an ideal dipole. One way to present it is in terms of spherical coordinates centered on the dipole and having the dipole moment vector as polar axis. Then the potential (2.25) can be written as

$$\phi_D = \frac{D \cos \vartheta}{r^2} \tag{2.26}$$

and the field components $\mathsf{E}_r \equiv -\partial\phi/\partial r$ and $\mathsf{E}_\vartheta \equiv -\partial\phi/(r\,\partial\vartheta)$ become

$$\mathsf{E}_r = \frac{2D \cos \vartheta}{r^3} \qquad \mathsf{E}_\vartheta = \frac{D \sin \vartheta}{r^3} \tag{2.27a}$$

The equivalent vector expression $\mathbf{E} = -\nabla\phi_D$ is also not difficult to calculate directly by using the fact that $\nabla(\mathbf{r} \cdot \mathbf{D}) = \mathbf{D}$ (Exercise A.8)

$$\mathbf{E} = \frac{3\hat{\mathbf{r}}(\hat{\mathbf{r}} \cdot \mathbf{D}) - \mathbf{D}}{r^3} = \frac{2\hat{\mathbf{r}}(\hat{\mathbf{r}} \cdot \mathbf{D}) + [\hat{\mathbf{r}} \times (\hat{\mathbf{r}} \times \mathbf{D})]}{r^3} \tag{2.27b}$$

with the last form following from the vector identity (A.53). The two vector terms of the last form correspond to the radial and transverse components shown in (2.27*a*).

The inquiry into dipole sources has made it evident what the largest effects are likely to be of a source distribution which is neutral in that its total charge vanishes ($q = 0$). If the equal positive and negative charge distributions are not exactly superposed, so that the positive and negative charge centers are relatively displaced, there will be a dipole effect. Whereas monopole effects, when they exist, fall off inversely as the square of the distance, that is, $\mathbf{E}_q = q/r^2$, dipole effects fall off as the cube of the distance, as the results (2.27) show. It is said that the higher-multipole force effects have the *shorter range* (whereas the monopole force is reduced by a factor one-fourth in a doubling of the distance, the dipole force is reduced to one-eighth).

It should still be explicitly pointed out that an analysis into multipole effects is defined only relative to some *chosen* center. Even a single point charge has higher (than monopole) force effects relative to a center not on the charge. Plainly, $\mathbf{D} = \oint dq\mathbf{r}_s \rightarrow q\mathbf{r}_s \neq 0$ for a point charge q at $\mathbf{r}_s \neq 0$. The choice of center is sometimes dictated by considerations other than those having to do with charge distributions alone; e.g., a point charge may simultaneously be subject to some *non*electric force center not coincident with the position of the charge. Notice, however, that the dipole moment of a *neutral* source is unchanged by shifts of the coordinate origin (Exercise 2.19); such a source has an *invariant* moment.

Electric Quadrupoles

A source may have equal amounts of positive and negative charge disposed so symmetrically that not only its monopole effect disappears ($q = 0$) but also its dipole effect, that is, $D = 0$, because its positive and negative charge centers coincide. It may still exhibit electrostatic effects through a nonvanishing quadrupole moment $\mathbf{Q}$.

An ideal quadrupole may be defined as having two equal and opposite ideal (point) dipoles $\mathbf{D}'$ and $\mathbf{D}'' = -\mathbf{D}'$ in positions which are relatively displaced by some $\mathbf{l} \rightarrow 0$. It is the involvement of two independent vector directions, of $\mathbf{D}' = -\mathbf{D}''$ and of $\mathbf{l}$, which makes it necessary to represent quadrupole effects with the help of second-rank *tensors* $\mathbf{Q}$. The entire configuration can be formed as a limit of one in which *four* equal point charges, two positive and two negative, are symmetrically disposed about some center; hence the name quadrupole (see Exercises 2.15 and 2.16).

Multipole Distributions

Just as the idealized point charges are generalizable to continuous charge-density distributions, so the ideal point multipoles can each be generalized to a continuous distribution of multipoles. For example, it will be found in Chap. D that certain dielectric materials can be represented as having a dipole moment $\mathbf{P}(\mathbf{r}_s)\,dV(\mathbf{r}_s)$ at each of a continuum of points $\mathbf{r}_s$, and so a *vector*, dipole-density, distribution $\mathbf{P}(\mathbf{r}_s)$ is defined. It will, by itself, give rise to the contributions [see (2.25)]

$$\phi_P(\mathbf{r}_f) = \int \frac{\hat{\mathbf{R}} \cdot \mathbf{P}(\mathbf{r}_s)\,dV(\mathbf{r}_s)}{R^2} \tag{2.28}$$

to the electrostatic potential when $\mathbf{R} = \hat{\mathbf{R}}R = \mathbf{r}_f - \mathbf{r}_s$, as before.

Some of the examples reviewed in the preceding sections dealt with sources which could be idealized as having charge densities σ per unit area distributed on surfaces of vanishing thickness, i.e., surface monopole distributions. Their contributions to a potential are

$$\phi_\sigma(\mathbf{r}_f) = \int \frac{\sigma\,dS}{R} \equiv \int \frac{\sigma(\mathbf{r}_S)\,dS_n(\mathbf{r}_S)}{|\mathbf{r}_f - \mathbf{r}_S|} \tag{2.29}$$

if $\sigma(\mathbf{r}_S) \gtrless 0$ at points of positive and negative charge, respectively. The last form has had the area element represented as the component of a vector simply to show the connection to the area-element vectors $d\mathbf{S}$, introduced earlier; they were defined as having directions normal to the surface, so that

$$dS_n \equiv \mathbf{n} \cdot d\mathbf{S} = dS \qquad \text{if } \mathbf{n} \equiv \frac{d\mathbf{S}}{dS} \tag{2.30}$$

is the unit vector in the direction chosen for $d\mathbf{S}$.

It was pointed out in connection with the example giving (2.5*a*) that the component of electric field normal to a charged surface should be expected to suffer a discontinuity there. This is true quite generally, as follows from the Gauss theorem, applied as suggested in Fig. 2.1. The rectangle indicates a *volume* of vanishing thickness, having dS as its inside and outside surface areas, at which the field may be $\mathbf{E}^i$ and $\mathbf{E}^o$, respectively. Then the flux out of the closed volume is

$$E_n^o\,dS - E_n^i\,dS = 4\pi\sigma\,dS \equiv 4\pi\sigma\mathbf{n} \cdot d\mathbf{S}$$

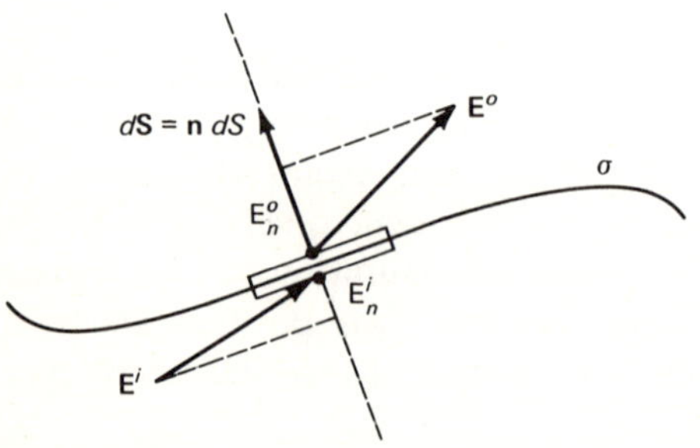

Figure 2.1

since the sides of the volume should be neglected for a surface of vanishing thickness. Thus

$$\Delta \mathbf{E}_n \equiv \mathbf{E}_n^o - \mathbf{E}_n^i \equiv \mathbf{n} \cdot \Delta \mathbf{E} = 4\pi\sigma \tag{2.31}$$

is the net efflux per unit area. An expression like $\mathbf{n} \cdot \Delta \mathbf{E} \equiv \mathbf{n} \cdot (\mathbf{E}^o - \mathbf{E}^i)$, where $\mathbf{n} = d\mathbf{S}/dS$, is called a *surface divergence* since it represents net flux per unit area at a point of a surface, just as $\nabla \cdot \mathbf{E}$ measures net flux per unit volume at a point in space.

It is also useful to represent some sources of field as having surface distributions of *dipole* moments, with

$$\tau(\mathbf{r}_S)\, d\mathbf{S}(\mathbf{r}_S) \tag{2.32}$$

the dipole moment *vector* contributed by a suitably chosen area element $d\mathbf{S}$. The area-density distribution $\tau(\mathbf{r}_S)$ may be positive or negative, but the vector is everywhere normal to the chosen surface. Here is an idealization of what are sometimes called double layers of charge, surfaces having equal and opposite charges on the two sides of each surface point $\mathbf{r}_S$, with a resultant dipole moment $\tau(\mathbf{r}_S)$ per unit area. Finding that the potential arising from (2.32) must have the discontinuities

$$\Delta\phi_\tau = 4\pi\tau(\mathbf{r}_S) \tag{2.33}$$

from one side of the surface to the other is left to Exercise 2.22. Plainly, $\Delta E_\tau = 0$ here since there is no *net* charge (monopole effect) to yield a net flux out of the surface elements.

2.4 ELECTROSTATIC FORCES AND ENERGIES

A survey of basic electrostatic theory is not complete before attention has been paid to its role as a theory of force, supplementing mechanics in the determination of some motions. The connections between the electrostatic field and mechanical concepts used in describing motion should be explored. The results obtainable at this point have a restricted range of validity (sometimes called the *nonrelativistic domain*), as will be found in later chapters, but they turn out to be completely adequate for a great variety of observations and should be considered first. (It can become absurd to take into account unobservably small corrections.)

Point Masses Subjected to Electrostatic Forces

The simplest case of motion subject to a fixed external field is that of a point charge q in a uniform electric field $\mathbf{E}_0$. The particle undergoes the uniform acceleration

$$\mathbf{a} = \frac{q\mathbf{E}_0}{m} \tag{2.34}$$

if its mass is m, and generally follows a parabolic orbit as under uniform gravitational acceleration. The particle can be attributed the potential energy

$$U = -q\mathbf{E}_0 \cdot \mathbf{r} \equiv q\phi(\mathbf{r}) \tag{2.35}$$

since the force $\mathbf{F} = q\mathbf{E}_0$ on it is derivable from this equation according to $\mathbf{F} = -\nabla U(\mathbf{r})$. The field $\mathbf{E}_0$ itself is derivable from the indicated electrostatic potential $\phi = -\mathbf{E}_0 \cdot \mathbf{r}$, which is a generalization (Exercise A.8) of (2.17) to an arbitrarily oriented frame, gauged by work from the frame origin wherever this may be put. The field will remain fixed as presumed here if its sources, perhaps the σ of (2.17), are massive enough to remain unperturbed by the motions of q.

The simplest case of a field internal to the moving system occurs for the relative motion of two point charges q_1 and q_2, a problem of fundamental interest to atomic physics. The Coulomb force $\mathbf{F} = \hat{\mathbf{r}} q_1 q_2 / r^2$ can be taken as the mutual force between the two particles, and this is derivable from the potential energy

$$U = \frac{q_1 q_2}{r} = q_1 \phi_2(r) = q_2 \phi_1(r) \tag{2.36}$$

where $\phi_{2(1)} = q_{2(1)}/r$ is the electrostatic potential at $q_{1(2)}$ due to $q_{2(1)}$. The orbits of the particles under the inverse-square law of force are well known† to be hyperbolas in general but may be closed elliptic orbits when the force is attractive and the relative kinetic energies are low enough to ensure that an escape velocity is never exceeded (then the particles are said to be bound together into an atom).

The potential energies (2.35) and (2.36) are examples of *coupling* or *interaction energies* between fields and particles. Such can be explored more systematically by first considering the work needed to construct any electrostatic system, as done in the next subsection.

The Energy of Electrostatic Systems

Setting up an electrostatic field requires work. A quite arbitrary field can be established by bringing together a suitable set of point charges $q_1, q_2, \ldots, q_i, \ldots$ into appropriate positions. Bringing the first two charges into definite positions a distance r_{12} apart entails the work $W_{12} = q_1 q_2 / r_{12}$. This may be positive or negative (in the latter case, whatever agency is responsible for stopping the attracting charges at the separation r_{12} has gained energy in the process). Bringing up the third charge now involves the further work $W_{13} + W_{23} = q_1 q_3 / r_{13} + q_2 q_3 / r_{23}$. Plainly the total work of thus establishing the system can be represented as

$$W = \sum_{i>j} \frac{q_i q_j}{r_{ij}} = \frac{1}{2} \sum_{i \neq j} \frac{q_i q_j}{r_{ij}} \tag{2.37}$$

† See, for example, Ref. 17, secs. 4.3 and 4.4.

In the first summation here there is a term for each *pair* of charges, whereas for the last expression both the summation indices i, j range over all their unequal values, so that each pair is counted twice. This permits rewriting the expression as

$$W = \frac{1}{2} \sum_i q_i \phi_i \qquad \text{with } \phi_i = \sum_{j \neq i} \frac{q_j}{r_{ij}} \tag{2.38}$$

the electrostatic potential at q_i arising from all its fellow charges.

Expressions (2.37) and (2.38) do not include any work that might be needed to construct each of the individual charges by itself; the charges were taken to be preexistent. All the work of constructing the electrostatic system is taken into account when the total charge is treated as being made up of *vanishingly* small elements $\rho(\mathbf{r})\, dV(\mathbf{r})$. Then (2.38) should be replaced by

$$W = \frac{1}{2} \oint dV(\mathbf{r}) \rho(\mathbf{r}) \phi(\mathbf{r}) \tag{2.39}$$

and (2.37) by

$$W = \frac{1}{2} \oint\oint dV(\mathbf{r})\, dV(\mathbf{r}') \frac{\rho(\mathbf{r})\rho(\mathbf{r}')}{|\mathbf{r} - \mathbf{r}'|} \tag{2.40}$$

which is just (2.39) with $\phi(\mathbf{r})$ replaced by the integral over $\mathbf{r}'$. That work of constructing charges themselves is now included can be seen after dividing the entire charge distribution into two separate parts, as described by $\rho(\mathbf{r}) = \rho_1(\mathbf{r}) + \rho_2(\mathbf{r})$, so that

$$q_1 = \oint dV \rho_1(\mathbf{r}) \qquad \text{and} \qquad q_2 = \oint dV \rho_2(\mathbf{r}) \tag{2.41}$$

Then the total energy (2.40) of the system can be written

$$W = \frac{1}{2} \oint\oint dV(\mathbf{r})\, dV(\mathbf{r}') \frac{[\rho_1(\mathbf{r}) + \rho_2(\mathbf{r})][\rho_1(\mathbf{r}') + \rho_2(\mathbf{r}')]}{|\mathbf{r} - \mathbf{r}'|}$$

$$= W_1 + W_2 + U_{12} \tag{2.42}$$

where

$$W_1 = \frac{1}{2} \oint\oint dV(\mathbf{r})\, dV(\mathbf{r}') \frac{\rho_1(\mathbf{r})\rho_1(\mathbf{r}')}{|\mathbf{r} - \mathbf{r}'|} \tag{2.43}$$

and W_2 is the same except that ρ_2 replaces ρ_1 for it. Clearly W_1 and W_2 respectively represent the energies of the individual systems of charges q_1 and q_2, each considered in isolation, on the same basis that W represented the energy of the entire charge $q = q_1 + q_2$. The remainder

$$U_{12} = \oint\oint dV(\mathbf{r})\, dV(\mathbf{r}') \frac{\rho_1(\mathbf{r})\rho_2(\mathbf{r}')}{|\mathbf{r} - \mathbf{r}'|} \tag{2.44}$$

is called the *interaction energy* between the charges q_1 and q_2; expression (2.36) is a specialization of (2.44) to a pair of *point* charges. Since the electrostatic potentials generated by the respective charges are

$$\phi_{1,2}(\mathbf{r}) = \oint dV(\mathbf{r}') \frac{\rho_{1,2}(\mathbf{r}')}{|\mathbf{r} - \mathbf{r}'|} \tag{2.45}$$

their interaction energy can also be written

$$U_{12} = \oint dV(\mathbf{r})\rho_1(\mathbf{r})\phi_2(\mathbf{r}) = \oint dV(\mathbf{r})\rho_2(\mathbf{r})\phi_1(\mathbf{r}) \tag{2.46}$$

comparable to the point-charge expressions in (2.36).

The significance of these results can best be appreciated if the energy of a system consisting of finite uniformly charged *spheres* is considered, in place of the point charges for which the evaluations (2.37) and (2.38) were performed. Suppose that none of the spheres overlap in the final configuration. Then the electrostatic potential at the center of the sphere of charge q_i due to the charge q_j is just $\phi_j = q_j/r_{ij}$ if r_{ij} is the distance between centers, exactly as if q_j were a point charge. Then the symmetries make it clear that the interaction energy contributed by the pair q_i, q_j will be just $U_{ij} = q_i q_j/r_{ij}$, equal to the quantity called W_{ij} in the development of the point-charge system. (This conclusion is borne out by the more detailed evaluation left to Exercise 2.26.) Now, however, the evaluation of the total energy of the system requires adding the so-called self-energies W_i, W_j, ... of the type (2.43), contributed by each individual charge q_i, q_j, ... If ϵ_i is the radius of the sphere with q_i, then

$$W_i = \frac{1}{2} \oint d\Omega \int_0^{\epsilon_i} d\epsilon \epsilon^2 \rho_i(\epsilon)\phi_i(\epsilon)$$

where

$$\phi_i(\epsilon) = \frac{q_i}{2\epsilon_i^3}(3\epsilon_i^2 - \epsilon^2)$$

is the potential at the radius ϵ inside the sphere due to the entire sphere of charge, according to (2.14). Since $\rho_i = q_i/\frac{4}{3}\pi\epsilon_i^3$ for the uniform charge distribution, the integration is easy to bring to the conclusion

$$W_i = \frac{3}{5}\frac{q_i^2}{\epsilon_i} \tag{2.47}$$

Thus, the total energy of the system of spherical charges consists not only of the interaction energies (2.37) or (2.38) but also of self-energies

$$\frac{3}{5}\sum_i \frac{q_i^2}{\epsilon_i}$$

that arise from internal interactions between the elements making up each sphere. It is clear that such internal self-energies were not counted in the expressions (2.37) and (2.38) for the point-charge system but are included in the more general energy expressions (2.39) and (2.40).

Evaluating the self-energies of point charges presents the difficulty that each is infinite, as can be seen by letting the radius ϵ_i of the spherical charge in (2.47) approach zero. It would take an infinite amount of energy to squeeze any finite amount of charge into a vanishingly small radius against the enormous mutual repulsions of charge elements a vanishing distance apart. The conclusion might be that ideal point charges cannot be constructed—that the concept is an over-idealization—a fact that continues to cause difficulties for theories of the so-called *elementary particles*. However, it does not prevent using point charges as models when dealing with interactions between separate ones as long as each self-energy can be regarded as being conserved at a constant value, no matter how large.

The separation of a system into parts, each having a conserved charge q_1 or q_2 or ..., is advantageous when dealing with processes during which the self-energies W_1 or W_2 or ... of the individual parts can be held constant and only their interaction energies, like U_{12}, are drawn upon in the processes. A simple example is provided by the two point-charge particles of (2.36); the process in this case is a motion during which exchanges occur between the interaction energy and the kinetic energy of the relative motion. It may be appropriate to refer to any of the partial charge distributions ρ_1 or ρ_2 or ... as a particle—an extended one rather than a point particle—for processes in which the individual charge and internal self-energy remain conserved. Of course, the internal electrostatic interactions between the charge elements that constitute an extended particle can continue to be identified with its constant self-energy, as in (2.43), only if those elements of charge remain in rigid relative positions whenever the center of mass of the whole particle moves.

In the motion of the two point particles of (2.36), their interaction energy $U(r) = q_1 q_2/r$ is drawn upon through serving as the potential energy from which the Coulomb force between them is derivable, as $\mathbf{F} = -\nabla U$. Motions of the centers of mass of extended particles can similarly draw on their interaction energies, which serve as potential energies from which forces on the centers of mass are derivable. However, only among such spherical particles as were discussed above and only while these remain external to each other in space will those forces have the simple Coulomb form, inversely proportional to the square of the distance between centers. A resultant force

$$\mathbf{F}_{12} = \oint\oint dV(\mathbf{r})\, dV(\mathbf{r}')\rho_1(\mathbf{r})\rho_2(\mathbf{r}')\frac{\mathbf{r}-\mathbf{r}'}{|\mathbf{r}-\mathbf{r}'|^3} = -\mathbf{F}_{21} \tag{2.48}$$

is to be anticipated between static charge distributions with the interaction energy U_{12} of (2.44), and this does not reduce to an inverse-square law of force between just two points, in general. Other forms of force that are derivable from the interaction energies will be explored in the next subsection.

Multipole Interactions

First consider a single finitely extended particle of localized charge $q = \oint dV(\mathbf{s})\rho(\mathbf{s})$, interacting with an external electrostatic field described by a potential distribution ϕ. [The situation can be fitted into the considerations of the preceding subsection by identifying q with a part q_1 of a system of charges consisting, in addition, of the sources q_2 of $\phi \equiv \phi_2$. Then the potential distribution ϕ can be fixed in space if its sources are so massive as to be immovable, despite perturbations by possible motions of $q \equiv q_1$. Compare the similar consideration about the sources of the uniform field in (2.35).] Let $\mathbf{r}$ be the center-of-mass (CM) position of the localized charge q and $\mathbf{s}$ a position of any one of its elements $dq \equiv \rho(\mathbf{s})\, dV(\mathbf{s})$ relative to the center of mass as origin. Then the interaction energy of q with the external field (or its sources) is calculable as in (2.46):

$$U = \oint dV(\mathbf{s})\rho(\mathbf{s})\phi(\mathbf{r}+\mathbf{s}) = U(\mathbf{r}) \tag{2.49}$$

More detailed conclusions become possible if the variations of ϕ across q from the CM value $\phi(\mathbf{r})$ are represented by a Taylor expansion [according to the generalization to three component variables of an expression given in the footnote to Eq. (2.18)]

$$\phi(\mathbf{r}+\mathbf{s}) = \phi(\mathbf{r}) + \mathbf{s}\cdot\nabla\phi(\mathbf{r}) + \frac{1}{2}\sum_{i,j} s_i s_j \frac{\partial^2\phi}{\partial x_i\,\partial x_j} + \cdots \tag{2.50}$$

where $i, j = 1, 2, 3$ and $s_{1,2,3} \equiv s_{x,y,z}$, $x_{1,2,3} \equiv x, y, z$. Substitution into the integral (2.49) for $U(\mathbf{r})$ now yields

$$U = q\phi(\mathbf{r}) + \nabla\phi\cdot\oint dq\mathbf{s} + \frac{1}{2}\sum_{i,j}\frac{\partial^2\phi}{\partial x_i\,\partial x_j}\oint dq s_i s_j + \cdots \tag{2.51}$$

Comparison with the definition (2.20) shows that the integral $\oint dq\mathbf{s} \equiv \mathbf{D}$ is just the dipole moment of the charge distribution constituting the localized particle. Similarly, comparison with the definitions (2.21) of quadrupole moments shows that

$$\oint dq s_i s_j = \tfrac{1}{3}Q_{ij} + \tfrac{1}{3}\delta_{ij}\oint dq s^2$$

Substitution of the last part of this into the summations occurring in the expression (2.50) has a result proportional to

$$\sum_{i,j}\frac{\partial^2\phi}{\partial x_i\,\partial x_j}\delta_{ij} = \sum_i \frac{\partial^2\phi}{\partial x_i^2} \equiv \nabla^2\phi = 0$$

vanishing as long as the particle remains outside the sources of ϕ. Finally, it can be recognized that $\mathsf{E} = -\nabla\phi$ is just the external vector field incident on the particle, and so the interaction energy (2.51) can be given the expression

$$U(\mathbf{r}) = q\phi(\mathbf{r}) - \mathbf{D}\cdot\mathsf{E} - \frac{1}{6}\sum_{i,j} Q_{ij}\frac{\partial \mathsf{E}_i}{\partial x_j} + \cdots \tag{2.52}$$

The last summation is invariant to an interchange of the indices i, j since the quadrupole moment tensors are always symmetric: $Q_{ji} = Q_{ij}$. The significance of each of the terms shown in (2.52) will be discussed next.

The first term, $U_q(\mathbf{r}) \equiv q\phi$, is just the interaction energy between field and particle in a monopole approximation of the particle, i.e., treating it as if all its charge were concentrated at a point $\mathbf{r}$. It leads to an acceleration of this point by the force $\mathbf{F}_q = -\nabla U_q = q\mathbf{E}(\mathbf{r})$; the case (2.35) of a point charge in a uniform field provides the simplest example.

As analyzed in terms of the CM position $\mathbf{r}$, the energy must be corrected by the dipole interaction with the applied field

$$U_D(\mathbf{r}) = -\mathbf{D} \cdot \mathbf{E}(\mathbf{r}) \tag{2.53}$$

whenever the localized particle possesses a dipole moment relative to the center of mass. The dipole moment may arise because q is a surplus over equal positive and negative charges within the particle that have noncoincident charge centers, or it may merely mean that the center of the net charge distribution $\mathbf{D}/q$ does not coincide with the center of mass. In any case the center of mass is accelerated by the force

$$\mathbf{F}_D = -\nabla[-\mathbf{D} \cdot \mathbf{E}(\mathbf{r})] = \mathbf{D} \times (\nabla \times \mathbf{E}) + (\mathbf{D} \cdot \nabla)\mathbf{E}(\mathbf{r}) = (\mathbf{D} \cdot \nabla)\mathbf{E}(\mathbf{r}) \tag{2.54}$$

since an electrostatic field is curlless (1.23). The center of mass of the particle is subject to this force, by virtue of its possession of a dipole moment, only in a nonuniform field. A dipole always behaves like a pair of equal and opposite point charges some distance apart, and it is easy to understand that there will be a resultant force at the center if the field $\mathbf{E}(\mathbf{r})$ is different at the sites of the positive and negative charges.

Even in a uniform field $\mathbf{E} \equiv \mathbf{E}_o$, when there is no resultant force like (2.54), there is still the contribution $U_D = -\mathbf{D} \cdot \mathbf{E}_o$ to the interaction energy. This is because even a uniform field can serve to *reorient* the direction of $\mathbf{D}$ while maintaining consistency with the rigid internal charge distribution the particle must have for a constant electrostatic internal (self-) energy (only the interaction energy with the external field is being drawn upon in these considerations). The reorientations are the result of a torque about the center of mass, or couple,

$$\boldsymbol{\Theta} \equiv \oint dq\mathbf{s} \times \mathbf{E}_o = \mathbf{D} \times \mathbf{E}_o \tag{2.55}$$

that acts on the particle whenever it has a dipole moment relative to the center of mass. The situation can be understood from the idealized picture presented in Fig. 2.2, in terms of which there is a torque about the dipole center having the magnitude

$$\Theta = (q_D \mathrm{E}_o)\tfrac{1}{2}l(\sin \vartheta)2 = |q_D \mathbf{l} \times \mathbf{E}_o|$$

and a direction tending to align the dipole moment vector $\mathbf{D} \equiv q_D \mathbf{l}$ with the imposed field $\mathbf{E}_o$. If any agency were to force an angular increase $\delta\vartheta$, it would have to do work $\delta W = q_D \mathrm{E}_o(\tfrac{1}{2}l\, \delta\vartheta)(\sin \vartheta)2 = |\mathbf{D} \times \mathbf{E}_o|\, \delta\theta$ against that torque; it would thus increase the potential energy of the system by $\delta U_D = \delta W = D\mathrm{E}_o \sin \vartheta \delta\vartheta = -D\mathrm{E}_o\, \delta(\cos \vartheta) = -\delta(\mathbf{D} \cdot \mathbf{E}_o)$, exactly as expected from $U_D = -\mathbf{D} \cdot \mathbf{E}_o$.

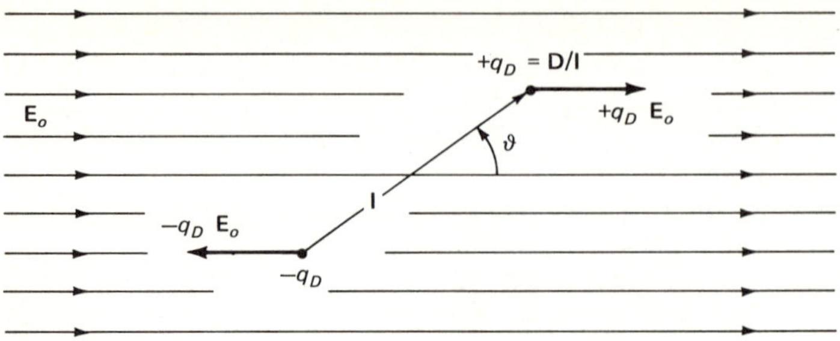

Figure 2.2

The sum $U_q + U_D = -q(\mathbf{r} \cdot \mathbf{E}_0) - \mathbf{D} \cdot \mathbf{E}_0$ is the entire interaction energy of the particle with a *uniform* field, since all higher-order multipole moments, like the quadrupole moment in (2.52), contribute only in a nonuniform field because moments of higher order than the dipole are closer to isotropy.

Even a point particle may require more than just the monopole term of (2.52) for describing its interactions with an electrostatic field. The point particle may be endowed with "ideal" (point) multipole moments of the type mentioned in Sec. 2.3.

The electrostatic interactions between *two* particles having complex internal structures can be obtained from the potential-energy expression (2.52) by substituting for the field in it the field $\mathbf{E}_2 = -\nabla\phi_2$ arising from the second localized particle $q_2 \equiv \oint dV(\mathbf{s}_2)\rho_2(\mathbf{s}_2)$. With this source of finite extent, the field ϕ_2 can be analyzed into its multipole effects according to (2.19), in terms of moments q_2, $\mathbf{D}_2$, $\mathbf{Q}_2$, ... ascribed to the second particle. It also becomes appropriate now to denote the moments of the first particle: q_1, $\mathbf{D}_1$, $\mathbf{Q}_1$, ... The result (2.19) for

$$\phi_2(\mathbf{r}) = \frac{q_2}{r} + \frac{\hat{\mathbf{r}} \cdot \mathbf{D}_2}{r^2} + \frac{\hat{\mathbf{r}} \cdot \mathbf{Q}_2 \cdot \hat{\mathbf{r}}}{2r^3} + \cdots$$

permits systematic expansion of the interaction $U(\mathbf{r})$ only through terms of order $1/r^3$, so that only

$$\mathbf{E}_2 = -\nabla\phi_2 = \frac{q_2}{r^2}\hat{\mathbf{r}} + \frac{3\hat{\mathbf{r}}(\hat{\mathbf{r}} \cdot \mathbf{D}_2) - \mathbf{D}_2}{r^3} + \cdots$$

obtainable with the help of the dipole field (2.27), and

$$\frac{\partial^2\phi_2}{\partial x_i\,\partial x_j} = \frac{\partial}{\partial x_i}\left(-\frac{q_2}{r^2}\frac{x_j}{r}\right) = q_2\left(\frac{3x_i x_j}{r^5} - \frac{\delta_{ij}}{r^3}\right) \tag{2.56}$$

are to be used. Substitution into $U(\mathbf{r})$ of (2.52) can now be found to yield

$$U = \frac{q_1 q_2}{r} + \frac{\hat{\mathbf{r}} \cdot (q_1\mathbf{D}_2 - q_2\mathbf{D}_1)}{r^2} + \frac{\hat{\mathbf{r}} \cdot (q_1\mathbf{Q}_2 + q_2\mathbf{Q}_1) \cdot \hat{\mathbf{r}}}{2r^3} + \frac{\mathbf{D}_1 \cdot \mathbf{D}_2 - 3(\hat{\mathbf{r}} \cdot \mathbf{D}_1)(\hat{\mathbf{r}} \cdot \mathbf{D}_2)}{r^3} + \cdots \tag{2.57}$$

Terms proportional to the Kronecker delta in (2.56)

$$-\frac{q_2}{3r^3}\sum_{i,j} Q_{ij}\delta_{ij} = -\frac{q_2}{3r^3}\sum_i Q_{ii} = 0 \tag{2.58}$$

have vanished because quadrupole moments have vanishing *traces*, $\sum_i Q_{ii} = Q_{xx} + Q_{yy} + Q_{zz} = 0$, as is readily confirmed from the definition (2.21). The two monopole-dipole interaction terms are actually symmetric in the two particles, despite the opposite signs, because the direction $\hat{\mathbf{r}}$ is defined antisymmetrically, as that of $\mathbf{r}_1 - \mathbf{r}_2$ (from q_2 to q_1). The first term is just the simple Coulomb interaction between q_1 and q_2 treated as point charges, the monopole approximation of both. All the terms in the first line of (2.57) come from treating at least one of the particles in the monopole approximation and the other as presenting a potential field in the form (2.19). The last term in (2.57), known as the *dipole-dipole interaction*, is the largest one whenever the others shown disappear. That happens, for example, in the interaction of certain *neutral* ($q_1 = q_2 = 0$ but $\mathbf{D}_{1,2} \neq 0$) molecules.

The Electrostatic Field Energy

The energy invested in constructing an electrostatic system was given in (2.39) as an integral over the parts of space containing material charge, with $\rho \neq 0$, the energy thus being treated as residing in the sources of the field. It will now be shown that it can equally well be considered to be spread out over the entire field, outside as well as inside the sources. For this purpose, it must be recognized that the charges exist at points of divergence of the field: $\rho = \nabla \cdot \mathbf{E}/4\pi$. Then (2.39) can be rewritten as

$$\begin{aligned} W &= \frac{1}{8\pi}\oint dV(\mathbf{r})\phi(\nabla \cdot \mathbf{E}) \\ &= \frac{1}{8\pi}\oint dV[\nabla \cdot \phi\mathbf{E} - (\mathbf{E} \cdot \nabla)\phi] \end{aligned} \tag{2.59}$$

Since the first of the volume integrals in the last line has only a divergence in its integrand, it can be expressed as an integral over the enclosing surface, according to the Gauss theorem. The surface is a very remote one that can be put completely outside the field whenever the energy of an *entire* field is being evaluated and hence the surface integral vanishes. There remains only an integral which, after a substitution of $\mathbf{E} = -\nabla\phi$, can be written

$$W = \oint dV(\mathbf{r})\frac{\mathbf{E}^2(\mathbf{r})}{8\pi} \tag{2.60}$$

Here the energy is evaluated as a sum of contributions by every element of space in which field $\mathbf{E}$ exists. It is being treated as a possession of the field, rather than its sources, existing in the density

$$w(\mathbf{r}) = \frac{\mathbf{E}^2(\mathbf{r})}{8\pi} \tag{2.61}$$

at the point $\mathbf{r}$.

Consider the implications for a system consisting of just one of the uniform spheres of charge for which (2.47) was calculated; that gave the energy $W = \frac{3}{5}q^2/a$ for a sphere of charge q and radius a when calculated as resident in the charge. The field inside the sphere is just $\mathbf{E}(r \le a) = qr/a^3$, according to (2.6), and when the field energy density (2.61) $w = q^2r^2/8\pi a^6$ is integrated over the interior of the sphere ($r \le a$), the result is $W_{\text{in}} = \frac{1}{6}W$ only. Outside the sphere, the field energy exists in the density $w(\mathbf{r}) = q^2/8\pi r^4$, and its integral yields exactly the remaining five-sixths of the total energy W. This energy in the outer part of the field $W_{\text{out}} = q^2/2a$ is the entire energy of the system when all the charge is uniformly concentrated at the surface of the sphere, for then $\mathbf{E} = 0$ inside. The conclusion checks with the result calculable from (2.39): $W = \frac{1}{2}\oint dq\phi = \frac{1}{2}q\phi(r = a)$ and $\phi(r = a) = q/a$.

The phenomenon of the energy being spread out over the field might be regarded as the result of coming to a stable equilibrium, in the process of which the energy has been propagated out of the source as far as its influence reaches.

The analysis into pairs of interacting systems that was introduced in (2.42) by splitting the sources as $\rho = \rho_1 + \rho_2$ can now be implemented through a corresponding splitting of the field $\mathbf{E} = \mathbf{E}_1 + \mathbf{E}_2$. This separates the total-energy expression (2.60) into the self-energies

$$W_1 = \oint dV \frac{\mathbf{E}_1^2}{8\pi} \quad \text{and} \quad W_2 = \oint dV \frac{\mathbf{E}_2^2}{8\pi} \tag{2.62}$$

and the interaction energy

$$U_{12} = \oint dV \frac{\mathbf{E}_1 \cdot \mathbf{E}_2}{4\pi} \tag{2.63}$$

The same expressions can be obtained directly from the forms (2.43) and (2.46) of these energies, in the same way that (2.60) was obtained from (2.39), after discarding integrals over surfaces enclosing the entire fields. Notice that the interaction, or coupling, energy density at a point

$$w_{12}(\mathbf{r}) = \frac{\mathbf{E}_1 \cdot \mathbf{E}_2}{4\pi} \tag{2.64}$$

may be positive or negative, and so can U_{12} in the forms (2.44) or (2.36). However, the total energy density of the composite system

$$w = w_1 + w_2 + w_{12} \tag{2.65}$$

is always positive definite since $\mathbf{E}_1^2 + \mathbf{E}_2^2 \ge 2|\mathbf{E}_1 \cdot \mathbf{E}_2|$ (a so-called *Schwarz inequality*).

Such forms as (2.64) for the interaction energy densities of distinct fields are known as *contact interactions*, depending as they do upon the values of the fields at a common point. Contact interactions play an important role in the field-theoretic formulations of all kinds of systems (including mesonic, baryonic, and leptonic ones).

The findings in this section lead to a fundamental physical interpretation of electrostatic potential distributions $\phi(\mathbf{r})$ that lends them quite as substantial a reality of presence as is often accorded only to $\mathbf{E}(\mathbf{r})$. Just as $\mathbf{E}(\mathbf{r})$ gains its primary physical meaning from the detectability of a force $q\mathbf{E}(\mathbf{r}_q)$ whenever a test charge q is put at any point $\mathbf{r}_q$ in the field, so the corresponding potential energy of the test charge $q\phi(\mathbf{r}_q)$ can be used for a primary, operational definition of $\phi(\mathbf{r})$.

What the findings in this section add is that the potential energy is actually present in the form of an electrostatic *field energy*. This follows from considering that no test charge can be put into a preexistent field $\mathbf{E} = -\nabla\phi$ without altering the field energy that is present, a fact essential to any detectability of a field even if the test procedure does not alter that field itself appreciably.

A system composed of a point charge q at a fixed position $\mathbf{r}_q$ within a field due to sources external to q has a resultant that can be described as $\mathbf{E}(\mathbf{r}) + \mathbf{E}_q(\mathbf{r} - \mathbf{r}_q)$, where $\mathbf{E} = -\nabla\phi$ and $\mathbf{E}_q$ is just the Coulomb field of the point charge. The *total* field energy then present can be analyzed into parts $W + W_q + U(\mathbf{r}_q)$, as in (2.62) and (2.63). Here, W is the self-energy of $\mathbf{E} = -\nabla\phi$ itself, independent of $\mathbf{r}_q$ and to be left untouched in a proper test procedure. W_q is a Coulomb field energy, "attached" to q and invariant in amount. $U(\mathbf{r}_q)$ is the interaction energy arising from the interference (2.63) of $\mathbf{E}(\mathbf{r})$ with q as represented by its self-field $\mathbf{E}_q(\mathbf{r} - \mathbf{r}_q)$. It is the part of the field energy present that changes with $\mathbf{r}_q$ and is drawn upon when q is disturbed, as in any test procedure. It is also the part that forms the potential energy

$$U(\mathbf{r}_q) = q\phi(\mathbf{r}_q) \tag{2.66}$$

the version† of (2.46) adapted to a single point of charge at $\mathbf{r}_q$. This field energy exists only after the work of introducing q has been done, although it is determined by a $\phi(\mathbf{r})$ that describes the preexistent field alone.

The outcome is a physical meaning for $\phi(\mathbf{r})$ quite as concrete as that of $\mathbf{E}(\mathbf{r})$. Either represents the field equally well since either is derivable from the other, as $\mathbf{E} = -\nabla\phi$ and

$$\phi = -\int_{\mathbf{r}_o}^{\mathbf{r}} d\mathbf{r} \cdot \mathbf{E}(\mathbf{r}) \tag{2.67}$$

the version of (1.29) gauged relative to some $\phi(\mathbf{r}_o) = 0$. Each describes the field through a property of it. Whereas $\mathbf{E}(\mathbf{r})$ determines a force per unit charge that the field can exert on charged matter, $\phi(\mathbf{r})$ describes instead the energy per unit charge it can transfer to the charged matter.

† How the result follows directly from the expression (2.63) instead is the subject of Exercise 3.2.

The detailed emphasis being put here on a physical meaning of $\phi(\mathbf{r})$ seems required because it is still quite often held that potentials are no more than mathematical conveniences, useful in solving the Maxwell equations for $\mathbf{E}(\mathbf{r})$. This in spite of the fact that every observation of $\mathbf{E}(\mathbf{r})$ can be interpreted equally well as a detection of a nonuniformity in a potential distribution $\phi(\mathbf{r})$. Indeed, electrostatic fields are perhaps most often mapped by obtaining $\phi(\mathbf{r})$ directly, as from detecting currents like those of (1.30) connecting probes with some fixed gauge point $\mathbf{r}_o$.

It is perhaps the variety of choices open for representing a potential—the arbitrariness with which $\mathbf{r}_o$ can be chosen and the potential gauged—that leads some to deny it as concrete an existence as that of $\mathbf{E}(\mathbf{r})$. However, quantities that require some kind of reference point for their expression are common in physical descriptions. Properly, one should discriminate between a physical observable and how one chooses to represent it; the "word" is not the "thing," as the semanticists insist.

This chapter reviewed the features of electrostatics most essential to the general electromagnetic field theory, principally the vector and scalar field descriptions and their relation to charges, both as sources and as subjects of force and work. The charges have been treated as practically disembodied. Chapters B and D pay more attention to the fact that charge actually resides in matter. They are designed to be read after Chap. 3, which will continue with developments of electrostatic formulations from which much can be learned about techniques for dealing with fields in general.

EXERCISES

2.1 (*a*) Obtain the result (2.2*a*) by integrating (1.5)

(*b*) Show how (2.2*b*) can be used to find the field anywhere in the free space of a plane that contains an infinitely long flat strip of charge λ per unit length and distributed uniformly within the breadth a of the strip. Is your result correct for $a \to 0$?

2.2 An appreciation of the superposition principle as it applies to vector fields can be gained from evaluating the integral (1.5) in the following sequence of cases. Obtain:

(*a*) The result (2.3) for the charged ring

(*b*) The field (2.4) by treating a disk as a superposition of rings

(*c*) The field inside and outside a uniformly charged spherical surface by treating it as a superposition of rings

(*d*) The field inside and outside a homogeneously charged spherical volume by treating it as a superposition of disks and also as a superposition of spherical shells

2.3 Show how results of Exercise 2.2 enable one to put down expressions for the field $\mathbf{E}$ everywhere on the axis of a uniformly charged spherical shell when the extension of its equatorial plane everywhere *outside* the sphere bears the same uniform density of charge. Check your results against your expectations for the field infinitely far away and also at the center of the sphere.

2.4 Suppose $\mathbf{E}$ radial lines per unit area are used to represent the field outside an homogeneously charged sphere and that just those lines are continued inward to the sphere center. They then give the directions of the inner field correctly, but how does their density at any radius inside the sphere compare to the magnitude of $\mathbf{E}$ there? (Note the relevance to remarks attending Fig. A.2.)

2.5 If **E** is the field at a distance l from a uniformly charged plane, within what radius is contained the *nearest* charge that can account for $\frac{1}{2}\mathbf{E}$? What happens to that radius when the inquiry is about the $\frac{1}{2}\mathbf{E}$ at $l \to 0$? (Compare Fig. B.3.)

2.6 A uniformly charged *disk* of negative charge is encircled by a coplanar ring of positive charge just sufficient to neutralize the field at a point above the disk's center by a distance equal to its radius. What is the least ratio of the positive to the negative charge that can do this? *Ans.* 1.656 ....

2.7 (*a*) Obtain the result (2.9) in appropriate cylindrical coordinates.

(*b*) Use your result (*a*) to find the field $\mathbf{E}(s = 0,\ z > \frac{1}{2}L)$ on the same line as that of the charge but beyond its end.

(*c*) Show how (2.10) reduces to (2.15) as $L \to \infty$.

2.8 An appreciation of the superposition principle as it applies to scalar fields can be gained from evaluating integral (1.26) in the following sequence.

(*a*) Obtain (2.12) by treating the disk as a superposition of such rings as yield (2.11*b*).

(*b*) Obtain (2.13) by treating the spherical shell as a superposition of rings.

(*c*) Obtain (2.14) by treating the sphere both as a superposition of disks and as a superposition of spherical shells.

2.9 Show how the results of Exercise 2.8 enable you to set down expressions for the potential at the field points of Exercise 2.3. Check your results against those of the earlier exercise.

2.10 (*a*) Find an approximation of (2.11*a*) valid for $\phi(r \gg a,\ \vartheta)$ through terms proportional to a^2/r^3 by first forming an appropriate power-series expansion of the integrand and then integrating.

(*b*) Obtain an approximation for $\phi(r \ll a,\ \vartheta)$ through terms proportional to r^2/a^3.

2.11 Charge is distributed on a plane in a surface density

$$\sigma(s) = \frac{-ql}{2\pi(s^2 + l^2)^{3/2}}$$

where s is the distance from a central point on the plane and l is a length parameter associated with the spatial concentration of the charge.

(*a*) What is the total charge on the plane?

(*b*) Find the potential $\phi(z \gtrless 0)$ everywhere on the *axis* of the cylindrically symmetric distribution.

(*c*) Specify the magnitude and position of a point charge that could give the same potential as (*b*) on one side of the plane and another point charge for the potential on the remaining side.

(*d*) Assume that the potential $\phi(s,\ z > 0)$ *everywhere* on one side of the plane can be represented by the appropriate point charge of (*c*), and similarly for $\phi(s,\ z < 0)$. Show that these representations are correct because they give the right *net* surface divergences at $z = 0$.

2.12 The centrosymmetric potential distributions

$$\phi(r) = \frac{q}{r} e^{-2r/a}$$

can exist with various degrees of concentration as measured by the length parameter a. Find charge distributions $\rho(r)$ that can produce them. Are your results consistent with what you must expect when $a \to \infty$? What is the total charge for any $a < \infty$?

2.13 Unlimited numbers of ions with mass M and charge q can enter through a plane grid at $x = 0$ into a space in which they are accelerated toward a plate at $x = l$ by a potential difference ϕ_o. They form a steady and uniform current density j across the space, but their charge density $\rho(x) = j/u(x)$ is nonuniform because of space-charge limitations [representable through an equilibrium condition $\mathbf{E}(x = 0) = 0$]. Find $u(x)$ in terms of $\phi(x)$ and solve the appropriate Poisson equation for $\phi(0 < x < l)$. Show that *Childs' equation*

$$j = \frac{\phi_o^{3/2}}{9\pi l^2}\left(\frac{2q}{M}\right)^{1/2}$$

for the space-charge-limited current is a result.

2.14 Two coaxial, uniformly charged disks, each of radius a and bearing total charges $\pm q$, respectively, are a distance L apart.

(*a*) Find ϕ and $\mathbf{E}$ everywhere on the common (z) axis of the disks.

(*b*) Show that for $z \to \pm\infty$, the fields approach an ideal dipole form.

(*c*) What is the dipole moment of the system?

2.15 Form a linear quadrupole having a point charge q at each end of a length l and a charge $-2q$ in the middle of it. Find the resultant quadrupole field distribution $\mathbf{E}_Q(r, \vartheta)$ in terms of appropriate spherical coordinates. On what cones coaxial with the line of charges is there a radial field component only, and on what cones does this component vanish?

2.16 Form a square quadrupole of side l having point charges q at the two ends of one diagonal and charges $-q$ at the remaining corners. Evaluate the components of the quadrupole moment (*a*) on x, y, z axes parallel to the sides of the square and (*b*) on axes that pass through the charges.

(*c*) Find the general relation between cartesian quadrupole components Q_{ij} and Q'_{ij} referred to frames that are relatively rotated by some arbitrary angle α about a common $z \equiv z'$ axis.

(*d*) Find the field $\mathbf{E}_Q(r, \vartheta)$ of the square quadrupole in terms of appropriately chosen spherical coordinates.

2.17 (*a*) Find the dipole and quadrupole moments of a uniformly charged ring relative to the ring center.

(*b*) Compare your results for the dipole and quadrupole effects with those in Exercise 2.10(*a*).

2.18 Find the dipole and quadrupole effects outside a sphere of charge (radius a) containing the distribution

$$\rho(z) = \frac{4qz}{\pi a^4}$$

at $-a < z < a$ relative to an equatorial plane.

2.19 (*a*) Show that a dipole moment of a neutral (zero net charge) distribution is invariant to the center relative to which it is evaluated.

(*b*) Show that if the neutral distribution also has no dipole moment, its quadrupole moment is invariant to displacements of the reference center.

2.20 An indefinitely long line of uniformly distributed and oriented dipole moment $\boldsymbol{\delta}$ per unit length of the line and perpendicular to it is formed. Show that the resultant $\mathbf{E}$ has the same magnitude at all points of any circle having the line as axis and that its direction rotates twice in one circling about the line (as against a single such rotation about a monopole line).

2.21 Suppose that a line of dipoles as in Exercise 2.20 is formed instead into a circular ring of radius a with $\boldsymbol{\delta}$ perpendicular to the plane of the ring.

(*a*) Find the potential on the axis. Need it be known off axis to yield $\mathbf{E} = -\nabla\phi$ on the axis?

(*b*) Check your result by integrating the elements $d\mathbf{E}$ on the axis that arise from the ring of source elements $\boldsymbol{\delta}\, a d\varphi$, as in (2.27*a*).

2.22 (*a*) Show that the contribution to the potential at $\mathbf{r}$ by the element (2.32) is just $d\phi(\mathbf{r}) = \tau(\mathbf{r}_S)\, d\Omega$, where $d\Omega$ is the element of solid angle subtended at $\mathbf{r}$ by the area $d\mathbf{S}$.

(*b*) Suppose that a uniform dipole distribution, of constant τ, extends over an infinite plane. Find the potential everywhere. Why should this be expected to have opposite signs on opposite sides of the plane? (Give a physical argument, as well as the formal one implicit in the discussion of Fig. 1.1.)

(*c*) Give arguments that the discontinuity (2.33) should be expected even for a nonuniform τ spread over an arbitrary (but smoothly connected) surface.

2.23 (*a*) A double layer with a uniform density τ of dipole moment is distributed on the half-infinite plane $x > 0$. Describe the resultant equipotentials and field lines.

(*b*) Instead, distribute the double layer over the curved surface of a cylinder of radius a and length L. Find the potential everywhere on the cylinder's axis.

2.24 A neutral particle of mass m and a permanent electric dipole moment D (like a water molecule) is placed at the center of a uniformly charged ring (radius a, charge $q > 0$). What is the minimum initial velocity u_m that the particle must be given if it is to escape to infinity, tail first along the axis of the ring? What happens if the particle is disturbed with less than that minimum velocity?

Ans. $u_m = 2(qD/3\sqrt{3}ma^2)^{1/2}$

2.25 Suppose that the particle of Exercise 2.24 has no dipole moment but instead a linear quadrupole moment, of the type considered in Exercise 2.15, with $Q_{zz} \equiv Q_o$. Assume that it is lined up with the axis but placed at a small distance ($z \ll a$) from the ring center and on its axis. Show that it will start oscillating and find the frequency of this small oscillation. *Ans.* $\omega = (3qQ_o/ma^5)^{1/2}$.

2.26 *Show* that the interaction energy (2.46) of two non-overlapping, homogeneous spheres of charge q_1 and q_2 is just $q_1 q_2/r$ if r is the distance between their centers.

2.27 A point nucleus of charge $+Ze$ is surrounded by an electron cloud forming the charge distribution

$$\rho(r) = -\frac{Ze}{\pi a^2 r} e^{-2r/a}$$

centered on $r = 0$ at the nucleus. What is the minimum energy it would cost to strip the resultant "atom" of its electrons?

2.28 The stability of charged fluid spheres against division into smaller spheres is relevant to understanding thunderstorms and nuclear fission. Suppose that an energy S per unit area must be invested in surface tension to maintain a sphericity of a fluid drop. Any charge Q can be assumed to be uniformly distributed on the surface in the case of water but homogeneously distributed throughout nuclear fluid. In each case, what is the critical charge for which an isolated initial sphere of radius a can be divided into two equal, isolated, and equally charged spheres of the incompressible fluids in terms of a and S? What was the electrostatic potential, relative to ground, of the initial, critically charged sphere?

Ans. $2^{5/3}(\pi a S)^{1/2}$. In the case of water, it can amount to some thousands of volts for droplets of the order of a millimeter in radius.

2.29 Two ideal dipoles $\mathbf{D}_{1,2}$ are respectively centered at points apart by a distance r. $\mathbf{D}_1$ is perpendicular to the line joining the dipole centers and $\mathbf{D}_2$ is parallel to this line.

(*a*) Find the translational force on each dipole and show that each force is a reaction from the other.

(*b*) Show that each force in (*a*) is derivable from the same dipole-dipole interaction energy, U.

(*c*) By thinking of a dipole as suggested by Figure 2.2, argue that a torque like $\mathbf{D} \times \mathbf{E}$ is the same relative to *any* momental center.

(*d*) Find the *total* torque on each dipole.

(*e*) Show that relative to *any one* momental center, all clockwise torques are equal and opposite reactions from counterclockwise torques.

CHAPTER

THREE

LAPLACE FIELDS

Electrostatic fields in free space and the necessity of their attachment to sources · Boundary conditions as replacements for source descriptions · Fourier and spherical-harmonic analyses · The electrostatic multipole expansion to all orders

In regions containing no sources the Poisson equation of (2.7) reduces to the *Laplace* equation, $\nabla^2\phi = 0$. Since satisfying this demands only a certain smoothness of its solutions [see the discussion of (A.37)], it has an enormous variety of them, called *Laplace fields* here. They are important to electromagnetic theory because various superpositions of them describe various ways in which static fields can span the spaces between sources.

The Laplace fields can be classified into alternative *orthogonal function sets*, introduced in Secs. 3.2 and 3.3 (more can be found in Chap. C). The first section will be concerned with how they fit in with solutions of the Poisson equation, which is the one that must be satisfied in the regions containing the sources of the Laplace fields.

3.1 GENERAL FORMS FOR THE POTENTIAL

To prepare for situations in which the source distributions may not all be given a priori and to find what pieces of information may suffice in their stead, other forms than just $\phi = \oint dV\rho/R$ [Eq. (1.26)] for a potential satisfying the Poisson equation should be considered. The general solution will now be discussed.

The Poisson differential equation is classified as *linear* and *inhomogeneous* by mathematicians. A differential equation for a dependent variable ϕ is linear if each term that contains ϕ is linear in it or in one of its derivatives. Terms not containing ϕ at all are the inhomogeneities; the Poisson equation has just the inhomogeneity $-4\pi\rho(\mathbf{r})$. The general solution of such an equation can always be expressed as a sum of any *special* solution of it (one specified completely and not containing any arbitrary integration constants) plus the *general* solution of the *homogeneous* equation (one formed by eliminating the inhomogeneities). Thus, the general solution of the Poisson equation can be written

$$\phi(\mathbf{r}) = \phi_\rho(\mathbf{r}) + \phi_o(\mathbf{r}) \tag{3.1}$$

where $$\nabla^2\phi_\rho = -4\pi\rho \qquad \text{and} \qquad \nabla^2\phi_o = 0 \tag{3.2}$$

That the sum (3.1) satisfies the Poisson equation if only ϕ_ρ does is obvious. That adding on the general solution ϕ_o of the *Laplace* equation (A.37) is sufficient to make the sum the general solution of the *Poisson* equation follows from the fact that both equations have the same number of independent variables (three) and the same degree (second order).

Possibilities for the Laplace fields ϕ_o will be left to succeeding sections. It will turn out that the many solutions can be classified into alternative complete sets of orthogonal solutions. Each set is complete in the sense that, if $\phi_1, \phi_2, \ldots, \phi_n, \ldots$ represent the members of the set, any linear combination of them

$$\phi_o = \Sigma_n c_n \phi_n \tag{3.3}$$

can represent the most general solution (including other complete orthogonal sets!). The $c_1, c_2, \ldots, c_n, \ldots$ are arbitrary integration constants, such as characterize a general solution. Various choices of values for them yield descriptions of various specific circumstances. It is just to gain this adaptability to a variety of specific situations that ϕ_o is added to the solution of the Poisson equation.

The part ϕ_o leads only to a divergenceless part $\mathbf{E}_o = -\nabla\phi_o$ of the total vector field $\mathbf{E} = \mathbf{E}_\rho + \mathbf{E}_o$, since $\nabla \cdot \mathbf{E}_o = -\nabla^2\phi_o = 0$. It is the remaining part ϕ_ρ that must represent any divergences $\nabla \cdot \mathbf{E} = -\nabla^2\phi = 4\pi\rho$ specified by the Poisson equation. As long as it satisfies the full Poisson equation, the solution ϕ_ρ may be one obtained by any means that present themselves. Different ϕ_ρ's need not engender any difference in the final result $\phi = \phi_\rho + \phi_o$ as adapted to any specific situation; they will merely lead to different choices for the supplementation by ϕ_o, with no difference to the total result. This emphasizes the importance of checking that any conditions on the solution which may be taken to represent a specific situation are *sufficient* to delimit a properly *unique* description of it.

Findings about any solution of the Poisson equation as obtained from the Green theorem, in Sec. A.5, will now be analyzed in the terms $\phi = \phi_\rho + \phi_o$. Those findings will here be discussed in the more physical way made possible by the definite interpretation of ϕ as an electrostatic potential.

The Delta Distribution

First, the process of extracting a value $\phi(\mathbf{r}_f)$ at any chosen field point $\mathbf{r}_f$ out of the integral

$$\int_V \phi(\mathbf{r})\, \nabla^2\chi(\mathbf{r})\, dV(\mathbf{r}) \tag{3.4}$$

in the Green's theorem expression (A.42) will be discussed in the electrostatic context.

The extraction was performed with the help of a *Green's function*

$$\chi(\mathbf{r}) = \frac{1}{|\mathbf{r} - \mathbf{r}_f|} \tag{3.5}$$

which can now be recognized as just the electrostatic potential at various points $\mathbf{r}$ arising from a *unit test charge* placed at the point $\mathbf{r}_f$, where the field value is desired. It must be a solution of the Poisson equation

$$\nabla^2\chi(\mathbf{r}) = -4\pi\delta(\mathbf{r} - \mathbf{r}_f) \tag{3.6}$$

if δ, as a function of $\mathbf{r}$, is the density distribution of the unit of test charge.

Distributions like δ play an important role in field-theoretic descriptions and deserve some attention. Plainly, for $\delta(\mathbf{r} - \mathbf{r}_f)$ to represent the density of a *point* charge at $\mathbf{r}_f$ it must be defined to vanish everywhere except at the zero value of the indicated argument of it. Moreover, its value at $\mathbf{r} = \mathbf{r}_f$ must be such that

$$\int_{V(\mathbf{r}_f)} dV(\mathbf{r})\delta(\mathbf{r} - \mathbf{r}_f) = 1 \tag{3.7}$$

upon integration over any volume $V(\mathbf{r}_f)$ which contains the singular point of δ, in order that this properly equal the whole unit of test charge. It is apparent that δ must tend to infinity at the zero point of its argument, in such a way that $\delta \to 1/dV(\mathbf{r}_f)$ as $dV(\mathbf{r}_f) \to 0$. The entire volume of integration $V(\mathbf{r}_f)$ of (3.7) may have any size, as long as it contains the singular point, since $\delta = 0$ at all other points and they contribute nothing. The behavior required of δ does not conform to conventional mathematical definitions of functions that exist, but after its usefulness was demonstrated, it was granted some kind of existence as a distribution rather than a function. However, it is sometimes classed as an improper function, and physicists continue calling it either a *needle function* or, more often, a *Dirac delta function.*

For any function $\phi(\mathbf{r})$ which has a definite value $\phi(\mathbf{r}_f)$ at the singular point of $\delta(\mathbf{r} - \mathbf{r}_f)$

$$\int_{V(\mathbf{r}_f)} dV(\mathbf{r})\phi(\mathbf{r})\delta(\mathbf{r} - \mathbf{r}_f) = \phi(\mathbf{r}_f) \int dV\delta(\mathbf{r} - \mathbf{r}_f) = \phi(\mathbf{r}_f) \tag{3.8}$$

since every value of ϕ except $\phi(\mathbf{r}_f)$ is multiplied by zero in the course of the integration. As a consequence, the substitution of $\nabla^2\chi = -4\pi\delta$ (3.6) into the integral (3.4) yields

$$\int_V \phi\nabla^2\chi\, dV = -4\pi\phi(\mathbf{r}_f) \tag{3.9}$$

when V contains the field point $\mathbf{r}_f$, exactly as in (A.46). This procedure appears to bypass the steps (A.43) and (A.45) formerly used, but this is not actually so, since equivalents of those steps are needed in the formal demonstration that $\chi = 1/|\mathbf{r} - \mathbf{r}_f|$ [Eq. (3.5)] is indeed a solution of (3.6): $\nabla^2\chi = -4\pi\delta(\mathbf{r} - \mathbf{r}_f)$. Such a demonstration would start from the fact that no particular directions from $\mathbf{r}_f$ are singled out by the isotropic charge distribution $\delta(\mathbf{r} - \mathbf{r}_f)$ and hence a solution $\chi(|\mathbf{r} - \mathbf{r}_f|)$, depending only on the radial distance $R \equiv |\mathbf{r} - \mathbf{r}_f|$ from $\mathbf{r}_f$, should suffice to represent the resultant potential. Then $\nabla^2\chi(R) = [d^2/(R\, dR^2)] \times (R\chi) = 0$, as in (A.43*a*), everywhere except at $R = 0$, and hence the solution must be proportional to $\chi \sim 1/R \equiv 1/|\mathbf{r} - \mathbf{r}_f|$ in the suitable gauge. The proportionality constant determining the magnitude of χ must be such that integrating (3.6) over volume yields $\int dV \nabla^2 \chi = -4\pi$. This is where the step (A.45) is needed, with the result $\chi = 1/|\mathbf{r} - \mathbf{r}_f|$.

The delta distributions are useful for representing systems of point charges in general. For example, the two-point configuration discussed as having the dipole moment $\mathbf{D} = q\mathbf{l}$ [Eq. (2.24)] can be said to have the charge distribution $\rho(\mathbf{r}_s) = q\delta(\mathbf{r}_s - \frac{1}{2}\mathbf{l}) - q\delta(\mathbf{r}_s + \frac{1}{2}\mathbf{l})$, and then the defining integral $\mathbf{D} \equiv \oint dV \rho \mathbf{r}_s$ [Eq. (2.20)] yields just $\mathbf{D} = q\mathbf{l}$.

The Special Solution

The final result obtained from Green's theorem was that any solution, the *general* one, of the Poisson equation satisfies the relation (A.47). The relation is expressed only for field values in a certain volume of definition, one in which a Poisson equation is defined and so the sources $\rho(\mathbf{r})$ known. It can include regions between sources where ρ is definitely known to be zero (with the Poisson equation reducing to a Laplace equation there), and hence includes the main regions of interest. Moreover, as will be seen, provision is made for the effects of any unknown sources there might be outside the integration volume.

The first volume integral in (A.47) can serve as the special solution needed for the expression of the general solution (3.1),

$$\phi_\rho(\mathbf{r}_f) = \int_V \frac{dV(\mathbf{r}_s)\rho(\mathbf{r}_s)}{|\mathbf{r}_f - \mathbf{r}_s|} \tag{3.10}$$

It does satisfy the Poisson equation, in its region of definition, all by itself. This can be rechecked by applying the operation ∇_f^2 and using (3.6), with the result

$$\nabla_f^2 \phi_\rho(\mathbf{r}_f) = \int_V dV(\mathbf{r}_s)\rho(\mathbf{r}_s)[-4\pi\delta(\mathbf{r}_s - \mathbf{r}_f)] = -4\pi\rho(\mathbf{r}_f)$$

Expression (3.10) is a special solution since it contains no arbitrariness, with $\rho(\mathbf{r}_s)$ definite in V. Moreover, it does not include effects at $\mathbf{r}_f$ of any sources there might be outside V. It differs from $\phi = \oint dV\rho/R$ [Eq. (1.26)] in that the latter was meant to stand for an *entire* field, with ρ *all* its sources and hence the integration $\oint dV$ extensible to all space (as denoted by the symbol $\oint$).

The inverse distance occurring in (3.10) had its origin in the Green's function χ, which must be mathematically the same as the test-charge potential (3.5) in the integral (3.9). However, its *physical* role in ϕ_ρ (3.10) is different. Here it represents the potential at the field point due to each unit of charge $\rho(\mathbf{r}_s)\, dV(\mathbf{r}_s)$ at a source point. It measures the propagation of the influence to $\mathbf{r}_f$ of the charge at $\mathbf{r}_s$ in the form of a contribution to the potential. For such reasons, Green's functions like $\chi(|\mathbf{r}_f - \mathbf{r}_s|)$ are frequently called *propagators*.

Boundary Conditions

The remainder of the Green's theorem result (A.47), added to the special solution ϕ_ρ (3.10) to yield a full value $\phi(\mathbf{r}_f)$ of the general solution, must be identifiable with the addition ϕ_o in the general-solution form (3.1). It will now be written

$$\phi_o(\mathbf{r}_f) = (4\pi)^{-1} \oint dS(\mathbf{r}_S) \left[\phi(\mathbf{r}_S)\nabla_n\left(\frac{1}{R}\right) - \frac{1}{R}\nabla_n\phi\right] \tag{3.11}$$

with the ∇_n standing for *inward* (into V) components of the gradient derivative. It will prove desirable, for the considerations here, to thus redefine the area-element vectors $d\mathbf{S}$ and the normal unit vectors $\mathbf{n} \equiv d\mathbf{S}/dS$ on the enclosing surface of the integration volume so that they are directed *in*ward, into the volume of definition. As before, $R = |\mathbf{r}_S - \mathbf{r}_f|$ is the distance to the field point from the point $\mathbf{r}_S$ on the surface of the area element $dS(\mathbf{r}_S)$. The subscript n attached to the gradient operators serves as a reminder that, for the fixed field point at which $\phi(\mathbf{r}_f)$ is desired, the derivatives arise from variations in the surface-point positions $\mathbf{r}_S$ of $d\mathbf{S} = \mathbf{n}\, dS$ and that only their components ∇_n normal to the surface are needed for the scalar products with $d\mathbf{S}$.

For the equation of the integral (3.11) to the part ϕ_o of the general solution form (3.1) to be a valid one, the integral as a function of $\mathbf{r}_f$ must satisfy the Laplace equation $\nabla_f^2 \to 0$ of (3.2). An application of the Laplace operator to the integral does indeed yield zero because $\nabla_f^2(1/R) = 0$ everywhere except for $\mathbf{r}_f$'s right on the surface, according to (A.43*b*).

The result is a useful statement about the determinability of a field within some volume V when only those sources $\rho(\mathbf{r}_s)$ of it which lie within V have been given. It shows that the entire field is determinable everywhere within V if it is already known on the enclosing surface only. Giving just the sources within V need only be supplemented by *boundary conditions* on its surface. In more detail, the contents of the integral (3.11) indicate that it is *sufficient* to supply information about the field values $\phi(\mathbf{r}_S)$ everywhere on the surface, together with the normal derivatives $\nabla_n\phi = -\mathbf{E}_n$. It remains to be seen whether as much information as this is really *necessary*.

The outcome so far is easiest to understand in terms of a one-dimensional illustration, in which the equation to be satisfied by some field $\phi(x)$ is given as $d^2\phi/dx^2 = s(x)$, say. (The x here might be a distance along a line normal to such a surface as was considered above, with constant values for y and z on it.) The general solution in this simple example is

$$\phi(x) = \int^x dx \left[\int^x dx' s(x') + a\right] + b$$

where a and b are the two arbitrary integration constants to be expected of a second-order equation in just one independent variable. The two constants can have definition only relative to whatever lower limit $x = x_0$ might be chosen for the integrations over the inhomogeneity $s(x)$. Let x_0 be an edge point of a region in which $s(x)$ has been given explicitly, and then

$$\phi_s(x) = \int_{x_0}^{x} dx' \int_{x_0}^{x'} dx'' s(x'')$$

can be given the role of the *special* solution. The remainder, $\phi_0(x) \equiv a(x - x_0) + b$, is the *general* solution of the *homogeneous* equation.

Now it is clear that $b = \phi(x_0)$ and $a = (d\phi/dx)_{x_0}$. Thus, giving just these starting values, or boundary conditions, is sufficient to determine a unique $\phi(x)$ over the region in which $s(x)$ is known. The solution $\phi(x)$ may be thought of as a continuation of the given $\phi(x_0)$ at the prescribed initial rate $(d\phi/dx)_{x_0}$. The continuation to points x in the immediate neighborhood of x_0 can be represented as a Taylor expansion [introduced in the footnote to Eq. (2.18)]

$$\begin{aligned}\phi(x) &= \phi(x_0) + (x - x_0)\phi'(x_0) + \tfrac{1}{2}(x - x_0)^2\phi''(x_0) + \tfrac{1}{6}(x - x_0)^3\phi'''(x_0) + \cdots \\ &= \phi(x_0) + (x - x_0)\left(\frac{d\phi}{dx}\right)_{x_0} + \tfrac{1}{2}(x - x_0)^2 s(x_0) + \tfrac{1}{6}(x - x_0)^3 s'(x_0) + \cdots\end{aligned} \tag{3.12}$$

This outcome, that a starting value and its initial rate of change determine a unique solution, is familiar in newtonian mechanics. There, only second rates of change (accelerations) are given directly by the forces, and so initial velocities and positions must be supplied before a unique motion is determined.

More can be learned from the one-dimensional example. Suppose $s(x)$ is given only in a closed region, $x_0 \le x \le x_1$. A unique solution in this region can be determined by prescribing just $\phi(x_0)$ and $\phi(x_1)$, for then $b = \phi(x_0)$, as before, and

$$a = \frac{\phi(x_1) - \phi_s(x_1) - \phi(x_0)}{x_1 - x_0}$$

The implication of this finding for the three-dimensional case is that when boundary conditions are prescribed everywhere on a *closed* surface, it should be sufficient to give just $\phi(\mathbf{r}_S)$. Giving the rates of change $(\nabla_n\phi)_{\mathbf{r}_S}$ in addition should be superfluous. This conclusion is borne out by the uniqueness considerations connected with (A.49), which can be rewritten

$$\int_V dV\,|\mathbf{E}_1 - \mathbf{E}_2|^2 = \oint dS(\phi_1 - \phi_2)\nabla_n(\phi_1 - \phi_2) \tag{3.13}$$

According to this, a unique solution $\mathbf{E}_1 \equiv \mathbf{E}_2$ is determined as soon as definite values $\phi_1 = \phi_2$ are prescribed for $\phi(\mathbf{r}_S)$ everywhere on the enclosing surface. A consequence is that the form (3.11) is no longer useful for calculating the part ϕ_o of the field inside the space, since it also requires preknowledge of $\nabla_n\phi$ on the surface. Other approaches should exist [(3.3), for example], which will allow calculation of the solution when only

$$\phi(\mathbf{r}_S) = \text{Dirichlet boundary condition} \tag{3.14}$$

is prescribed everywhere on a surface enclosing a volume of definition. Examples will be seen in the next sections and in Chaps. B to E, as well as in Exercises 3.3 to 3.5, 3.8, and 3.10.

The uniqueness relation (3.13) also offers alternatives to the boundary conditions (3.14). For example, it shows that prescribing

$$\mathbf{E}_n(\mathbf{r}_S) = -\nabla_n \phi = \text{Neumann boundary condition} \tag{3.15}$$

in place of $\phi(\mathbf{r}_S)$ everywhere on the closed surface should be sufficient to determine a unique solution $\mathbf{E}_1 \equiv \mathbf{E}_2$ within it (Exercises 3.7, 3.11, and 3.12). Mixed conditions should also be possible, i.e., prescribing $\phi(\mathbf{r}_S)$ on some parts of the surface and $\mathbf{E}_n(\mathbf{r}_S)$ on the remaining ones. Still another alternative will be shown in Chap. B, applicable when the surface is known to be an equipotential but no value is prescribed for that potential.

The reader may be finding the conclusions here distressingly untidy in comparison to the explicit prescriptions for calculating the field when *all* its sources are given, as in Secs. 2.1 and 2.2. This must be regarded as a natural cost of generality, of applicability to an enormous variety of situations which might be presented by nature. It is symptomatic of the necessity for keeping physical theories open-ended, to ensure that coping with unexpected problems will not be unnecessarily circumscribed.

The Boundary of All Space

The boundary conditions (3.14) are stated for enclosed spaces, but open spaces extending to infinity can be represented by letting at least part of the enclosing boundary approach infinity in its relative positions. This works out unambiguously except when sources are overidealized by being treated as themselves extending to infinity. The overidealized representations may still be very useful ones (in situations where edge effects are negligible) but are best left to individual (ad hoc) treatment, since the conclusions may depend on the special respects in which various finite situations can be idealized [compare (2.15) with (2.16).]

The most noteworthy general conclusion about open spaces is implicit in the case where the volume of definition consists of all space, with the charge distributions $\rho(\mathbf{r}_s)$ given everywhere and held to be nonvanishing only in some finite portion of all space. Then the special solution ϕ_ρ of (3.10) becomes the entire one $\phi = \oint dV\rho/R$ of (1.26), since the volume which encloses the given sources may be taken to be all space ($V \to \infty$, as symbolized by $\oint dV$). This means that for the Poisson equation to be consistent with the basis (1.26) from which it was in effect generalized, the addition ϕ_o as given by (3.11) must vanish.

The supplement ϕ_o now consists of an integral over an extremely remote surface as $V \to \infty$. Such integrals are presumed to vanish in obtaining the result (A.48), but the presumption may seem a bit cavalier in view of the fact that the area of integration itself approaches infinity. The field at a surface so remote from the sources has a net monopole effect $\phi = q/R$ as its largest possible survival, with $R \to \infty$ the distance to points on the surface from some origin among the sources. The normal gradients $\nabla_n \phi$ will be fractions of q/R^2, and $R \equiv |\mathbf{r}_S - \mathbf{r}_f|$ will approach infinity as well, since only field points in the vicinity of the source will be in any finite portion of the space. Then the contribution by each element of the integral will be less than $dS(q/R^3) = d\Omega R^2(q/R^3) \to 0$ as $R \to \infty$. Thus, the Poisson equation does indeed lead to consistency with $\phi = \oint dV\rho/R$ as the entire field due to all sources $\rho(\mathbf{r}_s)$ when these are given everywhere, and the treatment which led to (A.48) is justified.

Meanwhile, the rather obvious expectation that $\mathbf{E} \to 0$ and $\phi \to 0$ in a suitable gauge sufficiently far from all sources can be regarded as the formal boundary condition at infinity, at least for finite sources. Some expectations about sources which can be idealized as infinite in extent have been indicated in (2.16) and (2.17).

External Effects

It has now been checked that the general physical solution of the Poisson equation, like any real electrostatic field, becomes representable by the integral over all space, $\phi = \oint dV\rho/R$, once *all* sources of the field become known. Thus confirmed is the interpretation of $\phi_o = \phi - \phi_\rho$ as the field arising from unknown external sources outside the volume V in which ϕ_ρ is defined whenever only the sources within V are actually known. This warrants further discussion since the mathematical finding (3.11) about ϕ_o appears to require using the boundary values of the *entire* field for its evaluation and not only the field arising from the external sources. It seems to imply that even when the given sources within V are all the sources with any effect at the field point, so that $\phi = \phi_\rho$ is to be expected, an evaluation of ϕ_o on the finite boundary of V should still be carried out, since the entire field certainly does not vanish there, as it does on the boundary at infinity. The point is that the contributions to the surface integral (3.11) of the field ϕ_ρ by itself actually cancel out. This can be confirmed quite directly, after substituting $\phi \equiv \phi_\rho$ into (3.11) (see Exercise 3.1).

3.2 FOURIER ANALYSIS

Expressions of Laplace fields as superpositions, like (3.3), of mutually orthogonal functions are particularly useful because of their comparatively straightforward adaptability to boundary conditions when a suitable set of orthogonal functions is used as a basis. How a suitable set can be defined by the needs of a given problem will be illustrated by first considering the kinds of electrostatic fields that can exist between two grounded planes. The latter are equipotentials with $\phi = 0$ on them (how they are physically supplied by conductor boundaries that are connected to the earth is reviewed in Chap. B).

The Sine-Wave Modes

First consider two-dimensional situations, ones in which nothing varies appreciably along one (z) direction of space and so description by potentials $\phi(x, y)$ is adequate. These are to satisfy the Laplace equation (over source-free regions) and also boundary conditions at grounded planes some distance L apart: $\phi(0, y) = \phi(L, y) = 0$.

A frequently effective approach to finding solutions of partial differential equations starts with looking for separable ones, having the dependence on each independent variable in a separate factor, as in $\phi(x, y) = X(x)Y(y)$. For such forms, the Laplace equation yields

$$\frac{\nabla^2\phi}{\phi} = \frac{X''(x)}{X} + \frac{Y''(y)}{Y} = 0 \tag{3.16}$$

Here, a term independent of x is being required to cancel a term which can vary only with x, if at all, and this can be so for all x, y only if each term is constant. The conclusion is that

$$X'' = -k^2X \qquad Y'' = +k^2Y \tag{3.17}$$

with some arbitrary separation constant k^2 (not necessarily positive or even real, so far). The problem is thus reduced to a matter of solving two *ordinary* (rather than partial) differential equations, and the first of them has solutions proportional to

$$X_k(x) \sim \sin\,(kx + \alpha_k) \tag{3.18}$$

The consequent special solutions $\phi_k(x, y) = X_k\,Y_k$ will satisfy the boundary condition $\phi_k(0, y) = 0$, for any y and k, if $\alpha_k = 0$ is chosen, and then the second condition, $\phi_k(L, y) = 0$ for every y, requires choosing k so that $\sin kL = 0$ or $k = n\pi/L$ with any integer n. The definite modes thus found can be labeled by n in place of k, and all the distinct ones, each proportional to a

$$\sin\frac{n\pi x}{L} \qquad \text{with } n = 1, 2, 3, \ldots \tag{3.19}$$

are enumerable by just the positive integers. (The use of just such modes for describing standing waves† in strings stretched between fixed points is probably familiar to the reader.) If advantage is taken of the fact that arbitrary linear combinations of solutions will also satisfy the Laplace equation, a highly general solution

$$\phi(x, y) = \sum_{n=1}^{\infty} Y_n(y)\sin\frac{n\pi x}{L} \tag{3.20}$$

† See Ref. 17, secs. 12.1 and 12.2.

with Y_n's such that

$$\nabla^2 Y_n = Y_n''(y) = +\left(\frac{n\pi}{L}\right)^2 Y_n \tag{3.21}$$

will have been found. Since the general Y_n will contain two arbitrary integration constants, a double infinity of such constants is implicit in (3.20).

Despite the special character of the steps taken in finding the separate terms of (3.20), the sum nevertheless constitutes a fully general solution, in the sense that it can represent, to arbitrarily high accuracy, any functional dependence on $0 < x < L$ that is not too pathological to be used for physical description. Even rather highly idealized x dependences, like those of only piecewise continuous functions (continuous except for abrupt jumps of value in passages through isolated points), can be represented by such superpositions of the sines as (3.20). All this stems from the fact that the sine functions (3.19) form a complete orthogonal set on the interval $0 < x < L$.

Any set of functions $u_1(x), u_2(x), \ldots, u_n(x), \ldots$ is said to have orthogonality within a given domain of the independent variable if, for any pair of its members,

$$\oint dx u_n^*(x) u_{n'}(x) = 0 \qquad n \neq n' \tag{3.22}$$

The symbol $\oint dx$ here stands for integration over the entire domain of the definition. In the case of the sines (3.19), the domain is $0 < x < L$, and the orthogonality of those functions is readily proved by direct integration over it.

Orthogonalities of function sets make resolutions on them (or decompositions into their members), the formation of expressions like

$$f(x) = \sum_n c_n u_n(x) \tag{3.23}$$

for various functions $f(x)$, as useful and convenient as resolutions of ordinary vectors into their components on some chosen frame. The properties of a description by a function $f(x)$ becomes deducible from knowledge about each of the component modes $u_n(x)$ it contains. Moreover, the coefficient c_n with which any given mode $u_n(x)$ enters the description is simple to determine, for any given $f(x)$. It is only necessary to multiply $f(x)$ by $u_n^*(x)$ and integrate, a process aptly called *projection* of $f(x)$ onto $u_n(x)$. The complete formula is

$$c_n = \frac{\oint dx u_n^*(x) f(x)}{\oint dx \, |u_n|^2} \tag{3.24}$$

as follows quite readily from carrying out the indicated operations on an expression like (3.23) and using the orthogonalities (3.22). The denominator here can be replaced by unity if the set has been normalized to unity beforehand, through multiplying each member function by an appropriate constant when defining the set. The sines (3.19) become such an *orthonormal* set when each is defined with an amplitude factor $(2/L)^{1/2}$.

The essential characteristic of the mutually orthogonal modes is their linear independence of each other; each can be said to describe a component behavior not representable by its fellow modes. Formally, this manifests itself through the inexpressability of any one mode as a linear combination of the others. Any attempt to construct an expression for $u_{n'}(x)$, say, as $\sum_{n \neq n'} c_n u_n(x)$ would result in $c_n \equiv 0$ for every n ($\neq n'$). This also emphasizes how essential it is to leave out no mode of a complete set when undertaking to analyze arbitrary functional dependences. That would be comparable to using a two-dimensional frame for the resolution of general three-dimensional vectors! Proof of the sufficient completeness of any set to be used is therefore desirable, but the formal aspects of this question may be left to mathematicians. In physical applications, it is usually clear from physical considerations just what conditions on a solution should be expected to yield a properly unique description [see discussions of (3.14), (3.15), and (B.2)]. Then any set of functions adopted is complete enough for the situation in question if only it permits satisfying all those conditions.

The analysis into the superposition (3.20) reduces the problem of how electrostatic fields confined between grounded planes may behave to an investigation of the component modes $Y_n(y) \sin (n\pi x/L)$ that may be present. The amplitude Y_n of any one mode may vary in accordance with the general solution of Eq. (3.21)

$$Y_n(y) = a_n e^{+n\pi y/L} + b_n e^{-n\pi y/L} \tag{3.25}$$

The Laplace equation therefore demands that bounded variations in one direction, like those proportional to sin $(n\pi x/L)$, be accompanied by unbounded, here *exponential*, variations along an orthogonal direction. This accords with the expectation that to have any field at all in a sourceless region, there must be sources in some direction—with effects that will inevitably decrease with distance from them, in some fashion. For any finite a_n, b_n at distances sufficiently far in the direction of increasing y, the decreasing exponential $b_n e^{-n\pi y/L}$ will have become negligible and only the exponential increase proportional to $a_n e^{+n\pi y/L}$ will remain. Such a behavior can represent an approach to sources of some kind; the sourceless region must eventually end and a region of sources begin. If there actually are no sources in that direction, the arbitrary constants a_n must be taken equal to zero, for otherwise $\phi(x, y \to \infty) \to \infty$ in a region with no sources, and this accords with no physical situation there is to describe. Thus the representation of the field is reduced to

$$\phi(x, y) = \sum_{n=1}^{\infty} b_n e^{-n\pi y/L} \sin \frac{n\pi x}{L} \tag{3.26}$$

whenever the sourceless region extends to $y \to \infty$.

Similar considerations naturally apply to $y \to -\infty$ also. If there are no sources to interrupt the sourceless region in either direction, not only is every $a_n = 0$ but also every $b_n = 0$, and there is no field at all. However, consider cases in which sources do exist somewhere in the direction of $y \to -\infty$. Their effects become most directly amenable to such an analysis as (3.26) when they are represented by giving a potential distribution $\phi_o(x)$ they generate on some boundary plane of given y. A frame in which this is the $y = 0$ plane can always be adopted, and then the form (3.26) must reduce to

$$\phi(x, 0) = \sum_{n=1}^{\infty} b_n \sin \frac{n\pi x}{L} = \phi_o(x) \tag{3.27}$$

at $y = 0$. The coefficients in this decomposition of the given $\phi_o(x)$ are determined by

$$b_n = \int_0^L \frac{2\,dx}{L} \phi_o(x) \sin \frac{n\pi x}{L} \tag{3.28}$$

and so an expression in the form (3.26) for the resultant field $\phi(x, y)$ is completely defined.

Take a case in which the applied potential $\phi_o(x)$ is simply a *constant*, $\phi_o \neq 0$. This "function" can be given the representation (3.27) despite the fact that the series of sines vanishes at $x = 0$ and L while $\phi_o \neq 0$ there; by summing the series sufficiently far, it can be made to yield the finite ϕ_o at points arbitrarily close to $x = 0$ and L (where insulation from the grounded planes must in any case be placed). Then (3.28) yields

$$b_n = \frac{2\phi_o}{n\pi}[1 - (-1)^n] \tag{3.29}$$

vanishing for every even n and having the value $4\phi_o/n\pi$ for any odd n. The sine functions with even n will thus be missing from the series representations, as might have been expected from the fact that these sines are odd about the middle of the interval (they change sign upon replacements $x \leftrightarrow L - x$), whereas the excitation $\phi_o(x) \equiv \phi_o$ is even. The result for the field anywhere between the grounded planes $(0 < x < L)$ and outside the "source plane" $(y > 0)$ can then be written

$$\phi(x, y) = \frac{4\phi_o}{\pi} \sum_{N=0}^{\infty} \frac{\sin [(2N + 1)\pi x/L]}{2N + 1} e^{-(2N+1)\pi y/L} \tag{3.30}$$

since $n = 2N + 1$ with $N = 0, 1, 2, \ldots$ in the nonvanishing terms of (3.26). The part of this field that penetrates farthest in the y direction is $(4\phi_o/\pi)e^{-\pi y/L} \sin(\pi x/L)$.

That the infinite series (3.30) converges uniformly at all points in the domain of definition becomes evident from the fact that it is summable into a closed functional form. The summation can be carried out by using the familiar result for geometric series†:

$$\sum_{N=0}^{\infty} \epsilon^N = \frac{1}{1 - \epsilon} \qquad \text{if } |\epsilon| < 1$$

† If $1 + \epsilon + \epsilon^2 + \cdots + = S$, then $S - 1 = \epsilon S$.

When the sine functions in (3.30) are decomposed into the complex exponentials, it becomes simple to see that $\phi = \text{Im}\,[f(\zeta)]$, where

$$f(\zeta) = \frac{4\phi_o}{\pi} \sum_{N=0}^{\infty} \frac{\epsilon^{2N+1}}{2N+1} \qquad \text{with } |\epsilon \equiv e^{i\pi\zeta/L}| < 1 \tag{3.31}$$

and $\zeta \equiv x + iy$ (see Exercise 3.16). Taking a derivative yields

$$f'(\zeta) = \frac{4i\phi_o \epsilon}{L} \sum_{N=0}^{\infty} \epsilon^{2N} = \frac{4i\phi_o}{L} \frac{\epsilon}{1-\epsilon^2} = -\frac{2\phi_o/L}{\sin(\pi\zeta/L)} \tag{3.32}$$

Reintegration gives

$$f(\zeta) = \frac{2\phi_o}{\pi} \ln \cot \frac{\pi\zeta}{2L} \tag{3.33}$$

within an arbitrary additive constant, and its imaginary part yields

$$\phi(x, y) = \frac{2\phi_o}{\pi} \tan^{-1} \frac{\sin(\pi x/L)}{\sinh(\pi y/L)} \tag{3.34}$$

in a gauge which is suitable if the range 0 to $\frac{1}{2}\pi$ is used for the arctangent. Then ϕ vanishes properly at $x = 0$ and $x = L$ and $y \to \infty$, while $\phi = \phi_o$ at $y = 0$. It is doubtful, however, whether the closed form (3.34) yields as direct insight into the important features of the field as the superposition of simple modes (3.30).

If the space for the field is closed off at some finite $y = L_y$, both the exponentials in (3.25) are needed to satisfy boundary conditions even if $y = L_y$ is taken to be a third *grounded* plane. In that case the exponentials increasing in the y direction must be given the coefficients $a_n = -b_n \exp[-2\pi n(L_y/L_x)]$. The expression (3.29) for b_n becomes an expression for

$$a_n + b_n = b_n\left[1 - \exp\left(-2\pi n \frac{L_y}{L_x}\right)\right]$$

instead. The consequent occurrence of the increasing exponential may be regarded as a result of reflection from the grounded $y = L_y$ plane of the field transmitted to it from the $y = 0$ source plane.

A finite interval, $0 < y < L_y$, along the y direction permits using the complete orthogonal set

$$\sin \frac{n_y \pi y}{L_y} \qquad \text{with } n_y = 1, 2, 3, \ldots \tag{3.35}$$

for analyzing the functional dependences on y. There is no profit in doing this in the problems considered above, since it would only lead to a replacement of the simple exponentials, $\exp[\pm(n_x \pi y/L_x)]$, by infinite series of the sines (3.35), valid only over an interval $0 < y < L_y$. However, joint use of both the orthogonal sets (3.19) and (3.35) can become quite profitable when variations along the third (z) direction are also to be described. For instance, there may be interest in the transmission of an electrostatic field down a rectangular channel with grounded walls. The field within such a tube is clearly obtainable in the form

$$\phi(x, y, z) = \sum_{\alpha} (a_\alpha e^{k_\alpha z} + b_\alpha e^{-k_\alpha z}) \sin \frac{n_x \pi x}{L_x} \sin \frac{n_y \pi y}{L_y} \tag{3.36}$$

where α is shorthand for the *pair* of indices n_x, $n_y = 1, 2, 3, \ldots$ and

$$k_\alpha = \left[\left(\frac{n_x}{L_x} \right)^2 + \left(\frac{n_y}{L_y} \right)^2 \right]^{1/2} \pi \tag{3.37}$$

is necessary if the three-dimensional Laplace equation is to be satisfied by ϕ. When an arbitrary potential distribution $\phi(x, y, 0) \equiv \phi_o(x, y)$ is applied to the $z = 0$ plane,

$$a_\alpha + b_\alpha = \frac{4}{L_x L_y} \int_0^{L_x} dx \int_0^{L_y} dy \phi_o(x, y) \sin \frac{n_x \pi x}{L_x} \sin \frac{n_y \pi y}{L_y} \tag{3.38}$$

The coefficients a_α, b_α are not determined individually until sufficient boundary conditions on the z dimension have been satisfied. For example, all $a_\alpha = 0$ if the tube extends to $z \to +\infty$ without interruption by sources, thus satisfying the boundary condition $\phi(x, y, \infty) = 0$.

Fourier Series

Each member of the sine-wave set (3.19) vanishes at the endpoints of the interval of definition, and so the superpositions of those sines are particularly suited to the efficient representation of functions that also vanish at the endpoints. More general functions are better represented, i.e., with more rapid convergence of the series, by sets that include cosines of the same arguments as members, since such cosines have finite values where the sines vanish (and vice versa). However, to have the cosines orthogonal to (linearly independent of) the sines, the interval of orthogonality must be doubled (whole wavelengths, rather than integral numbers of half wavelengths, must fit into the interval). It is a joint set defined by

$$\sin \frac{2\pi n x}{L}, \cos \frac{2\pi n x}{L} \qquad n = 0, 1, 2, \ldots \tag{3.39}$$

that will be found to have orthogonality on an interval of length L. Moreover, the interval here can begin at any x, unlike the interval for the sines alone. The definition (3.39) constitutes one of the standard forms used for producing Fourier-series representations.

The standard Fourier set (3.39) could also have been used for the problems of the preceding subsection but would have produced the same series representations only if the distance between the grounded planes was taken to be *half* the interval for the Fourier set and the other (unused) half of the Fourier interval was allowed to produce odd extensions of the functions that can be arbitrary only between the grounded planes. These assertions do not warrant extensive discussion but serve to emphasize that the representations can sometimes be improved by judicious choice of the interval of orthogonality relative to the interval in which the representation is actually used.

Briefer expressions for Fourier series can be written when the trigonometric functions (3.39) are decomposed into the complex exponentials

$$e_n(x) = L^{-1/2} e^{2\pi i n x/L} \qquad n = 0, \pm 1, \pm 2, \ldots \tag{3.40}$$

which form an ortho*normal* set on any interval of length L. The orthonormality property

$$\oint dx e_n^* e_{n'} \equiv \int_{x_0}^{x_0+L} \frac{dx}{L} e^{2\pi i(n'-n)x/L} = \delta_{n'n} \tag{3.41}$$

is quite obvious, in view of the periodicity of the exponentials for $n' \neq n$. It makes possible the determination of the coefficients for decompositions like

$$f(x) = \sum_{n=-\infty}^{\infty} c_n e_n(x) \qquad x_0 < x < x_0 + L \tag{3.42}$$

from

$$c_n = \oint dx e_n^*(x) f(x) = L^{-1/2} \int_{x_0}^{x_0+L} dx f(x) e^{-2\pi i n x/L} \tag{3.43}$$

If a real function $f(x)$ is to be represented, the series will have $c_{-n} = c_n^*$. Expression in terms of the real Fourier set (3.39) is always obtainable from (3.42) through the substitutions $\exp(2\pi i n x/L) = \cos(2\pi n x/L) + i \sin(2\pi n x/L)$.

Fourier Transforms

The sets considered so far can furnish representations of quite arbitrary functional dependences only over some interval of *finite* length L; if the same representations were evaluated outside that interval, they would merely repeat the function values within L. To provide for the representation of functions that can be arbitrary over the full span $-\infty < x < +\infty$, the set (3.40) can be defined for an interval $-\frac{1}{2}L < x < +\frac{1}{2}L$, by choosing $x_0 = -\frac{1}{2}L$ in (3.42) and (3.43), and then the limit $L \to \infty$ approached.

For the finite interval, a succession of *discrete* wave numbers $k_n \equiv 2\pi n/L$ ($n = 0, \pm 1, \ldots, \pm \infty$) is used, differing by steps $\Delta k = 2\pi/L$. When $L \to \infty$, the steps approach $\Delta k \to 0$ and so a *continuum* of wave numbers, $-\infty < k < +\infty$, is needed. The consequences for a representation like (3.42), with (3.40), can be seen upon substituting k for $2\pi n/L$ and inserting $L\, \Delta k/2\pi \equiv 1$ into the series expression

$$f(x) = \sum_{k=-\infty}^{\infty} \left(\frac{L^{1/2}\, \Delta k}{2\pi} \right) c_n e^{ikx}$$

As $\Delta k \to 0$, the summation here approaches an integration which can be expressed as

$$f(x) = (2\pi)^{-1/2} \int_{-\infty}^{\infty} dk c(k) e^{ikx} \tag{3.44a}$$

if $(L/2\pi)^{1/2}c_n \to c(k)$ is defined, to be obtained from (3.43) as

$$c(k) = (2\pi)^{-1/2} \int_{-\infty}^{\infty} dx f(x) e^{-ikx} \tag{3.44b}$$

The two expressions (3.44) together constitute the *Fourier integral theorem*. The factors used in defining $c(k)$ were chosen to make plainer a reciprocal symmetry between the two expressions, one that has led to their being called *each other's Fourier transforms*. A correspondence between a mapping on a space of $-\infty < x < +\infty$ and on a k-space, of $-\infty < k < +\infty$, is thus furnished.

The *Fourier amplitudes* $c(k)$ may be regarded as projections of $f(x)$ onto the orthonormal basis

$$e(kx) \equiv (2\pi)^{-1/2} e^{ikx} \tag{3.45}$$

consisting of a continuum of members enumerated by all the possible real ($\gtrless 0$) values of k. The orthonormality property can be represented as

$$\oint dx e^*(kx) e(k'x) \equiv \int_{-\infty}^{\infty} \frac{dx}{2\pi} e^{i(k'-k)x} = \delta(k' - k) \tag{3.46}$$

This follows from the $L \to \infty$ limit of (3.41), which (before the limit is reached) yields $(L/2\pi)\delta_{nn'} = \delta_{n'n}/\Delta k$ for the integrals (3.46). If $\delta(k' - k)$ is to represent the limit of this result, it must vanish for $k' \neq k$ and for $k' = k$ it must approach $1/\Delta k \to \infty$ as $L = 2\pi/\Delta k \to \infty$, in such a way that integration over k' becomes equivalent to multiplication by Δk and so

$$\int dk' \delta(k' - k) = 1 \tag{3.47}$$

over any integration range containing $k' = k$. Thus, the δ here is just a one-dimensional version of a delta distribution like that of (3.7). The result (3.47) is real, and so $\delta(k' - k) \equiv \delta(k - k')$ in (3.46).

Interchanging the symbols k and x in (3.46) yields

$$\int_{-\infty}^{\infty} \frac{dk}{2\pi} e^{ik(x-x')} = \delta(x - x') \tag{3.48}$$

amounting to a Fourier decomposition, like (3.44*a*), of a delta function in x. It follows that

$$\oint dk e(kx) e^*(kx') = \delta(x - x') \tag{3.49}$$

and this is said to constitute a *completeness condition* on the set (3.45) because it is just the property needed to ensure that superpositions like (3.44*a*) of the set members will be able to reproduce any value $f(x')$ that a quite arbitrary function $f(x)$ may take on. That becomes clear upon substitution of the amplitudes as given by (3.44*b*) into $f(x')$ as given by (3.44*a*):

$$\begin{aligned} f(x') &= \int_{-\infty}^{\infty} dk c(k) e(kx') = \int_{-\infty}^{\infty} dk e(kx') \int_{-\infty}^{\infty} dx f(x) e^*(kx) \\ &= \int_{-\infty}^{\infty} dx f(x) \int_{-\infty}^{\infty} dk e(kx') e^*(kx) \end{aligned} \tag{3.50}$$

If $f(x)$ is quite arbitrary, the integral over k here must be such as to multiply by zero every value $f(x \neq x')$ except $f(x')$, to produce just the value on the left. The delta function (3.49) does this and yields

$$\int_{-\infty}^{\infty} dx f(x)\delta(x - x') = f(x') \int_{-\infty}^{\infty} dx \delta(x - x') = f(x') \qquad (3.51)$$

because the last integral has a unit value by definition of the delta function [compare (3.8)].

If the analysis here is applied to the x dependence of a Laplace field $\phi(x, y)$, the Fourier transform will be a function $c(k, y)$ of y, as well as k. Satisfaction of the Laplace equation will entail an exponential behavior of $c(k, y)$, so that the result can be written

$$\phi(x, y) = \int_{-\infty}^{\infty} dk[a_+(k)e^{+ky} + a_-(k)e^{-ky}]e^{ikx} \qquad (3.52)$$

For a *real* field, (3.44*b*) makes it obvious that $c(-k, y) = c^*(k, y)$, and it follows that $a_\pm(-k) = a^*_\mp(k)$ because the exponential variations $e^{\pm ky}$ are linearly independent [in the same sense that any two distinct powers y^{p_1} and y^{p_2} are linearly independent, despite the inexpressibility of this as an orthogonality like (3.22) or (3.49)].

When the Laplace field also varies with z, it can become profitable to Fourier-analyze not only the x, but also the y, dependence, much as in (3.36). Later chapters (see Sec. 7.1) will deal with electromagnetic wave fields that are not restricted to the static Laplace field behavior, and then the analysis of all three degrees of freedom in functions $f(x, y, z)$ will become desirable. The analysis will entail introducing three independent wave numbers k_x, k_y, k_z, one for each degree of freedom, and then expression in terms of vectors $\mathbf{r}(x, y, z)$ and $\mathbf{k}(k_x, k_y, k_z)$ becomes possible. The Fourier integral (3.44) is generalized to

$$f(x, y, z) = (2\pi)^{-3/2} \iiint_{-\infty}^{\infty} dk_x\, dk_y\, dk_z\, c(k_x k_y k_z)e^{i(k_x x + k_y y + k_z z)}$$

which can be written more economically as

$$f(\mathbf{r}) = (2\pi)^{-3/2} \oint dV(\mathbf{k})c(\mathbf{k})e^{i\mathbf{k}\cdot\mathbf{r}} \qquad (3.53)$$

where $dV(\mathbf{k}) \equiv dk_x\, dk_y\, dk_z$ is a volume element in the space of wave vectors $\mathbf{k}$. The Fourier transform (3.44) is generalized to

$$c(\mathbf{k}) = (2\pi)^{-3/2} \oint dV(\mathbf{r}) f(\mathbf{r})e^{-i\mathbf{k}\cdot\mathbf{r}} \qquad (3.54)$$

The representation (3.53) amounts to a resolution of $f(\mathbf{r})$ onto the basis

$$e(\mathbf{k}, \mathbf{r}) = (2\pi)^{-3/2} e^{i\mathbf{k}\cdot\mathbf{r}} \qquad (3.55)$$

enumerated by all possible directions and real magnitudes of the vector $\mathbf{k}$. The set has the orthonormality property

$$\oint dV(\mathbf{r})e^*(\mathbf{k}, \mathbf{r})e(\mathbf{k}', \mathbf{r}) = \oint \frac{dV(\mathbf{r})}{(2\pi)^3} e^{i(\mathbf{k}'-\mathbf{k})\cdot\mathbf{r}} = \delta(\mathbf{k}' - \mathbf{k}) \tag{3.56}$$

and a completeness expressible as

$$\oint dV(\mathbf{k})e(\mathbf{k}, \mathbf{r})e^*(\mathbf{k}, \mathbf{r}') = \oint \frac{dV(\mathbf{k})}{(2\pi)^3} e^{i\mathbf{k}\cdot(\mathbf{r}-\mathbf{r}')} = \delta(\mathbf{r} - \mathbf{r}') \tag{3.57}$$

Three-dimensional delta functions

$$\begin{aligned} \delta(\mathbf{k}' - \mathbf{k}) &\equiv \delta(k_x' - k_x)\delta(k_y' - k_y)\delta(k_z' - k_z) \\ \delta(\mathbf{r} - \mathbf{r}') &\equiv \delta(x - x')\delta(y - y')\delta(z - z') \end{aligned} \tag{3.58}$$

are used here. Expression (3.57) affords a useful decomposition of such unit point-charge distributions as in (3.7).

3.3 SPHERICAL REPRESENTATIONS

Perhaps the most fundamentally important orthogonal basis, after the Fourier sets, is provided by the *spherical harmonics* $Y_{lm}(\vartheta, \varphi)$, on which arbitrary *directional* dependences, in three-dimensional space, can be resolved. The basis will be developed here as a means for representing Laplace fields in terms of spherical coordinates $\phi(r, \vartheta, \varphi)$.

The dependence of a field on any one of its variables can be Fourier-analyzed, as dependences on rectangular coordinates were in the preceding section, but it will soon appear that there is profit in doing this here only for the azimuthal dependence, on φ. This can be decomposed on a discretely enumerable basis, like the complex Fourier set (3.40), basically because only a finite interval $0 \le \varphi < 2\pi$ is needed for locating any point in the space. The substitutions $x \to \varphi$ and $L \to 2\pi$ into (3.40) provide the complete orthonormal basis

$$(2\pi)^{-1/2} e^{im\varphi} \qquad m = 0, \pm 1, \pm 2, \ldots \tag{3.59}$$

for representing the azimuthal dependence of

$$\phi(r, \vartheta, \varphi) = \sum_{m=-\infty}^{\infty} c_m(r, \vartheta) e^{im\varphi} \tag{3.60}$$

Subjecting this form to the Laplace condition, with the Laplacian expressed in spherical coordinates (A.39), will lead to a partial differential equation for each of the linearly independent coefficients $c_m(r, \vartheta)$. For solutions of a separable form $c_m \to R(r)\Theta_m(\vartheta)$, it is readily found that each equation can be written

$$\frac{1}{R}\frac{d}{dr}\left(r^2 \frac{dR}{dr}\right) + \frac{1}{\Theta_m \sin\vartheta}\frac{d}{d\vartheta}\left(\sin\vartheta \frac{d\Theta_m}{d\vartheta}\right) - \frac{m^2}{\sin^2\vartheta} = 0$$

On a basis like that discussed for the cartesian coordinate separation in (3.16), the radially dependent term must be set equal to some *separation constant* λ, and then the two ordinary differential equations

$$\frac{d}{dr} r^2 \frac{dR}{dr} = \lambda R \tag{3.61}$$

and

$$\left(-\frac{1}{\sin \vartheta} \frac{d}{d\vartheta} \sin \vartheta \frac{d}{d\vartheta} + \frac{m^2}{\sin^2 \vartheta} \right) \Theta_m = \lambda \Theta_m \tag{3.62}$$

are the result. Solutions of the latter will be discussed next.

The Legendre Polynomials

Solutions of (3.62) for the special case $m = 0$ are plainly sufficient for finding azimuthally symmetric Laplace fields $\phi(r, \vartheta)$ independent of φ. Attention here will be restricted to fields that are bounded in their ϑ dependence on the full interval $0 \le \vartheta \le \pi$.

The equation produced by setting $m = 0$ in (3.62) is well known as the *Legendre differential equation.* It is usually written in terms of the substitute $\mu \equiv \cos \vartheta \, [\to d/d\mu = -d/(\sin \vartheta \, d\vartheta)]$ for the independent variable, as

$$\left[\frac{d}{d\mu} (1 - \mu^2) \frac{d}{d\mu} + \lambda \right] \Theta_o = 0 \tag{3.63}$$

This is found† to have solutions that are finite everywhere on the sphere ($-1 \le \mu \equiv \cos \vartheta \le +1$) only for certain discrete values of the constant λ,

$$\lambda = l(l + 1) \qquad \text{with } l = 0, 1, 2, 3, \ldots \tag{3.64}$$

The solution for each of these values of l is known as a *Legendre polynomial*, $P_l(\mu)$, because it consists of a power series in μ with μ^l as a largest power. The simplest examples are

$$\begin{aligned} &P_0 = 1 \qquad P_1 = \mu \equiv \cos \vartheta \qquad P_2 = \tfrac{1}{2}(3\mu^2 - 1) \\ &P_3 = \tfrac{1}{2}(5\mu^3 - 3\mu) \qquad P_4 = \tfrac{1}{8}(35\mu^4 - 30\mu^2 + 3) \end{aligned} \tag{3.65}$$

and any one can be generated from the *Rodrigues formula*

$$P_l(\mu) = \frac{1}{2^l l!} \left(\frac{d}{d\mu} \right)^l (\mu^2 - 1)^l \tag{3.66}$$

The normalization is such that $P_l(1) = +1$ at the "north pole," $\vartheta = 0$. $P_l(-1) = (-1)^l$ at the "south pole," $\vartheta = \pi$, since the polynomials are respectively even and

† See Ref. 17, sec. 13.3, or any of numerous other references.

odd, in $\mu \equiv \cos \vartheta$, for l even and odd. At the "equator," $\vartheta = \frac{1}{2}\pi$, $P_l(0) = 0$ for odd l and

$$P_l(0) = \frac{(l-1)!!}{(-2)^{(1/2)l}(\frac{1}{2}l)!} \qquad \text{for even } l \geq 2 \tag{3.67}$$

The double factorial symbol used here is defined for any odd integer, $2n + 1$, as

$$(2n+1)!! \equiv 1 \cdot 3 \cdot 5 \ldots (2n-1)(2n+1) = \frac{(2n+1)!}{2^n n!} \tag{3.68}$$

just an ordinary factorial, like $n! = 1 \cdot 2 \cdot 3 \ldots (n-1)n$, with the even integers divided out.

The Legendre polynomials form a complete orthogonal set on the interval $-1 \leq \mu \leq 1$, having the property

$$\int_{-1}^{1} d\mu P_l(\mu) P_{l'}(\mu) = \delta_{ll'} \frac{2}{2l+1} \tag{3.69}$$

Any, not too pathological functional dependence on $0 \leq \vartheta = \cos^{-1} \mu \leq \pi$ can be resolved on the set.

Azimuthally Symmetric Laplace Fields

It should now be clear that a potential distribution independent of φ can be given the representation

$$\phi(r, \vartheta) = \sum_{l=0}^{\infty} R_l(r) P_l(\cos \vartheta) \tag{3.70}$$

if R_l satisfies (3.61) with $\lambda \equiv l(l+1)$. The general solution of this radial equation follows most simply from trying a power $R_l \sim r^{\alpha}$. It is then easy to find that either of the two roots of the quadratic $\alpha(\alpha + 1) = l(l+1)$ will do for α, these being $\alpha = +l$ and $\alpha = -(l+1)$. Thus, the general radial behavior required of a Laplace field is

$$R_l = A_l r^l + \frac{B_l}{r^{l+1}} \tag{3.71}$$

unbounded as $r \to \infty$ if $A_l \neq 0$ and as $r \to 0$ if $B_l \neq 0$. There must be sources of the field, to interrupt the Laplace field behavior, in one or both of the radial directions, orthogonal to the directions of increasing or decreasing ϑ, along which the field is bounded by the everywhere finite $P_l(\mu)$. The result is

$$\phi(r, \vartheta) = \sum_{l=0}^{\infty} \left(A_l r^l + \frac{B_l}{r^{l+1}} \right) P_l(\cos \vartheta) \tag{3.72}$$

for the general azimuthally symmetric Laplace field that is finite everywhere on $0 \leq \vartheta \leq \pi$ (and $r > 0$, $r < \infty$).

The form (3.72) is most immediately adaptable to boundary conditions given on some sphere, say an azimuthally symmetric potential distribution $\phi_a(\vartheta)$ on a sphere of radius a. If there are no other sources of field, either inside or outside the sphere, $\phi(r, \vartheta)$ must remain finite both as $r \to 0$ and as $r \to \infty$. This can be represented as in (3.72) only in the two pieces

$$\phi(r \le a, \vartheta) = \sum A_l r^l P_l \qquad \phi(r \ge a, \vartheta) = \sum A_l \frac{a^{2l+1}}{r^{l+1}} P_l \tag{3.73}$$

Coefficients $B_l = A_l a^{2l+1}$ have been adopted for the last sum in order to have the two pieces agree at $r = a$ for all ϑ. The remaining constants A_l are to be so determined that

$$\phi(a, \vartheta) = \sum A_l a^l P_l(\cos \vartheta) = \phi_a(\vartheta) \tag{3.74}$$

Finding the A_l from the given $\phi_a(\vartheta)$ is again a problem of the type (3.24), as follows from the orthogonality (3.69), so that the expressions

$$A_l = \frac{2l+1}{2a^l} \int_{-1}^{1} d(\cos \vartheta) P_l(\cos \vartheta) \phi_a(\vartheta) \tag{3.75}$$

complete defining the resultant field (3.73). More explicit applications are left to Exercises 3.10 to 3.12.

Representation by the infinite series of Legendre polynomials (3.72) is sometimes preferable to expression by some closed function, like the elliptic functions sometimes used for the field (2.11) arising from a uniform ring of charge. The coefficients needed for the series can be obtained by taking advantage of the fact that the potential on the axis of the ring, $\phi = q/(r^2 + a^2)^{1/2}$ with $q = 2\pi a \lambda$ the total charge, is elementary to find, and the further fact† that its power-series development must coincide with expression (3.72) evaluated at $\vartheta = 0$, where $P_l(1) = +1$:

$$\phi(r, 0) = \sum_{l=0}^{\infty} \left(A_l r^l + \frac{B_l}{r^{l+1}} \right) \tag{3.76}$$

Power series for $\phi = q/(r^2 + a^2)^{1/2}$ can be obtained from the binomial expansion [see the footnote to Eq. (2.18)]

$$\phi(r < a, 0) = \frac{q}{a}\left(1 + \frac{r^2}{a^2}\right)^{-1/2} = \frac{q}{a} \sum_{s=0}^{\infty} \frac{(-\frac{1}{2})!}{s!\,(-\frac{1}{2} - s)!} \left(\frac{r}{a}\right)^{2s}$$

and $\phi(r > a, 0)$ the same except that r and a are interchanged. This application of the binomial expansion to the noninteger power $-\frac{1}{2}$ can be justified if the factorials are treated as in

$$(-\tfrac{1}{2})! = [(-\tfrac{1}{2}) \cdot (-\tfrac{1}{2} - 1) \ldots (\tfrac{3}{2} - s) \cdot (\tfrac{1}{2} - s)] \cdot (-\tfrac{1}{2} - s)!$$

† This is equivalent to giving ϕ on a surface enclosing all space, the Dirichlet condition (3.14), with the surface a cylinder of vanishing diameter, centered on the $\vartheta = 0$ axis.

The final factorial here is canceled out in the above series coefficients and the first s factors can have their signs reversed to produce

$$\frac{(-\frac{1}{2})!}{(-\frac{1}{2}-s)!} = (-1)^s(s-\tfrac{1}{2})(s-\tfrac{3}{2})\dots(\tfrac{3}{2})\cdot(\tfrac{1}{2})$$
$$= \frac{(-1)^s(2s-1)(2s-3)\dots(3)(1)}{2^s} \qquad s \geq 1 \qquad (3.77)$$

A double factorial of the type (3.68), $(2s-1)!!$, occurs here, and if the substitution $l \equiv 2s$ is now made, the above power series can be written as

$$\phi(r<a, 0) = \frac{q}{a}\left[1 + \sum \frac{(l-1)!!}{(-2)^{(1/2)l}(\frac{1}{2}l)!}\left(\frac{r}{a}\right)^l\right]$$

with the summation restricted to $l = 2, 4, 6 \dots$. Expressed in this way, the coefficients are recognizable as just the equator values $P_l(0)$ of (3.67) for the Legendre polynomials of even $l \geq 2$. Moreover, $P_l(0) = 0$ for odd l's and $P_0 \equiv 1$ everywhere. It thus turns out that all $B_l = 0$, $A_l = qP_l(0)/a^{l+1}$ in $\phi(r<a, 0)$ and all $A_l = 0$, $B_l = qP_l(0)a^l$ in $\phi(r>a, 0)$. Then at *any* position $r \neq a$,

$$\phi(r<a, \vartheta) = \frac{q}{a}\sum_{l=0}^{\infty}\left(\frac{r}{a}\right)^l P_l(0)P_l(\cos\vartheta)$$
$$\phi(r>a, \vartheta) = \frac{q}{r}\sum_{l=0}^{\infty}\left(\frac{a}{r}\right)^l P_l(0)P_l(\cos\vartheta) \qquad (3.78)$$

Just why coefficients have turned out to be proportional to equator values of the Legendre polynomials will become obvious in connection with (3.105). Neither of the expressions (3.78) can be used in the immediate vicinity of $r = a$ $[\mathsf{E}(a, \frac{1}{2}\pi) \to \infty]$.

Multipole Expansion of a Point-Charge Field

Representation of the effects of arbitrary charge distributions on situations of spherical symmetry can be prepared by expressing the field contributed by any one charge element, treated as a point charge q, in a series of the Legendre polynomials like (3.72). If the charge element is situated at some source point $\mathbf{r}_s$,

$$\phi_q(\mathbf{r}_f) = \frac{q}{|\mathbf{r}_f - \mathbf{r}_s|} = \frac{q}{(r_f^2 - 2r_f r_s\cos\vartheta + r_s^2)^{1/2}} \qquad (3.79)$$

where ϑ is the angle between the radius vector $\mathbf{r}_f$ to the field point and the position vector $\mathbf{r}_s$, which is thus employed as the polar axis. The expression can be expanded into a power series in the ratio $\rho \equiv r_f/r_s$ for $r_f < r_s$ (and in $\rho \equiv r_s/r_f$ for $r_f > r_s$), as already seen in (2.18). The coefficients in the power series will have to be functions of $\mu \equiv \cos\vartheta$, and it turns out that these are exactly the Legendre polynomials! Indeed, $P_{0,1,2}$ of (3.65) can be recognized as coefficients in the terms of the expansion as presented in (2.18). The power-series expansion thus becomes a generator of the polynomials, alternative to the Rodrigues formula (3.66).

The general identification of the coefficients in (3.72) for the azimuthally symmetric point-charge field (3.79) can be approached as for the charged-ring field (3.78) through analytic continuation from the field values on the axis. The field (3.79) is everywhere proportional to $(1 - 2\rho \cos \vartheta + \rho^2)^{-1/2}$, and on the axis, where $\cos \vartheta = +1$, this becomes

$$\frac{1}{(1 - 2\rho + \rho^2)^{1/2}} = \frac{1}{1 - \rho} = \sum_{l=0}^{\infty} \rho^l \tag{3.80}$$

just a geometric series like that used on page 65 (it is also a special case of the binomial expansion). This shows that the expression for $\phi_q(r_f < r_s, 0)$ is given by (3.76) with $A_l = q/r_s^{l+1}$, all $B_l = 0$, and that $\phi(r_f > r_s, 0)$ has all $A_l = 0$, $B_l = qr_s^l$. At a general field position, the field arising from each *unit* of charge at r_s, $\vartheta = 0$ is the special case of (3.72),

$$\frac{1}{(r_f^2 - 2r_f r_s \cos \vartheta + r_s^2)^{1/2}} = \sum_{l=0}^{\infty} \frac{r_<^l}{r_>^{l+1}} P_l(\cos \vartheta) \tag{3.81}$$

where $r_<$ and $r_>$ are respectively the lesser and greater of the magnitudes r_f and r_s. This result amounts to a multipole expansion of a point unit-charge field [or of the propagator (3.5)] relative to an axis passing through the charge. This aspect will be enlarged upon in an ensuing subsection. Applications to point charges introduced into spherically symmetric electrostatic situations are taken up in Chap. D and Exercise 3.15.

Spherical Harmonics

A Laplace field $\phi(r, \vartheta, \varphi)$ that is not azimuthally symmetric is described with the help of solutions of the associated Legendre equation (3.62) with $m \neq 0$. The solutions that are bounded at all angles $0 \leq \vartheta \leq \pi$ again exist only for the discrete eigenvalues $\lambda = l(l + 1)$ and, moreover, only for $l \geq |m|$. They are the associated Legendre functions

$$P_l^m(\vartheta) = \frac{(1 - \mu^2)^{(1/2)m}}{2^l l!} \left(\frac{d}{d\mu}\right)^{l+m} (\mu^2 - 1)^l \qquad \text{with } \mu \equiv \cos \vartheta \tag{3.82}$$

defined for each $l = 0, 1, 2 \ldots$ and for $m = 0, \pm 1, \pm 2, \ldots, \pm l$ only. Clearly $P_l^0 \equiv P_l(\cos \vartheta)$ of (3.66), and for $m > 0$

$$P_l^m(\vartheta) = (1 - \mu^2)^{(1/2)m} \left(\frac{d}{d\mu}\right)^m P_l(\mu) \tag{3.83}$$

The associated Legendre functions with $m < 0$ are not linearly independent of those with $m > 0$, since the connection

$$P_l^{-m}(\vartheta) = (-1)^m \frac{(l - m)!}{(l + m)!} P_l^m \tag{3.84}$$

can be deduced† from (3.82); this conforms to the fact that the associated Legendre equation is independent of the sign of m.

The associated Legendre functions appear only in $e^{im\varphi}P_l^m(\vartheta)$ when the decompositions like (3.60) are formed. Moreover, the product functions with $m \gtrless 0$ are linearly independent of each other because of the factors $e^{\pm im\varphi}$. For such reasons it becomes appropriate to define *spherical harmonics* $Y_{lm}(\vartheta, \varphi)$, each of which is proportional to one of the product functions and is directly calculable from

$$Y_{lm}(\vartheta, \varphi) \equiv \frac{(-)^{l+m}}{2^l l!}\left[\frac{2l+1}{4\pi}\frac{(l-m)!}{(l+m)!}\right]^{1/2} \sin^m \vartheta \left[\frac{d}{d(\cos\vartheta)}\right]^{l+m} \sin^{2l}\vartheta\, e^{im\varphi} \tag{3.85}$$

These form an orthogonal set, and the factors adopted in their definition normalize them to

$$\oint d\Omega\, Y_{lm}^*(\vartheta, \varphi) Y_{l'm'}(\vartheta, \varphi) = \delta_{ll'}\delta_{mm'} \tag{3.86}$$

where $d\Omega$ is the element of solid angle (1.9) and $\oint d\Omega = 4\pi$.

The most important low-order spherical harmonics are $Y_{00} = (4\pi)^{-1/2}$,

$$Y_{10} = \left(\frac{3}{4\pi}\right)^{1/2} \cos\vartheta \qquad Y_{1,\pm 1} = \mp\left(\frac{3}{8\pi}\right)^{1/2} \sin\vartheta\, e^{\pm i\varphi} \tag{3.87}$$

and

$$\begin{aligned} Y_{20} &= \left(\frac{5}{16\pi}\right)^{1/2} (3\cos^2\vartheta - 1) \\ Y_{2,\pm 1} &= \mp\left(\frac{15}{8\pi}\right)^{1/2} \sin\vartheta\cos\vartheta\, e^{\pm i\varphi} \\ Y_{2,\pm 2} &= \left(\frac{15}{32\pi}\right)^{1/2} \sin^2\vartheta\, e^{\pm 2i\varphi} \end{aligned} \tag{3.88}$$

Quite generally,

$$Y_{l0} \equiv \left(\frac{2l+1}{4\pi}\right)^{1/2} P_l(\cos\vartheta) \tag{3.89}$$

$$Y_{l,-m} = (-1)^m Y_{lm}^* \tag{3.90}$$

$$Y_{ll} = (-1)^l Y_{l,-l}^* = \frac{(-1)^l}{2^l l!}\left[\frac{(2l+1)!}{4\pi}\right]^{1/2} \sin^l\vartheta\, e^{il\varphi} \tag{3.91}$$

Cases with $m < 0$ are simpler to calculate from (3.85) than cases with $m > 0$.

† As in Ref. 17, sec. 13.3, for example.

The products $r^l Y_{lm}(\vartheta, \varphi)$ are sometimes called *solid spherical harmonics*. With the help of

$$x = r \sin\vartheta \cos\varphi, \qquad y = r \sin\vartheta \sin\varphi \qquad z = r\cos\vartheta$$

the three of order $l = 1$ can be written

$$rY_{10} = \sqrt{\frac{3}{4\pi}}\, z \qquad rY_{1,\pm 1} = \mp\sqrt{\frac{3}{8\pi}}\,(x \pm iy) \tag{3.92}$$

Implicit here are resolutions of the vector **r** into its so-called† *helicity components* $r_{\pm 1} \equiv \mp 2^{-1/2}(x \pm iy)$ and its *axial component* $r_0 \equiv z$. The $l = 1$ spherical harmonics are themselves respectively proportional to components $\hat{r}_{m=0,\pm 1} \equiv r_m/r$, of the unit vector $\hat{\mathbf{r}} \equiv \mathbf{r}/r$ that points out the (ϑ, φ) direction:

$$Y_{1m}(\vartheta, \varphi) \equiv \sqrt{\frac{3}{4\pi}}\,\hat{r}_m = Y_{1m}(\hat{\mathbf{r}}) \tag{3.93}$$

The last form here introduces an appropriate notation for the direction, using $\hat{\mathbf{r}}$ in place of (ϑ, φ). The higher-order spherical harmonics correspondingly form resolutions of reduced tensors having a higher rank than the vector (rank 1). An example of a rank 2 reduced tensor was introduced in Sec. 2.3, to represent quadrupole moments, and its connection with $Y_{2m}(\hat{\mathbf{r}})$ will be shown in the next subsection.

In terms of the complex component numbers $r_{\pm 1} = \mp 2^{-1/2}(x \pm iy)$, the scalar product of the vector **r** with itself can be written

$$\mathbf{r} \cdot \mathbf{r} \equiv r^2 = x^2 + y^2 + z^2 = \sum_{m=0,\,\pm 1} r_m^* r_m \tag{3.94}$$

as is easy to verify. For the unit vector,

$$\hat{\mathbf{r}} \cdot \hat{\mathbf{r}} \equiv 1 = \sum_{m=-1}^{+1} \hat{r}_m^* \hat{r}_m = \frac{4\pi}{3} \sum_{m=-1}^{1} |Y_{1m}|^2 \tag{3.95}$$

according to (3.93), and this calls attention to a general property possessed by all the spherical harmonics, of any order l,

$$\sum_{m=-l}^{l} |Y_{lm}|^2 = \frac{2l+1}{4\pi} \tag{3.96}$$

This can be verified for $l = 1$ from (3.87) and for $l = 2$ from (3.88).

The result (3.95) stemming from the scalar product of a unit vector direction with itself can be generalized to one that follows from the scalar product of the two different unit vectors $\hat{\mathbf{r}}_1$, $\hat{\mathbf{r}}_2$, respectively pointing out two distinct directions (ϑ_1, φ_1) and (ϑ_2, φ_2):

$$\hat{\mathbf{r}}_1 \cdot \hat{\mathbf{r}}_2 \equiv \sum_{m=-1}^{+1} \hat{r}_{1m}^* \hat{r}_{2m} = \frac{4\pi}{3} \sum_{m=-1}^{+1} Y_{1m}^*(\vartheta_1, \varphi_1) Y_{1m}(\vartheta_2, \varphi_2) \tag{3.97}$$

† See Ref. 17, sec. 10.5 and p. 4.26.

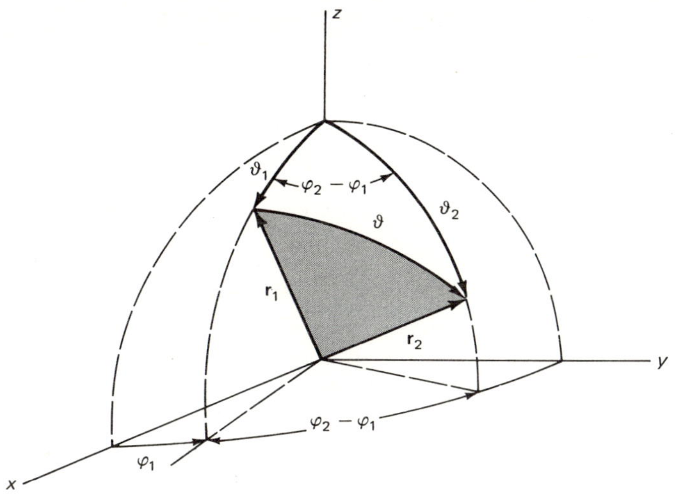

Figure 3.1

This can be easily seen to amount to the well-known relation between the "sides" of a spherical triangle† indicated in Fig. 3.1,

$$\cos \vartheta = \cos \vartheta_1 \cos \vartheta_2 + \sin \vartheta_1 \sin \vartheta_2 \cos (\varphi_2 - \varphi_1) \tag{3.98}$$

with the left side equivalent to $P_1(\cos \vartheta)$. This and its generalization to

$$P_l(\cos \vartheta) = \frac{4\pi}{2l+1} \sum_{m=-l}^{l} Y_{lm}^*(\vartheta_1, \varphi_1) Y_{lm}(\vartheta_2, \varphi_2) \tag{3.99}$$

turn out to be very useful properties of the spherical harmonics. They permit changing from an initial polar axis, $\hat{\mathbf{r}}_1$ say, relative to which the point at $\mathbf{r}_2$ has angle ϑ, to another (z) axis relative to which $\mathbf{r}_2$ has ϑ_2. Relation (3.99), known as the *spherical-harmonic addition theorem*, is actually symmetrical in (ϑ_1, φ_1) and (ϑ_2, φ_2); the complex-conjugation operation can be put on either factor in the terms of the sum, since the result is real and the switch amounts to no more than an interchange of the order of the terms, in view of the $m \gtrless 0$ interrelationship (3.90). Notice that (3.96) is a special case of (3.99) with $\vartheta = 0$.

Multipoles of Order 2^l

Any finite electrostatic source can be decomposed into multipoles, as suggested in Sec. 2.3. There this was done by expanding the propagator $|\mathbf{r}_f - \mathbf{r}_s|^{-1}$, as in (2.18), into successive powers of r_s/r_f. Then only monopole, dipole, and quadru-

† Discussed more fully in Ref. 17, fig. 13.8.

pole moments were defined explicitly because the development into cartesian tensor components is awkward to carry on to higher orders. Now at hand, in (3.81), is a development of the propagator to arbitrarily high order. Inserting this into the expression (1.26) for the field arising from an entire source of charge $q = \oint dV(\mathbf{r}_s)\rho(\mathbf{r}_s)$ produces

$$\phi(\mathbf{r}_f) = \sum_{l=0}^{\infty} \frac{1}{r_f^{l+1}} \oint dV(\mathbf{r}_s)\rho(\mathbf{r}_s) r_s^l P_l(\hat{\mathbf{r}}_s \cdot \hat{\mathbf{r}}_f) \tag{3.100}$$

where the notation $\hat{\mathbf{r}}_s \cdot \hat{\mathbf{r}}_f$ for $\cos \vartheta$ emphasizes that the angle ϑ of (3.81) is the one between the source and field point directions. The successive terms, inversely proportional to the powers 1, 2, 3, ..., $l + 1$, ... of r_f, respectively represent the monopole, dipole, quadrupole, ..., 2^l pole, ... effects of the source.

The integral coefficients in (3.100) still depend on the field-point direction $\hat{\mathbf{r}}_f$ from which the source is being viewed. To find moments characteristic of the source itself and independent of the viewpoint, the dependence on $\hat{\mathbf{r}}_f$ must be factored out of the integrals; this can be done with the help of the spherical-harmonic addition theorem (3.99). It transforms the integrals into

$$\oint dV(\mathbf{r}_s)\rho(\mathbf{r}_s) r_s^l P_l(\hat{\mathbf{r}}_f \cdot \hat{\mathbf{r}}_s) = \frac{4\pi}{2l+1} \sum_{m=-l}^{l} Y_{lm}(\hat{\mathbf{r}}_f) \oint dV(\mathbf{r}_s)\rho(\mathbf{r}_s) r_s^l Y_{lm}^*(\hat{\mathbf{r}}_s)$$

Now $2l + 1$ components of a 2^l-pole moment can be defined:

$$q_{lm} \equiv \oint dV(\mathbf{r}_s)\rho(\mathbf{r}_s) r_s^l Y_{lm}^*(\hat{\mathbf{r}}_s) \qquad m = 0, \pm 1, \ldots, \pm l \tag{3.101}$$

each a (generally complex) number that depends only on the distribution $\rho(\mathbf{r}_s)$ of the charge in the source. The resultant potential (3.100) becomes

$$\phi(\mathbf{r}_f) = 4\pi \sum_{l=0}^{\infty} \sum_{m=-l}^{l} \frac{1}{2l+1} \frac{q_{lm}}{r_f^{l+1}} Y_{lm}(\hat{\mathbf{r}}_f) \tag{3.102}$$

This is a real expression, despite the complexity of $q_{lm} = (-)^m q_{l,-m}^*$ and $Y_{lm} = (-)^m Y_{l,-m}^*$.

Clearly $(4\pi)^{1/2} q_{00} = q$ is the total charge in the source. The connection of q_{1m} and q_{2m} to the cartesian components of the dipole moment vector $\mathbf{D}$ and the quadrupole moment tensor $\mathbf{Q}$, defined in (2.20) and (2.21), can be seen from the explicit expressions (3.92) and (3.88) for the spherical harmonics:

$$\begin{aligned} q_{10} &= \oint dq \sqrt{\frac{3}{4\pi}}\, z_s = \sqrt{\frac{3}{4\pi}}\, D_z \\ q_{1,\pm 1} &= \mp \sqrt{\frac{3}{8\pi}} \oint dq (x_s \pm i y_s)^* = \mp \sqrt{\frac{3}{8\pi}} (D_x \mp i D_y) \end{aligned} \tag{3.103}$$

where the abbreviation $dq \equiv \rho(\mathbf{r}_s)\, dV(\mathbf{r}_s)$ has been used for the charge element, as before, and

$$\begin{aligned} q_{20} &= \sqrt{\frac{5}{16\pi}} \oint dq(3z_s^2 - r_s^2) = \sqrt{\frac{5}{16\pi}}\, Q_{zz} \\ q_{2,\pm 1} &= \mp \sqrt{\frac{15}{8\pi}} \oint dq z_s(x_s \mp iy_s) = \mp \frac{1}{3}\sqrt{\frac{15}{8\pi}}\,(Q_{xz} \mp iQ_{yz}) \\ q_{2,\pm 2} &= \sqrt{\frac{15}{32\pi}} \oint dq(x_s \mp iy_s)^2 = \frac{1}{3}\sqrt{\frac{15}{32\pi}}\,(Q_{xx} - Q_{yy} \mp 2iQ_{xy}) \end{aligned} \tag{3.104}$$

The combinations $D_{\pm 1} \equiv \mp 2^{-1/2}(D_x \pm iD_y)$ occurring here form helicity components of the vector $\mathbf{D}$ on the same basis as the resolutions $r_{\pm 1}$ of $\mathbf{r}$ discussed in the preceding subsection. The results, for $m = 0, \pm 1$,

$$D_m = \oint dq r_m = \left(\frac{4\pi}{3}\right)^{1/2} q_{1m}^* = \left(\frac{4\pi}{3}\right)^{1/2} (-)^m q_{1,-m}$$

show that the definitions of the components q_{1m} correspond to resolutions on a helicity rather than a cartesian basis. The five components q_{2m} likewise correspond to resolutions of the rank 2 quadrupole tensor on a helicity basis. Notice that the quadrupole tensor also really has only five independent components on the cartesian basis of (2.21). A general rank 2 tensor has nine cartesian components $(xx, xy, xz, yx, yy, \ldots)$, but the quadrupole tensor is first a symmetrical one, $Q_{ij} = Q_{ji}$, and this means it has only three, rather than six, independent off-diagonal components $Q_{xy} = Q_{yx}$, Q_{yz}, Q_{zx}. Moreover, it has a zero trace, $Q_{xx} + Q_{yy} + Q_{zz} \equiv 0$, as noted in (2.58), so that only two of the three diagonal components are independent for *any* source. Such a five-component entity is known as a rank 2 *reduced symmetrical tensor*. For like reasons, the rank l tensor needed to represent a 2^l-pole moment has only $2l + 1$ rather than 3^l, independent components, and this accounts for the felicity of representing the 2^l-pole effect of the source through the $2l + 1$ spherical harmonics $Y_{lm}(\hat{\mathbf{r}}_f)$ of (3.102).

For a source with an axis of symmetry, like the ring of charge for which the field (3.78) was found, it is naturally advantageous to choose the symmetry axis as the polar axis for the definition of the m values that distinguish the components of each moment q_{lm}. Then the charge elements $dq = \lambda a\, d\varphi_s = q\, d\varphi_s/2\pi$ of the ring all have the same $r_s = a$ and $\vartheta_s = \frac{1}{2}\pi$, so that (3.101) becomes

$$q_{lm} = \frac{q}{2\pi} a^l \int_0^{2\pi} d\varphi_s\, Y_{lm}^*(\tfrac{1}{2}\pi, \varphi_s)$$

The integral here vanishes except for $m = 0$ because Y_{lm}^* is proportional to $e^{-im\varphi_s}$ in its φ_s-dependence. With $Y_{l0}^*(\frac{1}{2}\pi, \varphi) = [(2l + 1)/4\pi]^{1/2} P_l(0)$ according to (3.89), the moments of the ring source become

$$q_{lm} = \delta_{m0}\left(\frac{2l + 1}{4\pi}\right)^{1/2} qa^l P_l(0) \tag{3.105}$$

Only the $m = 0$ terms of the multipole expansion (3.102) survive, and the second of the two expressions (3.78) for the field is the result. The first of the two expressions, applying inside the source, follows from the symmetry of the propagator (3.81) in r_s and r_f. The considerations here clarify the occurrence of the equator values of the Legendre polynomials in the field expression (3.78); they measure the moments of a source lying on the equator of the reference frame being used.

EXERCISES

3.1 (*a*) Show that the contribution by ϕ_o of (3.10) to the surface integral (3.11) vanishes, as asserted on page 61. (Why is it actually sufficient to show this only for a single point charge?)

(*b*) By (*a*), the fields ϕ and $E_n = -\nabla_n \phi$ appearing in (3.11) need only be ones generated by sources external to the volume of definition. Show that the two terms of ϕ_o can be understood as arising from a surface-charge distribution $\sigma = E_n/4\pi$ superposed on a double layer $\tau = \phi/4\pi$, consistent with the discontinuities (2.31) and (2.33) (and hence consequences of ignoring fields outside the surface and their actual sources).

3.2 A point charge q is fixed at $\mathbf{r}_q$ in a field $\mathbf{E} = -\nabla\phi$ due to sources external to q. Show (*a*) that the interaction energy (2.63) can here be expressed as

$$U(\mathbf{r}_q) = -\oint dV(\mathbf{r})\mathbf{E}_q \cdot \frac{\nabla\phi(\mathbf{r})}{4\pi}$$

where $\mathbf{E}_q(\mathbf{r} - \mathbf{r}_q)$ is the Coulomb field of q, and (*b*) how this expression can be integrated to the result (2.66) with the help of $\nabla \cdot \mathbf{E}_q = 4\pi q\delta(\mathbf{r} - \mathbf{r}_q)$.

3.3 Suppose that the space described by the potential distribution (3.30) is cut off from $y \to \infty$ by a grounded ($\phi = 0$) plane at $y = l$.

(*a*) Show that the exponentials in (3.30) are then to be replaced by the ratio

$$\frac{\sinh[(2N+1)\pi(l-y)/L]}{\sinh[(2N+1)\pi l/L]}$$

and that the result reverts to (3.30) for $l \to \infty$.

(*b*) Show that $\phi(x = \frac{1}{2}L, l \to 0) \to \phi_o(1 - y/l)$, as to be expected.

3.4 Suppose that, as suggested by Exercise 3.3(*b*), a series representation is sought for only the *difference* $\phi(x, y) - \phi_o(1 - y/l)$. Show that enough generality to satisfy the boundary conditions on that difference is provided by a series in the set (3.35).

3.5 Demonstrate how the expressions found in either Exercises 3.3 or 3.4 can be used to write down (without calculation) the field $\phi(x, y)$ in a space enclosed by walls at $x = 0$, $y = 0$ that are *both* at a potential ϕ_o and by grounded walls at $x = L$ and $y = l$ ($-\infty < z < +\infty$, as before).

3.6 (*a*) Find expressions for the *vector* field distribution $\mathbf{E}(x, 0)$ on the $y = 0$ surface of the situation described by (3.30) and (3.34).

(*b*) Do the same for the situation of Exercise 3.3.

(*c*) Decide whether the interruption of the space by the grounded plane at $y = l$ has enhanced or diminished the field strengths at $y = 0$.

3.7 Suppose that the source of field between two infinite grounded planes at $x = 0$ and $x = L$ is provided by a uniform charge distribution, σ per unit area, on the part of the $y = 0$ plane ($0 < x < L$, $-\infty < z < +\infty$) enclosed by the grounded walls. The situation is to be presumed symmetrical in $y \leftrightarrow -y$. [See (3.15).] Show that the field component $\phi \sim \exp(-\pi y/L)$ is the same as in (3.30) if $\phi_o = 2\sigma L$ is chosen but that it is subject to smaller corrections.

3.8 Four sides of a cubical space, defined by the planes $x = y = 0$, $x = y = L$, are all grounded to $\phi = 0$. Find $\phi(x, y, z)$ within the space for each of the following cases:

(*a*) The bottom $z = 0$ is also grounded, but the $z = L$ top is maintained at the uniform potential ϕ_o.

(*b*) *Both* top and bottom are maintained at ϕ_o.

(*c*) Show that if the top is kept at ϕ_o but the $z = 0$ bottom is given the equal and opposite potential $-\phi_o$, then $\phi(x, y, z)$ of case (*b*) need only be modified by replacing the ratios

$$\frac{\cosh k_{mn}(z - \frac{1}{2}L)}{\cosh (k_{mn} L/2)}$$

with ratios of hyperbolic *sines* having the same arguments. Here $k_{mn} \equiv (m^2 + n^2)^{1/2}\pi/L$ if $m, n = 1, 3, 5, \ldots$.

3.9 Useful relations between the Legendre polynomials can be obtained from (3.81), expressed for $\rho \equiv r_< / r_> < 1$ and $\mu \equiv \cos \vartheta$.

(*a*) Show that a derivative $d/d\rho$ of the expression yields

$$(\mu - \rho)\Sigma\rho^l P_l = (1 - 2\mu\rho + \rho^2)\Sigma l\rho^{l-1}P_l$$

(*b*) If (*a*) is to hold for any $0 < \rho < 1$, coefficients of each different power of ρ on the two sides must match. Show how this leads to the recurrence relations

$$(2l + 1)\mu P_l = (l + 1)P_{l+1} + lP_{l-1}$$

(*c*) Show that a derivative $d/d\mu$ of (3.81) yields

$$P_l = P'_{l+1} - 2\mu P'_l + P'_{l-1}$$

where $P'_l(\mu) \equiv dP_l/d\mu$.

(*d*) Show how combining (*c*) with *derivatives* of (*b*) leads to

$$P'_{l+1} - P'_{l-1} = (2l + 1)P_l \qquad P'_l = \frac{l(P_{l-1} - \mu P_l)}{1 - \mu^2}$$

(*e*) Use (*b*) to add $P_5(\mu)$ to the list in (3.65) and then check (*d*) for $l = 1, 2, 3, 4$. Is the last expression in (*d*) consistent with the original differential equation (3.63) for $\lambda \equiv l(l + 1)$?

3.10 Apply (3.73) to the case of a sphere $r = a$ having a uniform potential ϕ_o maintained on one hemispherical part of its surface and $-\phi_o$ on the other half.

(*a*) Evaluate the field $\mathbf{E}_o$ at the sphere center.

(*b*) You should expect the field *outside* the sphere to take on an ideal dipole form sufficiently far away. Express the dipole moment in terms of $\mathbf{E}_o$ and a.

(*c*) What is the *quadrupole* effect of the sphere?

3.11 Suppose that the sphere of Exercise 3.10 instead has *charges* $+q$ and $-q$ distributed uniformly on its respective hemispherical surfaces.

(*a*) Find all the moments (3.101) of the distribution here.

(*b*) Modify (3.102) so that it represents $\phi(r > a, \vartheta)$ and provide another modification to represent $\phi(r < a, \vartheta)$.

(*c*) Find $\mathbf{E}_o$ at the sphere center and express the dipole moment $\mathbf{D}$ in terms of it. [Compare Exercise 3.10(*b*).]

3.12 (*a*) Adapt the procedures of Exercise 3.11 to finding $\phi(r \gtrless a, \vartheta)$ *everywhere* about a *single* hemispherical surface $\vartheta < \frac{1}{2}\pi$, $r = a$, that has a charge q distributed uniformly on it. (Pay special attention to the $l = 0$ component!)

(*b*) Express the radial field components $\mathsf{E}_r(r \gtrless a, \vartheta)$.

(*c*) Show that for all $\vartheta < \frac{1}{2}\pi$, E_r has the proper discontinuity $\Delta\mathsf{E}_r = 4\pi(q/2\pi a^2)$ at $r = a$ and that E_r is properly *continuous* at all $\vartheta > \frac{1}{2}\pi$ despite the different expressions for $\mathsf{E}_r(r > a, \vartheta)$ and $\mathsf{E}_r(r < a, \vartheta)$. (Results of Exercise 3.10 can help here.)

(*d*) Evaluate the field strength E_o at $r = 0$ and show that this agrees with a result of an integration like (1.5).

3.13 Compare the results (3.78) with those of Exercise 2.17.

3.14 Charge is distributed in the density $\rho(r, \vartheta) = 4qr \cos \vartheta/\pi a^4$ inside a sphere of radius a. Find *all* its moments and compare your results with those of Exercise 2.18.

3.15 Consider the field $\phi(r > a, \vartheta)$ outside a grounded ($\phi = 0$) sphere of radius a that has a point charge q at a distance $r_o > a$ from its center.

(*a*) Determine the coefficients for such a representation as (3.72) for the *difference* $\phi_o = \phi - q/|\mathbf{r} - \mathbf{r}_o|$, with the help of (3.81) in seeing how the condition $\phi(a, \vartheta) = 0$ can be satisfied.

(*b*) Show that the difference ϕ_o is the same as the field of a point charge $q' = -qa/r_o$ placed at a position $r' < a$; find r'.

3.16 (*a*) Show that any pair of differentiable functional forms $f(\zeta)$ and $g(\zeta)$ can be used to produce the general solution of the Laplace equation in two dimensions

$$\phi(x, y) = f(x + iy) + g(x - iy)$$

where $i \equiv \sqrt{-1}$.

(*b*) Show that the solution in (*a*) can be represented by (3.52) if $(2\pi)^{1/2}a_{\pm}(k)$ are respectively the Fourier transforms of $g(\zeta)$ and $f(\zeta)$.

The following exercises, together with Exercises C.8 to C.15, introduce elements of the complex representations of fields like $\phi(x, y)$, independent of z, as functions of a complex plane of points $\zeta \equiv x + iy$ which can be used to represent field points in place of the two-component position *vector* $\mathbf{r} = \mathbf{i}x + \mathbf{j}y$. The complex-number calculus thus becomes deducible from vector calculus specialized to just two dimensions.

3.17 (*a*) Show that any differentiable expression $f(\zeta)$ analytic in ζ, that is, containing x, y only in the combination $\zeta = x + iy$, automatically satisfies the Laplace equation.

(*b*) In general, $f(\zeta)$ will be complex-valued, i.e., have

$$f(\zeta) = u(x, y) + iv(x, y)$$

where u and v are both real. These are called the *real* and *imaginary* parts of $f(\zeta)$, respectively, and are denoted $u(x, y) = \text{Re}\,[f(\zeta)]$ and $v = \text{Im}\,[f(\zeta)]$. Show that u and v *individually* satisfy the Laplace equation because two complex numbers are equal only if their real and imaginary parts are separately equal. (Thus, setting down any expression analytic in ζ immediately presents two real functions, *each* of which can represent an electrostatic potential in some situation!)

3.18 (*a*) To illustrate the results of Exercise 3.17 find the real and imaginary parts of $f(\zeta) = (\phi_o/x_o^2)\zeta^2$, where ϕ_o and x_o are real constants; then check that $\nabla^2 \equiv 0$ for *each*.

(*b*) Show that an electrostatic field $\phi(x, y) = (\phi_o/x_o^2)\,\text{Re}\,(\zeta^2)$ has as its equipotentials the rectangular hyperbolas asymptotic to $y = \pm x$.

(*c*) Consider, instead, the field $\phi = (\phi_o/x_o^2)\,\text{Im}\,(\zeta^2)$. Through finding its equipotentials, show that it differs from the field in (*a*) only in being bodily reoriented at 45° to ϕ of (*a*). [Such fields are called *quadrupole fields* because central parts of them can be generated by replacing an equipotential in each *quadrant* with an hyperbolically shaped conducting surface at the corresponding potential (see page 495). They are useful in ways indicated in Exercises 5.3 and 5.4.]

3.19 Complex conjugates (cc) like $\zeta^* \equiv x - iy$ and $f^*(\zeta)$ are formed by reversing the sign of i wherever it occurs, explicitly or implicitly. Then, when $x = (\zeta + \zeta^*)/2$ and $y = (\zeta - \zeta^*)/2i$ are substituted into an arbitrary $F(x, y)$, it will contain *both* ζ and ζ^* and will *not* be analytic in ζ alone.

(*a*) Plainly $u(x, y) + iv(x, y) = f(\zeta)$, as in Exercise 3.17(*b*), will contain $\zeta = x + iy$ but *not* ζ^* only if u and v bear some relation to each other. Show how it follows from $(\partial f/\partial \zeta^*)_\zeta = 0$ that

$$\frac{\partial u}{\partial x} = \frac{\partial v}{\partial y} \qquad \text{and} \qquad \frac{\partial u}{\partial y} = -\frac{\partial v}{\partial x}$$

(These are the famous *Cauchy-Riemann conditions* for the analyticity of $u + iv$.)

(*b*) Show from (*a*) that curves, e.g., equipotentials, on which $u(x, y)$ is constant are orthogonal to curves of constant $v(x, y)$. [Thus, when $u = \phi(x, y)$ represents an electrostatic potential, the curves of constant v are force-field lines of $\mathbf{E} = -\nabla\phi$. See (A.9).]

3.20 (*a*) Using the Cauchy-Riemann conditions of Exercise 3.19(*a*), show that when $\phi = \text{Re}\,[f(\zeta)] \equiv \text{Im}\,[+if(\zeta)]$, the components of $\mathbf{E} = -\nabla\phi$ are given by

$$-f'(\zeta) \equiv -\frac{df}{d\zeta} = \mathsf{E}_x - i\mathsf{E}_y$$

(*b*) By applying the Cauchy-Riemann conditions to the analytic function $\mathsf{E}(\zeta) \equiv \mathsf{E}_x(x, y) - i\mathsf{E}_y(x, y)$, like $-f'(\zeta)$ of (*a*), show that they are equivalent to the Maxwell equations (2.1) in a sourceless region ($\rho \equiv 0$).

3.21 Demonstrate how (3.32) can be used as in Exercise 3.20(*a*) to provide $\mathbf{E}(x, 0)$ of Exercise 3.6(*a*).

3.22 (*a*) Let the analytic function $\mathsf{E}(\zeta) = \mathsf{E}_x - i\mathsf{E}_y$ represent an electrostatic field $\mathbf{E}(x, y)$ with $\mathsf{E}_z = 0$, and form the integral on the left side of

$$\int \mathsf{E}(\zeta)\, d\zeta = \int \mathbf{E} \cdot d\mathbf{r} + i \int \mathbf{E} \cdot d\mathbf{S}_1$$

where $d\zeta \equiv dx + i\,dy$ stands for an element $d\mathbf{r}(dx, dy)$ of path (contour) in the plane. Show the equivalence of the contour integral to the right side, in which $d\mathbf{S}_1(dy, -dx)$ stands for an area element perpendicular to the (xy) plane and having unit width in that (third) direction.

(*b*) From your knowledge of electrostatic fields representable by analytic functions $\mathsf{E}(\zeta)$, that is, in sourceless regions, as in Exercise 3.20(*b*), argue that

$$\oint \mathsf{E}(\zeta)\, d\zeta = 0$$

is to be expected when the integration is extended over a closed contour (loop). [This is a well-known *Cauchy theorem*, valid for all loop integrals enclosing regions in which the integrand is analytic. It is the basis for the deformability of paths in the complex plane across regions of analyticity only (therefore sourceless ones).]

CHAPTER

FOUR

THE MAGNETOSTATIC FIELD

Vector potential descriptions · The Biot-Savart Law · Magnetic dipoles as the simplest elements of magnetic sources; the gyromagnetic ratio · Magnetic scalar potentials · The magnetostatic multipole series to all orders · Magnetic forces on and between currents · The Lorentz force and galilean relativity · Magnetostatic field energy and circuit inductances · The nonrelativistic mass renormalization of charged particles

The basis provided by the Maxwell equations (1.40) for deducing expectations about magnetostatic fields is

$$\nabla \cdot \mathbf{B} = 0 \qquad \nabla \times \mathbf{B} = \frac{4\pi \mathbf{j}(\mathbf{r})}{c} \tag{4.1}$$

Since the divergence of any curl vanishes, this applies only to divergenceless sources $\mathbf{j}(\mathbf{r})$, such as are characteristic of *steady-current* distributions, without charge accumulations: $\partial\rho/\partial t = -\nabla \cdot \mathbf{j} = 0$. Naturally the sources must be steady if the resultant fields are to be static. The first concern here will be to find the fields that arise from *given* steady-current distributions, just as the initial explorations of the electrostatic field were carried out for sources presumed given.

4.1 THE GENERATION OF MAGNETOSTATIC FIELDS BY CURRENTS

Electrostatic fields could be obtained as superpositions (1.5) of contributions by charge elements because the Coulomb law was available to give the field from any single element. No comparable law has so far been promulgated here for the magnetostatic field of a current element, but one can now be derived as a solution of Eqs. (4.1), just as the Coulomb law is derivable from the electrostatic Maxwell equations (2.1).

The Magnetostatic Vector Potential

Both the electrostatic and magnetostatic sets of Maxwell equations, (2.1) and (4.1), include equations that are independent of what the sources may be and can be solved once and for all by introducing suitable potentials. In the magnetostatic case, the source-independent equation merely expresses the divergencelessness of the field $\mathbf{B}(\mathbf{r})$, and so, as in (A.26) or (A.50), it has the general solution

$$\mathbf{B}(\mathbf{r}) = \nabla \times \mathbf{A}(\mathbf{r}) \tag{4.2}$$

where $\mathbf{A}(\mathbf{r})$ is some *vector* potential. As discussed in connection with (A.29), there is a range of choices for a vector potential that can represent a given field, and a gauge can always be chosen in which $\mathbf{A}$ is divergenceless, as in the development of the vector version (A.54) of the Poisson equation. A divergenceless vector potential that will represent the field arising from a given steady-current distribution $\mathbf{j}(\mathbf{r})$ must then be a solution of

$$\nabla^2 \mathbf{A} = -\frac{4\pi \mathbf{j}(\mathbf{r})}{c} \tag{4.3}$$

if the remaining one of the Maxwell equations (4.1) is to be satisfied.

A solution of an equation like (4.3) that represents the field arising solely from a given divergenceless source distribution takes the form (A.55):

$$\mathbf{A}(\mathbf{r}_f) = \oint \frac{dV(\mathbf{r}_s)\mathbf{j}(\mathbf{r}_s)}{c|\mathbf{r}_f - \mathbf{r}_s|} \tag{4.4}$$

According to this, each element $\mathbf{j}\, dV(\mathbf{r}_s)$ contributes to the magnitude of $\mathbf{A}(\mathbf{r}_f)$ in proportion to its projection on the resultant direction of $\mathbf{A}$ at $\mathbf{r}_f$.

An illustration can be provided by a rigidly homogeneous sphere of charge that becomes a source of magnetostatic field by being set spinning around one of its diameters with some uniform angular velocity ω. Let the vector $\boldsymbol{\omega}$ have the direction of the spin axis, and use the center of the sphere as the origin for the source-point vectors $\mathbf{r}_s$. Then the charge at $\mathbf{r}_s$ will have the velocity $\mathbf{u} = \boldsymbol{\omega} \times \mathbf{r}_s$, making it circle about $\boldsymbol{\omega}$, and the consequent current-density distribution (1.13) is

$$\mathbf{j}(\mathbf{r}_s) = \rho\mathbf{u} = \frac{3q}{4\pi a^3}\boldsymbol{\omega} \times \mathbf{r}_s \tag{4.5}$$

if q is the total charge and $4\pi a^3/3$ is the sphere volume (see Exercise A.8 for a demonstration that $\nabla_s \cdot \mathbf{j} = 0$). Expression (4.4) for the vector potential generated becomes

$$\mathbf{A}(\mathbf{r}) = \frac{3q}{4\pi a^3 c}\boldsymbol{\omega} \times \oint dV(\mathbf{r}_s)\frac{\mathbf{r}_s}{|\mathbf{r} - \mathbf{r}_s|}$$

at a field point $\mathbf{r}_f \equiv \mathbf{r}$. The volume integration here is to extend over a sphere, and because of the symmetry of the integrand about the field-point direction $\hat{\mathbf{r}} \equiv \mathbf{r}/r$, the vector resultant of the integration can only have the direction of $\hat{\mathbf{r}}$. In more detail, for every contribution proportional to a component $(\mathbf{r}_s)_\perp = \mathbf{r}_s - \hat{\mathbf{r}}(\hat{\mathbf{r}} \cdot \mathbf{r}_s)$ perpendicular to $\hat{\mathbf{r}}$, there is an exactly canceling contribution from $(\mathbf{r}_s')_\perp = -(\mathbf{r}_s)_\perp$ that has the same distance from the field point. Consequently the vector potential can be written in terms of the *scalar* integral in the expression

$$\mathbf{A}(\mathbf{r}) = \frac{3q}{4\pi a^3 c}(\boldsymbol{\omega} \times \hat{\mathbf{r}}) \oint \frac{dV(\mathbf{r}_s)(\hat{\mathbf{r}} \cdot \mathbf{r}_s)}{[r^2 + r_s^2 - 2(\mathbf{r} \cdot \mathbf{r}_s)]^{1/2}}$$

With spherical coordinates having the direction $\hat{\mathbf{r}}$ to the field point as polar axis, the volume integral is

$$2\pi \int_0^a dr_s r_s^2 \int_{-1}^1 d(\cos \vartheta_s) \frac{r_s \cos \vartheta_s}{(r^2 + r_s^2 - 2rr_s \cos \vartheta_s)^{1/2}}$$

and the angular integration here can be carried out most expeditiously by taking advantage of the fact that $\cos \vartheta_s = P_1$ is orthogonal to all but the $l = 1$ term of an expansion like (3.81) into Legendre polynomials. As a result

$$\mathbf{A}(\mathbf{r}) = \frac{q}{a^3 c} (\boldsymbol{\omega} \times \hat{\mathbf{r}}) \int_0^a dr_s r_s^3 \frac{r_<}{r_>^2}$$

where $r_<$ and $r_>$ are, respectively, the lesser and greater of r and r_s in the integration range $0 < r_s < a$. For $r \geq a \geq r_s$, the radial integration gives

$$\mathbf{A}(r \geq a) = \frac{qa^2 \boldsymbol{\omega}}{5c} \times \frac{\hat{\mathbf{r}}}{r^2} \tag{4.6}$$

as the field outside the sphere. For $r < a$, the radial integration must be carried out in two pieces: over $0 < r_s < r$, where $r_< = r_s$ and $r_> = r$, and between the limits of $r < r_s < a$, where $r_< = r$ and $r_> = r_s$. The result is

$$\mathbf{A}(r < a) = \frac{qa^2 \boldsymbol{\omega}}{5c} \times \mathbf{r} \frac{5a^2 - 3r^2}{2a^5} \tag{4.7}$$

for the field inside the sphere. Notice that this makes $\mathbf{A}$ continuous through the sphere surface $r = a$.

The abbreviation

$$\frac{1}{5} \frac{qa^2 \boldsymbol{\omega}}{c} \equiv \boldsymbol{\mu} \tag{4.8}$$

will be convenient in calculating the result for $\mathbf{B} = \nabla \times \mathbf{A}$. Then, outside the sphere,

$$\mathbf{B}(r \geq a) = \nabla \times \left(\boldsymbol{\mu} \times \frac{\hat{\mathbf{r}}}{r^2} \right) = \boldsymbol{\mu} \left(\nabla \cdot \frac{\hat{\mathbf{r}}}{r^2} \right) - (\boldsymbol{\mu} \cdot \nabla) \frac{\hat{\mathbf{r}}}{r^2} \tag{4.9}$$

follows from the triple vector identity (A.53). The divergence derivative here vanishes for $r \geq a$ since $\hat{\mathbf{r}}/r^2 \equiv \mathbf{E}_1$ is *formally* the same as the electrostatic field due to a unit point charge at $\mathbf{r} = 0$, and this is divergenceless everywhere except right at $r = 0$. Moreover, since $\mathbf{E}_1 = \hat{\mathbf{r}}/r^2 = -\nabla(1/r)$,

$$\mathbf{B}(r \geq a) = \nabla(\boldsymbol{\mu} \cdot \nabla) \frac{1}{r} = -\nabla \frac{\boldsymbol{\mu} \cdot \hat{\mathbf{r}}}{r^2} \tag{4.10}$$

Thus $\mathbf{B}(r > a)$ is the negative gradient of a scalar field which has the same form as the electrostatic potential (2.25) due to an ideal electric dipole of moment

$\mathbf{D} = \boldsymbol{\mu}$. For this reason the resulting *magnetic* field outside the sphere of spinning charge, which will have the form (2.27),

$$\mathbf{B} = \frac{3\hat{\mathbf{r}}(\hat{\mathbf{r}} \cdot \boldsymbol{\mu}) - \boldsymbol{\mu}}{r^3} = \frac{2\hat{\mathbf{r}}(\hat{\mathbf{r}} \cdot \boldsymbol{\mu}) + [\hat{\mathbf{r}} \times (\hat{\mathbf{r}} \times \boldsymbol{\mu})]}{r^3} \tag{4.11}$$

is said to be an *ideal magnetic dipole field.*

Inside the sphere

$$\mathbf{B}(r \leq a) = [3\hat{\mathbf{r}}(\hat{\mathbf{r}} \cdot \boldsymbol{\mu}) - \boldsymbol{\mu}]\frac{r^2}{a^5} + \frac{5\boldsymbol{\mu}(a^2 - r^2)}{a^5} \tag{4.12}$$

This is clearly continuous with the outside, dipole field (4.11) at the sphere surface. Near the center of the sphere, where $r^2 \ll a^2$, the field becomes parallel to the spin axis, and

$$\left|\mathbf{B}(r \ll a) \approx \frac{5\boldsymbol{\mu}}{a^3}\right| = \frac{2\pi I}{ca} \tag{4.13}$$

where $I = q\omega/2\pi$ is the total current in the sphere (Exercise 1.7).

The Vector Potential of Filamentary Currents

Among the most important of macroscopic sources are ones in which current is conducted along *filaments* of negligible cross section, like those discussed in connection with the flux expression $I = \int d\mathbf{S} \cdot \mathbf{j}$ of (1.18). If $d\mathbf{r}_s$ is a line-element vector along a filament, the volume element enclosing $d\mathbf{r}_s$ can be written $dV(\mathbf{r}_s) = d\mathbf{r}_s \cdot d\mathbf{S}$ and

$$\int_S dV(\mathbf{r}_s)\, \mathbf{j}(\mathbf{r}_s) = d\mathbf{r}_s \int_S d\mathbf{S} \cdot \mathbf{j}(\mathbf{r}_s) = I\, d\mathbf{r}_s \tag{4.14}$$

for the integral over just the filament cross section $S(\mathbf{r}_s) \to 0$ at $\mathbf{r}_s$. Use has been made of the fact that a divergenceless current density $\mathbf{j}$ must be taken entirely parallel to the filamentary segments $d\mathbf{r}_s$ as their cross sections $S(\mathbf{r}_s)$ approach the vanishing point, since there can be no flux of current through the sides. Moreover, the flux I that enters one end of each segment $d\mathbf{r}_s$ must emerge unchanged from the other end in a divergenceless, steady current source. Thus I must be uniform along the filament.

It may be remarked that electric-circuit theory deals with cases in which a filament carries some current I to a junction point beyond which *several* filaments may emerge. These must carry away currents $I_1, I_2, \ldots, I_s, \ldots$ (positive or negative) that add up to

$$I = I_1 + I_2 + \cdots + I_s + \cdots \equiv \Sigma_s I_s \tag{4.15}$$

when there are no charge accumulations at the junction. This is known as a *Kirchhoff law* in elementary physics, and so the expression $\nabla \cdot \mathbf{j} = -\partial\rho/\partial t = 0$ may be regarded as a generalization, to a continuous distribution, of the elementary Kirchhoff law.

When the vector potential (4.4) is expressed for the idealized filamentary sources, the integrations over the vanishing filament cross sections $S(\mathbf{r}_s)$ at each $\mathbf{r}_s$ can be carried out as in (4.14). Then

$$\mathbf{A}(\mathbf{r}_f) = \Sigma_s \frac{I_s}{c} \int_s \frac{d\mathbf{r}_s}{|\mathbf{r}_f - \mathbf{r}_s|} \tag{4.16}$$

if there are several stretches of filament, carrying currents $I_1, I_2, \ldots, I_s, \ldots$, respectively.

The result (4.16) will now be illustrated for a single filament that forms a circular loop carrying a steady current I. [The ostensibly simpler case of an indefinitely long *straight* filament has already been treated in (1.20) by means of

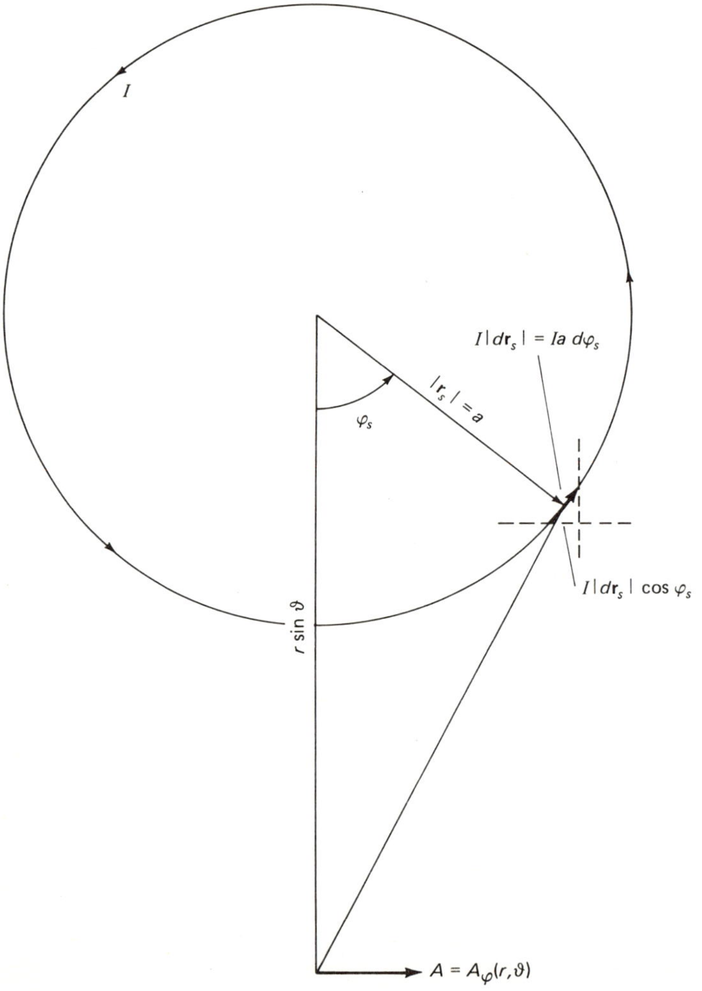

Figure 4.1

the Ampère law, and other treatments are left to Exercise 4.2.] Figure 4.1 indicates a top view of the circular loop, given the radius a. The evaluation is to be made for an arbitrary field point $\mathbf{r}_f$, having spherical coordinates r, ϑ, (φ) relative to the ring center as origin and the ring axis as polar axis. Then the field point lies at a distance $|z \equiv r \cos \vartheta|$ from the plane of the ring. In terms indicated in the figure, the distance from a current element to the field point is

$$\begin{aligned}|\mathbf{r}_f - \mathbf{r}_s| &= \{z^2 + [a^2 + (r \sin \vartheta)^2 - 2a(r \sin \vartheta)\cos \varphi_s]\}^{1/2} \\ &= (r^2 + a^2 - 2ar \sin \vartheta \cos \varphi_s)^{1/2}\end{aligned}$$

It should be apparent that only a vector-potential component A_φ, normal to the (meridional) plane containing the axis and the field point, will survive the integration needed for evaluating the resultant vector potential (4.16). The projections of the current elements onto the meridional plane itself, which contribute toward components A_r and A_ϑ, come in canceling pairs from elements to the left and right of the plane and equidistant from the field point. There thus remains just the one component A_φ to evaluate, and this constitutes the entire magnitude of the vector potential

$$A(r, \vartheta) = A_\varphi = \frac{Ia}{c} \int_0^{2\pi} \frac{d\varphi_s \cos \varphi_s}{(r^2 + a^2 - 2ar \sin \vartheta \cos \varphi_s)^{1/2}} \tag{4.17}$$

Putting this expression into terms of the parameter defined by

$$\epsilon \equiv \frac{2ar \sin \vartheta}{r^2 + a^2} \qquad (\leq 1) \tag{4.18a}$$

so that

$$A = \frac{Ia}{c(r^2 + a^2)^{1/2}} f(\epsilon) \tag{4.18b}$$

where

$$f(\epsilon) \equiv \int_0^{2\pi} \frac{d\varphi_s \cos \varphi_s}{(1 - \epsilon \cos \varphi_s)^{1/2}} \tag{4.18c}$$

shows that the integral remaining is a function of just the one parameter. The integral is of an elliptic type, like the integral in (2.11a), describing the electrostatic potential of a charged ring. Again, it is about as convenient to work with the integral expressions (4.17) or (4.18c) as they stand as it is to introduce the standardized elliptic functions defined in the literature (see Exercise 4.4).

It is quite clear that the result for the vector potential is properly divergenceless, since there is just the one, azimuthal, component A_φ and it is independent of the azimuth φ [see (A.5) for the appropriate expression of the divergence]. The lines of the vector-potential field form rings coaxial with the ring of current. They become rarefied ($A \to 0$) near the axis $\vartheta = 0$ or π, where $f(\epsilon \to 0) \to 0$, and tend to an infinite density at the filament itself,† where $f(\epsilon \to 1) \to \infty$.

† The result to be expected is one that corresponds to $\mathbf{B} \to 2I/cs$ [Eq. (1.20)] as the distance s from the filament approaches zero, since that is the field circling a long, straight filament. Formal investigation of (4.17) for $s \to 0$ shows that it does yield the proper potential, $A \to -(2I/c) \ln s$ (Exercise 4.2), but only within an infinite additive constant, much as in the case (2.15) of the electrostatic potential near a charged line.

The integration for the vector potential becomes simple to complete as a power series in the parameter ϵ of (4.18). The series can be introduced by expanding the inverse square root of the binomial that appears in the integral, as demonstrated when (3.78) was produced:

$$(1 - \epsilon \cos \varphi)^{-1/2} = 1 + \sum_{s=1}^{\infty} \frac{(2s-1)!!}{2^s s!} \epsilon^s \cos^s \varphi$$

Only the terms with odd $s \equiv 2N + 1 = 1, 3, 5, \ldots$ survive the integrations, which yield (Exercise C.15)

$$\int_0^{2\pi} d\varphi \cos^{2N+2} \varphi = \frac{\pi}{2^N} \frac{(2N+1)!!}{(N+1)!}$$

As a result

$$A = \frac{\pi a I}{c(r^2 + a^2)^{1/2}} \sum_{N=0}^{\infty} \frac{(4N+1)!!}{2^{4N+1} N!\,(N+1)!} \left(\frac{2ar \sin \vartheta}{r^2 + a^2} \right)^{2N+1} \tag{4.19}$$

This converges everywhere except on the filament itself (where $r = a$, $\vartheta = \frac{1}{2}\pi$, and $\epsilon = 1$).

The first term of the series,

$$A \approx \frac{\pi a^2 I}{c} \frac{r \sin \vartheta}{(r^2 + a^2)^{3/2}} \tag{4.20}$$

by itself furnishes an adequate picture of the field except too near the ring of current. The corrections are small wherever $\epsilon \ll 1$, and this is so all along the axis ($\vartheta \approx 0$ or π), in all directions from and near the center ($r \ll a$), and also in all directions sufficiently far away ($r \gg a$).

The magnetic field itself, $\mathbf{B} = \nabla \times \mathbf{A}$, can now be obtained by applying the curl operation as given in (A.6). The azimuthal component $\mathbf{B}_\varphi$ vanishes because $A_r = A_\vartheta = 0$; thus the magnetic field vectors everywhere lie in meridional planes orthogonal to the ring of current (and to the parallel rings forming the vector-potential field lines). Moreover, because of the azimuthal symmetry there remain only the operations

$$\mathbf{B}_r = (r \sin \vartheta)^{-1} \frac{\partial}{\partial \vartheta} \sin \vartheta\, A \qquad \mathbf{B}_\vartheta = -\frac{\partial}{r\, \partial r} r A \tag{4.21}$$

In the approximation of (4.20), the results are

$$\begin{aligned} \mathbf{B}_r &\approx \frac{\pi a^2 I}{c} \frac{2 \cos \vartheta}{(r^2 + a^2)^{3/2}} \\ \mathbf{B}_\vartheta &\approx -\frac{\pi a^2 I}{c} \frac{\sin \vartheta}{(r^2 + a^2)^{5/2}} (2a^2 - r^2) \end{aligned} \tag{4.22}$$

Near the center ($r \ll a$) this field becomes parallel to the axis

$$\mathbf{B}_z \equiv \mathbf{B}_r \cos \vartheta - \mathbf{B}_\vartheta \sin \vartheta \rightarrow \frac{2\pi I}{ca} \approx \mathbf{B} \tag{4.23}$$

Its form far away ($r \gg a$) will be discussed in Sec. 4.2.

The Biot-Savart Law

An expression for an element $d\mathbf{B}$ of magnetostatic field arising from an element of source, comparable to the expression for $d\mathbf{E}$ in (1.4), can be obtained by applying the curl operation to the expression (4.4) for the vector potential

$$\mathbf{B}(\mathbf{r}_f) = \nabla_f \times \mathbf{A}(\mathbf{r}_f) = \oint \frac{dV(\mathbf{r}_s)}{c}[-\mathbf{j}(\mathbf{r}_s) \times \nabla_f]\frac{1}{|\mathbf{r}_f - \mathbf{r}_s|}$$

$$= \oint \frac{dV(\mathbf{r}_s)j(\mathbf{r}_s) \times \hat{\mathbf{R}}}{cR^2} \tag{4.24a}$$

when $\mathbf{R} \equiv \mathbf{r}_f - \mathbf{r}_s$ is the vector from source to field point. Thus

$$d\mathbf{B} = \frac{\mathbf{j}\, dV \times \hat{\mathbf{R}}}{cR^2} \tag{4.24b}$$

varies inversely with the square of the distance from the source element, as in the Coulomb law for $d\mathbf{E}$. However, whereas $d\mathbf{E}$ has the direction $\hat{\mathbf{R}}$ from source to field point, $d\mathbf{B}$ is orthogonal to that direction, as well as to the direction of the current element $\mathbf{j}\, dV$. The outcome is consistent with the familiar right-hand rule mentioned in the discussion of (1.21). Expressions (4.24) are usually referred to as constituting the *Biot-Savart law*.

Another form of the Biot-Savart law expresses it for filamentary elements of source like $I\, d\mathbf{r}_s$ of (4.14):

$$d\mathbf{B} = I\, d\mathbf{r}_s \times \frac{\hat{\mathbf{R}}}{cR^2} \tag{4.25}$$

This form makes it most immediately clear why the field (4.23) at the center of the circular loop turned out to be $\mathbf{B} = \mathbf{B}_z = 2\pi a I/ca^2$, an *exact* result (as might have been expected with $\epsilon = 0$).

The Biot-Savart law, like the Coulomb law, is valid only in *static* situations, when any moving charges form steady, divergenceless currents; it was found as a solution of the magneto*static* Maxwell equations. It is therefore not strictly applicable to finding the magnetic field of a single moving point charge, which cannot by itself form an everywhere-divergenceless current. However, the point charge can be an element of a current, and it is worth considering what follows from the Biot-Savart law for a current density $\rho\mathbf{u}$ when $\rho(\mathbf{r}_s) = q\delta[\mathbf{r}_s - \mathbf{r}_q(t)]$, as is characteristic of a point charge passing through positions $\mathbf{r}_q(t)$ with a velocity $\mathbf{u} = \dot{\mathbf{r}}_q(t)$. When this $\mathbf{j} = \rho\mathbf{u}$ is substituted into (4.24*a*) and the integration over the δ distribution is carried out for an instantaneous value of $\mathbf{r}_q(t)$, the result (see Exercise 1.12) is

$$\mathbf{B}_q(\mathbf{r}_f) \approx q\mathbf{u} \times \frac{\hat{\mathbf{R}}}{cR^2} \approx \frac{\mathbf{u}}{c} \times \mathbf{E}_q \tag{4.26}$$

where $\mathbf{R} = \mathbf{r}_f - \mathbf{r}_q(t)$ and $\mathbf{E}_q = \hat{\mathbf{R}}q/R^2$ is the electric field of the temporarily static point charge. The fields here are not actually static, but the result might be expected to be *approximately* valid if the motion of the charge is slow enough, with $u \ll c$. It will indeed prove to be an excellent approximation when the solution of the appropriate nonstatic Maxwell equations is found in (10.12). The treatment of a situation as temporarily static, as here, is known as a *quasi-static* or *adiabatic approximation*.

4.2 THE MAGNETIC DIPOLE APPROXIMATION

Far enough from the loop of current considered in the preceding section, its field (4.22) attains the form (for $r \gg a$)

$$\mathbf{B}_r = \frac{2\mu \cos \vartheta}{r^3} \qquad \mathbf{B}_\vartheta = \frac{\mu \sin \vartheta}{r^3} \tag{4.27}$$

where $\mu \equiv \pi a^2 I/c$. Comparison with (2.27) shows that this has again the form, described in spherical coordinates, of an ideal dipole field, like the field found outside the spinning sphere of charge in (4.11). The result suggests that current loops (and the spheres) behave much like the bipolar magnetic needles discussed in connection with the initial mappings and measurements of magnetic fields in Sec. 1.2. Each needle is characterized by equal and opposite pole strengths $\pm q_M$, and since the poles generate force fields of the point-charge form (1.15), the needle can be treated in the same way as the electric dipoles of (2.23). The north pole has magnetic field lines emanating from it, while the south pole has lines entering it, like the two sides of the current loop, as the continuous field lines circle the filament forming the loop.

The strength and orientation of any ideal magnetic dipole field is always entirely determined by some magnetic dipole moment vector $\boldsymbol{\mu}$. [Examples of such moments were given in (4.8) and Exercise 4.1 for the spinning spheres, and the magnitude $\mu = \pi a^2 I/c$ for the circular loop was quoted in connection with (4.27).] The field $\mathbf{B}(\mathbf{r})$ due to any given $\boldsymbol{\mu}$ was shown in (4.11) and (4.27). It is derivable as $\mathbf{B} = \nabla \times \mathbf{A}$ from the vector potential

$$\mathbf{A} = \boldsymbol{\mu} \times \frac{\hat{\mathbf{r}}}{r^2} \tag{4.28}$$

as demonstrated in the steps from (4.9) to (4.11). The spherical coordinate form (4.27) is obtainable, by the curl operation (A.6) or (4.21), from

$$A = A_\varphi = \frac{\mu \sin \vartheta}{r^2} \tag{4.29}$$

the appropriate version of (4.28). The same field forms $\mathbf{B}$ are *also* derivable as $\mathbf{B} = -\nabla\phi_\mu$, from the *scalar* field

$$\phi_\mu = \boldsymbol{\mu} \cdot \frac{\hat{\mathbf{r}}}{r^2} = \frac{\mu \cos \vartheta}{r^2} \tag{4.30}$$

as shown in (4.10).

The result (4.30) is an example of a *magnetic scalar* potential ϕ_M. Such scalar potentials are always definable for magnetic fields *over regions in which they are curlless*, in accordance with (A.28). Thus, the simple field $\mathbf{B} = \mathbf{B}_\varphi = 2I/cs$ (1.20) is derivable as $\mathbf{B} = -\nabla\phi_M$ from $\phi_M = -2I\varphi/c$ [the form also used as the *electro*static potential of a double layer in Exercise C.9(*b*)]. It is characteristic for a magnetic scalar potential to be multiple-valued, like $\phi_M \sim \varphi$, along circuits that enclose current (nonvanishing $\nabla \times \mathbf{B}$). This must be so, since loop integrals of single-valued gradient fields vanish everywhere, yet magnetomotive force loops must not vanish when they enclose current.

For an Arbitrary Loop of Current

The field of the circular loop was found to attain the ideal dipole form (4.27) only at distances sufficiently remote from its source, and this suggests that the result is obtainable by the approximation procedure of (2.18), used there to develop the electric multipoles. The same procedure will now be applied to a vector-potential field (4.16) arising from a single *arbitrarily shaped* loop of current, by expanding the propagator $|\mathbf{r}_f - \mathbf{r}_s|^{-1}$ into inverse powers of the distance $r_f = r$ to the field point

$$\mathbf{A}(\mathbf{r}) = \frac{I}{c}\oint \frac{d\mathbf{r}_s}{|\mathbf{r} - \mathbf{r}_s|} \approx \frac{I}{cr}\oint d\mathbf{r}_s\left(1 + \frac{\hat{\mathbf{r}} \cdot \mathbf{r}_s}{r} + \cdots\right) \tag{4.31}$$

The sum $\oint d\mathbf{r}_s$ of line-element vectors forming a closed loop has a vanishing resultant, and so

$$\mathbf{A}(\mathbf{r}) \approx \frac{I}{cr^2}\oint d\mathbf{r}_s(\hat{\mathbf{r}} \cdot \mathbf{r}_s) \tag{4.32}$$

in the first approximation. The integrand here can be reexpressed in the equivalent alternative ways

$$(\hat{\mathbf{r}} \cdot \mathbf{r}_s)\, d\mathbf{r}_s = d[\mathbf{r}_s(\hat{\mathbf{r}} \cdot \mathbf{r}_s)] - \mathbf{r}_s(\hat{\mathbf{r}} \cdot d\mathbf{r}_s)$$

and

$$(\hat{\mathbf{r}} \cdot \mathbf{r}_s)\, d\mathbf{r}_s = \hat{\mathbf{r}} \times (d\mathbf{r}_s \times \mathbf{r}_s) + \mathbf{r}_s(\hat{\mathbf{r}} \cdot d\mathbf{r}_s)$$

$$= \tfrac{1}{2}\{d[\mathbf{r}_s(\hat{\mathbf{r}} \cdot \mathbf{r}_s)] + \hat{\mathbf{r}} \times (d\mathbf{r}_s \times \mathbf{r}_s)\}$$

The complete differential term in the last line integrates to zero around the closed loop because $\mathbf{r}_s(\hat{\mathbf{r}} \cdot \mathbf{r}_s)$ returns to its starting value in the completion of such an integration, and so

$$\mathbf{A} = \frac{I}{cr^2}\,\hat{\mathbf{r}} \times \oint \tfrac{1}{2}(d\mathbf{r}_s \times \mathbf{r}_s) \equiv \boldsymbol{\mu} \times \frac{\hat{\mathbf{r}}}{r^2}$$

with

$$\boldsymbol{\mu} \equiv \frac{I}{c}\oint \tfrac{1}{2}\mathbf{r}_s \times d\mathbf{r}_s \tag{4.33}$$

The result has the form (4.28) of an ideal magnetic dipole field, and the expression for $\boldsymbol{\mu}$ defines the magnetic dipole moment of an arbitrary loop.

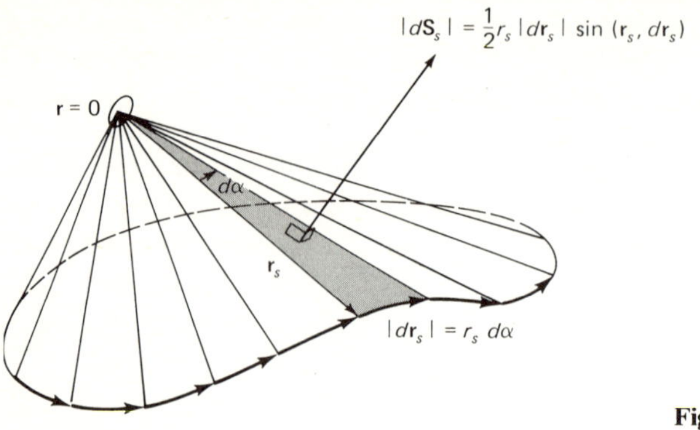

Figure 4.2

The elements $\frac{1}{2}\mathbf{r}_s \times d\mathbf{r}_s \equiv d\mathbf{S}_s$ of the integral for $\boldsymbol{\mu}$ are recognizable as directed triangular-area elements, each having a vertex at the chosen momental center $\mathbf{r} = 0$ and an element $d\mathbf{r}_s$ of the loop as the triangle side opposite that central vertex, all as indicated in Fig. 4.2. The choice of the reference center indicated in the figure produces a construction that might suggest a lack of uniqueness in the value of the moment. However, consider the effect of shifting the center by some $\mathbf{r}_o$:

$$\mathbf{S} = \oint \tfrac{1}{2}(\mathbf{r}_s + \mathbf{r}_o) \times d\mathbf{r}_s = \oint \tfrac{1}{2}\mathbf{r}_s \times d\mathbf{r}_s$$

because $\oint d\mathbf{r}_s = 0$, as has already been seen. Thus the directed area enclosed by a given loop, arising as a vector resultant of area elements, is after all unique (it is a minimal surface), and the magnetic moment vector

$$\boldsymbol{\mu} = \frac{I\mathbf{S}}{c} \tag{4.34}$$

is *invariant* to the choice of the center relative to which it is gauged (like the electric dipole moment of a *neutral* source, as seen in Exercise 2.19, and the monopole of an unneutral one, in each case the lowest-order moment). Notice that the magnetic moment $\mu = \pi a^2 I/c$ found for the circular loop and quoted in connection with (4.27) is a special case of (4.34).

For an Arbitrary Steady-Current Distribution

The dipole effect of the arbitrary steady-current distribution for which the vector potential (4.4) is expressed can be obtained from

$$\mathbf{A}(\mathbf{r}) = \oint \frac{dV(\mathbf{r}_s)\mathbf{j}(\mathbf{r}_s)}{c|\mathbf{r} - \mathbf{r}_s|} \approx \frac{1}{cr}\oint dV\mathbf{j}\left(1 + \frac{\hat{\mathbf{r}} \cdot \mathbf{r}_s}{r} + \cdots\right) \tag{4.35}$$

Perhaps the best way to acquire insight into this case is to recognize that any steady-current distribution can be regarded as a superposition of such closed loops as were treated in the preceding subsection. Any divergenceless field like the steady-current density $\mathbf{j}(\mathbf{r}_s)$ consists of lines forming closed loops. A sheaf of the $\mathbf{j}$ lines passing through any cross-sectional element $d\mathbf{S}$ will form a closed, tubelike circuit that may have a *variable* geometric cross section but will nevertheless continue to carry a steady current $(dI) = \mathbf{j} \cdot d\mathbf{S}$ that is constant all along the closed tube. There is no transfer of charge from any one tubular element to any contiguous one simply because the flow directions of $\mathbf{j}$ are everywhere parallel to the tube walls by the way they are constructed.† The tube elements (dI) can be introduced into the volume integral (4.35) by defining line-element vectors $d\mathbf{r}_s$ that are segments of the tube circuits (of vanishing cross section $d\mathbf{S}$) and so are parallel to $\mathbf{j}$ at each $\mathbf{r}_s$. Then

$$\mathbf{j}\,dV = \mathbf{j}(d\mathbf{S} \cdot d\mathbf{r}_s) = (\mathbf{j} \cdot d\mathbf{S})\,d\mathbf{r}_s = (dI)\,d\mathbf{r}_s \tag{4.36}$$

With this, the contribution of each element (dI) to the vector potential can be evaluated like the field (4.33) due to a filamentary circuit

$$d\mathbf{A} \approx \frac{(d\boldsymbol{\mu}) \times \hat{\mathbf{r}}}{r^2} \qquad \text{with } (d\boldsymbol{\mu}) = \frac{(dI)}{2c} \oint \mathbf{r}_s \times d\mathbf{r}_s$$

Notice that this entails the conclusion

$$\oint dV\mathbf{j} = (dI) \oint d\mathbf{r}_s = 0 \tag{4.37}$$

for a divergenceless current distribution (the current-element vectors, laid end to end, naturally have a zero vector resultant as the field lines $\mathbf{j}$ complete their closed loops). After a replacement of $(dI)\,d\mathbf{r}_s$ by $\mathbf{j}\,dV$, in accordance with (4.36), the fully integrated result is

$$\mathbf{A} = \frac{\boldsymbol{\mu} \times \hat{\mathbf{r}}}{r^2} \qquad \text{with } \boldsymbol{\mu} = \frac{1}{2c} \oint \mathbf{r}_s \times \mathbf{j}\,dV \tag{4.38}$$

The last expression constitutes the general definition of the magnetic dipole moment vector, a characteristic of the source independent of the field point $\mathbf{r}$. Illustrations are provided in Exercises 4.9 and 4.11.

The Gyromagnetic Ratio

Charges are borne by matter, and their motions have inertial mass associated with them. A definite mass is often associated with each unit of charge, representable by giving a charge-to-mass ratio $\equiv q/m$. A source consisting solely of matter with the given charge-to-mass ratio will have a mass $\rho_m(\mathbf{r}_s)\,dV(\mathbf{r}_s) = (m/q)\rho\,dV$ associated with each charge element $\rho\,dV$. If the matter is in motion with some velocity distribution $\mathbf{u}(\mathbf{r}_s)$, each element contributes $(2c)^{-1}\mathbf{r}_s \times \rho\mathbf{u}\,dV$ toward a magnetic moment (4.38) and, at the same time, a moment of momen-

† A *picture* of a divergenceless current source might resemble Fig. A.3.

tum $\mathbf{r}_s \times \rho_m \mathbf{u}\, dV$. This mechanical moment, usually called angular momentum, is important because its total

$$\mathbf{L} = \oint \mathbf{r}_s \times \rho_m \mathbf{u}\, dV(\mathbf{r}_s) \tag{4.39}$$

remains conserved at whatever starting value it may be given unless torques are imposed. Since

$$\boldsymbol{\mu} = (2c)^{-1} \oint \mathbf{r}_s \times \rho \mathbf{u}\, dV = \frac{q}{2mc} \oint \mathbf{r}_s \times \rho_m \mathbf{u}\, dV(\mathbf{r}_s)$$

in the situation described above, the source has a definite value for

$$g \equiv \frac{\mu}{L} = \frac{q}{2mc} \tag{4.40}$$

called† its *gyromagnetic ratio.*

Suppose that the spinning homogeneous sphere of charge, for which the magnetic moment $\boldsymbol{\mu} = \boldsymbol{\omega}(qa^2/5c)$ was found in (4.8), has a uniformly distributed total mass m. The angular momentum imparted to it by the spin can be evaluated from (4.39), but it is well known to be $\mathbf{L} = \mathscr{I}\boldsymbol{\omega}$, where $\mathscr{I} = \frac{2}{5}ma^2$ is its moment of inertia. As expected, the gyromagnetic ratio here is $g = q/2mc$.

Other sources may not have all their charge codistributed with the mass. There is then only a definite charge-to-mass ratio q/m of the entire charge q to the entire mass m. It is frequently the case that such a source has a single symmetry axis for both the charge and mass distributions, so that giving it an angular momentum $\mathbf{L}$ results in a magnetic moment $\boldsymbol{\mu}$ parallel to $\mathbf{L}$. Now a gyromagnetic ratio

$$g = \frac{\mu}{L} \neq \frac{q}{2mc} \tag{4.41}$$

is generally the result.‡

The spinning sphere of charge distributed on the surface only, for which the magnetic moment $\boldsymbol{\mu} = \boldsymbol{\omega}(qa^2/3c)$ was found in Exercise 4.1, may have its mass m distributed throughout its interior, so that its angular momentum is $\mathbf{L} = \frac{2}{5}ma^2\boldsymbol{\omega}$ again. In this case,

$$g = \frac{5}{3}\frac{q}{2mc} \tag{4.42}$$

with nearly twice the gyromagnetic ratio of the entirely homogeneous sphere.

† Sometimes it is μ/L *in units of* $q/2mc$ that is called the gyromagnetic ratio, and then $g = 1$ replaces (4.40).

‡ A notable example is the spinning neutron, which has $q = 0$ yet $\mu = g_n L \neq 0$ because it contains equal positive and negative charges that are not codistributed.

The Sequence of Magnetic Multipoles

A continuation of the expansion into inverse powers of the distance from source to field point, in (4.35), should be expected to yield higher-order magnetic multipoles, just as that procedure led to the higher-order electric multipoles in (2.19) and (3.102). However, properly covariant expressions for the multipole vector potentials call for representation in terms of so-called *vector spherical harmonics*, which will not be introduced until Chap. 9, for more essential uses. Much less mathematical erudition is needed for descriptions by magnetic *scalar* potentials—generalizations to the higher orders of the magnetic dipole scalar potential ϕ_μ of (4.30). Such descriptions are possible because multipole fields are, by definition, *exterior* to their sources, being distributed over regions in which curlless expressions $\mathbf{B} \equiv -\nabla\phi_M$ are valid.

There is not yet at hand an expression for the magnetic scalar potential of an arbitrary source, a counterpart of (4.35) for the vector potential, from which a multipole development could be started. It can be obtained from the Biot-Savart law (4.24) for $\mathbf{B}$ by a line integration of $\mathbf{B} = -\nabla\phi_M$ like that in (2.67)

$$\phi_M(\mathbf{r}) = -\int_C d\mathbf{r}_C \cdot \mathbf{B}(\mathbf{r}_C) \tag{4.43}$$

where the $d\mathbf{r}_C$ are elements of an integration path that must not enclose any part of the source (where $\mathbf{B}$ would acquire curl). A suitable path to any given field point $\mathbf{r}$ exterior to the source is provided by a straight line $\mathbf{r}_C = \hat{\mathbf{r}} r_C$ extended from $r_C = \infty$ to $r_C = r$, radially inward toward an origin in source. The result† will be a potential gauged from $\phi_M(\infty) = 0$ like the electrostatic potential (1.26).

Substituting the Biot-Savart expression (4.24) for $\mathbf{B}$ and the radial line elements $d\mathbf{r}_C = \hat{\mathbf{r}}\, dr_C$ into (4.43) yields

$$\phi_M(\mathbf{r}) = -\oint \frac{dV(\mathbf{r}_s)}{c} \int_\infty^r dr_C \frac{\hat{\mathbf{r}} \cdot [\mathbf{j}(\mathbf{r}_s) \times (\mathbf{r}_C - \mathbf{r}_s)]}{|\mathbf{r}_C - \mathbf{r}_s|^3}$$

Since $\hat{\mathbf{r}} \cdot (\mathbf{j} \times \mathbf{r}_C) = 0$ when $\mathbf{r}_C$ is parallel to $\mathbf{r}$, this can be rewritten as

$$\phi_M = \oint \frac{dV(\mathbf{r}_s)}{c} [\mathbf{j}(\mathbf{r}_s) \times \mathbf{r}_s] \cdot \hat{\mathbf{r}} \int_\infty^r \frac{dr_C}{|\mathbf{r}_C - \mathbf{r}_s|^3} \tag{4.44}$$

Notice that for sufficiently distant field points, when $|\mathbf{r}_C - \mathbf{r}_s| \approx r_C$ becomes a good approximation, the radial integral here is simply $-1/2r^2$ and the entire expression reduces to the dipole scalar potential (4.30), in terms of the general magnetic dipole moment (4.38). The radial integration is also elementary for closer field points, but the expansion into multipoles is eased when the propagator (3.81) is introduced through the use of

$$\nabla_s \frac{1}{|\mathbf{r}_C - \mathbf{r}_s|} = \frac{\mathbf{r}_C - \mathbf{r}_s}{|\mathbf{r}_C - \mathbf{r}_s|^3} \tag{4.45}$$

† This approach to developing the static magnetic multipoles seems to be relatively new to the literature. The author's attention was called to its publication by Bronzan (Ref. 6).

The expression on the right can be formed in (4.44) by replacing the path direction $\hat{\mathbf{r}} = \mathbf{r}_C/r_C$ with $(\mathbf{r}_C - \mathbf{r}_s)/r_C$, a valid step because $(\mathbf{j} \times \mathbf{r}_s) \cdot \mathbf{r}_s = 0$. Then

$$\phi_M = \oint \frac{dV(\mathbf{r}_s)}{c} [\mathbf{j}(\mathbf{r}_s) \times \mathbf{r}_s] \cdot \nabla_s \int_\infty^r \frac{dr_C}{r_C} \frac{1}{|\mathbf{r}_C - \mathbf{r}_s|} \tag{4.46}$$

Upon introducing the expansion (3.81) into Legendre polynomials the radial integral here becomes

$$\int_\infty^r \frac{dr_C}{r_C} \frac{1}{|\mathbf{r}_C - \mathbf{r}_s|} = \sum_{l=0}^{\infty} r_s^l P_l(\hat{\mathbf{r}} \cdot \hat{\mathbf{r}}_s) \int_\infty^r \frac{dr_C}{r_C^{l+2}}$$

The integration in the lth term has the simple result $-1/(l+1)r^{l+1}$. It should be plain that the arguments of the Legendre polynomials must be the cosines of the angles between source and field-point directions. To obtain source integrals independent of the direction to the field point, the Legendre polynomials are separated into spherical harmonics of the individual directions $\hat{\mathbf{r}}$ and $\hat{\mathbf{r}}_s$ through use of the addition theorem (3.99), just as this was done to obtain the electric multipole moments (3.101). As a consequence, the magnetic scalar potential (4.46) can be given a form analogous to (3.102)

$$\phi_M(\mathbf{r}) = \sum_{l=1}^{\infty} \sum_{m=-l}^{l} \frac{4\pi}{2l+1} \frac{\mu_{lm}}{r^{l+1}} Y_{lm}(\hat{\mathbf{r}}) \tag{4.47}$$

with magnetic multipole moments defined by

$$\mu_{lm} \equiv \oint dV(\mathbf{r}_s) \frac{\mathbf{r}_s \times \mathbf{j}(\mathbf{r}_s)}{(l+1)c} \cdot \nabla_s [r_s^l Y_{lm}^*(\hat{\mathbf{r}}_s)] \tag{4.48}$$

This does not exist for $l = 0$, since $Y_{00} = (4\pi)^{-1/2}$ has no gradients, corresponding to the fact that no magnetic monopoles (charges) exist. Whereas a point charge is the simplest element of an electrostatic source, the ideal magnetic dipole is the simplest element of sources of magnetism. Notice that whereas the electric multipole moments (3.101) consist of weighted integrations over charge elements $dq \equiv \rho\, dV$, the elements $d\boldsymbol{\mu} = (\mathbf{r}_s \times \mathbf{j})\, dV/2c$ of the dipole moment vector (4.38) are distinguishable in (4.48).

The *dipole* effects of a general localized source of magnetism are represented by the three $l = 1$ terms of (4.47). In the corresponding moments μ_{1m}, the gradient operations result in directional factors independent of the source integrations. The operands $r_s Y_{1m}^*(\hat{\mathbf{r}}_s) = (3/4\pi)^{1/2}(\mathbf{r}_s)_m^*$, according to (3.93), with $(\mathbf{r}_s)_0 = z_s$ and $(\mathbf{r}_s)_{\pm 1} = \mp 2^{-1/2}(x_s \pm iy_s)$ just the spherical components of the source-point position $\mathbf{r}_s$. Then

$$\nabla_s(\mathbf{r}_s)_0^* = \mathbf{k} \qquad \nabla_s(\mathbf{r}_s)_{\pm 1}^* = \mp 2^{-1/2}(\mathbf{i} \mp i\mathbf{j}) \tag{4.49}$$

in terms of fixed unit vectors $\mathbf{i}, \mathbf{j}, \mathbf{k}$ that point out directions of cartesian axes, as in (A.1). Each of the moments μ_{1m} thus becomes proportional to a component of the vector integral $\oint d\boldsymbol{\mu} = \boldsymbol{\mu}$, and the connections

$$\mu_{10} = \left(\frac{3}{4\pi}\right)^{1/2} \mu_z \qquad \mu_{1,\pm 1} = \mp \left(\frac{3}{8\pi}\right)^{1/2} (\mu_x \mp i\mu_y) \tag{4.50}$$

to the dipole moment vector $\boldsymbol{\mu}$ are the result. These are exactly like the connections (3.103) of the complex electric moments q_{1m} to the electric dipole vector $\mathbf{D}$, and, as in that case, the $l = 1$ terms of (4.47) reduce to the dipole-potential form $\phi_\mu = \boldsymbol{\mu} \cdot \hat{\mathbf{r}}/r^2$, as to be expected.

More conventional expressions than (4.48) for the magnetic multipole moments have the gradient operation transferred from the propagation factor $r_s^l Y_{lm}^*$ to the source factor $\mathbf{r}_s \times \mathbf{j}(\mathbf{r}_s)$ through the use of a *hermitean property* of the gradient operator. Quite generally

$$\oint dV(\mathbf{r}_s)\mathbf{f}(\mathbf{r}_s) \cdot \nabla_s g(\mathbf{r}_s) = -\oint dV g \nabla_s \cdot \mathbf{f} \tag{4.51}$$

The two sides here formally differ by

$$\oint dV(\mathbf{r}_s)\nabla_s \cdot (g\mathbf{f}) = \oint d\mathbf{S} \cdot g\mathbf{f}$$

but this vanishes for gaussian surfaces on which $f_n = 0$ or $g = 0$, a condition for the hermiteanship. In (4.48), $\mathbf{f}(\mathbf{r}_s) = \mathbf{r}_s \times \mathbf{j}$ vanishes on a surface placed just outside the source (and within the field points of interest). Thus

$$\mu_{lm} = -\frac{1}{(l+1)c}\oint dV(\mathbf{r}_s) r_s^l Y_{lm}^*(\hat{\mathbf{r}}_s)\nabla_s \cdot [\mathbf{r}_s \times \mathbf{j}(\mathbf{r}_s)] \tag{4.52}$$

is equivalent to (4.48) and constitutes the more conventional definition of the moments.

4.3 MAGNETIC FORCES

The measure of magnetic field $\mathbf{B}$ has so far been taken to be the force on a unit test pole. Meanwhile, the concept of magnetic pole has been relegated to a secondary status, as something *derivable* from properties of atomic currents in materials. On such a basis, the magnetic field should finally be defined in terms of measurable forces on test currents rather than test poles.

The connection between force on a pole and force on a current can be made by considering the effect of putting a pole having some strength q_M into the magnetic field arising from some current. An element $d\mathbf{B}'$ of the field arising from an element $\mathbf{j}\,dV$ of the current will be detected as the force

$$q_M\,d\mathbf{B}' = \frac{q_M(\mathbf{j}\,dV \times \hat{\mathbf{R}})}{cR^2}$$

on the pole, according to the Biot-Savart law (4.24*b*). The *reaction* to this will be a force on the element of current equal to

$$d\mathbf{F} = -q_M\,d\mathbf{B}' = \frac{\mathbf{j}\,dV}{c} \times \left(\frac{-q_M\hat{\mathbf{R}}}{R^2}\right)$$

The final parenthesis here can be recognized as just the magnetic field $\mathbf{B}$ at the current element, in this case arising from the pole, as in expression (1.15), with $\mathbf{R}$ reversed, of the force between poles. The conclusion to be reached is that a current element $\mathbf{j}\,dV$ is subjected to the force

$$d\mathrm{F} = \frac{\mathbf{j}\,dV}{c} \times \mathbf{B} \tag{4.53}$$

whenever a magnetic field $\mathbf{B}$ is incident on it, whatever its sources. It is this expression that, in principle, replaces the force per unit pole (1.16) as the operational definition of the magnetic field. All the findings about magnetic fields already reviewed have been most accurately confirmed on the basis of this definition.

The qualification "in principle" was appended to the statement above because development of the force expression (4.53) is obviously necessary before it can correspond to any practical way of measuring $\mathbf{B}$. In the first place, detecting force on any one element can yield information only about the field component normal to it. Current elements in *various orientations*, as well as various spatial positions, must be used to explore a field completely. Moreover, it must be recognized that isolated steady-current elements cannot be obtained, in the way that effectively isolated test poles could be for the secondary definition (1.16) of $\mathbf{B}$. In the redefinition, conclusions must be drawn from measuring force per unit length on small segments of *extended* test currents. Using long, straight filaments of steady current [see (4.63)] should yield the least ambiguous results since parts of such a filament outside a test segment of it do not contribute to the field at the segment, according to the Biot-Savart law.

Finally, it should be pointed out that the definition (4.53) yields directly only a value for the *ratio* $\mathbf{B}/c$. The presence of c in the definition stems from its development in consistency with all the earlier expressions of the findings about electromagnetic fields, so that they can continue to be used without change. The consequence is that a value for c must be available before a force measurement on a known test current can yield a definite value for the field strength $\mathbf{B}$ itself in some system of units.

How a measure of the constant c, independent of the units used for $\mathbf{B}$, can be obtained becomes clear upon referring to its first introduction in Chap. 1, to help represent the proportionality between a magnetomotive force and its current source, the Ampère law (1.21) or (1.22). There a *product* $c\mathbf{B} = c^2(\mathbf{B}/c)$ was taken to be proportional to the source current. If the field of a known source current is now measured through its force on a known test current, a value for c^2 follows, since the force measurement yields the factor $\mathbf{B}/c$. Here the magnetic field is merely serving as a mediating agent in the influence of two currents on each other and can be eliminated in expressions of forces between them, as done in the next subsection.

Forces between Currents

The magnetostatic field that arises from any given steady-current source in accordance with the Ampère law was made explicit in the Biot-Savart expression

(4.24*a*). Then the force $d^2\mathbf{F}$ on any current element $\mathbf{j}\,dV$ due to a second current element $\mathbf{j}'\,dV'$ is obtainable from the force expression (4.53) for any magnetic field $\mathbf{B}$ by replacing $\mathbf{B}$ with an element $d\mathbf{B}'$, as given by (4.24*b*):

$$d^2\mathbf{F} = \frac{dV\,dV'}{c^2}\mathbf{j} \times \frac{\mathbf{j}' \times \hat{\mathbf{R}}}{R^2} \tag{4.54}$$

where $\mathbf{R}$ is the separation vector extending from the source element $\mathbf{j}'\,dV'$, as origin, to the element $\mathbf{j}\,dV$ being subjected to the force $d^2\mathbf{F}$. Whereas the expression (4.53) is held generally valid, as a definition of $\mathbf{B}$, the result (4.54) is restricted to steady currents because the Biot-Savart law has been used in its construction.

The force expression (4.54) is the magnetostatic analog of the electrostatic Coulomb law as expressed for charge elements:

$$d^2\mathbf{F} = \frac{dV\,dV'\rho\rho'\hat{\mathbf{R}}}{R^2} \tag{4.55}$$

Recall now that measurements of the Coulomb force could be used to define the unit of charge, as discussed in connection with (1.1). A comparable definition of a unit for the current, through measurements on forces between currents, is not in principle necessary since a current can be produced by moving known charges with known velocities, as suggested by the representation $\mathbf{j} = \rho\mathbf{u}$. Instead, as anticipated in the preceding subsection, the force measurements between currents can be used to determine the constant c^2, since all other factors in the force expression (4.54) can be measured independently of it. The result

$$c = 2.9974 \times 10^{10} \text{ cm/s} \tag{4.56}$$

has already been quoted (somewhat less precisely) in connection with its introduction in the Ampère law (1.21).

Because it is comparable to the Coulomb law for static charges, the integrated form of the force (4.54) between steady currents

$$\mathbf{F} = \oint \frac{dV\mathbf{j}}{c} \times \left(\oint \frac{dV'\mathbf{j}'}{c} \times \frac{\hat{\mathbf{R}}}{R^2} \right) \tag{4.57}$$

is often regarded as the logical starting point for developing the theory of magnetism. A magnetic field $\mathbf{B}$ is then first introduced as the resultant of the integration in the factor in parentheses, consistent with the definition adopted in (4.53). The Biot-Savart form (4.24) for a magneto*static* field emerges first, just as the Coulomb law served only to introduce the simplest form (1.2) of an electrostatic field initially. Such a line of development has the advantage that it avoids the temporary use of poles to define a magnetic field and might well have become the way this concept originated had it not been for the earlier discovery of magnetic ores as sources of magnetism.

The force expression (4.57) is sometimes used to define currents $\mathbf{j}_{\text{emu}} \equiv \mathbf{j}/c$ in *electromagnetic units* (emu), as against the currents like $\mathbf{j} = \rho\mathbf{u}$ in the *electrostatic units* (esu) defined by the Coulomb law involving ρ. On this account, the constant $c = j/j_{\text{emu}} = \rho/\rho_{\text{emu}}$ is referred to as the ratio of electromagnetic to the electrostatic units. It is more appropriate to think of c as a natural unit for *velocities* as these influence electromagnetic phenomena, e.g., in such a ratio as $\mathbf{j}/c = \rho(\mathbf{u}/c)$, which helps determine forces like (4.57). It is in such connections that c has attained its importance as a fundamental constant of nature.

Expression (4.54) as it stands gives the force $d^2\mathbf{F}$ on the current element $\mathbf{j}\,dV$ due to the element $\mathbf{j}'\,dV'$, which serves as the origin of the vector $\mathbf{R}$. It should require merely an interchange of $\mathbf{j}\,dV$ and $\mathbf{j}'\,dV'$, together with a reversal of the sign with which $\mathbf{R}$ occurs, to produce an expression for the force on $\mathbf{j}'\,dV'$ due to $\mathbf{j}\,dV$:

$$d^2\mathbf{F}' = \frac{dV\,dV'}{c^2}\mathbf{j}' \times \frac{\mathbf{j} \times (-\hat{\mathbf{R}})}{R^2} \tag{4.58}$$

Notice that $d^2\mathbf{F}' \neq -d^2\mathbf{F}$ in general, since

$$d^2\mathbf{F} = \frac{dV\,dV'}{c^2}\frac{\mathbf{j}'(\mathbf{j}\cdot\hat{\mathbf{R}}) - \hat{\mathbf{R}}(\mathbf{j}\cdot\mathbf{j}')}{R^2} \tag{4.59a}$$

whereas

$$d^2\mathbf{F}' = \frac{dV'\,dV}{c^2}\frac{-\mathbf{j}(\mathbf{j}'\cdot\hat{\mathbf{R}}) + \hat{\mathbf{R}}(\mathbf{j}'\cdot\mathbf{j})}{R^2} \tag{4.59b}$$

follows from the triple-vector-product identity (A.53). This outcome need not be inconsistent with Newton's law of the equality of action and reaction (which is used basically to maintain a detailed conservation of momentum throughout interactions) since the expressions are supposed to hold only in static, equilibrium situations, with the current elements as parts of complete, divergenceless loops. The law of reactions should rather be expected to hold as $\mathbf{F}' = -\mathbf{F}$, for integrated resultant forces between loops that are isolated except from each other.†

The situation here contrasts with the electrostatic case (4.55), since

$$d^2\mathbf{F}' \equiv \frac{dV'\,dV\rho'\rho(-\hat{\mathbf{R}})}{R^2} = -d^2\mathbf{F} \tag{4.60}$$

is immediately (element by element) consistent with the law of reactions. This happens because there need be no charges present other than just $\rho\,\Delta V$ and $\rho'\,\Delta V'$ to have a static situation. An electrostatic equilibrium more comparable to the above magnetic one is discussed for (B.19), where force due to a surface charge element is not oppositely directed to the force on the element itself.

Still to be checked is the expectation that when the force elements (4.54) and (4.58) or (4.59*a*) and (4.59*b*) are integrated over complete circuits of current, such

† This does not invalidate the use that was made of the law of reactions to arrive at (4.53) since this yields the same force on a current element, *due to a pole alone*, whether the element is treated as isolated or as part of a divergenceless current.

as can exist in isolation and yet be divergenceless, then indeed the resultants **F** and **F**′ are related as action and reaction $\mathbf{F}' = -\mathbf{F}$. Integrations over divergenceless currents are best carried out after decompositions of the source volumes into the filamentary tubes of (4.36). It will be sufficient to consider only the forces F_{12} and F_{21} between two discrete idealized tubes of vanishing cross section, as in (4.14), respectively carrying currents I_1 and I_2. Such can be considered to exist in isolation and will suffice to demonstrate the point. The general case can always be obtained by superposition (integration) after replacements like $I_1\, d\mathbf{r}_1 \to dI\, d\mathbf{r} \equiv \mathbf{j}\, dV$ and $I_2\, d\mathbf{r}_2 \to \mathbf{j}'\, dV'$.

It follows from (4.57) that the resultant force $\mathbf{F}_{12}$ on a loop I_1 in the magnetic field of a loop I_2 is

$$\mathbf{F}_{12} = \frac{I_1 I_2}{c^2} \oint\oint \frac{d\mathbf{r}_1 \times [d\mathbf{r}_2 \times (\mathbf{r}_1 - \mathbf{r}_2)]}{|\mathbf{r}_1 - \mathbf{r}_2|^3} \tag{4.61}$$

and the force $\mathbf{F}_{21}$ on I_2 due to I_1 has an expression differing from this only by an interchange of the indices 1, 2. The triple vector product in the integrand can be expanded as in (4.59) to yield

$$d\mathbf{r}_1 \times [d\mathbf{r}_2 \times (\mathbf{r}_1 - \mathbf{r}_2)] = d\mathbf{r}_2[d\mathbf{r}_1 \cdot (\mathbf{r}_1 - \mathbf{r}_2)] - (\mathbf{r}_1 - \mathbf{r}_2)(d\mathbf{r}_1 \cdot d\mathbf{r}_2)$$

Now

$$\oint d\mathbf{r}_1 \cdot \frac{\mathbf{r}_1 - \mathbf{r}_2}{|\mathbf{r}_1 - \mathbf{r}_2|^3} = -\oint d\mathbf{r}_1 \cdot \nabla_1 \frac{1}{|\mathbf{r}_1 - \mathbf{r}_2|} = -\left[\frac{1}{|\mathbf{r}_1 - \mathbf{r}_2|}\right]_{\mathbf{r}_1 = \mathbf{r}_0}^{\mathbf{r}_1 = \mathbf{r}_0} = 0$$

if the integration starts and ends at the point $\mathbf{r}_1 = \mathbf{r}_0$ of the I_1 loop. Thus

$$\mathbf{F}_{12} = -\frac{I_1 I_2}{c^2} \oint\oint (d\mathbf{r}_1 \cdot d\mathbf{r}_2) \frac{\mathbf{r}_1 - \mathbf{r}_2}{|\mathbf{r}_1 - \mathbf{r}_2|^3} = -\mathbf{F}_{21} \tag{4.62}$$

and the reaction relation holds, as it should. Notice that the two loops attract each other when their nearest elements are predominantly parallel, with currents in the same direction ($I_1\, d\mathbf{r}_1 \cdot I_2\, d\mathbf{r}_2 > 0$); antiparallel currents repel each other. These conclusions are consistent with the fact that two magnets (dipoles) attract each other when their unlike poles are closest together, whereas they repel when their like poles are nearest each other.

Perhaps the most important example is provided by a pair of currents I_1, I_2, flowing along straight-line filaments that are parallel to each other and apart by some fixed distance s. Obtaining the magnitude f of the force on each unit length of either filament

$$f = \frac{2I_1 I_2}{c^2 s} \tag{4.63}$$

is elementary and so is left to Exercise 4.14. The force is a mutual attraction when the flow directions of I_1 and I_2 are the same and a repulsion when they are antiparallel.

The Lorentz Force

The force expression (4.53) is the magnetic counterpart of the expression $d\mathbf{F} = (\rho\, dV)\mathbf{E}$ for the force on a charge element in an *electric* field $\mathbf{E}$. Whenever both an electric field $\mathbf{E}$ and a magnetic field $\mathbf{B}$ are present, the resultant force per unit volume, the force density $\mathbf{f}$, is

$$\mathbf{f} \equiv \frac{d\mathbf{F}}{dV} = \rho\mathbf{E} + \mathbf{j} \times \frac{\mathbf{B}}{c} \tag{4.64}$$

The magnetic part of the force can also be described as proportional to the charge density ρ whenever the discrimination of elements is fine enough to permit assigning a unique velocity to each. Then the expression $\mathbf{j} = \rho\mathbf{u}$ of (1.13) can be used for the current density, and the force per unit test charge at a point becomes

$$\frac{\mathbf{f}}{\rho} = \mathbf{E} + \frac{\mathbf{u}}{c} \times \mathbf{B} \tag{4.65}$$

This is valid whether the fields are static or not, since it is constituted of the *definitions* of $\mathbf{E}$ and $\mathbf{B}$.

Because the last expression is not restricted to static, divergenceless currents, it can also be applied to a moving *point* charge q. If the charge takes positions $\mathbf{r}_q(t)$ as time goes on, it has the velocity $\mathbf{u}(t) = \dot{\mathbf{r}}_q(t)$, and then $\rho = q\delta[\mathbf{r} - \mathbf{r}_q(t)]$ and $\mathbf{j} = \rho\mathbf{u}$. Now the integration over the δ distribution of the point charge yields the instantaneous resultant force on it

$$\mathbf{F}_q = q\left(\mathbf{E} + \frac{\mathbf{u}}{c} \times \mathbf{B}\right) \tag{4.66}$$

where $\mathbf{E}$ and $\mathbf{B}$ are field values at the instantaneous position $\mathbf{r}_q(t)$ of the point charge. It is being presumed here that the point charge is a simple monopole, not endowed with higher-order point-multipole moments, subject to additional forces like those reviewed in connection with the interaction form (2.57). Expression (4.66) is called the *Lorentz force*, and several applications of it will be reviewed in the next chapter.

For the discussion to be carried on at this stage, it will be useful to point out some implications of the Lorentz-force expression for electromagnetic fields as detected by a moving test charge.

It is concluded that a field described by $\mathbf{E}$ and $\mathbf{B}$ is present whenever a static charge q detects a force $q\mathbf{E}$, and an extra force $q(\mathbf{u}/c) \times \mathbf{B}$ is detected upon giving the test charge a velocity $\mathbf{u}$. Now suppose the test charge is given a *uniform* velocity, and consider the force it detects relative to a reference frame that moves with the charge, the so-called *proper* frame of the point charge. The charge is static relative to this frame and so can detect only an electric field in it; yet the force is *unchanged* from that in the initial frame, according to the ideas of classical, galilean relativity.†

† See Ref. 17, sec. 2.1 and eq. (2.11).

Forces are measured by the accelerations they produce in any mass m to which they are applied, and accelerations are supposed to be the same relative to either frame of any pair in uniform relative motion, according to the classical ideas. A velocity $\mathbf{u}(t)$ in one of the frames is taken to differ from the velocity $\mathbf{u}'(t)$ measured in the other by just a constant, as in

$$\mathbf{u}(t) = \mathbf{u}'(t) + \mathbf{v} \tag{4.67}$$

if $\mathbf{v}$ is the constant relative velocity of the frames. Thus $\dot{\mathbf{u}} \equiv \mathbf{F}/m = \dot{\mathbf{u}}' \equiv \mathbf{F}'/m$ and $\mathbf{F}' = \mathbf{F}$. This galilean invariance leads to the conclusion that the electric field $\mathbf{E}' = \mathbf{F}'/q$ that a charge q will detect in its proper frame is related by

$$\mathbf{E}' = \mathbf{E} + \frac{\mathbf{v}}{c} \times \mathbf{B} \tag{4.68}$$

to the electromagnetic field $\mathbf{E}$, $\mathbf{B}$ which it detects in a frame relative to which it has the velocity $\mathbf{v}$. The same electromagnetic field is described differently when referred to frames in relative motion. What may be exclusively an electric field $\mathbf{E}'$ in one of the frames (whenever $\mathbf{B}' \equiv 0$) can at least partially be converted into a magnetic field relative to the other. Here again is an indication that electric and magnetic fields should be regarded as components of a single entity, an electromagnetic field.

That an exclusively electric field in one frame should lead to a magnetic field in a relatively moving frame is not difficult to understand. A purely electric field $\mathbf{E}'$ arises when all its sources are some charges ρ that are static in its frame. The same charges are in motion relative to the other frame, and their current densities $\rho\mathbf{v}$ must be expected to produce a magnetic field. The finding (4.26) for each point of charge indicates that, at least for small velocities, this magnetic field will be of order $\mathbf{B} \approx (\mathbf{v}/c) \times \mathbf{E}' \approx (\mathbf{v}/c) \times \mathbf{E}$.

More generally, there may be both electric and magnetic fields in both of a pair of relatively moving frames. Then the connection (4.68) ought be supplemented by another one, giving $\mathbf{B}'$ in terms of $\mathbf{E}$ and $\mathbf{B}$. Consider what can be said about this on the basis of the classical ideas that led to (4.68). The frame of $\mathbf{E}'$, $\mathbf{B}'$ has the velocity $\mathbf{v}$ with respect to the frame of $\mathbf{E}$, $\mathbf{B}$, and so the latter frame has the velocity $-\mathbf{v}$ relative to the former. A connection $\mathbf{E} = \mathbf{E}' - (\mathbf{v}/c) \times \mathbf{B}'$, reciprocal to (4.68), is thus implied. The sum of the reciprocal expressions leads to $\mathbf{v} \times (\mathbf{B} - \mathbf{B}') = 0$, and $\mathbf{B}' = \mathbf{B}$ is implied if the connections are to be valid however the direction of $\mathbf{v}$ is chosen. The outcome $\mathbf{B}' = \mathbf{B}$ cannot be accepted as generally valid, however. For the case of the purely electrostatic field $\mathbf{E}' \neq 0$, $\mathbf{B}' \equiv 0$ discussed in the preceding paragraph, $\mathbf{B} \neq \mathbf{B}' = 0$ had to be expected. The outcome $\mathbf{B} = \mathbf{B}'$ here can still be valid only as an *approximation* in which a magnitude of order $|\mathbf{B}| = |\mathbf{E}'|v/c$ can be considered negligible, so that $\mathbf{B} \approx \mathbf{B}' = 0$ only approximately. That is the situation for velocities so small in comparison to the enormous value c in (4.56) that $v \ll c$, and it will eventually be learned (in Chap. 11) that galilean relativity is valid only for the so-called nonrelativistic velocities $v \ll c$.

Considerations already introduced above suggest how the approximation $\mathbf{B}' \approx \mathbf{B}$ should be corrected to first order in powers of v/c, the degree to which

(4.68) has been found valid. The result already quoted, (4.26), suggests that when charges responsible for the field $\mathbf{E}'$ are given an *extra* and not too large velocity $\mathbf{v}$, an extra magnetic field $(\mathbf{v}/c) \times \mathbf{E}'$ arises. When, in the frame of $\mathbf{E}$ and $\mathbf{B}$, the sources of $\mathbf{E}'$, $\mathbf{B}'$ are given the frame velocity $\mathbf{v}$, the field $\mathbf{B}$ that will be detected consists not only of $\mathbf{B}'$ but is supplemented to $\mathbf{B} \approx \mathbf{B}' + (\mathbf{v}/c) \times \mathbf{E}'$. The reciprocal of this is

$$\mathbf{B}' \approx \mathbf{B} - \frac{\mathbf{v}}{c} \times \mathbf{E} \tag{4.69}$$

in the desired approximation. The connections (4.68) and (4.69) together constitute expectations from galilean relativity for velocities $v \ll c$. It can readily be seen that the two connections can be consistent with each other and with their reciprocals only if terms of order v^2/c^2 are always neglected.

Forces Induced in Moving Circuits

The development of the force expression (4.65) and of the Lorentz force (4.66) makes it possible to discuss the Faraday-induction phenomenon of (1.33) as a result of such forces. The discussions will show how the occurrence of the constant c in the induction law could have been expected, on grounds other than just the beautiful symmetry it helps maintain between the Faraday induction (1.36) and the Maxwell induction (1.38).

Suppose that a filamentary conducting loop, which naturally contains charges that are mobile along it, is made to move in some way relative to a *static*, purely magnetic field $\mathbf{B}(\mathbf{r})$, and let $\mathbf{u}(\mathbf{r}_s, t)$ be the velocity of the loop segment $d\mathbf{r}_s$ during the motion. Then each unit of charge carried along with the segment will be subject to a magnetic Lorentz force

$$\mathbf{E}'(\mathbf{r}_s, t) = \frac{\mathbf{u}}{c} \times \mathbf{B}(\mathbf{r}_s) \tag{4.70}$$

This force per unit charge can be recognized as an electric field like (4.68), defined in the rest-frame of the moving segment, for a case in which the entire field is a magnetostatic one relative to the frame in which the segment has the velocity $\mathbf{u}$; that is, $\mathbf{E} = 0$.

The Lorentz force per unit charge (4.70) is also identifiable with the type of electric field that was found by Faraday to introduce emf into circuits moving with respect to fixed magnets. This interpretation is supported by the fact that the total emf thus introduced into the complete loop

$$\oint d\mathbf{r}_s \cdot \mathbf{E}' = \oint d\mathbf{r}_s \cdot \left(\frac{\mathbf{u}}{c} \times \mathbf{B}\right) = \frac{1}{c} \oint (d\mathbf{r}_s \times \mathbf{u}) \cdot \mathbf{B} \tag{4.71}$$

can be formulated as proportional to the rate at which the magnetic flux enclosed by the circuit changes during its motion. Reference to Fig. 4.3 should help make it clear that $(\mathbf{u}\,\Delta t) \times d\mathbf{r}_s = \Delta\mathbf{S}(\mathbf{r}_s, t)$ is a vector-element contribution to the area enclosed by the circuit as it is changed by the motion of the segment $d\mathbf{r}_s$

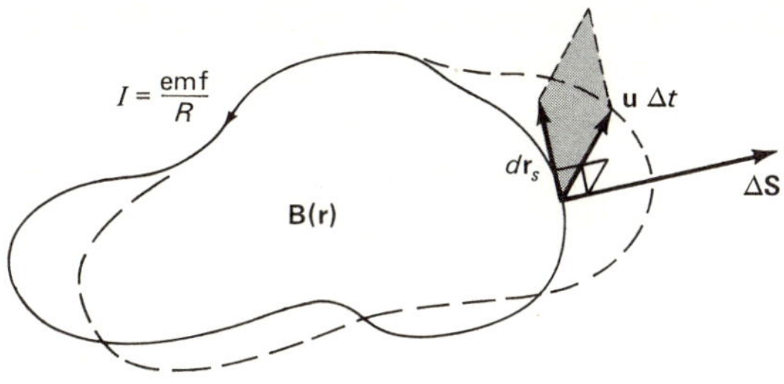

Figure 4.3

during a time increment Δt. Then $\Delta\Phi = \Delta t(\mathbf{u} \times d\mathbf{r}_s) \cdot \mathbf{B}(\mathbf{r}_s)$ is the corresponding change in the enclosed magnetic flux, and so

$$\frac{d\Phi}{dt} \equiv \frac{d}{dt}\oint d\mathbf{S} \cdot \mathbf{B}(\mathbf{r}) = -\oint (d\mathbf{r}_s \times \mathbf{u}) \cdot \mathbf{B}(r_s) \tag{4.72}$$

is just the rate of change of the flux enclosed due to the motion through a static field. Divided by c, this is identical with the negative of the emf (4.71), exactly as formulated in the Faraday law (1.33).

The effect in moving segments just described accounts for a usual statement that emf's can be generated by cutting magnetic field lines with a wire. The effect in each segment is present even in a *uniform* magnetostatic field, since the Lorentz forces plainly do not disappear in that situation. However, if all segments of a complete wire loop are translated together, with the same directed velocity, the flux of a *uniform* field $\mathbf{B}_o$ enclosed by the loop is not changing and there can be no net emf; the integral (4.71) reduces to $-c^{-1}\mathbf{B}_o \cdot (\mathbf{u} \times \oint d\mathbf{r}_s) = 0$ for the complete loop, indicating cancellations between the contributions to the emf by the individual segments. A *net* emf is generated in a uniform field when different parts of the circuit are given different velocities. A standard way used in electric power generators is to rotate the circuit rigidly with some angular velocity ω about an axis perpendicular to uniform field lines (as between the poles of a magnet). If $\Phi(0) = \mathrm{B}_o S$ is the magnetic flux through the circuit at a phase in the rotation at which it is maximal, then $\Phi(t) = \mathbf{B}_o \cdot \mathbf{S} = \mathrm{B}_o S \cos \omega t$ is the flux at a later time t. This yields an emf equal to $-\dot{\Phi}/c = (\omega/c)\mathrm{B}_o S \sin \omega t$ and an alternating current

$$I = \frac{\text{emf}}{R} = I_o \sin \omega t \tag{4.73}$$

with $I_o = (\mathrm{B}_o S/R)(\omega/c)$, if the resistance in the circuit is R (see Exercise 1.10).

It may not be immediately clear how the induction of a Faraday emf in a magneto*static* field can be consistent with the purportedly general point form $\nabla \times \mathbf{E} = -\mathring{\mathbf{B}}/c$ [Eq. (1.36)] of the Faraday law, in which $\mathring{\mathbf{B}} = 0$ for a magnetostatic field. In this relation, the fields on both sides must be referred to the same frame, and the electric field $\mathbf{E}$ does remain zero in the frame relative to which $\mathbf{B}$ is static and $\mathring{\mathbf{B}} = 0$. It is only in the frames of segments moving through the magnetostatic field that electric fields $\mathbf{E}' \neq 0$ are induced, and relative to these the magnetic field ($\mathbf{B}'$) at each segment does change in value. The result can be viewed as the motion of the magnetic field relative to the segment, whereas it was the segment that was regarded as moving and the field static in the initial frame. This bears out the statement, made in connection with (1.33), that Faraday induction arises whenever circuit and field move relative to each other regardless of which is viewed as being at rest.

4.4 MAGNETOSTATIC ENERGY

A magnetostatic field energy can be identified in steps paralleling those by which expression (2.60) for the electrostatic field energy was developed. The first step toward the latter was an evaluation of the work (2.37) needed to bring together point-charge sources of a field. The correspondingly idealized sources of a magnetostatic field are discrete, divergenceless loops of steady current. As for the point charges, it will be supposed that the individual current loops have already been formed, in their final magnitudes and shapes, while still far apart, i.e., isolated from each other. Thus self-energies of the currents are omitted initially. Only the work of bringing the prefabricated current loops from infinity into each other's presence—to their final static positions as sources of the field—is to be evaluated at first.

In the electrostatic case, expressions for the work were immediately available as potential energies per unit charge. No comparable expressions have so far been developed for work by magnetostatic forces. However, expressions are available for the forces between steady-current loops in arbitrary relative configurations. Let I_1 stand for the first current brought into position, at the cost of no work as long as self-energies are not counted, and then consider bringing up a second preformed current loop I_2. At any given stage of the process the force on I_2 due to I_1 is given by the expression for F_{21} in (4.62)

$$\mathbf{F}_{21} = \oint\oint d^2\mathbf{F}'_{21} \equiv \frac{I_1 I_2}{c^2} \oint\oint \frac{(d\mathbf{r}_1 \cdot d\mathbf{r}_2)\hat{\mathbf{R}}}{R^2} \tag{4.74}$$

if $\mathbf{R} \equiv R\hat{\mathbf{R}} \equiv \mathbf{r}_1 - \mathbf{r}_2$ is the relative position vector between the elements $(I_1\, d\mathbf{r}_1)$ and $(I_2\, d\mathbf{r}_2)$ of the two currents. The symbol $d^2\mathbf{F}'_{21}$ here is a temporarily convenient one for an element of the double integral (it is only the part of the force between current elements that survives integration and hence its distinction by the prime). While the separations R are being reduced from infinite ones to their final values, the work

$$W''_{12} = \oint\oint\int (d^2\mathbf{F}'_{21} \cdot d\mathbf{R}) = \frac{I_1 I_2}{c^2} \oint\oint (d\mathbf{r}_1 \cdot d\mathbf{r}_2) \int_\infty^R \frac{\hat{\mathbf{R}} \cdot d\mathbf{R}}{R^2}$$

is done. The element in the last integral is equivalent to $d(-1/R)$, and so

$$W''_{12} = -\frac{I_1 I_2}{c^2} \oint\oint \frac{(d\mathbf{r}_1 \cdot d\mathbf{r}_2)}{R} \tag{4.75}$$

The two apostrophes on the work symbol here are meant to indicate that, as will be found next, a second omission besides that of the self-energy has been made in the evaluation so far.

The negative sign on the expression for W''_{12} is to be expected because parallel loops of current, with $I_1\, d\mathbf{r}_1 \cdot I_2\, d\mathbf{r}_2 > 0$, are known to attract each other and so there is negative work in bringing them together; this conforms to the fact that two bar magnets having their opposite poles nearest each other tend to slide together *spontaneously*. A negative work was also found for bringing *charges* of opposite sign together. However, a close approach of opposite charges tends to *weaken* the resultant electric field around them, whereas the approach of parallel currents *strengthens* the magnetostatic field since the contributions $\mathbf{B}_1$ and $\mathbf{B}_2$ by the individual currents tend to become *co*directional. On this basis alone, the implication would be that magnetostatic field strengths can come into being spontaneously, instead of costing positive energy.

Actually, an essential part of the work needed to construct the field has not even been mentioned so far. Each of the pair of currents considered above is a source of magnetic field, and the amount of magnetic flux each sends through the area enclosed by the other changes as the two currents are being brought together. This will result in the induction of electric forces in the current-carrying loops, according to the Faraday law (1.33), and so work will have to be done against those induced forces just to maintain the currents at their prefabricated values during the establishment of the final configuration. The necessary work will be evaluated next.

The part to be played in these considerations by the Lenz law, mentioned in the discussion of the Faraday law (1.33), should be recognized. It makes explicit the fact that the field induced opposes the inducing field, and so the resultant field is *weakened* by the process unless work energy is supplied to counter the induction.

The Effects of the Induced Electromotive Forces

To see the effect of Faraday induction on the loop carrying the current I_1 as the second current loop I_2 is brought into its neighborhood, the flux through the first loop

$$\Phi_{12} \equiv \int_{(I_1)} d\mathbf{S}_1 \cdot \mathbf{B}_2(\mathbf{r}_1) \tag{4.76}$$

of magnetic field $\mathbf{B}_2$ generated by I_2 must be evaluated at any stage. The subscript (I_1) indicates that integration over elements $d\mathbf{S}_1$ of an area enclosed by I_1 is meant. When the derivability of the field $\mathbf{B}_2$ from the vector potential $\mathbf{A}_2(\mathbf{r}_1) = (I_2/c) \oint d\mathbf{r}_2/R$ is taken into account, the Stokes theorem yields

$$\Phi_{12} = \int_{(I_1)} d\mathbf{S}_1 \cdot (\nabla_1 \times \mathbf{A}_2) = \oint d\mathbf{r}_1 \cdot \mathbf{A}_2 = \frac{I_2}{c} \oint\oint \frac{d\mathbf{r}_1 \cdot d\mathbf{r}_2}{R} \tag{4.77}$$

The integral multiplying I_2 here depends only on the geometrical configurations of the loops, including their relative positions. This suggests defining a quantity

$$M_{12} = c^{-2} \oint\oint \frac{d\mathbf{r}_1 \cdot d\mathbf{r}_2}{R} = M_{21} \tag{4.78}$$

a property of the entire geometrical configuration known as its *mutual inductance*. In terms of it,

$$\Phi_{12} = cM_{12}I_2 \qquad \text{and} \qquad \Phi_{21} = cM_{12}I_1 \tag{4.79}$$

are the magnetic fluxes being sent through the loops of I_1 and I_2 by the currents I_2 and I_1, respectively.

During any *changes* of the magnetic flux Φ_{12} through the loop of I_1 there exists in the loop an electromotive force proportional to the *rate* at which the flux is changing,

$$\oint \mathbf{E}_1 \cdot d\mathbf{r}_1 = -\frac{d\Phi_{12}}{c\, dt} \tag{4.80}$$

according to Faraday's law (1.33). The induced force $\mathbf{E}_1$ per unit charge tends to generate new motions of any charges in the loop, and so the current I_1 in it will be changed unless the agency constructing the system supplies counter emf, $-\oint \mathbf{E}_1 \cdot d\mathbf{r}_1$, to cancel the forces in the induced emf. It is the energy that must be thus supplied, the work that must be done by the counter emf, that is wanted.

Whenever a charge element $\rho(\mathbf{r})\, dV(\mathbf{r})$ undergoes a displacement $\mathbf{u}(\mathbf{r})\, dt$ in the presence of an electric field $\mathbf{E}$, work $(\rho\, dV)\mathbf{E}(\mathbf{r}) \cdot \mathbf{u}\, dt$ is done by the electric force. Thus *power* (work per unit time)

$$\mathbf{E} \cdot \mathbf{j} \tag{4.81}$$

is being expended *per unit of volume* occupied by current in density $\mathbf{j}(\mathbf{r})$. The total power being expended in a circuit carrying a current I is then

$$\oint dV \mathbf{j} \cdot \mathbf{E} = I \oint d\mathbf{r} \cdot \mathbf{E} = (\text{emf})(I) \tag{4.82}$$

The last expression is familiar in elementary physics, as is also the expression I^2R for a circuit with resistance $R = \text{emf}/I$, as in (1.30) and (1.31). This remark serves as a reminder that an increment of emf equal to IR is also needed in a real wire, to maintain a steady current I against any resistance in it, quite aside from the emf counter to any induced emf's. However, work against resistance should plainly *not* be counted as part of the energy needed to establish a field. Power losses I^2R are irreversible heat losses, leading to the appearance of *thermal* energies in the material of the circuit, and not a part of any reversibly storable field energy.

It is now plain that the power needed to counter the induced emf (4.80) is just

$$-I_1 \oint \mathbf{E}_1 \cdot d\mathbf{r}_1 = +\frac{I_1\, d\Phi_{12}}{c\, dt} \tag{4.83}$$

Its magnitude at any given stage of bringing up the loop of I_2 that is supplying the flux Φ_{12} depends on the speed with which the stage is passed through, but the total work done, over whatever time is taken to increase Φ_{12} from the zero value it has when the loops are far apart, is independent of such details of the relative motion if this is carried out slowly enough (adiabatically), with field equilibria at each stage of the process. Integrating (4.83) over the time, with a constantly maintained I_1, yields simply

$$\frac{I_1 \Phi_{12}}{c} = M_{12} I_1 I_2 \tag{4.84}$$

where Φ_{12} and M_{12} are the values of the magnetic flux and the mutual inductance in the final configuration. The symmetry of the expression shows that an exactly equal amount of work is needed to counter the emf's induced in the loop of I_2 because of changes in the flux $\Phi_{21} = CM_{12} I_1$ through it during its approach to I_1.

It has now been found that the work of establishing a pair of preformed current loops in each other's neighborhood (still omitting the work of establishing either current by itself) consists of

$$W'_{12} = W''_{12} + \frac{I_1 \Phi_{12} + I_2 \Phi_{21}}{c} = W''_{12} + 2M_{12} I_1 I_2 = -W''_{12} \tag{4.85}$$

The last of these equalities follows from noticing that expression (4.75) for W''_{12} can be written $-M_{12} I_1 I_2$ when the definition (4.78) of the mutual inductance is used. Implicit in these results are the expressions

$$W'_{12} = -W''_{12} = \frac{I_1 I_2}{c^2} \oint\oint \frac{d\mathbf{r}_1 \cdot d\mathbf{r}_2}{R} = I_1 \oint d\mathbf{r}_1 \cdot \frac{\mathbf{A}_2}{c} = I_2 \oint d\mathbf{r}_2 \cdot \frac{\mathbf{A}_1}{c} \tag{4.86}$$

where $\mathbf{A}_1$ and $\mathbf{A}_2$ are appropriate vector potentials.

Perhaps the equal absolute magnitudes of the three contributions discriminated in the work expression (4.85) would seem less fortuitous if it were considered that the same system can also be established in other ways. For example, the initial situation might be taken to be one in which the loop circuits are already in place but with $I_1 = I_2 = 0$ in them. Then, as power is supplied for the increase of the currents to their eventual values, the loops carrying them need not be displaced and such a force between them as (4.74) does no work at all. Only work against emf's induced as the fluxes, $\Phi_{12} = cM_{12} I_2$ and $\Phi_{21} = cM_{12} I_1$ of (4.79), increase with the currents need be evaluated. The part countering the emf induced in the loop of I_1 is done at the rate (4.83) again, and a corresponding part is done in the loop of I_2, to give a total rate

$$\frac{dW'_{12}}{dt} = M_{12}\left(I_1 \frac{dI_2}{dt} + I_2 \frac{dI_1}{dt}\right) = \frac{d}{dt}(M_{12} I_1 I_2) \tag{4.87}$$

In the process here, the flux changes arise from the changes of the currents, rather than in the mutual inductance, which is constant in the fixed geometrical configuration. The final result

$$W_{12}^{\cdot} = M_{12} I_1 I_2 \tag{4.88}$$

is plainly the same as the one found in (4.85).

The last considerations suggest an approach to finding the *self-energy* of any one loop of current. The flux through its own enclosed area, of field $\mathbf{B}_1 = \nabla \times \mathbf{A}_1$ generated by a loop of current I_1, is *formally* representable as in (4.77) with $\mathbf{A}_2$ replaced by $\mathbf{A}_1$. Then

$$\Phi_{11} = c L_1 I_1 \tag{4.89}$$

if

$$L_1 \equiv M_{11} = c^{-2} \oint\oint \frac{d\mathbf{r}_1 \cdot d\mathbf{r}_1'}{R}$$

is a self-inductance, obtained by integrating twice over the same loop and with $R = |\mathbf{r}_1 - \mathbf{r}_1'|$. Replacing (4.87) is a power to be supplied in order to increase I_1 from zero to its final value against the *self*-inductances

$$\frac{dW_1}{dt} = L_1 I_1 \frac{dI_1}{dt} = \frac{d}{dt} \tfrac{1}{2} L_1 I_1^2 \tag{4.90}$$

This leads to an expression

$$W_1 = \tfrac{1}{2} L_1 I_1^2 \tag{4.91}$$

for the self-energy of the current I_1, which will turn out to be formally correct. However, the integral given above for evaluating the self-inductance actually diverges. In the course of the double integration over a single geometric curve, $R^{-1} = |\mathbf{r}_1 - \mathbf{r}_1'|^{-1}$ becomes infinite whenever the elements $d\mathbf{r}_1$ and $d\mathbf{r}_1'$ coincide. The infinite result should also have been anticipated from the finding (4.17) about the circular loop; the vector potential $\mathbf{A}_1$ should be expected to approach infinity right at current idealized as having infinite density. The self-energy of a *finite* current passing through the *vanishing* cross section of an idealized loop becomes infinite for basically the same reasons that the electrostatic self-energy of a *point* charge turned out to be infinite. It can also be anticipated that a finite self-inductance and self-energy will be calculable for currents spread with finite density over *nonvanishing* cross sections, just as a finite self-energy was found for charge spread over a sphere of nonvanishing radius.

Granting the definability [see (4.95)] of finite self-inductances for sources that are not overidealized, an expression for the total work of establishing any system of discrete currents $I_1, I_2, \ldots, I_i, \ldots, I_j, \ldots$ can be written as

$$W = \sum_{i<j} M_{ij} I_i I_j + \sum_i \tfrac{1}{2} L_i I_i^2 = \sum_{i,j} \tfrac{1}{2} M_{ij} I_i I_j \tag{4.92}$$

where the last expression is predicated on counting each pair twice, on using $M_{ji} = M_{ij}$ for $i \neq j$, and on adopting $M_{ii} \equiv L_i$ for $i = j$. When calculating *mutual* inductances, the currents *may* become adequately representable as having negligible lateral cross sections and integrations like (4.78) may be used. Then the interaction energies alone, W minus the self-energies, can be expressed as

$$W' = \sum_{i \neq j} \tfrac{1}{2} M_{ij} I_i I_j = \sum_{i \neq j} \frac{I_i I_j}{2c^2} \oint \oint \frac{d\mathbf{r}_i \cdot d\mathbf{r}_j}{r_{ij}} \qquad \text{with } r_{ij} \equiv |\mathbf{r}_i - \mathbf{r}_j| \tag{4.93}$$

Expressions for the Magnetostatic Field Energy

The work of constructing any sources of magnetostatic field, as described by arbitrary distributions $\mathbf{j}(\mathbf{r})$ of divergenceless current densities, can be obtained as a suitable generalization of the expression (4.93) for discrete current loops. The basis for the generalization is provided by the findings in connection with (4.36) that any divergenceless current distribution can be divided up into tube elements (dI), each of which forms a complete loop. Then the work of establishing any pair (dI) and (dI') of the tube elements is given by (4.86) with $I_1\, dr_1$ replaced by $(dI)\, d\mathbf{r} \equiv \mathbf{j}(\mathbf{r})\, dV(\mathbf{r})$ and $I_2\, dr_2$ by $(dI')\, d\mathbf{r}' \equiv \mathbf{j}(\mathbf{r}')\, dV(\mathbf{r}')$. The suitable generalization of (4.93) becomes

$$W = \frac{1}{2c^2} \oint \oint dV(\mathbf{r})\, dV(\mathbf{r}') \frac{\mathbf{j}(\mathbf{r}) \cdot \mathbf{j}(\mathbf{r}')}{|\mathbf{r} - \mathbf{r}'|} \tag{4.94}$$

Like its electrostatic analog (2.40), this can be held to omit no energy of establishing the entire source, since constructing a tube with $(dI) \to 0$ requires no supply of self-energy.

The general result (4.94) makes it obvious how to calculate inductances for the expression (4.92) when this is used for discrete loops having finite thicknesses and finite current densities

$$M_{ij} = \frac{1}{I_i I_j} \oint \oint \frac{dV(\mathbf{r}_i)\, dV(\mathbf{r}_j)}{c^2} \frac{\mathbf{j}(\mathbf{r}_i) \cdot \mathbf{j}(\mathbf{r}_j)}{|\mathbf{r}_i - \mathbf{r}_j|} \tag{4.95}$$

This can give finite results even for a self-inductance $L_i \equiv M_{ii}$. It is plainly independent of the absolute sizes of the currents unless their magnitudes affect current distributions. Steady and quasi-static currents are usually quite uniformly distributed over homogeneous conductors of constant cross section (see page 595), and the circuit inductances usually spoken of are ones calculated on that basis.

Just as the electrostatic energy (2.39) can be expressed in terms of the scalar-potential field generated by the entire charge distribution, so the magnetostatic energy (4.94) can be reexpressed in terms of the vector potential (4.4) generated by the entire current distribution

$$W = (2c)^{-1} \oint dV(\mathbf{r}) \mathbf{j}(\mathbf{r}) \cdot \mathbf{A}(\mathbf{r}) \tag{4.96}$$

Moreover, as in the electrostatic case, the source description can be eliminated from this expression entirely, here by using the Ampère law $\mathbf{j} = (c/4\pi)(\nabla \times \mathbf{B})$. Then

$$\begin{aligned} W &= (8\pi)^{-1} \oint dV \mathbf{A} \cdot (\nabla \times \mathbf{B}) \\ &= (8\pi)^{-1} \oint dV (\mathbf{A} \times \nabla) \cdot \mathbf{B} \\ &= (8\pi)^{-1} \oint dV [-(\nabla \times \mathbf{A}) \cdot \mathbf{B} + \mathbf{B} \cdot (\nabla \times \mathbf{A})] \end{aligned}$$

The gradient operator is being allowed to apply to all functions following it in each term of these expressions, hence the correction by the last term in the last line. The interchangeability of dot and cross has been used in going from the first to the second line, and when it is used again in the first term of the last line, this term becomes proportional to a volume integral of a divergence $\nabla \cdot (\mathbf{A} \times \mathbf{B})$. Such an integral is equivalent to a surface integral, according to the Gauss theorem, and here the surface is a remote one, since the volume integral was to extend over the *entire* field, with $\mathbf{A} \times \mathbf{B}$ reduced to zero on the enclosing surface. There remains only a volume integral of $\mathbf{B} \cdot (\nabla \times \mathbf{A}) = \mathbf{B} \cdot \mathbf{B}$, and so

$$W = \oint \frac{dV(\mathbf{r}) \mathbf{B}^2(\mathbf{r})}{8\pi} \tag{4.97}$$

comparable to the electrostatic-energy expression (2.60) in that there is merely a replacement of $\mathbf{E}$ by $\mathbf{B}$. As in the electrostatic case, the magnetostatic energy is now being treated as a possession of the field, existing in the density

$$w(\mathbf{r}) = \frac{\mathbf{B}^2(\mathbf{r})}{8\pi} \tag{4.98}$$

at the point $\mathbf{r}$. Whenever *both* electric *and* magnetic fields are present,

$$w(\mathbf{r}) = \frac{\mathbf{E}^2 + \mathbf{B}^2}{8\pi} \tag{4.99}$$

is the energy-density distribution in the electromagnetic field. Examples are treated in Exercises 4.22, 4.25, and 4.26.

An example of fundamental interest is concerned with the magnetic energy gained by the field of an uniformly charged spherical shell when it is set into translational motion with some velocity $\mathbf{u}$. For the present, only a quasi-static evaluation is possible, valid only for $u \ll c$, like the magnetic field evaluation (4.26) and the galilean field transformations (4.68) and (4.69). The latter present a significant way to obtain the field here (others, suggested in Exercise 4.8, yield the same results).

In the *instantaneous* frame with velocity $\mathbf{u}$, the sphere is at rest and the entire field is $\mathbf{E}'(R > a) = q\hat{\mathbf{R}}/R^2$, a Coulomb field, with $\mathbf{E}'(R < a) = 0$ and $\mathbf{B}' = 0$.

Thus, gauged relative to the rest frame, the field energy is entirely electric, and the value $W_o = q^2/2a$ for a sphere of radius a has already been quoted on page 48.

In the frame relative to which the velocity $\mathbf{u}$ is observed, $\mathbf{E} \approx \mathbf{E}'$ because $\mathbf{B}' = 0$, according to the reciprocal of (4.68), and then $\mathbf{B} \approx \mathbf{u} \times \mathbf{E}/c \approx \mathbf{u} \times \mathbf{E}'/c$ follows from (4.69). The electric part of the field energy is thus unchanged from $W_o = q^2/2a$, the field's rest energy, but its motion with the sphere adds *magnetic* field energy in the density $\mathbf{B}^2/8\pi = (\mathbf{E}^2/8\pi)(u \sin \vartheta/c)^2$ at field points outside the moving sphere, where ϑ is the angle between $\mathbf{u}$ and the direction $\hat{\mathbf{R}}$ from sphere center to field point. Integration over the entire field gives

$$\frac{q^2}{2a}\left(\frac{2}{3}\frac{u^2}{c^2}\right) = \frac{1}{2}\left(\frac{4}{3}\frac{W_o}{c^2}\right)u^2 \tag{4.100}$$

as the total magnetic field energy that is added to the electric part W_o when this is set into motion with a speed $u \ll c$.

What is being described here is a result of the fact that no point of charge can move unless it carries along bodily its (Coulomb) self-field, which must remain centered on it. Whenever a velocity $\mathbf{u}$ is imparted to a charge-bearing bare mass m_o, not only must the work of supplying kinetic energy $\frac{1}{2}m_o u^2$ be done but the field energy (4.100) must also be supplied, besides the self-field energy W_o of the charge at rest. The extra energy also has a kinetic-energy form, as if the field itself were endowed with a field mass $4W_o/3c^2$. Any force $\mathbf{F}$ responsible for imparting the motion will supply the needed energy only if it equals

$$\mathbf{F} = \frac{d}{dt}\left(m_o + \frac{4W_o}{3c^2}\right)\mathbf{u} \tag{4.101}$$

an equation of motion first advanced by Abraham and by Lorentz, aside from radiative corrections that are completely negligible for $u \ll c$. Besides the momentum $m_o\mathbf{u}$, the force must impart field momentum $(4W_o/3c^2)\mathbf{u}$. Note that nothing *observable* is thus changed; $\mathbf{F} = m\dot{\mathbf{u}}$ is still valid if it is supposed that endowment with charge renormalizes the bare mass m_o to the constant $m = m_o + 4W_o/3c^2$. Corrections of (4.101), arising when higher velocities than $u \ll c$ are considered, will be taken up in Chap. 14.

Development of expressions for magnetic *interaction* energies, paralleling the developments in the electrostatic case, is left to the next chapter.

EXERCISES

4.1 Suppose that the spinning sphere presenting the source (4.5) has all its charge q uniformly distributed on its surface instead, so that $\rho = (q/4\pi a^2)\delta(r_s - a)$.

(*a*) Check that $\oint dV\rho = q$ upon integration over all space and that the flux $I = \int d\mathbf{S} \cdot \mathbf{j}$ through any meridional half plane equals $I = q\omega/2\pi$ properly (see Exercise 1.7).

(*b*) Show that the field outside the sphere again has the magnetic dipole form, except that now $\boldsymbol{\mu} = \frac{1}{3}qa^2\omega/c$.

(*c*) Show that the magnetic field is uniform throughout the interior of the spinning shell [$\mathbf{B}(r < a) = 2\boldsymbol{\mu}/a^3$].

4.2 For a long (∞) straight line of steady current I:

(*a*) Integrate (4.25).

(*b*) Integrate (4.16) and determine $\mathbf{B}$ from it.

(*c*) Determine a magnetic scalar potential ϕ_M such that $\mathbf{B} = -\nabla\phi_M$.

4.3 Use the Biot-Savart law to find the exact result for $\mathbf{B}$ everywhere on the axis of the circular loop of current in Fig. 4.1. Compare it with the approximation (4.22). Is it sufficient to know $\mathbf{A}$ *only* on the axis in order to determine $\mathbf{B}$ on it?

4.4 A substitution $\alpha = (\pi - \varphi_s)/2$ in the integral (4.17) puts this into a form that is standard in the definitions of elliptic integrals (see almost any integral table).

(*a*) Show that

$$A = \frac{8Ia}{ck^2} \frac{(1 - \frac{1}{2}k^2)K(k) - E(k)}{(r^2 + a^2 + 2ar \sin \vartheta)^{1/2}}$$

where $k^2 \equiv (4ar \sin \vartheta)/(r^2 + a^2 + 2ar \sin \vartheta)$ and K and E are, respectively, elliptic integrals of the first and second kinds.

(*b*) It is now most natural to approximate by power series in $k^2 < 1$, in place of the ϵ of (4.18); the series expansions are also given in the integral tables. Find $\mathbf{B}_r$ and $\mathbf{B}_\vartheta$ to first order in k^2, as a replacement of (4.22). [The result is a somewhat different and slightly poorer approximation. It yields a $\mathbf{B}$ in the plane of the loop that reverses sign at $r = 2a$ instead of the proper $r = a$. In (4.22) the reversal takes place at $r = \sqrt{2}a$.]

4.5 (*a*) Find the vector potential $\mathbf{A}$ inside and outside an infinitely long cylindrical solenoid (see Fig. E.1) as a continuous coaxial superposition of such current loops as the one that led to (4.17), with n loops (turns) per unit length of their axis, each carrying the current I. [Reference to Exercise C.15 may help; see also (6.20).]

(*b*) Derive $\mathbf{B}$ from the result of (*a*).

(*c*) Show how the same $\mathbf{B}$ follows from the Biot-Savart law.

(*d*) Show that the axial field components at the open end of half-infinite solenoid are one-half of those far inside it.

4.6 Experimenters must be content with solenoids (Exercise 4.5) of *finite* length (say, $-\frac{1}{2}L < z_s < \frac{1}{2}L$). Then there is interest in departures from uniformity of the field near $s = 0$, $z = 0$.

(*a*) Find an *exact* expression for $\mathbf{B}(s = 0, z)$ everywhere on the solenoid's axis.

(*b*) Expand your result (*a*) into a power series in z to get the first-order correction factor

$$\frac{\mathbf{B}(0, z) - \mathbf{B}(0, 0)}{\mathbf{B}(0, 0)}$$

as a function of z.

(*c*) Find correspondingly corrected expressions for $\mathbf{B}(s, z)$ *off axis*.

$$\left[\mathbf{B}_s \approx \mathbf{B}(0, 0) \frac{24a^2 zs}{(L^2 + 4a^2)^2} \right]$$

(*d*) Which is stronger, $\mathbf{B}_z(s > 0, 0)$ or $\mathbf{B}_z(0, 0)$? Give qualitative reasons for expecting your result.

4.7 Show that the quasi-static $\mathbf{B}_q$ of (4.26) is derivable as the curl of

$$\mathbf{A}_q = \frac{\mathbf{u}}{c}\phi_q \qquad \text{where } \phi_q = \frac{q}{R}$$

[This vector potential can be made divergenceless (without altering its curl) by replacing half of it with only its longitudinal (parallel to $\hat{\mathbf{R}}$) component.]

4.8 A sphere of charge q homogeneously distributed within a radius a is being translated with a uniform speed $u \ll c$.

(a) Use a quasi-static approximation like that used for (4.26) to find $\mathbf{B}$ both inside and outside the sphere in terms of a radius vector $\mathbf{R}$ based on the center of the moving sphere.

(b) Show how the galilean transformation (4.69) yields the same results, starting from the known $\mathbf{E}$ in the frame in which the sphere is at rest (its rest frame).

(c) In a third approach, different from that of Exercise 1.12 only for $\mathbf{B}(R < a)$, use the Maxwell induction integral (1.39).

4.9 Evaluate the integral for $\boldsymbol{\mu}$ of (4.38) for the spinning spheres treated in connection with (4.8) and in Exercise 4.1.

4.10 Two coaxial circular loops, each of radius a, lie in planes normal to their common axis and a distance l apart. They carry equal but oppositely directed currents I.

(a) Show that at remote field points ($r \gg a, l$) the vector potential becomes

$$A(r, \vartheta) \approx \frac{3\mu l \sin\vartheta \cos\vartheta}{r^3}$$

where $\mu \equiv \pi a^2 I/c$ is the dipole moment of each loop.

(b) Show that the field $\mathbf{B}$ derived from $\mathbf{A}$ of (a) is also derivable from a scalar potential having the same form as that found for the linear electric quadrupole moment of Exercise 2.15.

4.11 The infinitely extended charge distribution

$$\rho(r, \vartheta) = -\frac{e}{64\pi a^5} r^2 \sin^2\vartheta e^{-r/a}$$

is rigidly maintained and set spinning about its polar axis with an angular velocity $\omega = \hbar/24ma^2$.

(a) What is the magnetic dipole moment in terms of the given constants e, a, $\hbar$, and m?

(b) The m is a total mass codistributed with the charge. What is the angular momentum of the motion? (The results here pertain to an electron bound in the lowest p state of an hydrogen atom.)

4.12 Suppose that the symbols δ and τ of Exercises 2.20, 2.21, and 2.23 stand for magnetic rather than electric dipole moment distributions. Find results for the magnetic fields corresponding to the electric ones found in the earlier exercises.

4.13 Evaluate the lowest-order moment (4.48) that does not vanish for the complete system of Exercise 4.10. [The more conventional definition (4.52) would here require treating derivatives of delta distributions.] Show that your result yields the same scalar potential as that found in Exercise 4.10.

4.14 (a) Obtain (4.63) by integrating (4.62).

(b) Give recognition to $\mathbf{B}$ as the transmitter of the force by using the appropriate modification

$$\mathbf{F} = \frac{I}{c}\oint d\mathbf{r}_s \times \mathbf{B}(\mathbf{r}_s)$$

of (4.53) and the fact that $\mathbf{B}_1 = 2I_1/cs$ is the field at I_2 due to I_1.

4.15 Two coaxial circular loops of current I_1 and I_2, having radii a_1 and a_2, lie in parallel planes a distance z apart.

(a) Show that the resultant force on each is determined solely by components $\mathbf{B}_s$ of their respective fields, i.e., radial components directed to or from the common axis.

(b) Show that the approximation of (4.22) yields

$$F = \frac{6\mu_1\mu_2 z}{(z^2 + a_1^2 + a_2^2)^{5/2}}$$

for the magnitude of the mutual force if $\mu_{1,2}$ are the magnetic dipole moments of the respective loops. [This force properly vanishes when the loops are copolanar ($z = 0$) even though the approximation becomes least trustworthy in that configuration!]

(*c*) Show that the result for $z \gg a_{1,2}$ is exactly what is to be expected for a force derivable as a negative gradient of the dipole-dipole potential in (2.57), with $\mathbf{D}_{1,2} \to \boldsymbol{\mu}_{1,2}$.

4.16 (*a*) In the situation of Exercise 1.10, show that the work rate against the magnetic force is exactly sufficient to make up for the instantaneous ohmic-loss rate I^2R.

(*b*) Suppose that the rotating loop serves as a generator that also provides energy used externally. How can such an extra load be represented in the formulation of (*a*), and how does throwing it in affect the current when the rotation rate is maintained?

4.17 An illustration frequently used concerns a rectangular loop having as two sides the parallel axles of locomotive wheel pairs and, for the other two, tie rods connecting the two wheels on each track. The locomotive is presumed to be moving with uniform speed through a uniform magnetic field (usually the earth's field). According to the discussion of Fig. 4.3, there will be no *net* emf in the loop circuit. Questions that frequently result in controversy are: Will a voltmeter joining opposite points on the tie rods and moving with them register a potential difference? Will a stationary voltmeter on the ground joining opposite points on the two tracks register a potential difference? Compare the situation here with a circuit that has in it two like batteries bucking each other.

4.18 (*a*) Show that the force (4.63) is derivable as a negative gradient of a potential energy per unit length $U_1(s)$ that could be calculated from the work expression (4.75).

(*b*) When a cylinder, radius a, carries a uniformly distributed current I, there is a mutual attraction between every pair of the parallel current elements, $dI = (I/\pi a^2)s\, ds\, d\varphi$, that form I. Show that, as a result, there is an inward force per unit *volume* on elements at a distance $s < a$ from the cylinder axis that is given by $2I^2 s/\pi a^4 c^2$. (Reference to Exercise C.15 can help with the integration. The force density here is known as a *pinch*.)

4.19 A square loop of side a and mass m is placed in a vertical plane with its uppermost side parallel to, and at a distance z below, a long horizontal wire fixed in space.

(*a*) If the loop is carrying a steady current I, what current I' must be sent through the wire to keep the otherwise unsupported loop from falling?

(*b*) Calculate the mutual inductance of wire and loop.

(*c*) Show how the mutual inductance is related to the work of reversing the current I in situation (*a*).

(*d*) Show how the work in (*c*) is related to the work of taking the initial current loop off to infinity and returning it with I reversed.

4.20 Suppose that the magnetic-bottle field

$$\mathbf{B}(s, z) = \alpha(z\mathbf{e}_z - \tfrac{1}{2}s\mathbf{e}_s)$$

of Exercise 1.9 has its axis parallel to the uniform gravitational field. Also suppose that the carefully centered wire loop of that exercise has a mass m and is allowed to fall from rest with the loop in an horizontal plane.

(*a*) Find the speed of fall as a function of time and in terms of the loop and field parameters.

(*b*) Show that if the fall lasts long enough, the current in the loop reaches a maximum value independent of the resistance in it.

(*c*) Compare the motion and the orientation of the induced magnetic moment with cases in which (1) the field direction (relative to gravity) is reversed and (2) the direction in which the field intensifies is reversed.

4.21 (*a*) Using the approximation of Exercise 4.15 for the mutual force between coaxial current loops, find the work needed to take the loops from a distance z apart to $z \to \infty$.

(*b*) What does the result of (*a*) imply about the mutual inductance between the loops?

(*c*) Starting from the expression

$$\Phi \equiv \oint d\mathbf{S} \cdot \mathbf{B} = \oint d\mathbf{r} \cdot \mathbf{A}$$

implicit in (4.77), for the magnetic flux through any circuit, show how an exact expression for the mutual inductance in (*b*) can be obtained, as a direct proportionality to the series (4.19), suitably evaluated, or as the equivalent expression in terms of the elliptic functions displayed in Exercise 4.4.

4.22 The definition (4.95) of a self-inductance $L = M_{ii}$ exhibits its purely geometric character for uniformly distributed currents, but it is generally more expeditious to derive it from equivalent expressions implicit in the self-energy

$$W = \tfrac{1}{2}LI^2 = \frac{1}{2c}\oint dV\mathbf{j}\cdot\mathbf{A} = \oint dV\frac{\mathbf{B}^2}{8\pi}$$

(*a*) Confirm these equivalences.

(*b*) Use the vector potential found in Exercise 4.5 to show that the self-inductance contributed by each unit length of the infinite solenoid is just $(2\pi na/c)^2$.

(*c*) Use the magnetic field value to obtain the same result.

(*d*) Show that a consideration of the magnetic flux (4.89) also gives the same result. [The solenoid can be made finite without changing the results appreciably by forming it into a closed circle with a very large radius $R \gg a$, thus producing a ring or toroidal solenoid. Its *total* self-inductance is then well approximated as $2\pi R(2\pi na/c)^2$.]

4.23 The solenoid of Exercise 4.22 becomes an air-core transformer when secondary windings of insulated wire, a total of N turns confined to some finite stretch along the outer surface, are wrapped around the tightly wound solenoid. Find the total mutual inductance between the primary and secondary circuits.

4.24 Suppose that a circuit carrying the current around the solenoid of Exercise 4.22 is suddenly interrupted at $t = 0$, so that I is no longer sustained against the ohmic-loss rate I^2R in the circuit.

(*a*) Considering the initial field energy available, construct an equation for the rate of decrease $\dot{I}(t > 0) < 0$.

(*b*) Show that the decay of $I(t)$ will be exponential, with a time constant L/R.

4.25 Repeat the considerations of (4.100) and (4.101) for a sphere that has the charge q homogeneously distributed within its radius a. How is the field-mass contribution to the mass related to the self-field energy $W_o = 3q^2/5a$ of (2.47) in this case?

4.26 Reconsider the spinning spheres of Exercise 4.1 and the moment (4.8). In each case:

(*a*) Find the total magnetic field energies from the integration over the source distribution (4.96).

(*b*) Find the fractions of the energies in (*a*) that are distributed over the fields inside and outside each sphere.

(*c*) In terms of the field masses discussed in connection with (4.100) and Exercise 4.25, calculate field moments of inertia for the hollow sphere and for its homogeneously charged counterpart.

Ans. $4W_o a^2/21c^2$.

CHAPTER

FIVE

NONRELATIVISTIC MOTIONS IN STATIC FIELDS

Newtonian equations in electromagnetic fields and their canonical reformulation in terms of potentials · Charged particles in purely magnetostatic fields and their guidance by field lines · $\mathbf{E} \times \mathbf{B}$ *drifts · Forces and torques on magnetic dipoles · Interaction energies between sources and fields · The magnetic mirror effect · Larmor precession and Zeeman effects on atoms · The vector potential arising from frame rotation*

The earliest discoveries about electricity and magnetism were associated with matter in bulk. Conclusions about point charges were drawn from experiments with macroscopic balls of material and about charges in motion from observations on currents in conducting wires. Such inferences were provided with a better foundation after discoveries about individual "atoms of electricity," the electrons and ions, which not only led to a deeper understanding of the macroscopic properties of matter, now analyzable into constituent atoms, but fostered the exploitation of special forms of matter, such as the highly ionized plasmas that can be created in gaseous discharges.

The behavior of electrons and ions in electromagnetic fields was, quite naturally, first treated as the motion of simple point charges in accordance with newtonian equations of motion,

$$m\dot{\mathbf{u}} = q\left(\mathbf{E} + \frac{\mathbf{u}}{c} \times \mathbf{B}\right) \tag{5.1}$$

subjecting a mass m associated with a point charge q to the Lorentz force (4.66) that arises in a given electromagnetic field $\mathbf{E}$, $\mathbf{B}$. Investigating the simplest cases, in given electrostatic and magnetostatic fields, will be a concern of this chapter.

Using the classical newtonian mechanics will generally restrict the validity of the results to nonrelativistic velocities $u \ll c$, as will be seen later. It is nevertheless valuable to review the classical formulations, not only because they remain valid in most actual applications, but also because the more comprehensive relativistic formulations arise as generalizations of the classical ones and it is important to understand just what requires modification. Moreover, the results for the purely magnetic fields, in particular, will be found to remain unchanged at relativistic velocities ($u \to c$) if the value attributed to the constant m is chosen appropriately.

Some general conclusions that follow from the equation of motion (5.1) will be useful for all the cases to be investigated here.

Consider first the work done by the Lorentz force during any motional displacement $d\mathbf{r}_q = \mathbf{u}\,dt$ of the particle. Since the magnetic part of the force, $q\mathbf{u} \times \mathbf{B}/c$, keeps acting perpendicularly to the particle's displacements in every phase of its motion, the magnetic force never does any work at all: $(q/c)\times(\mathbf{u} \times \mathbf{B}) \cdot \mathbf{u}\,dt \equiv 0$! The electromagnetic field can change the kinetic energy of the particle, according to

$$\frac{d}{dt}\,\tfrac{1}{2}mu^2 = \mathbf{u} \cdot (m\dot{\mathbf{u}}) = q\mathbf{E} \cdot \mathbf{u} \tag{5.2}$$

only through work by the electrical part $q\mathbf{E}$ of the Lorentz force.

In an electro*static* field, derivable from a potential as $\mathbf{E} = -\nabla\phi$, the work elements $q\mathbf{E} \cdot \mathbf{u}\,dt$ will have integrated effects like

$$q\int_{\mathbf{r}_o}^{\mathbf{r}_q} \mathbf{E} \cdot d\mathbf{r} = -q\phi(\mathbf{r}_q) + q\phi(\mathbf{r}_o) = \tfrac{1}{2}m(u^2 - u_o^2) \tag{5.3}$$

This makes it possible to define a total energy of the motion

$$E = \tfrac{1}{2}mu^2 + q\phi(\mathbf{r}_q) \tag{5.4}$$

that is conserved at whatever initial value it might be given and provides an expression for the kinetic energy that is not explicitly dependent on any magnetic field present.

The magnetic force is not an energy-changing agency but is a *momentum-changing* one. This can also be expressed in terms of a potential, the vector potential from which the magnetic field $\mathbf{B} = \nabla \times \mathbf{A}$ is derivable:

$$q\mathbf{u} \times \frac{\mathbf{B}}{c} = \frac{q}{c}\mathbf{u} \times (\nabla \times \mathbf{A}) = \frac{q}{c}[\nabla(\mathbf{u} \cdot \mathbf{A}) - (\mathbf{u} \cdot \nabla)\mathbf{A}] \tag{5.5}$$

Here only $\mathbf{A}(\mathbf{r}_q)$ as determined by the particle's position $\mathbf{r}_q$, and not $\mathbf{u}(t)$, is to be subject to the gradient operation. The change $d\mathbf{A}(\mathbf{r})$ during a displacement $d\mathbf{r}_q = \mathbf{u}\,dt$ is just $(\mathbf{u}\,dt \cdot \nabla)\mathbf{A}(\mathbf{r})$, so that $(\mathbf{u} \cdot \nabla)\mathbf{A} \equiv d\mathbf{A}/dt$. This allows writing the equation of motion (5.1) in the form

$$\frac{d}{dt}\left(m\mathbf{u} + \frac{q}{c}\mathbf{A}\right) = -q\nabla\left(\phi - \frac{\mathbf{u}}{c} \cdot \mathbf{A}\right) \tag{5.6}$$

entirely in terms of a potential description of the field. The combination

$$\mathbf{p}(\mathbf{r}_q, t) \equiv m\mathbf{u}(t) + \frac{q\mathbf{A}(\mathbf{r}_q)}{c} \tag{5.7}$$

is known as the *conjugate momentum* of the particle in the field. The part $q\mathbf{A}/c$ represents a store of field momentum available to q (as will be seen in Chap. 6), just as $q\phi$ serves as a store of field energy (2.66) available for exchange with the particle's kinetic energy in (5.4). Both the kinetic momentum $m\mathbf{u}$ and the amount of $q\mathbf{A}/c$ stored as field momentum can be changed wherever the particle meets gradients in an interaction energy $U(\mathbf{r}) = q(\phi - \mathbf{u} \cdot \mathbf{A}/c)$ of particle and field.

5.1 A POINT CHARGE IN A UNIFORM MAGNETIC FIELD

The simple uniformly accelerated motions a point charge will have in a uniform *electric* field $\mathbf{E}_o$, have already been pointed out in connection with the expression (2.34) for the acceleration $\mathbf{a} = q\mathbf{E}_o/m$ of the particle. The counterpart magnetostatic force problem is concerned with the motion of a point charge that enters a uniform magnetic field $\mathbf{B}_o$ with some initial velocity $\mathbf{u}_o$ (a particle initially at rest would merely remain at rest). The equation of motion (5.1) yields for this case the acceleration

$$\dot{\mathbf{u}}(t) = \frac{1}{m}\left(q\frac{\mathbf{u}}{c} \times \mathbf{B}_o\right) = -\boldsymbol{\omega}_o \times \mathbf{u} \tag{5.8}$$

where $\boldsymbol{\omega}_o \equiv q\mathbf{B}_o/mc$ is the so-called cyclotron frequency of the particle m, q in the field $\mathbf{B}_o$. (The name stems from the fact that ω_o can represent the revolution rate of a particle circulating in the magnetic field of a high-energy accelerator; see Exercise 5.7.)

The first point made evident by expression (5.8) is that every vector increment $d\mathbf{u} = \dot{\mathbf{u}}\,dt$ added to the instantaneous velocity vector $\mathbf{u}(t)$ is perpendicular to it at the instant; the magnetic force merely *rotates* the velocity vector without changing its magnitude so that $|\mathbf{u}| = |\mathbf{u}_o|$. Here the field is solely a momentum-*direction*-changing agency. This means that the particle's kinetic energy will be conserved by itself, as already expected from the energy expression (5.4) for any purely magnetic field.

The equation of motion (5.8) likewise supplies no acceleration along the direction of the field $\mathbf{B}_o$, and so any component $\mathbf{u}_\parallel$ the initial velocity $\mathbf{u}_o$ may have parallel to the field will remain unaltered throughout the ensuing motion. The result can be viewed as a uniform translation, with the speed $u_\parallel = \mathbf{u}_o \cdot \mathbf{B}_o/\mathrm{B}_o$, of a plane in which motions $\mathbf{u}_\perp(t)$ perpendicular to the field take place. Since $\mathbf{u} = \mathbf{u}_\parallel + \mathbf{u}_\perp$ and $\dot{\mathbf{u}}_\parallel = 0$,

$$\dot{\mathbf{u}}_\perp(t) = -\omega_o \times \mathbf{u} = -\boldsymbol{\omega}_o \times \mathbf{u}_\perp(t) \tag{5.9}$$

and $\mathbf{u}_\parallel$ can be completely ignored for the motion in the moving plane. Of course, the magnitude of $\mathbf{u}_\perp(t)$ is constant by itself.

Let $\mathbf{R}(t)$ be the position vector of the particle based on some suitably chosen origin, all in a plane perpendicular to $\mathbf{B}_o$. Then $\mathbf{u}_\perp = \dot{\mathbf{R}}(t)$ can be put into the right side of (5.9), with a result that is immediately integrable to $\mathbf{u}_\perp(t) = -\boldsymbol{\omega}_o \times (\mathbf{R} + \mathbf{C})$, where $\mathbf{C}$ is an integration constant (vector) that depends not only on what start $\mathbf{u}_\perp(0)$ is given the particle but also on how the origin for $\mathbf{R}$ is chosen relative to the initial position of the particle. For any given start, that origin can always be so chosen that $\mathbf{C} \equiv 0$, and then

$$\mathbf{u}_\perp = \dot{\mathbf{R}} = -\boldsymbol{\omega}_o \times \mathbf{R}(t) \tag{5.10}$$

describes a uniform rotation of the resulting position vector $\mathbf{R}$ with the angular velcocity $-\boldsymbol{\omega}_o$. The motion is a circling at a constant radius R, which, since $\mathbf{R}$

and $\boldsymbol{\omega}_o \sim \mathbf{B}_o$ are perpendicular, is given by

$$R = \frac{u_\perp}{\omega_0} = \frac{mcu_\perp}{q\mathsf{B}_o} \tag{5.11}$$

This outcome is well known in elementary physics, where it is usually derived from balancing a centrifugal repulsion against the centripetal magnetic force

$$\frac{mu_\perp^2}{R} = \frac{qu_\perp \mathsf{B}_o}{c} \tag{5.12}$$

The *sense* of the circulation about the direction of $\mathbf{B}_o$, as given by (5.10), is clearly opposite for oppositely charged particles, since $-\boldsymbol{\omega}_o = -q\mathbf{B}_o/mc$ is parallel to $\mathbf{B}_o$ for $q < 0$ and antiparallel for $q > 0$. Thus (5.10) shows how particles of opposite charge choose centers about which to circle, in opposite directions from their initial positions and on a perpendicular to both $\mathbf{B}_o$ and $\mathbf{u}_\perp(0)$.

Consider the sense of the circulation in relation to the new magnetic field that is being generated by the current of the circulating charge. A *time-averaged* magnetic flux can be understood as arising from the current $I = q/(2\pi R/u_\perp) = q\omega_o/2\pi$ averaged over an orbital period. The direction of the generated flux through the part of the plane of motion enclosed by the circular orbit is *anti*parallel to the direction of the applied field $\mathbf{B}_o$. This holds for either sign of the charge since a change of sign is accompanied by a reversal in the sense of the circulation. The imposition of a field on the moving charge has induced in it a diamagnetic behavior regardless of the sign of the charge!

As a source of field, the averaged current produces a dipole moment $\boldsymbol{\mu}$ of a magnitude calculable by multiplying I/c with the area of the orbit πR^2

$$\mu = \frac{|q|\omega_o}{2\pi c}\pi\left(\frac{mcu_\perp}{q\mathsf{B}_o}\right)^2 = \frac{\frac{1}{2}mu_\perp^2}{\mathsf{B}_o} \tag{5.13}$$

The moment is aligned antiparallel to $\mathbf{B}_o$ for either sign of the charge, so that the projection

$$\boldsymbol{\mu} \cdot \mathbf{B}_o = -\tfrac{1}{2}mu_\perp^2 \tag{5.14}$$

is always negative.

Finally, consider the orbit described in space when a component of initial velocity $\mathbf{u}_\parallel$ parallel to the field also exists. This component persists unchanged, as already noted, and so the circular motion opens up into a uniform *helical* orbit. Some one field line coincides with the axis of the helix, and the particle winds around this central field line as if guided by it. Such a behavior can be expected to persist even when the guiding line begins to curve in space, as in a nonuniform field, if the field strength is great enough for the particle to complete a circuit within a small region of comparative uniformity. The tight spiraling about strong field lines is responsible for the observed fact that the electrons in an ionized gas (a plasma) tend to follow lines of magnetic field wherever they may lead; the electron constituents of the plasma exhibit this behavior most readily because they have the smallest masses, and the circulation radius (5.11) is proportional to the mass, as well as being inversely proportional to the field strength.

5.2 THE $\mathbf{E} \times \mathbf{B}$ DRIFT

Now suppose that a uniform electric field $\mathbf{E}_o$ is imposed on the point charge circulating in the uniform magnetic field $\mathbf{B}_o$, as described in the preceding section. If the field $\mathbf{E}_o$ is established *parallel* to $\mathbf{B}_o$, only the motion $\mathbf{r}_{\parallel}(t)$ parallel to both fields is disturbed. It becomes the uniformly accelerated one

$$\mathbf{r}_{\parallel}(t) = \mathbf{u}_{\parallel}(0)t + \tfrac{1}{2}\mathbf{a}t^2 \qquad \text{with } \mathbf{a} = \frac{q\mathbf{E}_o}{m} \tag{5.15}$$

while $\mathbf{r}_{\perp}(t)$ continues to describe the circle of radius (5.11). The uniformly helical orbit followed in the absence of $\mathbf{E}_o$ is now distorted into a steadily stretching or contracting (for $\mathbf{u} \cdot \mathbf{a} \gtrless 0$) spiral. In the special case of $\mathbf{u}_{\perp}(0) = \mathbf{u}_o - \mathbf{u}_{\parallel}(0) = 0$, this becomes a uniformly accelerated motion in a straight line parallel to the fields, as if the magnetic field were absent. The magnetic field has no effect except on a $\mathbf{u}_{\perp} \neq 0$ (while a parallel electric field leaves $\mathbf{u}_{\perp}$ undisturbed).

In the more general case of fields not parallel to each other, any component of electric field in the direction of $\mathbf{B}_o$ will merely superpose a uniformly accelerated component, in that direction, on motion taking place in a plane perpendicular to $\mathbf{B}_o$. It will therefore be sufficient to examine a case of *crossed* fields, i.e., a situation in which the electric and magnetic field lines cross each other at right angles, with the motion confined to a plane normal to $\mathbf{B}_o$ by *starting* it with an initial velocity $\mathbf{u}_o$ perpendicular to $\mathbf{B}_o$. To be considered now is the modification of such planar motions as are described by (5.9), with $\mathbf{u}_{\perp}(0) = \mathbf{u}_o$, and generated by lines of $\mathbf{E}_o$ lying in that plane.

The equation of motion (5.1) for the case of the crossed fields can be written in terms of the electric acceleration $\mathbf{a} = q\mathbf{E}_o/m$ and the angular velocity $\boldsymbol{\omega}_o = q\mathbf{B}_o/mc$ as

$$\dot{\mathbf{u}} = \mathbf{a} - \boldsymbol{\omega}_o \times \mathbf{u} = -\boldsymbol{\omega}_o \times \left(\mathbf{u} - \frac{\mathbf{a} \times \boldsymbol{\omega}_o}{\omega_o^2}\right) \tag{5.16}$$

The equivalence to the last form follows from the fact that $\boldsymbol{\omega}_o \times (\mathbf{a} \times \boldsymbol{\omega}_o) = \omega_o^2 \mathbf{a}$ because $\mathbf{E}_o \cdot \mathbf{B}_o = 0$ in the crossed fields. The result (5.16) suggests using a frame that *moves* with the uniform velocity

$$\bar{\mathbf{u}} \equiv \frac{\mathbf{a} \times \boldsymbol{\omega}_o}{\omega_o^2} = c\frac{\mathbf{E}_o \times \mathbf{B}_o}{\mathbf{B}_o^2} \tag{5.17}$$

for then the particle velocity $\mathbf{U} \equiv \mathbf{u} - \bar{\mathbf{u}}$ relative to the moving frame† satisfies $\dot{\mathbf{U}} = -\boldsymbol{\omega}_o \times \mathbf{U}$, an equation like (5.9) for a particle in the field $\mathbf{B}_o$ alone. Thus, relative to the *moving* frame, the particle will have a circular orbit with the radius

$$R = \frac{U_o}{\omega_o} = \frac{|\mathbf{u}_o - \bar{\mathbf{u}}|}{\omega_o} \tag{5.18}$$

as indicated in Fig. 5.1*a* for both signs of charge. Relative to the fixed initial frame, the orbit is a cycloid, i.e., a curve traced out by a point on a circle rotating with angular velocity ω_o and being translated with the uniform drift velocity $\bar{\mathbf{u}}$. The direction of the drift is that of $\mathbf{E}_o \times \mathbf{B}_o$, *not* in the direction of the electric acceleration $\mathbf{a} = q\mathbf{E}_o/m$ but at right angles to it. Moreover this drift direction is independent of the sign of the charge.

The striking nature of this result is perhaps most evident in the case of a particle starting from *rest* ($\mathbf{u}_o = 0$ and $\mathbf{U}_o = -\bar{\mathbf{u}}$) when the orbit becomes the special cusped cycloid indicated in Fig. 5.1*b* for $q > 0$. Such a cycloid is traced out by a point on the *rim* of a wheel, radius $R = \bar{u}/\omega_o = a/\omega_o^2$, that is rolling with the speed $\bar{u} = \omega_o R$ on the negative y axis of the indicated cartesian frame (having $\mathbf{B}_o$ in its z direction). The coordinates of the particle at any moment t are

$$\begin{aligned} x &= R - R\cos\omega_o t = \frac{a}{\omega_o^2}(1 - \cos\omega_o t) \\ y &= -\bar{u}t + R\sin\omega_o t = -\frac{a}{\omega_o^2}(\omega_o t - \sin\omega_o t) \end{aligned} \tag{5.19}$$

also obtainable directly from the equation of motion (5.16) as resolved on the cartesian frame. In the earliest moments ($t \ll 1/\omega_o$), $x \approx \frac{1}{2}at^2$ and $y \approx 0$, as to be expected since the magnetic field cannot begin to have effect until the particle acquires some velocity $u \approx \dot{x} \approx at$ from acceleration in the direction of the electric field. The magnetic field then deflects the particle sideways to the electric field and even reverses its direction so that the electric field decelerates the particle to zero velocity at $x = 0$ again. The initial conditions are restored at $t = 2\pi/\omega_o$, except for the sideways displacement by $|y| = 2\pi a/\omega_o^2$, and so these sideways displacements are repeated in periods $2\pi/\omega_o$. The particle never penetrates farther than the distance $x(t = \pi/\omega_o) = 2a/\omega_o^2$ into the direction of the electric field despite the steady force $q\mathbf{E}_o$ on it. The sideways progression with the mean drift velocity $\bar{u} = (\mathsf{E}_o/\mathsf{B}_o)c$ may seem quite startling since a naïve expectation might have been that the circular orbits characteristic of the motion in the

† Notice that the absence of an electric force in this frame is consistent with the galilean field transformation (4.68). According to this, the electric force per unit charge will be

$$\mathbf{E}' = \mathbf{E}_o + \frac{\bar{\mathbf{u}}}{c} \times \mathbf{B}_o$$

in the frame moving with the velocity $\bar{\mathbf{u}} = c(\mathbf{E}_o \times \mathbf{B}_o)/\mathsf{B}_o^2$. Substitution reduces the result to $\mathbf{E}' = \mathbf{E}_o - \mathbf{E}_o = 0$.

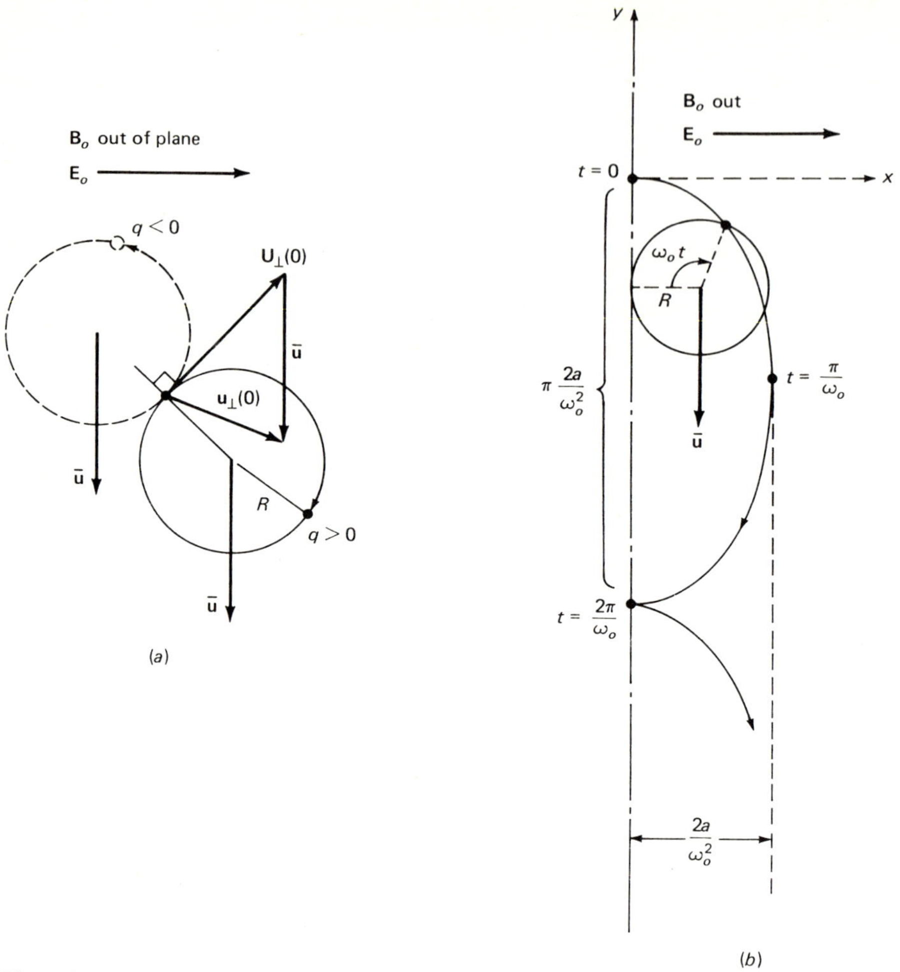

Figure 5.1

magnetic field alone would be translated into the direction of the electric force, as they are when the electric field is parallel to the magnetic field.

The $\mathbf{E} \times \mathbf{B}$ drift has been discussed here only as a motion of charged particles in free space, but its effects have been detected in a variety of circumstances. For example, it has an influence on currents in conductors known as the *Hall effect*. A typical way of observing that effect is to impose a magnetic field $\mathbf{B}$ normal to a conducting *sheet* across which a potential difference is applied. Then the sheet of current in the direction of the applied potential gradient $\nabla\phi = -\mathbf{E}$ is found to have superposed on it a transverse current† in just the direction of the drift velocity $\bar{\mathbf{u}} = c\mathbf{E} \times \mathbf{B}/B^2$ and calculable from it.

† If a closed circuit is provided for it, as through a voltmeter measuring the equivalent emf.

5.3 MAGNETIC DIPOLES IN MAGNETIC FIELDS

The investigation of forms the magnetic forces can take has not yet been carried as far as for the electrostatic forces in Sec. 2.4. Magnetic interactions with complex distributions of moving charge can be analyzed as the electrostatic interactions were in (2.52). For charge localized within any volume of finite extent, the electrostatic interactions could be expressed as forces on the electric multipole moments of the charge distribution. It can also be expected that the interactions of a localized steady-current distribution with a magnetostatic field can be expressed as forces on the *magneto*static multipoles of the distribution. Whereas the electrostatic interactions were developed up to quadrupole effects, it is comparably important to consider only magnetic *dipole* effects.

Forces on Magnetic Dipoles

First consider a steady, and hence divergenceless, current distribution localized about some point $\mathbf{r}$, fixed in space, that can serve as the origin of position vectors $\mathbf{s}$ to elements $\mathbf{j}(\mathbf{s})\,dV(\mathbf{s})$ of the distribution. The divergencelessness is then described by $\nabla_s \cdot \mathbf{j}(\mathbf{s}) = 0$, and it has the consequence $\oint dV(\mathbf{s})\mathbf{j}(\mathbf{s}) = 0$ as in (4.37). What is wanted is the resultant magnetostatic force on the center $\mathbf{r}$ of this distribution,

$$\mathbf{F} = c^{-1} \oint dV(\mathbf{s})\mathbf{j}(\mathbf{s}) \times \mathbf{B}(\mathbf{r}+\mathbf{s}) = \mathbf{F}(\mathbf{r}) \tag{5.20}$$

in an *external* field $\mathbf{B}$, one that has none of its sources overlapping on the distribution. As for the electrostatic interaction (2.49), the effect of the incident field is represented by its value $\mathbf{B}(\mathbf{r})$ at the center and by its rates of variation there, as introduced in a Taylor expansion of

$$\begin{aligned}\mathbf{B}(\mathbf{r}+\mathbf{s}) &= \mathbf{B}(\mathbf{r}) + \mathbf{s}\cdot[\nabla_s \mathbf{B}(\mathbf{r}+\mathbf{s})]_{\mathbf{s}=0} + \cdots \\ &= \mathbf{B}(\mathbf{r}) + (\mathbf{s}\cdot\nabla)\mathbf{B}(\mathbf{r}) + \cdots\end{aligned} \tag{5.21}$$

The gradient derivative in the last line operates on the $\mathbf{r}$ dependence of $\mathbf{B}(\mathbf{r})$ since $\nabla_s \mathbf{B}(\mathbf{r}+\mathbf{s}) = \nabla\mathbf{B}(\mathbf{r}+\mathbf{s})$ before $\mathbf{s}$ is set equal to zero, as required for the coefficients in the Taylor expansion. It is generally useful to note alternative forms that can be given to the first correction term by taking advantage of the divergencelessness of any magnetic field ($\nabla\cdot\mathbf{B} = 0$) or its curllessness ($\nabla\times\mathbf{B} = 0$) in regions not overlapped by its sources. Judicious applications of the triple vector identity (A.53) show that

$$(\mathbf{s}\cdot\nabla)\mathbf{B}(\mathbf{r}) = \mathbf{s}(\nabla\cdot\mathbf{B}) - \nabla\times(\mathbf{s}\times\mathbf{B}) = -\nabla\times(\mathbf{s}\times\mathbf{B}) \tag{5.22}$$

and

$$(\mathbf{s}\cdot\nabla)\mathbf{B}(\mathbf{r}) = \nabla(\mathbf{s}\cdot\mathbf{B}) - \mathbf{s}\times(\nabla\times\mathbf{B}) = \nabla(\mathbf{s}\cdot\mathbf{B}) \tag{5.23}$$

with the second equation holding *only* where $\nabla\times\mathbf{B} = 0$.

Substituting the Taylor expansion into the force integral (5.20) yields the expression

$$\mathbf{F}(\mathbf{r}) = c^{-1}\left[\oint dV(\mathbf{s})\mathbf{j}(\mathbf{s})\right] \times \mathbf{B}(\mathbf{r}) + c^{-1}\oint dV\mathbf{j} \times (\mathbf{s}\cdot\nabla)\mathbf{B} + \cdots \tag{5.24}$$

having an integral over **s** in the first term that vanishes for divergenceless current distributions (4.37). Thus in the first approximation, which treats the field as the same at all elements of the current, the overall resultant force is zero. An immediate conclusion is that in a *uniform* field, for which also all the other terms vanish, there is no force at all on a steady-current distribution that would *translate* it bodily. However, a bodily *rotation* about the direction of the uniform field can still be started, and attention will be paid to this in Sec. 5.5.

The remaining term exhibited in (5.24) exists only in nonuniform fields and is usually the largest one in such circumstances. It can be evaluated most expeditiously by making use of formal results already obtained in developing the vector-potential field due to a magnetic dipole (4.38) from the power-series expansion in (4.35). The applicability of those results here becomes most evident when it is noted that the vector $-\hat{\mathbf{r}}/r^2$, which is fixed in the integration over the current distribution in (4.35), is formally the same as a magnetic field $\mathbf{B}_1(\mathbf{r})$ at the source, due to a pole of *unit* strength at $-\mathbf{r}$. Then the development from (4.35) to (4.38) can be represented as

$$\mathbf{A}(\mathbf{r}) = -c^{-1}\oint dV(\mathbf{s})\mathbf{j}(\mathbf{s})(\mathbf{s}\cdot\mathbf{B}_1) \rightarrow \mathbf{A} = \mathbf{B}_1 \times \boldsymbol{\mu} \tag{5.25}$$

An integral of just the type here occurs in (5.24) when the form (5.23) is used, so that

$$\mathbf{F}(\mathbf{r}) \approx c^{-1}\oint dV(\mathbf{s})\mathbf{j}(\mathbf{s}) \times \nabla(\mathbf{s}\cdot\mathbf{B}) = -c^{-1}\nabla \times \oint dV\mathbf{j}(\mathbf{s}\cdot\mathbf{B}) \tag{5.26}$$

The minus sign comes from interchanging the order of the factors in a cross product, permissible because the gradient operator applies only to the field $\mathbf{B}(\mathbf{r})$, fixed in the integration over the current distribution. Thus the force here is representable as the curl of a vector field

$$\mathbf{F}_\mu(\mathbf{r}) = -\nabla \times \boldsymbol{\Theta}(\mathbf{r}) \qquad \text{if } \boldsymbol{\Theta} = \boldsymbol{\mu} \times \mathbf{B}(\mathbf{r}) \tag{5.27}$$

with $\boldsymbol{\mu}$ just the dipole moment vector of the current distribution. The significance of the vector field $\boldsymbol{\Theta}(\mathbf{r})$ will be seen in Sec. 5.5.

Use of triple-vector-product identities, as in (5.22) and (5.23), yields

$$\mathbf{F}_\mu = -\nabla \times (\boldsymbol{\mu} \times \mathbf{B}) = (\boldsymbol{\mu}\cdot\nabla)\mathbf{B}(\mathbf{r}) = \nabla(\boldsymbol{\mu}\cdot\mathbf{B}) \tag{5.28}$$

This shows that, just as the field *arising from* a magnetic dipole is derivable both from a vector potential, as in (4.28), and from a scalar potential $\phi_\mu = -\boldsymbol{\mu}\cdot\mathbf{B}_1$, as in (4.30), so the force *acting on* a dipole in an external field is derivable both from a vector field $-\boldsymbol{\Theta}(\mathbf{r})$ and from a scalar field $U_\mu(\mathbf{r})$ such that

$$\mathbf{F}_\mu = -\nabla U_\mu(\mathbf{r}) \qquad \text{with } U_\mu = -\boldsymbol{\mu} \cdot \mathbf{B}(\mathbf{r}) \tag{5.29}$$

The relationship of this scalar field to the force is like that of a potential energy drawn upon by work $\mathbf{F}_\mu \cdot d\mathbf{r} = -dU_\mu(\mathbf{r})$ done by the force here, just as the electric dipole interaction energy $U_D = -\mathbf{D} \cdot \mathbf{E}$ of (2.53) serves as a potential from which is derivable the force $\mathbf{F}_D = (\mathbf{D} \cdot \nabla)\mathbf{E}$ of (2.54) on an electric dipole in a nonuniform electrostatic field. It should be clear from their development that the force expressions (5.27) to (5.29) become exact in the limit of an *ideal* magnetic dipole, extending only over some one point $\mathbf{r}$ in space.

The existence of a magnetic force $\mathbf{F}_\mu$ that does work on the charges in $\mathbf{j}(\mathbf{s}) = \rho\mathbf{u}(\mathbf{s})$ is not necessarily inconsistent with the basic finding leading to (5.2), that the complete magnetic force per unit charge $\mathbf{u} \times \mathbf{B}(\mathbf{r} + \mathbf{s})/c$ does *no* work during displacements $\mathbf{u}\, dt$ of the charge. In the first place, $\mathbf{F}_\mu \cdot d\mathbf{r} = -dU_\mu(\mathbf{r})$ represents only a *part* of that work, serving to transfer energy† to a newly activated external degree of freedom $\mathbf{r}$. Moreover, the internal motions described by $\mathbf{u}(\mathbf{s})$ are being assumed to be steady ones, maintained with the help of some agency (perhaps even a nonelectromagnetic one) that must cancel out some of the magnetic forces, and then $\mathbf{F}_\mu$ represents an unbalanced remainder. More insight into such situations can be expected from considering the interaction *energies* between dipole and field, as in the next subsection.

Examples of systems in which steady internal motions are somehow maintained are atoms or molecules that are found to possess permanent magnetic dipole moments. There is, in connection with them, an important application of such forces as $\mathbf{F}_\mu = (\boldsymbol{\mu} \cdot \nabla)\mathbf{B}$; they are used for *sorting* the dipoles according to their orientations in space. A beam of the electrically neutral particles is sent through a region (past a magnetized iron wedge, for example) in which magnetic field normal to the beam has a nonvanishing gradient. Then the particles with orientations parallel and antiparallel to the field gradient will receive oppositely directed "kicks," $\pm\mu|\nabla\mathbf{B}|$, out of the initial beam direction. They are separated into two beams and thus sorted out. It might be expected that a randomly prepared (unsorted) initial beam would contain dipoles with all possible orientations and so would be dispersed into a *continuous* distribution between the two extremes. An actual finding about *some* types of particles (in the famous Stern-Gerlach experiment) is that a randomly prepared beam behaves as if only the two opposite orientations exist in it. Some other types of atoms are found in more than two orientations but still in only a discrete number of them. These findings are understood as quantum phenomena, outside the purview of this book, but the laboratory procedures that reveal those phenomena are still subject to classical analysis. Even the classically expected quantitative results for the beam deflections turn out to be valid measures of the magnetic moments.

† Even with a vanishing work element like $F_x\,dx + F_y\,dy = 0$, for example, there may still be nonvanishing *partial* work $F_y\,dy = -F_x\,dx \neq 0$ of energy transfer from one degree of freedom to another.

Interaction Energies of Magnetic Dipoles with Fields

The forces on *electric* multipoles obtained in Sec. 2.4 were derived from interaction *energies* between the multipoles and the sources providing the field, equivalent to interactions between the field of the multipole and the source field. A more direct evaluation of the force on a *magnetic* dipole was employed in the preceding subsection mostly because the role of energy exchanges in a magnetic field is less straightforward. However, it ought to be understood, and so an analysis of the magnetostatic field energy comparable to that carried out for the electrostatic field will be undertaken now.

The electrostatic situations were treated by making distinctions between the self-energies of electrostatic sources, or of the fields they generate, and the interaction energies arising when two or more of the fields are superposed. The distinctions were found to be valuable whenever each of the self-energies formed remains constant throughout some process and can be regarded as belonging to an extended particle (the source of the field possessing the self-energy) having some fixed internal structure.

The analogous investigation of a magnetostatic field energy can begin with dividing its steady current source into two parts, represented through $\mathbf{j} = \mathbf{j}_1 + \mathbf{j}_2$, each divergenceless and respectively generating a vector-potential field $\mathbf{A}_1$ and $\mathbf{A}_2$. Putting $\mathbf{A} = \mathbf{A}_1 + \mathbf{A}_2$ in the energy expression (4.96) allows one to write it as

$$W = W_1 + W_2 + W_{12} \tag{5.30}$$

with
$$W_1 = (2c)^{-1} \oint dV \mathbf{j}_1 \cdot \mathbf{A}_1 \qquad W_2 = (2c)^{-1} \oint dV \mathbf{j}_2 \cdot \mathbf{A}_2 \tag{5.31}$$

and
$$W_{12} = c^{-1} \oint dV \mathbf{j}_1 \cdot \mathbf{A}_2 = c^{-1} \oint dV \mathbf{j}_2 \cdot \mathbf{A}_1 \tag{5.32}$$

formal analogs of the electrostatic self-energy expressions W_1, W_2 (2.43) and of the interaction energy U_{12} (2.46). It is also clear that, at least as long as each of the distributions $\mathbf{j}_1$ and $\mathbf{j}_2$ remains fixed in space, the energies W_1 and W_2 can be regarded as being spread out over their respective fields in densities $\mathsf{B}_1^2/8\pi$ and $\mathsf{B}_2^2/8\pi$, on the same basis that the entire energy W is distributed in the density $(\mathsf{B}_1 + \mathsf{B}_2)^2/8\pi$ of (4.98).

The object of such an analysis as (5.30) is the identification of a part of the field energy that can be drawn upon, like a potential energy $U_{12}(\mathbf{r})$, for a relative motion in $\mathbf{r}$ of the two partial distributions when each of these behaves like a particle with a stable internal structure. The identification of the electrostatic potential energy $U_{12}(\mathbf{r})$ of (2.44) was relatively simple because rigid internal charge distributions can be maintained without drawing upon the field energy; any nonelectromagnetic forces† responsible for the rigidity do no work during

† Such must exist if the individual charges are not to "explode"! (For the considerations here, effects of the Pauli principle and quantization are classed as nonelectromagnetic forces.)

any relative motions because there need be no internal motion in the process. However, maintaining the internal currents as needed for the stability of the individual current distributions $\mathbf{j}_1$ and $\mathbf{j}_2$ requires fresh work against inductive effects of relative motions, and this can draw on the field energy, as was seen in the development of the expression (4.85) for the various pieces of work done during relative displacements of currents. A self-energy like W_1 or W_2 of (5.31), calculated for the static field of a steady-current distribution, has that constant value only relative to a frame in which the distribution is stationary in the proper frame of the distribution. A different evaluation can be expected to be necessary in a frame relative to which the distribution is being displaced bodily, just as the kinetic energy of a particle is zero in its own rest frame yet equals $\frac{1}{2}mv^2 \neq 0$ relative to a frame in which it has the velocity v. It is significant that such facts require explicit attention particularly for magnetic fields, which have charges in motion as their sources.

This discussion indicates that a complete treatment would require dealing with the *non*static fields that arise in frames relative to which the steady sources are being displaced. This actually applies to the electric as well as the magnetic sources. However, the evaluations used so far, which treat each momentary phase of the bodily motion as a static situation, should be valid in what is called an *adiabatic approximation*, applicable to slow enough motions. Inquiry into the nonstatic field descriptions should be expected to be necessary in dealing with more rapid bodily motions (as with relativistic velocities), and that will eventually indicate the range of validity of the adiabatic treatment more precisely.

For the considerations of the magnetic energies here, the adiabatic treatment corresponds to a generalization of the development of (4.86) from $I_{1,2}\, d\mathbf{r}_{1,2}$ to $\mathbf{j}_{1,2}\, dV(\mathbf{r}_{1,2})$. Then the expression (5.30) should be rewritten as

$$W = (W_1 + W_{12}) + (W_2 + W_{21}) + U_{12} \tag{5.33}$$

where $W_{21} = W_{12}$ of (5.32) and $U_{12} = -W_{12}$. A part like $W_1 + W_{12}$ must be regarded as an adiabatic approximation of a self-energy as evaluated in a frame defined with the help of a second distribution, and it is

$$U_{12} = -c^{-1}\oint dV\, \mathbf{j}_1 \cdot \mathbf{A}_2 = -c^{-1}\oint dV\, \mathbf{j}_2 \cdot \mathbf{A}_1 \tag{5.34}$$

that represents energy drawn upon by relative motions of the two distributions $\mathbf{j}_1$ and $\mathbf{j}_2$. Its individual deduction from considering the work done by the mutual force between the currents, an element of which is given by (4.54), would be a suitably generalized ($I\, d\mathbf{r} \to \mathbf{j}\, dV$) repetition of the development of (4.75).

The result (5.34) helps account for the occurrence of the interaction energy

$$U(\mathbf{r}_q) = q\left(\phi - \frac{\mathbf{u}}{c} \cdot \mathbf{A}\right) \tag{5.35}$$

between a point charge and a field ϕ, $\mathbf{A}$ in (5.6). The part $q\phi$, the only one present in a purely electrostatic field, has already been discussed in connection with (2.66). The remainder is an obviously suitable specialization of the magnetic interaction energy (5.34), as is shown by integration over a point-charge current density $\mathbf{j} = \rho\mathbf{u}$, where $\rho = q\delta(\mathbf{r} - \mathbf{r}_q)$. The force $-\nabla U$ thus discriminated is one that can change not only the kinetic momentum $m\mathbf{u}$ but also field momentum $q\mathbf{A}/c$, with results the same as those of (5.1), since (5.6) was found equivalent to (5.1).

It can next be checked that the force on a magnetic dipole (5.28) is indeed derivable from a potential energy like (5.34). Written for a distribution $\mathbf{j}(\mathbf{s})$ centered at $\mathbf{r}$ (where $\mathbf{s} = 0$) in an arbitrary field $\mathbf{A}$ external to it,

$$U(\mathbf{r}) = -c^{-1} \oint dV(\mathbf{s})\mathbf{j}(\mathbf{s}) \cdot \mathbf{A}(\mathbf{r} + \mathbf{s}) \tag{5.36}$$

the analog of the electrostatic interaction expression (2.49). A Taylor expansion like (5.22), now applied to the vector potential, can be seen to yield

$$\begin{aligned} U(\mathbf{r}) &\approx -c^{-1} \oint dV(\mathbf{s})\mathbf{j}(\mathbf{s}) \cdot [(\mathbf{s} \cdot \nabla)\mathbf{A}(\mathbf{r})] \\ &= -c^{-1} \sum_{k,l} \nabla_k A_l(\mathbf{r}) \oint dV(\mathbf{s}) s_k j_l(\mathbf{s}) \end{aligned} \tag{5.37}$$

in first approximation. A divergencelessness $\nabla_s \cdot \mathbf{j}(\mathbf{s}) = 0$ of the distribution affects the integration in a way that can be seen by splitting the tensor (outer product) $s_k j_l$ into its symmetric and antisymmetric parts, as in

$$s_k j_l = \tfrac{1}{2}(s_k j_l + s_l j_k) + \tfrac{1}{2}(s_k j_l - s_l j_k) \tag{5.38}$$

The effect of $\nabla_s \cdot \mathbf{j} = 0$ on this can be obtained through the trick of considering

$$\nabla_s \cdot (s_k s_l \mathbf{j}) = (\mathbf{j} \cdot \nabla_s) s_k s_l = j_k s_l + j_l s_k \tag{5.39}$$

This shows that according to the Gauss theorem, the volume integral of the symmetric part in (5.38) can be replaced by a surface integral of $\frac{1}{2}s_k s_l \mathbf{j}$ and this must vanish upon integration over the entire distribution, enclosable by a surface on which $j_n = 0$. For the antisymmetric part of (5.38),

$$\begin{aligned} \sum_{k,l} \nabla_k A_l[\tfrac{1}{2}(s_k j_l - s_l j_k)] &= \frac{1}{2} \sum_{k<l} (\nabla_k A_l - \nabla_l A_k)(s_k j_l - s_l j_k) \\ &= \tfrac{1}{2}(\nabla \times \mathbf{A}) \cdot (\mathbf{s} \times \mathbf{j}) = \tfrac{1}{2}\mathbf{B} \cdot (\mathbf{s} \times \mathbf{j}) \end{aligned}$$

Thus

$$U(\mathbf{r}) = -(2c)^{-1}\mathbf{B} \cdot \oint dV \mathbf{s} \times \mathbf{j} = -\boldsymbol{\mu} \cdot \mathbf{B}(\mathbf{r}) \tag{5.40}$$

in terms of the magnetic moment as defined in (4.38). This is the magnetostatic analog of the electric dipole interaction energy $U_D(\mathbf{r}) = -\mathbf{D} \cdot \mathbf{E}$ of (2.53), and the derivability from it of the force on a magnetic dipole as $\mathbf{F}_\mu = -\nabla U_\mu(\mathbf{r})$ was exhibited in (5.29). Indeed, that identification of the energy in question might be regarded as sufficient, but its development from (5.33) and (5.34) shows how it is to be associated with the total magnetostatic field energy.

An example in which the potential energy $U_\mu = -\boldsymbol{\mu} \cdot \mathbf{B}(\mathbf{r})$ is drawn upon for imparting kinetic energy to a new degree of freedom is provided by the kick sorting of permanent magnetic dipoles, mentioned on page 131, in which kinetic energy is transferred to the deflected motions, transverse to the initial beam.

The finding (5.14) about the point charge circling in a uniform field can now be recognized as a special case of the interaction energy $U_\mu = -\boldsymbol{\mu} \cdot \mathbf{B} = \frac{1}{2}mu_\perp^2$. It is a consequence of having replaced the actual motion, of energy $\frac{1}{2}mu_\perp^2$ by a steady current, through an averaging typical of the adiabatic approximation. Reference back to the way results for the magneto*static* field energies were found in Sec. 4.4 shows that the equilibrium of the field with the source makes them equivalents to mechanical energy in the source itself. Here this is just the simple, transverse kinetic energy $\frac{1}{2}mu_\perp^2$ being reexpressed as a field interaction energy $-\boldsymbol{\mu} \cdot \mathbf{B}$. How the interaction thus treated can be drawn upon for transfers of energy between kinetic energies in various degrees of freedom will be illustrated in the next section.

5.4 THE MAGNETIC MIRROR EFFECT

Motion as guided by a line of nonuniform magnetostatic field, of a *point charge* spiraling about the line, can be treated, in a first approximation, as the motion of a *magnetic dipole*, subject to the force $\mathbf{F}_\mu = (\boldsymbol{\mu} \cdot \nabla)\mathbf{B}(\mathbf{r})$ of (5.28). That the circulating charge behaves, in an average over each cycle, like a dipole antiparallel to the guiding field line was first pointed out in connection with effective magnetic moment magnitude (5.13). The approximation thus suggested arises quite naturally in treating the point-charge equation of motion itself, as will now be demonstrated in a special case of some interest.

Suppose that a field line guides a point charge q spiraling about it into a region of stronger field. This happens in field configurations called magnetic bottles, formed so that a region of uniform field is terminated by a convergence of the lines toward a central axis, as indicated in Fig. 5.2. The central line can be used as the axis of a cylindrical coordinate system, and then the field is represented by some vector function $\mathbf{B}(s, z)$ with $\mathbf{B}_\varphi \equiv 0$. As the figure suggests, $\mathbf{B}_z$ remains the largest component in the region of intensification, where $\partial \mathbf{B}_z/\partial z > 0$ exists, but this variation is necessarily accompanied by a growth in the negative *radial* component, with $-\mathbf{B}_s = 0$ on the axis $s = 0$ but with $-\mathbf{B}_s > 0$ away from it, as required for the divergencelessness of the field

$$\frac{\partial}{s\,\partial s} s\mathbf{B}_s = -\frac{\partial \mathbf{B}_z}{\partial z} \rightarrow \mathbf{B}_s \approx -\tfrac{1}{2}s\frac{\partial \mathbf{B}_z}{\partial z} \tag{5.41}$$

The derivative expression here comes from $\nabla \cdot \mathbf{B} = 0$ in the cylindrical coordinate representation of (A.5), and the result quoted for $\mathbf{B}_s$ follows from the simple integration that is valid near the axis, where the variation with s of the gradient $\partial \mathbf{B}_z/\partial z$ can be treated as negligibly small. Since this simple approximation is best for the effect on particles that do not stray too far from the central axis, the

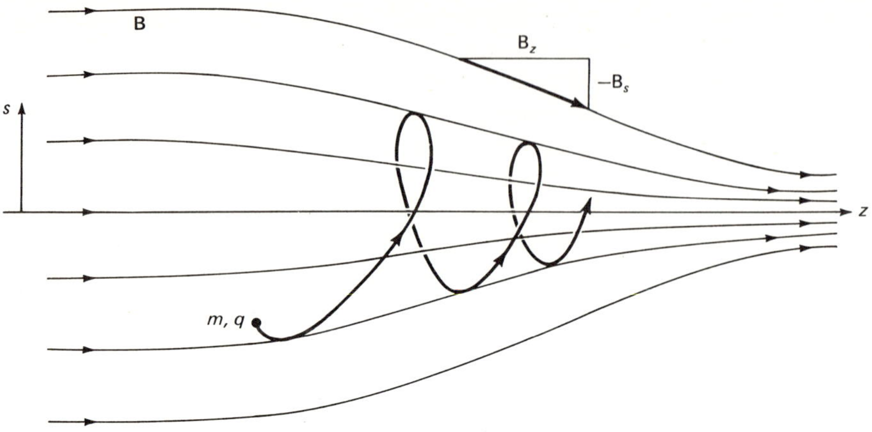

Figure 5.2

specific case treated will be concerned with a particle spiraling about the central field line itself as axis. The case will not be completely atypical; particles guided by *noncentral* field lines will merely be subject to small additional drifts that are best studied separately (as in Exercise 5.19). If the field is strong enough, the considerations can be restricted to $B_z \approx B(0, z) \equiv B_o(z)$ and $\partial B_z/\partial z = B_o'(z)$.

The small radial field component B_s furnishes all the force in the equation of motion for the velocity component $u_{\parallel} \equiv \dot{z}$ of the point charge (or of its instantaneous orbit center)

$$m\ddot{z} = \frac{q}{c}(\mathbf{u} \times \mathbf{B})_z = \frac{q}{c}(-u_\varphi B_s) \approx +\frac{q}{2c}s^2\dot{\varphi}\frac{\partial B_z}{\partial z} \tag{5.42}$$

The coefficient of $\partial B_z/\partial z$ here can be identified as just the negative of an effective magnetic moment $\mu = -\mu_z$, so that the force guiding the orbit center is indeed being evaluated as the appropriate (z) component, $-\mu\,\partial B_z/\partial z$, of $\mathbf{F}_\mu = (\boldsymbol{\mu} \cdot \nabla)\mathbf{B}$.

That $-(q/2c)s^2\dot{\varphi} = -\mu_z = +\mu$ is to be identified as the magnitude of an effective magnetic moment directed antiparallel to the central field line can be seen in various ways. For example, the quantity $ms^2\dot{\varphi} \equiv L_z$ can be recognized as just the angular momentum $\mathbf{s} \times m\mathbf{u}$ about the z direction; then a magnetic moment $\boldsymbol{\mu} = (q/2mc)\mathbf{L}$ is only to be expected, from the gyromagnetic ratio (4.40). In (5.14), essentially the expression $\mu = \frac{1}{2}mu_\varphi^2/B_o$ was given, on the basis of a circulation velocity $u_\varphi = s\dot{\varphi} = -s\omega_o$ with $\omega_o \equiv qB_o/mc$; this can also be written $\mu = \frac{1}{2}ms^2\dot{\varphi}^2/B_o = -(q/2c)s^2\dot{\varphi}$, as above.

Multiplying Eq. (5.42) for $m\dot{u}_{\parallel}$ by $(2/m)\,\Delta z \equiv (2/m)u_{\parallel}\,\Delta t \to 0$ yields

$$2u_{\parallel}\dot{u}_{\parallel}\,\Delta t \approx \frac{q}{mc}s^2\dot{\varphi}B_o'(z)\,\Delta z \equiv \Delta(u_{\parallel}^2) \approx \frac{q}{mc}s^2\dot{\varphi}\,\Delta B_o(z) \tag{5.43}$$

since $2u_{\parallel}\dot{u}_{\parallel} = d(u_{\parallel}^2)/dt$. Thus given is the decrement (recall that $q\dot{\varphi} < 0!$) in $u_{\parallel}^2$ that takes place while the orbit center moves through an interval $\Delta z \to 0$ on the central field line. The factor $s^2\dot{\varphi}$ here must generally be expected to vary during the motion, in a way that should properly be found from the radial and azimuthal components of the vector equation of motion having (5.42) for its z component. Actually, one of the three component equations can be replaced by a first integral that follows from all three of them; this expresses the kinetic-energy conservation in the motion of a charge through any purely magnetic field, as found in (5.4) and here it is representable as a constancy of $u^2 = u_\parallel^2 + u_\perp^2 = \dot{z}^2 + (s^2\dot{\varphi}^2 + \dot{s}^2)$.

Detailed solution of the three simultaneous differential equations for $\ddot{z}(t)$, $\ddot{s}(t)$, and $\ddot{\varphi}(t)$ would, in principle, give the instantaneous changes taking place in the radius $s(t)$ and angular velocity $\dot{\varphi}(t)$ within each cycle of the spiraling motion, but the main features of the behavior become plainer in the adiabatic approximation. This widely useful approach treats a particle's orbit over any individual cycle as unperturbed by whatever changes of circumstance it may encounter during that cycle; only secular effects, i.e., averages over each cycle, are taken into account. In the problem here, each cycle of the orbit can be approximated as such a circular helix, of constant s and $\dot{\varphi}$, as would be followed if the field remained uniform at its average value over that cycle; the change of the field along each cycle is thus treated as a small perturbation, with an appreciable net effect only from cycle to cycle. The average field in a given cycle can be represented as $\mathsf{B}_o(\bar{z})$, where $\bar{z}$ is some coordinate within the span of the cycle. Then the constants specifying the unperturbed helical orbit can be taken to be $\dot{\varphi}(\bar{z}) \approx -q\mathsf{B}_o(\bar{z})/mc$ and $s(\bar{z}) \approx mcu_\perp(\bar{z})/q\mathsf{B}_o(\bar{z})$, as in the uniform field of (5.11), when approximating the actual orbital cycle at $\bar{z}$. Radial velocities are neglected in this approximation. The entire transverse velocity is taken to be $u_\perp(\bar{z}) \approx -s\dot{\varphi}$, and so the equation expressing the constancy of the total velocity u yields $u_\perp^2(\bar{z}) \approx u^2 - u_\parallel^2(\bar{z})$. Now Eq. (5.43) can be used to evaluate changes that occur over the span of an entire cycle as

$$\Delta(u_\perp^2) = -\Delta(u_\parallel^2) \approx \frac{u_\perp^2(\bar{z})}{\mathsf{B}_o(\bar{z})} \Delta\mathsf{B}_o(\bar{z}) \tag{5.44}$$

for the cycle at $\bar{z}$. The secular changes (in the average over many cycles) can then be obtained by treating $\bar{z}$ as a continuous variable and the relation as a differential equation in $\bar{z}$. This procedure yields $\Delta(\ln u_\perp^2/\mathsf{B}_o) = 0$, and so

$$\frac{u_\perp^2(z)}{\mathsf{B}_o(z)} \approx \text{const} \approx \frac{u_\perp^2(0)}{\mathsf{B}_o(0)} \tag{5.45}$$

is expected, from cycle to cycle. A constant so obtained is called an *adiabatic invariant*. The constancy should plainly hold best if $\Delta\mathsf{B}_o/\mathsf{B}_o \ll 1$, where $\Delta\mathsf{B}_o$ is the change of field during any one cycle.

The result (5.45) amounts to a statement that the effective magnetic moment, the $\mu \approx \frac{1}{2}mu_\perp^2/\mathsf{B}_o$ discussed above, stays the same in succeeding cycles of the motion. It implies that the spiral orbit continues to enclose the same flux

$$\pi s^2(z)\mathsf{B}_o(z) \approx \frac{\pi u_\perp^2 \mathsf{B}_o}{\dot{\varphi}^2} \approx \pi\left(\frac{mc}{q}\right)^2 \frac{u_\perp^2}{\mathsf{B}_o} \tag{5.46}$$

as indicated in Fig. 5.2 by having the orbit continue to touch the same tube wall of field lines throughout.

The proportionality $u_\perp^2(z) \sim \mathsf{B}_o(z)$ shows that the transverse motion speeds up as the particle penetrates farther into the region of strengthening field. At the same time, the longitudinal translation is slowed, in accordance with

$$u_\parallel^2(z) \approx u^2 - \frac{u_\perp^2(0)}{\mathsf{B}_o(0)}\mathsf{B}_o(z) \tag{5.47}$$

as the constant resultant velocity $u^2 = u_\parallel^2(z) + u_\perp^2(z)$ is maintained. The deceleration of the translation is directly evident from its equation of motion (5.42), also expressible as $m\dot{u}_\parallel \approx -\mu\, \partial \mathsf{B}_o/\partial z$ now that the approximate constancy of the effective magnetic moment μ has been established. The deceleration will continue until the translation is brought to rest ($u_\parallel \to 0$) at a turning point $z_{\max}$ determinable from $\mathsf{B}_o(z_{\max}) \approx [u^2/u_\perp^2(0)]\mathsf{B}_o(0)$. The negative z component of force $-\mu \mathsf{B}_o'(z)$ reaches its greatest value at the turning point [if $\mathsf{B}_o'(z) > 0$ continues to hold to $z_{\max}$] and continues to decrease the longitudinal velocity component from $u_\parallel(z_{\max}) = 0$ to negative values. There is a *reflection* of the translational motion and a spiraling back into the region of weaker field. The reflected motion does not retrace the original orbit but follows a complementary spiral, having the same absolute sense of rotation ($q\dot{\varphi} < 0$ and $\mu = -\mu_z$ still) but with a reverse-accelerated translation that opens up the spans between cycles as the reflected motion progresses in the negative z direction. As an overall result there has been a bottling up of the motion at the neck of the magnetic bottle.

The behavior just reviewed is known as the *magnetic mirror effect*. Charges of any sign are thus reflected from regions of strengthening field. The arrowheads attached to the spiral drawn in Fig. 5.2, indicating $\dot{\varphi} < 0$, make it an orbit of a positive charge *entering* the bottleneck. A reversal of the arrows would make the same spiral an orbit of a *negative* charge being *reflected* from the bottleneck, so that $q\dot{\varphi} < 0$ and $\mu_z = -\mu$ still. It is the complementary spiral referred to above, with $\dot{\varphi} > 0$ on it, that would be followed by a negative charge *entering* the bottleneck.

The magnetic mirror effect and the magnetic bottle are used to contain high-temperature plasmas in the hope of keeping them out of contact with material walls that might be melted down by the high temperature or might cool the plasma down. This is important to attempts at inducing thermonuclear reactions (releasing nuclear-fusion power) between the ions in the plasma.

Aside from its intrinsic interest, the example here has lent insight into the role of the magnetostatic field energies during motions of a distribution $\mathbf{j}_1$ in an external field $\mathbf{B}_2$. The distribution here was a particularly simple one, consisting of a spiraling point charge that can be treated as forming a steady current, with a constant magnetic moment, only in an adiabatic approximation. The energy density $\mathsf{B}_2^2/8\pi$ has not been drawn upon at all, consistent with the finding (5.2) that a magnetic field does no work on charges moving in it. In toto, there have only been kinetic-energy exchanges between longitudinal and transversal degrees of freedom, as represented by (5.44). Forces derivable from an interaction energy

$U = -\boldsymbol{\mu} \cdot \mathbf{B}_2$ provided the mechanism for the momentum transfers $(mu_{\parallel} \leftrightarrow mu_{\perp})$; the instantaneous values of $U = \frac{1}{2}mu_{\perp}^2$ merely represented the internal (transverse) kinetic energies, while the internal motions were treated as effective current distributions giving the magnetic moment. From the latter point of view, the mirror effect arises from a minimization of an effective potential energy† $U = -\boldsymbol{\mu} \cdot \mathbf{B}_2 > 0$ that is lowest where the external field is weakest,‡ so that the particles are repelled from the regions of strongest field.

5.5 THE ROTATION OF DIPOLES

The translations of magnetic dipole moments, as reviewed in the preceding sections, are generated by *nonuniform* magnetic fields. That a *uniform* field can impart no translational (linear) momentum to a dipole is part of the conclusion that was reached in the immediate discussion of (5.24). However, a uniform field can impart rotational (angular) momentum to a dipole, through a torque like that found for electric dipoles in electric fields and illustrated by Fig. 2.2.

The Magnetic Torque

The resultant of the torques exerted by the magnetic force elements in (5.20) about a momental center at the position **r** in space will be evaluated as

$$\boldsymbol{\Theta} = c^{-1} \oint dV(\mathbf{s})\mathbf{s} \times [\mathbf{j}(\mathbf{s}) \times \mathbf{B}(\mathbf{r})] \tag{5.48}$$

Only the first term, $\mathbf{B}(\mathbf{r})$, of the Taylor expansion (5.21) is being used here because that will give the main effect—and the *only* one in a uniform field. Expansion of the triple vector product yields

$$\boldsymbol{\Theta} = c^{-1} \oint dV(\mathbf{s})[\mathbf{j}(\mathbf{s})(\mathbf{s} \cdot \mathbf{B}) - (\mathbf{s} \cdot \mathbf{j})\mathbf{B}(\mathbf{r})]$$

and the integral in the last term

$$\oint dV(\mathbf{s})(\mathbf{s} \cdot \mathbf{j}) = \sum_l \oint dV s_l j_l$$

can be seen to vanish from such considerations as those attending (5.39), which show that

$$\oint dV s_k j_l = -\oint dV s_l j_k \to 0 \qquad \text{for } k = l \tag{5.49}$$

† Since $U = \frac{1}{2}mu_{\perp}^2$ is a kinetic energy of rotational motion, the situation here is quite comparable to the use of a centrifugal potential for representing the role of rotational kinetic energies in the radial motions taking place in a central force field, as discussed in Ref. 17, sec. 4.2 and figs. 4.3 and 4.4.

‡ Or when $\boldsymbol{\mu}$ is *parallel* to $\mathbf{B}_2$, in situations where this is at all possible (see page 140).

The remainder of the expression for $\boldsymbol{\Theta}$ is identical with (5.26), and hence

$$\boldsymbol{\Theta} = \boldsymbol{\mu} \times \mathbf{B} \tag{5.50}$$

of (5.27), a result analogous to the electrostatic torque $\mathbf{D} \times \mathbf{E}$ found in (2.55).

The torque (5.50) is also derivable from the same interaction energy $U_\mu = -\boldsymbol{\mu} \cdot \mathbf{B}$ as the translation force $\mathbf{F}_\mu$ of (5.29) on a dipole in a nonuniform field $\mathbf{B}(\mathbf{r})$. The interaction energy U_μ is drawn upon for work $\mathbf{F}_\mu \cdot d\mathbf{r} = -dU_\mu = -\boldsymbol{\mu} \cdot d\mathbf{B}$, with $d\mathbf{B} = (d\mathbf{r} \cdot \nabla)\mathbf{B}(\mathbf{r})$, whenever there is a translation $d\mathbf{r}$ of a dipole $\boldsymbol{\mu}$ having a fixed orientation. The same interaction energy can change even when there is no translation of the dipole to new points in space or when the field is uniform so that $d\mathbf{B} = 0$ in a translation. It can also change (be drawn upon for work) whenever there is a rotational displacement† $d\boldsymbol{\mu} = d\boldsymbol{\vartheta} \times \boldsymbol{\mu}$ (a rotation through angle $|d\boldsymbol{\vartheta}|$ about the direction of the vector $d\boldsymbol{\vartheta}$ at a point in space). Then a torque $\boldsymbol{\Theta}$ does work $-dU_\mu = \mathbf{B} \cdot d\boldsymbol{\mu}$, such that

$$\boldsymbol{\Theta} \cdot d\boldsymbol{\vartheta} = -dU_\mu = (d\boldsymbol{\vartheta} \times \boldsymbol{\mu}) \cdot \mathbf{B} = d\boldsymbol{\vartheta} \cdot (\boldsymbol{\mu} \times \mathbf{B}) \tag{5.51}$$

(Compare the *electric* dipole rotations discussed in connection with Fig. 2.2.)

The torque of (5.50) or (5.51) tends to align the dipole with the field, and this makes dipoles (either in the form of short magnetic needles or tiny coils of current) useful for mapping and measuring magnetic fields [in place of the poles at ends of *long* needles discussed in connection with the preliminary definition (1.16) of a magnetic field]. The alignment of a dipole with a field can be viewed as a transition to a more stable lower-energy configuration having a least value for $U_\mu = -\boldsymbol{\mu} \cdot \mathbf{B}$. Energy conservation will then demand that the dipole not be isolated from other energy-possessing mechanisms (such as thermal motions to which the energy is transferable by friction or ohmic resistance).

Larmor Precession

The conservation of dynamical quantities characteristic of isolated systems, other than just the energy, may play a role in preventing the alignment of a dipole with a field, even though the external field disturbs the isolation, since that depends on just what perturbation by the magnetic torque (5.50) can do. Examples are provided by systems that owe their magnetic moments to charges rotating with conserved *angular momenta* $\mathbf{L}$. The magnetic moment is then

† More generally, *simultaneous* translations and rotations of a dipole in a nonuniform field might be considered. Then $-dU_\mu = \boldsymbol{\mu} \cdot d\mathbf{B} + \mathbf{B} \cdot d\boldsymbol{\mu}$, and the joint motions are subject to such *coupled* equations of motion as

$$m_\mu \ddot{\mathbf{r}} = (\boldsymbol{\mu} \cdot \nabla)\mathbf{B}, \qquad \frac{d\mathbf{L}}{dt} = \boldsymbol{\mu} \times \mathbf{B}$$

usually entailing energy exchanges between the translational and rotational motions. However, the atomic systems that have greatest interest are so small (the dipole is so nearly ideal) that nonuniformities in externally applied fields are usually negligible over spans that are large compared with atomic dimensions.

supposed to be $\boldsymbol{\mu} = g\mathbf{L}$ with some gyromagnetic ratio $g = L/\mu$, according to the discussion of (4.41). To align such a moment (when it is not initially parallel to the imposed magnetic field) requires the torque to change the angular momentum, i.e., to overcome inertial effects of mass motion that are responsible for *conserving* the angular momentum while the system is left to itself.

A torque $\boldsymbol{\Theta}$ changes angular momentum at the rate $d\mathbf{L}/dt = \boldsymbol{\Theta}$ just as force $\mathbf{F}$ changes linear momentum at the rate $d\mathbf{p}/dt = \mathbf{F}$. Thus, the equation of motion for a dipole $\boldsymbol{\mu} = g\mathbf{L}$ at a point in a magnetic field $\mathbf{B}$ can be written

$$\frac{d\mathbf{L}}{dt} = g\mathbf{L} \times \mathbf{B} = -\boldsymbol{\omega}_L \times \mathbf{L} \qquad \text{with } \boldsymbol{\omega}_L \equiv g\mathbf{B} \tag{5.52}$$

This means that the magnetic torque rotates the angular-momentum vector (and the magnetic moment) about the direction of the field at the point as axis, just as the field was found to rotate the velocity, or linear-momentum, vector of a moving point charge in (5.8). The circular frequencies of the two types of rotation are generally different; even for equal charge-to-mass ratios, $\omega_o = q\mathsf{B}/mc$ in (5.8) and $\omega_L = q\mathsf{B}/2mc$, half as great, according to the gyromagnetic ratio (4.40). Just as the velocity magnitude was left unchanged by the rotation in (5.8), so here the magnitude of the angular momentum is *conserved* during the motion; $\mathbf{L} \cdot (d\mathbf{L}/dt) = d(\frac{1}{2}L^2)/dt = 0$ follows from the perpendicularity of the torque in (5.52) to the moments. Moreover, the projection $(\boldsymbol{\omega}_L \cdot \mathbf{L})/\omega_L$ of the angular momentum on the field direction likewise stays constant, since $\boldsymbol{\omega}_L \cdot (d\mathbf{L}/dt) = 0$ also follows from the equation of motion. Thus only the component of $\mathbf{L}$ normal to the field can be changed by the magnetic torque, and the rotation of this component results in a *precession* of the angular momentum vector about the field direction, as indicated in Fig. 5.3. Such a result is called *Larmor precession*, and $\omega_L = g\mathsf{B}$ is called the *Larmor precession frequency*. The magnetic torque is here unable to align the dipole with the field because it cannot change the angular-momentum component parallel to the field, and this must remain conserved. The resultant precession is a type of gyroscopic effect familiar from experience with spinning tops in the uniform gravitational field.†

The Larmor precession is important in the interpretation of *Zeeman effects* on atoms, the results of probing them with external magnetic fields. Each atom is supposed to contain orbiting charges with some angular momentum $\mathbf{L}$ helping define a state of motion in which the probing process may find it. This process is just the imposition of an external field that will start various Larmor precessions in atoms found with various magnitudes and orientations of a resultant $\mathbf{L}$. These precessions become detectable in the form of energy differences expressable as interaction energies $U_\mu = -\boldsymbol{\mu} \cdot \mathbf{B} = -g\mathbf{L} \cdot \mathbf{B}$ (usually through effects on electro-

† In that case (see Ref. 17, sec. 10.4) there is generally, besides the precession, a *nutation* of the top, i.e., an oscillation of the *spin*-angular-momentum component parallel to the field, with compensating oscillations of the *precession* angular momentum. That happens because the gravitational moment of the top is independent of the spin angular momentum instead of being proportional to it as the magnetic moment is in the Larmor precession.

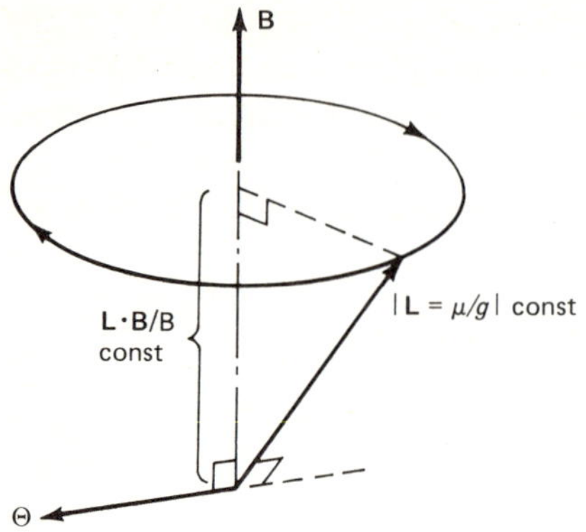

Figure 5.3

magnetic radiation† emitted or absorbed by the atom). Then conclusions can be drawn about states of motion in the atom and the gyromagnetic ratios they possess. More details of Zeeman effects can be understood with the help of the idealized example in the next subsection.

Zeeman Effects on an Orbiting Point Charge

Treating the effect of an external magnetic field on a system of orbiting charges as simply inducing a Larmor precession of it is plainly adequate only insofar as the dipole approximation of the magnetic interaction is sufficient. The complete effect is simplest to investigate for a case in which the motion is an orbiting of a single point charge. The system will be taken to consist of a negative point charge $q = -e$ moving in the electrostatic field $\mathbf{E} = \hat{\mathbf{r}}Ze/r^2$ of a second, fixed point charge $+Ze$ at $\mathbf{r} = 0$. That constitutes a classical model of an hydrogenlike atom, as used in early explorations of atomic dynamics. The motion will have some conserved energy $E = \frac{1}{2}mu^2 - Ze^2/r$, as in (5.4), and, in the absence of any external field, also a constant angular momentum $\mathbf{L} = \mathbf{r} \times m\mathbf{u}$, since $\dot{\mathbf{L}} = \mathbf{r} \times m\dot{\mathbf{u}} = \mathbf{r} \times (-eE\hat{\mathbf{r}}) = 0$ when only the central force operates. These conservation principles can by themselves yield a complete description‡ of the motion for any given values of E and $\mathbf{L}$. When the energy constant is negative ($|-Ze^2/r| > \frac{1}{2}mu^2$ in every phase of the motion), the radial distance r is restricted within finite bounds and an elliptic orbit is followed, as already remarked in connection with (2.36). The area enclosed by the ellipse can be expressed as Kepler's *sectorial velocity* $\frac{1}{2}r^2\dot{\varphi} = L/2m$ (the area being swept by the radius vector per unit time) times the period $2\pi/\omega$ of one revolution, and it faces into the direction of $\mathbf{L}$. The orbiting of the charge $-e$ produces a current $-e\omega/2\pi$,

† To be discussed in connection with Fig. 8.4.

‡ See Ref. 17, sec. 4.3, for example. It adds to an understanding of what will follow to have in mind a well-known result from quantum mechanics: the energies E permitted to an atomic electron are actually restricted to certain discrete values, $E = \epsilon_1, \epsilon_2, \epsilon_3, \ldots$, and $\mathbf{L}$ is restricted to discrete values for each E.

averaged over the orbit, and so† a magnetic moment $\boldsymbol{\mu} = -e\mathbf{L}/2mc$, as is also expected from (4.40). This magnetic moment is just of the type that motivated the Larmor equation of motion (5.52) and is ready for interaction with any magnetic field that may now be imposed.

The effect of a magnetic field is more properly taken into account in an equation of motion like (5.1)

$$m\dot{\mathbf{u}} = -eE\hat{\mathbf{r}} - e\mathbf{u} \times \frac{\mathbf{B}}{c} = -\frac{Ze^2}{r^2}\hat{\mathbf{r}} - 2m\boldsymbol{\omega}_L \times \mathbf{u} \tag{5.53}$$

where $\boldsymbol{\omega}_L = -e\mathbf{B}/2mc$ is a Larmor frequency like that of (5.52). An energy expressed as in (5.4),

$$E = \tfrac{1}{2}mu^2 - \frac{Ze^2}{r} \tag{5.54}$$

is still conserved because the steady magnetic field does no work, but this may not be the same energy as the system had before the imposition of the field because of the intermediate flux changes. The angular momentum $\mathbf{L} = \mathbf{r} \times m\mathbf{u}$ is no longer conserved, as before the imposition of the field, because

$$\dot{\mathbf{L}} = \mathbf{r} \times m\dot{\mathbf{u}} = -2m\mathbf{r} \times (\boldsymbol{\omega}_L \times \mathbf{u}) \neq 0 \qquad \text{for } \mathbf{B} \neq 0 \tag{5.55}$$

follows from the equation of motion (5.53). Nor is the rate of this angular-momentum change given by a Larmor equation $\dot{\mathbf{L}} + \boldsymbol{\omega}_L \times \mathbf{L} = 0$ of (5.52), being different by

$$\begin{aligned}\dot{\mathbf{L}} + \boldsymbol{\omega}_L \times (\mathbf{r} \times m\mathbf{u}) &= -m[\mathbf{r} \times (\boldsymbol{\omega}_L \times \mathbf{u}) + \mathbf{u} \times (\boldsymbol{\omega}_L \times \mathbf{r})] \\ &= -\frac{d}{dt}[m\mathbf{r} \times (\boldsymbol{\omega}_L \times \mathbf{r})]\end{aligned} \tag{5.56}$$

since $\dot{\mathbf{r}} \equiv \mathbf{u}$. Basically this happens because a rigid magnetic moment magnitude is not being maintained here, as presumed for the Larmor equation.

Next consider the moment $\mathbf{l} = \mathbf{r} \times \mathbf{p}$ of the canonical momentum $\mathbf{p}(\mathbf{r}, t)$ of (5.7). The vector potential in its definition can be taken to be $\mathbf{A} = \mathbf{B} \times \mathbf{r}/2$ (Exercise A.8), since a field $\mathbf{B} = \nabla \times \mathbf{A}$ that is uniform over the system is derivable from it, and so

$$\mathbf{p}(\mathbf{r}, t) = m(\mathbf{u} + \boldsymbol{\omega}_L \times \mathbf{r}) \tag{5.57}$$

Its moment differs from the kinetic angular momentum $\mathbf{L} = \mathbf{r} \times m\mathbf{u}$ as in

$$\mathbf{l} = \mathbf{r} \times \mathbf{p} = \mathbf{L} + m\mathbf{r} \times (\boldsymbol{\omega}_L \times \mathbf{r}) \tag{5.58}$$

† Using the expression (4.34) for the connection between magnetic moment and area.

It follows, most directly from the equation of motion (5.53) [or its equivalent (5.6)], that the rate of change of $\mathbf{l}$ is given by

$$\begin{aligned}\dot{\mathbf{l}} &= \frac{d}{dt}(\mathbf{r} \times \mathbf{p}) = \mathbf{u} \times \mathbf{p} + \mathbf{r} \times m(\dot{\mathbf{u}} + \boldsymbol{\omega}_L \times \mathbf{u}) \\ &= m\mathbf{u} \times (\boldsymbol{\omega}_L \times \mathbf{r}) - m\mathbf{r} \times (\boldsymbol{\omega}_L \times \mathbf{u}) \\ &= -m\boldsymbol{\omega}_L \times (\mathbf{r} \times \mathbf{u}) \equiv -\boldsymbol{\omega}_L \times \mathbf{L} \end{aligned} \tag{5.59}$$

differing from the Larmor equation in that $\dot{\mathbf{l}}$ replaces $\dot{\mathbf{L}}$. It also differs from the Larmor type of equation $\dot{\mathbf{l}} + \boldsymbol{\omega}_L \times \mathbf{l} = 0$, by

$$\dot{\mathbf{l}} + \boldsymbol{\omega}_L \times \mathbf{l} = \boldsymbol{\omega}_L \times [m\mathbf{r} \times (\boldsymbol{\omega}_L \times \mathbf{r})] = -m(\boldsymbol{\omega}_L \times \mathbf{r})(\boldsymbol{\omega}_L \cdot \mathbf{r}) \tag{5.60}$$

However, the extra torque here is only *quadratic* in the field strength $\mathbf{B} = -2mc\boldsymbol{\omega}_L/e$, whereas the angular momentum $\mathbf{L}$ fails to execute Larmor precession by effects *linear* in the field, as (5.56) shows. The external field strengths that are practically imposable have far smaller effects on atomic motions than the natural electromagnetic forces within atoms, and the so-called *quadratic Zeeman effects* require a much greater accuracy of observation than those linear in the field strength (the square of a fractional number is much smaller than the number). In evaluating linear Zeeman effects, terms quadratic in $\boldsymbol{\omega}_L$ like those in (5.60) are quite completely negligible. Thus it is $\mathbf{l}$, the moment of the conjugate momentum, rather than $\mathbf{L} = \mathbf{r} \times m\mathbf{u}$, that executes Larmor precession the more accurately.†

A physical interpretation of the conjugate momentum (5.57) becomes plain when it is recognized that the factor $\mathbf{u} + \boldsymbol{\omega}_L \times \mathbf{r}$ in its expression is just the velocity of the particle relative to a frame that is rotating about $\mathbf{r} = 0$ (the electric force center) with an angular velocity $\boldsymbol{\Omega} = -\boldsymbol{\omega}_L$. Much greater insight into the entire problem is obtained when it is considered relative to such a rotating frame.

The position vector of the electron relative to the rotating frame is $\mathbf{R} = \mathbf{r}$ at all times, but the velocity $\mathbf{U} \equiv \mathring{\mathbf{R}}$ calculated as if the rotating frame were fixed differs from the absolute velocity $\mathbf{u} = \dot{\mathbf{r}}$ according to

$$\mathbf{u} = \mathbf{U} + \boldsymbol{\Omega} \times \mathbf{R} \qquad \text{and} \qquad \mathbf{U} = \mathbf{u} - \boldsymbol{\Omega} \times \mathbf{r} \tag{5.61}$$

since a velocity $\boldsymbol{\Omega} \times \mathbf{R} = \boldsymbol{\Omega} \times \mathbf{r}$ is plainly imparted by the relative frame rotation. The relation between the two types of time derivative, $\mathbf{u} \equiv \dot{\mathbf{r}}$ and $\mathbf{U} \equiv \mathring{\mathbf{R}} = \mathring{\mathbf{r}}$, that is implicit here is valid not only for the vector $\mathbf{r} = \mathbf{R}$ but for any vector‡ $\mathbf{V}$:

$$\dot{\mathbf{V}} = \mathring{\mathbf{V}} + \boldsymbol{\Omega} \times \mathbf{V} \tag{5.62}$$

whenever $\dot{\mathbf{V}}$ and $\mathring{\mathbf{V}}$ refer to changes with respect to relatively rotating frames.

† The same conclusions are obtained from the hamiltonian formulation of classical mechanics in Ref. 17, exercises 8.4 to 8.6.

‡ If this does not appear obvious, see Ref. 17, eq. (1.46). More details pertinent to the ensuing discussion can also be seen there, in eq. (2.37) and exercise 8.4.

When this is applied to the velocity $\mathbf{u}$, it leads to the reciprocal relations

$$\begin{aligned} \dot{\mathbf{u}} &= \mathring{\mathbf{U}} + 2\boldsymbol{\Omega} \times \mathbf{U} + \boldsymbol{\Omega} \times (\boldsymbol{\Omega} \times \mathbf{R}) \\ \mathring{\mathbf{U}} &= \dot{\mathbf{u}} - 2\boldsymbol{\Omega} \times \mathbf{u} + \boldsymbol{\Omega} \times (\boldsymbol{\Omega} \times \mathbf{r}) \end{aligned} \tag{5.63}$$

between the absolute acceleration $\dot{\mathbf{u}}$ and the acceleration $\mathring{\mathbf{U}}$ calculated as if the rotating frame were fixed. They differ not only by a centrifugal acceleration

$$\boldsymbol{\Omega} \times (\boldsymbol{\Omega} \times \mathbf{r}) = -\Omega^2[\mathbf{r} - \hat{\boldsymbol{\Omega}}(\hat{\boldsymbol{\Omega}} \cdot \mathbf{r})]$$

with the bracket just the perpendicular distance from the rotation axis, but also by a Coriolis acceleration $2\boldsymbol{\Omega} \times \mathbf{U}$ or its reciprocal $-2\boldsymbol{\Omega} \times \mathbf{u}$ that is needed to carry any given velocity vector with the rotation of a frame.

When the frame angular velocity $\boldsymbol{\Omega} = -\boldsymbol{\omega}_L$ is now adopted, the linear and angular momentum relative to the rotating frame

$$\mathbf{p} = m\mathbf{U} \qquad \text{and} \qquad \mathbf{l} = \mathbf{R} \times m\mathbf{U} \tag{5.64}$$

are identical with the conjugate momenta (5.57) and (5.58), as (5.61) shows. Substitution into (5.63) of the acceleration given by the equation of motion (5.53) shows that the trajectory as viewed from the rotating frame has

$$m\mathring{\mathbf{U}} = -\frac{Ze^2}{R^2}\hat{\mathbf{R}} + m\boldsymbol{\omega}_L \times (\boldsymbol{\omega}_L \times \mathbf{R}) \tag{5.65}$$

on it. The magnetic force $-e\mathbf{u} \times \mathbf{B}/c = 2m\mathbf{u} \times \boldsymbol{\omega}_L$ has here provided just sufficient acceleration to cancel out the inertial Coriolis force $-2m\boldsymbol{\Omega} \times \mathbf{u}$ that follows from (5.63). Consequently, when the quadratic effects remaining in (5.65) are negligible, the orbit relative to the rotating frame is the same as in the inertial frame with the magnetic field absent. Moreover, the quadratic effects can now be viewed as arising from the centrifugal force $m\boldsymbol{\omega}_L \times (\boldsymbol{\omega}_L \times \mathbf{R})$ imparted by the frame rotation. These will merely distend and distort the orbit slightly because the Larmor precessions in any practical external field are slow in comparison to the orbital motions of the electrons.

Because the particle's trajectory relative to the rotating frame is the same as in the absence of a magnetic field, insofar as the quadratic effects are negligible, the system's internal angular momentum $\mathbf{l} = \mathbf{R} \times m\mathbf{U}$ will be a nearly conserved vector in that frame: $\mathring{\mathbf{l}} = \mathbf{R} \times m\mathring{\mathbf{U}} \approx 0$ follows from $m\mathring{\mathbf{U}} \approx -(Ze^2/R^2)\hat{\mathbf{R}}$ as in the field-free case. Thus the internal angular momentum $\mathbf{l}$ will be viewed in the initial, inertial, frame as rotating with the Larmor precession frequency quite accurately—the conclusion already reached from (5.60).

When the quadratic effects in (5.65) are taken into account,

$$\mathring{\mathbf{l}} = m(\mathbf{R} \times \boldsymbol{\omega}_L)(\boldsymbol{\omega}_L \cdot \mathbf{R}) \tag{5.66}$$

follows, in agreement with (5.60), when the connection between $\mathring{\mathbf{l}}$ and $\dot{\mathbf{l}}$ given by (5.62) is recognized. The internal angular momentum is not quite conserved even in the system's proper frame, because of the centrifugal distortion of the orbit. However, $\boldsymbol{\omega}_L \cdot \mathring{\mathbf{l}} = \boldsymbol{\omega}_L \cdot \dot{\mathbf{l}} = 0$, and so the projection $\boldsymbol{\omega}_L \cdot \mathbf{l}$ does remain constant in the distorted orbits. This is important for expressions of the energies, considered next.

When the quadratic effects in the equation of motion (5.65) are neglected, it yields a conservation of the energy

$$\epsilon = \tfrac{1}{2}mU^2 - \frac{Ze^2}{R} \tag{5.67}$$

exactly as in the absence of the magnetic field. This is energy characteristic of the unperturbed system, arising from the internal motion of the electron in it, as when uninfluenced by any external field. When the magnetic field is imposed, it becomes the energy (aside from the quadratic effects) relative to a frame rotating with the Larmor precession frequency, as viewed from a laboratory frame. The energy in this frame is thereby supplemented to the value given by (5.54). The connection between this energy and the unperturbed energy ϵ can be found by reexpressing it in terms of the variables **R**, **U**, and **l** referring to the rotating frame

$$E = \tfrac{1}{2}m(\mathbf{U} - \boldsymbol{\omega}_L \times \mathbf{R})^2 - \frac{Ze^2}{R} = \epsilon - \boldsymbol{\omega}_L \cdot \mathbf{l} + \tfrac{1}{2}m(\boldsymbol{\omega}_L \times \mathbf{R})^2 \tag{5.68}$$

because $m(\boldsymbol{\omega}_L \times \mathbf{R}) \cdot \mathbf{U} = \boldsymbol{\omega}_L \cdot (\mathbf{R} \times m\mathbf{U})$. The supplementations of ϵ here are rotational increments of kinetic energy arising as **U** is changed to $\mathbf{u} = \mathbf{U} - \boldsymbol{\omega}_L \times \mathbf{R}$ by the magnetic field. The pure field increment $\tfrac{1}{2}m(\boldsymbol{\omega}_L \times \mathbf{R})^2$, uncoupled to **U**, is a quadratic effect such as is being neglected in the present considerations. The remainder is the system's energy, in the laboratory frame, as perturbed by the magnetic field. It is most often expressed as

$$E \approx \epsilon - \boldsymbol{\mu} \cdot \mathbf{B} \tag{5.69}$$

after recalling that $\boldsymbol{\omega}_L = -e\mathbf{B}/2mc$ and putting $-e\mathbf{l}/2mc \equiv \boldsymbol{\mu}$, a magnetic dipole moment characteristic of the motion described by the internal angular momentum **l**. It is the perturbed laboratory energy (5.69) that becomes available for radiations by the system, as detected in the laboratory frame. That the perturbation consists of just the interaction energy $U_\mu = -\boldsymbol{\mu} \cdot \mathbf{B}$ of the dipole with the field has already been mentioned on page 141.

It is not difficult to take the quadratic effects into account as well. Then the energy ϵ of (5.67) is no longer quite conserved; instead, the conservation of

$$\epsilon + \tfrac{1}{2}m(\boldsymbol{\omega}_L \times \mathbf{R})^2 \tag{5.70}$$

follows from the full equation of motion (5.65). This can be seen by multiplying the equation by the velocity **U**, as in the procedure of (5.2), and identifying the resulting time derivatives or by finding the consequences of the equation for the time derivative ($\rightarrow 0$) of (5.70). It does not matter for time derivatives of scalar expressions, as it does for the vector in (5.62), whether they are taken as if the rotating frame were fixed or not, both yield the same result. The reason that both the different quantities of energy (5.70) and (5.68) can be simultaneously conserved is that their difference $\boldsymbol{\omega}_L \cdot \mathbf{l}$ is conserved by itself, as seen in the discussion of (5.66).

The difference from ϵ of the laboratory energy in (5.68) can also be expressed in terms of the laboratory variables, through the use of the connection (5.58) between the internal angular momentum $\mathbf{l}$ and the laboratory moment $\mathbf{L} = \mathbf{r} \times m\mathbf{u}$. Then

$$E = \epsilon - \boldsymbol{\omega}_L \cdot \mathbf{L} - \tfrac{1}{2}m(\boldsymbol{\omega}_L \times \mathbf{r})^2 \tag{5.71}$$

Comparison to (5.68) shows that both the expressions in

$$E \approx \epsilon - \boldsymbol{\omega}_L \cdot \mathbf{l} \approx \epsilon - \boldsymbol{\omega}_L \cdot \mathbf{L} \tag{5.72}$$

are equally accurate, being subject to equal and opposite quadratic corrections $\pm\frac{1}{2}m(\boldsymbol{\omega}_L \times \mathbf{r})^2$. This is so despite the fact that the internal angular momentum $\mathbf{l}$ was found to execute Larmor precession more accurately than $\mathbf{L}$ does. On the other hand, it is the $\boldsymbol{\omega}_L \cdot \mathbf{l} = \boldsymbol{\mu} \cdot \mathbf{B}$ as in (5.69) that is proportional to the *intrinsic* magnetic moment, which, like ϵ, characterizes the unperturbed atom.

EXERCISES

All speeds are to be assumed $\ll c$.

5.1 Results of imposing a uniform electric field $\mathbf{E}_o$ on oscillating or orbiting charges are called *Stark effects*. Consider such effects on the motions of a point charge $q = -e$, m that is normally subject only to the central force $\mathbf{F} = -m\omega_o^2\mathbf{r}(t)$ whenever it is displaced from $\mathbf{r} = 0$ (the *isotropic oscillator*).

(*a*) Arbitrary linear oscillations can be started from some initial displacement $\mathbf{r}_o$, and $\dot{\mathbf{r}}(0) \equiv \mathbf{u}_o = 0$. Find how the motion ensuing from an $\mathbf{r}_o$ parallel to $e\mathbf{E}_o$ differs from the case in which $\mathbf{E}_o$ is absent.

(*b*) Do the same for an $\mathbf{r}_o$ normal to $e\mathbf{E}_o$ and also for an $\mathbf{r}_o = 0$ but $\mathbf{u}_o \neq 0$ and normal to $e\mathbf{E}_o$.

(*c*) The most general possible orbit the particle can describe is started when it is given some $\mathbf{u}_o \neq 0$ not parallel to an $\mathbf{r}_o \neq 0$. While $\mathbf{E}_o \equiv 0$, the orbit will be an ellipse centered on $\mathbf{r} = 0$ and in the plane of $\mathbf{r}_o$ and $\mathbf{u}_o$ (Ref. 17, sec. 4.1). Directly from the vector equation of motion, show that in the presence of an $\mathbf{E}_o \neq 0$ the orbit will again be an ellipse but with its center shifted from $\mathbf{r} = 0$, with its plane differently oriented, and with a newly centered conserved angular momentum.

5.2 As mentioned on page 142, the energy E of an electron bound in an atom with a fixed nucleus (as in the surface of a metal) and having an effective charge $+Ze$ is usually represented by a negative value. Find an (approximate) expression for the minimal strength E_o of an imposed electric field, gauged relative to $\phi = 0$ at the atomic nucleus, that can release an electron of the energy E from the atom (on a classical basis; quantum-mechanical barrier penetration can reduce the necessary field strength; such considerations help one understand the cold emission mentioned in the footnote to page 491). How can a negative energy value for the unbound electron, in the gauge here, be interpreted?

5.3 Something can be learned about the operation of electrostatic lens systems by considering the penetration of charged particles through a region $|z| < l/2$ containing the quadrupole lens field

$$\phi\left(|z| < \frac{l}{2}\right) = \alpha(x^2 - y^2) \qquad \phi\left(|z| > \frac{l}{2}\right) = 0$$

where α ($\gtrless 0$) is a field-strength parameter (see Exercise 3.18). Compare normally incident rays, consisting of particles with $\dot{z} = u_\parallel$ but $u_\perp = 0$ at incidence $z = -l/2$, one in the xz plane, the other in the yz plane.

(*a*) Show that one of the above rays can be made to converge toward the z axis after its transmission with a proper choice of the lens parameters α and l but that the other is necessarily divergent.

(*b*) Normally incident rays define *focal lengths*, i.e., distances to points at which projections of the straight emergent rays intersect the axis. Show that the focal lengths of both the convergent and divergent lens actions are properly independent of the part of the lens on which the rays are incident. (Convergence of both rays is achieved by making them pass through more than one lens, some distance apart, with the sign of the lens fields *alternated*.)

5.4 The magnetostatic counterpart of the quadrupole lens treated in Exercise 5.3 is best represented by a field $\mathbf{B} = -\nabla\phi_M$ with $\phi_M = \beta xy$, a scalar potential formally different from the electrostatic potential only in being referred to a cartesian frame rotated through $\pi/4$ [compare Exercise 3.18(*c*) and (b)].

(*a*) The principal difference in the resulting motions is that the velocity magnitude u must remain constant in a purely magnetic field. Show that enforcing this requires an equation $\ddot{z} \neq 0$ that changes $\dot{z}$ when $\dot{x}$, $\dot{y} \neq 0$ are generated.

(*b*) For paraxial rays, when terms quadratic in x and y are negligible, $\dot{z} \approx u$ is an adequate approximation. Show that the equations of motion for $\ddot{x}$, $\ddot{y}$ then become identical with those of the electrostatic case except for the representation of effective field strengths $2\alpha \to \pm\beta u/c$.

5.5 In a 180° spectrometer (β) electrons ($q = -e$) from a source placed at $s = r_o$, $\varphi = 0$, $z = 0$ pass through a uniform magnetic field $\mathbf{B} = \mathbf{B}_z$ and are detected as they arrive at $\varphi = \pi$, $z = 0$ within a narrow range Δs at $s = r_o$. The spectrometer is focused on electrons of momentum p_o when the field strength is set at $\mathbf{B}_o = cp_o/er_o$. Intermediate collimation allows only electrons that emerge in the $z = 0$ plane within a small angular range $\Delta\alpha$, centered on the direction $\mathbf{e}_\varphi$, to reach the detector. Show that the electrons with p_o are consequently spread over the radial range $\Delta s = r_o\,(\Delta\alpha)^2/4$ at the detector and that this leads to $\Delta p/p = (\Delta\alpha)^2/8$ as the resolving power of the spectrometer.

5.6 In the instrument of Exercise 5.5 any electrons that emerge from the source with a velocity component parallel to the uniform $\mathbf{B} = \mathbf{B}_z$ field lines tend to be defocused from the central $z = 0$ plane of source and detector. Therefore consider a field that is made the function of radius $\mathbf{B}_z(s) \approx \mathbf{B}_o(r_o/s)^n$, with some power n, at least in the neighborhood of the central $s = r_o$, $z = 0$ orbit (through shaping the pole faces of the electromagnet that provides the field).

(*a*) Show that $\nabla \times \mathbf{B} = 0$ requires a $\mathbf{B}_s \approx -n\mathbf{B}_z(z/s)$ to exist. (The resulting approximations of $\mathbf{B}_{z,s}$ violate $\nabla \cdot \mathbf{B} = 0$ only negligibly for $z \ll r_o/n^2$ near $s = r_o$.)

(*b*) By approximating the equation of motion for $\ddot{z}$ in a way suitable for orbits near $s = r_o$ and with $\dot{\varphi} \approx u/r_o = \omega_o$ show that, for $n > 0$, any electrons straying from the $z = 0$ plane are sent back to it in a half-period $\pi/\sqrt{n}\,\omega_o$, where $\omega_o \equiv e\mathbf{B}_o/mc$.

(*c*) The choice $n = \frac{1}{2}$ is made for the Siegbahn type of spectrometer. At what angle from the source (instead of $\varphi = 180°$) should the detector in this type of spectrometer be placed in order to take maximum advantage of the z focusing provided when $n = \frac{1}{2}$?

5.7 A cyclotron, like the spectrometers of the preceding exercises, also provides a $\mathbf{B} \approx \mathbf{B}_z$ field, between cylindrical pole faces of some large radius R, in which charged particles describe orbits near a $z = 0$ plane. There is a natural weakening of $\mathbf{B}_z$ as $s \to R$, where the space between pole faces ends, in a way approximately representable as in Exercise 5.6, with $n > 0$. However, the cyclotron is primarily used for *positively* charged particles.

(*a*) Does the weakened fringe field near $s = R$ tend to focus or defocus the positive charges?

(*b*) The ion source in a cyclotron starts low-velocity particles near the center and then accelerates them by an alternating potential difference as they pass across a narrow gap along a diameter. Show that a *fixed* frequency of alternating the emf will make the accelerations in a fixed field $\mathbf{B}_z$ cumulative. (The result is called a *cyclotron resonance*.)

(*c*) In terms of q, M, $\mathbf{B}$, and R, express the approximate energy an ion that reaches the largest possible orbit, $s \approx R$, will have.

5.8 Despite its absence from (5.4), a magnetic field *can* influence the kinetic energy a particle will have because it helps determine its positions in a coexistent electric potential distribution.

(*a*) Find the kinetic energy of the motion (5.19) at any instant, showing that it fluctuates with a frequency determined by $\mathbf{B}_o$.

(*b*) Show that the net energy (5.4) of the motion in (*a*) vanishes at all times *if* the potential is gauged as $\phi = -\mathbf{E}_o \cdot \mathbf{r}$. Is this consistent with the start given to the motion?

5.9 A magnetic field confined to a finite space can influence the motion of charges *outside* that space if it is allowed to change with time. Let a small bead bearing the charge q be free to move on a circular fiber of radius s outside of, and coaxial with, an infinite solenoid of radius $a < s$.

(*a*) After an instant $t = 0$ at which the bead is at rest, the current in the solenoid is allowed to decrease to zero at some rate $\dot{I}(t) < 0$, perhaps as in Exercise 4.24, made slow enough (as by self-inductance in the solenoid) to permit a quasi-static field equilibrium to be presumed for every stage. What is the force on the bead in terms of $\dot{I}(t)$?

(*b*) Find the eventual angular momentum imparted to the bead in terms of $I(0)$, a, s, and the turns n per unit length of the solenoid.

(*c*) Express the eventual *linear* momentum gain in terms of the initial vector potential at the radius $s > a$ (see Exercise 4.5).

5.10 In a betatron, electrons are accelerated as they follow a circular orbit of radius $r_0 = mcu(t)/e\mathsf{B}_o(t)$, kept fixed by increasing the field $\mathsf{B}_o(t)$ at that radius as the electron's velocity $u(t)$ increases. The emf for the acceleration is provided by increasing the magnetic flux enclosed by the orbit together with $\mathsf{B}_o(t)$, slowly enough for each instant to be treated as an equilibrium (quasi-static) situation.

(*a*) Show that r_0 can be maintained if the flux has the value $2\pi r_o^2 \mathsf{B}_o(t)$ at each instant.

(*b*) Plainly, the average magnetic field within the orbit must be greater in magnitude than $\mathsf{B}_o(t)$ is at r_0. Suppose that the field $\mathsf{B}(s, t) = \mathsf{B}_z$ everywhere in the plane of the orbit is represented as

$$\mathsf{B}(s < r_1 < r_0, t) = \left(\frac{r_0}{r_1}\right)^n \mathsf{B}_o(t), \ \mathsf{B}(r_1 < s < r_0) = \left(\frac{r_0}{s}\right)^n \mathsf{B}_o(t)$$

Show that the nonuniform, outer part of this field must be weakened with radius $s > r_1$ as given by some power $n \geq 2$, whatever $r_1 < r_0$ is chosen.

5.11 A particle m, q moves on a circle of radius r_0, with speed u_o, in a plane normal to a uniform field $\mathbf{B}_o$. The field magnitude is then *very slowly* increased at a rate $\dot{\mathsf{B}} > 0$, so that the particle completes many cycles during each increment of field and an equilibrium orbit can be assumed at every stage.

(*a*) Find the relation between $\dot{\mathsf{B}}$, $\dot{r}$, $\dot{u}$ that can be assumed at every stage.

(*b*) Find the average acceleration $\dot{u}$ that the particle can be assumed to undergo during each cycle.

(*c*) Show that after the field has increased, the orbit will have contracted and the speed increased in just such a way that the flux through the orbit and its effective magnetic moment (5.13) remain unchanged (they are *adiabatic invariants*).

5.12 For the loop falling in a magnetic bottle, treated in Exercise 4.20, show that in the steady terminal phases of the motion:

(*a*) The gravitational force is exactly balanced by a magnetic force $\mathbf{F} = (\boldsymbol{\mu} \cdot \nabla)\mathbf{B}$.

(*b*) The steady loss of gravitational potential is offset by a steady gain in the magnetic potential energy $-\boldsymbol{\mu} \cdot \mathbf{B}$ and that this corresponds to energy going into I^2R losses.

5.13 As a magnetostatic counterpart of the problem in Exercise 2.24, consider a point particle of mass m and magnetic moment $\boldsymbol{\mu}$ placed at the center of a fixed circular loop of current I on radius a. The particle is given an initial impulse mu_o, directed along the axis of the loop. Describe the ensuing motions for various initial orientations of the particle.

5.14 (*a*) Starting from the expression $U_\mu = -\boldsymbol{\mu} \cdot \mathbf{B}$, derive an expression for the interaction energy of a *pair* of dipoles at a distance r apart and compare your result with (2.57) [and see Exercise 4.15(*c*)]. Why are there no magnetic interactions corresponding to all but the one part of (2.57)?

(*b*) Find the signs of the mutual forces between the dipoles for various relative orientations of them while they are at some fixed distance apart.

5.15 (*a*) A charged particle is dropped (from rest and subject to uniform gravity) into a uniform horizontal magnetic field. Describe the ensuing motions in each of the two cases of oppositely charged particles. How do these differ from the case in which gravity is replaced by a uniform electric field?

(*b*) Now suppose that the particle had been started with some initial *horizontal* velocity $\mathbf{u}_o$ perpendicular to the direction of the magnetic field. Specify the rolling concentric circles that can generate the orbits for each of opposite directions of $\mathbf{u}_o$ and for both $|\mathbf{u}_o| \gtrless g/\omega$ in each case ($\omega \equiv q\mathsf{B}/mc$). Make sketches that show the contrasting features of the various orbits, including cases of $u_o = g/\omega$ and $2g/\omega$.

(*c*) How are the orbits (*a*) and (*b*) modified by an initial velocity component parallel to the magnetic field?

(*d*) Suppose that the starting velocity $\mathbf{u}_o$ had been directed upward, against gravity. Would it be correct to say that the ensuing orbit is the same as in (*a*) *if* referred to a frame moving upward with the velocity $\mathbf{u}_o$? If not, why not?

5.16 (*a*) Suppose it possible to establish a uniform magnetic field $\mathsf{B}_o = \mathsf{B}_z$ in an $x > 0$ part of space and an exactly reversed field in $x < 0$. Then start a particle m, q with just such a velocity that it begins a circular arc in the xy plane that is centered on the boundary. Sketch the ensuing motion. How would it be altered if the starting arc were centered off the boundary by a distance less than its radius?

(*b*) Now suppose that the uniform fields in $x \gtrless 0$ have like directions parallel to the boundary but that one is stronger than the other. Again start a particle circling about a boundary point. Characterize the direction taken by the mean drift of the particle's gyrations and express the ratio $\bar{u}/u$ of the drift velocity to the particle's speed u in terms of the two field magnitudes. Compare the directions of drift for opposite charges on the particle.

5.17 Exercise 5.16 suggests how a charged particle will behave in a region of magnetostatic field having a gradient of the field *magnitude* in a direction perpendicular to nearly parallel field lines. Most often the interest is in gradients $\nabla\mathsf{B} \perp \mathbf{B}$ sufficiently gradual for the change in the field to be small ($\Delta\mathsf{B} \ll \mathsf{B}_o$) over a radius of gyration $r_o = mcu_\perp/q\mathsf{B}_o$ like (5.11). Then in the terms of the figure,

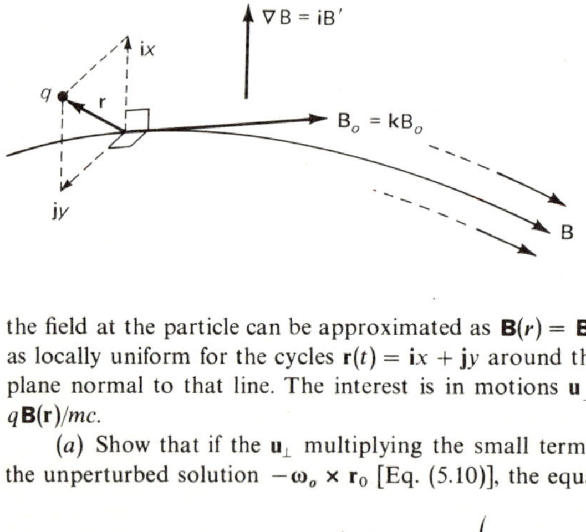

Exercise 5.17

the field at the particle can be approximated as $\mathbf{B}(r) = \mathbf{B}_o(1 + x\mathsf{B}'/\mathsf{B}_o)$ where B_0 and B' can be treated as locally uniform for the cycles $\mathbf{r}(t) = \mathbf{i}x + \mathbf{j}y$ around the guiding field line in the (generally moving) plane normal to that line. The interest is in motions $\mathbf{u}_\perp = \dot{\mathbf{r}}(t)$ yielded by (5.9) with ω_o replaced by $q\mathbf{B}(\mathbf{r})/mc$.

(*a*) Show that if the $\mathbf{u}_\perp$ multiplying the small term proportional to $x\mathsf{B}'/\mathsf{B}_o$ is approximated by the unperturbed solution $-\boldsymbol{\omega}_o \times \mathbf{r}_0$ [Eq. (5.10)], the equation of motion can be written as

$$\dot{\mathbf{u}}_\perp \approx -\boldsymbol{\omega}_o \times \left(\mathbf{u}_\perp - \boldsymbol{\omega}_o \times \frac{\langle \mathbf{r}_o x_o \rangle \mathsf{B}'}{\mathsf{B}_o}\right)$$

where it is appropriate to use the average over a cycle: $\langle \mathbf{r}_o x_o \rangle = \mathbf{i}r_o^2/2$.

(*b*) By comparison with (5.16) and (5.17), show that the cycles have an average drift velocity $\bar{\mathbf{u}} = (qr_o^2/2mc)[(\mathbf{B} \times \nabla)\mathsf{B}]/\mathsf{B}$, which reverses with the sign of the charge.

(*c*) Show that the results (*b*) agree with those of Exercise 5.16(*b*) if $|\nabla\mathsf{B}| \equiv \Delta\mathsf{B}/\Delta x$ is used, with $\Delta\mathsf{B}$ equal to the difference of the two given fields and $\Delta x = \pi r_o/2$, where r_o is the gyration radius in the average of the two fields.

5.18 Reliance on magnetic mirrors, like that of Fig. 5.2, at each end of a stretch of field lines to confine ions spiraling about the field lines can be avoided by making the field lines endless, through forming them into a toroidal sheaf of large circles. This can be done within a ring solenoid like that mentioned in Exercise 4.22. Thus consider the motions of a charged particle spiraling about a field line $\mathsf{B} = \mathsf{B}_\varphi$ at some large radius $s = R$, this being a distance from the axis of the toroidal ring, which can also serve as the z axis for cylindrical coordinates s, φ, z of the particle.

(*a*) Set up the equation of motion for $\ddot{s}(t)$ of q, m in $\mathsf{B} = \mathsf{B}_\varphi$.

(*b*) For tight spiraling about a field line at $s = R$, the approximations $s(t) \approx R(\dot{s},\ \ddot{s} \approx 0)$ and $u_\parallel \approx R\dot{\varphi}$ can be adopted for the particle orbit. Show that the existence of a drift $\dot{z} \neq 0$ is necessitated by the finite curvature ($R < \infty$) of the field lines.

(*c*) Show that this can be understood as an $\mathbf{E} \times \mathbf{B}$ drift (5.17), with the centrifugal force $\mathbf{e}_\varphi m u_\parallel^2/R$ replacing the electric force $q\mathbf{E}$ (as $m\mathbf{g}$ did in Exercise 5.15). (The magnetic mirror of Fig. 5.2 will also allow such curvature drifts of spiralings about off-axis field lines.)

5.19 Molding a long solenoid into a ring, as in Exercise 5.18, necessarily introduces a small gradient in the field $\mathsf{B} = \mathsf{B}_\varphi$.

(*a*) Show that $s\mathsf{B}(s)$ must be constant for a field $\mathsf{B} = \mathsf{B}_\varphi$, and so for a particle spiraling about a ring of a large radius $s = R$ the field can be approximated as $\mathsf{B}(s) \approx \mathsf{B}_o(1 - x/R)$ if $x \equiv s - R \ll R$ and $\mathsf{B}_o = \mathsf{B}(R)$.

(*b*) The field gradient $\mathsf{B}'(s) \approx -\mathsf{B}_o/R$ will cause a gradient drift of the particle cycles like that of Exercise 5.17. Evaluate the gradient-drift velocity for this case.

(*c*) Show that the resultant of the curvature (Exercise 5.18) and gradient drifts can be represented by

$$\bar{\mathbf{u}} = \frac{\mathbf{e}_z(u_\parallel^2 + \frac{1}{2}u_\perp^2)}{\omega R} \qquad \text{where } \omega = \frac{q\mathsf{B}}{mc}$$

[To avoid the losses to the walls of the solenoid caused by these first-order drifts, the solenoid ring is given a half-twist, making it a figure eight, in the Princeton stellarator device. Particles making circuits of the twisted field lines meet opposite signs of curvature ($\sim 1/R$) in each circuit.]

5.20 A *magnetron* is a device that has a hot filament (a thin cylinder of some radius a) stretched along a central (z) axis and ejecting thermoelectrons (m, $q = -e$) of negligible kinetic energy into a uniform magnetic field $\mathbf{B}_o$ parallel to the filament. The electrons are accelerated to an enclosing cylindrical plate of some radius $b \gg a$ by maintaining the plate at a sufficient potential difference ϕ_o from the filament.

(*a*) Show that in its motion, an electron will have a conserved angular momentum relative to a frame rotating with the Larmor frequency $\omega_L = e\mathsf{B}_o/2mc$ about the filament as axis.

(*b*) Show that if ϕ_o is decreased sufficiently, the current of electrons to the plate can be stopped and that the critical potential difference needed to do this is

$$\phi_o = \frac{e\mathsf{B}_o^2}{8mc^2} b^2 \left(1 - \frac{a^2}{b^2}\right)$$

5.21 For effects of the earth's magnetic field on charged particles, approaching from far reaches of space, it can be treated as an ideal dipole field centered on the earth's core and oriented so that the north magnetic pole is near the south geographical pole.

(*a*) Show that the moment can be estimated as $\mu \approx \mathsf{B}_o R^3$, where B_o is a mean value ($\approx \frac{1}{3}$ gauss) at the earth's surface ($r = R$) in latitudes near the equator.

(*b*) Can you suggest why the auroral phenomena, supposed due to atmospheric ionization by streams of electrons from the hot sun, should be largely confined to the polar regions?

(*c*) The primary cosmic rays consist largely of protons. The first hint that they consist predominantly of positive charges came from observations that more arrive from the west. Do the expected orbits with various velocities suggest the basis for that conclusion?

(*d*) The explanation of (*b*) is related to an observed latitude effect on cosmic rays: greater intensities are found in both polar latitudes than near the equator. Where do you expect the mean energy of the arriving particles to be greater (after corrections for atmospheric interceptions)?

(*e*) Consider protons (M, $q = +e$) that come from $r \approx \infty$ with velocities $\mathbf{u}$ lying in the equatorial plane and directed at the earth's center. Show that those with speeds $u < e\mathsf{B}_o R/Mc$ will miss the earth's surface (unless intercepted by the atmosphere).

5.22 The earth's field is found to trap charged particles in regions encircling the earth's axis, far beyond any appreciable remnants of atmosphere. The regions are known as *van Allen belts* and extend from the equatorial plane toward both poles.

(*a*) Explain how the field as characterized in Exercise 5.21 can act as a magnetic bottle for the confinement of charges spiraling about field lines.

(*b*) Consider a charged particle that is being guided by a field line passing through the equatorial plane at a distance $a \gg R$ from the earth (see Exercise A.5). Find the field magnitudes $\mathsf{B}(\lambda)$ along the line as a function of the latitude angle ($\lambda = |\vartheta - \pi/2|$ if ϑ is the angle with the earth's axis; thus $\lambda = 0$ at the equator).

(*c*) $\mathsf{B}(\lambda)$ plays the role of $\mathsf{B}_o(z)$ in (5.47) if curvature drifts are ignored. Use the approximation to show that a particle having a gyration radius r_o in the equatorial plane must have a total kinetic energy proportional to

$$u^2 = \frac{\omega^2 r_o^2 (4 - 3\cos^2\lambda)^{1/2}}{\cos^6\lambda}$$

to reach a maximum latitude λ in its bottled oscillations about the equatorial plane. Here $\omega = q\mathsf{B}(0)/mc$, where $\mathsf{B}(0)$ is the field magnitude on the guiding line at the equatorial plane (thus, ωr_o is the velocity of gyration $u_\perp$ there).

CHAPTER
SIX

DESCRIBING THE GENERAL ELECTROMAGNETIC FIELD

Identifications of field-energy, momentum, and angular-momentum distributions and their flows · The transmission of energy to currents · The physical interpretation of the vector potential and its measurability relative to gauge · The Maxwell stress tensor and the transmission of force by fields · The field angular-momentum tensor and the transmission of torque · Lorentz potentials and gauge invariance · Monochromatic representatives of fields

Nonstatic electromagnetic fields are to be described by vector distributions $\mathbf{E}(\mathbf{r}, t)$ and $\mathbf{B}(\mathbf{r}, t)$, as defined by the complete set of Maxwell equations (1.40). Directly from them, some of the results obtained for the static fields can at once be generalized to nonstatic situations, and that will be the first concern of this chapter.

6.1 FIELD ENERGY AND ITS FLUX

The conservation of energy, momentum, and angular momentum constitutes principles that are among the most generally† tenable throughout all physical description. The quantities in question were first identified for matter in motion, and representing their conservation requires considering exchanges between various forms they can take. Then the shares contributed by electromagnetic fields can be identified through their interactions with matter. Identifications of energy distributed over electrostatic and magnetostatic fields, in static equilibrium with their sources, have already been made in Secs. 2.4 and 4.4. Energy that can be attributed to a generally nonstatic electromagnetic field will be identified in this section, and momentum-conservation principles will be developed in succeeding sections.

The Energy of an Isolated Field

Electromagnetic forces perform work during displacements of charges; these motions are represented by current densities $\mathbf{j}(\mathbf{r}, t)$. Moreover, only the electrical part $\mathbf{E}(\mathbf{r}, t)$ of the field does the work directly, since the magnetic part $\mathbf{j} \times \mathbf{B}/c$ of

† A basis for this great generality, residing in the invariances (symmetries) that any description in space-time must be expected to have, will be discussed in Chap. 13.

the force density (4.64) has no component along the direction of the net charge displacement at a point. Thus the expression $\mathbf{E} \cdot \mathbf{j}$ found in (4.81) represents all the work being done by the field at a point per unit time and per unit volume.

The work elements $\mathbf{j} \cdot \mathbf{E}\, dV\, dt$ may be negative at some points, indicating energy transfers to the field there—from, rather than to, the charged matter. In the entire situation under consideration, $\mathbf{j}(\mathbf{r}, t)$ and $\rho(\mathbf{r}, t)$ are supposed to represent all the sources responsible for the field present, in accordance with the Maxwell equations (1.40). Thus $(c/4\pi)(\nabla \times \mathbf{B} - \mathring{\mathbf{E}}/c)$ can be substituted for $\mathbf{j}$ in the evaluation of the entire power transfer to the field

$$-\oint \mathbf{j} \cdot \mathbf{E}\, dV = (4\pi)^{-1} \oint dV \mathbf{E} \cdot [\mathring{\mathbf{E}} - c(\nabla \times \mathbf{B})]$$

with the volume integration $\oint dV$ to cover the space of the entire field, regarded as an entity distributed over the space. The term $\mathbf{E} \cdot (\nabla \times \mathbf{B})$ can be reevaluated as in

$$\begin{aligned} \nabla \cdot (\mathbf{E} \times \mathbf{B}) &= (\nabla \times \mathbf{E}) \cdot \mathbf{B} = -(\nabla \times \mathbf{B}) \cdot \mathbf{E} \\ &= \mathbf{B} \cdot (\nabla \times \mathbf{E}) - \mathbf{E} \cdot (\nabla \times \mathbf{B}) \end{aligned} \tag{6.1}$$

where the operation by ∇ applies to all the field factors that follow it in each term of the expressions. Since $\nabla \times \mathbf{E} = -\mathring{\mathbf{B}}/c$ according to the Maxwell equations,

$$\begin{aligned} -\oint \mathbf{j} \cdot \mathbf{E}\, dV &= (4\pi)^{-1} \oint dV [\mathbf{E} \cdot \mathring{\mathbf{E}} + \mathbf{B} \cdot \mathring{\mathbf{B}} + c\nabla \cdot (\mathbf{E} \times \mathbf{B})] \\ &= \frac{d}{dt} \oint dV \frac{\mathsf{E}^2 + \mathsf{B}^2}{8\pi} + \frac{c}{4\pi} \oint d\mathbf{S} \cdot (\mathbf{E} \times \mathbf{B}) \end{aligned} \tag{6.2}$$

With the integration to be extended over the entire field, $\mathbf{E} \times \mathbf{B}$ can be taken to vanish on the remote surface outside the field, and the remainder describes a growth of field energy in the density

$$w(\mathbf{r}, t) = \frac{\mathsf{E}^2 + \mathsf{B}^2}{8\pi} \tag{6.3}$$

whenever positive work, at the rate $-\oint \mathbf{j} \cdot \mathbf{E}\, dV$, is done by the source motions. Such an expression for the energy density has already been obtained in (4.99) for the special case of static fields.

As in (2.60) and (4.97), the symbol

$$W(t) \equiv \int_V dV(\mathbf{r})\, w(\mathbf{r}, t) \tag{6.4}$$

will be used to represent the field energy in any given volume V. It can be time-independent even for a nonstatic field† when the field is an *isolated* one,

† Explicit representations of such will be seen in (7.11).

enclosable by a surface on which it vanishes and having no charge-bearing matter within the volume it occupies. Then $dW/dt = 0$ in (6.2), and a conservation of the isolated field energy to itself is described.

The Poynting Vector

Suppose now that the integrations in (6.2) are restricted to just some part of the field, contained in a finite volume V. Then

$$\frac{dW}{dt} = -\int_V dV\,\mathbf{j}\cdot\mathbf{E} - \oint d\mathbf{S}\cdot\frac{c}{4\pi}(\mathbf{E}\times\mathbf{B}) \tag{6.5}$$

with the surface integral no longer vanishing, in general. The expression now describes the fact that the field energy $W(t)$ contained in V can diminish whenever the field is doing net positive work (at the rate $\mathbf{j}\cdot\mathbf{E}$ per unit volume) on charges in V and *also* if there is a net efflux of field energy out through the surface enclosing V. The flux-density expression

$$\mathbf{N} \equiv \frac{c}{4\pi}(\mathbf{E}\times\mathbf{B}) \tag{6.6}$$

is known as the *Poynting vector*. It should plainly be interpreted as an amount of field energy flowing at the point of its evaluation per unit time and per unit of cross-sectional area transverse to the direction of $\mathbf{N}$.

By reexpressing the flux integral in (6.5) as the volume integral of a divergence the result can be written

$$\int_V dV\left(\frac{\partial w}{\partial t} + \nabla\cdot\mathbf{N} + \mathbf{j}\cdot\mathbf{E}\right) = 0$$

for any volume V. The usual reduction to a point volume produces

$$\frac{\partial w}{\partial t} + \nabla\cdot\mathbf{N} = -\mathbf{j}\cdot\mathbf{E} \tag{6.7}$$

describing conservation of the field energy in its flux, except at points where $\mathbf{j} \neq 0$ and work is being expended on charges. The mode of expression is comparable to that used for charge conservation in (1.14).

The expression for the Poynting vector shows that propagation of electromagnetic field energy requires a coexistence of both electric and magnetic fields. That arises naturally in nonstatic situations because time variations in either field generate the other, according to the Maxwell equations.

Static electric and magnetic fields can also coexist in a space, and then there must be a *steady* flow of field energy at every point where the two fields are not parallel to each other. For illustration, consider the situation indicated in Fig. 6.1. Two plane parallel conducting plates, a distance Δl apart, are maintained at a potential difference $\Delta\phi$, and so the space between them is occupied by a uniform electric field $\mathsf{E} = |-\nabla\phi| = \Delta\phi/\Delta l$. Since the two plates are connected by a uniformly conducting cylinder of radius a, a uniformly distributed current $I = \Delta\phi/R = \mathsf{E}\,\Delta l/R$ can be expected to flow in the cylinder (see page 595) if R is

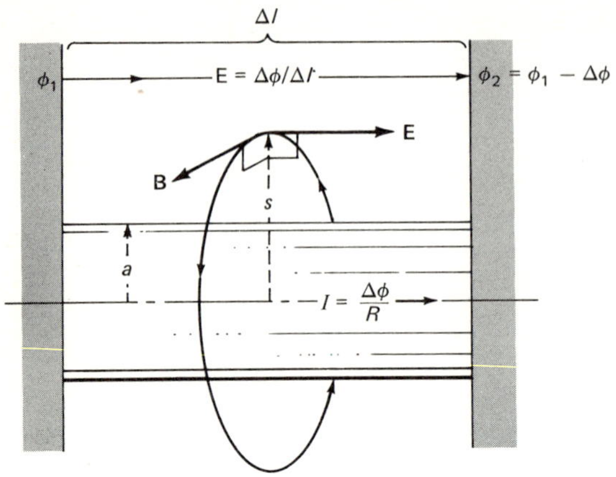

Figure 6.1

the resistance of the cylinder according to the Ohm finding (1.30). It is such a uniformity of response, made possible by the existence of the resistance, that permits the uniformity of the electric field under the steady potential difference to persist within the cylinder.

The current will be encircled by magnetic field lines in strength $\mathsf{B} = 2I/cs$ [Eq. (1.20)] at distance s from the cylinder axis and outside the cylinder ($s > a$). The crossed electric and magnetic fields are plainly transmitting the energy flux

$$N = \frac{c}{4\pi}\,\mathsf{EB} = \frac{\mathsf{E}I}{2\pi s} \tag{6.8}$$

directed radially inward, through each unit of area on the cylindrical surface at radius s. The total flux impinging on the cylindrical conductor itself is

$$(2\pi a\ \Delta l)N = I\mathsf{E}\ \Delta l = I^2 R \tag{6.9}$$

and this supplies just enough power to offset such ohmic losses as were alluded to in connection with (4.82). The middle expression in (6.9) is just

$$\oint dV\,\mathbf{j}\cdot\mathbf{E} = (\pi a^2\ \Delta l)\,\frac{I}{\pi a^2}\,\mathsf{E} \tag{6.10}$$

of (6.5), the rate of energy loss by the field to the charged matter. Of course, there is no diminution of the field energy that is present, because an equilibrium is being maintained by the agency supplying the steady potential difference $\Delta\phi = \mathsf{E}\ \Delta l = IR$.

6.2 FIELD MOMENTUM AND STRESS

The directed electromagnetic forces at a point impart momentum to any charged matter there, and the rate at which such mechanical momentum $\mathbf{p}$ is imparted to all the matter in some volume V is

$$\frac{d\mathbf{p}}{dt} = \mathbf{F} = \int_V dV\,\mathbf{f} = \int_V dV\left(\rho\mathbf{E} + \mathbf{j}\times\frac{\mathbf{B}}{c}\right) \tag{6.11}$$

according to the expression for the force density **f** in (4.64). A reaction $-\mathbf{F}$ on the field in the volume can be held to impart field momentum $\mathbf{P}(t)$ to it at a rate $d\mathbf{P}/dt = -\mathbf{F}$, as given by

$$-\int_V dV\,\mathbf{f} = -(4\pi)^{-1}\int_V dV\left\{\mathbf{E}(\nabla\cdot\mathbf{E}) + \left[(\nabla\times\mathbf{B}) - \frac{\mathring{\mathbf{E}}}{c}\right]\times\mathbf{B}\right\} \tag{6.12}$$

obtained by substitutions for ρ and **j** from the Maxwell equations. To be considered first is the rate of momentum transfer to an entire field. The final term in (6.12) suggests investigating the difference of the result for this from the expression

$$-\oint dV\,\mathbf{f} = \frac{d}{dt}\oint dV\,\frac{\mathbf{E}\times\mathbf{B}}{4\pi c} \tag{6.13}$$

The investigation, left to a later subsection, will show that the difference between (6.13) and the result of the corresponding integration in (6.12) can be expressed as the volume integral of a divergence. As usual, the Gauss theorem will make this equivalent to an integration over an enclosing surface, and what is meant by an entire field is one having all effects vanish on a sufficiently remote surface. In this way, (6.13) becomes the complete result for a field as a whole.

The result shows that the force of reaction on the field imparts to it a momentum that is distributed over the field in the density

$$\mathbf{g} = \frac{\mathbf{E}\times\mathbf{B}}{4\pi c} = \frac{\mathbf{N}}{c^2} \tag{6.14}$$

As indicated, this field-momentum density at a point differs from the *energy-flux* density (6.6) there only by a constant factor $1/c^2$. This is consistent with the dimensions of the two quantities since N is an energy density multiplied by a velocity and so has the same dimensionality as a momentum density multiplied by the square of a velocity like c^2.

The resultant field momentum contained within any finite volume V is now understood as

$$\mathbf{P}(t) \equiv \int_V dV\,\frac{\mathbf{E}\times\mathbf{B}}{4\pi c} \tag{6.15}$$

Its value for an entire field, with $\int_V dV \to \oint dV$, occurs in the reaction expression (6.13), and when the force **F** in this is put in terms of the mechanical momentum **p** as in (6.11),

$$\frac{d}{dt}(\mathbf{P} + \mathbf{p}) = 0 \tag{6.16}$$

is the result for the entire field, together with all the sources with which it interacts. Such a conservation of the total momentum of an isolated system is expected quite generally, and was here implemented through the identification of $d\mathbf{P}/dt$ as a reaction to $\mathbf{F} = d\mathbf{p}/dt$, in accordance with Newton's law of reactions.

The important conclusion to be emphasized is that momentum *must* be attributed to electromagnetic fields if there is to be a general conservation of momentum.

The Physical Meaning of the Vector Potential

The identification of (6.15) as electromagnetic field momentum makes it possible now to justify an assertion made in connection with (5.6) that $q\mathbf{A}(\mathbf{r}_q)/c$ represents field momentum available for transfer to motions of charge q at $\mathbf{r}_q$.

In the correspondent demonstration that $q\phi(\mathbf{r}_q)$ of (2.66) measures the field *energy* available at $\mathbf{r}_q$ from the entire field, the superposition $\mathbf{E} + \mathbf{E}_q$ was considered, describing the system composed of a static field with $\mathbf{E}(\mathbf{r}) = -\nabla\phi$ and a static point charge that contributes a Coulomb field $\mathbf{E}_q(\mathbf{r} - \mathbf{r}_q)$. It was then found that only the interaction part of the total field energy, i.e., that having the density $\mathbf{E}_q \cdot \mathbf{E}/4\pi$ and arising from interference between the superposed fields, goes into forming potential energy $q\phi$ for motions of q (see Exercise 3.2).

To have field *momentum* present together with a static charge, the applied field must have a magnetic component $\mathbf{B}(\mathbf{r}) = \nabla \times \mathbf{A}$, according to (6.15). The system will then contain a momentum density $\mathbf{g} = (\mathbf{E} + \mathbf{E}_q) \times \mathbf{B}/4\pi c$, and only the interference part of the total field momentum

$$\mathbf{P}_q(\mathbf{r}_q) = (4\pi c)^{-1} \oint dV(r)\, \mathbf{E}_q(\mathbf{r} - \mathbf{r}_q) \times \mathbf{B}(\mathbf{r}) \tag{6.17}$$

will take on different values when the position $\mathbf{r}_q$ of q is changed. The difference must be imparted to the mechanical momentum in the motion changing $\mathbf{r}_q$, at least whenever q and $\mathbf{B}(\mathbf{r})$ are isolated except from each other.

The integration in (6.17) leads to a simple result that is easiest to obtain after introducing the sources $\mathbf{j}(\mathbf{r}) = (c/4\pi)\nabla \times \mathbf{B}$, that a magnetostatic field must have, according to Ampère's law, and letting $\mathbf{E}_q = -\nabla\phi_q$ with $\phi_q = q/|\mathbf{r} - \mathbf{r}_q|$. Then

$$\mathbf{P}_q = \oint \frac{dV}{4\pi c}\, \mathbf{B} \times \nabla\phi_q = \oint \frac{dV}{4\pi c}\, \phi_q \nabla \times \mathbf{B} = \frac{q}{c} \oint \frac{dV\mathbf{j}}{c|\mathbf{r} - \mathbf{r}_q|}$$

The first and second integral expressions here may appear to differ in proportion to [see (A.15)]

$$\oint dV \nabla \times \phi_q \mathbf{B} \equiv \oint d\mathbf{S} \times \phi_q \mathbf{B}$$

but this must vanish on remote surfaces able to enclose an entire field if $\mathbf{P}_q$ is to represent field momentum made available by an entire field. The last integral for $\mathbf{P}_q$ can be recognized as just the vector potential (4.4) from which $\mathbf{B} = \nabla \times \mathbf{A}$ is derivable. Thus

$$\mathbf{P}_q(\mathbf{r}_q) = \frac{q\mathbf{A}(\mathbf{r}_q)}{c} \tag{6.18}$$

and the assertion in question is confirmed. It can be said that just as $q\phi$ is potential energy, so $q\mathbf{A}/c$ serves as potential momentum for motions of charge.

The physical interpretation of vector potentials† found here requires emphasis because it has been persistently asserted that **A** has *no* physical meaning. All the remarks about the scalar potential ϕ that attend (2.66) could be closely paralleled for **A**. The outcome is that potential descriptions, by ϕ and **A**, of electromagnetic fields are at least as meaningful as descriptions by **E** and **B**. Whereas **E**, **B** describe a field in terms of forces the field can exert on charged matter, ϕ and **A** describe the same field in terms of energies and momenta that the entire field makes available to the matter.

A situation in which it becomes crucial to recognize the physical effectiveness of having vector potentials present is one having a space in which $\mathbf{E} \equiv 0$ and $\mathbf{B} \equiv 0$ but $\mathbf{A} \neq 0$ as measured in some gauge. This happens outside an effectively infinite cylindrical solenoid, like that treated in Exercise 4.5, which is well known to carry a magnetic flux $\Phi = \pi a^2(4\pi nI/c)$ within its radius $s = a$. The magnitude of the azimuthally directed vector potential follows most directly from the equivalence of the flux expressions

$$\Phi \equiv \oint d\mathbf{S} \cdot \mathbf{B} = \oint d\mathbf{S} \cdot (\nabla \times \mathbf{A}) \equiv \oint d\mathbf{r} \cdot \mathbf{A} \tag{6.19}$$

[see also (4.77) or Exercise 4.21]. Since $A = A_\varphi(s)$ must be constant along any circle concentric with the solenoid, the *total* flux above can also be equated to $\Phi = 2\pi s A(s)$ for any $s > a$ outside the solenoid. Then

$$A(s > a) = \frac{2\pi a^2 nI}{cs} \tag{6.20}$$

consistent with $\mathsf{B}(s > a) = \partial(sA)/(s\,\partial s) \equiv 0$ for its curl.

According to the interpretation found above, the vector potential (6.20) makes azimuthally directed (circling) field momentum $q\mathbf{A}(\mathbf{r}_q)/c$ transferable to any charge q placed at a point $\mathbf{r}_q$ *outside* the field **B** of the solenoid. A complete transfer can be made detectable by putting a charged bead on a circular fiber concentric with the solenoid and along which the bead is free to slide, then letting the field momentum $q\mathbf{A}/c$ be reduced to zero by interrupting the solenoid current I that has been maintaining the static equilibrium field (6.20). The rate of decrease $\dot{I}(t) < 0$ of the current can be sufficiently slowed, as by the self-inductance in the solenoid (Exercise 4.24), for a quasi-static field equilibrium to be established for each moment. Then a relation like (6.20) holds between all $A(t \geq 0)$ and the corresponding $I(t \geq 0)$, and there is a rate of transfer of momentum from the field to the motion given by

$$m\dot{u} \equiv m\dot{u}_\varphi = -\frac{q}{c}\frac{\partial A}{\partial t} = -\frac{2\pi q a^2 n\dot{I}}{c^2 s} \tag{6.21}$$

† See Refs. 6a and 18. A physical significance of **A** in quantum-mechanical contexts was pointed out by Bohm and Aharonov (Ref. 5), who identified $q\mathbf{A}/ch$ as a detectable phase shift per unit path length in a De Broglie wave, now to be regarded as a *result* of exchanges between mechanical and field momenta. In a similar way, a physical basis is provided for requiring local gauge invariance, a principle from which electromagnetism, chromodynamics, and other boson fields can be derived (developments due to C. N. Yang, R. L. Mills, S. Weinberg, A. Salam, and others).

This is just the azimuthal component of the equation of motion (5.6) and actually expresses a *conservation* of the *canonical* momentum $p = p_\varphi = mu + qA/c$ (5.7) along the circular trajectory, on which the gradients on the right side of (5.6) vanish. Integrating (6.21) for a start from $u(t = 0) = 0$ and $I(0) \neq 0$ yields an ultimate observable momentum

$$mu(t \to \infty) = \frac{+\,2\pi q a^2 n I(0)}{c} = \frac{qA(0)}{c} \tag{6.22}$$

replacing the initial *field* momentum and equal to it.

Kinetic-momentum gains obtained in similar ways are well known; they form the basis of the betatron principle (Exercise 5.10). More usually they are calculated from the newtonian form (5.1) of the equations of motion, as in Exercise 5.9, again using the quasi-static approximation, without recognizing the role of the vector potential as the transmitter of field momentum to the site of the charge.

The preceding example demonstrates that field momentum can be made locally available at points where $\mathbf{g} = \mathbf{E} \times \mathbf{B}/4\pi c = 0$. It is only necessary that a vector potential extend to those points, and then charge cannot be introduced there without accumulating field momentum $q\mathbf{A}/c$, just as putting a charge q into a field $\mathbf{E}$ calls forth a force $q\mathbf{E}$.

Although a charge need not be put inside a field $\mathbf{B}$ for its motion to be influenced by it, reference to the integral (6.18) for $\mathbf{P}_q = q\mathbf{A}/c$ shows that it *is* essential for the (Coulomb) self-field of the charge to overlap on the space of $\mathbf{B} \neq 0$. This indicates that the self-field attached to the charge and moving with it not only contributes to the mass of the charge, as discussed in connection with (4.101), but also allows that integral entity to begin interaction with fields $\mathbf{E}$, $\mathbf{B}$ at points not reached by the charge itself, through interferences $\mathbf{E}_q \cdot \mathbf{E}/4\pi$ and $\mathbf{E}_q \times \mathbf{B}/4\pi c$ of the self-field $\mathbf{E}_q$ with $\mathbf{E}$, $\mathbf{B}$. It is the latter effects that are representable by potential $(\phi, \mathbf{A})$ values at the location of the charge itself.

It should still be affirmed that $w = (E^2 + B^2)/8\pi$ and $\mathbf{g} = \mathbf{E} \times \mathbf{B}/4\pi c$ remain the most economical expressions for *isolated* systems, including ones with charged matter present if only $\mathbf{E}$ and $\mathbf{B}$ include the self-fields of the charges. It is only when these are analyzed into interacting (interfering) parts that representation by ϕ and $\mathbf{A}$ becomes important, as was already evident in the interaction-energy expressions (2.46) and (5.34). The importance stems from the fact that isolated systems do not become observable until the isolation is disturbed by interactions.

The Flux of Field Momenta

The assertions following (6.13), used to justify accepting (6.15) as the momentum in an entire field, are still to be confirmed. For this, consider what (6.12) yields, instead, when the rate of momentum transfer to the field in some *finite* volume V is evaluated from it:

$$\frac{d\mathbf{P}}{dt} = \frac{d}{dt}\int_V dV \frac{\mathbf{E} \times \mathbf{B}}{4\pi c} = -\int dV\,\mathbf{f} + \int \frac{dV}{4\pi}\left[\mathbf{E} \times \frac{\mathring{\mathbf{B}}}{c} + \mathbf{E}(\nabla \cdot \mathbf{E}) - \mathbf{B} \times (\nabla \times \mathbf{B})\right] \tag{6.23}$$

The time derivative $\mathring{\mathbf{B}}$ can be replaced by $-c(\nabla \times \mathbf{E})$ according to a Maxwell equation, and then the last integrand can be expressed as the sum of the two contributions

$$\mathbf{E}(\nabla \cdot \mathbf{E}) - \mathbf{E} \times (\nabla \times \mathbf{E})$$

by the electrical part of the field and

$$\mathbf{B}(\nabla \cdot \mathbf{B}) - \mathbf{B} \times (\nabla \times \mathbf{B})$$

by the magnetic part. An extra term, $\mathbf{B}(\nabla \cdot \mathbf{B}) = 0$, vanishing because $\mathbf{B}$ is divergenceless, has been inserted here, to make evident a formal symmetry between the two contributions. An ith cartesian component of the electric contribution can be written

$$\mathsf{E}_i(\nabla \cdot \mathbf{E}) - [\textstyle\sum_j \mathsf{E}_j \nabla_i \mathsf{E}_j - (\mathbf{E} \cdot \nabla)\mathsf{E}_i]$$

after use of a triple vector identity like (A.53), with due care for the order in which the factors must appear when the gradient operator ∇ is one of them. The result is equivalent to

$$\begin{aligned} \textstyle\sum_j(\mathsf{E}_i \nabla_j \mathsf{E}_j - \tfrac{1}{2}\nabla_i \mathsf{E}_j^2 + \mathsf{E}_j \nabla_j \mathsf{E}_i) &= [\textstyle\sum_j \nabla_j(\mathsf{E}_i \mathsf{E}_j)] - \tfrac{1}{2}\nabla_i \mathsf{E}^2 \\ &= \textstyle\sum_j \nabla_j(\mathsf{E}_i \mathsf{E}_j - \tfrac{1}{2}\delta_{ij} \mathsf{E}^2) \end{aligned}$$

since $\mathsf{E}^2 = \mathbf{E} \cdot \mathbf{E} = \sum_j \mathsf{E}_j^2$ and $\sum_j \delta_{ij} \nabla_j = \nabla_i$. The quantity in the last parentheses is the i, j component of a tensor, $\mathbf{EE} - \frac{1}{2}(\mathbf{1}\mathsf{E}^2)$. The formal symmetry noted above makes it plain that the magnetic contribution can be similarly expressed, in terms of a tensor $\mathbf{BB} - \frac{1}{2}(\mathbf{1}\mathsf{B}^2)$. It will therefore be useful for expressing results of the relation (6.23) to define the tensor

$$\mathbf{T} = \frac{\mathbf{EE} + \mathbf{BB} - \frac{1}{2}[\mathbf{1}(\mathsf{E}^2 + \mathsf{B}^2)]}{4\pi} = \frac{\mathbf{EE} + \mathbf{BB}}{4\pi} - \mathbf{1}w \tag{6.24a}$$

having nine cartesian components like

$$T_{ij} = (4\pi)^{-1}[\mathsf{E}_i \mathsf{E}_j + \mathsf{B}_i \mathsf{B}_j - \tfrac{1}{2}\delta_{ij}(\mathsf{E}^2 + \mathsf{B}^2)] \tag{6.24b}$$

The tensor is plainly a symmetrical one, with $T_{ji} = T_{ij}$, and has the trace $\sum_j T_{jj} = -w$, just the negative of the energy density (6.3). The tensor is also symmetric in the interchange of $\mathbf{E}$ and $\mathbf{B}$ and is separable into a sum of contributions from each field individually, $\mathbf{T} = \mathbf{T}(\mathbf{E}) + \mathbf{T}(\mathbf{B})$.

The electric and magnetic contributions to the last integrand of (6.23) are together proportional to the sum of spatial derivatives

$$\textstyle\sum_j \nabla_j T_{ji} \equiv (\nabla \cdot \mathbf{T})_i$$

each a component of a vector quantity $\nabla \cdot \mathbf{T}$, called the divergence of the tensor. Its volume integrals are subject to a Gauss theorem as in (A.15), with an outcome for the relation (6.23):

$$\frac{d\mathbf{P}}{dt} = -\mathbf{F} + \oint d\mathbf{S} \cdot \mathbf{T} \tag{6.25}$$

The flux integral here vanishes when the surface can be put outside the entire field, and this has already been used to obtain the version (6.13) of (6.25).

The result (6.25) makes it clear that $\mathbf{T} \cdot d\mathbf{S}$ should be interpreted as an efflux out of the volume V, through $d\mathbf{S}$, of field momentum having the direction of the vector $-\mathbf{T} \cdot d\mathbf{S}$, so that $-\mathbf{T}$ itself represents a current *density* of field momentum. Representation by a tensor is necessary because the direction of this flux density through a point may be quite independent of the direction possessed by the vector field momentum that is being transported. A tensor component $-T_{ij}$ is the ith component of the flux density of jth component of momentum and also the jth component of the flux density of ith component of momentum.

Reanalyzing (6.25) into the volume-element contributions out of which it has been formed leads to

$$\int_V dV\left(\frac{\partial \mathbf{g}}{\partial t} + \mathbf{f} - \nabla \cdot \mathbf{T}\right) = 0$$

and the usual reduction to a point yields

$$\frac{\partial \mathbf{g}}{\partial t} + \nabla \cdot (-\mathbf{T}) = -\mathbf{f} \tag{6.26}$$

This describes a conservation of the vector field momentum except at points where there are charges or currents to which momentum can be transferred by the electromagnetic forces. The expression is comparable to the continuity equation for the energy (6.7) but generalized to a vector equation concerned with a directed momentum.

The Maxwell Stress Tensor

Equation (6.25) for the rate of change of field momentum can also be put into the form

$$\frac{d}{dt}(\mathbf{P} + \mathbf{p}) = +\oint d\mathbf{S} \cdot \mathbf{T} = \oint dS\, \mathbf{n} \cdot \mathbf{T} \tag{6.27}$$

where $\mathbf{n} \equiv d\mathbf{S}/dS$ is a direction facing out of the volume V containing $\mathbf{P} + \mathbf{p}$, since the connection of $d\mathbf{S}$ to V was introduced through a Gauss theorem formulated on that basis. Expressed in this way, relation (6.27) gives its right side the role of a resultant force on *both* matter and field, contained in a region demarked by the surface chosen and behind it. The force is being transmitted through the surface in the form of a stress represented by the vector $\mathbf{n} \cdot \mathbf{T}$ evaluated on the surface elements, a force per unit area like the one considered in† (B.19). It is such an interpretation that led to the name Maxwell stress tensor for $\mathbf{T}$. This is not inconsistent with viewing $-\mathbf{T}$ as a current density of field momentum, the interpretation emphasized in the preceding subsection, since any force can be regarded as a rate of momentum transfer.

† There the vector $\mathbf{n} \cdot \mathbf{T}$ is denoted $\mathbf{T}$.

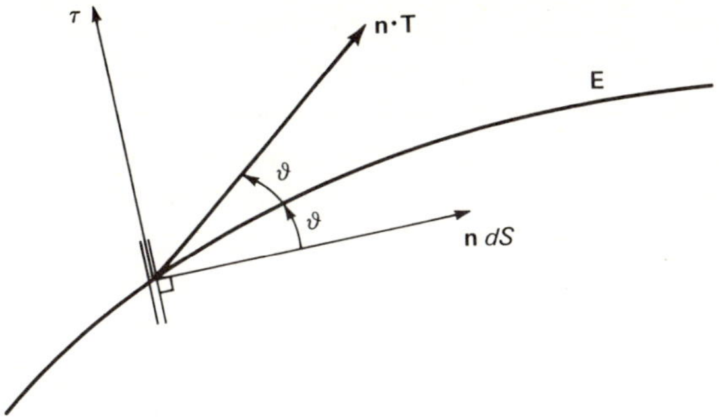

Figure 6.2

The separability of the electric and magnetic contributions to the stress tensor makes it possible to discuss each separately. Moreover, the symmetry in **E** and **B** makes it possible to take over any conclusions about stresses in an electric field as conclusions about a magnetic field that has the same field-line representation, with **B** lines merely replacing the **E** lines. When both are present, their effects are simply additive: $\mathbf{T} = \mathbf{T}(\mathbf{E}) + \mathbf{T}(\mathbf{B})$. Consider therefore the stress $\mathbf{n} \cdot \mathbf{T}(\mathbf{E})$ being transmitted through an arbitrarily directed element of geometric surface by just the electric field, as indicated in Fig. 6.2. The surface element chosen for consideration faces into some direction **n** that generally makes some angle ϑ with the field line passing through the element. Then the direction of the field line can be represented as in

$$\mathbf{E} = \mathsf{E}(\mathbf{n} \cos \vartheta + \boldsymbol{\tau} \sin \vartheta)$$

where $\boldsymbol{\tau}$ is a unit vector lying in the surface element and coplanar with **E** and **n**. The magnitude E can be taken either positive or negative, since the *sense* of the **E** line makes no difference to the stress tensor, as is immediately evident. The consequent stress vector can now be expressed as

$$\begin{aligned}\mathbf{n} \cdot \mathbf{T}(\mathbf{E}) &= (8\pi)^{-1}[2(\mathbf{n} \cdot \mathbf{E})\mathbf{E} - \mathbf{n}\mathsf{E}^2] \\ &= \frac{\mathsf{E}^2}{8\pi}[2 \cos \vartheta\, (\mathbf{n} \cos \vartheta + \boldsymbol{\tau} \sin \vartheta) - \mathbf{n}] \\ &= \frac{\mathsf{E}^2}{8\pi}(\mathbf{n} \cos 2\vartheta + \boldsymbol{\tau} \sin 2\vartheta) \qquad (6.28)\end{aligned}$$

As the figure should make clear, this vector also makes an angle ϑ with the field line but to the side opposite that of the vector **n**. It always has the absolute magnitude $|\mathbf{n} \cdot \mathbf{T}(\mathbf{E})| = \mathsf{E}^2/8\pi$.

Consider first the stress being transmitted into a region behind a surface element transverse to the field line, so that $\vartheta = 0$. (The surface element is then part of an equipotential surface in electrostatic situations.) The corresponding stress vector $\mathbf{n} \cdot \mathbf{T} = +\mathbf{n}(\mathsf{E}^2/8\pi)$ describes a pulling tension, or attraction, irrespective of the sense of the field line. Such situations have already been met in Sec. B.6, where an electrostatic field incident on the (equipotential) surface of a conductor was found to exert a pull, of just the magnitude $E^2/8\pi$ per unit area, on surface charges of either sign, induced by the field and thereby consistent with the sense of the incident field. In those situations the force was being transmitted to charged matter immediately behind the surface; there is no field inside a conductor to carry field momentum. When instead the surface element considered is placed somewhere in the field where there is no charged matter, it is force along a continuous field line that is being transmitted. Such a tension along a line is characteristic of a string that is being stretched under a tension applied to its ends. For this reason, the field lines themselves are sometimes said to be stretched under tension. Faraday, who had no formal training in mathematical description, formed intuitive pictures of fields with the help of such notions. Maxwell introduced the stress tensor just to formulate Faraday's intuitions.

Another aspect of Faraday's intuitive pictures of fields was a mutual repulsion between neighboring field lines, tending to disperse them as far as is consistent with the tension in each. For this, surfaces of separation parallel to the field lines should be considered, as represented by $\vartheta = \frac{1}{2}\pi$ in (6.28). The corresponding stress vector $\mathbf{n} \cdot \mathbf{T}(\mathbf{E}) = -\mathbf{n}(E^2/8\pi)$ does describe repulsion. Such mutual repulsions between contiguous regions are characteristic of fluids under pressure, and the distributed stresses are called pressures. It is then unsurprising that when dense ionized plasmas are treated as continuous fluids, their behavior in magnetic fields becomes deducible from a hydrodynamic formalism (sometimes called *magnetohydrodynamics*) in which $|\mathbf{n} \cdot \mathbf{T}(\mathbf{B})| = B^2/8\pi$ is additive to the purely mechanical pressures in the system.

Finally, whenever it becomes significant† to form a surface of separation on which electric or magnetic field lines are incident at $\vartheta = \pi/4 = 45°$, a purely shearing stress is transmitted through it. In such a case, $\mathbf{n} \cdot \mathbf{T}(\mathbf{E}) = \tau(E^2/8\pi)$ is parallel to the surface. The direction of the shearing force makes the acute (45°) angle with the incident field line, regardless of the sense this has.

Transmissions of Force in Electrostatic Fields

Perhaps the most important implication of the stress-tensor formulation for general electromagnetic theory is that the resultant forces on charges and currents are not actions at a distance, as the Coulomb law might have seemed to imply, but are transmitted through interactions of *contiguous* volume elements of the intervening field, propagated in the form of field momenta. This will now be investigated for several simple examples of already familiar static electric fields. Some of the results may seem only academic, but the importance of their implications makes them worth attention.

† Such situations can arise at the surfaces of separation of different dielectrics. However, for reasons like those discussed in connection with the expression $w = \mathbf{E} \cdot \mathbf{D}/8\pi$ for the energy density (D.30) in a dielectric, the appropriate stress tensor must have one of its two factors $\mathbf{E}$ replaced by $\mathbf{D}$. Similarly, one of the factors $\mathbf{B}$ in the magnetic contribution must be replaced by $\mathbf{H}$ in magnetically permeable media. See Exercise E.10 and Eq. (F.7).

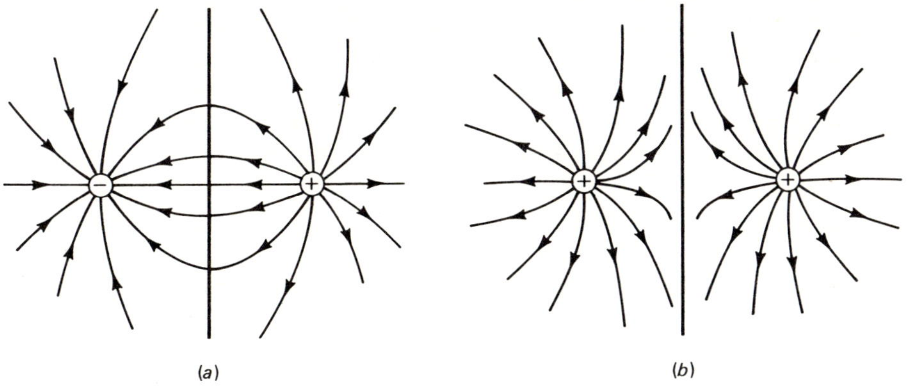

Figure 6.3

A comparison of the two situations indicated in Fig. 6.3 is instructive. In Fig. 6.3*a* field lines under tension are transmitting the force between two equal and opposite charges. The lines to the right of the median plane are just the same as those found in connection with Fig. B.1, representing the field due to a point charge and its image in a conducting plane. The tension they transmit, $\mathbf{E}^2/8\pi$ in magnitude according to (6.28), has already been found to integrate to just the Coulomb attraction, in the discussions of Fig. B.1. When the two charges have the same sign, as in Fig. 6.3*b*, the lines at the median plane are parallel to it and so transmit pressure or repulsion. The magnitude of the pressure, $\mathbf{E}^2/8\pi$ again, is readily found to integrate to the expected Coulomb repulsion (Exercise 6.5).

Stresses of variable direction on a surface are involved in a simple case of fundamental interest, showing how electrostatic force is transmitted through successive surfaces between the sources of a uniform field $\mathbf{E}_o = \mathbf{k}\mathsf{E}_o$ and a static point charge q. Let the succession of surfaces be spheres centered on q, with progressively smaller radii r. The stress tensor $\mathbf{T}(\mathbf{E})$ describing the transmissions must be evaluated for the superposition of fields

$$\mathbf{E} = \mathbf{k}\mathsf{E}_o + \frac{\hat{\mathbf{r}}q}{r^2} \tag{6.29}$$

since the same tensor can describe either the transmission to q or the reactions transmitted to the sources of $\mathbf{E}_o$.

The net force $\oint d\mathbf{S} \cdot \mathbf{T}$ that is transmitted into a sphere of radius r is to be calculated with $d\mathbf{S} = \hat{\mathbf{r}}r^2\, d\Omega$, and then the part $\mathbf{T}(\mathbf{E})$ of (6.24) yields for (6.29)

$$\hat{\mathbf{r}} \cdot \mathbf{T}(\mathbf{E}) = (8\pi)^{-1}\left[2\left(\hat{\mathbf{r}} \cdot \mathbf{E}_o + \frac{q}{r^2}\right)\left(\mathbf{E}_o + \frac{\hat{\mathbf{r}}q}{r^2}\right) - \hat{\mathbf{r}}\left(\mathsf{E}_o^2 + \frac{q^2}{r^4} + 2q\hat{\mathbf{r}} \cdot \frac{\mathbf{E}_o}{r^2}\right)\right] \tag{6.30a}$$

The contributions proportional to one factor $\hat{\mathbf{r}}$ average to zero on the entire sphere, and the two terms proportional to $\hat{\mathbf{r}}(\hat{\mathbf{r}} \cdot \mathbf{E}_o)$ cancel out. Then the resultant becomes

$$\oint d\mathbf{S} \cdot \mathbf{T} = r^2 \oint \frac{d\Omega}{4\pi} \frac{\mathbf{E}_o q}{r^2} = q\mathbf{E}_o \qquad (6.30b)$$

just as expected for $r \to 0$. Actually the result is independent of r and so represents the net force transmitted through all surfaces enclosing q. This also should have been expected since $\nabla \cdot \mathbf{T}(\mathbf{E}) = 0$ (6.26) in the electrostatic field everywhere outside the charge at $r = 0$. No difficulty has arisen from the fact that the charge density $\rho = q\delta(\mathbf{r})$ and its self-field $\hat{\mathbf{r}}q/r^2$ becomes infinite at $\mathbf{r} = 0$. The corresponding interaction of the charge, through its self-field, with itself is represented by the term proportional to $\hat{\mathbf{r}}q^2/r^4$ in (6.30*a*), and this averaged to zero because of its opposite signs on opposite sides of charge, essentially the basis of the Laplace argument quoted in connection with Fig. B.3. This outcome has confirmed that proper evaluations require leaving out the force of any charge element on itself, just as practiced in all the force evaluations so far.

The elimination of self-field effects just found does *not* mean that explosive forces to be expected from the mutual repulsions between elements of any finite charge do not exist. Suppose that the charge q is homogeneously distributed over a spherical volume of radius a, generating the field $\mathbf{E} = \hat{\mathbf{r}}qr/a^3$ (2.6) at any radius $r \leq a$. This by itself, independently of any external field that might be incident, leads to an inward transmission $\hat{\mathbf{r}} \cdot \mathbf{T} = \hat{\mathbf{r}}(qr/a^3)^2/8\pi$ through each point inside the sphere. It is true that this averages to zero like the self-field effects considered above, but it results in outward radial forces on *individual* charge elements. The net force on each volume element dV amounts to a difference of opposing tensions transmitted through the inner and outer surfaces of dV, given by $\oint d\mathbf{S} \cdot \mathbf{T} = dV(\nabla \cdot \mathbf{T})$, and leads to a force per unit volume,

$$\nabla \cdot \mathbf{T} = \mathbf{f} \to \rho\mathbf{E} \qquad (6.31)$$

as follows from (6.26) specialized to the electrostatic situation here. The result $\rho\mathbf{E} = \hat{\mathbf{r}}(3q^2r/4\pi a^6)$, at radius r, has the expected, explosive radial direction and tends to infinity for a point charge with $a \to 0$. These forces on the individual elements must be somehow counterbalanced—perhaps by some kind of nonelectromagnetic attractions—if the finite charge is to be held together.

Transmissions of Force in Magnetostatic Fields

The point-to-point transmission of force described by the stress tensor is particularly important to making the expression (4.54), for the force between current elements, consistent with a detailed conservation of momentum. Interpreted as action at a distance, that expression was found *not* to be consistent with an element-by-element equality of action and reaction, i.e., with equal and opposite momentum increments of the current elements themselves. However, the force is now viewed as a resultant of transmissions via chains of actions and reactions

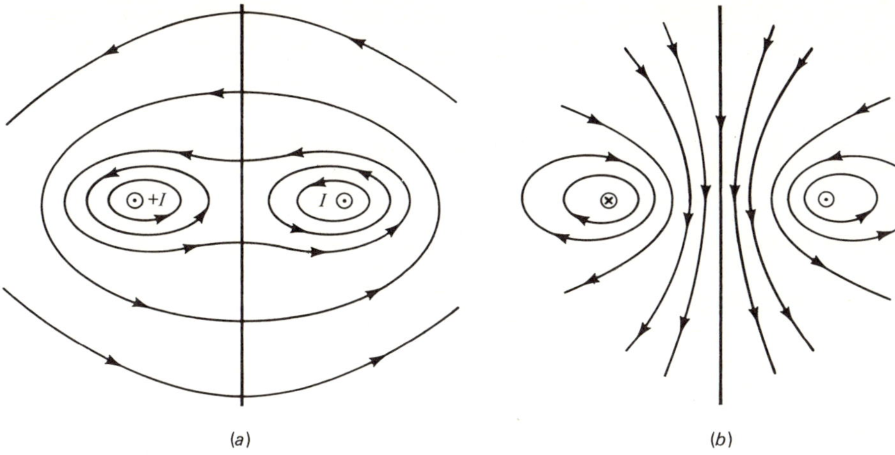

Figure 6.4

between contiguous elements in the field intervening between the currents. Field momentum plays a mediating role, and this was just so defined as to obtain reactions of the field equal and opposite to its actions on charged matter. Transmissions of force thus mediated will now be traced in some simple examples.

The magnetostatic counterparts of the electrostatic situations of Fig. 6.3 may be taken to be interactions of parallel and antiparallel currents like those indicated in Fig. 6.4. In Fig. 6.4*a*, with equal codirectional currents I, the field has the magnitude $\mathsf{B} = 4Iy/c(y^2 + \frac{1}{4}l^2)$ on the median plane at a distance y from its midline if l is the distance apart of the two currents. Its lines are normal to the plane, and so a tension attracting the currents to each other is being transmitted. It is easy to find (Exercise 6.5) that the integral $\int_{-\infty}^{\infty} dy(\mathsf{B}^2/8\pi)$ over unit-wide strips on the plane just equals the force per unit length of current already quoted in (4.63). For the antiparallel currents in Fig. 6.4*b*, the field lines are parallel to the median plane, and so a repelling pressure is being transmitted through it. The field is represented by $\mathsf{B} = 2Il/c(y^2 + \frac{1}{4}l^2)$ in this case, and the force integral yields the same absolute values as it does for the codirectional currents.

Considering the stresses in the case indicated by Fig. 6.1 calls attention to some interesting phenomena. Both the electric and magnetic fields in that situation are parallel to the surface of the conducting cylinder and so transmit pressures through it. However, the electrostatic pressures are distributed uniformly throughout the interior of the cylinder and so exert no net force, $\nabla \cdot \mathbf{T}(\mathbf{E}) = \rho\mathbf{E} = 0$, on any of its volume elements; they merely represent the equilibrium of mutual repulsions between the static field lines that also exist without the presence of the conducting cylinder. It is otherwise with the magnetostatic field, which weakens as the central axis of the cylinder is approached, resulting in pressure gradients that can move the charged matter of the current inward unless they are counterbalanced.

The decreasing magnitude of the magnetic field as a function of the radius $s < a$ inside the cylinder of uniform current is calculable from the Ampère law. This gives the magnetomotive force $2\pi s \mathsf{B}(s < a)$ encircling the fraction $(s/a)^2 I$ of the current as $4\pi s^2 I/ca^2$, and so $\mathsf{B}(s < a) = 2Is/ca^2$. Thus the inner magnetic pressures, $\mathsf{B}^2/8\pi = I^2 s^2/2\pi c^2 a^4$, decrease as $s \to 0$ in proportion to the square of the distance s. There is a net inward force on each unit of current-containing volume, given by $|\nabla \cdot \mathbf{T}(\mathbf{B})| = |\mathbf{j} \times \mathbf{B}/c| = 2I^2 s/\pi c^2 a^4$ and tending to constrict the current to the axis. Obtained in this way, the effect can be seen to be arising from the attractions that parallel filaments of current have for each other [see Exercise 4.18(*b*)]. It can also be understood as a variety of the Hall effect, mentioned in Sec. 5.2, due to a magnetic field arising from the current itself rather than to an externally imposed one. An $\mathbf{E} \times \mathbf{B}$ drift directed inward (for both signs of charge) is to be expected in view of the relative directions of the mutually perpendicular field lines as they exist here.

No actual constriction of the current in a metallic conductor is normally observable. Only the negative charges have the necessary mobility, since the positive ions form the rigid lattice of the solid metal. A constriction of the current of negative charges by itself is prevented by the strong electric fields that arise whenever charge separation is attempted (the ordinary, divergenceless conductor currents require no such charge separations; they always have moving charge passing stationary charge of equal density). This interpretation is borne out by the fact that very heavy currents *have* been observed to cause a constriction of the entire body of metal itself, a so-called *pinch phenomenon*.

Pinch effects occur more readily in plasma currents, where only interionic collisions or mechanical fluid pressures put any restriction on the mobility of the positive ions as they follow the attracting negative electrons inward. Such constrictions of the plasma can raise its temperature substantially, and attempts are being made to develop the effect as a means of igniting thermonuclear reactions (releases of nuclear-fusion power).

6.3 FIELD ANGULAR MOMENTUM

Any linear momentum at a point will have a moment about any momental center, off its line, that might be adopted. Thus a space V in which field momentum is distributed with a density $\mathbf{g}(\mathbf{r}, t)$ has a resultant field *angular* momentum

$$\mathbf{J}(t) \equiv \int_V dV\, \mathbf{r} \times \mathbf{g}(\mathbf{r}, t) \tag{6.32}$$

relative to the momental center taken as the origin for $\mathbf{r}$. Conditions under which it is conserved, whatever the momental center chosen, can be found from the continuity equation (6.26) for the linear field momentum:

$$\frac{d\mathbf{J}}{dt} = \int dV(\mathbf{r})\mathbf{r} \times \frac{\partial \mathbf{g}}{\partial t} = -\int dV\, \mathbf{r} \times \mathbf{f} + \int dV\, \mathbf{r} \times (\nabla \cdot \mathbf{T}) \tag{6.33}$$

Consider some cartesian component of the vector integrand in the last integral, such as

$$[\mathbf{r} \times (\nabla \cdot \mathbf{T})]_x = y \sum_i \nabla_i T_{iz} - z \sum_i \nabla_i T_{iy}$$

This is equal to

$$-\nabla_x(T_{xy} z - T_{xz} y) - \nabla_y(T_{yy} z - T_{yz} y) - \nabla_z(T_{zy} z - T_{zz} y)$$

since the indicated derivatives of the factors y and z in the fourth and fifth terms have results, T_{yz} and $-T_{zy} = -T_{yz}$ respectively, that cancel each other out. Thus

$$\mathbf{r} \times (\nabla \cdot \mathbf{T}) = -\nabla \cdot \mathbf{M} \qquad \text{with } \mathbf{M} \equiv \mathbf{T} \times \mathbf{r} \tag{6.34}$$

if it is understood that

$$M_{ix} \equiv T_{iy} z - T_{iz} y \qquad M_{iy} = T_{iz} x - T_{ix} z \qquad M_{iz} \equiv T_{ix} y - T_{iy} x$$

for the nine components of a (pseudo) tensor $\mathbf{M}$, thus defined. The remaining integral, $\int dV\, \mathbf{r} \times \mathbf{f}$, is plainly a resultant torque being imparted by the electromagnetic forces $\mathbf{f}(\mathbf{r}, t) \equiv \rho\mathbf{E} + \mathbf{j} \times \mathbf{B}/c$ to the charges and currents in the volume, and so

$$\int dV\, \mathbf{r} \times \mathbf{f} = \frac{d\mathbf{L}}{dt} \tag{6.35}$$

where $\mathbf{L}$ is the total *mechanical* angular momentum of the masses associated with the charges and currents. As a result, (6.33) can be written

$$\frac{d}{dt}(\mathbf{J} + \mathbf{L}) = -\int dV\, \nabla \cdot \mathbf{M} = -\oint d\mathbf{S} \cdot \mathbf{M} \tag{6.36}$$

a counterpart of the linear momentum relation (6.27). There is plainly a conservation of the total angular momentum $\mathbf{J} + \mathbf{L}$, distributed throughout the volume as both mechanical and field angular momentum, whenever the net flux represented on the right side of the equation vanishes. When this flux does not vanish, it measures angular momentum being transported by the field through the enclosing surface; $\mathbf{M} \equiv \mathbf{T} \times \mathbf{r}$ must be interpreted as a current density of field angular momentum, just as $-\mathbf{T}$ of (6.25) was found to be a current density of linear field momentum.

A continuity equation for the field angular momentum can be constructed by introducing the volume integrals (6.32) and (6.35) into the left side of (6.36) and reducing them to a point volume, in the usual way

$$\frac{\partial}{\partial t}(\mathbf{r} \times \mathbf{g}) + \nabla \cdot \mathbf{M} = -\mathbf{r} \times \mathbf{f} \tag{6.37}$$

This describes the conservation of the field angular momentum in its flux through every point except where there is matter to which the field imparts torque.

For a field-momentum distribution $\mathbf{g}(\mathbf{r}, t)$ and an angular-momentum density $\mathbf{r} \times \mathbf{g}$ to exist in a space, it is necessary that there be a flux $\mathbf{N} = c^2\mathbf{g}$

[Eq. (6.14)] of field energy in it. Since $\mathbf{N} \equiv (c/4\pi)\mathbf{E} \times \mathbf{B}$, this requires a coexistence of electric and magnetic fields. Examples are provided by the spinning charged spheres introduced in Sec. 4.1 and Exercise 4.1. The central Coulomb field and the ideal magnetic dipole field, which exist outside any one of the spheres, together produce the angular-momentum density distribution

$$\mathbf{r} \times \mathbf{g} = \frac{q}{4\pi c r^4} \hat{\mathbf{r}} \times (\boldsymbol{\mu} \times \hat{\mathbf{r}}) \tag{6.38}$$

There is none inside the sphere when the charge is concentrated on the surface, for then $\mathbf{E} = 0$ inside, but additional field angular momentum is distributed throughout the interior of the homogeneously charged sphere. In either case, the integrated resultant $\mathbf{J}$ is easy to find (Exercise 6.7), and, as might be expected, it has the direction of the spin: $\mathbf{J} \sim \boldsymbol{\omega}(\sim \boldsymbol{\mu})$.

The tensor $-\mathbf{M}$ measures a transmission of torque by the field, just as $\mathbf{T}$ was found to measure the transmission of force. That is particularly evident from (6.36), which gives the rate of change of angular momentum (equivalent to torque) as

$$\boldsymbol{\Theta} = -\oint d\mathbf{S} \cdot \mathbf{M} \tag{6.39}$$

An example is provided by a magnetic counterpart of the force transmission (6.30*b*) to a point charge in a uniform electric field, namely, the transmission of torque to an ideal dipole $\boldsymbol{\mu}$ in a uniform magnetic field $\mathbf{B}_o$. The total field in this case is

$$\mathbf{B} = \mathbf{B}_o + \frac{3\hat{\mathbf{r}}(\boldsymbol{\mu} \cdot \hat{\mathbf{r}}) - \boldsymbol{\mu}}{r^3}$$

and the inward radial flux density of angular momentum is

$$-\hat{\mathbf{r}} \cdot \mathbf{M} = \frac{(\hat{\mathbf{r}} \cdot \mathbf{B})(\mathbf{r} \times \mathbf{B})}{4\pi} \tag{6.40}$$

(see Exercise 6.8). Integrated as in (6.39), over any sphere centered on the dipole, this yields the expected result, $\boldsymbol{\Theta} = +r^2 \oint d\Omega\, \hat{\mathbf{r}} \cdot \mathbf{M} = \boldsymbol{\mu} \times \mathbf{B}_o$, for the torque transmitted to the dipole.

6.4 NONSTATIC POTENTIALS

The finding that the static fields can be represented by potentials as well as they are by $\mathbf{E}$ and $\mathbf{B}$ carries over to all electromagnetic fields, including nonstatic ones.

The potentials can again be introduced as general solutions of those Maxwell equations which are independent of sources. Since $\nabla \cdot \mathbf{B} = 0$ always, any magnetic field can be represented as the curl of some vector potential $\mathbf{A}(\mathbf{r}, t)$.

However, a nonstatic electric field is not curlless, in general, and so is no longer simply derivable from a scalar potential. On the other hand, when $\mathbf{B} = \nabla \times \mathbf{A}$ is introduced into the general Maxwell equation for the curl of $\mathbf{E}(\mathbf{r}, t)$, the result is

$$\nabla \times \left(\mathbf{E} + \frac{\dot{\mathbf{A}}}{c}\right) = 0 \tag{6.41}$$

Any such curlless quantity is derivable from some scalar potential $\phi(\mathbf{r}, t)$, and so any electromagnetic field can be described by potentials $\mathbf{A}(\mathbf{r}, t)$, $\phi(\mathbf{r}, t)$ *jointly*, with

$$\mathbf{E} = -\nabla\phi - \frac{\dot{\mathbf{A}}}{c} \qquad \mathbf{B} = \nabla \times \mathbf{A} \tag{6.42}$$

These expressions automatically satisfy the sourceless Maxwell equations, whatever the (differentiable) distributions $\mathbf{A}(\mathbf{r}, t)$, $\phi(\mathbf{r}, t)$ may be.

The necessity of the contribution to $\mathbf{E}$ additional to the curlless $-\nabla\phi$ in nonstatic situations is evident from the fact that Faraday induction can produce $\mathbf{E}$ with curl, having field lines that can form uninterrupted, divergenceless loops. That the supplementation should be just $-\dot{\mathbf{A}}/c$ is most directly evident when the Faraday-integral law (1.33) is valid. In this, the magnetic flux can be represented by $\Phi = \oint d\mathbf{r} \cdot \mathbf{A}$ [Eq. (6.19)], and then

$$\oint d\mathbf{r} \cdot \mathbf{E} = \oint d\mathbf{r} \cdot (\mathbf{E} + \nabla\phi) = -\frac{d}{c\,dt}\oint d\mathbf{r} \cdot \mathbf{A}$$

The insertion of $\nabla\phi$ here changes nothing since loop integrals of gradient fields vanish, and so the usual reduction of the arbitrary loop toward a point yields $\mathbf{E} + \nabla\phi = -\dot{\mathbf{A}}/c$, exactly the result in (6.42). An example with $\phi \equiv 0$ and $\mathbf{E} = -\dot{\mathbf{A}}/c$ has been met in (6.21), which exhibits a Faraday-induced electric force $q\mathbf{E} = -q\dot{\mathbf{A}}/c$ (see also Exercise 5.9).

The remaining Maxwell equations impose the restrictions needed if the potentials are to describe the field arising from given sources, $\rho(\mathbf{r}, t)$ and $\mathbf{j}(\mathbf{r}, t)$. Thus, the substitution of the relations (6.42) into the Maxwell equation for the magnetic curl yields

$$\nabla \times (\nabla \times \mathbf{A}) = \frac{4\pi\mathbf{j}}{c} + \frac{\partial}{c\,\partial t}\left(-\nabla\phi - \frac{\partial\mathbf{A}}{c\,\partial t}\right)$$

and since the left side can be replaced by $\nabla(\nabla \cdot \mathbf{A}) - \nabla^2\mathbf{A}$ according to (A.52), the result can be written

$$\left(\nabla^2 - \frac{\partial^2}{c^2\,\partial t^2}\right)\mathbf{A} = -\frac{4\pi}{c}\mathbf{j} + \nabla\left(\nabla \cdot \mathbf{A} + \frac{\dot{\phi}}{c}\right) \tag{6.43}$$

In a similar way, the equation $\nabla \cdot \mathbf{E} = 4\pi\rho$ becomes

$$\left(\nabla^2 - \frac{\partial^2}{c^2\,\partial t^2}\right)\phi = -4\pi\rho - \frac{\partial}{c\,\partial t}\left(\nabla \cdot \mathbf{A} + \frac{\dot{\phi}}{c}\right) \tag{6.44}$$

when $-\mathring{\mathring{\phi}}/c^2$ is added to both sides just to make evident a symmetry with (6.43). Now both equations can be simplified in a manner analogous to the way $\nabla \times (\nabla \times \mathbf{A}) = \mathbf{c}$ of (A.51) was reduced to $\nabla^2\mathbf{A} = -\mathbf{c}$ of (A.54) by making use of the gauge invariance of any potential description.

The Lorentz Potentials

The potential gauge invariance refers to the freedom with which $\mathbf{A}$, ϕ can be chosen when they are to represent a given electromagnetic field $\mathbf{E}$, $\mathbf{B}$ through the relations (6.42). It is first clear that, as in the static case, the vector potential $\mathbf{A}$ can be altered by the addition to it of any gradient field $\nabla\chi$; its curl $\mathbf{B}$ is not changed thereby because the curl of any gradient vanishes identically. However, if a time-dependent regauging by $\chi(\mathbf{r}, t)$ is chosen, the expression for the electric field in (6.42) is changed by $-\nabla\mathring{\chi}/c$ *unless* the gauging of the scalar potential ϕ is changed simultaneously by adding $-\mathring{\chi}/c$ to it. Thus, the same $\mathbf{E}$, $\mathbf{B}$ that are derivable from potentials $\mathbf{A}$, ϕ are also derivable from new potentials $\mathbf{A}'$, ϕ' if these are results of a gauge transformation,

$$\phi \to \phi' = \phi - \frac{\mathring{\chi}}{c} \qquad \mathbf{A} \to \mathbf{A}' = \mathbf{A} + \nabla\chi \tag{6.45}$$

by an arbitrary $\chi(\mathbf{r}, t)$. This constitutes a formal description of the gauge invariance possessed by potential descriptions of electromagnetic fields.

The arbitrariness in the choice of potentials starts from the fact that the requirement $\nabla \times \mathbf{A} = \mathbf{B}$ restricts what the curl of the vector potential must be without specifying its divergence. In the static case, it was found most convenient to work with divergenceless† vector potentials because they could be formed as solutions of an equation $\nabla \times (\nabla \times \mathbf{A}) = 4\pi\mathbf{j}/c$ simplified to one of the Poisson type, $\nabla^2\mathbf{A} = -4\pi\mathbf{j}/c$. Inspection of the nonstatic equations (6.43) and (6.44) indicates that they can be most simplified by using a gauge in which

$$\nabla \cdot \mathbf{A} + \frac{\mathring{\phi}}{c} = 0 \tag{6.46}$$

Potentials thus restricted are called *Lorentz potentials*, and it is said that a Lorentz gauge is being used in descriptions by them.

Lorentz potentials arising from given sources are thus to be found as solutions of

$$\left(\nabla^2 - \frac{\partial^2}{c^2\,\partial t^2}\right)\begin{Bmatrix}\mathbf{A}\\ \phi\end{Bmatrix} = -4\pi\begin{Bmatrix}\mathbf{j}/c\\ \rho\end{Bmatrix} \tag{6.47}$$

Perhaps the greatest simplification here, compared with Eqs. (6.43) and (6.44) for the potentials in an arbitrary gauge, is that the determinations of $\mathbf{A}$ and ϕ are

† A vector potential with $\nabla \cdot \mathbf{A}(\mathbf{r}, t) = 0$ has also been found convenient in some nonstatic situations. It is then said that a *solenoidal gauge* is being used. See Exercises 6.11 and 6.12 and Eqs. (7.13), (9.110), and (10.42) for examples.

decoupled in (6.47); each field can be solved for separately. On the other hand, whenever any solutions of (6.47) are obtained, it must be seen to that they are indeed Lorentz potentials related as in (6.46), a requirement that does couple $\mathbf{A}$, ϕ so that they describe an electromagnetic field arising from conserved charges, with

$$\frac{\partial \rho}{\partial t} + \nabla \cdot \mathbf{j} = -\frac{c}{4\pi}\left(\nabla^2 - \frac{\partial^2}{c^2\,\partial t^2}\right)\left(\frac{\mathring{\phi}}{c} + \nabla \cdot \mathbf{A}\right) \equiv 0$$

The supplementary condition (6.46) is comparable to that represented in (A.56) for the static case, where it had to be checked that the solution of the Poisson equation was divergenceless before it could be held to represent an actual (magnetostatic) field.

Equations (6.47) form the generalization of the Poisson equations (2.7) and (4.3) suitable for nonstatic sources, ρ, $\mathbf{j}(\mathbf{r}, t)$. The laplacian has been supplemented by derivatives with respect to time, and the resulting operator

$$\nabla^2 - \frac{\partial^2}{c^2\,\partial t^2} \equiv \Box^2 \tag{6.48}$$

is called a d'alembertian, sometimes symbolized as $\Box^2$ (in a generalization of the laplacian symbol ∇^2) but more often as simply $\Box$. (Mathematicians, in particular, also seem to prefer Δ in place of $\nabla^2 \equiv \nabla \cdot \nabla$.)

It has been presumed in the preceding development that the Lorentz condition is not too restrictive, i.e., that Lorentz potentials can be found for any electromagnetic field **E**, **B** whatever. That this is indeed so cannot be satisfactorily demonstrated until solutions of the d'alembertian equations (6.47) are obtained for arbitrary sources, as will be done in Chap. 8. Meanwhile it can at once be made apparent that the Lorentz conditions are not even restrictive enough to define unique potentials for any field.

Any Lorentz potentials $\mathbf{A}$, ϕ can be changed by the gauge transformation (6.45) and yet remain Lorentz potentials if the choice of gauge field $\chi(\mathbf{r}, t)$ is restricted to any of the numerous solutions of

$$\left(\nabla^2 - \frac{\partial^2}{c^2\,\partial t^2}\right)\chi = 0 \tag{6.49}$$

This condition is sufficient to make $\nabla \cdot \mathbf{A}' + \mathring{\phi}'/c = 0$ follow from $\nabla \cdot \mathbf{A} + \mathring{\phi}/c = 0$ when $\mathbf{A}' = \mathbf{A} + \nabla\chi$ and $\phi' = \phi - \mathring{\chi}/c$.

Equation (6.49) is known as the *wave equation* and all the fields† **E**, **B**, $\mathbf{A}$, ϕ are also propagated in free space as it requires. That is immediately obvious from (6.47) for the Lorentz potentials, and then it must also be true of derivatives of $\mathbf{A}$, ϕ like **E**, **B**. Thus the condition (6.49) on the gauge field corresponds to maintaining a Lorentz character of the gauge for the fields, as they are propagated in space and time.

† The same propagation equation also applies to mechanical disturbances, e.g., sounds that are not too loud, in elastic media if c is chosen appropriately. See Ref. 17, chaps. 12 to 15.

The word gauge was adopted in connection with the property (6.45) of descriptions by potentials because it has the meaning of a chosen standard of measurement. A consequence of the property is that results of measuring potentials must be expressed not only relative to choices of units and reference frame but also relative to a choice of gauge. The additional arbitrariness of definition here applies to all† energies and momenta, and not only to the electromagnetic field energies and momenta that, like (2.66) and (6.18), are represented with the help of potentials.

The above requires emphasis because it has sometimes been asserted that the arbitrariness with which they can be gauged renders potentials physically meaningless, despite their measurability relative to one gauge or another and the useful knowledge thus provided about actually existing field energies and momenta.

The assertions in question seem to stem from failures to appreciate how gauge transformations like (6.45) can be understood. It is simplest to consider an example like that of the electrostatic potential $\phi(\mathbf{r}) = -\mathbf{E}_o \cdot (\mathbf{r} - \mathbf{r}_o)$ of a uniform field $\mathbf{E}_o = -\nabla\phi$ measured relative to $\phi(\mathbf{r}_o) = 0$ at a fixed zero point $\mathbf{r}_0$. According to (6.45), it can be replaced by generally *nonstatic* potentials $\phi'(\mathbf{r}, t) = \phi(\mathbf{r}) - \dot{\chi}/c$, $\mathbf{A}' = \nabla\chi$ with $\mathbf{E}_o = -\nabla\phi' - \dot{\mathbf{A}}'/c$. In particular, $\chi(\mathbf{r}, t) = -\frac{1}{2}(\mathbf{E}_o \cdot \mathbf{u})ct^2$ may be chosen, with $\mathbf{u}$ some constant vector. Then $\phi' = -\mathbf{E}_o \cdot (\mathbf{r} - \mathbf{r}_o - \mathbf{u}t)$, which simply means that the fixed zero point has been replaced with a uniformly moving one. This may seem an unnatural choice, but it cannot be excluded except by special (and unnecessarily restrictive) convention. Quite generally, the gauge invariance arises from the arbitrariness with which zero points can be adopted; a new one can be chosen for every position $\mathbf{r}$ at which ϕ, $\mathbf{A}$ may be measured and for every moment in time, as well. Similar considerations also apply to the vector potentials measuring potential momenta instead of energies.‡ Such points seem too academic to warrant more space for their exploration here; this is left to any reader who is interested.

6.5 MONOCHROMATIC FIELDS

One of the most widely followed lines of generalization from static to nonstatic fields can be described as passing from zero-frequency to nonvanishing-frequency Fourier components of the time dependence. Any time-dependent field $f(\mathbf{r}, t)$, scalar or nonscalar, can be Fourier-analyzed according to

$$f(\mathbf{r}, t) = \int_{-\infty}^{\infty} d\omega f_\omega(\mathbf{r}) e^{-i\omega t} \qquad \text{with} \qquad f_\omega(\mathbf{r}) = \int_{-\infty}^{\infty} \frac{dt}{2\pi} f(\mathbf{r}, t) e^{+i\omega t} \tag{6.50}$$

† This refers to arbitrary canonical transforms that can be interpreted as redefinitions relative to new zeros in phase spaces and hence are even more general than the gauge transforms of electromagnetic potentials. See Ref. 17, sec. 8.2.

‡ That must be so, since energies and momenta are merely different (and intertransformable) components of a single entity, the energy-momentum 4-vector, as will be learned in Sec. 11.2 and in connection with (13.21).

as follows from the theorem (3.44). Attention can then be concentrated on any one of the Fourier amplitudes f_ω, or *monochromatic representatives*,

$$f_\omega(\mathbf{r})e^{-i\omega t} \tag{6.51}$$

of the field in question, with the idea that results obtained for this can be superposed as in (6.50) to obtain conclusions about an arbitrarily time-dependent situation as described by some $f(\mathbf{r}, t)$.

It should immediately be clear why each such an element as (6.51) is said to have a definite, nonvanishing frequency (when $\omega \neq 0$). It returns to exactly the same value after each increment of time by $T = 2\pi/\omega$, the period of the oscillations in value. The number of oscillations per unit time is the definite frequency, $\nu \equiv 1/T = \omega/2\pi$. It has become customary to refer to $\omega = 2\pi\nu$ itself, the number of oscillations in 2π units of time, as the frequency, although this is sometimes distinguished as the *angular* or *circular frequency*. The component is also called *monochromatic* because light of a definite color can be identified with an electromagnetic disturbance of definite frequency.

Each such representative as (6.51) is complex-valued and by itself cannot describe a real field. Whenever the Fourier analysis (6.50) is applied to a real field $f(\mathbf{r}, t) = f^*(\mathbf{r}, t)$, then $f^*_\omega(\mathbf{r}) = f_{-\omega}(\mathbf{r})$ and the superposition can be written

$$\begin{aligned} f &= \int_0^\infty d\omega\ f_\omega e^{-i\omega t} + \int_{-\infty}^0 d\omega\ f_\omega e^{-i\omega t} \\ &= \int_0^\infty d\omega\ f_\omega e^{-i\omega t} - \int_{+\infty}^0 d\omega'\ f_{-\omega'} e^{+i\omega' t} \end{aligned}$$

after a substitution $\omega' = -\omega$. Thus

$$f = \int_0^\infty d\omega\,(f_\omega e^{-i\omega t} + f^*_\omega e^{+i\omega t}) \tag{6.52}$$

for which only positive frequencies need be used. However, there is no compelling reason to exclude† linearly independent negative frequencies (with the same oscillation periods as their positive counterparts), and since (6.52) can equally well be written as

$$f = -\int_0^{-\infty} d\omega\,(f_{-\omega} e^{+i\omega t} + f^*_{-\omega} e^{-i\omega t}) = \int_{-\infty}^0 d\omega\,(f^*_\omega e^{i\omega t} + f_\omega e^{-i\omega t})$$

the real field can be treated as the superposition

$$f = \frac{1}{2}\int_{-\infty}^\infty d\omega (f_\omega e^{-i\omega t} + \text{cc}) = \int_{-\infty}^\infty d\omega\ \text{Re}\,(f_\omega e^{-i\omega t}) \tag{6.53}$$

† In quantum mechanics, which deals with particles as quanta of *complex* fields, the linearly independent negative-frequency components are very much needed to describe antiparticles. This is not essential to the quanta ≡ photons of the *real* electromagnetic fields since the antiphoton is identical with the photon (the quantum is said to be *self-conjugate*).

where cc stands for complex conjugate. Thus, whenever it becomes important to recognize the reality of the field, it can be regarded as a superposition of just the real parts of the complex representatives (6.51), having both positive and negative frequencies. It also continues to be possible to think of the complex representatives as having only positive frequencies, each accompanied by its complex conjugate, as in (6.52), in the representation of any real field.

A lemma that is useful in dealing with the complex representatives (6.51) follows from

$$\frac{\partial}{\partial t} f(\mathbf{r}, t) = \int_{-\infty}^{\infty} d\omega(-i\omega f_\omega)e^{-i\omega t} \tag{6.54}$$

This means that for each monochromatic representative, the time-derivative operation is replaced by multiplication with $-i\omega$:

$$\frac{\partial}{\partial t} \rightarrow -i\omega \tag{6.55}$$

Because operation on the representative with $i\partial/\partial t$ is equivalent to multiplying it by a real constant ω, the representative is said to be a (*frequency*) *eigenfunction* of the operator $i\partial/\partial t$, for the (*frequency*) *eigenvalue*, or *eigenfrequency*, ω.

The Fourier-Transform Equations

Fourier transforms of all the Maxwell equations (1.40) can be obtained simply by applying the integration operation

$$(2\pi)^{-1} \int_{-\infty}^{\infty} dt e^{i\omega t} \ldots \tag{6.56}$$

to them. The rather straightforward results are

$$\begin{aligned} \nabla \cdot \mathbf{E}_\omega(\mathbf{r}) &= 4\pi\rho_\omega(\mathbf{r}) \qquad & \nabla \times \mathbf{E}_\omega &= ik\mathbf{B}_\omega(\mathbf{r}) \\ \nabla \cdot \mathbf{B}_\omega(\mathbf{r}) &= 0 \qquad & \nabla \times \mathbf{B}_\omega &= \frac{4\pi\mathbf{j}_\omega}{c} - ik\mathbf{E}_\omega \end{aligned} \tag{6.57}$$

where

$$k \equiv \frac{\omega}{c} \tag{6.58}$$

The spatial distributions $\mathbf{E}_\omega$, $\mathbf{B}_\omega$, ρ_ω, and $\mathbf{j}_\omega(\mathbf{r})$ are Fourier transforms of $\mathbf{E}$, $\mathbf{B}$, ρ, and $\mathbf{j}(\mathbf{r}, t)$ defined on the model of (6.50). The source components are always related as

$$\nabla \cdot \mathbf{j}_\omega(\mathbf{r}) = i\omega\rho_\omega(\mathbf{r}) \tag{6.59}$$

because of the charge-conservation relation (1.14).

The potential descriptions can be transformed similarly. Thus the Fourier-component fields $\mathbf{E}_\omega(\mathbf{r})$ and $\mathbf{B}_\omega(\mathbf{r})$ are derivable from potentials according to the transforms

$$\mathbf{E}_\omega = -\nabla\phi_\omega + ik\mathbf{A}_\omega \qquad \mathbf{B}_\omega = \nabla \times \mathbf{A}_\omega \tag{6.60}$$

of (6.42). The fields $\mathbf{A}_\omega(\mathbf{r})$, $\phi_\omega(\mathbf{r})$ will be Fourier amplitudes of Lorentz potentials if they satisfy the transform

$$\nabla \cdot \mathbf{A}_\omega(\mathbf{r}) = ik\phi_\omega(\mathbf{r}) \tag{6.61}$$

of the Lorentz condition (6.46). For given sources, they are to be obtained as solutions of the equations

$$(\nabla^2 + k^2)\begin{Bmatrix}\mathbf{A}_\omega \\ \phi_\omega\end{Bmatrix} = -4\pi\begin{Bmatrix}\mathbf{j}_\omega/c \\ \rho_\omega\end{Bmatrix} \tag{6.62}$$

that follow from (6.47). These are generalizations to $\omega = ck \neq 0$ of the Poisson type of equations. In free spaces, where $\mathbf{j}_\omega, \rho_\omega \equiv 0$, they are generalizations of the *Laplace* equations ($\nabla^2 \equiv 0$) known as *Helmholtz equations* ($\nabla^2 + k^2 \equiv 0$).

The Energy in a Monochromatic Field

The simplest superposition of Fourier representatives that can stand for a real electromagnetic field describes the *ideally monochromatic* field presentable as

$$\mathbf{E}(\mathbf{r}, t) = \text{Re}\,[\mathbf{E}(\mathbf{r})e^{-i\omega t}] \qquad \mathbf{B}(\mathbf{r}, t) = \text{Re}\,[\mathbf{B}(\mathbf{r})e^{-i\omega t}] \tag{6.63}$$

Then attention can be confined to the complex amplitudes $\mathbf{E}(\mathbf{r})$, $\mathbf{B}(\mathbf{r})$ in all such linear relationships as those of the preceding subsection [with $\mathbf{E}_\omega$, $\mathbf{B}_\omega \rightarrow \mathbf{E}$, $\mathbf{B}(\mathbf{r})$], but explicit recognition must be given to the real combinations (6.63) in forming expressions nonlinear in the fields, like the energy.

The contribution to the field-energy density (6.3) at a point by just the electrical part of the above monochromatic field is

$$\begin{aligned} w(\mathbf{E}) &= \frac{\mathrm{E}^2(\mathbf{r}, t)}{8\pi} = \frac{1}{8\pi}\frac{1}{4}[\mathbf{E}(\mathbf{r})e^{-i\omega t} + \mathbf{E}^*(\mathbf{r})e^{+i\omega t}]^2 \\ &= (32\pi)^{-1}\{2\mathbf{E}^* \cdot \mathbf{E}(\mathbf{r}) + [\mathrm{E}^2(\mathbf{r})e^{-2i\omega t} + \text{cc}]\} \end{aligned} \tag{6.64}$$

This consists of a steady, time-independent part and a part that fluctuates in time, about a zero average. Such fluctuations occur only in a field idealized as in (6.63) and are cancelled out in any *entire* field because of energy conservation, as will be explicitly checked in the next chapter. For this reason, only the time-averaged contribution, $|\mathbf{E}(\mathbf{r})|^2/16\pi$, is retained, and this can be seen to have required considering not only the complex representative $\mathbf{E}(\mathbf{r}) \exp(-i\omega t)$ but also its complex conjugate (cc). The magnetic contribution is treated similarly, with the result

$$\langle w(\mathbf{r})\rangle = \frac{\frac{1}{2}(\mathbf{E}^* \cdot \mathbf{E} + \mathbf{B}^* \cdot \mathbf{B})}{8\pi} \tag{6.65}$$

for the time average, as denoted by the angular brackets.

The current density of the energy, as described by the Poynting vector (6.6), can also be expressed in terms of the complex field amplitudes $\mathbf{E}$, $\mathbf{B}(\mathbf{r})$. The time-averaged flux density will be

$$\langle \mathbf{N}(\mathbf{r}) \rangle = \frac{c}{4\pi} \cdot \tfrac{1}{4}(\mathbf{E} \times \mathbf{B}^* + \mathbf{E}^* \times \mathbf{B})$$

$$= \tfrac{1}{2}\,\mathrm{Re}\left[\frac{c}{4\pi}(\mathbf{E} \times \mathbf{B}^*)\right] = \tfrac{1}{2}\,\mathrm{Re}\left[\frac{c}{4\pi}(\mathbf{E}^* \times \mathbf{B})\right] \tag{6.66}$$

at any given point $\mathbf{r}$.

Illustrations of the general results obtained in Secs. 6.4 and 6.5 will be left to ensuing chapters. The ranges of application are very broad and require several chapters for adequate review.

EXERCISES

6.1 Extend the discussion of Fig. 6.1 to the *interior* of the current-carrying cylinder. Show that the difference of the energy fluxes through the outer and inner surfaces of a cylindrical shell, having radii differing by Δs, is exactly enough to supply the I^2R losses within the shell. As $\Delta s \to 0$, your result should become consistent with $\Delta V(\nabla \cdot \mathbf{N})$ for the energy deposited in the shell.

6.2 Reconsider the uniformly moving charged spherical shell discussed in connection with (4.100).

(*a*) Show that the quasi-static approximations of the energy flux and of the field-momentum densities around the sphere have no radial components but do have components in the direction of the sphere's velocity. Where do these have their maxima and minima?

(*b*) Find the total field momentum $\mathbf{P}$ and its relation to the field mass $4W_o/3c^2$.

6.3 Find the field-momentum distribution outside the spinning spherical shell of charge treated in Exercise 4.1, and its relation to the vector-potential distribution. What is the resultant (integrated) field momentum in this case?

6.4 A point charge q is placed at some distance s from an infinite straight line of steady current I. Carry out the integration (6.17) in this special case and express the result in terms of the vector potential of I (Exercise 4.2) in some gauge.

6.5 Carry out the integrations of the momentum fluxes through the midplanes of the configurations discussed in connection with Figs. 6.3 and 6.4.

6.6 Consider the transmission of stresses by a uniform field $\mathbf{E}_o$ to a point charge q.

(*a*) Show that the force transmitted through a *plane* transverse to $\mathbf{E}_o$ and at some distance l short of the position of q is just $\frac{1}{2}q\mathbf{E}_o$ minus an infinite amount $\mathrm{E}_o^2/8\pi$ per unit area.

(*b*) Supplement (*a*) with transmissions through a plane on the opposite side of q. How are the results here related to the fact that a uniform field can be produced between *two* remote planes bearing equal and opposite fixed charged densities?

6.7 (*a*) Evaluate the total angular momenta $\mathbf{J}$ in the fields generated by the spinning spheres considered in Sec. 4.1 and Exercise 4.1.

(*b*) Find radii a of the spheres for which the field angular momentum exceeds the mechanical angular momentum however the mass is distributed inside the spheres.

(*c*) Find coefficients $\mathscr{I}$ such that $\mathbf{J} = \mathscr{I}\boldsymbol{\omega}$ in each of the cases in (*a*) and compare your results to those of Exercise 4.26(*c*).

6.8 Confirm (6.40) and from it find the torque transmitted to an ideal magnetic dipole by a uniform magnetic field.

6.9 Suppose that a particle endowed with some magnetic monopole strength q_M is eventually found (see the footnote to page 18).

(*a*) Show that a test point charge q with some velocity $\mathbf{u}$ relative to the monopole should detect it through a force on it given by $\mathbf{F}_q = qq_M\mathbf{u} \times \hat{\mathbf{r}}/cr^2$ if $\mathbf{r}$ is centered on q_M.

(*b*) By considering the reaction on the monopole from the force in (*a*) show that in an electromagnetic field **E**, **B** the force on the monopole should be

$$\mathbf{F}_M = q_M \mathbf{B} + q_M \mathbf{u} \times \frac{\mathbf{E}}{c}$$

when it has a velocity **u**.

6.10 Suppose that the magnetic monopole q_M of Exercise 6.9 is placed at some distance r from a point charge $q = -e$.

(*a*) Show that the total field angular momentum **J** present in the static situation is independent of the distance r, existing unchanged even when $-e$ is remote from q_M.

(*b*) According to quantum-mechanical principles, every conserved quantity of angular momentum must be an integer or half-integer multiple of $\hbar = h/2\pi$, where h is Planck's constant, and thus $|\mathbf{J}| = n\hbar/2$. Show that this implies that the possible pole strengths of monopoles are restricted to $q_M = (n/2)(\hbar c/e) \approx 137ne/2$, if e is the charge on an electron. (These are just the Dirac quantized monopoles referred to in the footnote to page 18.)

6.11 Suppose that the solenoidal gauge $\nabla \cdot \mathbf{A}(\mathbf{r}, t) = 0$, mentioned in the footnote to page 172, is adopted for the potential description by ϕ, **A** in place of the Lorentz gauge (6.46).

(*a*) Show that in the new gauge ϕ determines a curlless part $\mathbf{E}_\rho$ and **A** a divergenceless part $\mathbf{E}_M$ of the total electric field $\mathbf{E} = \mathbf{E}_\rho + \mathbf{E}_M$ as

$$\mathbf{E}_\rho = -\nabla\phi \qquad \mathbf{E}_M = -\frac{\mathring{\mathbf{A}}}{c}$$

[See (A.32) for such a splitting of vector fields in general.]

(*b*) Show that the Maxwell equations lead to a pair

$$\nabla \cdot \mathbf{E}_\rho = 4\pi\rho \qquad \nabla \times \mathbf{E}_\rho = 0$$

for the curlless $\mathbf{E}_\rho$ alone and that (6.44) reduces to $\nabla^2\phi = -4\pi\rho(\mathbf{r}, t)$.

(*c*) The equations in (*b*) can only help determine spatial distributions at any one moment t. The strictures on the time dependence that should follow from charge conservation are best formulated after also splitting the current density in a general source into curlless and divergenceless parts: $\mathbf{j} = \mathbf{j}_\rho + \mathbf{j}_M$ with $\partial\rho/\partial t = -\nabla \cdot \mathbf{j}_\rho$. Show how a time derivative of the Gauss law in (*b*) leads to

$$\nabla \cdot (\mathring{\mathbf{E}}_\rho + 4\pi\mathbf{j}_\rho) = 0 \rightarrow \mathring{\mathbf{E}}_\rho = -4\pi\mathbf{j}_\rho(\mathbf{r}, t)$$

for any physical field, one that vanishes at surfaces remote from the sources.

6.12 Complete the development started in Exercise 6.11 by showing that:

(*a*) The Maxwell equations relate **B** to $\mathbf{E}_M$ only, according to

$$c\nabla \times \mathbf{E}_M = -\mathring{\mathbf{B}} \qquad c\nabla \times \mathbf{B} = 4\pi\mathbf{j}_M + \mathring{\mathbf{E}}_M$$

(*b*) Equation (6.43) reduces to the same form as (6.47), for the Lorentz-gauged vector potential, except that **j** is replaced by $\mathbf{j}_M$.

6.13 In view of Exercises 6.11 and 6.12, show that the splitting $\mathbf{j} = \mathbf{j}_\rho + \mathbf{j}_M$ of any finite source can, in principle, be found from

$$4\pi\mathbf{j}_\rho(\mathbf{r}, t) = -\nabla \oint dV \, \frac{\nabla_s \cdot \mathbf{j}(\mathbf{r}_s, t)}{R} \qquad 4\pi\mathbf{j}_M(\mathbf{r}, t) = \nabla \times \left[\nabla \times \oint dV \frac{\mathbf{j}(\mathbf{r}_s, t)}{R}\right]$$

where $R = |\mathbf{r} - \mathbf{r}_s|$. (To derive the latter, it is perhaps simplest to first recognize that the laplacian of the last integral is just $-4\pi\mathbf{j}$ and then use the result for $\mathbf{j}_\rho$.)

6.14 Supplement (6.65) and (6.66) by finding $\langle \mathbf{T} \rangle$ in terms of the complex representatives $\mathbf{E}(\mathbf{r})$, $\mathbf{B}(\mathbf{r})$.

6.15 The findings discussed in Secs. 2.4 and 6.2, that the potentials are detectable as are **E** and **B**, indicates that Eqs. (6.47) and (6.46) for the potentials can replace the Maxwell equations at the basis of electromagnetic theory, with $\mathbf{E} = -\nabla\phi - \mathring{\mathbf{A}}/c$ and $\mathbf{B} = \nabla \times \mathbf{A}$ as derived concepts. To support this notion, show how the Maxwell equations *follow* from the potential equations.

CHAPTER

SEVEN

PLANE ELECTROMAGNETIC WAVES

Self-sustaining fields in free space and the oscillations of their energies between electric and magnetic forms · The field momentum carried by a plane wave · Complementarity in wave packets · Analyses into linearly or circularly polarized components · Field-energy transmission by ideal wave guides · Resonant cavities and resonance breadths

When isolated electromagnetic fields were envisioned in the preceding chapter, it was being presumed that the Maxwell equations possess solutions that represent self-sustaining fields—out in free space and divorced from sources. That this might be expected, as a joint result of the Faraday and Maxwell inductions described by the equations, has already been suggested in connection with Fig. 1.4. Now explicit solutions of the anticipated type will be developed, as superpositions of plane waves.

7.1 WAVE FIELDS IN FREE SPACE

In spaces devoid of material, where ρ, $\mathbf{j} \equiv 0$, the general Maxwell equations (1.40) reduce to

$$\nabla \cdot \mathbf{E} = 0 \qquad \nabla \times \mathbf{E} = -\frac{\mathring{\mathbf{B}}}{c} \qquad \nabla \cdot \mathbf{B} = 0 \qquad \nabla \times \mathbf{B} = \frac{\mathring{\mathbf{E}}}{c} \tag{7.1}$$

The magnetic field can be eliminated from the equation for $\nabla \times \mathbf{E}$ by applying a second curl operation to it and using the equation for $\nabla \times \mathbf{B}$, with $\nabla \times (\nabla \times \mathbf{E}) = -\mathring{\mathring{\mathbf{E}}}/c^2$ as the result. For the divergenceless field here, $\nabla \times (\nabla \times \mathbf{E}) = -\nabla^2 \mathbf{E}$ follows from the mathematical identity (A.52). A like procedure can also be used to eliminate $\mathbf{E}$ in favor of $\mathbf{B}$. The results,

$$\left(\nabla^2 - \frac{\partial^2}{c^2\,\partial t^2}\right)\begin{Bmatrix}\mathbf{E}\\ \mathbf{B}\end{Bmatrix} = 0 \tag{7.2}$$

are called *wave* or *propagation equations*. They reduce to Laplace equations in static situations. Whereas Laplace equations demand a "smoothness" described by $\nabla^2 = 0$ for static equilibria, the wave equations give second rates of variation in time that are generated by any departures from a spatial smoothness, $\nabla^2 \neq 0$. All this has only to do with distributions over space and time of each field component separately. The initial Maxwell equations also couple different field components to each other. Thus, the wave equation, although a necessary consequence of the Maxwell equations, is not sufficient to replace them. Any solutions of the wave equation that are found must also be made to satisfy the original Maxwell conditions before they can represent possible electromagnetic fields.

Resolutions into Plane-Wave Components

Each of the vector field components, like any bounded distribution in space, can be resolved on the complete orthonormal Fourier basis (3.55), to form a superposition like (3.53). Thus any nonsingular electric field can be represented as

$$\mathbf{E}(\mathbf{r}, t) = (2\pi)^{-3/2} \oint dV(\mathbf{k})\, \mathbf{C}(\mathbf{k}, t) e^{i\mathbf{k}\cdot\mathbf{r}}$$

with some vector function $\mathbf{C}(\mathbf{k}, t)$. This must be restricted in its time dependence if the resultant field is to be propagated across free spaces in accordance with the wave equation (7.2). Substitution of the integral into the equation leads to the requirement $\overset{\circ\circ}{\mathbf{C}} = -c^2k^2\mathbf{C}$ for each $\mathbf{k}$, and so each $\mathbf{C}(\mathbf{k}, t)$ must be a linear combination of $\exp(\pm i\omega t)$ with

$$\omega \equiv c|\mathbf{k}| \equiv ck > 0 \tag{7.3}$$

Thus

$$\mathbf{E}(\mathbf{r}, t) = \frac{1}{2}\frac{1}{(2\pi)^{3/2}} \oint dV(\mathbf{k})\, [\mathbf{a}(\mathbf{k})e^{-i\omega t} + \mathbf{b}(\mathbf{k})e^{+i\omega t}] e^{i\mathbf{k}\cdot\mathbf{r}} \tag{7.4}$$

with some vector coefficients $\mathbf{a}$, $\mathbf{b}$ for each $\mathbf{k}$. There has emerged here a Fourier analysis of the time dependence, like that discussed in Sec. 6.5. The frequencies have been restricted by the wave equation to just the two linearly independent ones $\pm(\omega \equiv ck)$ for each $|\mathbf{k}| = k$. The same results could have been obtained by starting with the frequency analysis, and then the wave equation would have demanded spatial dependences for which $\nabla^2 + (\omega/c)^2 \to 0$, a Helmholtz equation [see the statements about (6.62)]. In that procedure, the exponentials $\exp i\mathbf{k}\cdot\mathbf{r}$ would be introduced as solutions of the Helmholtz equation, which they obviously satisfy if $|\mathbf{k}| = \omega/c$, but these do not furnish the *only* complete orthogonal set of solutions, as the experience with the Laplace ($\omega = 0$) equation in Chap. 3 should suggest. In this connection it should be noticed that, whereas the Laplace equation always demands an unbounded behavior for indefinite extensions along at least one dimension, here bounded periodic variations have been found for all extensions, including that in time. Mathematically this outcome is traceable to the occurrence of the negative sign in the d'alembertian operator, which makes the wave equation a hyperbolic one, as against the elliptic Laplace differential equation. It is important to physical description since it makes representable isolated fields that can remain bounded without being connected to sources, as Laplace fields must be. The possibility arises from the negative relative sign of the Faraday and Maxwell inductions, as anticipated.

To see how the coefficients in the superposition (7.4) must be restricted for the resultant to describe a *real* field, first make the substitution $\mathbf{k} \to -\mathbf{k}$ for this dummy integration variable in the term containing $\mathbf{b}(\mathbf{k})$:

$$\mathbf{E} = \frac{1}{2}\oint \frac{dV(\mathbf{k})}{(2\pi)^{3/2}} [\mathbf{a}(\mathbf{k})e^{i(\mathbf{k}\cdot\mathbf{r}-\omega t)} + \mathbf{b}(-\mathbf{k})e^{-i(\mathbf{k}\cdot\mathbf{r}-\omega t)}] \tag{7.5}$$

This makes it quite plain that the choices $\mathbf{b}(-\mathbf{k}) = \mathbf{a}^*(\mathbf{k})$ must be made for the (generally) complex coefficients. Another way to reach the same conclusion is to consider the complex conjugate of a Fourier-transform expression like (3.54) for the square bracket in (7.4) and duly recognize that the two exponentials exp $(\pm i\omega t)$ are linearly independent.

Expression (7.5) has now become a superposition of just the representatives

$$\mathbf{a}(\mathbf{k})e^{i(\mathbf{k}\cdot\mathbf{r}-\omega t)}$$

and their linearly independent complex conjugates. This satisfies the wave equation, but the Maxwell conditions (7.1) must also be satisfied before the result can represent an electric field in free space. For the field to be properly divergenceless, each of the linearly independent representatives must be divergenceless:

$$\nabla \cdot [\mathbf{a}(\mathbf{k})e^{i(\mathbf{k}\cdot\mathbf{r}-\omega t)}] = i\mathbf{k}\cdot\mathbf{a}e^{i(\mathbf{k}\cdot\mathbf{r}-\omega t)} = 0$$

It is thus found that the vector amplitude $\mathbf{a}(\mathbf{k})$ for each $\mathbf{k}$ must be perpendicular to $\mathbf{k}$, transverse to the direction $\hat{\mathbf{k}} \equiv \mathbf{k}/k$. This still leaves to arbitrary choice just two component amplitudes in the transverse plane.

Let some conveniently chosen direction in the planes transverse to a given $\mathbf{k}$ be pointed out by a unit vector $\boldsymbol{\epsilon}_{\mathbf{k}}^1$ and the third orthogonal direction by $\boldsymbol{\epsilon}_{\mathbf{k}}^2 \equiv \hat{\mathbf{k}} \times \boldsymbol{\epsilon}_{\mathbf{k}}^1$. A right-handed frame $\hat{\mathbf{k}}$, $\boldsymbol{\epsilon}_{\mathbf{k}}^{1,2}$, with the formal properties

$$\begin{aligned} &\boldsymbol{\epsilon}_{\mathbf{k}}^{\lambda}\cdot\boldsymbol{\epsilon}_{\mathbf{k}}^{\lambda'} = \delta_{\lambda\lambda'} \qquad &(\lambda, \lambda' = 1, 2)\\ &\hat{\mathbf{k}}\cdot\boldsymbol{\epsilon}_{\mathbf{k}}^{\lambda} = 0 \qquad &\boldsymbol{\epsilon}_{\mathbf{k}}^1 \times \boldsymbol{\epsilon}_{\mathbf{k}}^2 = \hat{\mathbf{k}} \end{aligned} \tag{7.6}$$

is thus chosen for each $\mathbf{k}$. Now the general, divergenceless electric field in free space can be represented as

$$\mathbf{E}(\mathbf{r}, t) = \text{Re}\left[\sum_{\lambda=1,2}\oint \frac{dV(\mathbf{k})}{(2\pi)^{3/2}} a_\lambda(\mathbf{k})\boldsymbol{\epsilon}_{\mathbf{k}}^{\lambda} e^{i(\mathbf{k}\cdot\mathbf{r}-\omega t)}\right] \tag{7.7}$$

with arbitrary *scalar* amplitudes $a_\lambda(\mathbf{k})$.

The remaining Maxwell equation that involves the electric field, $\mathring{\mathbf{B}} = -c(\nabla \times \mathbf{E})$, requires magnetic variations to accompany it. Each of the component representatives must be accompanied by a component of $\mathring{\mathbf{B}}$ given by

$$-c\,\nabla \times (a_\lambda \boldsymbol{\epsilon}_{\mathbf{k}}^{\lambda} e^{i(\mathbf{k}\cdot\mathbf{r}-\omega t)}) = -ica_\lambda \mathbf{k} \times \boldsymbol{\epsilon}_{\mathbf{k}}^{\lambda} e^{i(\mathbf{k}\cdot\mathbf{r}-\omega t)}$$

A simple time integration divides this by the factor $-i\omega = -ick$, and so the expression

$$\mathbf{B}(\mathbf{r}, t) = \text{Re}\left[\sum_{\lambda}\oint \frac{dV(\mathbf{k})}{(2\pi)^{3/2}} a_\lambda(\mathbf{k})(\hat{\mathbf{k}} \times \boldsymbol{\epsilon}_{\mathbf{k}}^{\lambda})e^{i(\mathbf{k}\cdot\mathbf{r}-\omega t)}\right] \tag{7.8}$$

describes a magnetic field that must coexist with the electric field (7.7) in free space. This field is divergenceless, as any magnetic field must be, since $i\mathbf{k} \cdot (\hat{\mathbf{k}} \times \boldsymbol{\epsilon}_{\mathbf{k}}^{\lambda}) = 0$. The joint fields (7.7) and (7.8) likewise satisfy the final Maxwell equation, $\dot{\mathbf{E}} = c\, \nabla \times \mathbf{B}$, without further strictures on the arbitrary amplitudes $a_\lambda(\mathbf{k})$ (because of the prior satisfaction of the wave equation).

The superpositions (7.7) and (7.8), with real **k**'s, can represent only the most general free fields that are *bounded* everywhere, and so exclude† the Laplace ($\omega = ck = 0$) field forms of Chap. 3, which become unbounded in at least one direction of space and thereby depend on the existence of sources in that direction.

It has now been found that any bounded electromagnetic field in free space can be decomposed into a superposition of the orthonormal transverse-plane-wave representatives labeled by **k** and λ,

$$\mathbf{e}_{\mathbf{k}\lambda}(\mathbf{r}, t) \equiv \frac{\boldsymbol{\epsilon}_{\mathbf{k}}^{\lambda} e^{i(\mathbf{k}\cdot\mathbf{r} - \omega t)}}{(2\pi)^{3/2}} \tag{7.9}$$

each with some amplitude $a_\lambda(\mathbf{k})$. The magnetic field superposition (7.8) is constituted of these representatives because $\hat{\mathbf{k}} \times \mathbf{e}_{\mathbf{k}1,2} = \pm \mathbf{e}_{\mathbf{k}2,1}$. The set is orthonormal in that

$$\oint dV(\mathbf{r})\, e_{\mathbf{k}\lambda}^{*} \cdot e_{\mathbf{k}'\lambda'} = \delta_{\lambda\lambda'}\delta(\mathbf{k} - \mathbf{k}') \tag{7.10}$$

follows from (3.56) and (7.6). Another integration, over the volume elements $dV(\mathbf{k})$ of the **k**-space, yields unity for $\lambda = \lambda'$ and, in that sense, each of the representatives (7.9) is normalized to unity in a phase space, the joint six-dimensional space of **r** and **k**. Each basis function (7.9) is said‡ to be a *simultaneous eigenfunction* of the operation $i\, \partial/\partial t$ (since this is equivalent to multiplication by the real constant ω) and of the operation $-i\nabla$ (since this multiplies the function by **k**).

A complex representative (7.9) has a phase that can be written as $k(\hat{\mathbf{k}} \cdot \mathbf{r} - ct)$ since $\omega = ck$. Any function of $\hat{\mathbf{k}} \cdot \mathbf{r} - ct$ forms a plane wave§ propagated in the direction $\hat{\mathbf{k}}$ with the velocity c. It is called a propagated wave because it forms a distribution in which each value, say the one characterized by the specific phase $\hat{\mathbf{k}} \cdot \mathbf{r}_1 - ct_1$, continues to exist at later times, like $t_1 + \Delta t$, at advanced positions $\mathbf{r}_1 + \Delta\mathbf{r}$ such that $\hat{\mathbf{k}} \cdot \Delta\mathbf{r} - c\,\Delta t = 0$. It is a plane wave because any given value exists, at any given moment, on a whole plane of points, transverse to the direction $\hat{\mathbf{k}}$, all having position vectors with the same projection

† The situation may be confusing because any bounded section of a Laplace field can be Fourier-analyzed into components with real k's, as in (3.52) for example. However, all this variety of k values corresponds to a *single* frequency ($\omega = 0$), rather than having the relations $k = \omega/c$ to the frequencies, as in the solutions (7.7) and (7.8).

‡ See Ref. 17, sec. 14.3.

§ For simple pictures corresponding to the discussion here see Ref. 17, sec. 14.1 and figs. 12.4 and 14.1.

$\hat{\mathbf{k}} \cdot \mathbf{r}$ on that direction. Since a given phase plane advances at the rate $(\hat{\mathbf{k}} \cdot \Delta\mathbf{r})/\Delta t = c$, this is the propagation velocity of the wave.

The particular wave functions (7.9) are periodic, both in time, with the frequency $\omega/2\pi$, and in space, with the wavelength $2\pi/k$. Specifying the wave vector $\mathbf{k}$ gives not only the direction of the propagation but also its wavelength and frequency.

Thus, electromagnetic wave fields can be propagated across spaces devoid of matter. The propagations have the constant velocity c, first introduced as a mere conversion factor between electric and magnetic units, as discussed in connection with the citation of its value (4.56). This value coincides very precisely with the velocity that has been directly measured for light. Moreover, light can also be propagated across vacua. These are just the first indications that light can indeed be identified with electromagnetic waves.

The Energy of an Isolated Field

The plane-wave superpositions (7.7), with (7.8), can represent any field in free space and consist of components that remain bounded without interruption by sources. They can therefore represent *isolated* fields, and such should have energies that are conserved to themselves at every instant, as discussed in connection with (6.4). The integrated energy should be time-independent despite the variations in time of every individual ($\omega \neq 0$) component. There should not even be such temporal fluctuations as were found associated with the ideally monochromatic field in (6.64).

To evaluate the electrical contribution $\mathsf{E}^2/8\pi$ to the field energy density, the square of expression (7.7) must be formed:

$$(2\pi)^{-3} \sum_{\lambda, \lambda'} \oint\!\!\oint dV(\mathbf{k})\, dV(\mathbf{k}')\, \boldsymbol{\epsilon}_{\mathbf{k}}^{\lambda} \cdot \boldsymbol{\epsilon}_{\mathbf{k}'}^{\lambda'}$$

$$\times \tfrac{1}{4}[a_\lambda(\mathbf{k})a_{\lambda'}(\mathbf{k}')e^{i(\mathbf{k}+\mathbf{k}')\cdot\mathbf{r}}e^{-i(\omega+\omega')t} + \mathrm{cc}$$

$$+ a_\lambda^*(\mathbf{k})a_{\lambda'}(\mathbf{k}')e^{-i(\mathbf{k}-\mathbf{k}')\cdot\mathbf{r}}e^{+i(\omega-\omega')t} + \mathrm{cc}]$$

This must be integrated over all space to obtain the total electric field energy $W(\mathsf{E})$, and the process will, according to (3.56), introduce the delta functions $\delta(\mathbf{k}+\mathbf{k}')$ and $\delta(\mathbf{k}-\mathbf{k}')$. Each of these is replaced by unity after the integration $\oint dV(\mathbf{k}')$, and since $\omega' = \omega = ck$ for both $\mathbf{k}' = \pm\mathbf{k}$, the result is

$$W(\mathsf{E}) = (32\pi)^{-1} \sum_{\lambda, \lambda'} \oint dV(\mathbf{k})\{\boldsymbol{\epsilon}_{\mathbf{k}}^{\lambda} \cdot \boldsymbol{\epsilon}_{\mathbf{k}}^{\lambda'} a_\lambda^*(\mathbf{k})a_{\lambda'}(\mathbf{k}) + \mathrm{cc}$$

$$+ \boldsymbol{\epsilon}_{\mathbf{k}}^{\lambda} \cdot \boldsymbol{\epsilon}_{-\mathbf{k}}^{\lambda'}[a_\lambda(\mathbf{k})a_{\lambda'}(-\mathbf{k})e^{-2i\omega t} + \mathrm{cc}]\}$$

This still contains such temporal fluctuations as were found for ideally monochromatic fields in (6.64).

The magnetic contribution must be similarly evaluated from the expression (7.8) for the magnetic field, which differs from the electrical expression (7.7) only in the replacement of each unit vector $\boldsymbol{\epsilon}_{\mathbf{k}}^{\lambda}$ by a $\hat{\mathbf{k}} \times \boldsymbol{\epsilon}_{\mathbf{k}}^{\lambda}$. Thus, the expression for $W(\mathbf{B})$ will be exactly like the expression for $W(\mathbf{E})$ except for the replacements

$$\boldsymbol{\epsilon}_{\mathbf{k}}^{\lambda} \cdot \boldsymbol{\epsilon}_{\mathbf{k}}^{\lambda'} \to (\hat{\mathbf{k}} \times \boldsymbol{\epsilon}_{\mathbf{k}}^{\lambda}) \cdot (\hat{\mathbf{k}} \times \boldsymbol{\epsilon}_{\mathbf{k}}^{\lambda'}) = \hat{\mathbf{k}} \cdot [\boldsymbol{\epsilon}_{\mathbf{k}}^{\lambda} \times (\hat{\mathbf{k}} \times \boldsymbol{\epsilon}_{\mathbf{k}}^{\lambda'})]$$
$$= \boldsymbol{\epsilon}_{\mathbf{k}}^{\lambda} \cdot \boldsymbol{\epsilon}_{\mathbf{k}}^{\lambda'} - (\hat{\mathbf{k}} \cdot \boldsymbol{\epsilon}_{\mathbf{k}}^{\lambda'})(\boldsymbol{\epsilon}_{\mathbf{k}}^{\lambda} \cdot \hat{\mathbf{k}}) = \boldsymbol{\epsilon}_{\mathbf{k}}^{\lambda} \cdot \boldsymbol{\epsilon}_{\mathbf{k}}^{\lambda'}$$

equivalent to $\delta_{\lambda\lambda'}$ because of the orthonormality (7.6), and

$$\boldsymbol{\epsilon}_{\mathbf{k}}^{\lambda} \cdot \boldsymbol{\epsilon}_{-\mathbf{k}}^{\lambda'} \to (\hat{\mathbf{k}} \times \boldsymbol{\epsilon}_{\mathbf{k}}^{\lambda}) \cdot (-\hat{\mathbf{k}} \times \boldsymbol{\epsilon}_{-\mathbf{k}}^{\lambda'}) = -\boldsymbol{\epsilon}_{\mathbf{k}}^{\lambda} \cdot \boldsymbol{\epsilon}_{-\mathbf{k}}^{\lambda'}$$

The latter result signifies that the temporal fluctuations found in $W(\mathbf{E})$ merely constitute exchanges with the magnetic energy, since everywhere and at every moment they cancel out in the total $W = W(\mathbf{E}) + W(\mathbf{B})$, which is then given by

$$W = \sum_{\lambda} \oint dV(\mathbf{k}) \frac{|a_{\lambda}(\mathbf{k})|^2}{8\pi} \tag{7.11}$$

This is properly constant in time and will be finite for appropriately chosen functional dependences of $a_{\lambda}(\mathbf{k})$ on $\mathbf{k}$.

The detailed (moment-by-moment) constancy in time of the total field energy is necessary to its conception as an isolatable entity. It accounts for the greater interest in the time-averaged energy expression like (6.65) than in the temporal fluctuations exhibited by the ideally monochromatic fields. The latter extend to infinity, as will be discussed in the next section, and remain useful only as field *components*, or as idealizations of finite, nearly monochromatic wave packets. The latter are superpositions like (7.7) with (7.8) and will not contain the temporal fluctuations, as the outcome (7.11) attests.

Potential Descriptions of the Free Fields

The form (7.8) for the magnetic vector field makes it plain that it is derivable as $\mathbf{B} = \nabla \times \mathbf{A}_{\tau}$ from a vector potential

$$\mathbf{A}_{\tau} = \mathrm{Re}\left[\sum_{\lambda=1,2} \oint \frac{dV(\mathbf{k})}{(2\pi)^{3/2}} \frac{a_{\lambda}(\mathbf{k})}{ik} \boldsymbol{\epsilon}_{\mathbf{k}}^{\lambda} e^{i(\mathbf{k}\cdot\mathbf{r}-\omega t)}\right] \tag{7.12}$$

The subscript τ is meant to indicate that this contains only *transverse* plane waves, each divergenceless so that $\nabla \cdot \mathbf{A}_{\tau} = 0$. The result remains in the family of Lorentz gauged potentials (6.46) if $\phi_{\tau} \equiv 0$ is adopted. Then the electric field must be derivable as in

$$\mathbf{E} = -\frac{\dot{\mathbf{A}}_{\tau}}{c} \qquad \mathbf{B} = \nabla \times \mathbf{A}_{\tau} \tag{7.13}$$

and this does yield (7.7) properly. It is thus found sufficient to use only a vector potential, with $\phi \equiv 0$, for gauging fields in free space. The gauge here is variously referred to as solenoidal (see the footnote to page 172), divergenceless, or transverse.

Since it only needs to satisfy the wave equation, (6.47) with $\mathbf{j} \equiv 0$, a general Lorentz vector potential in free space can have, added to $\mathbf{A}_\tau$ of (7.12), an arbitrary superposition of longitudinal plane waves $\hat{\mathbf{k}} a_o e^{i(\mathbf{k} \cdot \mathbf{r} - \omega t)}$. These have divergences and so, to conform to the Lorentz condition (6.46) or (6.61), each must be accompanied by a component $a_o e^{i(\mathbf{k} \cdot \mathbf{r} - \omega t)}$ of scalar potential. The resulting supplementations, which may be denoted $\mathbf{A}_l$ and ϕ_l, are called *longitudinal potentials*. However, $\mathbf{B} = \nabla \times (\mathbf{A}_\tau + \mathbf{A}_l) = \nabla \times \mathbf{A}_\tau$ still, and also

$$\mathbf{E} = -\nabla\phi_l - \frac{\mathring{\mathbf{A}}_\tau + \mathring{\mathbf{A}}_l}{c} = -\frac{\mathring{\mathbf{A}}_\tau}{c}$$

since

$$\left(-\nabla - \frac{\hat{\mathbf{k}}}{c}\frac{\partial}{\partial t}\right) a_o e^{i(\mathbf{k} \cdot \mathbf{r} - \omega t)} = 0$$

Thus, longitudinal supplementations play no role in determining the fields $\mathbf{E}$, $\mathbf{B}$ in the source-free parts of space. They do turn out to be useful for maintaining the continuity of potential fields that overlap regions containing sources, where the existence of scalar potentials becomes quite essential, as shown by Eq. (6.47) for ϕ with $\rho \neq 0$.

7.2 PLANE MONOCHROMATIC WAVES

Because any free-space field can be regarded as a superposition of plane monochromatic waves, it is often sufficient to confine attention to an *ideal* plane-wave field†:

$$\mathbf{E}(\mathbf{r}, t) = \text{Re}\,(\mathbf{a} e^{i(\mathbf{k} \cdot \mathbf{r} - \omega t)}) \qquad \mathbf{B} = \hat{\mathbf{k}} \times \mathbf{E} \tag{7.14}$$

This is a special case of an ideally monochromatic field (6.63) characterized not only by a definite frequency $\omega = ck$ but also by a unique propagation direction $\hat{\mathbf{k}}$. The generally complex amplitude vector $\mathbf{a}$ can be chosen arbitrarily except that it must be transverse to the wave vector: $\mathbf{k} \cdot \mathbf{a} = 0$. The electric and magnetic vectors have equal magnitudes ($|\mathbf{B}| = |\hat{\mathbf{k}} \times \mathbf{E}| = |\mathbf{E}|$ since $\hat{\mathbf{k}} \cdot \mathbf{E} = 0$) in every phase of the propagation, and they are normal to each other ($\mathbf{E} \cdot \mathbf{B} = 0$) as well as to the propagation direction, all as indicated in Fig. 7.1.

The equal magnitudes of the electric and magnetic fields mean that the field energy is shared equally by them, and its time-averaged density (6.66) can be written

$$\langle w \rangle = \left\langle \frac{\mathbf{E}^2(\mathbf{r}, t)}{4\pi} \right\rangle = \left\langle \frac{\mathbf{B}^2(\mathbf{r}, t)}{4\pi} \right\rangle = \frac{|a|^2}{8\pi} \tag{7.15}$$

† Here, $\mathbf{E}(\mathbf{r}) = \mathbf{a} e^{i\mathbf{k} \cdot \mathbf{r}}$ in the notation of (6.63). This is derivable as $\mathbf{E}(\mathbf{r}) = ik\mathbf{A}_\tau(\mathbf{r})$ from a potential in the transverse gauge of (7.13), having $\phi_\tau \equiv 0$. Thus, the potential field representative is simply $\mathbf{A}_\tau = \mathbf{E}/ik$.

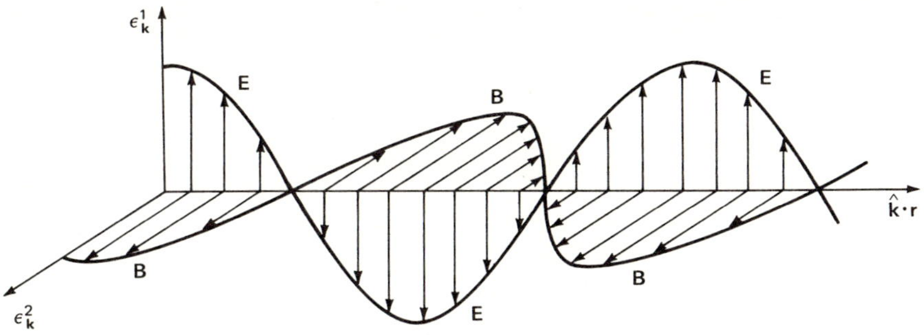

Figure 7.1 A linearly polarized plane wave.

describing a distribution that is uniform over all space. The Poynting flux (6.66) is

$$\mathbf{N} = \frac{c}{4\pi}\,\mathbf{E} \times (\hat{\mathbf{k}} \times \mathbf{E}) = \hat{\mathbf{k}}\,\frac{\mathsf{E}^2}{4\pi}\,c = \hat{\mathbf{k}}wc \tag{7.16}$$

as to be expected for energy of density w propagated with velocity c in the direction $\hat{\mathbf{k}}$. The momentum density (6.14) in the wave is

$$\mathbf{g} = \hat{\mathbf{k}}\frac{w}{c} \tag{7.17}$$

It is characteristic of the monodirectional plane wave to have a definitely directed field momentum $\hat{\mathbf{k}}/c = \mathbf{k}/\omega$ for each unit of energy that it carries.†

Complementarity in Plane-Wave Representations

The uniformity of the energy distribution (7.15) leads to an infinite energy for the ideally monochromatic plane-wave field (7.14), extending throughout all space. No field of finite extent can have the ideal monochromaticity but must form a wave packet, i.e., some superposition like (7.7) and (7.8) of the plane waves.

Such an effect of interrupting the extension of a field will here be illustrated for one described as in (7.14) but with the amplitude **a** defined to vanish outside some interval in one of the dimensions transverse to the propagation direction. Take the latter to be the z direction, so that the $\mathbf{k}\cdot\mathbf{r}$ in the exponentials of (7.14) is replaced by $k_z z$, and adopt for **a** any magnitude proportional to the function of x:

$$f(x) = \begin{cases} 1 & \text{for } |x| < \frac{1}{2}\,\Delta x \\ 0 & \text{for } |x| > \frac{1}{2}\,\Delta x \end{cases} \tag{7.18}$$

† In the contexts of quantum theory, this ratio would be written $\hbar\mathbf{k}/(\hbar\omega)$ or $h\mathbf{k}/2\pi(h\nu)$, where $h \equiv 2\pi\hbar$ is *Planck's constant*. Then $h\nu$ is called a quantum of energy and $\hbar\mathbf{k}$ is the momentum of one photon of light.

The Fourier decomposition of this function is given [see (3.44)] by

$$f(x) = \int_{-\infty}^{\infty} dk_x \, c(k_x) e^{ik_x x} \tag{7.19}$$

with

$$c(k_x) = \int_{-\Delta x/2}^{+\Delta x/2} \frac{dx}{2\pi} e^{-ik_x x} = \frac{\sin(k_x \, \Delta x/2)}{\pi k_x} \tag{7.20}$$

As a consequence, the field restricted to $\Delta x < \infty$ has plane-wave constituents with a variety of propagation directions $\mathbf{k}(k_x, k_z)$, besides the initial one with $k_x = 0$. It is such an introduction of new propagation directions that allows waves to turn corners—to be diffracted— upon passing beyond a space in which they are restricted in lateral extent.

The distribution (7.20) approaches $\delta(k_x)$ of (3.46) as $\Delta x \to \infty$, and so the new components with $k_x \neq 0$ appear only when $\Delta x < \infty$. Then the distribution of k_x values still has its maximum at $k_x = 0$, where $c(0) = \Delta x/2\pi$, and falls off symmetrically for $k_x \gtrless 0$. There is a finite *range* Δk_x of the k_x values that it is most essential to include in order to have the field largely confined within the spatial extent $\Delta x < \infty$. A suitable definition for Δk_x is suggested when it is recognized that, just as for the δ distribution obtained in the limit $\Delta x \to \infty$, the integral of the distribution over k_x has the unit value

$$\int_{-\infty}^{\infty} dk_x \frac{\sin(k_x \, \Delta x/2)}{\pi k_x} = 1 \tag{7.21}$$

independently of the size Δx. A *uniformly* distributed range Δk_x of k_x values would also have this invariant unit integral if $\Delta k_x c(0) = 1$. Thus

$$\Delta k_x = \frac{2\pi}{\Delta x} \tag{7.22a}$$

can be used to describe a complementarity that exists between spatial extent and wave-number range. The narrower the space, the wider the variety of wavelengths it must contain with appreciable amplitudes.

Such a complementarity as (7.22*a*) applies to confinements in any direction†:

$$\Delta k_y = \frac{2\pi}{\Delta y} \qquad \Delta k_z = \frac{2\pi}{\Delta z} \tag{7.22b}$$

Restrictions of both lateral dimensions are assumed whenever a ray picture of a field having extremely short wavelengths, like those of visible light, with $\lambda = (4 - 7) \times 10^{-5}$ cm, is used. The ray must be many wavelengths broad if it is to have negligible directional spreading confined to angles $\Delta k_x/k = 2\pi/(k \, \Delta x) = \lambda/\Delta x \ll 1$.

† In Ref. 17, sec. 12.4, this is illustrated for a wave train of finite length parallel to the propagation direction. A picture of such a distribution as (7.20) is also shown.

The complementarity

$$\Delta\omega = \frac{2\pi}{\Delta t} \tag{7.23}$$

follows similarly† for any field of finite duration, $\Delta t < \infty$. Thus, whenever the field participating in some process is assumed to be essentially monochromatic, it is being presumed that its duration is much longer than the time needed for the completion of the process. Perfect monochromaticity is an idealization that, in practice, can be achieved for an entire field only approximately.

The Polarizations of the Plane Waves

Any plane-wave field has a definite propagation direction $\hat{\mathbf{k}}$ and so can be resolved on a frame like that defined in (7.6). In general the result will have a form (7.14) with an amplitude vector

$$\mathbf{a} = \sum_{\lambda=1,2} a_\lambda \boldsymbol{\epsilon}_{\mathbf{k}}^\lambda = \sum_\lambda \boldsymbol{\epsilon}_{\mathbf{k}}^\lambda |a_\lambda| e^{i\alpha_\lambda} \tag{7.24a}$$

where a_1, a_2 are the generally complex components of $\mathbf{a}$ and α_1, α_2 are their phases. Now

$$\mathbf{E}(\mathbf{r}, t) = \sum_\lambda \boldsymbol{\epsilon}_{\mathbf{k}}^\lambda |a_\lambda| \text{ Re } (e^{i(\mathbf{k}\cdot\mathbf{r} - \omega t + \alpha_\lambda)}) \tag{7.24b}$$

Each of the two terms superposed here describes what is called a *linearly polarized* wave because the electric vector in each has a steady direction, $\boldsymbol{\epsilon}_{\mathbf{k}}^1$ and $\boldsymbol{\epsilon}_{\mathbf{k}}^2$, respectively, in all phases of the propagation. The latter unit vectors are accordingly called *polarization vectors*. The conclusion is that every well-defined plane wave is resolvable into two linearly polarized component waves orthogonal to each other (but with an absolute orientation that need be restricted only by the requirement of transversality).

The *resultant* wave (7.24) can be studied by considering its behavior as a function of time in any one of its transverse-phase planes (having some given value for $\mathbf{k}\cdot\mathbf{r}$). Since the plane, the frame orientation in it, the origin of $\mathbf{r}$ and the zero of time are all matters of choice, there will be no real loss of generality if the particular plane at $\mathbf{k}\cdot\mathbf{r} = -\alpha_1$ is chosen for examination. The representation of the two orthogonal components of $\mathbf{E}$ in the chosen plane is then simplified to

$$\mathsf{E}_1 = |a_1| \cos \omega t \qquad \mathsf{E}_2 = |a_2| \cos (\omega t + \delta) \tag{7.25}$$

where $\delta = \alpha_1 - \alpha_2$ is the phase difference between the components. It is well known‡ that a point executing simultaneous simple harmonic oscillations at right angles to each other follows a closed elliptical orbit when the two frequencies are equal, as here. For this reason, the most general well-defined plane wave is said to be *elliptically polarized*.

† See footnote on page 188.

‡ The reader is probably familiar with the Lissajous figures that can be formed by the superposition of simple harmonic oscillations at right angles to each other, as on an oscillograph screen. The Lissajous figure becomes a simple, closed ellipse for equal frequencies (Ref. 17, sec. 4.1 and exercises 4.3 and 4.4).

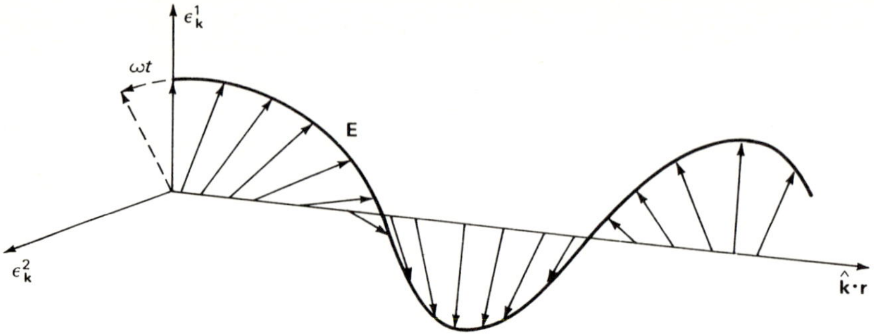

Figure 7.2 The electric vector in a left circularly polarized wave.

The ellipse degenerates into a straight-line segment whenever the phases of a_1, a_2 are such that δ is an integer multiple of π, for then (7.25) yields

$$\frac{\mathsf{E}_2}{\mathsf{E}_1} = \pm\frac{|a_2|}{|a_1|}$$

a constant slope made by the resultant electric vector with the axis $\boldsymbol{\epsilon}^1_{\mathbf{k}}$. This only means that the resultant plane wave can itself be linearly polarized, without having this polarization parallel to either $\boldsymbol{\epsilon}^1_{\mathbf{k}}$ or $\boldsymbol{\epsilon}^2_{\mathbf{k}}$ as initially chosen.

It can also happen† that the phase difference is 90°: $\delta = -\frac{1}{2}\pi$ or $+\frac{1}{2}\pi$. If, in addition, the amplitudes are equal, with $|a_1| = |a_2| = |a|$, then

$$\mathsf{E}_1 = |a| \cos \omega t \qquad \mathsf{E}_2 = \pm |a| \sin \omega t \tag{7.26}$$

and

$$\mathsf{E}_1^2 + \mathsf{E}_2^2 = \mathsf{E}^2 = |a|^2$$

The electric vector now describes a circle, and the resultant wave is said to be *circularly polarized.*

The upper sign in (7.26) corresponds to $\delta = -\frac{1}{2}\pi$ in (7.25) and helps describe a *counterclockwise* rotation of the electric vector in a transverse plane through which the wave is passing as viewed head on *from* the direction into which the wave is progressing. It is from this point of view that the wave is conventionally said to be *left* circularly polarized. Figure 7.2 indicates the distribution in space of such a wave; the arrows indicate the electric vector in phases that have occurred *earlier* at the origin. The phase difference $\delta = +\frac{1}{2}\pi$ similarly describes a right-handedly polarized plane wave.

The vector expression (7.24*b*) becomes

$$\mathbf{E}_{\pm}(\mathbf{r}, t) = \text{Re}\left[a(\boldsymbol{\epsilon}^1_{\mathbf{k}} \pm i\boldsymbol{\epsilon}^2_{\mathbf{k}})e^{i(\mathbf{k}\cdot\mathbf{r}-\omega t)}\right] \tag{7.27}$$

† $\delta = \pm\frac{1}{2}\pi$ will occur whenever the directions $\boldsymbol{\epsilon}^{1,2}_{\mathbf{k}}$ have been so chosen that they turn out to lie along the principal axes of the ellipse, as is easily verified (Exercise 7.3).

for the left and right circularly polarized waves, respectively. In the terms of (7.25) and (7.26), this expression has $\alpha_1 = \arg a$ and $\alpha_2 = \alpha_1 \pm \frac{1}{2}\pi$, and hence $\delta = \alpha_1 - \alpha_2 = \mp \frac{1}{2}\pi$ properly.

The expression (7.27) suggests using a helicity resolution of the polarization vectors, like that illustrated for the position vector in (3.92), in place of the cartesian frame of (7.6). Accordingly, define

$$\boldsymbol{\epsilon}_{k,\,m=\pm 1} \equiv \mp 2^{-1/2}(\boldsymbol{\epsilon}_{\mathbf{k}}^1 \pm i\boldsymbol{\epsilon}_{\mathbf{k}}^2) \qquad \boldsymbol{\epsilon}_{k0} \equiv \hat{\mathbf{k}} \tag{7.28a}$$

with the properties

$$\boldsymbol{\epsilon}_{km}^* \cdot \boldsymbol{\epsilon}_{km'} = \delta_{mm'} \qquad \boldsymbol{\epsilon}_{km}^* = (-1)^m \boldsymbol{\epsilon}_{k,\,-m} \tag{7.28b}$$

On this basis, the *left* circularly polarized wave in (8.27) is said to have *positive* helicity ($m = +1$), and the right circularly polarized wave has helicity $m = -1$.

The connections (7.28) can be used to eliminate $\boldsymbol{\epsilon}_{\mathbf{k}}^{1,2}$ in favor of $\boldsymbol{\epsilon}_{k,\,\pm 1}$ in the expression (7.24) for the *general* plane wave. The result can be written

$$\mathbf{E}(\mathbf{r}, t) = \sum_{m=\pm 1} \operatorname{Re}\,(\boldsymbol{\epsilon}_{km} a_m e^{i(\mathbf{k}\cdot\mathbf{r}-\omega t)}) \tag{7.29}$$

where $\qquad a_{+1} = -2^{-1/2}(a_1 - ia_2) \qquad a_{-1} = 2^{-1/2}(a_1 + ia_2)$

or

$$a_m = \boldsymbol{\epsilon}_{km}^* \cdot \mathbf{a} \tag{7.30}$$

projections of the amplitude vector $\mathbf{a}$ of (7.24*a*) onto the helicity basis (7.28). Thus, just as the general plane wave can be decomposed into orthogonal, *linearly* polarized waves as in (7.24), it can also be expressed as a superposition of the linearly independent left and right circularly polarized waves.

Light exhibits polarizability, and this furnishes one of the ways in which its identification with electromagnetic waves was confirmed.

There exist crystals† which are ordinarily transparent to light but which block it, by scattering and absorption, when two of them in a certain relative orientation are placed in the path of light. The distinguishability of a certain relative orientation testifies to an anisotropy of the crystals. Their behavior can then be interpreted by supposing that each has in it a preferred direction, an *optic axis*, and that only electric oscillations perpendicular to this axis are transmitted. It can then be expected that light which has been transmitted through one of the crystals (called the *polarizer*) will be linearly polarized and therefore not transmitted by a second crystal (called the polarization *analyzer*) having its optic axis at right angles to the optic axis of the polarizer. The fraction of the light transmitted by the analyzer as a function of angle, when this is not restricted to a right angle, is exactly what is expected from such an analysis as represented by (7.24) (see Exericse 7.4.)

† Tourmaline is one example. More familiar nowadays are Polaroid films which exhibit like properties. Specially constructed crystalline devices called *Nicol prisms* are also used to exhibit the properties discussed here.

Most essential to these interpretations is the *transverse* character of the light waves. It is one of the most distinctive characteristics found for electromagnetic waves and stems directly from their constitution of vector fields that are divergenceless between sources.

As suggested in the course of the preceding discussion, light from ordinary, macroscopic sources is usually *unpolarized*, so that only a part transmitted by a polarizer is extinguishable by an analyzer. Unpolarized light itself, even if it is monochromatic and monodirected, cannot be described by any of the plane waves (7.24) with a well-defined amplitude vector $\mathbf{a}$. Whatever the phases of the components a_1, a_2 may be, if they are definite constants, a resultant wave is described that is at least elliptically polarized. Unpolarized light must be considered a *mixture* of plane waves in all possible phases, i.e., a *random* superposition of them. This must be expected of light originating from sources spread over space and time by any more than a wavelength and period. Such a spread must generally be expected to produce phases $\hat{\mathbf{k}} \cdot (\mathbf{r} - \mathbf{r}_s) - c(t - t_s)$ with $\mathbf{r}_s$, t_s distributed at random over a wavelength and period, respectively.

7.3 WAVE FIELDS WITHIN PERFECTLY REFLECTING BOUNDARIES

The plane-wave representations are particularly useful in meeting boundary conditions at plane surfaces the propagated fields might encounter. Boundaries separating dielectric spaces are treated in Chap. F and imperfect-conductor surfaces in Chap. G. To be considered here are boundaries beyond which the fields cannot penetrate, like those used to confine electrostatic fields in Sec. 3.2. Such perfectly reflecting boundaries are, in practice, approximated by good conductor surfaces and so are also called perfect conductors. As is confirmed in Chap. G, any good low-resistance conductor excludes nonstatic electric fields with high enough frequencies of variation quite as thoroughly as in electrostatic situations. Moreover, the varying magnetic fields that must accompany the electric ones are also excluded, unlike the independent magnetostatic fields. As a consequence, good-conductor boundaries can be used to *shape* the free-space fields, and it is this fact that gives the attendant problems general interest.

The Boundary Conditions on Perfect Conductors

Conditions at surfaces of separation can always be derived as degenerated forms of the Maxwell equations applying in a transition layer of vanishing thickness $\delta \to 0$. Here the transition is to be from a free-space field to a completely fieldless region.

What happens in a transition layer to equations for field divergences was found in (2.31), where $\nabla \cdot \mathbf{E}$ was shown to degenerate to the surface divergence $\mathbf{n} \cdot \Delta\mathbf{E} = \Delta E_n$ at a surface element $d\mathbf{S} = \mathbf{n}\, dS$. The ΔE_n stands for a discontinuous jump in the value of the field component normal to the surface and, since $\mathbf{E} = 0$

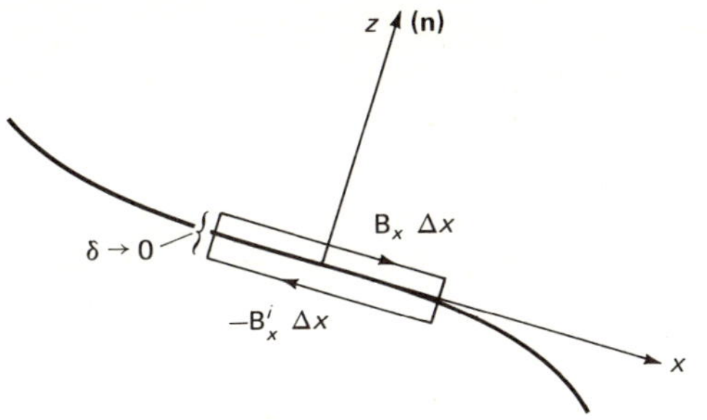

Figure 7.3

on one side of it here, $\Delta \mathsf{E}_n = \mathsf{E}_n$, just the normal component of the field on the free-space side. Thus, as in (2.31), $\nabla \cdot \mathbf{E} = 4\pi\rho$ is reduced to $\mathsf{E}_n = 4\pi\sigma$, where $\sigma = \lim \rho\delta$ represents a surface distribution of charge that is induced in the conductor by the electric field. In a similar way, the equation $\nabla \cdot \mathbf{B} = 0$ leads to $\mathsf{B}_n = 0$ at the surface.†

The curl equations require more elaborate consideration here. How they connect field values on the two sides of a surface of separation is found by applying the Stokes theorem to a rectangular area that encloses a line segment in the surface within its breadth $\delta \to 0$, as indicated in Fig. 7.3. The flux of the curl $\nabla \times \mathbf{B} = 4\pi\mathbf{j}/c + \mathring{\mathbf{E}}/c$ passing through the rectangle will have a y direction in the terms of the figure, and the equivalent line integral about the perimeter equals

$$(\mathsf{B}_x - \mathsf{B}_x^i)\,\Delta x = \frac{(4\pi j_y + \mathring{\mathsf{E}}_y)\,\Delta x\delta}{c}$$

A flux $\mathring{\mathsf{E}}_y\,\Delta x\delta$ of finite field values with finite rates of variation will vanish as $\delta \to 0$, but, just as $\rho\delta \to \sigma$ for charge induced on the conductor, a generation of *surface current* $j_y\delta \to J_y$ must be allowed for on a perfect conductor. A current $\mathbf{J}$, directed along the surface, will represent flux of charge per unit-wide strip of the surface. The factor $\mathsf{B}_x - \mathsf{B}_x^i = \Delta\mathsf{B}_x$ is the y component of the cross product

$$\mathbf{n} \times \Delta\mathbf{B} = \frac{4\pi\mathbf{J}}{c} \tag{7.31}$$

representing [compare (B.4)] the surface curl to which $\nabla \times \mathbf{B}$ has degenerated in the transition layer. Of course $\mathbf{B}^i = 0$ on the fieldless side, and $\Delta\mathbf{B} \to \mathbf{B}$ at the free surface. The remaining curl equation, $\nabla \times \mathbf{E} = -\mathring{\mathbf{B}}/c$, involves no charges or currents and so reduces to $\mathbf{n} \times \Delta\mathbf{E} = \mathbf{n} \times \mathbf{E} = 0$. This is equivalent to a vanishing of the component $\mathbf{E}_{\parallel}$ parallel to the surface.

In summary, the field in a free space outside a perfectly conducting boundary must adjust itself to having

$$\mathbf{E}_{\parallel} = 0 \qquad \text{and} \qquad \mathsf{B}_n = 0 \tag{7.32}$$

† Those who have read Chaps. B and E will recognize the same conditions listed in (B.2) and (E.9).

at the boundary. The remaining conditions put no such restrictions on the fields that can exist. They instead represent adjustments made in the conductor to any boundary values of E_n and $\mathbf{B}_{\|}$ that may be incident. Charges are induced and currents generated on the surface, as given by

$$\sigma = \frac{\mathsf{E}_n}{4\pi} \qquad \text{and} \qquad \mathbf{J} = \frac{c}{4\pi}\mathbf{n} \times \mathbf{B} \tag{7.33}$$

These help exclude the field from the perfectly conducting interior.

Waves Guided by Conductors

Electromagnetic wave fields that are confined to channels through lateral enclosure by perfectly conducting walls will now be considered. Such channels are called *wave guides* and are useful for transmitting power in the form of high-frequency (especially microwave) oscillations, which tend to be excluded from conducting lines (see page 596). The problem may be regarded as a generalization to nonstatic fields of one treated in Sec. 3.2, where (3.36) was obtained as an electro*static* field that can exist in a channel.

The channel to which the free-space field is confined will be taken to be indefinitely long and to have a uniform cross section. A z axis parallel to the channel and to its perfectly conducting walls will be adopted, and field representatives of the form

$$\mathbf{E} = \mathbf{f}(x, y)e^{i(k_z z - \omega t)} \qquad \mathbf{B} = \mathbf{g}(x, y)e^{i(k_z z - \omega t)} \tag{7.34}$$

will be sought. Such can be expected in the composition of arbitrary fields as results of Fourier-analyzing their z dependence and their time dependence. Fourier analysis of the x and y dependences would yield results like

$$\left.\begin{matrix}\mathbf{f}\\ \mathbf{g}\end{matrix}\right\} = \sum_{k_x, k_y} \mathbf{a}_{k_x k_y} e^{i(k_x x + k_y y)} \tag{7.35}$$

but other forms can be more convenient, depending on the shape of the channel cross section. If solutions of the type (7.34) can be found, they will describe the propagation of electromagnetic waves down the channel.

The forms (7.34) will properly obey the wave-propagation equations (7.2) if the modulating functions $\mathbf{f}(x, y)$ and $\mathbf{g}(x, y)$ satisfy

$$[\nabla^2 + (k')^2]\left\{\begin{matrix}\mathbf{f}(x, y)\\ \mathbf{g}(x, y)\end{matrix}\right\} = 0 \qquad \text{with } (k')^2 \equiv \left(\frac{\omega}{c}\right)^2 - k_z^2 \tag{7.36}$$

The forms must also obey the monochromatic Maxwell equations (6.57) as adapted to free space ($\rho_\omega \equiv 0$ and $\mathbf{j}_\omega \equiv 0$). It is sufficient for the curl equations, $ik\mathbf{B} = \nabla \times \mathbf{E}$ and $ik\mathbf{E} = -\nabla \times \mathbf{B}$, to be satisfied since their solutions will be divergenceless automatically. Those curl equations yield conditions on the components of $\mathbf{f}$ and $\mathbf{g}$ that are simple to form when it is recognized that solutions of the type (7.34) are eigenfunctions of the operation $-i\,\partial/\partial z$ ($\equiv k_z$), equivalent to multiplication by the eigenvalue k_z

$$
\begin{aligned}
ikg_x &= \frac{\partial f_z}{\partial y} - ik_z f_y & ikf_x &= -\frac{\partial g_z}{\partial y} + ik_z g_y \\
ikg_y &= ik_z f_x - \frac{\partial f_z}{\partial x} & ikf_y &= -ik_z g_x + \frac{\partial g_z}{\partial x} \\
ikg_z &= (\nabla \times \mathbf{f})_z & ikf_z &= -(\nabla \times \mathbf{g})_z
\end{aligned}
\tag{7.37}
$$

The transverse components can be put entirely in terms of the longitudinal components f_z, g_z through substitutions of the expressions for $f_{x,y}$ into the equations for $g_{x,y}$ and vice versa:

$$
\begin{aligned}
g_x &= \frac{i}{(k')^2}\left(k_z \frac{\partial g_z}{\partial x} - k\frac{\partial f_z}{\partial y}\right) & f_x &= \frac{i}{(k')^2}\left(k_z \frac{\partial f_z}{\partial x} + k\frac{\partial g_z}{\partial y}\right) \\
g_y &= \frac{i}{(k')^2}\left(k_z \frac{\partial g_z}{\partial y} + k\frac{\partial f_z}{\partial x}\right) & f_y &= \frac{i}{(k')^2}\left(k_z \frac{\partial f_z}{\partial y} - k\frac{\partial g_z}{\partial x}\right)
\end{aligned}
\tag{7.38}
$$

where $(k')^2 \equiv k^2 - k_z^2$ of (7.36). These automatically satisfy the final line of the requirements (7.37), $(\nabla \times \mathbf{f})_z = ikg_z$ and $(\nabla \times \mathbf{g})_z = -ikf_z$, for any choices of f_z and g_z that obey the Helmholtz equations (7.36). They then also yield $\nabla \cdot \mathbf{f} = -ik_z f_z$ and $\nabla \cdot \mathbf{g} = -ik_z g_z$, in consistency with $\nabla \cdot \mathbf{E} = 0$ and $\nabla \cdot \mathbf{B} = 0$.

To see the effect of the boundary conditions (7.32), choose a *temporary* cartesian frame with origin at any one wall element $d\mathbf{S} = \mathbf{n}\,dS$, as indicated in Fig. 7.4, with the y axis taken parallel to $\mathbf{n}$, normal to the surface element $dS = dx\,dz$. Then the components $\mathbf{E}_\parallel$, parallel to the surface, and the normal component $\mathbf{B}_n$ properly vanish if only

$$
f_z(x, 0) = 0 \qquad f_x(x, 0) = 0 \qquad g_y(x, 0) = 0
$$

relative to the temporary frame. Since $f_z(x, 0) = 0$ is to hold as x varies, $\partial f_z/\partial x = 0$ and it then follows from (7.38) that both f_x and g_y will vanish at the

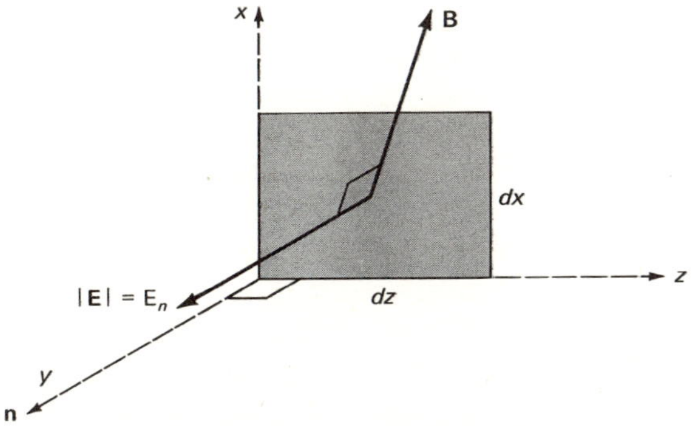

Figure 7.4

wall element if just the condition $\partial g_z/\partial y = 0$ is imposed. Thus, the conditions on **f** and **g** at all the lateral boundaries can be represented as

$$f_z = 0 \qquad \text{and} \qquad \nabla_n g_z \equiv (\mathbf{n} \cdot \nabla) g_z = 0 \tag{7.39}$$

if the z axis is taken along the channel direction.

It should now be clear that field representatives of the form (7.34) can be generated by making arbitrary and independent choices of the longitudinal modulators $f_z(x, y)$ and $g_z(x, y)$; each only needs to be a solution of the Helmholtz equation (7.36) satisfying the boundary conditions (7.39). An arbitrary field in the channel can be constructed as a superposition of independent modes of two types, one derived from choices of f_z with $g_z \equiv 0$ and a second type derivable from g_z with $f_z \equiv 0$. Fields of the first type are called *transverse magnetic* (TM) *modes*, and those of the second type are the *transverse electric* (TE) *modes*. Neither type has *both* its electric and magnetic fields transverse† to the channel direction, despite previous findings about the transversalities of waves in free spaces, because k_z is not the only component of the multiply reflected wave vectors $\mathbf{k}(k_x, k_y, k_z)$.

The description of each *TM mode* ($g_z \equiv 0$) is derivable from a solution $f_z(x, y)$ through the two-component (x, y) equations

$$\mathbf{f} = \frac{ik_z}{(k')^2} \nabla f_z(x, y) \qquad \mathbf{g} = -\frac{ik}{(k')^2} (\nabla \times \mathbf{f}) \tag{7.40}$$

according to (7.38). Both expressions yield only transverse (x, y) components because f_z does not vary with z, and $(\nabla \times \mathbf{f})_z = 0 = ikg_z$ follows from the gradient expressions for $f_{x,y}$. The curl expressions for **g** show that it is divergenceless, in keeping with $\nabla \cdot \mathbf{g} = -ik_z g_z = 0$ for a TM wave. Whereas only the z component of $\nabla \times \mathbf{f}$ vanishes, $(\nabla \times \mathbf{g})_{x,y} = 0$ and $(\nabla \times \mathbf{g})_z = -ikf_z \neq 0$. Note that $\mathbf{f} \cdot \mathbf{g} = 0$ and hence $\mathbf{E} \perp \mathbf{B}$.

Each *TE mode* ($f_z \equiv 0$) is similarly derivable from the two-component equations:

$$\mathbf{g} = \frac{ik_z}{(k')^2} \nabla g_z(x, y) \qquad \mathbf{f} = +\frac{ik}{(k')^2} (\nabla \times \mathbf{g}) \tag{7.41}$$

Here $\nabla \cdot \mathbf{f} = -ik_z f_z = 0$, $(\nabla \times \mathbf{f})_{x,y} = 0$, and $(\nabla \times \mathbf{g})_z = -ikf_z = 0$. Again **E** and **B** are mutually perpendicular.

Solutions of the form (7.35) will now be used for illustrative purposes, and since they are most readily adapted to the boundary conditions in a channel of rectangular cross section, one defined by perfectly conducting planes at $x = 0$, $x = L_x$, $y = 0$, and $y = L_y$ will be considered. To describe a TM mode in this

† Fields with both $f_z = 0$ and $g_z = 0$ (TEM modes) can exist when there is a second conductor occupying the central axis of the channel, as in coaxial cables, investigated in Exercise 7.15.

channel, a form for $f_z(x, y)$ that vanishes on all the walls is needed (7.39); any one of the complete set

$$f_z(x, y) = a \sin \frac{n_x \pi x}{L_x} \sin \frac{n_y \pi y}{L_y} \qquad n_x, n_y = 1, 2, 3, \ldots \tag{7.42}$$

may be used. This set has already been developed for (3.36), to represent electro*static* potentials required to vanish on the walls of a rectangular tube. Describing a TE mode in the same channel requires a form for $g_z(x, y)$ that has vanishing normal gradients at the walls. The requisite forms are obviously to be obtained by replacing (7.42) with the equally complete set

$$g_z(x, y) = b \cos \frac{n_x \pi x}{L_x} \cos \frac{n_y \pi y}{L_y} \qquad n_x, n_y = 0, 1, 2, 3, \ldots \tag{7.43}$$

Each component gradient $\nabla_{x,y} g_z$ has one of the cosines replaced by a sine that vanishes at the walls normal to the gradient.

Any one of the forms (7.42) or (7.43) properly satisfies the Helmholtz equation (7.36) if

$$(k')^2 \equiv \left(\frac{\omega}{c}\right)^2 - k_z^2 = \left[\left(\frac{n_x}{L_x}\right)^2 + \left(\frac{n_y}{L_y}\right)^2\right]\pi^2 \tag{7.44}$$

as in (3.37). In a mode labeled by a specific pair of integers n_x, n_y, a number n_x of transverse half wavelengths $\pi/k_x = L_x/n_x$ must fit into the dimension L_x of the channel and n_y half-wavelengths $\pi/k_y = L_y/n_y$ must fit into L_y. If the applied frequency ω has a value less than that of

$$\omega_{n_x n_y} = \left[\left(\frac{n_x}{L_x}\right)^2 + \left(\frac{n_y}{L_y}\right)^2\right]^{1/2} \pi c \tag{7.45}$$

k_z becomes imaginary and the representative (7.34) decays exponentially along the channel, in proportion to $\exp(-|k_z|z)$; the corresponding mode is not propagated indefinitely far even though no resistive losses have been taken into account. In the static ($\omega = 0$) limit, the exponential attenuation is proportional to $\exp(-\omega_{n_x n_y} z/c)$, exactly as found in connection with the electrostatic evaluations stemming from (3.36).

The result (7.45) is an example of a modal *cutoff frequency* characteristic of a channel with given dimensions. It is a value that must be surpassed by the applied frequency if the mode in question is to be propagated down the channel. It represents properties of wave guides that must be expected whatever their cross-sectional shape. There will always be maximal wavelengths that can fit into a finite cross section and hence a minimum of frequency that can be propagated indefinitely far.

A given mode (n_x, n_y) will be propagated by any of the *continuum* of applied frequencies $\omega > \omega_{n_x n_y}$; increasing ω primarily shortens the longitudinal wavelength

$$\frac{2\pi}{k_z} = \frac{2\pi c}{(\omega^2 - \omega_{n_x n_y}^2)^{1/2}} \tag{7.46}$$

of the propagation. If ω_0 is the lowest cutoff that can be given by (7.45) for given channel dimensions and ω_1 is the next lowest, any of the continuum of frequencies $\omega_0 < \omega < \omega_1$ will be propagated by just the modes that yield ω_0.

The lowest frequency with which a TM mode (7.42) can be propagated is

$$\omega_{11} = \frac{\pi c}{L_x}\left[1 + \left(\frac{L_x}{L_y}\right)^2\right]^{1/2} \tag{7.47}$$

The cutoff frequency of a TE mode (7.43) can be lower since a mode can exist for $n_x = 0$ *or* $n_y = 0$ [both cannot vanish, since that corresponds to $(k')^2 = 0$ and infinite results for the transverse components fields (7.41), requiring an infinite excitation energy]. Suppose the x axis has been chosen to lie parallel to the longer of the cross-sectional dimensions, so that† $L_x > L_y$. Then

$$\omega_0(\text{TE}) = \omega_{10} = \frac{\pi c}{L_x} \qquad \text{and} \qquad \omega_1(\text{TE}) = \omega_{01} = \frac{\pi c}{L_y} \tag{7.48}$$

are lowest and next to the lowest cutoff frequencies (it is being supposed for the latter that $L_y > \frac{1}{2}L_x > \frac{1}{2}L_y$). Even $\omega_1(\text{TE})$ is lower than $\omega_0(\text{TM})$, since the ratio $\omega_{11}/\omega_{01} = [(L_y/L_x)^2 + 1]^{1/2}$ is greater than unity. Thus a single TE mode is propagated when frequencies in the range $\pi c/L_x < \omega < \pi c/L_y$ are applied. (Usually it is the propagation of a given frequency ω that is desired, and channel dimensions are chosen so that the given ω will lie in a desired range.)

The complete description of the dominant ($n_x = 1$, $n_y = 0$) TE mode is implicit in the form (7.34) as generated by $g_z(x, y)$ of (7.43). The corresponding real field can be expressed as

$$\mathsf{B}_z = \text{Re}\,(g_z e^{i(k_z z - \omega t)}) = |b| \cos\frac{\pi x}{L_x} \sin\,(k_z z - \omega t) \tag{7.49a}$$

if the phase of the arbitrary constant b is so chosen ($\arg b = -\frac{1}{2}\pi$) that this has a node at $z = 0$, $t = 0$. Then the transverse components that follow from (7.41) are

$$\mathsf{B}_x = -\frac{k_z L_x}{\pi}|b| \sin\frac{\pi x}{L_x}\cos\,(k_z z - \omega t) \tag{7.49b}$$

$$\mathsf{E}_y = +\frac{k L_x}{\pi}|b| \sin\frac{\pi x}{L_x}\cos\,(k_z z - \omega t) \tag{7.49c}$$

with $\mathsf{B}_y = 0$ and $\mathsf{E}_x = 0$ because g_z is independent of y (a result of $n_y = 0$). The entire field is uniform along the short dimension ($L_y < L_x$). The transverse components are propagated out of phase with the longitudinal one by 90° [formally the result of the factors i in (7.41)].

An instantaneous ($t = 0$) field-line distribution, following from Eqs. (7.49), is sketched in Fig. 7.5 as "viewed" in a yz plane. The density of the vertical **E** lines

† Theoretically interesting degeneracies can occur when $L_x = L_y$. Compare the vibrations in a *square* membrane as discussed, for example, in Ref. 17, sec. 13.1.

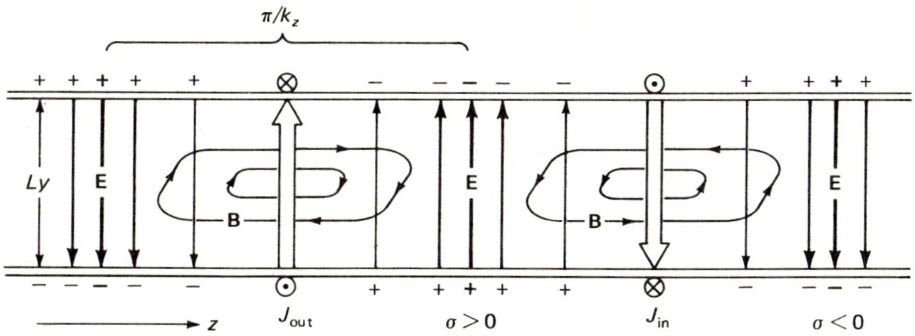

Figure 7.5

is supposed to be proportional to cos $k_z z$ in each yz plane, but only some lines in their regions of greatest concentration are shown. Their densities actually rarefy to zero [$\sim \sin(\pi x/L_x)$] toward front and back, away from the midplane at $x = \frac{1}{2}L_x$. Only a central xz plane of **B** loops, actually of density proportional to $\sin k_z z$, is indicated (in some perspective) since there are no variations with y. Also indicated are the directions of the maximal surface currents on the "front" ($x = 0$) face—as they follow from the Ampère law $\mathbf{J} = c(\mathbf{n} \times \mathbf{B})/4\pi$. The view here was chosen for display because of its resemblance to the chain of Faraday and Maxwell inductions shown in Fig. 1.4. The **E** loops of that chain are here interrupted where they intersect the conductor walls, by the induced charges $|\sigma| = \mathsf{E}/4\pi$, having the signs indicated. Note that the everywhere mutually perpendicular **E** and **B** lines yield a right-directed energy flow $c\mathbf{E} \times \mathbf{B}/4\pi$ everywhere along the channel when the conduction is perfect.

Cavity Resonators

A nice example of isolating an energy-conserving field is presented when modes like those described in the preceding subsection are trapped in a length of the channel by perfectly conducting planar end walls, placed at $z = 0$ and $z = L_z$, say. The result is an idealization of a *cavity resonator*.

Establishing a field in the cavity will require adaptation to the additional boundary conditions $\mathsf{E}_{x,y} = 0$ and $\mathsf{B}_z = 0$ at $z = 0$ and $z = L_z$, according to (7.32). The TM modes already have $\mathsf{B}_z \equiv 0$ everywhere, but their components $\mathsf{E}_{x,y}$ will have to be restricted to such linear combinations of forms (7.34) as have z dependences proportional to some $\sin(n_z \pi z/L_z)$ with integer n_z, a standing-wave combination of incidences on, and reflections from, the end walls (with $k_z = \pm n_z \pi/L_z$). The suitable choice should be introduced as an expression for $\mathsf{E}_z \sim f_z$, since the remaining TM field components are derivable from it, and what that may be becomes evident from just the divergencelessness $\nabla \cdot \mathbf{E} = 0$. This requires

$$\frac{\partial \mathsf{E}_z}{\partial z} = -(\nabla_x \mathsf{E}_x + \nabla_y \mathsf{E}_y) \sim \sin\frac{n_z \pi z}{L_z}$$

and hence any one of the complete set

$$\mathsf{E}_z = f_z(x, y) \cos \frac{n_z \pi z}{L_z} \cos \omega t \tag{7.50a}$$

may be chosen for a possible TM mode in the cavity; $f_z(x, y)$ is some standing wave like (7.42) and so can always be taken real ($\arg a = 0$). A given choice will lead to corresponding linear combinations of the transverse components (7.40), with

$$\mathsf{E}_{x,y} = -\frac{n_z \pi}{(k')^2 L_z} (\nabla_{x,y} f_z) \sin \frac{n_z \pi z}{L_z} \cos \omega t \tag{7.50b}$$

$$\mathbf{B} = -\frac{k}{(k')^2} (\nabla \times \mathbf{f}) \cos \frac{n_z \pi z}{L_z} \sin \omega t \tag{7.50c}$$

as the results. The switches from sines to cosines, and vice versa, are dictated by the phase factors $i = \exp \frac{1}{2} i\pi$ appearing in (7.40).

A TE mode may be excited in the cavity instead, and its description is derivable from

$$\mathsf{B}_z = g_z(x, y) \sin \frac{n_z \pi z}{L_z} \cos \omega t \tag{7.51a}$$

which properly vanishes at the end walls. The corresponding transverse components are

$$\mathsf{B}_{x,y} = \frac{n_z \pi}{(k')^2 L_z} (\nabla_{x,y} g_z) \cos \frac{n_z \pi z}{L_z} \cos \omega t \tag{7.51b}$$

$$\mathbf{E} = \frac{k}{(k')^2} (\nabla \times \mathbf{g}) \sin \frac{n_z \pi z}{L_z} \sin \omega t \tag{7.51c}$$

with the electric components properly vanishing at the end walls.

Satisfaction of the wave equation by any one of the above field components obviously demands that

$$\omega = c[(k')^2 + k_z^2]^{1/2} = c\left[(k')^2 + \left(\frac{n_z \pi}{L_z}\right)^2\right]^{1/2} \tag{7.52a}$$

and this has the symmetrical† expression

$$\omega = [\omega_{n_x n_y}^2 + (ck_z)^2]^{1/2} = \pi c\left[\left(\frac{n_x}{L_x}\right)^2 + \left(\frac{n_y}{L_y}\right)^2 + \left(\frac{n_z}{L_z}\right)^2\right]^{1/2} \tag{7.52b}$$

in the special case (7.44) of a rectangular cross section. It is evident in any case that the fields established in a cavity (by multiple reflections from the enclosing

† This calls attention to the fact that the classifications into TM and TE modes can be made relative to any one of the directions in a rectangular cavity. For example, a TM(x) mode must be some linear combination of phase-shifted TM(z) *and* TE(z) modes.

walls) will be superpositions of modes that have their frequencies restricted to *discrete* values. A single mode, such as one labeled by integers n_x, n_y, n_z in the case of a rectangular cavity, will be excited by itself only if it is in resonance with an applied frequency having the corresponding discrete value. To achieve resonance with a given frequency ω, the cavity must have its dimensions altered (the cavity must be tuned to ω, as by using a movable plunger for one of the end walls) until (7.52) is satisfied.

Notice that TM modes with $n_z = 0$, uniform in z, exist, having only a longitudinal component (7.50*a*) of electric field and magnetic components (7.50*c*) transverse to it. The TE modes must have $n_z \geq 1$ since (7.51*a*) vanishes for $n_z = 0$. In a rectangular cavity, the lowest TM resonance frequency is just $\omega_0(\text{TM}) = \omega_{11}$ of (7.47), while the lowest TE resonance has the fundamental

$$\omega_0(\text{TE}) = \frac{\pi c}{L_x}\left[1 + \left(\frac{L_x}{L_z}\right)^2\right]^{1/2} \qquad \text{if } L_x > L_y$$

greater than the cutoff frequency of (7.48). If $L_z < L_y$, the fundamental, $\omega_0(\text{TM}) = \omega_{11}$, of the TM mode is lower than $\omega_0(\text{TE})$!

Breadths of Resonances

The absolute discreteness (7.52) of the cavity resonance frequencies is an idealization that is the more nearly valid the longer the energy stored in the oscillations can be absolutely conserved. To see the connection, consider the field energy in the cavity while it houses some one of the modes and then the effect of energy losses that might occur.

The time-averaged energy (6.65) in just the longitudinal field, (7.50*a*) or (7.51*a*), is the simplest to calculate, with the use of $\langle \cos^2 \omega t \rangle = \frac{1}{2}$:

$$\left\langle W \begin{Bmatrix} \mathsf{E}_z \\ \mathsf{B}_z \end{Bmatrix} \right\rangle = \frac{A}{8\pi}\frac{1}{2}\oint \frac{dA}{A}\begin{Bmatrix} f_z^2 \\ g_z^2 \end{Bmatrix}\int_0^{L_z} dz \begin{Bmatrix} \cos^2 \\ \sin^2 \end{Bmatrix} \frac{n_z \pi z}{L_z}$$

where A is the cross-sectional area of the cavity, whatever its lateral shape. The integration over the length L_z yields $\frac{1}{2}L_z$, and so

$$\left\langle W \begin{Bmatrix} \mathsf{E}_z \\ \mathsf{B}_z \end{Bmatrix} \right\rangle = \frac{V}{32\pi}\left\langle \begin{matrix} f_z^2 \\ g_z^2 \end{matrix} \right\rangle_A$$

where $V \equiv AL_z$ is the volume of the cavity and the indicated average is over the cross section A. The energy contribution $W(\mathsf{E}_z)$ is doubled for the special TM mode $n_z = 0$, for then there is uniformity in z. Evaluation of the energy in the transverse components will require integrations like

$$\oint dA(\nabla f_z)^2 = \iint dx\, dy\left[\left(\frac{\partial f_z}{\partial x}\right)^2 + \left(\frac{\partial f_z}{\partial y}\right)^2\right] = \oint dA\,(-f_z\, \nabla^2 f_z)$$

$$= (k')^2 \oint dA\, f_z^2 = (k')^2 A \langle f_z^2 \rangle_A$$

where $(\nabla f_z)\cdot(\nabla f_z) = \nabla\cdot(f_z \nabla f_z) - f_z \nabla^2 f_z$ could be replaced by just $-f_z \nabla^2 f_z$ because f_z vanishes on the side walls (7.39). The area integral of $(\nabla g_z)^2$ yields $(k')^2 A\langle g_z^2\rangle_A$ because the normal gradients of g_z vanish on the side walls. Area integrals of $(\nabla \times \mathbf{f})^2$ and $(\nabla \times \mathbf{g})^2$ give exactly the same results because, for example, $(\nabla \times \mathbf{f})_x = \partial f_z/\partial y$, $(\nabla \times f)_y = -\partial f_z/\partial x$, and $(\nabla \times \mathbf{f})_z = 0$ in a TM mode. The proportionality constants with which the gradients and curls appear in (7.40) and (7.41) then lead to totals

$$\left\langle W\begin{Bmatrix}\mathrm{TM}\\ \mathrm{TE}\end{Bmatrix}\right\rangle = \frac{V}{32\pi}\left\langle\begin{matrix}f_z^2\\ g_z^2\end{matrix}\right\rangle_A \left\{1+\frac{1}{(k')^2}\left[\left(\frac{n_z\pi}{L_z}\right)^2+k^2\right]\right\}$$

equivalent to
$$\left\langle W\begin{Bmatrix}\mathrm{TM}\\ \mathrm{TE}\end{Bmatrix}\right\rangle = \frac{V}{16\pi}\frac{k^2}{(k')^2}\left\langle\begin{matrix}f_z^2\\ g_z^2\end{matrix}\right\rangle_A \tag{7.53a}$$

In the special case of the rectangular cavity, (7.42) and (7.43) yield for any one mode

$$\langle W\rangle = \frac{V}{64\pi}\left(\frac{k}{k'}\right)^2 |a|^2 \tag{7.53b}$$

where a is an arbitrary longitudinal field amplitude that may be excited and $V = L_x L_y L_z$.

A given energy W stored in the cavity field will remain conserved (time-independent) only if the surface currents $\mathbf{J}$ that help contain it meet no resistance on the cavity walls. However, even the best conductors offer some resistance; I^2R losses of the type mentioned in connection with (6.9) and (6.10) must be expected, and so any given initial cavity energy $\langle W\rangle_o$ must be expected to become a decreasing function of time: $\langle W\rangle_t$. The rate of energy loss

$$\mathscr{P} \equiv -\frac{d}{dt}\langle W\rangle_t$$

must be proportional to the flux of the energy incident on the walls, and since the fluxes will have magnitudes $\langle W\rangle_t c$ according to (7.16), a proportionality of $\mathscr{P}$ to $\langle W\rangle_t$ itself must be expected. A constant of proportionality†

$$\tau \equiv \frac{\langle W\rangle_t}{\mathscr{P}} = \frac{\langle W\rangle_t}{-d\langle W\rangle_t/dt} \tag{7.54}$$

should be determined by the configuration and dimensions of the cavity, by the wall resistance, and by the character of the field modes in which the energy is

† In the technology of cavity resonators, it is customary to define instead the quantity $Q = (\omega_o/2\pi)\langle W\rangle_t/\mathscr{P}$, where ω_o is the resonance frequency excited, as a measure of the quality of the cavity. Q^{-1} is plainly a fractional energy loss in one period $T = 2\pi/\omega_o$ and is related to the mean time (7.54) of the energy decay via $\tau = QT$.

invested. Plainly, τ serves as a mean time in which any given initial energy $\langle W \rangle_o$ decays in value, since

$$\langle W \rangle_{t>0} = \langle W \rangle_o e^{-t/\tau} \tag{7.55}$$

follows from (7.54).

The decreasing field energy must be represented by appropriately decreasing field amplitudes $a(t > 0)$. A proportionality $\langle W \rangle_t \sim |a(t)|^2$ is to be expected according to (7.53). This presumes that the field *distributions* derived on the basis of steady oscillations, with some given frequency, hold at each instant of the secular decreases. The assumption is quite accurate; since the fields are propagated with the enormous velocity c, they cross the cavity in negligible times and so there is an almost instantaneous adjustment to each changed situation; each can well be treated as a steady field for purposes of distributing the field in space. The treatment is recognizable as an adiabatic approximation, in which $\langle W \rangle_t$ at each t is calculated from the equilibrium fields yielding $\langle W \rangle$ of (7.53).

Thus, during the exponential attenuation (7.55) of the field energy, there is a corresponding attenuation of every field-component representative, as in

$$\mathsf{E}(t) \sim a(t > 0)e^{-i\omega_r t} = a(0)e^{-i\omega_r t - t/2\tau} \tag{7.56}$$

where ω_r is one of the discrete resonance frequencies (7.52). The exponential aperiodicity introduces a continuous spectrum of new frequencies, deviating from ω_r, as can be seen from an analysis like (6.50) of

$$a(t > 0)e^{-i\omega_r t} = \int_{-\infty}^{\infty} d\omega \, a_\omega e^{-i\omega t}$$

where $$a_\omega = a(0) \int_0^\infty \frac{dt}{2\pi} e^{i(\omega - \omega_r)t - t/2\tau} = -\frac{a(0)}{2\pi} \frac{1}{i(\omega - \omega_r) - 1/2\tau} \tag{7.57}$$

There will consequently be field oscillations with various frequencies $\omega \neq \omega_r$, in which is stored energy proportional to

$$|a_\omega|^2 = \left[\frac{a(0)}{2\pi}\right]^2 \frac{1}{(\omega - \omega_r)^2 + (1/2\tau)^2} \tag{7.58}$$

A plot of this against ω, shown in Fig. 7.6, has what is called a *Lorentz line shape* with a *breadth at half-maximum* given by

$$\Delta\omega = \frac{1}{\tau} \tag{7.59}$$

Such an inverse relationship between the mean life of the energy of oscillations and the breadth of their resonance with exciting agencies is a very general characteristic† of states of oscillation. Narrow resonances characterize long-lived

† See Ref. 17, formula (3.56), for an example of a simple mechanical oscillation with the same property.

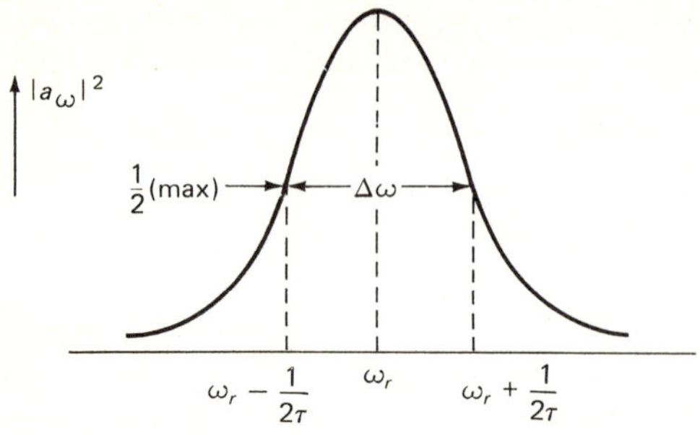

Figure 7.6

states. The lifetime τ represents a duration of oscillations such as was considered in the complementarity (7.23), and so a relation like (7.59) might have been expected.

EXERCISES

7.1 Because of unavoidable (random or statistical) errors of control, a pulse timed to take place at t_0 has an intensity *distributed* over time in proportion to

$$I(t) \sim e^{-(t-t_0)^2/2(\Delta t)^2}$$

a *normal* or *Gaussian distribution*, expected to arise from randomness about t_0.

(*a*) Show that the constant Δt is the rms value of the spread in time defined by

$$(\Delta t)^2 = \frac{\int_{-\infty}^{\infty} dt(t - t_0)^2 I(t)}{\int_{-\infty}^{\infty} dt\, I(t)} \qquad [\equiv \langle (t - t_0)^2 \rangle]$$

(*b*) The moment t_0 can always be chosen as the zero of the time scale. Find the monochromatic representative I_ω of $I(t)$ with $t_0 = 0$.

(*c*) Evaluate the rms spread $\Delta\omega$ about $\omega = 0$ (at which I_ω has its peak value) defined by

$$(\Delta\omega)^2 = \frac{\int_{-\infty}^{\infty} d\omega\, \omega^2 I_\omega}{\int_{-\infty}^{\infty} d\omega\, I_\omega}$$

(*d*) What is the product $(\Delta\omega)(\Delta t)$ according to the definitions adopted here? [Note that each ω here is *double* the frequency of the corresponding *amplitude* constituent the pulse might have, as in (6.64).]

7.2 A superposition like (7.7) but consisting of codirectionally polarized plane waves

$$\mathsf{E}(z, t) = \text{Re}\left[(2\pi)^{-1/2} \int_{-\infty}^{\infty} dk\; a(k) e^{i(kz - \omega t)}\right]$$

can be established at all $t > 0$ from the initial concentration near $z = 0$ described by

$$\mathsf{E}(z, 0) = \text{Re}\left[2\left(\frac{\sqrt{2\pi}}{\Delta z}\right)^{1/2} e^{-z^2/4(\Delta z)^2} e^{ik_o z}\right]$$

with $\Delta z = \text{const} \gg \lambda_o = 2\pi/k_o$.

(a) Show that the normalization factors adopted here correspond to 1 unit of total field energy per unit of the wave cross section. [For greater concentrations there would be an additional fraction, $\exp[-2(k_o \Delta z)^2]$, which doubles the total when $\Delta z \ll \lambda_o$.]

(b) Show that Δz is just the mean (rms) spread of the field-*energy* distribution, as defined by

$$(\Delta z)^2 = \frac{\int_{-\infty}^{\infty} dz\, z^2 \langle w(z) \rangle}{\int_{-\infty}^{\infty} dz\, \langle w(z) \rangle}$$

where $\langle w(z) \rangle$ is the energy density averaged over $\lambda_o \ll \Delta z$ via $\langle \cos^2 k_o z \rangle = \frac{1}{2}$.

(c) Determine $a(k)$ and show that $\int_{-\infty}^{\infty} dk\, |a(k)|^2/8\pi = 1$, a one-dimensional (monodirectional) counterpart of (7.11) representing the same unit of energy as that of (a).

(d) Evaluate the spread Δk of the energy as distributed over wave numbers, according to the definition

$$(\Delta k)^2 = \frac{\int_{-\infty}^{\infty} dk\, (k - k_o)^2 |a(k)|^2}{\int_{-\infty}^{\infty} dk\, |a(k)|^2}$$

What is the product $(\Delta k)(\Delta z)$ that replaces (7.22) according to the definitions here? [This is supposed to represent a *simultaneously* maximal definability of wavelength ($\lambda_o = 2\pi/k_o$) and position (initially $z = 0$) since the distributions in *both* k and z are normal, or gaussian.]

(e) Show that the $E(z, t > 0)$ above describes a propagation of the initial wave packet with its shape (spread Δz) unchanged but peaked at $z = ct$. [Thus there is no dispersion in free space. Compare (F. 47).]

7.3 Show that if the orthogonal field components (7.25) satisfy $(E_1^2/|a_1|^2) + (E_2^2/|a_2|^2) = 1$, a standard form for an ellipse referred to its principal axes, then $\delta = \pm \frac{1}{2}\pi$, as asserted in the footnote to page 190.

7.4 Suppose that one of the circularly polarized beams (7.27) first passes through a polarizer and then through an analyzer, both having their optic axes transverse to the beam but making some angle α with each other's directions. Find the fraction of the incident intensity $\langle N_o \rangle$ that is finally transmitted, as a function of α. Can observations of results like this be used to distinguish a circularly polarized beam from an unpolarized one?

7.5 Let unpolarized light of intensity $\langle N_o \rangle$ pass through a polarizer and be extinguished by an analyzer. Now place a second analyzer in the beam between the polarizer and the first analyzer, with its optic axis oriented at 45° relative to that of the polarizer. Find the result in quantitative terms.

7.6 Suppose that a plane wave represented by $\mathbf{E}'(\mathbf{r}) = \mathbf{a}' e^{i\mathbf{k}' \cdot \mathbf{r}}$ is incident from $z = \infty$ ($\equiv k_z' < 0$) on a perfectly conducting plane $z = 0$; $\mathbf{k}'$ and the normal to the plane define a plane of incidence. Suppose that the incident field has been analyzed into linearly polarized components as in (7.24) and that $\mathbf{E}'$ is a component polarized parallel to the plane of incidence. (All this can be pictured as shown in Fig. F.1.)

(a) Show that satisfying the boundary conditions (7.32) for all x, y, t requires that, with $\mathbf{E}'$, a second reflected field $\mathbf{E}'' = \mathbf{a}'' e^{i\mathbf{k}'' \cdot \mathbf{r}}$ exist in the space $z > 0$, that $k'' = k' = \omega/c$, and that $k_z'' = -k_z' > 0$ (the law of reflection).

(b) Show that the reflected wave must likewise be polarized in the plane of incidence.

(c) Show that the energy flux $\langle N'' \rangle$ in (the intensity of) the reflected wave has the same magnitude as the incident intensity $\langle N' \rangle$. (That is why perfect conductors are called perfect reflectors.)

(d) Show that like results hold for incident polarizations perpendicular to the plane of incidence and hence for unpolarized light.

7.7 For the reflections in Exercise 7.6 find the charge distribution σ and surface currents $\mathbf{J}$ induced on the conducting plane. Show that they are related as required by the charge-conservation condition (6.59).

7.8 Show that the reflections of Exercise 7.6 lead to the radiation pressure

$$|\langle \mathbf{n} \cdot \mathbf{T} \rangle| = 2\langle g' \rangle c \cos^2 i'$$

on the reflecting plane irrespective of how the incident radiation is polarized. Here i' is the angle of incidence made by $\mathbf{k}'$ with the normal $\mathbf{n}$ to the surface, and $\langle g' \rangle = \langle N' \rangle / c^2$ is the density of momentum in the incident radiation. Construct a simpler derivation of the same result (not requiring the evaluation of $\mathbf{T}$) by considering the reflections as an elastic bouncing of the field momentum off the plane.

7.9 (*a*) Sketch a field-line picture of the same field as that of Fig. 7.5 but in an xz plane as viewed from the positive y direction (use $+$ and $-$ appropriately to indicate ends of $\mathbf{E}$ lines).

(*b*) Sketch the field-line picture as seen in the xy plane at $z = 0$, $t = 0$.

(*c*) Repeat (*b*) at $z = 0$, $t = \pi/2\omega$.

7.10 (*a*) Form a real description analogous to (7.49) but for the lowest-frequency TM mode in a rectangular channel.

(*b*) Make three sketches of the field lines in xy, xz, and yz planes at some one moment t.

7.11 For a wave guide of *cylindrical* cross section and radius R:

(*a*) Show that TM and TE modes can be formed from choices of f_z, $g_z(s, \varphi)$ proportional to $e^{im\varphi} J_m(k's)$ [see (C.27)].

(*b*) Find the lowest frequency with which a TM mode can be propagated.

(*c*) Show that the cutoff frequencies of the TE modes are determined by the roots of $J'_m(\zeta'_{mn}) = 0$, where J'_m is the *derivative* of J_m with respect to its argument. (Evaluations give $\zeta'_{11} \approx 1{\cdot}84$ as the lowest root, corresponding to $m = 1$. Thus, as in the rectangular wave guides, a TE mode has the lowest cutoff frequency.)

7.12 Show that in the dominant ($m = n = 1$) TE mode of Exercise 7.11 the field lines in any given cross section form an orthogonal network which can rotate in time. (It requires a polarized excitation.) Find the maximum angular velocity this rotation can have. (Such rotations, with a monochromatic frequency, can also be made to occur in *square* wave guides. See the footnote to page 198.)

7.13 (*a*) Show that an ordinary transverse plane wave $\mathbf{E} = \mathbf{a}e^{ikz}$, $\mathbf{B} = \hat{\mathbf{k}} \times \mathbf{E}$ of *any* frequency $\omega = ck$ can be propagated without distortion by reflections between two infinite perfectly conducting planes at $x = 0$ and $x = L$.

(*b*) How must the wave be polarized?

(*c*) Show that at every instant there must be an emf, equal to $\mathbf{E}L$, between opposite points on the two conductors. (It is basically the necessity of supporting the purely transverse waves by potential differences between unconnected conductors that prevents their propagation within a single conductor like a hollow wave guide.)

7.14 Purely transverse waves of the type (7.34) with $f_z = g_z = 0$ are called TEM modes.

(*a*) Show that the Maxwell equations (7.37), when adapted to the TEM modes, lead to the conclusions

$$k_z = k \qquad \mathbf{g} = \hat{\mathbf{k}} \times \mathbf{f} \qquad (\nabla \times \mathbf{f})_z = (\nabla \times \mathbf{g})_z = 0$$

$$\nabla \cdot \begin{Bmatrix} \mathbf{f} \\ \mathbf{g} \end{Bmatrix} = 0 \qquad \nabla^2 \begin{Bmatrix} \mathbf{f} \\ \mathbf{g} \end{Bmatrix} = 0$$

and hence, the modes have no cutoff frequency.

(*b*) Show that over any closed path in the space of the mode

$$\oint \mathbf{f} \cdot d\mathbf{r} = \int dS_z\, (\nabla \times \mathbf{f})_z = 0$$

and hence $\mathbf{f} = -\nabla\phi(x, y)$ is derivable from a scalar potential that satisfies the Laplace equation. (TEM modes can be generated by choosing ϕ, just as TM and TE modes are generated by choices of f_z, g_z.)

(*c*) Show that the propagation in Exercise 7.13 is a particularly simple example of a TEM mode, one adapted to a space between parallel and unconnected conducting planes.

7.15 The results of Exercise 7.14 can be applied to transmissions by a coaxial cable. Let this consist of a conducting cylindrical shell of some inner radius a and an inner cylindrical core with an outer radius $b < a$. Assume the conductors perfect.

(*a*) Show that the fundamental (circularly symmetric) TEM mode propagated along the cable, in $b < s < a$, is

$$\mathbf{E} = \mathbf{E}_s(s, z, t) = \frac{\Delta\phi}{\ln(a/b)}\frac{1}{s}\cos(kz - \omega t) = \mathbf{B}_\varphi = \mathbf{B}$$

where $\Delta\phi$ is the peak potential difference between core and shell.

(*b*) Show that the alternating current I transmitted by the cable is related to $\mathbf{B}$ as to be expected from the Ampère law.

(*c*) Show how the results lead to a self-inductance $L = (2/c^2)\ln(b/a)$ per unit length of the cable. (See Exercise 4.22.)

(*d*) Show that the time-averaged power being transmitted is just $\frac{1}{2}LI_o^2$, where I_o is the peak value of the current.

7.16 In a cylindrical cavity of circular cross section, radius R, and length L_z, show that:

(*a*) Standing TE modes can be represented as in (7.51) with (see Exercise 7.11)

$$g_z = A\cos m\varphi J_m\left(\frac{\zeta'_{mn} s}{R}\right)$$

(*b*) Construct a corresponding description of TM modes.

(*c*) Find a ratio R/L_z for which the lowest resonant frequencies of the TE and TM modes will degenerate to the same value.

CHAPTER

EIGHT

THE GENERATION OF ELECTROMAGNETIC WAVES

The retarded responses of fields to variations in their sources and the detachment of radiation field energies in the form of electromagnetic waves · Antenna waves · The nonstatic electric dipole, magnetic dipole, and electric quadrupole fields · Electric dipole radiation from an oscillating charge · The radiation of energy and angular momentum by an orbiting charge · Zeeman splittings of radiated frequencies · Quadrupole radiation from an orbiting charge and from oscillating nuclei

The wave fields discussed in Chapter 7 can be expected to originate in sources that can be described most generally by charge and current distributions $\rho(\mathbf{r}, t)$ and $\mathbf{j}(\mathbf{r}, t)$. Fields as they are radiated by given source distributions will be the principal concern of this chapter.

8.1 RETARDED POTENTIALS

Each of the four components ρ, $\mathbf{j}$ of the source determines a corresponding component of the Lorentz potential ϕ, $\mathbf{A}$ in Eqs. (6.47). This makes it simplest to find the *potential* description of the field generated by a given source. The description by **E**, **B** can always be obtained afterward, by applying the derivative operations (6.42).

Retarded and Advanced Solutions

A special solution of the equation

$$\Box^2\phi \equiv \left(\nabla^2 - \frac{\partial^2}{c^2\,\partial t^2}\right)\phi = -4\pi\rho(\mathbf{r}, t) \tag{8.1}$$

that is the appropriate generalization of the static solution (3.10) will be sought first. Mathematically, any viable special solution should suffice for obtaining the general solution, which only requires supplementation with any of the solutions ϕ_o of the homogeneous wave equation $\Box^2\phi_o = 0$.

There will be a minimum of formal presumptions if a field contribution $d\phi(\mathbf{r}, t)$ by an individual source element at some fixed source point $\mathbf{r}_s$, treated as a point charge

$$q(t) \equiv \rho(\mathbf{r}_s, t)\, dV(\mathbf{r}_s) \tag{8.2}$$

is considered first. Because the charge at $\mathbf{r}_s$ can vary with time, the quantity $q(t)$ is not conserved by itself; it may even disappear or switch sign at some times during the motions of the charges in the source. When Eq. (8.1) is adapted for $\phi_q(\mathbf{r}, t) \equiv d\phi(\mathbf{r}, t)$, the point charge $q(t)$ enters in the isotropic density distribution $q(t)\delta(\mathbf{r} - \mathbf{r}_s)$. Because of the isotropy in $R \equiv |\mathbf{r} - \mathbf{r}_s|$, a centrosymmetric solution $\phi_q(R, t)$ having spatial dependence on $R = |\mathbf{r} - \mathbf{r}_s|$ only must be expected to exist, satisfying the equation

$$\frac{1}{R}\frac{\partial^2}{\partial R^2} R\phi_q - \frac{\partial^2}{c^2\,\partial t^2}\phi_q = -4\pi q(t)\delta(\mathbf{R}) \tag{8.3}$$

Everywhere except at $\mathbf{R} = 0$, it amounts to the one-dimensional wave equation

$$\frac{\partial^2}{\partial R^2} R\phi_q = \frac{\partial^2}{c^2\,\partial t^2} R\phi_q \tag{8.4}$$

for $R\phi_q(R, t)$, having the general solution†

$$R\phi_q(R, t) = f_r\left(t - \frac{R}{c}\right) + f_a\left(t + \frac{R}{c}\right) \tag{8.5}$$

where f_r and f_a are each an arbitrary twice differentiable functional form, containing t and R in only the linear combinations indicated. That is readily checked by forming $\partial(R\phi_q)/\partial R = -f'_r/c + f'_a/c$ and

$$\partial^2(R\phi_q)/\partial R^2 = f''_r/c^2 + f''_a/c^2 = \partial^2(R\phi_q)/(c^2\,\partial t^2)$$

Either of the forms $f_{r,a}$ can be adapted to give the inhomogeneity in (8.3), which requires that

$$\int dV(\mathbf{r})\left(\nabla^2\phi_q - \frac{\ddot{\phi}_q}{c^2}\right) = -4\pi q(t)$$

after integration over any volume, no matter how small, that contains $\mathbf{R} = 0$. Take the spherical volume $V_\epsilon = 4\pi\epsilon^3/3$, of radius $\epsilon \to 0$. Then

$$\begin{aligned}\int_{V_\epsilon} dV\,\ddot{\phi}_q &= \int_0^\epsilon 4\pi R^2\,dR\,\frac{f''_r(t - R/c) + f''_a(t + R/c)}{R} \\ &= 4\pi c^2\left[\int_t^{t-\epsilon/c} d\xi(\xi - t)f''_r(\xi) + \int_t^{t+\epsilon/c} d\eta(\eta - t)f''_a(\eta)\right] \\ &\to 0 \qquad \text{as } \epsilon \to 0\end{aligned}$$

† This can be correlated with similar findings about the two-dimensional Laplace equation by noting that the one-dimensional wave equation for any $f(R, t)$ is equivalent to a laplacian in x, y for $f(x \equiv R,\ y \equiv ict)$ having the general solution pointed out in Exercise 3.16,

$$f_1(x + iy) + f_2(x - iy) = f_1(R - ct) + f_2(R + ct)$$

and, as in the static case (A.45),

$$\int_{V_\epsilon} dV\nabla^2\phi_q = \oint d\mathbf{S}\cdot\nabla\phi_q = 4\pi\epsilon^2\left(\frac{\partial\phi_q}{\partial R}\right)_{R=\epsilon}$$

$$= 4\pi\epsilon^2\left(-\frac{f_r+f_a}{R^2}+\frac{-f'_r+f'_a}{cR}\right)_{R=\epsilon}$$

$$\rightarrow -4\pi[f_r(t)+f_a(t)] \qquad \text{as } \epsilon\rightarrow 0$$

Thus, Eq. (8.3) can be satisfied everywhere, including $R \rightarrow 0$, by any choice of the functional forms in (8.5) such that $f_r(t) + f_a(t) = q(t)$. The alternative choices $f_a \equiv 0$ and $f_r \equiv 0$ respectively yield

$$\phi_q(\mathbf{r}, t) = \frac{q(t - R/c)}{R} \qquad \text{and} \qquad \phi_q^{(a)}(\mathbf{r}, t) = \frac{q(t + R/c)}{R}$$

called *retarded* and *advanced solutions.* Rewritten in terms of the expression (8.2) for $q(t)$, the retarded solution is

$$\phi_q \equiv d\phi(\mathbf{r}, t) = \rho\left(\mathbf{r}_s, t - \frac{R}{c}\right)\frac{dV(\mathbf{r}_s)}{R}$$

and integration over all contributing source points yields

$$\phi(\mathbf{r}, t) = \oint \frac{\rho(\mathbf{r}_s, t_s)\, dV(\mathbf{r}_s)}{|\mathbf{r} - \mathbf{r}_s|}\Bigg|_{t_s \equiv t - |\mathbf{r}-\mathbf{r}_s|/c} \tag{8.6}$$

as a special solution of the original equation (8.1). Another special, advanced, solution, with $t_s \equiv t + |\mathbf{r} - \mathbf{r}_s|/c$, is also available.

Evaluating the retarded field (8.6) for a given moment t requires knowledge of the history of the source up to that moment. The contribution to it by any one source point $\mathbf{r}_s$ is proportional to the charge present there at a retarded time, earlier than t by an interval $|\mathbf{r} - \mathbf{r}_s|/c$, just such as would be required by a signal propagated with velocity c over the distance from the source point in question to the field point $\mathbf{r}$. The different charges that might occupy the volume element $dV(\mathbf{r}_s)$ at times earlier or later than the moment t_s are not needed to evaluate the effect at exactly the specified $\mathbf{r}$, t, as far as the solution (8.6) is concerned. In the course of integrating contributions from all possible source points, all having $\rho(\mathbf{r}_s, t_s) \neq 0$ at an appropriate moment, not only is $\mathbf{r}_s$ varied but also t_s together with $\mathbf{r}_s$, in the prescribed way. The consequence is that the totality of charge influencing $\phi(\mathbf{r}, t)$ cannot generally be expected to be a conserved quantity; it is rather an integration of $q \equiv \oint \rho(\mathbf{r}_s, t_s)\, dV(\mathbf{r}_s)$ keeping t_s *fixed* that must yield a conserved-charge value, independent of the moment t_s of the evaluation [if $\oint dV(\mathbf{r}_s)$ envelops the *entire* source at t_s].

The advanced special solution could be discussed in an analogous way, but it has little practical interest since it requires specifying a given source behavior for *all future* times. It depends on a sequence of source values $\rho(\mathbf{r}_s, t_s > t)$ that might be chosen quite independently of $\rho(\mathbf{r}_s, t_s \leq t)$ and should be ignored when the interest is in the effects of the latter. Thus putting $f_a \equiv 0$ in the general solution, with its twofold arbitrariness, is a supplementation of the second-order (in time) differential equation with an appropriate initial condition, always needed in delimiting specific physical circumstances.

Adopting the condition $f_a \equiv 0$ might be questioned when considering the *uniqueness* of the retarded solution as representative of a field that will *actually be observed* at a given time t. The formal possibility of taking some fraction of the retarded solution and replacing the rest with an advanced contribution also satisfies Eq. (8.1), but this is merely a result of having treated the function $\rho(\mathbf{r}, t)$ in the equation as given for all $\mathbf{r}$ and t and of the fact that representations by the d'alembertian $\Box^2$ are formally invariant to time reversal (the substitution $t \to -t$). Another essential part of the formalism is a supplementation by appropriate boundary conditions, and putting $f_a \equiv 0$ is *always* appropriate because actual observations have never shown effects attributable to an observer's plans for the future. Actual observations might be said to conform to a *principle of causality*, by which present observables are exclusively determined by past events, and the condition $f_a \equiv 0$ can be said to be a *causality requirement.*

A formal proof of uniqueness is not complete until attention has been paid to possible effects of removing the restriction to the centrosymmetric forms in (8.3) and (8.5). Such a question was bypassed in the static case by using a Green's theorem (A.47) to justify the final conclusion. That much elaboration is not really essential since it is known that the most general solution can be obtained just by supplementing the retarded solution (8.6) with solutions of the homogeneous wave equation $\Box^2\phi_o = 0$. This is entirely independent of the given sources $\rho(\mathbf{r}_s, t_s)$. Such effects were formulated in terms of boundary surface conditions, determined by exterior sources, in the static case, but here it is evident that they can be regarded to be arriving as electromagnetic waves from external sources that are not included in $\rho(\mathbf{r}_s, t_s)$. In any case, the retarded solution must be expected to represent the entire field arising from just the given sources $\rho(\mathbf{r}_s, t_s \leq t)$ by themselves.

The Retarded Propagator

The restrictions to $t_s = t - R/c$ in the retarded potential expression (8.6) can be enforced with the help of the delta function in the integration

$$\phi(\mathbf{r}, t) = \int_{-\infty}^{\infty} dt_s \oint dV(\mathbf{r}_s)\rho(\mathbf{r}_s, t_s)\frac{\delta(t_s - t + R/c)}{R} \tag{8.7}$$

with $R \equiv |\mathbf{r} - \mathbf{r}_s|$. Here the entity

$$\chi(\mathbf{r}, t; \mathbf{r}_s, t_s) \equiv \frac{\delta(t_s - t + R/c)}{R} \tag{8.8}$$

is serving as a propagator from an event at $\mathbf{r}_s$, t_s to another at $\mathbf{r}$, t, in the same way that $\chi \equiv 1/R$ served as a propagator in the static potential expression (3.10).

The initial event here is the occurrence of source charge $q(t_s)$ of (8.2) in the volume element $dV(\mathbf{r}_s)$, and the ultimate one is the generation of a field contribution $\phi_q(\mathbf{r}, t) \equiv d\phi$ at $\mathbf{r}$ and $t = t_s + R/c$. Sometimes χ of (8.8) is referred to as a *signal wave* (a pulse with a spherical front) originating at $\mathbf{r}_s$, t_s.

Just as $q(t - R/c)/R$ was found to be the solution of (8.3), so $\chi(\mathbf{r}_f, t_f; \mathbf{r}_s, t_s)$ satisfies

$$\Box_f^2 \chi = \Box_s^2 \chi = -4\pi\delta(t_f - t_s)\delta(\mathbf{r}_f - \mathbf{r}_s) \tag{8.9}$$

the generalization to the nonstatic case of Eq. (3.6) for the static Green's function. Applying the operation $\Box_f^2$ to

$$\phi(\mathbf{r}_f, t_f) = \int dt_s \oint dV(\mathbf{r}_s)\rho(\mathbf{r}_s, t_s)\chi(\mathbf{r}_f, t_f; \mathbf{r}_s, t_s) \tag{8.10}$$

and using (8.9) at once produces a corresponding solution of (8.1). Starting with the version (8.9) of the general equation (8.1), which amounts to a specialization for an instantaneously existing unit of test charge at $\mathbf{r}_s$ or $\mathbf{r}_f$, is a standard way of finding Green's-function solutions (8.10) of the general equation. From the point of view of a test charge at $\mathbf{r}_f$, the solution (8.8) can be said to represent a collector wave, converging on $\mathbf{r}_f$ with contributions from various $\mathbf{r}_s$, $t_s = t_f - R/c$. Such a visualization of the formalism will be found helpful in obtaining the field of a moving point charge in Chap. 10.

The Full Lorentz-Potential Field

Retarded solutions of Eqs. (6.47) for each of the three components of the vector potential $\mathbf{A}(\mathbf{r}, t)$ can be developed in the same way as (8.6) was for $\phi(\mathbf{r}, t)$. Thus, the retarded scalar potential $\phi(\mathbf{r}, t)$ should be supplemented with

$$\mathbf{A}(\mathbf{r}, t) = \oint dV(\mathbf{r}_s) \left. \frac{\mathbf{j}(\mathbf{r}_s, t_s)}{cR} \right|_{t_s = t - R/c} \tag{8.11}$$

to yield the complete Lorentz-potential field arising from the sources $\rho(\mathbf{r}_s, t_s \leq t)$ and $\mathbf{j}(\mathbf{r}_s, t_s \leq t)$.

It is still necessary to verify that the solutions (8.6) and (8.11) together satisfy the supplementary condition $\nabla \cdot \mathbf{A} + \mathring{\phi}/c = 0$ [Eq. (6.46)] that must be met by any Lorentz potentials arising from conserved charges [see the discussion following (6.47)]. At any given field point $\mathbf{r}_f$

$$\frac{\partial}{\partial t_f}\phi(\mathbf{r}_f, t_f) = \oint \frac{dV(\mathbf{r}_s)}{R}\frac{\partial}{\partial t_f}\rho(\mathbf{r}_s, t_s) = \oint \frac{dV}{R}\frac{\partial \rho}{\partial t_s}$$

since the $t_s = t_f - R/c$ at any one source point $\mathbf{r}_s$, having the fixed distance $R \equiv |\mathbf{r}_s - \mathbf{r}_f|$ from the field point, must be varied in proportion to t_f. To evaluate the field divergence

$$\nabla_f \cdot \mathbf{A}(\mathbf{r}_f, t_f) = \oint \frac{dV(\mathbf{r}_s)}{c}\left[\frac{1}{R}\nabla_f \cdot \mathbf{j}(\mathbf{r}_s, t_s) + \mathbf{j} \cdot \nabla_f \frac{1}{R}\right]$$

at any given moment t_f, use the fact that $\mathbf{j}(\mathbf{r}_s, t_s)$ depends on the field position $\mathbf{r}_f$ only through the dependence of $t_s(R) = t_f - R/c$ on $R = |\mathbf{r}_f - \mathbf{r}_s|$, and that $\nabla_f R = -\nabla_s R$. Thus

$$\nabla_f \cdot \mathbf{j}(\mathbf{r}_s, t_s) = \frac{\partial \mathbf{j}}{\partial t_s} \cdot \nabla_f t_s = -\frac{\partial \mathbf{j}}{\partial t_s} \cdot \nabla_s t_s$$

Comparison with the operation

$$\nabla_s \cdot \mathbf{j}(\mathbf{r}_s, t_s) = (\nabla_s \cdot \mathbf{j})_{t_s,\,\text{fixed}} + \frac{\partial \mathbf{j}}{\partial t_s} \cdot \nabla_s t_s$$

shows that $\nabla_f \cdot \mathbf{j}$ just equals the difference of the source divergences, $(\nabla_s \cdot \mathbf{j})_{t_s} - \nabla_s \cdot \mathbf{j}(\mathbf{r}_s, t_s)$, evaluated at a fixed t_s and when t_s has the requisite variation with R, respectively. Then

$$\nabla_f \cdot \mathbf{A} = \oint \frac{dV}{cR} (\nabla_s \cdot \mathbf{j})_{t_s} - \oint \frac{dV}{c} \left(\frac{1}{R} \nabla_s \cdot \mathbf{j} + \mathbf{j} \cdot \nabla_s \frac{1}{R} \right)$$

with the last two terms reducible to

$$\oint dV(\mathbf{r}_s) \nabla_s \cdot \frac{\mathbf{j}}{cR} = \oint d\mathbf{S} \cdot \frac{\mathbf{j}}{cR} = 0$$

after integration over the entire source. The result is

$$\nabla_f \cdot \mathbf{A} + \frac{\partial \phi}{c\, \partial t_f} = \oint \frac{dV(\mathbf{r}_s)}{cR} \left(\nabla_s \cdot \mathbf{j} + \frac{\partial \rho}{\partial t_s} \right)_{t_s} \tag{8.12}$$

consisting of elements that are each independent of the field variables at a fixed moment t_s and vanishing if the charge is conserved in its flux through every point at every moment. The retarded fields of conserved charges do indeed satisfy the Lorentz condition.

The results (8.6) and (8.11) answer a question that was posed in connection with generalizing the Maxwell equation (1.12) to nonstatic situations. They show that changes in remote sources do not instantaneously affect a field but that a finite time of propagation, at just the light velocity c, must be allowed for.

8.2 RADIATION FROM MONOCHROMATIC SOURCES

More explicit results are obtained when attention is concentrated on some one monochromatic representative of the source variations. That is described with the help of such distributions $\rho(\mathbf{r}_s) \equiv \rho_\omega$ and $\mathbf{j}(\mathbf{r}_s) \equiv \mathbf{j}_\omega$ as were introduced in the Fourier-transform equations (6.57). An ideally monochromatic source is represented by expressions like

$$\rho(\mathbf{r}_s, t_s) = \text{Re}\,[\rho(\mathbf{r}_s) e^{-i\omega t_s}] \qquad \mathbf{j}(\mathbf{r}_s, t_s) = \text{Re}\,[\mathbf{j}(\mathbf{r}_s) e^{-i\omega t_s}] \tag{8.13}$$

The consequence (6.59) of charge conservation is to be expressed as

$$\nabla_s \cdot \mathbf{j}(\mathbf{r}_s) = i\omega\rho(\mathbf{r}_s) \tag{8.14}$$

and it can be seen at once that integration over the entire source leads to

$$i\omega \oint dV(\mathbf{r}_s)\rho(\mathbf{r}_s) \equiv i\omega q_\omega = \oint d\mathbf{S} \cdot \mathbf{j} = 0 \tag{8.15}$$

The result $q_\omega = 0$ when $\omega \neq 0$ shows that no monochromatic representative will have monopole effects, and that such effects are confined to any static component $q_{\omega=0}$ the source may have.

Substitution of the monochromatic representatives (8.13) into the retarded potential expressions (8.6) and (8.11) yields the field representatives

$$\begin{aligned} \phi(\mathbf{r})e^{-i\omega t} &= \oint \frac{dV(\mathbf{r}_s)\rho(\mathbf{r}_s)e^{i(kR-\omega t)}}{R} \\ \mathbf{A}(\mathbf{r})e^{-i\omega t} &= \oint \frac{dV(\mathbf{r}_s)\mathbf{j}(\mathbf{r}_s)e^{i(kR-\omega t)}}{cR} \end{aligned} \tag{8.16}$$

where $k \equiv \omega/c$ and $R \equiv |\mathbf{r} - \mathbf{r}_s|$. Here serving as propagators of source influence to the field point are the signal waves

$$\chi_k(R, t) = \frac{e^{i(kR-\omega t)}}{R} \tag{8.17}$$

sometimes called *Huyghens' wavelets*. Their surfaces of constant phase form spheres centered on $R = 0$, and they are propagated uniformly into all outgoing directions, i.e., toward $R \to \infty$, and so can be described as *isotropic* spherical waves. They satisfy a unit-point-source version of (6.62),

$$(\nabla^2 + k^2)\chi_k = -4\pi\delta(\mathbf{R})e^{-i\omega t} \tag{8.18}$$

a fact readily verified by the same means† as in the static ($\omega = ck = 0$) case (3.6). The results (8.16) could have been obtained directly as solutions of the Fourier-transform equations (6.62), from appropriately weighted integrations of (8.18) over the source elements, in the same way that (8.10) follows from (8.9).

The monochromatic potential distributions

$$\phi(\mathbf{r}) = \oint \frac{dV(\mathbf{r}_s)\rho(\mathbf{r}_s)}{R} e^{ikR} \qquad \mathbf{A}(\mathbf{r}) = \oint \frac{dV(\mathbf{r}_s)\mathbf{j}(\mathbf{r}_s)}{cR} e^{ikR} \tag{8.19}$$

differ in form from the static potentials (3.10) and (4.4) only through the occurrence of the so-called retardation factor exp ikR. An important function of this

† Using the simple fact that $R\chi \sim e^{ikR}$ is a solution of $d^2(R\chi)/dR^2 = -k^2(R\chi)$ for $k \neq 0$. The Huyghens' wavelet defined in (8.17) becomes equivalent to a Green's-function solution of (8.9), for a $\rho(\mathbf{r}_s, t_s) = \rho(\mathbf{r}_s)e^{-i\omega t_s}$ in (8.10), only after multiplication by a further factor $e^{+i\omega t_s}$.

factor can be seen from considering contributions from just a pair of source points $\mathbf{r}_{1,2}$,

$$\frac{\rho(\mathbf{r}_1)\,dV(\mathbf{r}_1)}{R_1}e^{ikR_1}\left[1+\frac{R_1\rho(\mathbf{r}_2)\,dV(\mathbf{r}_2)}{R_2\rho(\mathbf{r}_1)\,dV(\mathbf{r}_1)}e^{ik(R_2-R_1)}\right]$$

where $R_{1,2} \equiv |\mathbf{r}-\mathbf{r}_{1,2}|$. When the distances $R_{1,2}$ from the source points to the field point differ by a whole number of wavelengths, $R_2 - R_1 = n\lambda = 2\pi n/k$, the contributions add, but an extra half wavelength would lead to $\exp ik(R_2 - R_1) = -1$ and the contributions are then said to interfere destructively.

The Field in a Near Zone

If the effective extent of a source is small enough relative to the wavelength $\lambda = 2\pi/k = 2\pi c/\omega$, there may be a region in which $kR \ll 1$ can be assumed. Such a so-called *near zone* is restricted to field points lying well within a wavelength of the contributing source points, such that $R \ll 1/k = \lambda/2\pi$. For these field points the retardation factor in (8.16) can be replaced by unity, and the monochromatic representative fields are given by

$$\phi(\mathbf{r}, t) \approx e^{-i\omega t}\oint \frac{dV(\mathbf{r}_s)\rho(\mathbf{r}_s)}{R} \qquad \mathbf{A} \approx e^{-i\omega t}\oint \frac{dV(\mathbf{r}_s)\mathbf{j}(\mathbf{r}_s)}{cR} \tag{8.20}$$

with spatial distributions that are calculable in the same way as for static fields. They vary in time in proportion to $\exp(-i\omega t)$ but slowly in comparison to the times needed for signal waves to traverse the source and the near zone, since $T \equiv 2\pi/\omega \gg R/c$ when $kR \ll 1$. The fields can be said to be adapting themselves comparatively instantaneously to the variations in the source, over a near zone, quite as completely as in the indefinitely long times available when the source is static. These results are valuable mostly because they make it plain how the static results follow from $\omega \equiv ck \to 0$.

The preceding considerations call attention to the fact that there are three characteristic lengths to be considered in evaluations of a field from a source of finite extent and a given frequency of variation ω. The lengths in question are most simply described when the origin for $\mathbf{r}_s$ and $\mathbf{r}_{(f)}$ is chosen to lie somewhere within the finite source. There are first various distances r from source center to field point that can be considered relative to a source size measured by

$$a \equiv |\mathbf{r}_{s,\,\max}| \tag{8.21a}$$

a radius within which the entire source lies, and to an effective free space wavelength

$$\lambda\!\!\!^{-} \equiv \frac{1}{k} = \frac{c}{\omega} = \frac{\lambda}{2\pi} \tag{8.21b}$$

Various ratios $a/\lambda\!\!\!^{-} = ka$ of source radius to wavelength can also be considered. For a near zone, in which the evaluation (8.20) can be near valid, to exist, both the ratios $ka = a/\lambda\!\!\!^-$ and $kr = r/\lambda\!\!\!^-$ must be small, with the relative size of a and r left open.

The near-zone approximation (8.20) is about the only general simplification possible for finding the field *inside* the source, and it is valid only for the effects of sources much smaller than a wavelength in extent. Of course, some progress toward finding the field inside a larger source could be made by dividing it into parts within a near zone and parts farther from the field point and then evaluating the fields due to the near and far parts of the source separately. The task of evaluating field contributions from far parts of a source is the same as finding fields outside an entire source and will be considered next.

The Radiation Field

When the interest is in the field outside a monochromatic source, it simplifies matters somewhat to aim for a description by $\mathbf{E}$, $\mathbf{B}$ since then attention can be confined to just the vector-potential components of (8.19). These are sufficient for obtaining $\mathbf{B} = \nabla \times \mathbf{A}$ (6.60), and then (6.57) leads to

$$\mathbf{E} = \frac{\nabla \times \mathbf{B}}{-ik} \qquad \text{where } \rho, j \equiv 0 \tag{8.22}$$

Just what ϕ serves to describe can be investigated afterwards.

The most important properties of the outer field are best made evident by examining its form at field points remote from the source: $r \gg a$. This first makes possible the replacement of the factor $1/R$ by $1/r$ in the vector-potential expression (8.19), but the effect of an approximation $R \approx r$ on the retardation factor must be investigated more carefully. Consider that, for $r \gg a \equiv |\mathbf{r}_{s,\,\max}|$,

$$kR = k|\mathbf{r} - \mathbf{r}_s| = kr\left(1 - \frac{2\hat{\mathbf{r}} \cdot \mathbf{r}_s}{r} + \frac{r_s^2}{r^2}\right)^{1/2}$$

$$\approx kr - k\hat{\mathbf{r}} \cdot \mathbf{r}_s + kr_s\frac{r_s}{2r}[1 - (\hat{\mathbf{r}} \cdot \hat{\mathbf{r}}_s)^2] + \cdots$$

and so

$$e^{ikR} \approx e^{ikr}e^{-i\mathbf{k} \cdot \mathbf{r}_s}\left\{1 + ikr_s\frac{r_s}{2r}[1 - (\hat{\mathbf{r}} \cdot \hat{\mathbf{r}}_s)^2] + \cdots\right\} \tag{8.23}$$

where

$$\mathbf{k} \equiv k\hat{\mathbf{r}} \equiv k\hat{\mathbf{k}} \tag{8.24}$$

is a wave vector directed toward the field point. Plainly the phase variations $\mathbf{k} \cdot \mathbf{r}_s$ cannot be neglected just because $a \ll r$; they help describe interferences between various parts of the source, as determined by the wavelength and existing regardless of the distance r to the point of observation. The final term discriminated in (8.23) is proportional to the presumed small ratio $r_s/r < a/r$, but its effect can be neglected only under a condition $r \gg (a/\lambda\!\!\!^-)a$ that becomes more stringent than just $r \gg a$ whenever the source is much larger than the wavelength it produces. For the present, attention will be confined to field points at distances r much greater than the larger of a and $(a/\lambda\!\!\!^{-})a$, so that the integration over the source in (8.16) becomes independent of the distance to the field point

$$\mathbf{A}(\mathbf{r})e^{-i\omega t} \approx \mathbf{\Lambda}(\mathbf{k})\frac{e^{i(kr-\omega t)}}{r} \tag{8.25}$$

where
$$\mathbf{\Lambda}(\mathbf{k}) \equiv c^{-1}\oint dV(\mathbf{r}_s)\mathbf{j}(\mathbf{r}_s)e^{-i\mathbf{k}\cdot\mathbf{r}_s} \tag{8.26}$$

Expression (8.25) describes a spherical electromagnetic wave being radiated outward from source to field point. It is not isotropic in general since the amplitude, being proportional to $\mathbf{\Lambda}(\mathbf{k})$, can vary with the *direction* $\hat{\mathbf{k}} \equiv \hat{\mathbf{r}}$ [Eq. (8.24)] from the source. Some of these variations will arise from the orientation of the current elements $\mathbf{j}\,dV(\mathbf{r}_s)$ relative to the field-point direction and some from the interferences between their contributions as determined by the phase factors $\exp(-i\mathbf{k}\cdot\mathbf{r}_s)$.

The expression for the vector potential (8.25) applies to any wavelength $\lambda\!\!\!^{-}$ provided only that its use is restricted to field points much farther from the source than the larger of a and $(a/\lambda\!\!\!^{-})a$ for a source within the radius a. However, the description of the fields **B** and **E** derivable from it is much the simplest at field points *many wavelengths away* so that $kr = r/\lambda\!\!\!^{-} \gg 1$ can be assumed. This is not an additional restriction for sources larger than a wavelength, since $r \gg (a/\lambda\!\!\!^{-})a = (a/\lambda\!\!\!^{-})^2\lambda\!\!\!^{-} > \lambda\!\!\!^{-}$ in that case, but it *is* a new requirement for $a < \lambda\!\!\!^{-}$, when $r \gg a$ was the only restriction previously imposed. All the conditions to be assumed for what will be called the radiation zone can be summarized as $r \gg \lambda\!\!\!^{-} \gg a^2/r$ with any relative size of source and wavelength.

The evaluation of $\mathbf{E} = \nabla \times \mathbf{A}$ and $\mathbf{E} \sim \nabla \times \mathbf{B}$ will require forming the gradient of the spherical wavelet

$$\nabla\frac{e^{ikr}}{r} = \hat{\mathbf{r}}\frac{\partial}{\partial r}\frac{e^{ikr}}{r} = ik\hat{\mathbf{r}}\left(1 - \frac{1}{ikr}\right)\frac{e^{ikr}}{r} \tag{8.27a}$$

and this can be approximated as

$$\nabla\frac{e^{ikr}}{r} \approx i\mathbf{k}\frac{e^{ikr}}{r} \qquad \text{for } kr \gg 1 \tag{8.27b}$$

since $\hat{\mathbf{k}} \equiv \hat{\mathbf{r}}$ [Eq. (8.24)]. Only variations in the phase kr are being taken into account since they reverse the sign of the result for each $\Delta r = \lambda\!\!\!^{-}$ while the changes arising from the spherical-wave curvature proportional to $1/r$ are relatively negligible when $r \gg \lambda\!\!\!^{-}$. Variations in the source factor $\mathbf{\Lambda}(\mathbf{k} \equiv k\mathbf{r}/r)$ are also to be neglected, since (8.26) shows that they will be proportional to

$$\nabla e^{-ik\mathbf{r}\cdot\mathbf{r}_s/r} = -ik\left(\frac{\mathbf{r}_s}{r} - \frac{\mathbf{r}\cdot\mathbf{r}_s}{r^2}\hat{\mathbf{r}}\right)e^{-i\mathbf{k}\cdot\mathbf{r}_s}$$

terms of order $a/r \ll 1$ like those already neglected in producing (8.25). Thus the entire factor $\mathbf{\Lambda}/r$ of (8.25) is to be treated as constant in obtaining

$$\mathbf{B}(\mathbf{r}) = \nabla \times \mathbf{A}(\mathbf{r}) \approx i\mathbf{k} \times \mathbf{\Lambda}(\mathbf{k})\frac{e^{ikr}}{r} \approx i\mathbf{k} \times \mathbf{A} \tag{8.28}$$

for $r \gg \lambda\!\!\!^{-} \gg a^2/r$. The curl operation must be applied a second time to obtain $\mathbf{E} = \nabla \times \mathbf{B}/(-ik)$, and again only the variation of the phase factor exp ikr is to be taken into account; the new factor $i\mathbf{k}/(-ik) = -\hat{\mathbf{k}} = -\mathbf{r}/r$ has variations $\nabla_i(-x_j/r) = (-\delta_{ij}/r) + x_j x_i/r^3$ that yield terms of the relative order $1/ikr$, already neglected in (8.28). Thus, for $r \gg \lambda\!\!\!^{-} \gg a^2/r$,

$$\mathbf{E} = i\mathbf{k} \times (-\hat{\mathbf{k}} \times \mathbf{\Lambda})\frac{e^{ikr}}{r} = -\hat{\mathbf{k}} \times \mathbf{B} \tag{8.29}$$

It can be seen that both the electric and magnetic vectors are transverse to the propagation direction $\hat{\mathbf{k}}$ through the field point, have equal magnitudes, and are transverse to each other (with $\mathbf{B} = \hat{\mathbf{k}} \times \mathbf{E}$) exactly as in the free-space plane waves of (7.14). Moreover, the phase factor is exp $ikr \equiv$ exp $i\mathbf{k} \cdot \mathbf{r}$, exactly as in the plane wave, and has amplitudes that are constant over large sectors of the spherical front at the very remote distances. This is a way in which such plane waves as (7.14) can be said to originate from charge-current sources. Electromagnetic radiation has thus been generated, and the energy it carries away from the source is evaluated next.

The Intensity of the Radiation

An expression for the time-averaged flux density of energy (6.66) in the outgoing spherical waves (8.28) and (8.29) can be developed as for the plane waves in (7.16), with the result

$$\langle \mathbf{N} \rangle = \frac{\hat{\mathbf{k}}\,|\mathbf{k} \times \mathbf{\Lambda}(\mathbf{k})|^2 c}{8\pi r^2} \tag{8.30}$$

It is directed precisely outward, in the direction of $\hat{\mathbf{k}} \equiv \hat{\mathbf{r}}$, because of the asymptotic transversality to the radial direction of the fields (8.28) and (8.29). It measures the intensity of the radiation as energy per unit time passing through unit areas on a very large sphere of radius r, centered on the source. Thus the power $d\mathscr{P}$ (energy per unit time) being radiated into a solid-angle element $d\Omega$ that subtends an area $\hat{\mathbf{r}} \cdot d\mathbf{S} = r^2\, d\Omega$ on the sphere is given by

$$\frac{d\mathscr{P}}{d\Omega} = r^2\hat{\mathbf{k}} \cdot \langle \mathbf{N} \rangle = \frac{\omega^2}{8\pi c^3}\left|\oint dV(\mathbf{r}_s)\hat{\mathbf{k}} \times \mathbf{j}(\mathbf{r}_s)e^{-i\mathbf{k} \cdot \mathbf{r}_s}\right|^2 \tag{8.31}$$

after a substitution for $\mathbf{\Lambda}(\mathbf{k})$ of its definition (8.26). This is completely independent of the distance r once this lies beyond the radiation zone, $r \gg \lambda\!\!\!^{-} \gg a^2/r$, and becomes precise for any finite wavelength and finite source size as $r \to \infty$.

Every element $d\mathscr{P}$ of the energy efflux continues outward, with a conserved value, to distances without limit.† It must therefore be considered *permanently* radiated and dissociated from the source. Accordingly, formula (8.31) is used for calculating the radiation that will be generated by any source characterized by current representatives $\mathbf{j}(\mathbf{r}_s)$.

The formula yields angular distributions (intensity patterns) the radiation can be expected to have. A general conclusion it makes evident is that the source must have current elements $\mathbf{j}$ that are perpendicular to the line of sight $\hat{\mathbf{k}} \equiv \hat{\mathbf{r}}$ if radiation is to be visible in that direction. This finding is used in the discussion of Fig. F.2.

Antenna Theory

The problems of designing antennas for the production of electromagnetic waves fall into two parts, respectively dealing with the excitation of the antenna and with the radiation pattern to be expected from it. Techniques of excitation have by now proliferated into an enormous variety and mostly require dealing with highly specialized extensions of the transmission-line theory alluded to on pages 590 and 596. Here, attention is perforce confined to what the radiation formula (8.31) can be used to describe.

A standard (prototype) example of antenna radiation deals with an idealized center-fed linear antenna, consisting of a straight conducting filament stretched between endpoints, at $z_s = \pm\frac{1}{2}L$ say, and carrying a standing wave of alternating current with some frequency ω. Center feeding produces a current antinode at the middle, $z_s = 0$, and nodes at the endpoints, as might be expected when current is fed in at the center and reflected at each end. Any one of the modes

$$I = I_o(\cos \omega t) \cos \frac{(2n+1)\pi z_s}{L} \qquad \text{with } n = 0, 1, 2, \ldots \tag{8.32}$$

may be the result. A single one of these can be produced, rather than a superposition of many, when the antenna length L is so chosen that the wavelength $L/(n+\frac{1}{2})$ implicit in (8.32) just matches the free-space wavelength $\lambda = 2\pi/k = 2\pi c/\omega$ for the frequency ω radiated into the free space outside the wire. This choice will be made, and then

$$I = I_o(\cos kz_s) \operatorname{Re}(e^{-i\omega t}) \qquad \text{for } |z_s| \le \tfrac{1}{2}L = \frac{(n+\frac{1}{2})\pi}{k} \tag{8.33}$$

Applying the radiation formula (8.31) requires analyzing the source current into current *density* representatives like (8.13), and corresponding to (8.33) is

$$j(\mathbf{r}_s) = j_z = I_o \cos kz_s \delta(x_s)\delta(y_s) \tag{8.34}$$

† *Formally* the result of the existence of terms in **E**, **B** that vanish as $r \to \infty$ only in proportion to $1/r$, giving intensities $\langle N \rangle \sim 1/r^2$ conforming to an inverse-square law.

If the wire is now used as a polar axis for the radiation field, so that $\vartheta = 0$ at $z_s = \frac{1}{2}L$ and $\vartheta = \frac{1}{2}\pi$ in an equatorial plane passing normally through the wire center, the magnitude of the factor $\hat{\mathbf{k}} \times \mathbf{j}$ becomes $j \sin \vartheta$ for radiation directed outward at an angle ϑ. Then

$$\frac{d\mathscr{P}}{d\Omega} = \frac{\omega^2}{8\pi c} \sin^2 \vartheta \, |\Lambda(\mathbf{k})|^2$$

with

$$\Lambda(\mathbf{k}) = \Lambda_z = \frac{I_o}{c} \int_{-(1/2)L}^{(1/2)L} dz_s \cos kz_s e^{-ikz_s \cos \vartheta} = [\Lambda_+ + \Lambda_-]_{-(1/2)L}^{(1/2)L}$$

and

$$\Lambda_\pm \equiv \frac{1}{2c} I_o \int dz_s \, e^{\pm ikz_s(1 \mp \cos \vartheta)} = \pm \frac{I_o}{2ikc} \frac{e^{\pm ikz_s(1 \mp \cos \vartheta)}}{1 \mp \cos \vartheta} \Bigg| \tag{8.35}$$

The latter definitions, evaluated over various stretches of the source, are useful in connection with many types of source modes (see Exercise 8.1). For (8.33) the results are

$$\Lambda_\pm \Bigg|_{-(1/2)L}^{+(1/2)L} = (-)^n \frac{I_o}{\omega} \frac{\cos [(n + \frac{1}{2})\pi \cos \vartheta]}{1 \mp \cos \vartheta}$$

and

$$\frac{d\mathscr{P}}{d\Omega} = \frac{I_o^2}{2\pi c} \frac{\cos^2 [(n + \frac{1}{2})\pi \cos \vartheta]}{\sin^2 \vartheta} \tag{8.36}$$

This is independent of antenna length and frequency individually but presumes that $L = (n + \frac{1}{2})2\pi c/\omega$. It is symmetrical with respect to the equatorial plane, being invariant to the substitution $\vartheta \to \pi - \vartheta$. Moreover, for any n, it vanishes at the poles since it can be found to be proportional to ϑ^2 for $\vartheta \ll 1$. That is consistent with the general conclusion that the current must have elements perpendicular to directions in which radiation is to be seen.

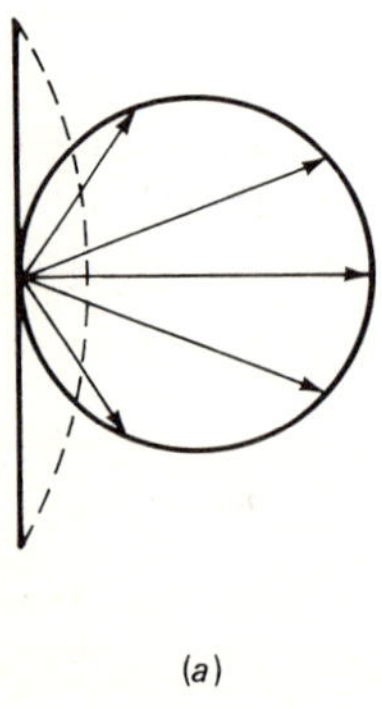

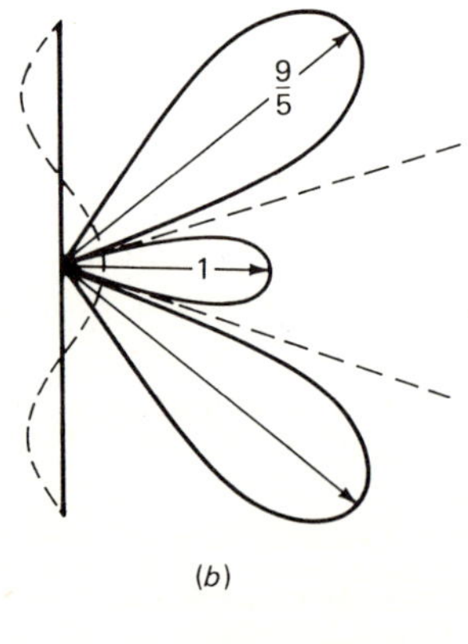

Figure 8.1 (*a*) $n = 0$; (*b*) $n = 1$.

Polar diagrams of the radiation pattern given by (8.36) are shown in Fig. 8.1 for a half-wave antenna ($n = 0$) and for the case $n = 1$. The smaller relative size of the central lobe in the latter case arises from a destructive interference of the outer half waves of source current with the central one. The radiation can be concentrated more into the equatorial plane by reversing the phases of the outer half waves of the current (see Exercise 8.1).

Half-wave antenna systems can be made practical only for relatively short-wave radiation. Dealing with radio waves of kilometer wavelengths makes it more essential to take into account the surroundings, like the earth's surface, treated as a conductor providing an image of the antenna.

8.3 FIELD DISTRIBUTIONS AROUND IDEALIZED SOURCES

The results for **E** and **B** in the radiation zone, (8.28) and (8.29), only describe the field distribution at distances many wavelengths away from the source. More complete descriptions, at $r \lesseqgtr \lambdabar$, are also comparatively simple for sources much smaller than a wavelength in extent: $a \ll \lambdabar$. Sources so small receive far the most attention in radiation theory because $\lambdabar \gg a$ characterizes the important radiations by molecular, atomic, and nuclear systems, as well as many macroscopic sources (Exercises 8.10 and 8.12).

The simplified expression (8.25) for the vector potential can be used for $r \lesseqgtr \lambdabar$ if only $r \gg a$, as noted on page 217; it can therefore include at least part of a region $r < \lambdabar$ when $\lambdabar \gg a$. With the source much smaller than a wavelength, the phase factor† $\exp(-i\mathbf{k} \cdot \mathbf{r}_s)$ in $\mathbf{\Lambda}(\mathbf{k})$ of (8.26) remains quite constant across the source and so can be replaced by unity ($\lambdabar \gg a$ implies $kr_s = r_s/\lambdabar \to 0$),

$$\mathbf{\Lambda}(k) \approx \oint \frac{dV(\mathbf{r}_s)\mathbf{j}(\mathbf{r}_s)}{c} \tag{8.37}$$

This vanishes in the static limit, as shown in (4.37), because steady currents must be divergenceless, but $\nabla \cdot \mathbf{j} = i\omega\rho$ (8.14) when $\omega \neq 0$. The effect of a condition on $\nabla \cdot \mathbf{j}$ in source integrations can always be obtained through a consideration like that exhibited in (5.39). Here consider the vector resulting from the divergence of the dyadic tensor (outer product) in

$$\nabla_s \cdot (\mathbf{j}\mathbf{r}_s) = \mathbf{r}_s(\nabla_s \cdot \mathbf{j}) + (\mathbf{j} \cdot \nabla_s)\mathbf{r}_s = i\omega\rho\mathbf{r}_s + \mathbf{j} \tag{8.38}$$

The complete divergence on the left has a zero resultant upon integration over an entire source, and so $\mathbf{j}(\mathbf{r}_s)$ can be replaced by $-i\omega\rho\mathbf{r}_s$ in (8.37):

$$\mathbf{\Lambda}(k) \approx -ik \oint dV(\mathbf{r}_s)\rho(\mathbf{r}_s)\mathbf{r}_s \equiv -ik\mathbf{D}(\omega) \tag{8.39}$$

† This factor plays the role of a photon wave function in quantum theory.

where $\mathbf{D}(\omega)$ is an *electric dipole moment* like the static one defined in (2.20). It must be kept in mind, however, that here $\rho(\mathbf{r})$ is a generally *complex* representative of a nonstatic quantity. That the first effect of a nonstatic source should be representable by a dipole moment might have been expected in view of the vanishing monopole (8.15).

Electric Dipole Fields

The fields $\mathbf{B}$ and $\mathbf{E}$ that are derivable from the vector potential

$$\mathbf{A}(\mathbf{r}) = -ik\mathbf{D}(\omega)\frac{e^{ikr}}{r} \tag{8.40}$$

as given by (8.25) for the case (8.39), will now be explored.

With $\mathbf{D}$ independent of the field point, the curl operation in $\mathbf{B} = \nabla \times \mathbf{A}$ yields

$$\mathbf{B} = ik\mathbf{D} \times \nabla\frac{e^{ikr}}{r} = k^2(\hat{\mathbf{k}} \times \mathbf{D})\left(1 - \frac{1}{ikr}\right)\frac{e^{ikr}}{r} \tag{8.41}$$

in view of the result (8.27*a*) for the gradient of a spherical wavelet. The more complete field thus differs from its radiation-zone form (8.28), for the case $\mathbf{\Lambda} = -ik\mathbf{D}$ [Eq. (8.39)], only in being radially modulated by a factor $1 + i/kr$ that reduces to unity in the radiation zone ($kr \gg 1$).

The second curl derivative needed for $\mathbf{E} = \nabla \times \mathbf{B}/(-ik)$ can be carried out through operations like

$$\nabla \times (\mathbf{r} \times \mathbf{D})f(r) = (\mathbf{D} \cdot \nabla)\mathbf{r}f - \mathbf{D}(\nabla \cdot \mathbf{r}f)$$

with the result

$$\mathbf{E} = k^2[\hat{\mathbf{k}} \times (\mathbf{D} \times \hat{\mathbf{k}})]\frac{e^{ikr}}{r} + \frac{3\hat{\mathbf{r}}(\hat{\mathbf{r}} \cdot \mathbf{D}) - \mathbf{D}}{r^3}(1 - ikr)e^{ikr} \tag{8.42}$$

It can be seen that the long-range term proportional to r^{-1} is just the radiation field (8.29) for the case of $\mathbf{\Lambda} = -ik\mathbf{D}$ and that the shorter-range terms reduce to the static form (2.27*b*) of a dipole field in the limit $k = \omega/c \to 0$. This indicates that $\mathbf{B}$ and $\mathbf{E}$ of (8.41) and (8.42) together† describe the entire field of an *ideal* electric dipole source ("extending" over one point only, so that $r/a \gg 1$ everywhere up to the point). This will be confirmed when the first corrections to the field of a small ($a \ll \lambda\!\!\!^{-}$) source are investigated in the next subsection.

† The same results, with $\mathbf{D}$ replaced by $\mathbf{\Lambda}(\mathbf{k})/(-ik)$ of (8.26), would have been obtained for a source of any size ($a \lessgtr \lambda\!\!\!^{-}$) and for field points farther out than the larger of a and $(a/\lambda\!\!\!^{-})a$ [so that (8.25) applies with $\mathbf{\Lambda}(\mathbf{k})$ invariant to field derivatives, as discussed in connection with (8.28)]. This shows that the field of a general source attains the dipole form sufficiently far away, just as the static field of the general finite charge distribution attains a monopole plus dipole form sufficiently far from it.

Evaluating the time-averaged energy flux density $\langle \mathbf{N} \rangle$ of (6.66) requires finding the real part of $\mathbf{E} \times \mathbf{B}^* = -\mathbf{B}^* \times \mathbf{E}$. For the electric dipole field (8.41) and (8.42) the longest-range terms are proportional to the vector factor

$$(\hat{\mathbf{k}} \times \mathbf{D}^*) \times [\hat{\mathbf{k}} \times (\hat{\mathbf{k}} \times \mathbf{D})] = \hat{\mathbf{k}}(\hat{\mathbf{k}} \times \mathbf{D}^*) \cdot (\hat{\mathbf{k}} \times \mathbf{D})$$

which is real by itself and has the radial (longitudinal) direction, $\hat{\mathbf{r}} \equiv \hat{\mathbf{k}}$. The real shorter-range terms have real scalar factors multiplying the vector

$$\text{Re}\,[2i(\hat{\mathbf{k}} \cdot \mathbf{D})\mathbf{D}^*] = i[(\hat{\mathbf{k}} \cdot \mathbf{D})\mathbf{D}^* - (\hat{\mathbf{k}} \cdot \mathbf{D}^*)\mathbf{D}] = i\hat{\mathbf{k}} \times (\mathbf{D}^* \times \mathbf{D})$$

which is transverse to the radial direction and can exist as a real quantity when $\mathbf{D}^* \neq \mathbf{D}$. The result is

$$\langle \mathbf{N} \rangle = \frac{ck^4}{8\pi} \frac{\hat{\mathbf{k}}\,|\hat{\mathbf{k}} \times \mathbf{D}|^2}{r^2} + \frac{ck^3}{8\pi r^3}\left(1 + \frac{1}{k^2 r^2}\right) i\hat{\mathbf{k}} \times (\mathbf{D}^* \times \mathbf{D}) \tag{8.43}$$

This yields an outward flux of radiation that is constant on every sphere of any radius $r > a$, as given by

$$\frac{d\mathscr{P}}{d\Omega} = r^2\hat{\mathbf{k}} \cdot \langle \mathbf{N} \rangle = \frac{\omega^4}{8\pi c^3}\,|\hat{\mathbf{k}} \times \mathbf{D}|^2 \tag{8.44}$$

agreeing with the radiation-zone result (8.31) for the case $\mathbf{\Lambda}(\mathbf{k}) = -ik\mathbf{D}$. The proportionality to the fourth power of the frequency is typical of dipole radiation when D is the same at each frequency [but see Exercise 8.10 and (9.116)].

The simplest monochromatic source is provided by a point charge that is *forced*† to oscillate linearly with some frequency ω_o and an amplitude $a \ll \lambda_o = 2\pi c/\omega_o$. It can be described as a charge distribution $\rho_q(\mathbf{r}_s, t_s) = q\delta[\mathbf{r}_s - \mathbf{r}_q(t_s)]$ with $\mathbf{r}_q(t_s) = \mathbf{a} \cos \omega_o t_s$. Then the dipole moment representative $\mathbf{D}(\omega)$ can be identified by considering the real moment defined by

$$\begin{aligned} \mathbf{D}(t_s) &\equiv \text{Re}\,[\mathbf{D}(\omega_o)e^{-i\omega_o t_s}] = \oint dV(\mathbf{r}_s)\mathbf{r}_s\,\text{Re}\,[\rho(\mathbf{r}_s)e^{-i\omega_o t_s}] \\ &= \oint dV(\mathbf{r}_s)\mathbf{r}_s q\delta[\mathbf{r}_s - \mathbf{r}_q(t_s)] = q\mathbf{r}_q(t_s) \\ &= q\mathbf{a} \cos \omega_o t_s = \text{Re}\,(q\mathbf{a}e^{-i\omega_o t_s}) \end{aligned} \tag{8.45}$$

Thus $\mathbf{D}(\omega_o) = q\mathbf{a}$ simply, and the monochromatic field it generates will have the forced frequency $\omega = \omega_o$. In such a case of a dipole with a constant direction,

† The forcing agency, whatever it is, will then offset the losses to radiation. The assumption of a monochromaticity always entails such a maintenance of a steady situation because of the complementarity of frequency and time span (7.23). The introduction of new frequencies by lapses from energy conservation with time was illustrated in connection with Fig. 7.6.

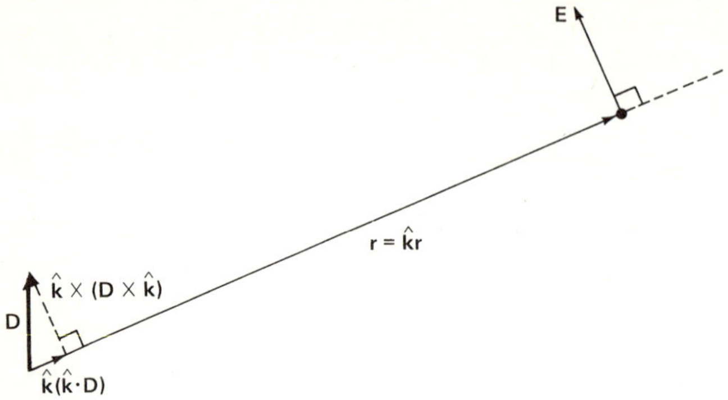

Figure 8.2

only the permanently radiated (long-range) part of the time-averaged energy flux (8.43) exists. If the steady direction of $\mathbf{D} = q\mathbf{a}$ is used as the polar axis,

$$\frac{d\mathscr{P}}{d\Omega} = \frac{q^2 a^2 \omega_o^4}{8\pi c^3} \sin^2 \vartheta \qquad \mathscr{P} = \frac{1}{3} \frac{(qa\omega_o^2)^2}{c^3} \tag{8.46}$$

having a pattern typical of steadily directed dipole sources.

Expression (8.42) for the electric vector shows that the entire field is polarized in meridional planes containing the dipole moment vector. The radiation emerging at infinity, into the direction of $\hat{\mathbf{k}}$, is linearly polarized in the direction of the vector $\hat{\mathbf{k}} \times (\mathbf{D} \times \hat{\mathbf{k}}) = \mathbf{D} - \hat{\mathbf{k}}(\hat{\mathbf{k}} \cdot \mathbf{D})$, as indicated in Fig. 8.2.

A source that provides a rotating electric dipole moment vector is formed by a point charge that is forced to follow a circular path with a constant angular velocity ω_o, as indicated in Fig. 8.3. Now (8.45) becomes

$$\mathbf{D}(t_s) = qa(\mathbf{e}_1 \cos \omega_o t_s + \mathbf{e}_2 \sin \omega_o t_s) = \mathrm{Re}\,[qa(\mathbf{e}_1 + i\mathbf{e}_2)e^{-i\omega_o t_s}] \tag{8.47a}$$

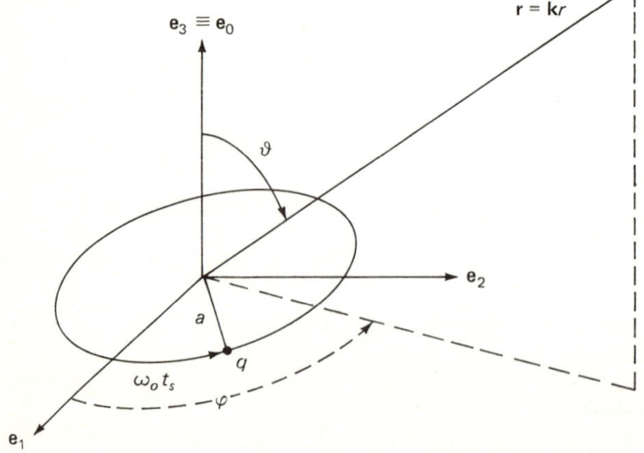

Figure 8.3

and its representative is

$$\mathbf{D}(\omega_o) = qa(\mathbf{e}_1 + i\mathbf{e}_2) \tag{8.47b}$$

Any rigid electric dipole moment that is rotated with a uniform angular velocity can provide such a representative. Since

$$\hat{\mathbf{k}} = \mathbf{e}_0 \cos \vartheta + (\mathbf{e}_1 \cos \varphi + \mathbf{e}_2 \sin \varphi) \sin \vartheta \tag{8.48}$$

in terms of the spherical coordinates indicated in the figure,

$$\hat{\mathbf{k}} \times \mathbf{D} = -iqa[(\mathbf{e}_1 + i\mathbf{e}_2) \cos \vartheta - \mathbf{e}_0 \sin \vartheta e^{+i\varphi}] \tag{8.49}$$

The radiation pattern (8.44) and the total intensity become respectively

$$\frac{d\mathscr{P}}{d\Omega} = \frac{(qa\omega_o^2)^2}{8\pi c^3}(1 + \cos^2 \vartheta) \qquad \text{and} \qquad \mathscr{P} = \frac{2}{3}\frac{(qa\omega_o^2)^2}{c^3} \tag{8.50}$$

In the plane of the orbit ($\vartheta = \frac{1}{2}\pi$) the intensity per unit solid angle equals the linear-oscillator result (8.46) because the motion has the aspect of a linear oscillation as viewed from field points in the orbital plane. There is twice as much radiation into the polar directions ($\vartheta = 0$ and π) because the circular motion can be regarded as a superposition of two linear oscillations at right angles to each other, as confirmed by the form (8.47) of the moment.

The polarizations of the plane-wave sectors [see the discussion of (8.29)] emerging at infinity are again determined by the vector factor $\hat{\mathbf{k}} \times (\mathbf{D} \times \hat{\mathbf{k}})$ in the electric vector expression (8.42). In the orbital plane, as to be expected from the linear aspect of the source motion, the polarization is linear in the plane of the orbit. In the polar directions $\hat{\mathbf{k}} = \pm\mathbf{e}_0$,

$$\mathbf{e}_0 \times (\mathbf{D} \times \mathbf{e}_0) = qa(\mathbf{e}_1 + i\mathbf{e}_2) \qquad (= \mathbf{D}) \tag{8.51}$$

and the radiation is *circularly* polarized, with a helicity [see the discussion of (7.27)] determined by the sense of the source rotation. In all other directions, the radiation is elliptically polarized in the way to be expected from the fact that a circle has the aspect of an ellipse as viewed from directions off axis.

Every electric dipole source yields the outward flux of permanently radiated energy represented by the first term of the expression for $\langle \mathbf{N} \rangle$ in (8.43). This component also helps describe a radially directed linear field momentum that exists in the density $\hat{\mathbf{r}} \cdot \langle \mathbf{g} \rangle = \hat{\mathbf{r}} \cdot \langle \mathbf{N} \rangle / c^2$ [Eq. (6.14)]. Since the radial momentum has equal magnitudes on opposite sides of the source, no net recoil is imparted to the source.

An electric dipole source like (8.47), arising from a rotational degree of freedom in the charge motion, produces, in addition, transverse fluxes of energy as described by the remainder of the expression for $\langle \mathbf{N} \rangle$. For $\mathbf{D} = qa(\mathbf{e}_1 + i\mathbf{e}_2)$,

$$i\hat{\mathbf{k}} \times (\mathbf{D}^* \times \mathbf{D}) = 2(qa)^2 \mathbf{e}_0 \times \hat{\mathbf{k}} = 2(qa)^2 (\sin \vartheta) \mathbf{e}_\varphi \tag{8.52}$$

where $\mathbf{e}_\varphi \equiv -\mathbf{e}_1 \sin \varphi + \mathbf{e}_2 \cos \varphi$ is a unit vector pointing out an azimuthal direction around the rotation axis, the direction of increasing φ in Fig. 8.3. The corresponding energy flux circles about the rotation axis in the same sense as the current of charge. The farthest portions of it decrease with distance r from the source as r^{-3}, and so the circulating energy is confined to the vicinity of the source. It is sometimes said to be only transiently or virtually radiated since it thus depends on the continued existence of the source and is not freed from it.

There are transverse components of linear field momentum proportional to the circulating flux of energy, and they have moments about the source center. There consequently exists a distribution of field *angular* momentum given by $\mathbf{r} \times \langle \mathbf{N} \rangle / c^2$ and having the density

$$\mathbf{r} \times \langle \mathbf{g} \rangle = \frac{k^3}{4\pi c r^2}\left(1 + \frac{1}{k^2 r^2}\right)(qa)^2 \sin \vartheta (\hat{\mathbf{k}} \times \mathbf{e}_\varphi) \tag{8.53}$$

Some of this diminishes with distance only in proportion to an inverse-square law and so a radiation of field angular momentum can be expected, permanently lost to the source and not confined to its vicinity. That can be calculated by integrating the current density† $(\mathbf{r} \times \langle \mathbf{g} \rangle)c$ over a sphere of radius $r \to \infty$:

$$\oint d\Omega r^2 c(\mathbf{r} \times \langle \mathbf{g} \rangle) = \frac{k^3 q^2 a^2}{4\pi} \oint d\Omega (\sin \vartheta)(\hat{\mathbf{k}} \times \mathbf{e}_\varphi)$$

Here, only the first term of

$$\hat{\mathbf{k}} \times \mathbf{e}_\varphi = \mathbf{e}_0 \sin \vartheta - (\mathbf{e}_1 \cos \varphi + \mathbf{e}_2 \sin \varphi) \cos \vartheta \qquad \text{where } \hat{\mathbf{k}} \times \mathbf{e}_\varphi \equiv -\mathbf{e}_\vartheta \tag{8.54}$$

survives the angular integration, and so the total angular momentum being radiated per unit time

$$\oint dS c(\mathbf{r} \times \langle \mathbf{g} \rangle) = \tfrac{2}{3} k^3 (qa)^2 \mathbf{e}_0 \tag{8.55}$$

† The expression $(\mathbf{r} \times \langle \mathbf{g} \rangle)c$ is valid only in the radiation zone ($r \gg \lambda\!\!\!^{-}$). The outward flux density of angular momentum everywhere should be calculated from the tensor $\mathbf{M} = \mathbf{T} \times \mathbf{r}$ [Eq. (6.34)] as

$$\begin{aligned}\hat{\mathbf{r}} \cdot \langle \mathbf{M} \rangle &= \hat{\mathbf{r}} \cdot \langle \mathbf{T} \rangle \times \mathbf{r} \\ &= \{(8\pi)^{-1} \operatorname{Re} [(\hat{\mathbf{r}} \cdot \mathbf{E})\mathbf{E}^* + (\hat{\mathbf{r}} \cdot \mathbf{B})\mathbf{B}^*] - \hat{\mathbf{r}}\langle w \rangle\} \times \mathbf{r}\end{aligned}$$

with the momentum flux tensor $\mathbf{T}$ [Eq. (6.24)] written in terms of the monochromatic field representatives of (6.63) (see Exercise 6.14). In the radiation zone $\mathbf{E}^* \times \hat{\mathbf{r}} = -\mathbf{B}^*$ and $\mathbf{B}^* \times \hat{\mathbf{r}} = \mathbf{E}^*$ [Eq. (8.29)], so that

$$\begin{aligned}\hat{\mathbf{r}} \cdot \langle \mathbf{M} \rangle &= (8\pi)^{-1} \operatorname{Re} [(\mathbf{r} \cdot \mathbf{B})\mathbf{E}^* - (\mathbf{r} \cdot \mathbf{E})\mathbf{B}^*] \\ &= \mathbf{r} \times \operatorname{Re}\left(\frac{\mathbf{E}^* \times \mathbf{B}}{8\pi}\right) = c(\mathbf{r} \times \langle \mathbf{g} \rangle)\end{aligned}$$

The radiation-zone fields are transverse, making $\mathbf{r} \cdot \mathbf{E} = \mathbf{r} \cdot \mathbf{B} = 0$, so that these projections must be evaluated to a higher order [in $(kr)^{-1} \ll 1$], as is justified by the extra power of r lent by the lever arm of the angular momentum. For the electric dipole field $\mathbf{r} \cdot \mathbf{B} = 0$ everywhere (8.41), but $(\mathbf{r} \cdot \mathbf{E})\mathbf{B}^* \to 0$ as $r \to \infty$ only in proportion to $1/r^2$.

has the vector direction of the source's rotation axis. It has been generated by a circulating field energy confined to the vicinity of the source but becomes dissociated from the source, i.e., permanently radiated, as attested by its conservation to distances without limit. Its ratio to the energy being radiated, $\frac{2}{3}ck^4(qa)^2$ [Eq. (8.50)], shows that $1/\omega$ units of angular momentum are radiated per unit energy.†

Considering the full *potential* description of the electric dipole field will permit understanding why the electric vector $\mathbf{E}$ of (8.42), derived from just the vector potential (8.40), has a near-zone part that reduces to the static description of a dipole field although in electrostatics this is derived from just a *scalar* potential $\phi = (\hat{\mathbf{r}} \cdot \mathbf{D})/r^2$ [Eq. (2.25)]. The point is that the Lorentz gauge on which (8.40) is predicated requires that the scalar potential $\phi = (\nabla \cdot \mathbf{A})/ik$ [Eq. (6.61)] coexist with the vector potential (8.40), and that yields

$$\phi = -\mathbf{D} \cdot \nabla \frac{e^{ikr}}{r} = \frac{\mathbf{D} \cdot \hat{\mathbf{r}}}{r^2}(1 - ikr)e^{ikr} \tag{8.56}$$

This survives as the static expression in the limit $k \to 0$, whereas the vector potential (8.40) vanishes. The limiting expression also follows from the near-zone ($kr \ll 1$) approximation (8.20), exactly as in the dipole approximation of electrostatics. The field derived as $\mathbf{E} = -\nabla\phi + ik\mathbf{A}$ [Eq. (6.60)] is precisely equivalent to (8.42), derived from $\mathbf{E} = \nabla \times (\nabla \times \mathbf{A})/(-ik)$, because $\nabla^2\mathbf{A} = -k^2\mathbf{A}$ [Eq. (6.62)] outside the source.

Magnetic Dipole Fields

Some sources may have variations so symmetrical that their electric dipole moments vanish.‡ For these sources the approximations that led to the electric dipole fields are insufficient, and first-order corrections to them should be considered. First the phase factor $\exp(-i\mathbf{k} \cdot \mathbf{r}_s)$, which was replaced by unity to produce the electric dipole (8.39), will now be replaced by $1 - i\mathbf{k} \cdot \mathbf{r}_s$, a small correction when $a \ll \lambda\!\!\!^{-}$ ($\mathbf{k} \cdot \mathbf{r}_s \ll 1$) but all-important if neglecting it leads to a vanishing field. Such corrections can be expected to produce higher-order multipoles as in electro- or magnetostatics, and then a simultaneous correction in the approximation of the static propagation factor $1/R \approx 1/r$, as used in producing the vector-potential expression (8.25), should also be made—by the factor $r/R \approx 1 + \hat{\mathbf{r}} \cdot \mathbf{r}_s/r$ of (2.18)—if the results are also to be valid in the near zone, where the static forms of the higher-order multipole fields should be anticipated. The product of the two correction factors must be taken to be

$$1 - i\mathbf{k} \cdot \mathbf{r}_s + \frac{\hat{\mathbf{r}} \cdot \mathbf{r}_s}{r} = 1 - i\mathbf{k} \cdot \mathbf{r}_s\left(1 - \frac{1}{ikr}\right)$$

† Compare the *linear* momentum $\mathbf{k}/\omega$ being carried by each unit of energy in a plane wave, as discussed in the footnote to page 187. It is a magnitude $\hbar$ of angular momentum that is carried by each photon of energy $\hbar\omega$ from the rotating electric dipole source.

‡ The symmetry thus introduces a *selection rule* against the emission of electric dipole radiation.

neglecting a second-order term of about the same size as the last term discriminated in (8.23), which should therefore continue to be neglected. Now the vector potential can be given the same form (8.25) as before, but with

$$\mathbf{A} \approx -ik\left[\mathbf{D} + \left(1 - \frac{1}{ikr}\right)\oint dV(\mathbf{r}_s)(\hat{\mathbf{r}} \cdot \mathbf{r}_s)\frac{\mathbf{j}(\mathbf{r}_s)}{c}\right] \tag{8.57}$$

replacing the electric dipole approximation (8.39).

To analyze the new source integral, the dyadic tensor $\mathbf{r}_s\mathbf{j}(\mathbf{r}_s)$ occurring in it can be split into antisymmetrical and symmetrical parts, as in (5.38), with the result

$$\begin{aligned}\hat{\mathbf{r}} \cdot \mathbf{r}_s\mathbf{j} &= \hat{\mathbf{r}} \cdot [\tfrac{1}{2}(\mathbf{r}_s\mathbf{j} - \mathbf{j}\mathbf{r}_s) + \tfrac{1}{2}(\mathbf{r}_s\mathbf{j} + \mathbf{j}\mathbf{r}_s)] \\ &= \tfrac{1}{2}[(\hat{\mathbf{r}} \cdot \mathbf{r}_s)\mathbf{j} - (\hat{\mathbf{r}} \cdot \mathbf{j})\mathbf{r}_s] + \tfrac{1}{2}\hat{\mathbf{r}} \cdot (\mathbf{r}_s\mathbf{j} + \mathbf{j}\mathbf{r}_s) \\ &= \tfrac{1}{2}\hat{\mathbf{r}} \times (\mathbf{j} \times \mathbf{r}_s) + \tfrac{1}{2}\hat{\mathbf{r}} \cdot (\mathbf{r}_s\mathbf{j} + \mathbf{j}\mathbf{r}_s)\end{aligned} \tag{8.58}$$

The substitution of this result into (8.57) and (8.25) yields the corrected vector potential as a superposition of three fields

$$\mathbf{A} = \mathbf{A}_D + \mathbf{A}_\mu + \mathbf{A}_Q + \cdots \tag{8.59}$$

Here $\mathbf{A}_D$ is just the electric dipole field (8.40),

$$\mathbf{A}_\mu \equiv \frac{\boldsymbol{\mu} \times \hat{\mathbf{r}}}{r^2}(1 - ikr)e^{ikr} \qquad \text{with } \boldsymbol{\mu} \equiv \frac{1}{2c}\oint dV\, \mathbf{r}_s \times \mathbf{j} \tag{8.60}$$

is obviously a *magnetic* dipole field arising from a dynamic moment $\boldsymbol{\mu}(\omega)$, defined on the model of the static one (4.38), and

$$\mathbf{A}_Q \equiv -i\mathbf{k} \cdot \left(\oint dV \frac{\mathbf{r}_s\mathbf{j} + \mathbf{j}\mathbf{r}_s}{2c}\right)\left(1 - \frac{1}{ikr}\right)\frac{e^{ikr}}{r} \tag{8.61}$$

is a contribution to be discussed in the next subsection.

The magnetic dipole field (8.60) plainly reduces to the static form (4.28) in the limit of $\omega = ck \to 0$, and for the evaluation of its curl $\mathbf{B}_\mu = \nabla \times \mathbf{A}_\mu$ the static result (4.8) can be taken over as the contribution of just the factor $\boldsymbol{\mu} \times \hat{\mathbf{r}}/r^2$. The entire result is

$$\mathbf{B}_\mu = k^2[\hat{\mathbf{k}} \times (\boldsymbol{\mu} \times \hat{\mathbf{k}})]\frac{e^{ikr}}{r} + \frac{3\hat{\mathbf{r}}(\hat{\mathbf{r}} \cdot \boldsymbol{\mu}) - \boldsymbol{\mu}}{r^3}(1 - ikr)e^{ikr} \tag{8.62}$$

The vector potential (8.60) is plainly divergenceless, as in the static case, and so the Lorentz potential has no scalar part (6.61): $\phi_\mu = \nabla \cdot \mathbf{A}_\mu/ik = 0$. Consequently the electric field $\mathbf{E}_\mu = ik\mathbf{A}_\mu - \nabla\phi_\mu$, according to (6.60), is simplest to obtain as

$$\mathbf{E}_\mu = ik\mathbf{A}_\mu = -k^2(\hat{\mathbf{k}} \times \boldsymbol{\mu})\left(1 - \frac{1}{ikr}\right)\frac{e^{ikr}}{r} \tag{8.63}$$

Comparison with (8.41) and (8.42) shows that the electric and magnetic dipole field *forms* differ only in the interchange of **E** and **B**. More precisely, the substitutions $\mathbf{B}_\mu = \mathbf{E}_D(\mathbf{D} \to \boldsymbol{\mu})$ and $\mathbf{E}_\mu = -\mathbf{B}_D(\mathbf{D} \to \boldsymbol{\mu})$ transform the electric into a magnetic dipole field. A ready consequence is that the energy-flux expression (8.43) is unchanged except for the substitution $\mathbf{D} \to \boldsymbol{\mu}$ [because $\mathbf{E}_\mu \times \mathbf{B}_\mu = (-\mathbf{B}_D \times \mathbf{E}_D)_{D=\mu} = (\mathbf{E}_D \times \mathbf{B}_D)_{D=\mu}$].

The foregoing discussion indicates that electric and magnetic dipole sources cannot be distinguished from each other simply by observing the radiation patterns they produce. A steadily directed dipole moment of either type will yield the angular distribution proportional to $\sin^2 \vartheta$, as in (8.46). However, observations on the polarization of the radiation relative to the polar axis of the pattern can distinguish the two types. Whereas the electric dipole radiation has polarizations as indicated in Fig. 8.2, the magnetic dipole radiation will be polarized perpendicularly to that since the electric vector of (8.63) has the direction of $\hat{\mathbf{k}} \times \boldsymbol{\mu}$.

The orbiting point charge of Fig. 8.3, which has already been seen to produce electric dipole radiation, also possesses a magnetic moment. Calculation paralleling (8.45) gives

$$\boldsymbol{\mu} = \frac{1}{2c} \oint dV\, \mathbf{r}_s \times q\delta[\mathbf{r}_s - \mathbf{r}_q(t_s)]\dot{\mathbf{r}}_q = \frac{q}{2c} \mathbf{r}_q \times \mathbf{u}_q(t_s) \tag{8.64}$$

with $\mathbf{r}_q(t_s) = a(\mathbf{e}_1 \cos \omega_o t_s + \mathbf{e}_2 \sin \omega_o t_s)$, as is implicit in (8.47a), and

$$\mathbf{u}_q \equiv \dot{\mathbf{r}}_q = -\omega_o a(\mathbf{e}_1 \sin \omega_o t_s - \mathbf{e}_2 \cos \omega_o t_s)$$

This yields a *constant* angular momentum $\mathbf{L} = m\mathbf{r}_q \times \mathbf{u}_q = \mathbf{e}_0\, ma^2\omega_o$, as might have been expected. Thus the magnetic moment, unlike the electric one, is *static*, with the expected value $\boldsymbol{\mu} = q\mathbf{L}/2mc$, and so will produce no magnetic dipole radiation. Such radiation *can* be induced by imposing an external magnetic field and thus starting a Larmor *precession* of the magnetic moment as in Fig. 5.3.

If the imposed magnetic field $\mathbf{B}_o$ is used as the polar axis $\mathbf{e}_0$ and the precessing magnetic moment makes the steady angle α with it, reference to the precession direction in Fig. 5.3 shows that it can be described by

$$\boldsymbol{\mu}(t_s) = \mathbf{e}_0 \mu \cos \alpha + \mu \sin \alpha(\mathbf{e}_1 \cos \omega_L t_s - \mathbf{e}_2 \sin \omega_L t_s)$$

where $\omega_L = q\mathsf{B}_o/2mc$ is the Larmor frequency. The consequent magnetic dipole radiation will therefore have this frequency, and its character is calculable from the monochromatic representative

$$\boldsymbol{\mu}(\omega_L) = \mu \sin \alpha(\mathbf{e}_1 - i\mathbf{e}_2) \tag{8.65}$$

As in the case of the electric dipole (8.47), a radiation pattern proportional to $1 + \cos^2 \vartheta$ is produced. The polarizations are determined by the vector $\hat{\mathbf{k}} \times \boldsymbol{\mu}$ in (8.63), and this yields circular polarizations of the radiation parallel or antiparallel to the imposed magnetic field $\mathbf{B}_o$. The radiation perpendicular to the field is linearly polarized, with its electric vector normal to the equatorial plane and hence parallel to the direction of $\mathbf{B}_o$. The total power being radiated

$$\mathscr{P} = \frac{2}{3}\frac{(\mu \sin \alpha)^2 \omega_L^4}{c^3} \tag{8.66}$$

is often too small to detect easily because the precession frequencies attainable with practical magnetic fields are low. The Larmor precession in atomic systems is therefore more often detected by a converse process, seeing what frequencies out of a superposed oscillating field are *absorbed* by the system precessing in the static field $\mathbf{B}_o$.

Such an orbiting charge as that of Fig. 8.3 continues to produce the *electric* dipole radiation after its angular momentum $|\mathbf{L}| = ma^2\omega_o$ is made to precess by imposing the external magnetic field $\mathbf{B}_o$. Its orbital rotation axis is then no longer steady, and it is relative to $\mathbf{B}_o$ as a polar axis that the radiation is now observed. It is primarily through the electric dipole radiation that the Zeeman effects discussed in the last subsection of Chap. 5 are detected. Since that discussion was concerned with an atomic model in which an electron of negative charge $q = -e$ orbits about a practically immovable nucleus, the considerations here will deal with a charge $q = -e$ following a circular orbit with frequency ω_o that itself precesses in the sense indicated in Fig. 8.4 (now $\boldsymbol{\mu} = -e\mathbf{L}/2mc$) with a Larmor frequency of magnitude $\omega_L \equiv e\mathbf{B}_o/2mc$.

The description of the orbital motion implicit in (8.47*a*) is now to be taken relative to the precessing axes (x' and y') of Fig. 8.4, and an instantaneous

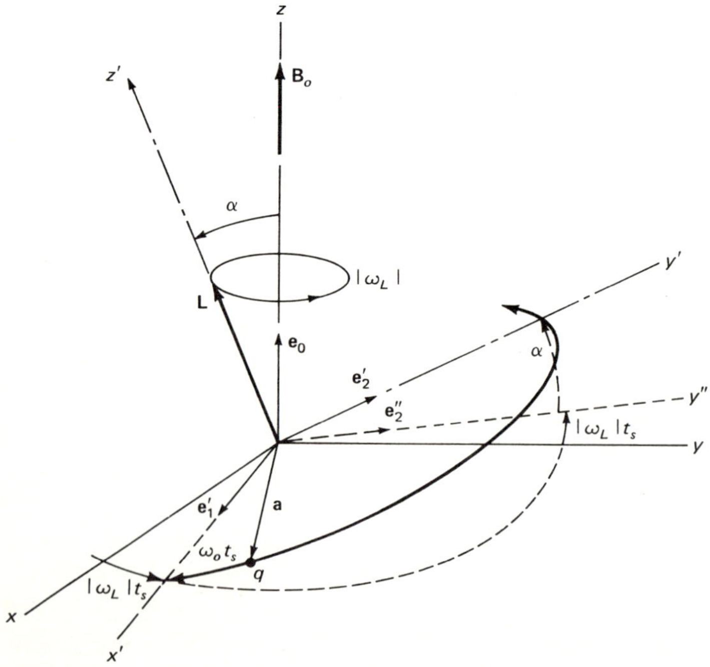

Figure 8.4

position of the charge is to be given by

$$\mathbf{a} = a(\mathbf{e}_1' \cos \omega_o t_s + \mathbf{e}_2' \sin \omega_o t_s) \qquad \text{where } \mathbf{a} \equiv \mathbf{r}_q$$

The figure makes it plain that $\mathbf{e}_1' = \mathbf{e}_1 \cos \omega_L t_s + \mathbf{e}_2 \sin \omega_L t_s$ relative to the fixed axes (x and y), while $\mathbf{e}_2'$ is simplest to resolve first as $\mathbf{e}_2' = \mathbf{e}_0 \sin \alpha + \mathbf{e}_2'' \cos \alpha$, where $\mathbf{e}_2''$ points out the axis (y'') in the xy plane that is perpendicular to the axis $\mathbf{e}_1'$ (along which the orbit intersects the xy plane). It is easy to see that $\mathbf{e}_2'' = -\mathbf{e}_1 \sin \omega_L t_s + \mathbf{e}_2 \cos \omega_L t_s$, and so substitutions yield

$$\begin{aligned}\mathbf{a}(t_s) = a[&\mathbf{e}_1(\cos \omega_o t_s \cos \omega_L t_s - \cos \alpha \sin \omega_o t_s \sin \omega_L t_s)\\ &+ \mathbf{e}_2(\cos \omega_o t_s \sin \omega_L t_s + \cos \alpha \sin \omega_o t_s \cos \omega_L t_s)\\ &+ \mathbf{e}_0 \sin \alpha \sin \omega_o t_s]\end{aligned}$$

This is Fourier-analyzed into monochromatic representatives of the electric dipole moment $\mathbf{D}(t_s) = -e\mathbf{a}(t_s)$ when it is written as in

$$\begin{aligned}\mathbf{D}(t_s) = -\tfrac{1}{2}ea\{&(1 + \cos \alpha)[\mathbf{e}_1 \cos (\omega_o + \omega_L)t_s + \mathbf{e}_2 \sin (\omega_o + \omega_L)t_s]\\ &+ (1 - \cos \alpha)[\mathbf{e}_1 \cos (\omega_o - \omega_L)t_s - \mathbf{e}_2 \sin (\omega_o - \omega_L)t_s]\\ &+ 2 \sin \alpha \cdot \mathbf{e}_0 \sin \omega_o t_s\}\end{aligned}$$

Thus there will still be electric dipole radiation with the orbital frequency ω_o but also radiation shifted in frequency by the Larmor precession to $\omega_o \pm \omega_L$. The corresponding electric dipole representatives are

$$\begin{aligned}\mathbf{D}(\omega_o) &= -ie(a \sin \alpha)\mathbf{e}_0\\ \mathbf{D}(\omega_o \pm \omega_L) &= -\tfrac{1}{2}ea(1 \pm \cos \alpha)(\mathbf{e}_1 \pm i\mathbf{e}_2)\end{aligned} \tag{8.67}$$

The radiation of frequency ω_o will be linearly polarized parallel to the imposed field $\mathbf{B}_o$, as from a steadily directed dipole along that direction. Those radiations with the shifted frequencies $\omega_o \pm \omega_L$ which emerge normally to $\mathbf{B}_o$ will likewise be linearly polarized, in the equatorial plane as from the representative (8.47*b*). Also as in the latter case (of $\mathbf{B}_o = 0$), the radiations in the polar directions (here parallel and antiparallel to $\mathbf{B}_o$) will be circularly polarized, with opposite helicities for the higher and the lower frequencies. Thus, in directions parallel to the imposed magnetic field there is a *doublet* of emitted frequencies $\omega_o \pm \omega_L$ with opposite circular polarizations, whereas perpendicularly to the field a *triplet* of frequencies is emitted; all are linearly polarized, with the central one (ω_o) in a plane perpendicular to that of the other two. Just these results are actually observed in Zeeman effects on atomic radiations.

Electric Quadrupole Radiation

There remains to be discussed the third contribution discriminated in the vector-potential expression (8.59). It is determined by the symmetrical source integral in (8.61), which can be related to a type of source characterization familiar in electrostatics by using the charge-conservation relation $\nabla_s \cdot \mathbf{j}(\mathbf{r}_s) = i\omega\rho(\mathbf{r}_s)$ to put it in terms of the charge-distribution representative, as in the step from the source integral (8.37) to the electric dipole moment in (8.39). Here the development in (5.39) is to be modified to the case of $\nabla_s \cdot \mathbf{j} \neq 0$ as in

$$\begin{aligned} \nabla_s \cdot (\mathbf{j}\mathbf{r}_s\mathbf{r}_s) &= \mathbf{r}_s\mathbf{r}_s\nabla_s \cdot \mathbf{j} + (\mathbf{j}\mathbf{r}_s + \mathbf{r}_s\mathbf{j}) \\ &= i\omega\rho\mathbf{r}_s\mathbf{r}_s + (\mathbf{j}\mathbf{r}_s + \mathbf{r}_s\mathbf{j}) \end{aligned} \tag{8.68}$$

Since, as usual, the complete divergence on the left will vanish upon integration over an entire source, the source integral of $\mathbf{A}_Q$ (8.61) can be expressed as

$$-\tfrac{1}{2}ik \oint dV\rho\mathbf{r}_s\mathbf{r}_s = -\tfrac{1}{6}ik(\mathbf{Q} + \mathbf{1}\oint dV\rho r_s^2) \tag{8.69}$$

in terms of a dynamic quadrupole moment $\mathbf{Q}(\omega)$, defined on the model of the static one (2.21). The vector potential $\mathbf{A}_Q$ is consequently said to describe an electric *quadrupole* field.

It can first be seen that the resultant field in the near zone ($kr \ll 1$), where

$$\mathbf{A}_Q \approx \frac{-ik(\hat{\mathbf{r}} \cdot \mathbf{Q} + \hat{\mathbf{r}} \oint dV\rho r_s^2)}{6r^2} \tag{8.70}$$

will attain just the electrostatic quadrupole form by examining the concomitant scalar potential $\phi_Q = \nabla \cdot \mathbf{A}_Q/ik$, as in the step to (8.56) of the electric *dipole* field description. Note that the factor $\hat{\mathbf{r}}/r^2$ is divergenceless (like the electrostatic field of a point charge, outside the point) and so the result for ϕ_Q will be proportional to

$$\begin{aligned} \nabla \cdot \left(\mathbf{Q} \cdot \frac{\mathbf{r}}{r^3}\right) &= \sum_{ij} Q_{ij}\nabla_i \frac{x_j}{r^3} = \sum_{ij} Q_{ij}\left(\frac{\delta_{ij}}{r^3} - \frac{3x_i x_j}{r^5}\right) \\ &= \frac{\sum_i Q_{ii}}{r^3} - 3\frac{\hat{\mathbf{r}} \cdot \mathbf{Q} \cdot \hat{\mathbf{r}}}{r^3} \end{aligned} \tag{8.71}$$

since $Q_{ij} = Q_{ji}$. Moreover the trace $\sum_i Q_{ii} = 0$, as already pointed out in (2.58). Thus

$$\phi_Q \approx \frac{\hat{\mathbf{r}} \cdot \mathbf{Q} \cdot \hat{\mathbf{r}}}{2r^3} \tag{8.72}$$

in the near zone, exactly like the static quadrupole term of (2.19). This persists in the static limit $k \to 0$, whereas the vector potential vanishes.

The complete vector potential $\mathbf{A}_Q$ that follows from substituting the source characterization (8.69) into (8.61) is

$$\mathbf{A}_Q = -\tfrac{1}{6}k^2(\hat{\mathbf{r}} \cdot \mathbf{Q} + \hat{\mathbf{r}} \oint dV\rho r_s^2)\left(1 - \frac{1}{ikr}\right)\frac{e^{ikr}}{r} \tag{8.73}$$

It depends on source integrals not reducible to $\mathbf{Q}$ alone, but the extra part is curlless and so does not contribute to the fields

$$\mathbf{B}_Q = -\frac{i}{6}k^3[\hat{\mathbf{k}} \times (\mathbf{Q} \cdot \hat{\mathbf{k}})]\left(1 - \frac{3}{ikr} - \frac{3}{k^2r^2}\right)\frac{e^{ikr}}{r} \tag{8.74}$$

and $\mathbf{E}_Q = \nabla \times \mathbf{B}_Q/(-ik)$ outside the source. In the radiation zone ($kr \gg 1$), these reduce to the forms (8.28) and (8.29) that are valid in the radiation zone of any source. With $\mathbf{E}_Q = -\hat{\mathbf{k}} \times \mathbf{B}_Q$ there, the power radiated per unit solid angle becomes

$$\frac{d\mathscr{P}}{d\Omega} = r^2 \frac{c}{8\pi} |\mathbf{B}_Q|^2 = \frac{\omega^6}{288\pi c^5} |\hat{\mathbf{k}} \times (\mathbf{Q} \cdot \hat{\mathbf{k}})|^2 \tag{8.75}$$

which is also directly derivable from the general formula (8.31) by replacing the retardation factor in it with $-i\mathbf{k} \cdot \mathbf{r}_s$ and retaining only the symmetrical part of the source integral. The integration over all directions needed to obtain the total power being radiated by the electric quadrupole

$$\mathscr{P} = \frac{\omega^6}{360c^5} \sum_{ij} |Q_{ij}|^2 \tag{8.76}$$

is left to the reader (Exercise 8.7). The *polarization* of the plane-wave sector emerging in the direction $\hat{\mathbf{k}}$ is determined by the direction of $\mathbf{E}_Q \sim \hat{\mathbf{k}} \times [\hat{\mathbf{k}} \times (\mathbf{Q} \cdot \hat{\mathbf{k}})] = \hat{\mathbf{k}}(\hat{\mathbf{k}} \cdot \mathbf{Q} \cdot \hat{\mathbf{k}}) - (\mathbf{Q} \cdot \hat{\mathbf{k}})$, as for the dipole in (8.42) with $\mathbf{D}$ replaced by the radiation-direction-dependent vector $\mathbf{Q} \cdot \hat{\mathbf{k}}$.

It is instructive to notice that such an orbiting charge as that of Fig. 8.3 emits electric quadrupole, in addition to the dipole, radiation. That becomes evident when its quadrupole moment is evaluated as the real tensor

$$\mathbf{Q}(t_s) = 3q[\mathbf{r}_q \mathbf{r}_q - \tfrac{1}{3}(\mathbf{1} r_q^2)] \tag{8.77a}$$

with $\mathbf{r}_q = a(\mathbf{e}_1 \cos \omega_o t_s + \mathbf{e}_2 \sin \omega_o t_s)$. The result has time-dependent terms quadratic in $\cos \omega_o t_s$ and $\sin \omega_o t_s$ and hence replaceable by terms linear in $\cos 2\omega_o t_s$ and $\sin 2\omega_o t_s$ (besides static ones). Thus the quadrupole radiation will have twice the frequency of the electric dipole radiation. The moment representative can easily be found as the tensor

$$\mathbf{Q}(2\omega_o) = \tfrac{3}{2} q a^2 (\mathbf{e}_1 + i\mathbf{e}_2)(\mathbf{e}_1 + i\mathbf{e}_2) \tag{8.77b}$$

which yields the vector

$$\mathbf{Q} \cdot \hat{\mathbf{k}} = \tfrac{3}{2} q a^2 \sin \vartheta e^{i\varphi}(\mathbf{e}_1 + i\mathbf{e}_2) = \tfrac{3}{2} a\mathbf{D} \sin \vartheta e^{i\varphi}$$

upon contraction with the radiation direction $\hat{\mathbf{k}}$ as decomposed in (8.48). The vector product $\hat{\mathbf{k}} \times (\mathbf{Q} \cdot \hat{\mathbf{k}})$ then has (8.49) as a factor, and so the radiation formula (8.75) yields

$$\frac{d\mathscr{P}}{d\Omega} = \frac{(2\omega_o)^6}{288\pi c^5} (\tfrac{3}{2} q a^2)^2 \sin^2 \vartheta (1 + \cos^2 \vartheta) = \frac{q^2 a^4 \omega_o^6}{2\pi c^5} (1 - \cos^4 \vartheta) \tag{8.78}$$

None of this emerges in the polar directions, whereas the electric dipole radiation has maxima in those directions. The total intensity is

$$\mathscr{P} = \frac{8}{5} \frac{q^2 a^4 \omega_o^6}{c^5} \tag{8.79}$$

differing from the dipole result by the factor $\frac{12}{5}(a\omega_o/c)^2$, and hence much weaker in view of the condition $ka = \omega_o a/c \ll 1$ that has been presumed. Atomic orbits have $a \approx 10^{-8}$ cm, and visible light has $c/\omega_o = \lambdabar \approx 10^{-5}$ cm (page 188), so that $(a/\lambdabar)^2 \approx 10^{-6}$ is typical of atomic radiations. Thus atomic radiation is almost completely dominated by electric dipole emissions; contributions by other multipoles are said to be forbidden, although accurate observations have detected them and forbidden spectral lines have great importance for the identification of atoms and excitation conditions in many sources of radiation.

Electric quadrupole radiations are relatively more prominent as γ rays from atomic *nuclei*, mostly because the simplest nuclear oscillations are so symmetrical that their electric dipole moments are suppressed. The simplest nuclei can be approximated as uniformly charged spheres of practically incompressible "nuclear fluid." A lowest-order excitation of such a structure is an azimuthally symmetric, volume-preserving oscillation between the shapes of a prolate (elongated) ellipsoid of revolution and an oblate (squashed) one. Now consider the symmetries between the positions of the charge elements at any one stage in the oscillation on a cartesian basis, with the axis of the azimuthal symmetry as z axis. Then for any one component of the dipole moment

$$D_i = \oint dV\rho x_{si} = \oint dV\rho(-x_{si}) = -D_i = 0$$

since replacing any one cartesian coordinate with its negative should make no difference to an integration over an ellipsoidal distribution. For the same reason, the off-diagonal components of the quadrupole moment will vanish

$$Q_{i \neq j} = 3\oint dV\rho x_{si} x_{sj} = 3\oint dV\rho x_{si}(-x_{sj}) = -Q_{ij} = 0$$

Because of the azimuthal (rotational) symmetry about the z axis, an interchange of x and y coordinates can also have no effect on the integrations, and since the trace $\sum_i Q_{ii} = 0$ always,

$$Q_{xx} = Q_{yy} = -\tfrac{1}{2}Q_{zz}$$

Thus all components of the quadrupole moment can be characterized by just one number, like Q_o in the representative

$$\mathbf{Q} = Q_o[\mathbf{e}_0\,\mathbf{e}_0 - \tfrac{1}{2}(\mathbf{e}_1\mathbf{e}_1 + \mathbf{e}_2\,\mathbf{e}_2)] \tag{8.80}$$

when the symmetry (z) axis above is taken as the polar axis for the radiation, as in the expression (8.48) for $\hat{\mathbf{k}}$. Now

$$\mathbf{Q} \cdot \hat{\mathbf{k}} = Q_o[\mathbf{e}_0 \cos \vartheta - \tfrac{1}{2} \sin \vartheta(\mathbf{e}_1 \cos \varphi + \mathbf{e}_2 \sin \varphi)]$$

$$\hat{\mathbf{k}} \times (\mathbf{Q} \cdot \hat{\mathbf{k}}) = \tfrac{3}{2}Q_o \sin \vartheta \cos \vartheta(\mathbf{e}_1 \sin \varphi - \mathbf{e}_2 \cos \varphi)$$

and

$$\frac{d\mathscr{P}}{d\Omega} = \frac{\omega^6}{128\pi c^5}\,|Q_o|^2 \sin^2 \vartheta \cos^2 \vartheta \qquad \mathscr{P} = \frac{\omega^6\,|Q_o|^2}{240c^5} \tag{8.81}$$

The pattern has a node in the equatorial plane as well as in the polar directions; its maxima occur at $\vartheta = \frac{1}{4}\pi$ and $\vartheta = \frac{3}{4}\pi$. The proportionality to the sixth power of the frequency, here and in (8.79), is typical of quadrupole radiations from small sources.

EXERCISES

8.1 The $n = 1$ antenna mode of Fig. 8.1 has its middle half wave of current flowing in directions opposite those in the outer half waves. There are means of excitation that change nothing except to reverse the middle phase, with three codirectional half waves as a result [three half-wave ($n = 0$) antennas lined up and oscillating in phase].

(*a*) Find the radiation pattern that replaces (8.36) and the locations of the nodes and maxima in it.

(*b*) Figure 8.1 indicates a ratio 5/9 of the equatorial intensity (per unit solid angle) to the maxima in the side lobes. Show that the new form of excitation increases such a ratio to almost 90. How does the equatorial intensity here compare with that of a half-wave ($n = 0$) antenna for equal wavelengths and peak currents I_o?

8.2 Particles m, q are maintained at a velocity $u \ll c$ while circling in a uniform magnetic field $\mathbf{B}_o$.

(*a*) Find the ratio of each particle's radiated energy loss per cycle to the energy with which it is maintained, in terms of the given quantities.

(*b*) Practical applied fields $\mathbf{B}_o$ cannot maintain orbits of less than macroscopic dimensions. Show that this makes the energy-loss fraction in (*a*) completely negligible for the motion because particles in nature have $q^2/mc^2 < e^2/m_e c^2 \approx 2.8 \times 10^{-13}$ cm. (The radiation here is sometimes called *cyclotron radiation*.)

8.3 The linearly oscillating point charge will emit not only the dipole radiation (8.46) but also electric *quadrupole* radiation. About the latter, show that:

(*a*) The radiation frequency is double the dipole frequency.

(*b*) None is emitted into directions perpendicular to the oscillation direction.

(*c*) Its total intensity is the small fraction $\frac{4}{5}(\omega_o a/c)^2$ of the dipole intensity.

(*d*) Like the dipole radiation, it is linearly polarized in meridional planes containing the emission and oscillation directions.

8.4 Two like point charges (q, m each) have equilibrium positions a distance l apart. Suppose that in the absence of the other each would oscillate with a simply harmonic natural frequency ω_o whenever displaced from its (then different) equilibrium position. Next suppose that, starting from the entire system in its equilibrium configuration, one particle is given a small impulse $mu_o \ll mc$ toward the other, insufficient to displace it by more than a small fraction of l.

(*a*) Neglecting the effects of losses to radiation on the ensuing motions, show that the *two* lowest frequencies that will be radiated if $q^2/l^2 \lesssim m\omega_o^2 l$ are $(\omega_o^2 + 4q^2/ml^3)^{1/2}$ and $2\omega_o$.

(*b*) Show that the energy loss to the radiation in each oscillation period is indeed a negligible fraction of the excitation energy $\frac{1}{2}mu_o^2$, even for frequencies high enough to be visible, because $q^2/mc^2 < e^2/m_e c^2 \approx 2.8 \times 10^{-13}$ cm for the largest chargings per unit mass that can be managed.

8.5 Two equal point charges at a rigid distance $2a$ apart rotate in a fixed plane and about the center between them with an uniform angular velocity $\omega_o \ll c/a$. What is the frequency of their most intense radiation (relative to the center):

(*a*) When the charges have opposite signs?

(*b*) When the charges have the same sign?

(*c*) Show how the angular distributions and intensities of those radiations can be obtained by simple modifications of results in the text.

8.6 Show that the angular momentum radiated during the quadrupole emissions (8.79) by the circling point charge amounts to $2\hbar$ units for each photon unit $\hbar\omega$ of the energy radiated.

8.7 Derive (8.76) for an arbitrary (symmetrical) quadrupole tensor.

8.8 (*a*) Form a linear quadrupole like that of Exercise 2.15 and set it spinning about its center with a constant angular velocity $\boldsymbol{\omega}_0$ normal to the line of charges. How will its radiation compare with that considered in Exercise 8.5(*b*)?

(*b*) Suppose, instead, that a square quadrupole of the type considered in Exercise 2.16 is set spinning uniformly about an axis through its center and normal to the plane of the charges. Find the angular distribution and the intensity of the radiation.

8.9 Observations on such radiations as are generated by the dipoles (8.67) must be made on assemblies of large numbers of the systems having a random distribution of the precession angles α $(0 < \alpha < \pi)$.

(*a*) Will the random orientations of the orbital planes wash out the observability of the characteristic angular distributions relative to the direction of the magnetic field?

(*b*) Do you expect the intensities of the $\omega_0 \pm \omega_L$ radiations to be the same?

(*c*) What is the ratio of each of the $\omega_0 \pm \omega_L$ intensities to the ω_0 intensity as detected in a direction perpendicular to the magnetic field? Parallel to it?

(*d*) Suppose that a polarization analyzer is placed in the path of the radiations into a direction perpendicular to the magnetic field. Compare the radiations it will transmit when the optic axis of the analyzer is oriented parallel to the magnetic field and perpendicular to it.

8.10 What is usually called a *dipole antenna* is a linear one that is short compared with the wavelength it radiates, so that $kL \ll 1$ and (8.32) is replaced by the *uniformly* distributed current $I = I_0 \cos \omega t$.

(*a*) Find its radiation pattern directly from (8.31).

(*b*) Dipole antennas meet the conditions for the definability of a dipole moment (8.39). Show that $D = iI_0 L/\omega$ and that (8.44) yields results agreeing with (*a*).

8.11 Two like dipole antennas (Exercise 8.10) are erected parallel to each other and at a (perpendicular) distance l apart.

(*a*) Find the radiation-zone approximation of the field (8.41) in each of the two cases respectively having the antenna currents parallel ($\uparrow\uparrow$) and antiparallel ($\downarrow\uparrow$) to each other, at every phase.

(*b*) Show that the radiation patterns are respectively representable as proportional to

$$\frac{dP}{d\Omega} \sim \sin^2 \vartheta \begin{Bmatrix}\cos^2\\ \sin^2\end{Bmatrix}(\tfrac{1}{2}kl \sin \vartheta \cos \varphi)\begin{Bmatrix}\uparrow\uparrow\\ \downarrow\uparrow\end{Bmatrix}$$

(*c*) For $l = \frac{1}{2}\lambda$ and $l = \lambda \equiv 2\pi/k$, make sketches of patterns in the $\vartheta = \frac{1}{2}\pi$ plane bisecting the antennas normally. Give qualitative arguments for expecting the directions in which the nodes and antinodes (lobe maxima) occur.

(*d*) Investigate the effects of putting the two antennas very close to each other ($kl \ll 1$), including rates of variation with frequency for given peak currents. How do the results compare with those from a single dipole and from a quadrupole?

8.12 Three dipole antennas lie in a straight line, equally spaced at distances $l = \frac{1}{2}\lambda$ apart. Compare radiation patterns with respect to node directions and relative intensities at maxima in the following cases.

(*a*) Let the outer dipoles oscillate in phase but the middle one in opposite phase. Compare results with the $n = 1$ case of Fig. 8.1.

(*b*) Now let all three dipoles oscillate in phase. Compare results with the case of Exercise 9.1.

CHAPTER
NINE

SPHERICAL WAVES†

Scalar spherical waves · Field displacement and rotation operators · Vector spherical harmonics and vector spherical waves · Nonstatic multipole fields of all orders and their angular momenta · Multipole sources and patterns of radiation · Circularly polarized beams as carriers of angular momentum · Photon spin

Each point element of a nonstatic source keeps emitting the spherical Huyghens' wavelets (8.17), a linearly independent one for each frequency needed to represent the source variations in time. The entire source emits a superposition of wavelets, centered on the various points of the source distribution, and their resultant cannot constitute a simply isotropic spherical wave. However, it will now be shown that *any* field distribution can be decomposed into concentric, though not isotropic, spherical waves centered on any desired point. Complete sets of the concentric waves can be defined so that they form a basis for the resolution of arbitrary distributions *alternative* to the plane-wave basis (7.9) used for the decompositions (7.7) and (7.8). The spherical-wave constituents can be expected to be the more directly relatable to a centrally localized source.

9.1 SCALAR SPHERICAL WAVES

The sets of spherical waves should be developed as solutions of the Helmholtz equation

$$(\nabla^2 + k^2)\phi(r, \vartheta, \varphi) = 0 \tag{9.1}$$

since this is just the wave equation expressed for representatives of frequency $\omega = ck$. It must be satisfied in the free spaces between sources not only by the scalar potential of (6.62) but also by each component of the vector potential and by derivatives of the potentials like the fields $\mathbf{E}$ and $\mathbf{B}$. Formally, the solutions constitute eigenfunctions of the operator $-\nabla^2$ for the real eigenvalue k^2 and are generalizations of the Laplace fields of Chap. 3, for which $k^2 \equiv \omega^2/c^2 = 0$ only.

The spherical waves centered on a point used as the origin of $\mathbf{r}$ are obtained when their dependence on the directions $\hat{\mathbf{r}}(\vartheta, \varphi)$ is separated off by taking advan-

† The sometimes complex developments in this chapter are not essential to understanding any other part of this book. They do have importance in many applications and for discriminating the role of photon spin in radiations.

tage of the resolvability of any such dependence on the orthonormal basis provided by the spherical harmonics (3.85). Thus, solutions of the form

$$\phi_{klm}(\mathbf{r}) = R(r)Y_{lm}(\hat{\mathbf{r}}) \tag{9.2}$$

are to be sought, with radial-wave factors $R(r)$ satisfying an equation to be obtained by substituting the form (9.2) into the Helmholtz equation.

The spherical coordinate representation (A.39*a*) of the Laplace operator ∇^2 is needed, and its effect on the direction dependence of (9.2) follows from the proportionality of the spherical harmonics to solutions of (3.62), with $\lambda = l(l+1)$ of (3.64):

$$-\left(\frac{1}{\sin\vartheta}\frac{\partial}{\partial\vartheta}\sin\vartheta\frac{\partial}{\partial\vartheta} + \frac{1}{\sin^2\vartheta}\frac{\partial^2}{\partial\varphi^2}\right)Y_{lm} = l(l+1)Y_{lm} \tag{9.3}$$

Then the Helmholtz equation $-\nabla^2(RY_{lm}) = k^2(RY_{lm})$ yields

$$\left[-\frac{d^2}{r\,dr^2}r + \frac{l(l+1)}{r^2}\right]R_l = k^2R_l(r) \tag{9.4}$$

as a *radial-wave* equation. The solutions of this for a given l are independent of m, and so the possible radial-wave factors will be the same for all the $2l+1$ differently oriented angular patterns that exist for each l.

The Radial Waves

Like the Bessel equation (C.7), the radial-wave equation (9.4) has its solutions depending on r only through the product $\zeta \equiv kr$, since this substitution yields

$$-\left[\frac{d^2}{d\zeta^2} - \frac{l(l+1)}{\zeta^2}\right](\zeta R_l) = (\zeta R_l) \tag{9.5}$$

Thus, the radial waves of various wavelengths $\lambda = 2\pi/k \equiv 2\pi\lambda\!\!\!^{-}$ differ only in the *scale* of their extensions into space, as measured by r in units of $\lambda\!\!\!^{-} \equiv 1/k$.

The simplest, $l = 0$, solutions are $\zeta R_0 \sim \exp(\pm i\zeta)$, describing radial waves that are propagated isotropically since $Y_{00} = (4\pi)^{-1/2}$ is independent of directions. To represent these, the function

$$h_0(\zeta) \equiv \frac{e^{i\zeta}}{i\zeta} \tag{9.6}$$

called a *spherical Hankel function of order zero*, is defined. The general isotropic ($l = 0$) solution is obviously $R_0 = a_o h_0 + b_o h_0^*$. Notice that the Huyghens' wavelets (8.17) are themselves just

$$\chi_k(R, t) = ikh_0(kR)e^{-i\omega t} \tag{9.7}$$

as befits their being isotropic solutions of the Helmholtz equation in (8.18).

A formula for generating higher-order solutions, $R_l \sim h_l(\zeta)$ with $l > 0$, from $h_0(\zeta)$ can be developed upon noticing that the operation in (9.5) can be factored into the product of the two operators in

$$-\left(\frac{d}{d\zeta}+\frac{l+1}{\zeta}\right)\left(\frac{d}{d\zeta}-\frac{l+1}{\zeta}\right)(\zeta h_l)=(\zeta h_l) \tag{9.8}$$

If still another operation by $d/d\zeta - (l+1)/\zeta$ is now applied to both sides, the result is

$$-\left[\frac{d^2}{d\zeta^2}-\frac{(l+1)(l+2)}{\zeta^2}\right]\left(\frac{d}{d\zeta}-\frac{l+1}{\zeta}\right)(\zeta h_l)=\left(\frac{d}{d\zeta}-\frac{l+1}{\zeta}\right)(\zeta h_l)$$

This shows that the function as derived on the right satisfies the radial equation for ζh_{l+1}. The conventional definitions put

$$\zeta h_{l+1}=-\left(\frac{d}{d\zeta}-\frac{l+1}{\zeta}\right)(\zeta h_l)=-\zeta^{l+1}\frac{d}{d\zeta}\zeta^{-(l+1)}(\zeta h_l)$$

so that

$$h_{l+1}(\zeta)=-\zeta^l\frac{d}{d\zeta}\zeta^{-l}h_l(\zeta) \tag{9.9}$$

a ladder operation, by which the one rung from h_l to h_{l+1} can be climbed. It is then quite simple to see that the repeated operations in

$$h_l=(-\zeta)^l\left(\frac{d}{\zeta\, d\zeta}\right)^l h_0 \equiv (-\zeta)^l\left(\frac{d}{\zeta\, d\zeta}\right)^l\frac{e^{i\zeta}}{i\zeta} \tag{9.10}$$

will yield solutions of any order l from h_0. Samples are

$$h_1=-\left(1-\frac{1}{i\zeta}\right)\frac{e^{i\zeta}}{\zeta} \qquad h_2=-\left(1-\frac{3}{i\zeta}-\frac{3}{\zeta^2}\right)\frac{e^{i\zeta}}{i\zeta} \tag{9.11}$$

Every h_l differs from h_0 only by a factor that is a polynomial in $1/\zeta$. The more highly inverse-powered terms become negligible as $\zeta \to \infty$, and

$$h_l(\zeta\to\infty)\to(-)^l i^l\frac{e^{i\zeta}}{i\zeta}=\frac{1}{\zeta}e^{i[\zeta-(l+1)\pi/2]} \tag{9.12}$$

is the asymptotic behavior. For this result, the factors $1/\zeta$ are treated as constants during the differentiations in (9.10), with only the phase variations $d^l e^{i\zeta}/d\zeta^l = i^l e^{i\zeta}$ being taken into account, just as in the radiation-zone ($kr \gg 1$) approximations of (8.27*b*) and (8.28).

The solutions $h_l(\zeta)$ are called *spherical Hankel functions* because of the relations

$$h_l(\zeta)=\left(\frac{\pi}{2\zeta}\right)^{1/2}H^{(1)}_{l+1/2}(\zeta) \tag{9.13}$$

they have to the Hankel functions of (C.14). The validity of these connections can readily be checked by substitution into the radial equation and seeing that this becomes the Bessel equation (C.7) with $m = l + \frac{1}{2}$. The identification with the particular cylindrical harmonics (C.14) follows from a comparison to the asymptotic behavior (9.12).

The asymptotic behavior also shows that a monochromatic representative like

$$\phi_{klm}(\mathbf{r})e^{-i\omega t} = h_l(kr)Y_{lm}(\hat{\mathbf{r}})e^{-i\omega t} \to (-i)^{l+1}Y_{lm}\frac{e^{i(kr-\omega t)}}{kr} \tag{9.14}$$

describes a spherical *outgoing* wave (propagated toward $r \to +\infty$). This is not only the asymptotic behavior but persists down to $r \to 0$, since closer in, there is only modulation of the amplitude by a polynomial in $1/kr$. Such outgoing waves have already appeared in the representations of the radiations in Chap. 8. The vector potential (8.25) is obviously proportional to $h_0(kr)$, and $h_1(kr)$ is introduced through (8.27*a*); $h_2(kr)$ is evident in the quadrupole field expression (8.74).

The *general* solution of the second-order radial-wave equation (9.5) can be written as

$$R_l(r) = a_l h_l(kr) + b_l h_l^*(kr) \tag{9.15}$$

since the phase reversal in $(e^{+i\zeta})^* = e^{-i\zeta}$ makes $h_l^*(kr)$ linearly independent of h_l. Using h_l^* in place of h_l when forming a representative like (9.14) would obviously produce an *incoming*, converging wave [but notice that the complex conjugate of the *entire* expression (9.14) still describes an outgoing wave, one proportional to $\exp(+i\omega t)$]. Thus the general solution (9.15) is expressed as an arbitrary superposition of outgoing and incoming waves.

A complex conjugation of the expression (9.13) introduces the Hankel function of the second kind, (C.15), differing from the first kind merely by the complex conjugation whenever the argument $\zeta \equiv kr$ is real. The real Bessel and Neumann functions (C.12) and (C.13) are introduced when the linear combinations

$$j_l(\zeta) \equiv \tfrac{1}{2}(h_l + h_l^*) = \left(\frac{\pi}{2\zeta}\right)^{1/2} J_{l+1/2}(\zeta) \tag{9.16a}$$

and

$$n_l(\zeta) \equiv \frac{1}{2i}(h_l - h_l^*) = \left(\frac{\pi}{2\zeta}\right)^{1/2} N_{l+1/2}(\zeta) \tag{9.16b}$$

are formed. Their asymptotic behavior is evident both from (9.12) and from (C.12) and (C.13):

$$j_l(kr \to \infty) = (kr)^{-1} \cos\left[kr - (l+1)\frac{\pi}{2}\right] \tag{9.17a}$$

$$n_l(kr \to \infty) = (kr)^{-1} \sin\left[kr - (l+1)\frac{\pi}{2}\right] \tag{9.17b}$$

They thus describe *standing* spherical waves, and the general solution (9.15) can, as an alternative, be expressed as the superposition

$$R_l(r) = (a_l + b_l)j_l(kr) + i(a_l - b_l)n_l(kr) \tag{9.18}$$

of standing waves that are asymptotically out of phase with each other by 90°.

The linearly independent parts discriminated in (9.18) are also referred to as respectively the *regular* and *irregular* solutions, on the basis of their behavior as $r \to 0$. It can be seen from the expression (9.10) for h_l together with (9.16*b*) for n_l, or from (C.10), that

$$h_{l\geq 1}(kr \to 0) \to -i\frac{(2l-1)!!}{(kr)^{l+1}} \qquad n_{l\geq 1} \to -1\frac{(2l-1)!!}{(kr)^{l+1}} \tag{9.19}$$

both types of solution becoming infinite at the origin (as also $h_0 \to 1/ikr$ and $n_0 \to -1/kr$). However, (9.16*a*) and (C.9) together show that

$$j_l(kr \to 0) \to \frac{(kr)^l}{(2l+1)!!} \tag{9.20}$$

is nonsingular and actually vanishes at $r = 0$, except for $j_0(0) = 1$. The existence of linearly independent solutions proportional to r^l and to $r^{-(l+1)}$ for $r \to 0$ should be expected from the static case (3.71), since $\zeta = kr \to 0$ for either $k \to 0$ or $r \to 0$.

The lowest-order radial standing waves are

$$j_0(kr) = \frac{\sin kr}{kr} \qquad \text{and} \qquad n_0 = -\frac{\cos kr}{kr} \tag{9.21}$$

These start their asymptotic behavior (9.17) right at the origin, as does $h_0 = j_0 + in_0$ of (9.6). The higher-order functions are plainly derivable from j_0 and n_0 by the operations in (9.9) and (9.10) again, as well as by forming the linear combinations (9.16). Some of the lower-order examples are

$$j_1(\zeta) = \frac{\sin \zeta}{\zeta^2} - \frac{\cos \zeta}{\zeta}$$

$$j_2 = \left(\frac{3}{\zeta^3} - \frac{1}{\zeta}\right) \sin \zeta - \frac{3}{\zeta^2} \cos \zeta \tag{9.22}$$

and

$$n_1(\zeta) = -\frac{\cos \zeta}{\zeta^2} - \frac{\sin \zeta}{\zeta}$$

$$n_2 = -\left(\frac{3}{\zeta^3} - \frac{1}{\zeta}\right) \cos \zeta - \frac{3}{\zeta^2} \sin \zeta \tag{9.23}$$

Spherical Wave Packets

An *orthogonal* set of spherical waves can be formed, to constitute a basis on which any everywhere-finite field can be resolved. The set is an alternative to the plane-wave basis (3.55), which is also restricted to representing fields where they are finite since a plane wave, being proportional to $\exp i\mathbf{k} \cdot \mathbf{r}$, plainly has no singularities anywhere. The spherical waves of the basis are accordingly restricted to the regular radial factors described by the spherical Bessel functions (9.16*a*)

$$u_{klm}(\mathbf{r}) \equiv j_l(kr) Y_{lm}(\hat{\mathbf{r}}) \tag{9.24}$$

Their orthogonality property† will be made explicit below.

That the regular spherical-wave set (9.24) forms as complete a basis as the plane waves follows from the fact that any one plane wave can be expressed as a superposition of the regular spherical waves. More widely known is the *partial wave analysis*

$$e^{i\mathbf{k}\cdot\mathbf{r}} = \sum_{l=0}^{\infty} (2l+1) i^l j_l(kr) P_l(\hat{\mathbf{k}} \cdot \hat{\mathbf{r}}) \tag{9.25}$$

usually derived by regarding the plane wave as a function of the cosine $\hat{\mathbf{k}} \cdot \hat{\mathbf{r}} \equiv \mu$ and resolving this dependence on the Legendre polynomial set of (3.69). The coefficients in this expansion must be functions of the remaining variable $\zeta \equiv kr$ and are proportional to integrals

$$\int_{-1}^{1} d\mu P_l(\mu) e^{i\mathbf{k}\cdot\mathbf{r}} = 2 i^l j_l(kr) \tag{9.26}$$

having the results indicated.‡ The integration is elementary in the $l = 0$ case, since $P_0 = 1$ and

$$\frac{1}{2} \int_{-1}^{1} d\mu e^{ikr\mu} = \frac{\sin kr}{kr} \equiv j_0(kr) \tag{9.27}$$

a result worth repeating because it shows that the lowest-order Bessel wave is just an average of plane waves into isotropically distributed directions. The resolution on the basis u_{klm}, as specified in (9.24), is made evident when the spherical-harmonic addition theorem (3.99) is used to reexpress (9.25) as

$$e^{i\mathbf{k}\cdot\mathbf{r}} = 4\pi \sum_{l=0}^{\infty} \sum_{m=-l}^{l} [i^l Y_{lm}^*(\hat{\mathbf{k}})] j_l(kr) Y_{lm}(\hat{\mathbf{r}}) \tag{9.28}$$

The spherical-wave set (9.24) has the orthogonality represented by

$$\oint dV(\mathbf{r}) u_{klm}^* u_{k'l'm'} = \delta_{ll'} \delta_{mm'} \frac{\pi}{2k^2} \delta(k - k') \tag{9.29}$$

where $\delta(k - k') \equiv \delta(k' - k)$ is just the *one*-dimensional delta function defined for $k, k' > 0$ so that $\int_0^\infty dk\delta = 1$. The proportionality to $\delta_{ll'} \delta_{mm'}$ results from the orthonormality (3.86) of the spherical harmonics, but the remaining factor

$$I_l(k, k') \equiv \int_0^\infty dr r^2 j_l(kr) j_l(k'r) = \frac{\pi}{2k^2} \delta(k - k') \tag{9.30}$$

† That an orthogonality over all space, as for the plane waves in (3.56), can be adduced only for the regular spherical waves is made clear in Ref. 17, page 446. The irregular solutions are linearly independent on other grounds.

‡ See, for example, Ref. 17, page 454.

may be less familiar. One way to obtain it is to substitute the plane-wave decomposition (9.28) into the orthogonality condition (3.56), with the result

$$\frac{(4\pi)^2}{(2\pi)^3}\sum_l\left[\sum_m Y_{lm}(\hat{\mathbf{k}}')Y^*_{lm}(\hat{\mathbf{k}})\right]I_l(k,k') = \delta(\mathbf{k}-\mathbf{k}')$$

The square-bracketed expression here,

$$\sum_m Y_{lm}(\hat{\mathbf{k}}')Y^*_{lm}(\hat{\mathbf{k}}) = \delta(\hat{\mathbf{k}}-\hat{\mathbf{k}}') \tag{9.31}$$

can be equated to the indicated delta function, which is defined to yield unity when integrated over all directions, $\oint d\Omega(\hat{\mathbf{k}})$ or $\oint d\Omega(\hat{\mathbf{k}}')$. That is just a completeness relation of the type (3.57) for the set $Y_{lm}(\hat{\mathbf{k}})$ in this instance; it follows from the decomposition

$$\delta(\hat{\mathbf{k}}'-\hat{\mathbf{k}}) = \sum_{lm} a_{lm}Y_{lm}(\hat{\mathbf{k}}')$$

which must have the coefficients

$$a_{lm} = \oint d\Omega(\hat{\mathbf{k}}')Y^*_{lm}(\hat{\mathbf{k}}')\delta(\hat{\mathbf{k}}'-\hat{\mathbf{k}}) = Y^*_{lm}(\hat{\mathbf{k}})$$

Now (9.30) is obtained when it is recognized that $\delta(\mathbf{k}-\mathbf{k}')$ must be equivalently representable as

$$\delta(\mathbf{k}-\mathbf{k}') = \delta(\hat{\mathbf{k}}-\hat{\mathbf{k}}')\cdot k^{-2}\delta(k-k') \tag{9.32}$$

to retain the property of yielding unity upon an integration $\oint dV(\mathbf{k}) \equiv \oint d\Omega(\hat{\mathbf{k}})\int_0^\infty dk k^2$.

Just a decomposition on the spherical-wave set is needed to represent an arbitrary, everywhere-finite, scalar wave field as a packet

$$\phi(\mathbf{r},t) = \mathrm{Re}\left[\sum_{lm}\int_0^\infty dk c_{lm}(k)u_{klm}(\mathbf{r})e^{-ickt}\right] \tag{9.33}$$

of components centered on whatever point has been chosen as the origin $\mathbf{r} = 0$.

The Association of the Scalar Potential with Its Sources

A development like that outlined in the opening paragraph of this chapter is simplest to carry out for the scalar-potential representative in (8.19). Unlike (9.33), this field can have singularities, at any points $\mathbf{r} = \mathbf{r}_s$ where finite charge exists. The various positions $\mathbf{r} = \mathbf{r}_s$ are points of origin for the Huyghens' wavelets, and a resolution into spherical waves centered on $\mathbf{r} = 0$ can be obtained by using a long-known generalization of the static propagator expression (3.81)

$$\frac{e^{ik|\mathbf{r}-\mathbf{r}_s|}}{ik|\mathbf{r}-\mathbf{r}_s|} = 4\pi\sum_{l,m}[h_l(kr_>)Y_{lm}(\hat{\mathbf{r}}_>)][j_l(kr_<)Y^*_{lm}(\hat{\mathbf{r}}_<)] \tag{9.34}$$

which can also be written

$$h_0(k|\mathbf{r}-\mathbf{r}_s|) = \sum_l (2l+1)h_l(kr_>)j_l(kr_<)P_l(\hat{\mathbf{r}}\cdot\hat{\mathbf{r}}_s) \tag{9.35}$$

with the help of the spherical-harmonic addition theorem (3.99).

The validity of the expressions can be seen by starting with the fact that the left side $[\equiv h_0(kR)]$ is a solution of the Helmholtz equation, except at $R \equiv |\mathbf{r} - \mathbf{r}_s| = 0$, regardless of the origin used in expressing the operation ∇^2. Thus the dependence on $\mathbf{r}$ for fixed $\mathbf{r}_s$ and also the dependence on $\mathbf{r}_s$ for fixed $\mathbf{r}$ must individually satisfy the Helmholtz equation [as in (8.9)!], and expression (9.34) is plainly a superposition of such solutions in each case. The coefficients can be checked by seeing that the superposition has the behavior required on the left side as $r \to 0$ and also as $r \to \infty$, a sufficient check because of the formal interchangeability of $\mathbf{r}$ and $\mathbf{r}_s$. For $r \to 0$, the behavior (9.20) of $j_l(kr \to 0)$ shows that

$$\lim_{r\to 0} \sum_{l=0}^{\infty} (2l+1)\frac{(kr)^l}{(2l+1)!!}h_l(kr_s)P_l(\hat{\mathbf{r}}\cdot\hat{\mathbf{r}}_s) \to h_0(kr_s)$$

properly, since only the $l = 0$ term survives and $P_0 = 1$. For the check at $r \to \infty$, the result (8.23) can be called on to represent the requirements of the left side in (9.35):

$$\frac{e^{ikr}}{ikr}e^{-i\mathbf{k}\cdot\mathbf{r}_s} \approx \sum_l (2l+1)j_l(kr_s)\frac{i^{-(l+1)}e^{ikr}}{kr}P_l(\hat{\mathbf{k}}\cdot\hat{\mathbf{r}}_s)$$

when the asymptotic behavior (9.12) of $h_l(kr \to \infty)$ is used. The spherical wave common to both sides can be canceled out, and what remains is just the complex conjugate of the partial wave analysis (9.25). It is also simple to check that expression (9.34) reduces to just the static propagator decomposition (3.81) as $k \to 0$.

A substitution of (9.34) into the scalar-potential expression (8.19) yields as the field outside the source

$$\phi(\mathbf{r}) = 4\pi i \sum_{l,m} \frac{k^{l+1}q_{lm}}{(2l+1)!!}h_l(kr)Y_{lm}(\hat{\mathbf{r}}) \tag{9.36}$$

for which the dynamic source moments

$$q_{lm} \equiv \frac{(2l+1)!!}{k^l}\oint dV(\mathbf{r}_s)\rho(\mathbf{r}_s)j_l(kr_s)Y^*_{lm}(\hat{\mathbf{r}}_s) \tag{9.37}$$

have been defined. These definitions are constructed so that they reduce to just the electro*static* moments (3.101) in the static limit, $k \to 0$, when $j_l(kr_s) \to (kr_s)^l/(2l+1)!!$. Indeed, the entire expression (9.36) reduces to the electrostatic potential (3.102) for $k \to 0$, as can be seen from the property (9.19) of $h_l(kr \to 0)$.

The simplified electrostatic moment *forms* of (3.101) can also be used for $\omega \neq 0$ whenever the source is much smaller than a wavelength, so that $kr_s \ll 1$, as in the cases discussed in Sec. 8.3. Then $q_{00} \approx 0$ because of (8.15), and the $l = 1$ contribution to the potential is

$$\phi_1 = \left[\frac{4\pi}{3} \sum_m q_{1m} Y_{1m}(\hat{\mathbf{r}})\right] \frac{e^{ikr}}{r^2} (1 - ikr) \tag{9.38}$$

with the square bracket replaceable by $\hat{\mathbf{r}} \cdot \mathbf{D}$ according to (3.93) and (3.103). The result then agrees with the expression (8.56) for the scalar potential of an ideal electric dipole. The higher-order terms of (9.36) help represent a whole sequence of electric multipole fields.

Basic Properties of Scalar Waves

To prepare for classifying the possible varieties of *vector* spherical waves, it will help to have prior acquaintance with a more primitive means of classifying any fields, i.e., according to their behavior in spatial displacements. The intrinsic importance of considering this behavior stems from the fact that any shift of coordinate frame relative to which a given, *isolated* field can be described should be entirely equivalent to keeping the frame fixed and displacing the field bodily instead. Space should be homogeneous and isotropic for any entire, isolated field, as for any complete system, uninfluenced by anything external to it.

A homogeneity of space is presumed for formulating a simple bodily *translation* of any given field $\phi(\mathbf{r})$ by some displacement $\delta\mathbf{r}$ having the same direction and magnitude at every point of the field. The displaced field is to be described by a new function $\phi'(\mathbf{r})$ related to the initial description as

$$\phi'(\mathbf{r}) = \phi(\mathbf{r} - \delta\mathbf{r}) \tag{9.39}$$

Particularly useful conclusions emerge when an infinitesimal displacement, by a $\delta\mathbf{r} \to 0$, is considered and there is the *change* in the field description by

$$\phi' - \phi \equiv \delta\phi = \phi(\mathbf{r} - \delta\mathbf{r}) - \phi(\mathbf{r}) = -\delta\mathbf{r} \cdot \nabla\phi(\mathbf{r}) \tag{9.40}$$

It is at once evident that finding $\phi'(\mathbf{r})$ for a translated field becomes simplest after the undisplaced field $\phi(\mathbf{r})$ has been resolved into its plane-wave constituents, as in (3.53), since

$$e^{i\mathbf{k} \cdot (\mathbf{r} - \delta\mathbf{r})} = e^{-i\mathbf{k} \cdot \delta\mathbf{r}} e^{i\mathbf{k} \cdot \mathbf{r}} \tag{9.41}$$

and each plane-wave component merely undergoes a uniform, overall, phase shift. For an infinitesimal translation, the phase factor is approximately equal to $1 - i\mathbf{k} \cdot \delta\mathbf{r}$, and its contribution to the change $\delta\phi$ is found by multiplying each plane-wave constituent by a factor $-i\mathbf{k} \cdot \delta\mathbf{r}$ that is uniform over all space, though it differs for the various components, with their different wavelengths and orientations as determined by $\mathbf{k}$. The simple behavior of each plane wave in the translations is due to its being an eigenfunction of the infinitesimal translation generator $-\delta\mathbf{r} \cdot \nabla$ of (9.40), in the sense that applying the operation

$$-i\nabla \; (\equiv \mathbf{k}) \tag{9.42}$$

to any plane wave is equivalent to multiplying it by a constant, its eigenvalue **k**. The definition of the infinitesimal translation operator (9.42) includes the factor $-i$ just so that its eigenvalues can have physically significant interpretations as real wave vectors.†

The three component operations $-i\nabla_{x,y,z}$ ($\equiv k_{x,y,z}$) are said to form a complete set of mutually commuting operators because giving their eigenvalues $k_{x,y,z}$ is sufficient to delimit an eigenfunction, a plane wave, that is unique apart from its arbitrarily choosable amplitude. The words mutually commuting signify that the operations can be successively‡ applied in any relative order with the same effect. It has already been pointed out [most recently for (9.1)] that the plane waves are also eigenfunctions of $-\nabla^2$ ($\equiv k^2$), but this property they share with all the great variety of solutions (spherical, cylindrical, paraboloidal waves, and many others) possessed by the Helmholtz equation; moreover $-\nabla^2$ ($\equiv k^2$) *follows* from the more stringent condition $-i\nabla$ ($\equiv \mathbf{k}$).

A field displacement that tests the isotropy of space is a bodily *rotation* of it about some fixed axis. It will be supposed that the given field $\phi(\mathbf{r})$ is described relative to a frame centered on the proposed rotation axis so that the displacement vector can be given the simple representation $\delta\mathbf{r} = \delta\boldsymbol{\alpha} \times \mathbf{r}$, where $|\delta\boldsymbol{\alpha}|$ is the magnitude of the rotation angle and the vector $\delta\boldsymbol{\alpha}$ has the direction of the rotation axis. The rotationally displaced field $\phi'(\mathbf{r})$ can be represented by $\phi(\mathbf{r} - \delta\mathbf{r})$, as in (9.39) again, but now $\delta\mathbf{r}$ is not uniform over space but is proportional to the distance of the field point from the rotation axis. In the limit of an infinitesimal rotation ($|\delta\boldsymbol{\alpha}| \to 0$), the change $\delta\phi \equiv \phi' - \phi$ in the field description has the form (9.40) with

$$\delta\phi = -(\delta\boldsymbol{\alpha} \times \mathbf{r}) \cdot \nabla\phi = -\delta\boldsymbol{\alpha} \cdot (\mathbf{r} \times \nabla)\phi(\mathbf{r}) \tag{9.43}$$

an effect determined by an infinitesimal rotation operator

$$\mathbf{L} \equiv -i\mathbf{r} \times \nabla \tag{9.44}$$

just as $-i\nabla$ of (9.42) generates infinitesimal translations of the field.

Whereas the effect of translating a field $\phi(\mathbf{r})$ was simplest to express for its plane-wave components, the effect of the rotation is simplest to find after the field has been decomposed into its *spherical*-wave constituents about the rotation center, as in (9.33). This is particularly so if the spherical-wave basis adopted is defined relative to the rotation axis of $\delta\boldsymbol{\alpha}$ as the polar (z) axis. Then the spherical coordinates of $\mathbf{r} - \delta\mathbf{r}$ will be just r, ϑ, $\varphi - \delta\alpha$ relative to the field point $\mathbf{r}(r, \vartheta, \varphi)$, and the dependence on r, ϑ is unchanged by rotating the field. A spherical wave like $\phi_{klm}(r, \vartheta, \varphi)$ of (9.2) is merely transformed to

$$\phi'_{klm}(\mathbf{r}) \equiv \phi_{klm}(r, \vartheta, \varphi - \delta\alpha) = e^{-im\delta\alpha}\phi_{klm}(\mathbf{r}) \tag{9.45}$$

† The operator $-i\nabla$ gains deeper physical significance from the fact, pointed out in connection with (7.17), that a plane wave carries a field *momentum* **k** for each ω units of energy it carries. As will be discussed in Chap. 13 (see also Ref. 17, page 476), momenta are most generally definable as quantities that are conserved because space must be homogeneous for isolated systems, as supposed when $\delta\phi = -i\delta\mathbf{r} \cdot (-i\nabla)\phi(\mathbf{r})$ is taken to be the effect of a uniform translation, by $\delta\mathbf{r}$, of the field.

‡ As in building up translations over finite distances from the infinitesimal ones.

because ϕ_{klm} depends on φ only in proportion to $e^{im\varphi}$, as reference to the spherical-harmonic expression (3.85) shows. There is here only a uniform phase shift, independent of the field point, just as in the translation (9.41) of a plane wave. For an infinitesimal rotation, $\delta\alpha \to 0$, the phase factor becomes approximately equal to $1 - im\delta\alpha$ and is the result of the operation in

$$\phi_{klm}(r, \vartheta, \varphi - \delta\alpha) \approx \phi_{klm}(r, \vartheta, \varphi) - \delta\alpha \frac{\partial \phi_{klm}}{\partial \varphi} \tag{9.46}$$

Thus, in this case of a $\delta\alpha \equiv \delta\alpha_z$ (a rotation about the z axis), when the more general operation in (9.43) reduces to $-i\delta\boldsymbol{\alpha} \cdot \mathbf{L} = -i\delta\alpha\mathbf{L}_z$, it must be true that

$$\mathbf{L}_z \equiv -i\left(x\frac{\partial}{\partial y} - y\frac{\partial}{\partial x}\right) = -i\frac{\partial}{\partial \varphi} \tag{9.47}$$

for the z component of the vector operator (9.44). It can indeed be checked that a straightforward substitution of spherical for cartesian coordinates in the operations here does result in this equivalence. A conclusion is that the spherical waves owe their simple behavior in rotations about their polar axis to the fact that they are *eigenfunctions* of the operation $\mathbf{L}_z = -i\partial/\partial\varphi$, this being equivalent to multiplying each by an eigenvalue constant m, that is, $\mathbf{L}_z$ $[\equiv m]$.

The operator components $\mathbf{L}_{x,y}$ of (9.44) would be needed to give the contributions $(-i\delta\alpha_x \mathbf{L}_x)\Phi$ and $(-i\delta\alpha_y \mathbf{L}_y)\phi$ to $\delta\phi$ of (9.43) if $\delta\alpha_{x,y} \neq 0$, as in rotations of the field about axes other than the polar axis of its spherical-wave constituents. Such rotations do *not* merely shift each component in phase. They have an effect only on the angular dependences, as can be found by transforming their obvious cartesian forms into operations on spherical coordinates. The results are simplest to express for the linear combinations

$$\mathbf{L}_\pm \equiv \mathbf{L}_x \pm i\mathbf{L}_y = e^{\pm i\varphi}\left(\pm\frac{\partial}{\partial\vartheta} + i\cot\vartheta\frac{\partial}{\partial\varphi}\right) \tag{9.48}$$

The effect of these on a spherical wave proportional to $Y_{lm}(\vartheta, \varphi)$ follows from the derivative of the associated Legendre function $P_l^m(\vartheta)$ of (3.82):

$$\begin{aligned}\frac{d}{d\vartheta}P_l^m &= \left(-\sqrt{1-\mu^2}\frac{d}{d\mu}\right)\frac{(1-\mu^2)^{m/2}}{2^l l!}\left(\frac{d}{d\mu}\right)^{l+m}(\mu^2-1)^l \\ &= -\frac{\sqrt{1-\mu^2}}{2^l l!}\left[-m\mu(1-\mu^2)^{(1/2)m-1}\left(\frac{d}{d\mu}\right)^{l+m}\right. \\ &\qquad \left. + (1-\mu^2)^{(1/2)m}\left(\frac{d}{d\mu}\right)^{l+m+1}\right](\mu^2-1)^l \\ &= m\cot\vartheta P_l^m - P_l^{m+1}\end{aligned} \tag{9.49}$$

The first of the terms here is canceled out in the full operation by $\mathbf{L}_+$ of (9.48), and when the coefficients in the definition (3.85) of the spherical harmonics are taken into account,

$$\mathbf{L}_+ Y_{lm} = [(l-m)(l+m+1)]^{1/2} Y_{l,m+1} \tag{9.50a}$$

a ladder operation in which m is raised by 1 unit. Because $\mathbf{L}_+^* = -\mathbf{L}_-$ follows from the definitions (9.44) and (9.48), and with the property (3.90) of the spherical harmonics,

$$(\mathbf{L}_+ Y_{lm})^* = (-\mathbf{L}_-)(-)^m Y_{l,\,-m} = [(l-m)(l+m+1)]^{1/2}(-)^{m+1} Y_{l,\,-m-1}$$

An interchange $m \leftrightarrow -m$ then shows that

$$\mathbf{L}_- Y_{lm} = [(l+m)(l-m+1)]^{1/2} Y_{l,\,m-1} \tag{9.50b}$$

an m-lowering operation. A conclusion from these findings is that the spherical waves are not simultaneous eigenfunctions of the three component operations $\mathbf{L}_{x,\,y,\,z}$ of (9.44) but only of $\mathbf{L}_z$ ($\equiv m$), in contrast to the three translation operators $-i\nabla_{x,\,y,\,z}$, which do have plane waves as simultaneous eigenfunctions of all three.

There is in fact *no* wave field that can be a simultaneous eigenfunction of the operators $\mathbf{L}_{x,\,y,\,z}$ because those do not commute with each other; it makes a difference in what order they are successively applied to any function. Commutability is essential† to the existence of a simultaneous eigenfunction because the individual operations on such a function must be equivalent to mere multiplications by eigenvalue constants. A useful measure of the lack of commutability of a pair of operators, $\mathbf{L}_x$ and $\mathbf{L}_y$ say, is given by the difference

$$\mathbf{L}_x \mathbf{L}_y - \mathbf{L}_y \mathbf{L}_x \equiv [\mathbf{L}_x, \mathbf{L}_y] \tag{9.51}$$

conventionally symbolized by the brackets and called the *commutator* of the operator pair. It is quite simple to find, by considering the effect of each operator as composed in (9.44) applied to any function, that

$$[\mathbf{L}_x, \mathbf{L}_y] = i\mathbf{L}_z \qquad [\mathbf{L}_y, \mathbf{L}_z] = i\mathbf{L}_x \qquad [\mathbf{L}_z, \mathbf{L}_x] = i\mathbf{L}_y$$

These three relations can be expressed by the single (axial) vector equation

$$\mathbf{L} \times \mathbf{L} = i\mathbf{L} \tag{9.52}$$

It is an equation satisfied by the generators of the infinitesimal rotations of any objects. Most fundamentally it expresses the fact that successive rotations about different axes lead to end configurations that depend on the order in which the individual rotations are performed and that any resultant of rotations about a center is also obtainable from a single rotation about some one axis.‡

Although the three operations $\mathbf{L}_{x,\,y,\,z}$ do not commute with each other, each of them commutes with the resultant $\mathbf{L}^2 \equiv \mathbf{L}_x^2 + \mathbf{L}_y^2 + \mathbf{L}_z^2$:

$$[\mathbf{L}^2, \mathbf{L}] = 0 \tag{9.53}$$

(Exercise 9.2). This means that there can exist simultaneous eigenfunctions of any *one* rectangular component of $\mathbf{L}$ and of $\mathbf{L}^2$. Indeed, the spherical harmonics

† For a more leisurely discussion, see Ref. 17, eq. (14.60), and the text attending it, or any one of numerous books on elementary quantum theory.

‡ See, for example, Ref. 17, page 239, and eqs. (8.49) and (8.50).

are simultaneous eigenfunctions of $\mathbf{L}_z = -i\partial/\partial\varphi \equiv m$ and of $\mathbf{L}^2$ [$\equiv l(l+1)$], since the operation in (9.3) is just the spherical coordinate representation of $\mathbf{L}^2 \equiv \mathbf{L}_x^2 + \mathbf{L}_y^2 + \mathbf{L}_z^2$, as is readily found by substituting the forms (9.47) and (9.48) into it. With $-\nabla^2$ becoming equivalent to just the radial derivative operation in (9.4) for any spherical wave proportional to a Y_{lm}, the three operators

$$\mathbf{L}_z\ (\equiv m) \qquad \mathbf{L}^2\ [\equiv l(l+1)] \qquad -\nabla^2\ (\equiv k^2) \tag{9.54}$$

form a complete mutually commuting set that defines a unique regular spherical wave for each choice of the eigenvalues k, l, m. This set plays the role for the scalar spherical waves that the properties $-i\nabla_{x,\,y,\,z}$ $(\equiv k_{x,\,y,\,z})$ (9.42) play in classifying the plane waves.

9.2 VECTOR SPHERICAL WAVES

Like the scalar waves just reviewed, the spherical waves into which any *vector* field can be decomposed are classifiable according to their behavior when rotated.

The Unit-Spin Operators

The bodily rotation of any vector field $\mathbf{A}(\mathbf{r})$ is complicated by the fact that not only is its dependence on the field point changed, as for a scalar field, but also the direction of the field vector at each field point is rotated. The effect is simplest to illustrate for a rotation about a z axis, when the z components of both $\mathbf{r}$ and $\mathbf{A}$ are temporarily ignored, as indicated in Fig. 9.1 for the rotated field vector $\mathbf{A}'(\mathbf{r})$ and the undisplaced one $\mathbf{A}(\mathbf{r})$. In the terms of the figure, at $\mathbf{r} - \delta\mathbf{r}$ the field $\mathbf{A}$ has components $A_x = A \cos\gamma$ and $A_y = A \sin\gamma$, and at $\mathbf{r}$

$$A'_x = |\mathbf{A}'| \cos(\gamma + \delta\alpha) \qquad A'_y = |\mathbf{A}'| \sin(\gamma + \delta\alpha)$$

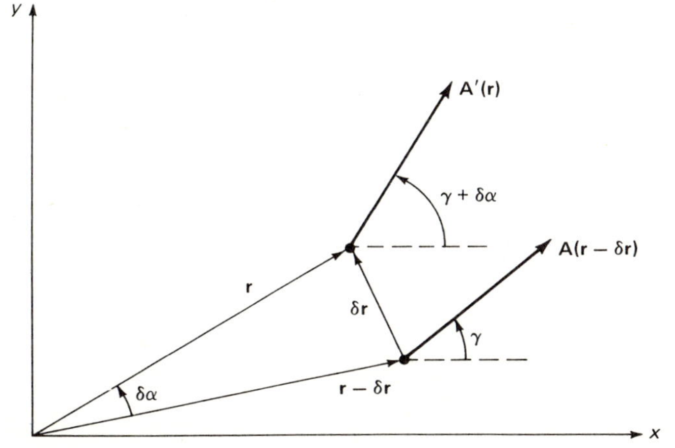

Figure 9.1

Since $|\mathbf{A}'(\mathbf{r})| = |\mathbf{A}(\mathbf{r} - \delta\mathbf{r})| \equiv A$,

$$A'_x = A(\cos\gamma\cos\delta\alpha - \sin\gamma\sin\delta\alpha) = A_x\cos\delta\alpha - A_y\sin\delta\alpha$$
$$A'_y = A(\sin\gamma\cos\delta\alpha + \cos\gamma\sin\delta\alpha) = A_x\sin\delta\alpha + A_y\cos\delta\alpha \qquad (9.55)$$

It will be sufficient to consider an infinitesimal rotation, by $\delta\alpha \to 0$, for which $\sin\delta\alpha \approx \delta\alpha$ and $\cos\delta\alpha \approx 1$:

$$A'_x(\mathbf{r}) \approx A_x - A_y\delta\alpha \approx A_x(\mathbf{r} - \delta\mathbf{r}) - \delta\alpha A_y(\mathbf{r})$$
$$A'_y(\mathbf{r}) \approx A_y + A_x\delta\alpha \approx A_y(\mathbf{r} - \delta\mathbf{r}) + \delta\alpha A_x(\mathbf{r}) \qquad (9.56)$$
$$A'_z(\mathbf{r}) \approx A_z(\mathbf{r} - \delta\mathbf{r})$$

with the last line the only supplementation needed when z components are no longer ignored. Notice that $A(\mathbf{r} - \delta\mathbf{r}) \approx A(\mathbf{r})$ is a sufficient approximation in the small terms proportional to $\delta\alpha$, since second-order differentials are being systematically neglected. Now, in the rotation about the z axis,

$$A_i(\mathbf{r} - \delta\mathbf{r}) \approx A_i(\mathbf{r}) - i\delta\alpha\mathsf{L}_z A_i(\mathbf{r})$$

for each of the components $A_{x,\,y,\,z}(\mathbf{r} - \delta\mathbf{r})$ in (9.56), just as for the scalar field in the preceding section. Then the complete change $\delta\mathbf{A} \equiv \mathbf{A}'(\mathbf{r}) - \mathbf{A}(\mathbf{r})$ of the vector field by the infinitesimal rotation about the z axis, as given by (9.56), can be represented as

$$\delta\mathbf{A} = -i\delta\alpha(\mathsf{L}_z + \mathsf{S}_z)\mathbf{A}(\mathbf{r}) \qquad (9.57a)$$

if S_z is *defined* as such an operator on any vector $\mathbf{A}$ that

$$(\mathsf{S}_z\mathbf{A})_x \equiv -iA_y \qquad (\mathsf{S}_z\mathbf{A})_y = +iA_x \qquad (\mathsf{S}_z\mathbf{A})_z = 0 \qquad (9.58)$$

are the components of the vector resultant of the operation. These are equivalent to the vector equation

$$\mathsf{S}_z\mathbf{A} \equiv i\mathbf{e}_z \times \mathbf{A} \qquad (9.59a)$$

if $\mathbf{e}_z \equiv \mathbf{e}_0$ is the unit vector pointing out the rotation (z) axis.

The finding (9.57a) is straightforwardly generalizable to an infinitesimal rotation $\delta\boldsymbol{\alpha}$ about an arbitrary axis, so that $\delta\alpha_{x,\,y} \neq 0$:

$$\delta\mathbf{A} = -i(\delta\boldsymbol{\alpha}\cdot\mathsf{J})\mathbf{A}(\mathbf{r}) \qquad \text{with } \mathsf{J} \equiv \mathsf{L} + \mathsf{S} \qquad (9.57b)$$

Infinitesimal effects, linear in $\delta\alpha_{x,\,y,\,z} \to 0$, are additive,† and so the results of operating with S_x and S_y are obtainable exactly as for the rotation about the z axis, above. The components (9.58) are replaced by ones obtainable by the cyclic interchanges $x \to y$, $y \to z$, $z \to x$, and the vector expression (9.59a) is generalized to

$$\mathsf{S}_j\mathbf{A} \equiv i\mathbf{e}_j \times \mathbf{A} \qquad \text{with } j = x, y, z \qquad (9.59b)$$

where $\mathbf{e}_{x,\,y,\,z} \equiv \mathbf{e}_{1,\,2,\,3(0)}$ are the rectangular directions.

† If there is any doubt, see Ref. 17, page 238. This is why angular *velocities* are vectorially additive but finite angular displacements are not.

Such tabulations as (9.58) can also be expressed in a matrix language,† as results of matrix multiplications into a column of vector components:

$$\mathbf{A} = \begin{bmatrix} A_x \\ A_y \\ A_z \end{bmatrix} \rightarrow \mathbf{S}_i \mathbf{A} = \begin{bmatrix} (\mathbf{S}_i \mathbf{A})_x \\ (\mathbf{S}_i \mathbf{A})_y \\ (\mathbf{S}_i \mathbf{A})_z \end{bmatrix} \tag{9.60}$$

On the basis here, the matrix-operator representatives of $\mathbf{S}_{x,\,y,\,z}$ are

$$\mathbf{S}_x = \begin{bmatrix} 0 & 0 & 0 \\ 0 & 0 & -i \\ 0 & i & 0 \end{bmatrix} \qquad \mathbf{S}_y = \begin{bmatrix} 0 & 0 & i \\ 0 & 0 & 0 \\ -i & 0 & 0 \end{bmatrix} \qquad \mathbf{S}_z = \begin{bmatrix} 0 & -i & 0 \\ i & 0 & 0 \\ 0 & 0 & 0 \end{bmatrix} \tag{9.61}$$

Such representations are more easily generalized to a greater variety of fields, but the vector representations (9.59) are sufficient for vector fields.

In either way it is straightforward to find that, like $\mathbf{L}_{x,\,y,\,z}$, the operators $\mathbf{S}_{x,\,y,\,z}$ do not commute with each other and obey the commutator relations

$$\mathbf{S} \times \mathbf{S} = i\mathbf{S} \tag{9.62}$$

which, like (9.52) for $\mathbf{L}$, characterize rotation generators in general. Moreover,

$$\mathbf{J} \times \mathbf{J} = i\mathbf{J} \tag{9.63}$$

since $\mathbf{L}$ and $\mathbf{S}$ commute with each other, each having effect only on entirely different degrees of freedom in the vector-field descriptions.

Even *uniform* vector fields, on which the operator $\mathbf{L}$ has a null effect, can be classified according to their properties under rotations of their directionalities by $\mathbf{S}$ as eigen*vectors* labeled by distinct eigenvalues, analogous to the scalar eigen*function* fields of the preceding section. Just as the directional dependencies proportional to $Y_{lm}(\hat{\mathbf{r}})$ of the spherical scalar fields could be classified as simultaneous eigenfunctions of $\mathbf{L}^2$ $[\equiv l(l+1)]$, $\mathbf{L}_z$ $(\equiv m)$, so the uniform vector fields are classifiable as simultaneous eigenvectors of $\mathbf{S}^2$, $\mathbf{S}_z$.

Actually *any* three-dimensional vector, no matter how directed, is an eigenvector of $\mathbf{S}^2$ $(\equiv 2)$. This invariant eigenvalue is obtainable in various ways, most directly by showing that a matrix $\mathbf{S}^2 = \mathbf{S}_x^2 + \mathbf{S}_y^2 + \mathbf{S}_z^2$ constructed from the representations (9.61) is just twice the unit matrix or by carrying out the operations vectorially

$$\sum_i \mathbf{S}_i(\mathbf{S}_i \mathbf{A}) = \sum_i [-\mathbf{e}_i \times (\mathbf{e}_i \times \mathbf{A})] = \sum_i [\mathbf{A} - \mathbf{e}_i(\mathbf{e}_i \cdot \mathbf{A})] = 3\mathbf{A} - \mathbf{A}$$

with the result $\mathbf{S}^2\mathbf{A} = 2\mathbf{A}$ for any $\mathbf{A}$.

Finding the distinct vector directions $\mathbf{e}_\mu$ that can be discriminated as eigenvectors of $\mathbf{S}_z$ $(\equiv \mu)$, with μ a distinct eigenvalue for each, is a problem of a conventional type‡ with the readily verifiable results

$$\mathbf{e}_{\pm 1} = \mp 2^{-1/2}(\mathbf{e}_1 \pm i\mathbf{e}_2) \qquad \mathbf{e}_0 = \mathbf{e}_3 \tag{9.64}$$

† See Ref. 17, secs. 9.3, 9.5, 9.6, and 10.5, or any book treating linear vector spaces.
‡ See Ref. 17, for example.

just such an orthonormal vector basis as was defined in (7.28) to help represent circular polarizations. Thus, the eigenvalues are quite naturally restricted to the three $\mu = 0, \pm 1$ and the isomorphism of the results

$$\mathbf{S}^2 \mathbf{e}_\mu = 2\mathbf{e}_\mu \qquad \mathbf{S}_z \mathbf{e}_\mu = \mu \mathbf{e}_\mu \tag{9.65}$$

to

$$\mathbf{L}^2 Y_{1m} = l(l+1) Y_{1m} = 2Y_{1m} \qquad \mathbf{L}_z Y_{1m} = mY_{1m} \tag{9.66}$$

might well have been anticipated in view of the discussion attending (3.93). The analogy to the $l = 1$ case has led to calling $\mathbf{S}$ a *unit-spin operator* and to the fact that $\mathbf{S}^2$ $(\equiv 2)$ for any vector field as a possession by it of a unit intrinsic (invariant) spin. The normalization $\mathbf{e}_\mu^* \cdot \mathbf{e}_{\mu'} = \delta_{\mu\mu'}$ makes the eigenvectors a convenient basis for the resolution of an arbitrarily directed vector field, just as the property $\oint d\Omega Y_{lm}^* Y_{l'm'} = \delta_{ll'} \delta_{mm'}$ made the spherical harmonics a convenient basis for resolving dependences on $\hat{\mathbf{r}}$. It will also be helpful to notice that just as

$$\mathbf{L}_\pm Y_{1m} \equiv (\mathbf{L}_x \pm i\mathbf{L}_y) Y_{1m} = \sqrt{(1 \mp m)(2 \pm m)}\, Y_{1,\, m \pm 1}$$

according to (9.50), so

$$\mathbf{S}_\pm \mathbf{e}_\mu \equiv (\mathbf{S}_x \pm i\mathbf{S}_y)\mathbf{e}_\mu = \sqrt{(1 \mp \mu)(2 \pm \mu)}\, \mathbf{e}_{\mu \pm 1} \tag{9.67}$$

follow from the matrices (9.61) or the vector products (9.59).

The Vector Spherical Harmonics

Nonuniform vector fields require applications of the entire operator $\mathbf{S} + \mathbf{L} = \mathbf{J}$ for their rotation, as in (9.57), and so can be classified into *eigenvector functions* of $\mathbf{J}^2$ $[= j(j+1)]$ and $\mathbf{J}_z$ $(\equiv m)$. The eigenvalues being presumed here will be found to be given by

$$j = 0, 1, 2, \ldots \qquad \text{and} \qquad m = 0, \pm 1, \pm 2, \ldots, \pm j \tag{9.68}$$

like† the values of l, m that label the spherical harmonics $Y_{lm}(\hat{\mathbf{r}})$ used to help represent the possible scalar fields. The corresponding functions for the vector fields are denoted‡ $\mathbf{T}_{jm}^l(\hat{\mathbf{r}})$ and called the *vector* spherical harmonics. The third index l, which will be found to be restricted to the values $l = j$ and $j \pm 1$, is needed because the operators $\mathbf{J}^2$, $\mathbf{J}_z$ turn out to be an insufficiently complete set for determining a unique directional dependence. An additional requirement $\mathbf{L}^2$ $[\equiv l(l+1)]$ will serve to complete the set [together with the automatic property $\mathbf{S}^2$ $(\equiv 2)$ of any vector field].

† This is no accident since it can be shown to follow from just the commutation relations $\mathbf{L} \times \mathbf{L} = i\mathbf{L}$ or $\mathbf{J} \times \mathbf{J} = i\mathbf{J}$, without regard for the degrees of freedom involved. That approach is not the one adopted in the text here because its full significance can be better appreciated in the context of quantum mechanics.

‡ Sometimes notations like $\mathbf{Y}_{jl}^m$ are used for the vector spherical harmonics.

The most explicit representation of the vector spherical harmonics gives them in terms of their components on the complex unit-vector basis (9.64) as

$$\mathbf{T}^l_{jm}(\hat{\mathbf{r}}) = \sum_{\mu=0,\,\pm 1} c_\mu(ljm)\mathbf{e}_\mu Y_{l,\,m-\mu}(\hat{\mathbf{r}}) \tag{9.69}$$

The proportionality of every component to a spherical harmonic of the order l makes the entire expression an eigenfunction of $\mathbf{L}^2$ $[\equiv l(l+1)]$, and the restriction to indices μ and $m - \mu$ that add up to m in every term make it simultaneously an eigenfunction of $\mathbf{J}_z = \mathbf{L}_z + \mathbf{S}_z$ $(\equiv m)$. The expression will be finally specified after the coefficients $c_{0,\,\pm 1}$ are so determined that operation with $\mathbf{J}^2$ becomes equivalent to multiplication by an eigenvalue constant, in completion of the requirements set down in the preceding paragraph. It is then said that the expression *couples* the description of the dependence on the *external* variable $\hat{\mathbf{r}}$ of the vector field, with its dependence on its *internal*, or *spin*, variable as represented by the values 0, ± 1 possible to the index μ.

The coupling coefficients $c_\mu(ljm)$ are to follow from making the expression for $\mathbf{T}^l_{jm}(\hat{\mathbf{r}})$ an eigenvector function of $\mathbf{J}^2 = (\mathbf{L} + \mathbf{S})^2 = \mathbf{L}^2 + \mathbf{S}^2 + 2\mathbf{L}\cdot\mathbf{S}$. Operation by $\mathbf{L}^2 + \mathbf{S}^2$ is equivalent to multiplication by $l(l+1) + 2$ regardless of what the c_μ's may be, and so these are in the end determined so that operation by $\mathbf{L}\cdot\mathbf{S}$ becomes equivalent to multiplication by some eigenvalue constant, say K, now to be found. An equation

$$\begin{aligned}(\mathbf{S}\cdot\mathbf{L} - K)\mathbf{T}^l_{jm}(\hat{\mathbf{r}}) &= 0 \\ &\equiv \sum_\mu c_\mu\left[\sum_i (\mathbf{S}_i\mathbf{e}_\mu)(\mathbf{L}_i Y_{l,\,m-\mu}) - K\mathbf{e}_\mu Y_{l,\,m-\mu}\right] = 0\end{aligned} \tag{9.70}$$

is to be solved for the possible values of K, as well as the corresponding coefficients c_μ. Here

$$\sum_i \mathbf{S}_i\mathbf{L}_i \equiv \mathbf{S}\cdot\mathbf{L} = \mathbf{S}_z\mathbf{L}_z + \tfrac{1}{2}(\mathbf{S}_+\mathbf{L}_- + \mathbf{S}_-\mathbf{L}_+) \tag{9.71}$$

where $\mathbf{S}_\pm \equiv \mathbf{S}_x \pm i\mathbf{S}_y$ and $\mathbf{L}_\pm \equiv \mathbf{L}_x \pm i\mathbf{L}_y$ have the known effects (9.67) and (9.50). The eigenvector equation (9.70) can be resolved into components on $\mathbf{e}_{+1}$, $\mathbf{e}_0$, $\mathbf{e}_{-1}$, and the respective results are

$$\begin{aligned}
&[(m-1) - K]c_1 + \tfrac{1}{2}[2(l+m)(l-m+1)]^{1/2}c_0 = 0 \\
&\tfrac{1}{2}[2(l+m)(l-m+1)]^{1/2}c_1 - Kc_0 + \tfrac{1}{2}[2(l-m)(l+m+1)]^{1/2}c_{-1} = 0 \\
&\tfrac{1}{2}[2(l-m)(l+m+1)]^{1/2}c_0 + [-(m+1) - K]c_{-1} = 0
\end{aligned} \tag{9.72}$$

Such linear homogeneous equations can only determine ratios like $c_{\pm 1}/c_0$ and have nonvanishing solutions only if the determinant of the coefficients vanishes. This yields the cubic

$$(K+1)[K(K+1) - l(l+1)] = 0 \tag{9.73}$$

Table 9.1 Coefficients $c_\mu(ljm)$ defining the vector spherical harmonics

j	c_{+1}	c_0	c_{-1}
$l+1$	$\left[\frac{(l+m)(l+m+1)}{2(l+1)(2l+1)}\right]^{1/2}$	$\left[\frac{(l-m+1)(l+m+1)}{(l+1)(2l+1)}\right]^{1/2}$	$\left[\frac{(l-m)(l-m+1)}{2(l+1)(2l+1)}\right]^{1/2}$
l	$-\left[\frac{(l+m)(l-m+1)}{2l(l+1)}\right]^{1/2}$	$\frac{m}{\sqrt{l(l+1)}}$	$\left[\frac{(l-m)(l+m+1)}{2l(l+1)}\right]^{1/2}$
$l-1$	$\left[\frac{(l-m)(l-m+1)}{2l(2l+1)}\right]^{1/2}$	$-\left[\frac{(l-m)(l+m)}{l(2l+1)}\right]^{1/2}$	$\left[\frac{(l+m)(l+m+1)}{2l(2l+1)}\right]^{1/2}$

for K, thus determining the possible eigenvalues as $K = +l$, -1 and $-(l+1)$ with $l \neq 0$. These correspond to $(\mathbf{J}^2 \equiv)\ l(l+1) + 2 + 2K \equiv j(j+1)$ with $j = (l+1)$, l, and $(l-1) \geq 0$, respectively, yielding just the variety of vector spherical harmonics anticipated above.

Putting, in turn, each of the three eigenvalues K into the secular equations (9.72) determines only the ratios between the three coefficients $c_{0,\pm 1}$ needed to represent each of the three varieties of vector spherical harmonics that exist for each j, m (or for each l, m). The determination of the coefficients can be completed within a conventional sign by requiring $\sum_\mu c_\mu^2 = 1$ for each set, so that the vector spherical harmonics will form an ortho*normalized* set, with the property

$$\oint d\Omega(\hat{\mathbf{r}})(\mathbf{T}^l_{jm})^* \cdot \mathbf{T}^{l'}_{j'm'} = \delta_{ll'}\delta_{jj'}\delta_{mm'} \tag{9.74}$$

The coefficients thus found are given in Table 9.1. They are examples of so-called *vector-addition*, *Wigner*, or *Clebsch-Gordan coefficients*,† much used in coupling many varieties of rotational descriptions. It is the readily confirmed property

$$\sum_\mu c_\mu(ljm)c_\mu(lj'm) = \delta_{jj'} \tag{9.75a}$$

of the tabulated coefficients that is responsible for the orthogonality when $j \neq j'$ in (9.74). Products of the tabulated columns can similarly be seen to have the sums

$$\sum_j c_\mu(ljm)c_{\mu'}(ljm) = \delta_{\mu\mu'} \tag{9.75b}$$

It is therefore said that the table constitutes a *unitary matrix* of coefficients.

† As such they have denotations like $c_\mu \equiv \langle l(m-\mu)1(\mu)|j(m)\rangle$, or $\langle l\ \ 1\ \ m-\mu, \mu|l\ \ 1\ \ j\ \ m\rangle$, or $C_{l1}(jm;\ m-\mu, m)$, or $C(l\ \ 1\ \ j;\ m-\mu, \mu)$ in various works. Sometimes they are replaced by Wigner's $3-j$ symbols:

$$c_\mu(ljm) \equiv (-)^{l+m-1}\sqrt{2j+1}\begin{pmatrix} l & 1 & j \\ m-\mu & \mu & -m \end{pmatrix}$$

The coefficients $c_{\pm 1}(llm)$ in the table can be recognized as proportional to the ones needed in (9.50) for expressing results of the operations $\mathbf{L}_{\pm} Y_{lm}$. Moreover, $c_0(llm)Y_{lm} \sim \mathbf{L}_z Y_{lm}$ in its m dependence. Consequently the particular vector spherical harmonics (9.69) having $j = l$ can be written

$$\mathbf{T}^l_{lm}(\hat{\mathbf{r}}) = \frac{\mathbf{L}Y_{lm}(\hat{\mathbf{r}})}{[l(l+1)]^{1/2}} \tag{9.76}$$

where $\mathbf{L} \equiv -i\mathbf{r} \times \nabla$. It is through this form that the vector spherical harmonics were earlier introduced into the theory of multipole radiations, but they are only a part of the complete set that also includes $j = l \pm 1$.

Some of the lowest-order vector spherical harmonics can be expressed more simply, in a manner analogous to (3.92) for $Y_{1m}(\hat{\mathbf{r}})$:

$$\mathbf{T}^1_{00}(\hat{\mathbf{r}}) = -\frac{\hat{\mathbf{r}}}{\sqrt{4\pi}} \qquad \mathbf{T}^0_{1m} = \frac{\mathbf{e}_m}{\sqrt{4\pi}} \qquad \mathbf{T}^1_{1m} = \sqrt{\frac{3}{8\pi}}\, i\mathbf{e}_m \times \hat{\mathbf{r}}$$
$$\mathbf{T}^2_{1m} = \frac{\mathbf{e}_m - 3\hat{\mathbf{r}}(\hat{\mathbf{r}} \cdot \mathbf{e}_m)}{\sqrt{8\pi}} \tag{9.77}$$

In these special forms, only $m = 0, \pm 1$ occur.

The Eigenfields and Their Derivatives

A general vector-field solution of the wave (or Helmholtz) equation will be an arbitrary linear combination of the vector spherical waves

$$\mathbf{A}^l_{kjm}(\mathbf{r}) = R_l(kr)\mathbf{T}^l_{jm}(\hat{\mathbf{r}}) \tag{9.78}$$

just as the general scalar-field solution can be constituted of the representatives $\phi_{klm}(\mathbf{r})$ of (9.2). The radial-wave factors R_l are again the superpositions (9.15) of outgoing and incoming waves or, equivalently, of the standing waves (9.18), since $\mathbf{T}^l_{jm}$ is just a linear combination of spherical harmonics all having the same order l. All three varieties ($j = l$ and $l \pm 1$) of the vector spherical harmonics are generally needed for expressing an arbitrary vector field. A unique one of the vector spherical waves (9.78) is singled out when it is required to be a simultaneous eigenfunction of the complete set of mutually commuting operators

$$\mathbf{L}^2\,[\equiv l(l+1)] \qquad \mathbf{S}^2\,(\equiv 2) \qquad \mathbf{J}^2\,[\equiv j(j+1)]$$
$$\mathbf{J}_z\,(\equiv m) \qquad -\nabla^2\,(\equiv k^2) \tag{9.79}$$

an augmentation of the set (9.54) that was found sufficient to define a unique *scalar* spherical wave.

The real part of the eigenfield (9.78), multiplied with $e^{-i\omega t}$, can constitute an entire vector-potential field since such is only required to satisfy the wave equation if it is accompanied by a scalar potential derivable from it as $\phi = \nabla \cdot \mathbf{A}/ik$. The corresponding fields $\mathbf{E}$ and $\mathbf{B}$ then follow as $\mathbf{E} = -\nabla\phi + ik\mathbf{A}$ and $\mathbf{B} = \nabla \times \mathbf{A}$.

Whatever **E** and **B** emerge, they must be expressible as superpositions of vector spherical waves like (9.78) since they also must satisfy the wave equation. However, **E** and **B** can only be *divergenceless* superpositions in the free spaces between sources, and, moreover, they must be interrelated as $\mathbf{E} = -\nabla \times \mathbf{B}/ik$ and $\mathbf{B} = +\nabla \times \mathbf{E}/ik$. It becomes important to examine the results of applying gradient, divergence, and curl operations to the spherical waves.

The purely formal work of applying the derivative operations to the spherical waves is sufficiently lengthy to have interest only for specialists, and so just the results will be quoted† here. (Their use can be avoided, as will be indicated at various junctures, if the reader can be satisfied with less explicit results.)

The vector field that arises as the *gradient* of a scalar spherical wave can be expressed as the linear combination of the vector spherical waves in

$$\nabla(R_l Y_{lm}) = k\left[\sqrt{\frac{l}{2l+1}}R_{l-1}\mathbf{T}_{lm}^{l-1} + \sqrt{\frac{l+1}{2l+1}}R_{l+1}\mathbf{T}_{lm}^{l+1}\right] \tag{9.80}$$

Formula (8.27*a*) is the special case of this with $l = 0$, $R_0 \equiv h_0$, and $\mathbf{T}_{00}^1 = -\hat{\mathbf{r}}/\sqrt{4\pi}$ (9.77). In the general formula, R_l may be as arbitrary as the general solution (9.15), but then $R_{l,\,l\pm1} = a_l h_{l,\,l\pm1} + b_l h^*_{l,\,l\pm1}$, with the same coefficients a_l, b_l for the linearly independent parts of all the radial waves that occur.

The *divergence* derivative of any vector spherical wave like (9.78) has the scalar result

$$\nabla\cdot R_l\mathbf{T}_{jm}^l = -k\left(\delta_{j,\,l+1}\sqrt{\frac{j}{2j+1}} + \delta_{j,\,l-1}\sqrt{\frac{j+1}{2j+1}}\right)R_j Y_{jm} \tag{9.81}$$

The field with $l = j$ is divergenceless, a fact most directly evident from the form (9.76), since $\nabla\cdot\mathbf{L} \equiv -i\nabla\cdot(\mathbf{r}\times\nabla) \equiv 0$ identically and also $\mathbf{L}\cdot\nabla \equiv 0$. Although the spherical waves with $l = j \pm 1$ are not individually divergenceless, a divergenceless *linear combination* of them,

$$\mathbf{U}_{kjm}(\mathbf{r}) \equiv \sqrt{\frac{j+1}{2j+1}}R_{j-1}\mathbf{T}_{jm}^{j-1} - \sqrt{\frac{j}{2j+1}}R_{j+1}\mathbf{T}_{jm}^{j+1} \tag{9.82}$$

is readily found by applying (9.81) to a linear combination with arbitrary coefficients to be determined so that the divergence does vanish. There is then a unique (within a normalization) second linear combination that is orthogonal to (9.82),

$$\mathbf{W}_{kjm}(r) \equiv \sqrt{\frac{j}{2j+1}}R_{j-1}\mathbf{T}_{jm}^{j-1} + \sqrt{\frac{j+1}{2j+1}}R_{j+1}\mathbf{T}_{jm}^{j+1} \tag{9.83}$$

† See Ref. 27 or Ref. 10. The calculations can also be carried out with less erudition than required in those books, but all derivations remain too lengthy to be given here. See Exercise 9.3.

readily found by substituting an arbitrary linear combination representing $\mathbf{W}_{kjm}$ into

$$\oint dV(\mathbf{r})\mathbf{U}^{*}_{kjm} \cdot \mathbf{W}_{kjm} = 0 \tag{9.84}$$

and using the orthonormality (9.74). Comparison with (9.80) shows at once that

$$\mathbf{W}_{kjm}(\mathbf{r}) = k^{-1}\nabla(R_j Y_{jm}) \tag{9.85}$$

is a gradient of the scalar spherical wave. There have thus been descried the three linearly independent vector spherical waves that must exist for each choice of j, m and are classified according to their divergence properties

$$R_j \mathbf{T}^j_{jm} \equiv \mathbf{L}\frac{R_j Y_{jm}}{\sqrt{j(j+1)}} \qquad \mathbf{U}_{kjm}(\mathbf{r}) \qquad \mathbf{W}_{kjm} \equiv \frac{1}{k}\nabla(R_j Y_{jm}) \tag{9.86}$$

with the first form following from (9.76) and the fact that operations by $\mathbf{L}$ require taking derivatives with respect to angles only. The first two of the set (9.86) are both divergence*less*, being different in having $l = j$ and $l \neq j$ respectively, whereas

$$\nabla \cdot \mathbf{W}_{kjm}(\mathbf{r}) = -kR_j(kr)Y_{jm}(\hat{\mathbf{r}}) \tag{9.87}$$

follows from (9.81) or (9.85) and the Helmholtz equation.

A perhaps more telling distinction between the two divergenceless spherical waves is that the first one, $R_j\mathbf{T}^j_{jm}$, is everywhere transverse to the field-point directions $\hat{\mathbf{r}}$, evident from the fact that $\hat{\mathbf{r}} \cdot \mathbf{L} \equiv -i\hat{\mathbf{r}} \cdot (\mathbf{r} \times \nabla) \equiv 0$, whereas $\mathbf{U}_{kjm}$ has a longitudinal component deducible from a result

$$\hat{\mathbf{r}} \cdot \mathbf{T}^l_{jm}(\hat{\mathbf{r}}) = \left(\delta_{j,\,l+1}\sqrt{\frac{j}{2j+1}} - \delta_{j,\,l-1}\sqrt{\frac{j+1}{2j+1}}\right)Y_{jm} \tag{9.88}$$

that is obtained by such a procedure as yielded the above divergence derivatives. The consequences (Exercise 9.1)

$$\hat{\mathbf{r}} \cdot \mathbf{U}_{kjm} = \frac{\sqrt{j(j+1)}}{2j+1}(R_{j-1} + R_{j+1})Y_{jm} = \sqrt{j(j+1)}\frac{R_j(kr)}{kr}Y_{jm}(\hat{\mathbf{r}}) \tag{9.89a}$$

$$\hat{\mathbf{r}} \cdot \mathbf{W}_{kjm} = \frac{jR_{j-1} - (j+1)R_{j+1}}{2j+1}Y_{jm} = \frac{dR_j}{d(kr)}Y_{jm}(\hat{\mathbf{r}}) \tag{9.89b}$$

follow. The latter is also obvious from (9.85), using $\hat{\mathbf{r}} \cdot \nabla \equiv \partial/\partial r$.

Finally, the *curl* operations have been found to yield, besides $\nabla \times \mathbf{W}_{kjm} = 0$ [as for any gradient field like (9.85)]

$$\nabla \times (R_j\mathbf{T}^j_{jm}) = ik\mathbf{U}_{kjm} \qquad \nabla \times \mathbf{U}_{kim} = -ik(R_i\mathbf{T}^j_{im}) \tag{9.90}$$

Naturally the curls must be divergenceless, and so they relate the two divergenceless waves only to each other. It can now be seen that just as $R_j\mathbf{T}^j_{jm}$ emerges as the result of the derivative operation $\mathbf{L}/\sqrt{j(j+1)}$ on the scalar spherical wave $R_j Y_{jm}$ and $\mathbf{W}_{kjm}$ is the result of the gradient operation $k^{-1}\nabla$ on it, so

$$\mathbf{U}_{kjm} \equiv [ik\sqrt{j(j+1)}]^{-1}\nabla \times \mathbf{L}(R_j Y_{jm}) \tag{9.91}$$

It is such less explicit forms, expressed as derivatives of the scalar spherical waves, that were earlier† used for formulating the multipole radiations in place of the eigenfields $R_l \mathbf{T}^l_{jm}$ of the displacement operations in (9.79).

The form (9.91) provides a way of seeing that the divergenceless field $\mathbf{U}_{kjm}(r)$ is everywhere perpendicular to its divergenceless companion $R_j \mathbf{T}^j_{jm}$ besides having the integrated orthogonality to it that follows from (9.74). The operator ∇ in the form can be separated into radial and angular derivatives via the identity

$$\nabla = \hat{\mathbf{r}}(\hat{\mathbf{r}} \cdot \nabla) - \hat{\mathbf{r}} \times (\hat{\mathbf{r}} \times \nabla) \equiv \hat{\mathbf{r}}\frac{\partial}{\partial r} - \frac{i}{r}\hat{\mathbf{r}} \times \mathbf{L} \tag{9.92a}$$

Then a little play with such commutation relations between the operations as (9.52) leads to the expression‡ (see also Exercise A.18)

$$\nabla \times \mathbf{L} = (\hat{\mathbf{r}} \times \mathbf{L})\left(\frac{\partial}{\partial r} + \frac{1}{r}\right) + \frac{i}{r}\hat{\mathbf{r}}\mathbf{L}^2 \tag{9.92b}$$

Using this in the form (9.91) yields

$$\mathbf{U}_{kjm} = \frac{1}{ik}\left(\frac{\partial}{r\,\partial r} rR_j\right)(\hat{\mathbf{r}} \times \mathbf{T}^j_{jm}) + \hat{\mathbf{r}}\sqrt{j(j+1)}\frac{R_j}{kr} Y_{jm} \tag{9.93}$$

both parts of which are obviously perpendicular to the transversely directed vector $\mathbf{T}^j_{jm}$ [Eq. (9.88)]. This result also indicates that asymptotically (as $r \to \infty$) the field $\mathbf{U}_{kjm}$ attains the direction of $\hat{\mathbf{r}} \times \mathbf{T}^j_{jm}$, transverse to both $\hat{\mathbf{r}}$ and $\mathbf{T}^j_{jm}$.

The foregoing discussion makes it possible to draw an important conclusion about the various derivative operations (gradient, divergence, and curl) that connect the electromagnetic-field descriptions ϕ, **A**, **E**, and **B** to each other. All those operations relate waves labeled by a given pair of j, m values only to each other, as for any linear operations on linearly independent expressions. The consequence is that a field characterized by a specific j, m pair can constitute an *entire* one which can exist in isolation, linearly independent of any field constituents having other j, m values.

† Thus, what are called the spin degrees of freedom were not given as explicit recognition; see Sec. 9.5.

‡ The radial derivative operator

$$-i\left(\frac{\partial}{\partial r} + \frac{1}{r}\right) \equiv -i\frac{\partial}{r\,\partial r}r \equiv \mathbf{p}_r$$

is known as the *radial-momentum operator* (in units of $\hbar$) in the contexts of quantum mechanics. Notice (Exercise A.18) that

$$-\nabla^2 \equiv \mathbf{p}_r^2 + \frac{\mathbf{L}^2}{r^2}$$

9.3 ISOLATED MULTIPOLE FIELDS

An electromagnetic field characterized by a given value of j is called a *multipole field of order* 2^j or a 2^j*-pole field*, for reasons that can be made plainer after it is related to the type of source that produces it, in the next section. A given multipolarity can be produced in any one of $2(2j + 1)$ linearly independent varieties. The multiplicity $2j + 1$ is due to the number of distinct m values (*orientations*), $m = 0, \pm 1, \ldots, \pm j$, that can be chosen for any given j. Even for a definite pair of j, m values there are two independent varieties, called an *electric-multipole* (Ej) *field* and a *magnetic multipole* (Mj) *field.*

The last distinctions are easiest to make plain for descriptions by $\mathbf{E}$ and $\mathbf{B}$. Since both these vectors are divergenceless in free space, either might be described by the function $R_j \mathbf{T}^j_{jm}$ or by $\mathbf{U}_{kjm}(\mathbf{r})$ of (9.82). Moreover, because of the curl connections between $\mathbf{E}$ and $\mathbf{B}$, when either is represented by $R_j \mathbf{T}^j_{jm}$ the other must be described by $\pm \mathbf{U}_{kjm}$, according to the curl properties (9.90) of these descriptions. The (Ej) field is then represented as

$$\mathbf{E}_m(Ej) = +\mathbf{U}_{kjm}(\mathbf{r}) \qquad \mathbf{B}_m(Ej) = \frac{\nabla \times \mathbf{E}_m}{ik} = -R_j \mathbf{T}^j_{jm} \tag{9.94}$$

and the linearly independent (Mj) field by

$$\mathbf{E}_m(Mj) = +R_j \mathbf{T}^j_{jm} \qquad \mathbf{B}_m(Mj) = +\mathbf{U}_{kjm}(\mathbf{r}) \tag{9.95}$$

In each case the electric and magnetic vectors are everywhere perpendicular to each other†, as follows from (9.93). Expressions (9.94) and (9.95) allow for arbitrary amplitudes of excitation if it is understood that any $R_j = a_j h_j + b_j h_j^*$, with arbitrary coefficients a_j, b_j, may be chosen. For a given R_j, the two types of field bear a duality relation to each other, like that pointed out for the electric and magnetic dipole fields of (8.63) and (8.62). An intrinsic distinction is provided by the fact that the electric multipole field has its magnetic vector everywhere transverse to the radial direction $\hat{\mathbf{r}}$, whereas it is the electric vector that has no longitudinal components in the magnetic multipole field.‡ All the longitudinal components tend to vanish asymptotically, relative to the transverse ones, as is evident from the extra power of $1/r$ in the expression (9.89) of $\hat{\mathbf{r}} \cdot \mathbf{U}_{kjm}$. A consequence is that sectors of the fields, as they emerge at infinity, take on the transverse characteristics of the electromagnetic plane waves, with polarizations determined by the direction of the vector $(\mathbf{U}_{kjm})_\perp \equiv \hat{\mathbf{r}} \times (\mathbf{U}_{kjm} \times \hat{\mathbf{r}})$ for an (Ej) field and of $\mathbf{T}^j_{jm}$ for an (Mj) field. The expression (9.93) for $\mathbf{U}_{kjm}$ shows that the polarization direction of an (Ej) field is parallel to $\hat{\mathbf{r}} \times \mathbf{T}^j_{jm}$ and therefore perpendicular to that of the (Mj) field.

† An example is evident in the dipole field descriptions (8.41) and (8.42).

‡ In these respects, the (Ej) fields are comparable to the TM modes propagated in wave guides (Sec. 7.3) and the (Mj) fields to the TE modes.

The dynamical characteristics of a given, isolated multipole field will be considered next.

The energy of a field does not endow it with any particular distinction, since it may be excited by any amount. As usual, only a finitely extended, and hence nonmonochromatic, wave packet will have a finite energy. Accordingly, a *packet* characterized by a specific pair of j, m values

$$\left.\begin{matrix}\mathbf{E}(\mathbf{r}, t)\\ \mathbf{B}(\mathbf{r}, t)\end{matrix}\right\} = \operatorname{Re}\left[\int_0^\infty dk c(k)\left\{\begin{matrix}\mathbf{U}_{kjm}(\mathbf{r})\\ -j_j(kr)\mathbf{T}^j_{jm}(\hat{\mathbf{r}})\end{matrix}\right\}e^{-ickt}\right] \tag{9.96}$$

will be considered. Only the regular radial waves can appear in a field isolated from sources, and hence $R_l \equiv j_l$ is taken. The resultant here is specifically an (Ej) field, but results for an (Mj) field are obtainable merely by the substitutions $\mathbf{E} \to \mathbf{B}$, $\mathbf{B} \to -\mathbf{E}$, which plainly have no effect on the energy density $(\mathbf{E}^2 + \mathbf{B}^2)/8\pi$. The integration for the total energy in the packet is quite as straightforward as it was for the plane-wave packet† on pages 184 and 185, if the orthonormalities (9.30) and (9.74) are used, together with their consequence

$$\oint dV(\mathbf{r})\mathbf{U}^*_{kjm} \cdot \mathbf{U}_{k'j'm'} = \delta_{jj'}\,\delta_{mm'}\frac{\pi}{2k^2}\,\delta(k - k') \tag{9.97}$$

The outcome is the constant

$$W = \int_0^\infty \frac{dk\,|c(k)|^2}{16k^2} \tag{9.98}$$

for the total field energy of a packet like (9.96) or its (Mj) counterpart.

It is the field *angular momentum*, per unit of the excitation energy, that provides the characterizing distinctions between fields of different j, m values. The total angular momentum $\mathbf{J}$ of (6.32) is to be calculated as a volume integral of the density $\mathbf{r} \times (\mathbf{E} \times \mathbf{B})/4\pi c$. Since $\mathbf{E} \times \mathbf{B} = \mathbf{B} \times (-\mathbf{E})$ is invariant to the substitutions that transform an (Ej) into an (Mj) field, it will suffice to evaluate $\mathbf{J}$ for the (Ej) packet of (9.96). In this, the magnetic field is everywhere transverse to $\mathbf{r}$, and so $\mathbf{r} \times (\mathbf{E} \times \mathbf{B}) = -(\mathbf{r} \cdot \mathbf{E})\mathbf{B}$. As a result

$$\mathbf{J} = \oint \frac{dV(\mathbf{r})}{8\pi c}\operatorname{Re}\left[\iint_0^\infty dk\,dk' c^*(k)c(k')(\mathbf{r} \cdot \mathbf{U}^*_{kjm})j_j(k'r)\mathbf{T}^j_{jm}\right]$$

For brevity, only time-averaged contributions are included here, since the general conservation of $\mathbf{J}$ assures that this changes nothing (explicit inclusion of the time factors also leads to the same ultimate result). The finding (9.89*a*) for the

† The temporal fluctuations of the individual electric and magnetic energies found in the plane-wave packet do not occur here except for $m = 0$, and in the latter case, the magnetic fluctuations again cancel the electric ones.

longitudinal component of $\mathbf{U}_{kjm}$ shows that the factor $\mathbf{r} \cdot \mathbf{U}^*_{kjm}$ can be replaced by†

$$\mathbf{r} \cdot \mathbf{U}^*_{kjm} = k^{-1}\sqrt{j(j+1)}\, j_j(kr) Y^*_{jm}$$

With this, the orthogonality (9.30) of the radial waves introduces a delta function that reduces the integral for $\mathbf{J}$ to

$$\mathbf{J} = \left[(16c)^{-1} \int_0^\infty dk k^{-3} |c(k)|^2\right] \mathrm{Re}\left[\sqrt{j(j+1)} \oint d\Omega Y^*_{jm} \mathbf{T}^j_{jm}\right]$$

Here, the representation (9.76) can be inserted for $\mathbf{T}^j_{jm}(\hat{\mathbf{r}})$, to yield an expression proportional to

$$\sqrt{j(j+1)} \oint d\Omega Y^*_{jm} \mathbf{T}^j_{jm} = \oint d\Omega Y^*_{jm} \mathbf{L} Y_{jm} \equiv \langle \mathbf{L} \rangle_{jm} \tag{9.99}$$

called a diagonal matrix element $\langle \mathbf{L} \rangle_{jm}$ of the operator $\mathbf{L}$. The effect of the components $\mathbf{L}_\pm$ of this operator is to respectively raise and lower the order m of Y_{jm} by a unit, as reference to (9.50) shows, and so the orthogonality of the spherical harmonics causes all components of the matrix element to vanish except for $\langle \mathbf{L}_z \rangle_{jm}$. Since $\mathbf{L}_z Y_{lm} = m Y_{lm}$, $\langle \mathbf{L}_z \rangle_{jm} = m$, which is just the eigenvalue of $\mathbf{L}_z$ in the field state (9.96). The total angular momentum in the packet has the magnitude

$$J = J_z = m \int_0^\infty \frac{dk\,|c(k)|^2}{16k^2\omega} \tag{9.100}$$

and has the direction about the polar axis relative to which m is defined. The amount resident per unit of a frequency range $\Delta\omega = \Delta(ck)$, of the linearly independent component frequencies, is $\Delta J/\Delta\omega = m|c(k)|^2/16ck^2\omega$, compared with the energy $\Delta W/\Delta\omega = |c(k)|^2/16ck^2$ of (9.98) in the same range. Thus, the ratio of angular momentum to energy at a given frequency ω is just‡

$$\frac{\Delta J}{\Delta W} = \frac{m}{\omega} \qquad \text{with } m = 0, \pm 1, \ldots, \pm j \tag{9.101}$$

A significant point about this is its variation by *discrete*§ steps, among the possible spherical-wave constituents of any field. Moreover, the eigenvalues m of the components of a given j vary from $-j$ to $+j$ (in the discrete steps) when

† Notice that the extra power of r, introduced as the lever arm of the momentum, is thus canceled out, a result for which the existence of an asymptotically vanishing longitudinal component is essential.

‡ In the contexts of quantum theory this ratio would be written $mh/h\omega$, and mh would be an angular-momentum component of a photon $h\omega$ having the extended "shape" given by the distribution of a field characterized by j, m. Compare the linear momentum in a plane-wave photon of given $\mathbf{k}$, alluded to in the footnote to page 187.

§ A general characteristic of wave-angular-momentum descriptions, illustrated for classical mechanical waves in an elastic medium in Ref. 17, sec. 15.3. There, too, can be found a discussion of the fact that characterizations by the integer m depend on the polar axis used for the analysis.

giving the angular-momentum component J_z. Since $j(j+1)$ is an eigenvalue of $\mathbf{J}^2 = \mathbf{J}_x^2 + \mathbf{J}_y^2 + \mathbf{J}_z^2$, the discrete values of j are said to measure the magnitudes (as against directions) of the angular momenta per unit energy in fields characterized by the j values.

9.4 MULTIPOLE SOURCES

The field generated by an arbitrary localized source is represented by the retarded vector potential (8.19). Its decomposition into its spherical-wave constituents can be prepared by resolving $\mathbf{A}(\mathbf{r})$ on the complex unit-vector basis $\mathbf{e}_{\mu=0,\,\pm1}$ of (9.64), having the orthonormality $\mathbf{e}_\mu^* \cdot \mathbf{e}_{\mu'} = \delta_{\mu\mu'}$, so that

$$\mathbf{A}(\mathbf{r}) = \sum_{\mu=0,\,\pm1} \mathbf{e}_\mu(\mathbf{e}_\mu^* \cdot \mathbf{A}) = \oint \frac{dV(\mathbf{r}_s)}{c} \frac{e^{ikR}}{R} \sum_\mu \mathbf{e}_\mu[\mathbf{e}_\mu^* \cdot \mathbf{j}(\mathbf{r}_s)] \tag{9.102}$$

where $\mathbf{j}(\mathbf{r}_s)$ is a current-density representative and $R \equiv |\mathbf{r} - \mathbf{r}_s|$, as usual. The integrand here contains a projection, onto the current-density vector, of a *tensor* $\sum \mathbf{e}_\mu \mathbf{e}_\mu^* \equiv \mathbf{1}$ having a dyadic form. As indicated, it is a unit tensor in that

$$\mathbf{1} \equiv \sum_{\mu=0,\,\pm1} \mathbf{e}_\mu \mathbf{e}_\mu^* = \mathbf{e}_x \mathbf{e}_x + \mathbf{e}_y \mathbf{e}_y + \mathbf{e}_z \mathbf{e}_z \tag{9.103}$$

has only diagonal cartesian components of magnitude $\mathbf{1}_{xx} = \mathbf{1}_{yy} = \mathbf{1}_{zz} = 1$, and $\mathbf{1}_{ij} = \delta_{ij}$ represents an arbitrary component.

The decomposition can now proceed by using a generalization of the Huyghens' wavelet decomposition (9.34) to

$$\mathbf{1}\frac{e^{ikR}}{R} = 4\pi ik \sum_{ljm} [h_l(kr)\mathbf{T}_{jm}^l(\hat{\mathbf{r}})][j_l(kr_s)\mathbf{T}_{jm}^l(\hat{\mathbf{r}}_s)]^* \tag{9.104}$$

here written for $r > r_s$ and sometimes called the *dyadic Green's function.* Its validity stems from a *vector* spherical harmonic addition theorem generalized from the scalar one (3.99). This is readily found by examining the dyadic product occurring in (9.104), for each l value,

$$\sum_{jm} \mathbf{T}_{jm}^l(\hat{\mathbf{r}})[\mathbf{T}_{jm}^l(\hat{\mathbf{r}}_s)]^* \equiv \sum_{jm\mu\mu'} c_\mu c_{\mu'} \mathbf{e}_\mu \mathbf{e}_{\mu'}^* Y_{l,\,m-\mu}(\hat{\mathbf{r}}) Y_{l,\,m-\mu'}^*(\hat{\mathbf{r}}_s)$$

after substitutions of the definition (9.69). The summation variable j occurs only in the factor $\sum_j c_\mu c_{\mu'} = \delta_{\mu\mu'}$ (9.75*b*). The summation over m can as well be carried out after substituting for m the variable $M \equiv m - \mu$, and then

$$\sum_{jm} \mathbf{T}_{jm}^l(\hat{\mathbf{r}})[\mathbf{T}_{jm}^l(\hat{\mathbf{r}}_s)]^* = \left(\sum_\mu \mathbf{e}_\mu \mathbf{e}_\mu^*\right)\left[\sum_M Y_{lM}(\hat{\mathbf{r}}) Y_{lM}^*(\hat{\mathbf{r}}_s)\right]$$

$$= \mathbf{1}\frac{2l+1}{4\pi} P_l(\hat{\mathbf{r}} \cdot \hat{\mathbf{r}}_s) \tag{9.105}$$

Now (9.104) follows merely from putting this into expression (9.35) after multiplication by the unit tensor.

Using expression (9.104) as it stands in (9.102) would obviously produce a decomposition of the vector potential on a basis of vector spherical-wave components like (9.78), classified according to $l = j$ and $j \pm 1$. It is rather a decomposition on the equally complete, alternative basis $R_j \mathbf{T}^j_{jm}$, $\mathbf{U}_{kjm}$, $\mathbf{W}_{kjm}$ of (9.86) that will permit finding the sources of each multipole field like (9.94) and (9.95). The requisite transformation of the basis merely requires noticing that, for just the $l = j \pm 1$ terms of (9.104),

$$\sum_{l=j\pm1} [h_l(kr)\mathbf{T}^l_{jm}(\hat{\mathbf{r}})][j_l(kr_s)\mathbf{T}^l_{jm}(\hat{\mathbf{r}}_s)]^* = \mathbf{U}^o_{kjm}(\mathbf{r})\mathbf{U}^*_{kjm}(\mathbf{r}_s) + \mathbf{W}^o_{kjm}(\mathbf{r})W^*_{kjm}(\mathbf{r}_s) \tag{9.106}$$

This becomes evident after some very simple algebra upon forming the expression on the right in terms of the definitions (9.82) and (9.83). The superscript o indicates that the corresponding fields are to be formed of the *outgoing* radial waves, $R_l = h_l(kr)$, whereas the starred fields must have the real standing waves, $R_l = j_l(kr_s)$, as constituents when results for $r > r_s$ are desired.

Multipole Potentials

The insertion of the tensor (9.104), with the modification (9.106), into the expression (9.102) for the vector potential produces an analysis of it into three linearly independent types of terms:

$$\mathbf{A}(\mathbf{r}) = \sum_j [\mathbf{A}(Ej) + \mathbf{A}(Mj)] + \mathbf{A}(\phi) \tag{9.107}$$

where

$$\mathbf{A}(Mj) \equiv h_j(kr) \sum_m \mathbf{T}^j_{jm}(\hat{\mathbf{r}}) \frac{a_m(Mj)}{ik} \tag{9.107M}$$

is the $l = j$ contribution,

$$\mathbf{A}(Ej) \equiv \sum_m \mathbf{U}^o_{kjm}(\mathbf{r}) \frac{a_m(Ej)}{ik} \tag{9.107E}$$

comes from the second divergenceless member of the set (9.86), and

$$\mathbf{A}(\phi) \equiv \sum_{jm} \mathbf{W}^o_{kjm}(\mathbf{r}) \frac{4\pi ik}{c} \oint dV(\mathbf{r}_s)\mathbf{W}^*_{kjm}(\mathbf{r}_s) \cdot \mathbf{j}(\mathbf{r}_s) \tag{9.107ϕ}$$

is the so-called† *longitudinal potential.* The amplitude factors a_m are determined by the scalar source integrals

$$a_m(Mj) \equiv ik \frac{4\pi ik}{c} \oint dV(\mathbf{r}_s) j_j(kr_s)[\mathbf{T}^j_{jm}(\hat{\mathbf{r}}_s)]^* \cdot \mathbf{j}(\mathbf{r}_s) \tag{9.108M}$$

$$a_m(Ej) \equiv ik \frac{4\pi ik}{c} \oint dV(\mathbf{r}_s)\mathbf{U}^*_{kjm}(\mathbf{r}_s) \cdot \mathbf{j}(\mathbf{r}_s) \tag{9.108E}$$

that have arisen in the decomposition of the integration in (9.102).

† Actually, the vector $\mathbf{W}_{kjm}$ becomes longitudinal only asymptotically; it has a component transverse to $\hat{\mathbf{r}}$ (proportional to $\hat{\mathbf{r}} \times \mathbf{T}^j_{jm}$) that dies out in comparison to its longitudinal component (9.89*b*) as $r \to \infty$. On the other hand, it is the longitudinal component (9.89*a*) of $\mathbf{U}_{kjm}$ that dies out relative to its transverse one, which becomes proportional to $\hat{\mathbf{r}} \times \mathbf{T}^j_{jm}$ [see (9.93)].

Since the gradient field $\mathbf{W}_{kjm}$ is the only one of the set (9.86) that possesses a divergence (9.87), $\mathbf{A}(\phi)$ is the only part of the vector potential that must be accompanied by a scalar potential

$$\phi = \frac{\nabla \cdot \mathbf{A}}{ik} = \frac{\nabla \cdot \mathbf{A}(\phi)}{ik} \tag{9.109}$$

This is exactly the scalar potential already found in (9.36), as can be seen from $\nabla \cdot \mathbf{W}^o_{kjm} = -kh_j Y_{jm}$ of (9.87) and using the charge-conservation condition $\nabla \cdot \mathbf{j}(\mathbf{r}_s) = i\omega\rho(\mathbf{r}_s)$ in the source integral of (9.107ϕ). The latter is done with the help of the gradient form (9.85) of $\mathbf{W}^*_{kjm}$, which makes

$$\oint dV(\mathbf{r}_s)\mathbf{j} \cdot \mathbf{W}^*_{kjm} = k^{-1} \oint dV\mathbf{j}(\mathbf{r}_s) \cdot \nabla_s[j_j(kr_s)Y^*_{jm}(\hat{\mathbf{r}}_s)]$$

$$= -k^{-1} \oint dV[j_j(kr_s)Y^*_{jm}]\nabla_s \cdot \mathbf{j}(\mathbf{r}_s)$$

just the source integral (9.37) over the charge-distribution representative $\rho(\mathbf{r}_s) = \nabla_s \cdot \mathbf{j}/i\omega$ that defines the electric multipole moments q_{jm} in (9.36).

Because $\mathbf{W}^o_{kjm}$ is a gradient field, $\mathbf{A}(\phi)$ is absolutely curlless [see the statement leading to (9.90)], and so $\mathbf{B}(\phi) = \nabla \times \mathbf{A}(\phi) \equiv 0$. Moreover, outside the source, $\mathbf{E}(\phi) = \nabla \times \mathbf{B}(\phi)/(-ik) \equiv 0$ also! Thus neither potential $\mathbf{A}(\phi)$ nor ϕ leads to any $\omega \neq 0$ electric or magnetic fields outside the source.† These potentials exist there only as continuations of the potentials inside the source needed for gauging inner electric fields that have divergence $\nabla \cdot \mathbf{E} = 4\pi\rho$ wherever there is charge. They are not needed for describing the outer fields and can be eliminated from their description by regauging the Lorentz potential to $\phi' = \phi + ik\chi \equiv 0$ outside the sources, where a gauge function $\chi = -\phi/ik$ satisfies the free-space wave equation (6.49). Then, also outside sources only,

$$\mathbf{A}'(\mathbf{r}) = \mathbf{A} + \nabla\chi = \mathbf{A} - \frac{\nabla\phi}{ik} = \mathbf{A} - \mathbf{A}(\phi) = \sum_j [\mathbf{A}(Ej) + \mathbf{A}(Mj)] \tag{9.110}$$

That $\mathbf{A}(\phi) = \nabla\phi/ik$ is most easily seen from $\phi = \nabla \cdot \mathbf{A}(\phi)/ik$ [Eq. (9.109)] and the fact that $\nabla(\nabla \cdot \mathbf{A}) = ik\mathbf{A}(\phi)$ because

$$\nabla(\nabla \cdot \mathbf{W}_{kjm}) = \nabla \times (\nabla \times \mathbf{W}_{kjm}) + \nabla^2\mathbf{W}_{kjm} = -k^2\mathbf{W}_{kjm}$$

for the curlless solution of the Helmholtz equation. The gauging by $\phi' \equiv 0$ and $\mathbf{A}' = \mathbf{A} - \mathbf{A}(\phi)$ can be recognized as a solenoidal gauging, of a type reviewed for plane waves in connection with (7.13).

Now $\mathbf{E} = -\nabla\phi + ik\mathbf{A} = ik[\mathbf{A} - \mathbf{A}(\phi)]$, and expressions (9.107) lead to superpositions of the multipole fields (9.94) and (9.95),

$$\mathbf{E}(\mathbf{r}) = \sum_{jm} [a_m(Mj)h_j(kr)\mathbf{T}^j_{jm}(\hat{\mathbf{r}}) + a_m(Ej)\mathbf{U}^o_{kjm}(\mathbf{r})]$$

$$= \sum_{jm} \left[a_m(Mj) + \frac{a_m(Ej)}{ik}\nabla \times\right]\mathbf{L}\frac{h_j(kr)Y_{jm}(\hat{\mathbf{r}})}{\sqrt{j(j+1)}} \tag{9.111}$$

† Just as for the plane-wave decompositions discussed in connection with (7.12).

according to (9.86) and (9.91). Because of the duality relations between the multipole fields, the comparable expressions for $\mathbf{B}$ differ only in having $a_m(Mj) \to -a_m(Ej)$ and $a_m(Ej) \to a_m(Mj)$. The last expression in (9.111) also indicates how explicit definition of the vector spherical harmonics $\mathbf{T}^l_{jm}$ could be avoided in the past by settling for the derivative expressions instead.

The Electric Multipole Radiation Moments

The finding (9.111) shows that a given source will produce the electric multipole field

$$\mathbf{E}(Ej) = \sum_m a_m(Ej)\mathbf{U}^o_{kjm} \qquad \mathbf{B}(Ej) = -\sum_m a_m(Ej)h_j\mathbf{T}^j_{jm} \tag{9.112}$$

having the potential description given by $\mathbf{A}(Ej)$ of (9.107E) and $\phi(Ej) \equiv 0$ in an intensity determined by the coefficients $a_m(Ej)$ of (9.108E). These coefficients are related to the electric multipole moments q_{jm} of (3.101) and (9.37), and that is responsible for the name given to the field.

The relationship can be found by using the representation (9.91) for the eigenfield $\mathbf{U}_{kjm}$ occurring in the source integral $a_m(Ej)$ of (9.108E). The operation $\nabla \times \mathbf{L}$ in the representation (9.91) is for this purpose developed from (9.92*b*) to the form

$$\nabla \times \mathbf{L} = -i\mathbf{r}\nabla^2 + i\nabla\left(1 + r\frac{\partial}{\partial r}\right) \tag{9.113}$$

obtainable by substituting $\mathbf{L} = -i\mathbf{r} \times \nabla$ [Eq. (9.44)] and $\mathbf{L}^2/r^2 = -\nabla^2 + (\partial/\partial r + 1/r)^2$ (see the footnote on page 258) into (9.92*b*). Since the scalar spherical wave operated on in (9.91) is an eigenfield of ∇^2 ($\equiv -k^2$), the source integral (9.108E) can now be written

$$a_m(Ej) = -\frac{4\pi k^2}{c\sqrt{j(j+1)}} \oint dV_s \left\{ \mathbf{j} \cdot \nabla_s \frac{\partial}{k\,\partial r_s}[r_s j_j(kr_s)] + k j_j(kr_s)(\mathbf{r}_s \cdot \mathbf{j}) \right\} Y^*_{jm}$$

Here, the charge-density representative $\rho(\mathbf{r}_s) = \nabla_s \cdot \mathbf{j}/i\omega$ can be introduced into the first term of the integrand by making use of the hermiteanship (4.51). The result will be represented as

$$a_m(Ej) = \frac{4\pi i k^{j+2}}{(2j+1)!!}\sqrt{\frac{j+1}{j}}\,(Q_{jm} + Q'_{jm}) \tag{9.114}$$

in terms of the definitions

$$Q_{jm} \equiv \frac{(2j+1)!!}{(j+1)k^j} \oint dV(\mathbf{r}_s)\rho(\mathbf{r}_s)\frac{\partial}{\partial r_s}[r_s j_j(kr_s)]Y^*_{jm}(\hat{\mathbf{r}}_s)$$

and

$$Q'_{jm} \equiv \frac{ik}{c}\frac{(2j+1)!!}{(j+1)k^j} \oint dV(\mathbf{r}_s)\mathbf{r}_s \cdot \mathbf{j}(\mathbf{r}_s) j_j(kr_s) Y^*_{jm}(\hat{\mathbf{r}}_s) \tag{9.115}$$

to be called electric multipole *radiation* moments. The definition of Q_{jm} is so constructed that it reduces to the form (3.101) of the electrostatic moment q_{jm} for sources so small that $j_j \approx (kr_s)^j/(2j+1)!!$ is an adequate approximation. For such sources, Q'_{jm} becomes negligible, being smaller than Q_{jm} by a factor kr_s. The moments $Q_{jm} + Q'_{jm}$ generally needed for determining the intensity of the (Ej) radiation constitute different generalizations of q_{jm} from the moments (9.37) which determine the nonradiating longitudinal field of the source [ϕ and $\mathbf{A}(\phi)$]. The forms (9.115) are valid for any relative size of source and wavelength, with Q'_{jm} representing contributions by longitudinal current elements, for which $\hat{\mathbf{r}} \cdot \mathbf{j} \neq 0$. In view of the complex of interferences that must be occurring between different parts of an arbitrarily extended source, the moments (9.115) are surprisingly simple.

It will now be seen that the $(E1)$ and $(E2)$ fields as found here can, even for sources not small in comparison to a wavelength, be reexpressed in just the forms found for small sources in (8.41), (8.42), and (8.74) with changes only in the way the moments $\mathbf{D}$ and $\mathbf{Q}$ are to be calculated for a given source ρ, $\mathbf{j}$ (now of arbitrary size).

The expression in (9.77) for $\mathbf{T}^1_{1m}(\hat{\mathbf{r}})$ shows that

$$\mathbf{B}(E1) = i\sqrt{\frac{3}{8\pi}}\, h_1(kr_s)\hat{\mathbf{r}} \times \sum_m \mathbf{e}_m a_m(E1)$$

follows from (9.112). The relation (9.114) of the multipole coefficient to the moments (9.115) leads to

$$i\sqrt{\frac{3}{8\pi}} \sum_m \mathbf{e}_m a_m = -\sqrt{3\pi}\, k^2 \oint dV \left[\rho \frac{\partial}{\partial r_s} r_s j_1 + \frac{ik}{c}(\mathbf{r}_s \cdot \mathbf{j}) j_1\right] \sum_m \mathbf{e}_m Y^*_{1m}(\hat{\mathbf{r}}_s)$$

The last factor in the integrand, formed of the summation over m, can be replaced by $\sqrt{3/4\pi}\,\hat{\mathbf{r}}_s$, according to (3.93), being a decomposition of this vector on the basis $\mathbf{e}_{0,\pm 1}$. It is now clear that the expression for $\mathbf{B}(E1)$ can be written as $-k^3 h_1 \hat{\mathbf{r}} \times \mathbf{D}$, just the form (8.41) according to the definition (9.11) of $h_1(kr_s)$, if $\mathbf{D}$ differs from the small source form $\oint dV \rho \mathbf{r}_s$ by the replacement of the charge-density representative $\rho(\mathbf{r}_s)$ with

$$\rho \to \frac{3}{2kr_s}\left[\rho(\mathbf{r}_s) \frac{\partial}{\partial r_s} r_s j_1 + \frac{ik}{c}(\mathbf{r}_s \cdot \mathbf{j}) j_1(kr_s)\right] \tag{9.116}$$

It is easy to see that this reduces to $\rho(\mathbf{r}_s)$ itself for the small sources in which $j_1(kr_s) \approx kr_s/3$ and the longitudinal current contribution is negligible.

The reduction of $\mathbf{E}(E1)$ in (9.112) to just the form (8.42) can be similarly demonstrated with the help of (9.82), giving $\mathbf{U}^o_{k1m}$ in terms of $\mathbf{T}^{0;2}_{1m}(\hat{\mathbf{r}})$ and the expressions in (9.77) for the latter. This supplementation of the above findings for $\mathbf{B}(E1)$ is not really necessary since the connection $\mathbf{E} = \nabla \times \mathbf{B}/(-ik)$ has been maintained in all the representations. It does call attention to the fact that the vector potential $\mathbf{A}(E1) = \mathbf{E}(E1)/ik$ of (9.107E) has the lengthy form following from (8.42) even in the small-source limit, not at all the simple form (8.40) from which the small-source results were derived. This is because the simple form is equivalent to $\mathbf{A}(E1) + \mathbf{A}_1(\phi)$, with $\mathbf{A}_1(\phi)$ the $j = 1$ term of the longitudinal potential (9.107ϕ), which contributes nothing to the fields $\mathbf{E}$, $\mathbf{B}(E1)$, as discussed in connection with (9.110).

The reexpression of $\mathbf{B}(E2)$ in terms of a quadrupole tensor $\mathbf{Q}$ as in (8.74) requires performing the summation

$$\sum_m Y^*_{2m}(\hat{\mathbf{r}}_s)\mathbf{T}^2_{2m}(\hat{\mathbf{r}}) = 6^{-1/2} \sum_m Y^*_{2m}(\hat{\mathbf{r}}_s)\mathbf{L}Y_{2m}(\hat{\mathbf{r}})$$

with the first spherical-harmonic factor coming from the integrand of $a_m(E2)$ in (9.115) and the second from the representation (9.76) of the vector spherical harmonic. Then the spherical-harmonic addition theorem (3.99) leads to a result proportional to $\mathbf{L}P_2(\hat{\mathbf{r}} \cdot \hat{\mathbf{r}}_s)$ with the operation $\mathbf{L} = -i\mathbf{r} \times \nabla$ applying only to the $\hat{\mathbf{r}}$ dependence in $P_2 = \frac{3}{2}(\hat{\mathbf{r}} \cdot \hat{\mathbf{r}}_s)^2 - \frac{1}{2}$. The operation is very simple to carry out and yields

$$\frac{4\pi}{5} \sum_m Y^*_{2m} \mathbf{T}^2_{2m} = -i\sqrt{\tfrac{3}{2}}\,(\hat{\mathbf{r}} \times \hat{\mathbf{r}}_s)(\hat{\mathbf{r}}_s \cdot \hat{\mathbf{r}}) = -i\frac{1}{\sqrt{6}}\hat{\mathbf{r}} \times (\Delta \cdot \hat{\mathbf{r}})$$

in terms of the dyad $\Delta \equiv 3\hat{\mathbf{r}}_s\hat{\mathbf{r}}_s - \mathbf{1}$ used for the expression of $\mathbf{Q}$ as in (2.21b) or (8.69). It should now be plain how (9.115) leads to the form $-\frac{1}{6}k^4[\hat{\mathbf{r}} \times (\mathbf{Q} \cdot \hat{\mathbf{r}})]h_2(kr_s)$, equivalent to (8.74). For this, the charge-density representative $\rho(\mathbf{r}_s)$ in the form (2.21b) of $\mathbf{Q}$ must be replaced by

$$\rho \rightarrow \frac{5}{k^2 r_s^2}\left[\rho(\mathbf{r}_s)\frac{\partial}{\partial r_s} r_s j_2 + \frac{ik}{c}(\mathbf{r}_s \cdot \mathbf{j})j_2(kr_s)\right] \tag{9.117}$$

which reduces to $\rho(\mathbf{r}_s)$ in the small sources for which the approximation $j_2 \approx (kr_s)^2/15$ is adequate.

Magnetic Multipole Moments

A given source will produce a magnetic multipole field

$$\mathbf{E}(Mj) = h_j(kr) \sum_m a_m(Mj)\mathbf{T}^j_{jm} \qquad \mathbf{B}(Mj) = \sum_m a_m(Mj)\mathbf{U}^o_{kjm} \tag{9.118}$$

determined by the source integrals $a_m(Mj)$ of (9.108M). It is customary to express them in terms of magnetic multipole moments μ_{jm} defined analogously to the electric ones $(Q_{jm} + Q'_{jm})$ of (9.114), so that

$$a_m(Mj) = \frac{4\pi i k^{j+2}}{(2j+1)!!}\sqrt{\frac{j+1}{j}}\,\mu_{jm} \tag{9.119}$$

Then (9.108M) yields

$$\mu_{jm} = -i\frac{(2j+1)!!}{(j+1)ck^j}\oint dV(\mathbf{r}_s)\mathbf{j}(\mathbf{r}_s) \cdot \mathbf{L}_s[j_j(kr_s)Y^*_{jm}(\hat{\mathbf{r}}_s)] \tag{9.120a}$$

when the representation (9.76) is introduced for the vector spherical harmonic. The gradient factor in the operation $\mathbf{L}_s = -i\mathbf{r}_s \times \nabla_s$ can be used to introduce the eigenfield $\mathbf{W}^*_{kjm}(\mathbf{r}_s)$ of (9.85),

$$\mu_{jm} = \frac{(2j+1)!!}{(j+1)ck^{j-1}}\oint dV(\mathbf{r}_s)(\mathbf{r}_s \times \mathbf{j}) \cdot \mathbf{W}^*_{kjm}(\mathbf{r}_s) \tag{9.120b}$$

a form of some significance because it has an explicit factor the magnetization density $(\mathbf{r}_s \times \mathbf{j})/2c$ of the dipole (8.60). It is more customary, however, to transfer the gradient operation to the magnetization factor by making use of the hermiteanship of $-i\boldsymbol{\nabla}_s$ (4.51), with the result

$$\mu_{jm} = -\frac{(2j+1)!!}{(j+1)ck^j}\oint dV j_j(kr_s)Y^*_{jm}\boldsymbol{\nabla}_s \cdot (\mathbf{r}_s \times \mathbf{j}) \tag{9.120c}$$

essentially a projection onto a scalar spherical wave of the divergence of the magnetization distribution.†

It will next be seen that the magnetic dipole field ($M1$) can again be represented as in (8.62) and (8.63), with the help of a vector moment $\boldsymbol{\mu}$ that is suitably generalized from the small-source form in (8.60). Inserting the expression for $\mathbf{T}^1_{1m}(\hat{\mathbf{r}})$ given in (9.77) and the moments μ_{jm} via (9.119) into the electric vector of (9.118) yields

$$\mathbf{E}(M1) = \sqrt{\frac{4\pi}{3}}\, k^3 h_1(kr)\hat{\mathbf{r}} \times \sum_m \mathbf{e}_m \mu_{1m}$$

This coincides with the form (8.63) upon the replacement (4.50)

$$\boldsymbol{\mu} \to \sqrt{\frac{4\pi}{3}} \sum_m \mathbf{e}_m \mu_{1m} \tag{9.121a}$$

Now put into the form (9.120b) of μ_{1m}, the result following from (9.83) and (9.77) for

$$\mathbf{W}^*_{k1m} = (12\pi)^{-1/2}[(j_0 + j_2)\mathbf{e}^*_m - 3j_2(kr_s)\hat{\mathbf{r}}_s(\hat{\mathbf{r}}_s \cdot \mathbf{e}^*_m)]$$

The longitudinal term here has no projection on the magnetization $(\mathbf{r}_s \times \mathbf{j})/2c$, and the above summation over m becomes proportional to the unit tensor $\sum_m \mathbf{e}_m \mathbf{e}^*_m = \mathbf{1}$ of (9.103). It follows that

$$\boldsymbol{\mu} \to \oint dV(\mathbf{r}_s)\frac{\mathbf{r}_s \times \mathbf{j}}{2c}[j_0(kr_s) + j_2(kr_s)] \tag{9.121b}$$

which reduces to the small-source form in (8.60) when the approximation $j_0 + j_2 \approx 1$ is adequate [see (9.20)].

The small-source limits of all the moment forms (9.120c)

$$\mu_{jm} = -\oint dV r_s^j Y^*_{jm}\boldsymbol{\nabla}_s \cdot \frac{\mathbf{r}_s \times \mathbf{j}}{(j+1)c} \tag{9.122}$$

become independent of the wavelength $2\pi/k$ because $j_j(kr_s) \approx (kr_s)^j/(2j+1)!!$ in that limit. It is therefore no surprise that (9.122) agrees with the *static* ($k \to 0$) moment forms already introduced in (4.52). There they helped form the magnetostatic *scalar*-potential description of the static multipole fields, by ϕ_M of

† It is sometimes profitable to include in the source description such magnetization distributions in matter as are denoted $\mathbf{M}(\mathbf{r}, t)$ in Chap. E and are not analyzed into current representatives $\mathbf{j}(\mathbf{r}_s)$. The inclusion of such descriptions is here left to Exercise 9.5.

(4.47). Here there is the option of representation by the divergenceless *vector* potentials $\mathbf{A}(Mj)$ of (9.107), with $\phi \equiv 0$, as discussed in connection with (9.109) and (9.110). The magnetostatic vector-potential forms are deducible from the $k \to 0$ limit of $\mathbf{A}(Mj) = \mathbf{E}(Mj)/ik$. In this static limit, the electric field in (9.118) properly vanishes, since $h_j(kr \to 0) \sim k^{-(j+1)}$, according to (9.19), whereas $a_m(Mj) \sim k^{+(j+2)}$ in (9.119). Since $\mathbf{E}(Mj) \to 0$ in proportion to just the first power of $k \to 0$, the vector potential $\mathbf{A} = \mathbf{E}/ik$ survives, with the value

$$\mathbf{A}(Mj) = -\frac{4\pi i}{2j+1}\sqrt{\frac{j+1}{j}}\sum_m \frac{\mu_{jm}}{r^{j+1}}\mathbf{T}^j_{jm}(\hat{\mathbf{r}}) \tag{9.123}$$

Summed over $j = 1, 2, \ldots$, this is the vector-potential description of the same magnetostatic multipole fields as are described by the scalar potential (4.47). Like the latter, (9.123) is a *real* field, as follows from the property

$$(\mathbf{T}^l_{jm})^* = (-)^{m+l+1-j}\mathbf{T}^l_{j,\,-m} \tag{9.124}$$

of the vector spherical harmonics that corresponds to $Y^*_{lm} = (-)^m Y_{l,\,-m}$ of the scalar spherical harmonics. For the special case of the $(M1)$ static field, a repetition of such a reduction as led to the dipole moment expression (9.121a) makes (9.123) yield $\mathbf{A}(M1) = \boldsymbol{\mu} \times \hat{\mathbf{r}}/r^2$, exactly as in (4.38).

The connection between the descriptions by $\mathbf{A} = \sum_j \mathbf{A}(Mj)$ and by ϕ_M is deducible from $\mathbf{B} = \nabla \times \mathbf{A} = -\nabla\phi_M$. An expression for $\mathbf{B} = \nabla \times \mathbf{A}$ will contain

$$\nabla \times \frac{\mathbf{T}^j_{jm}}{r^{j+1}} = \nabla \times \mathbf{L}\frac{r^{-(j+1)}Y_{jm}}{\sqrt{j(j+1)}} = i\nabla\frac{Y_{jm}}{\sqrt{j(j+1)}}\left(1 + r\frac{\partial}{\partial r}\right)r^{-(j+1)}$$

with the last result following from the expression (9.113) for $\nabla \times \mathbf{L}$ and the fact that $-\nabla^2 \; (\equiv k^2) \to 0$ for the static field. The operation by $1 + r\partial/\partial r$ is here equivalent to multiplication by the negative of the j value, and this makes

$$\mathbf{B} = \nabla \times \mathbf{A} = -\nabla 4\pi \sum_{jm}\frac{\mu_{jm}Y_{jm}}{(2j+1)r^{j+1}}$$

There is here a negative gradient of a scalar† expression that coincides with ϕ_M of (4.47).

The Radiated Energies and Angular Momenta

In the radiation zone ($kr \gg 1$), $h_j(kr) \to (-i)^{j+1}e^{ikr}/kr$ according to (9.12), and

$$\mathbf{U}^o_{kjm}(r) \to (-i)^{j+1}\frac{e^{ikr}}{kr}(\hat{\mathbf{r}} \times \mathbf{T}^j_{jm}) \tag{9.125}$$

† It is a procedure paralleling the one here by which it can be demonstrated that the static limits of the *electric* multipole fields (9.112) are representable as negative gradients of the scalar potential (3.102), despite their derivation from Lorentz potentials $\mathbf{A}(Mj)$ with $\phi = \nabla \cdot \mathbf{A}(Mj)/ik = 0$ before $k \to 0$. This would be a generalization of the finding for the electric dipole field discussed in connection with (8.56).

follows from (9.93). Then the general field (9.111), arising from an arbitrary source, becomes

$$\mathbf{E} \to (-i)^{j+1} \frac{e^{ikr}}{kr} \sum_{jm} [a_m(Mj)\mathbf{T}^j_{jm} + a_m(Ej)\hat{\mathbf{r}} \times \mathbf{T}^j_{jm}] \tag{9.126}$$

transverse to $\hat{\mathbf{r}} \equiv \hat{\mathbf{k}}$ and accompanied by $\mathbf{B} = \hat{\mathbf{r}} \times \mathbf{E}$. Sectors of it in any given direction take on the characteristics of plane waves, with polarizations determinable from the amplitude in (9.126). The energy-flux density is simply $\langle \mathbf{N} \rangle = \hat{\mathbf{r}} |\mathbf{E}|^2 c/8\pi$, and so the power radiated per unit solid angle, calculated as in (8.31) or (8.44), becomes

$$\frac{d\mathscr{P}}{d\Omega} = \frac{c}{8\pi k^2} \left| \sum_{jm} [a_m(Mj)\mathbf{T}^j_{jm} + a_m(Ej)\hat{\mathbf{r}} \times \mathbf{T}^j_{jm}] \right|^2 \tag{9.127}$$

This yields the radiation pattern (the angular distribution) from a completely arbitrary localized source, as characterized by its multipole moments.

For the angular distribution (9.127), the *amplitudes* contributed by each multipole source component must be added before forming the square of the total amplitude that determines the intensity. Interferences between the various multipole contributions, as represented by the cross terms, are thus taken into account. However, the interferences average to zero in the evaluation of the integrated intensity, radiated into all directions, because of the orthogonality (9.74) between amplitudes proportional to different vector spherical harmonics. Notice that

$$[\hat{\mathbf{r}} \times (\mathbf{T}^j_{jm})^*] \cdot (\hat{\mathbf{r}} \times \mathbf{T}^{j'}_{j'm'}) = \hat{\mathbf{r}} \cdot [(\mathbf{T}^j_{jm})^* \times (\hat{\mathbf{r}} \times \mathbf{T}^{j'}_{j'm'})] = (\mathbf{T}^j_{jm})^* \cdot \mathbf{T}^{j'}_{j'm'}$$

because of the transversality of $\mathbf{T}^j_{jm}$ to $\hat{\mathbf{r}}$. Moreover, the equation of (9.125) to the same vector $\mathbf{U}^o_{kjm}$ as given by the radiation-zone limit of its equivalent expression (9.82) shows that

$$\hat{\mathbf{r}} \times \mathbf{T}^j_{jm} = i\left(\sqrt{\frac{j+1}{2j+1}}\,\mathbf{T}^{j-1}_{jm} + \sqrt{\frac{j}{2j+1}}\,\mathbf{T}^{j+1}_{jm}\right) \tag{9.128}$$

which has the integrated orthogonality (9.74) to every $\mathbf{T}^{j'}_{j'm'}$. Thus,

$$\mathscr{P} = \frac{c}{8\pi k^2} \sum_{jm} [|a_m(Ej)|^2 + |a_m(Mj)|^2] \tag{9.129}$$

for which individual *intensity* contributions are added to each other.

The localized sources of greatest interest, having steadily maintained bound-state motions within a limited space, are restricted to essentially discrete frequencies,† as in the simple examples reviewed in Sec. 8.3, and for a given one

† Discrete spectra of frequencies are particularly well known for systems it is most essential to quantize, like atoms or nuclei. Classically described bounded motions, steadily maintained, also have discrete spectra, arising from the restriction of oscillations to wavelengths that fit within finite boundaries (Ref. 17, chaps. 11 to 15).

of these frequencies, some one type of multipole moment has a dominant magnitude (referred to as the operation of a *selection rule*). Then parts of the radiation expressions (9.127) and (9.129) referring to a given multipole order yield essentially all the radiation that can be observed with some one frequency. In terms of the moments introduced in (9.114) and (9.119)

$$\frac{d\mathscr{P}}{d\Omega}\begin{Bmatrix} Ej \\ Mj \end{Bmatrix} = \frac{2\pi c k^{2j+2}}{[(2j+1)!!]^2}\frac{j+1}{j}\left|\sum_m \begin{Bmatrix} Q_{jm} + Q'_{jm} \\ \mu_{jm} \end{Bmatrix}\mathbf{T}^j_{jm}\right|^2 \tag{9.130a}$$

$$\mathscr{P}\begin{Bmatrix} Ej \\ Mj \end{Bmatrix} = \frac{2\pi c k^{2j+2}}{[(2j+1)!!]^2}\frac{j+1}{j}\sum_m \left|\begin{Bmatrix} Q_{jm} + Q'_{jm} \\ \mu_{jm} \end{Bmatrix}\right|^2 \tag{9.130b}$$

These easily reduce to the special results (8.44), (8.75), and (8.66) when reexpressed in terms of moments $\mathbf{D}$, $\boldsymbol{\mu}$, and $\mathbf{Q}$, as illustrated in preceding subsections. In particular, the result (8.76) can be obtained through such connections as (3.104). The expressions here show that small sources, i.e., ones for which the moments become independent of the wavelength, radiate in proportion to the $(2j+2)$th power of the frequency when they have the multipolarity 2^j.

A sustained source frequently has a definite orientation as described by a single value of m and then has a radiation rate $\mathscr{P}_m$ given by just the one term of (9.130*b*). Its angular distribution is the more explicitly representable one

$$\frac{d\mathscr{P}_m}{d\Omega} = \mathscr{P}_m\,|\mathbf{T}^j_{jm}|^2 = \mathscr{P}_m \sum_\mu c^2_\mu\,|Y_{j,\,m-\mu}|^2$$

$$= \frac{\mathscr{P}_m}{2j(j+1)}[2m^2\,|Y_{jm}|^2 + (j+m)(j-m+1)\,|Y_{j,\,m-1}|^2$$

$$+ (j-m)(j+m+1)\,|Y_{j,\,m+1}|^2] \tag{9.131}$$

after substitution of the definition (9.69) of the vector spherical harmonic. Examples for small sources, with $Q_{jm} + Q'_{jm} \to q_{jm}$ [see the discussion of (9.115)], were given in Sec. 8.3. It is simple to verify that the dipole patterns (8.46) and (8.50) follow from (9.131) for cases in which $q_{1m} = \delta_{m0}(3/4\pi)^{1/2}qa$ and $q_{1m} = -\delta_{m1}(3/2\pi)^{1/2}qa$, respectively. The quadrupole patterns (8.78) and (8.81) arise in cases of $q_{2m} = \delta_{m2}(15/8\pi)^{1/2}qa^2$ and $q_{2m} = \delta_{m0}(5/16\pi)^{1/2}Q_o$, as is most easily verified with the help of the connections (3.104). Sources of any size will produce the same patterns for the same values of j, m since the moments $Q_{jm} + Q'_{jm}$ and μ_{jm} were so defined as to be independent of the viewing direction.

As might be expected from the dynamical properties of the multipole fields found in Sec. 9.3, a source producing a field characterized by given values j, m radiates m/ω units of angular momentum for each unit of energy it emits. This can be verified in quite the same way as for the dipole radiation when (8.55) was derived. A permanently radiated wave packet with ω units of energy, carrying one of the discrete numbers m of the angular momentum units, is formed in the shape described by the pattern (9.131). There is always such an association of a characteristic angular distribution with the definite angular momentum.

9.5 SPHERICAL-WAVE CONSTITUENTS OF VECTOR PLANE WAVES

The simple vector plane-wave descriptions, like the ones in (7.24) or (7.29), were found best adaptable to boundary conditions on plane surfaces, as in the rectangular wave guides of Sec. 7.3 and the reflection problems of Chaps. F and G. To deal with *finite* reflecting bodies, especially when the dimensions of the entire body may be only of the order of a wavelength, it is more effective first to decompose any incident plane wave into spherical waves centered on the body. Just as each such constituent was found more directly relatable to centrally localized primary sources of radiation, so it has better adaptability to the secondary sources presented by centrally localized reflecting bodies.

It is the individual *circularly* polarized plane-wave components of any incident field, represented as in (7.29), that can be expected to be simplest to express as superpositions of the vector spherical waves. The circular polarization vectors $\boldsymbol{\epsilon}_{\mathbf{k},\pm 1}$ of (7.28) can be identified with the spherical basis vectors $\mathbf{e}_{\pm 1}$ of (9.64) by choosing the plane wave's propagation direction $\hat{\mathbf{k}}$ as the polar (z) axis of the spherical-wave constituents to be discriminated in the plane waves. Then the transversality, $\mathbf{e}_z \cdot \mathbf{e}_{\pm 1} = 0$, of each wave can be made explicit by writing $\mathbf{k} \cdot \mathbf{r} = kz$ for the plane-wave phase in the expressions

$$\mathbf{E}_{\pm 1}(\mathbf{r}) = \sqrt{8\pi}\,\mathbf{e}_{\pm 1} e^{ikz} \qquad \mathbf{B}_{\pm 1}(\mathbf{r}) = \hat{\mathbf{k}} \times \mathbf{E}_{\pm 1} = \mp i\mathbf{E}_{\pm 1}(\mathbf{r}) \tag{9.132}$$

for the monochromatic plane-wave representatives of the incident field. That the magnetic vectors differ from the circularly polarized electric ones by just the phase factors $\mp i = \exp(\mp i\pi/2)$ follows directly from crossing $\hat{\mathbf{k}} \equiv \mathbf{e}_z$ into the $\mathbf{e}_{\pm 1}$ of (9.64); the always orthogonal electric and magnetic vectors rotate together (see Fig. 7.2), and so each comes into coincidence with the direction taken by the other a quarter cycle earlier or later. The normalization factor $\sqrt{8\pi}$ adopted for the amplitudes makes each of the representatives (9.132) correspond to a unit field energy density $\langle w \rangle \to 1$, in (7.15), for each.

Since the plane waves are divergenceless, each can contain as linearly independent spherical-wave constituents only the divergenceless members $j_j(kr)\mathbf{T}^j_{jm}(\hat{\mathbf{r}})$ and $\mathbf{U}_{kjm}(\mathbf{r})$ of the complete set (9.86). Moreover, the radial-wave factors that can occur are restricted to the regular ones, $R_l = j_l(kr)$, as in the scalar plane wave (9.25), because a plane-wave field has no singularities anywhere. Thus, only the coefficients for a decomposition like

$$\mathbf{E}_{\pm 1}(\mathbf{r}) = \sum_{jm} [A_{jm} j_j(kr)\mathbf{T}^j_{jm}(\hat{\mathbf{r}}) + B_{jm}\mathbf{U}_{kjm}(\mathbf{r})] \tag{9.133}$$

need be determined. That is simple to do when advantage is taken of the decomposition (9.25) of the scalar plane-wave factor in (9.132), which makes

$$\mathbf{E}_{\pm 1}(\mathbf{r}) = 4\pi\mathbf{e}_{\pm 1} \sum_{l=0}^{\infty} \sqrt{2(2l+1)}\, i^l j_l(kr) Y_{l0}(\hat{\mathbf{r}}) \tag{9.134}$$

Expression (9.133) could also be made an explicit superposition of products like $\mathbf{e}_\mu j_l Y_{l,m-\mu}$, as reference to the definitions (9.82) and (9.69) of $\mathbf{U}_{kjm}$ and the $\mathbf{T}^l_{jm}$

shows. The circularly polarized waves (9.134) are plainly eigenfunctions of $S_z[\pm 1]$ and $J_z[\pm 1]$, and so only $m = +1$ or $m = -1$ can appear in (9.133) for the respective cases. To find the coefficients that will make the form (9.133) equivalent to (9.134), it is perhaps simplest to project the expression (9.133) onto $\mathbf{T}^l_{jm}$, with $l = j,\ j \pm 1$ in turn, and use the orthonormality (9.74) of the vector spherical harmonics

$$\oint d\Omega(\mathbf{T}^j_{jm})^* \cdot \mathbf{E}_{\pm 1} = A_{jm} j_j(kr)$$

$$\oint d\Omega(\mathbf{T}^{j\pm 1}_{jm})^* \cdot \mathbf{E}_{\pm 1} = B_{jm} \begin{Bmatrix} -\sqrt{j} \\ \sqrt{j+1} \end{Bmatrix} \frac{j_{j\pm 1}(kr)}{\sqrt{2j+1}}$$

There are actually two ways to determine B_{jm} offered here; they *must* give the same results because $\mathbf{E}$ can contain only divergenceless linear combinations of the spherical waves. Now the definitions (9.69) and the orthonormalities of the scalar spherical harmonics in them to those of (9.134) lead to

$$A_{jm} = \mp iB_{jm} = \mp \delta_{m,\,\pm 1} 4\pi \sqrt{2j+1}\, i^j$$

as can be found after consulting Table 9.1 for the appropriate vector-addition coefficients.

The desired decompositions of the circularly polarized plane waves (9.132) turn out to be

$$\begin{aligned} \mathbf{E}_{\pm 1}(\mathbf{r}) &= \mp 4\pi \sum_{j=1}^{\infty} \sqrt{2j+1}\, i^j [j_j(kr)\mathbf{T}^j_{j,\,\pm 1}(\hat{\mathbf{r}}) \pm i\mathbf{U}_{kj,\,\pm 1}(\mathbf{r})] \\ \mathbf{B}_{\pm 1}(\mathbf{r}) &= \mp 4\pi \sum_{j=1}^{\infty} \sqrt{2j+1}\, i^j [\mathbf{U}_{kj,\,\pm 1}(\mathbf{r}) \mp i j_j(kr)\mathbf{T}^j_{j,\,\pm 1}(\hat{\mathbf{r}})] \end{aligned} \tag{9.135}$$

Reference to (9.94) and (9.95) shows that the plane waves contain equal parts of the (Mj) and (Ej) *standing* waves for each j value and in different relative phases for the opposite helicities ($\mu = \pm 1$). The superpositions are restricted to $j \geq 1$ because the transverse waves can have no longitudinal component proportional to $\mathbf{T}^1_{00}(\hat{\mathbf{r}}) \sim \hat{\mathbf{r}}$ of (9.77). Nevertheless, the important $l = 0$ component† of (9.134), the only one that does not vanish at the origin, is properly contained in $\mathbf{U}_{k1,\,\pm 1}(\mathbf{r})$. It is easy to check that both expressions (9.134) and (9.135) reduce to $\sqrt{8\pi}\,\mathbf{e}_{\pm 1}$, at $\mathbf{r} = 0$, as they should according to (9.132).

An illustration of how the result (9.135) can be applied is left to Exercises 9.10 to 9.13. Here, a more fundamental point brought out by the result will be discussed.

† Called the *s wave*. It is the one that will be influenced most by any small reflecting object that may be placed at the origin used for the decomposition, as in Exercise 9.13.

Angular Momenta Carried by Plane Waves

Every one of the spherical-wave constituents in the plane waves (9.135) was found, in Sec. 9.3, to contain an angular momentum $(m = \pm 1)/\omega$ per unit of field energy in the constituent. Here, the angular momentum $m = +1$ is parallel to the propagation direction of the circularly polarized plane-wave component of helicity $\mu = +1$, and the wave of helicity $\mu = -1$ carries $m = \mu = -1$, a unit of angular momentum antiparallel to the propagation direction. It must therefore be expected that circularly polarized plane waves and linear combinations of them should be able to carry nonvanishing, conserved, field angular momentum to distances indefinitely far from any point $\mathbf{r} = 0$ that might be taken as momental center. Only linearly polarized linear combinations, like those having polarization vectors $\mathbf{e}_x = -2^{-1/2}(\mathbf{e}_{+1} - \mathbf{e}_{-1})$ or $\mathbf{e}_y = 2^{-1/2}i(\mathbf{e}_{+1} + \mathbf{e}_{-1})$ containing equal parts of the two opposing helicities, can be expected to bear no net angular momentum.

These conclusions present a puzzle in that *any* ideal plane wave, however polarized, bears a linear momentum density $\langle \mathbf{g} \rangle = \hat{\mathbf{k}}\langle w \rangle / c$ [Eq. (7.17)] that is uniformly distributed about any axis parallel to the propagation direction that might be chosen. Then every contribution $\mathbf{r} \times \langle \mathbf{g} \rangle$ to the angular-momentum density should be canceled by a contribution from an element at $-\mathbf{r}$. On this account, no net angular momentum should be expected in any idealized plane wave, however polarized.

The resolution of the puzzle lies in the circumstance that the absolutely monodirectional, infinitely extended plane wave is an *over*idealization of any physically realizable finite field, in that it does not permit identification of any conserved angular momentum this field may have, just as the same idealized representation prevents identification of a finite *total* energy, like (7.11), that an actual, physically realizable, finite field must have. For evaluating total angular momenta in actual fields, it must be recognized that it is departures from the infinitely extended *spatial* uniformity of the ideal plane waves that have effect on conclusions from momentum distributions. The well-defined total angular momenta of Sec. 9.3 were identifiable only for finitely extended wave *packets* of the spherical waves, constructed by superposing waves with a continuous range of radial wavelengths that can interfere destructively† at great distances from the momental centers. The outward transport of angular momenta was found to depend essentially on the existence of longitudinal field components parallel to the radial propagation directions. This was also found earlier, in the radiated waves of Chap. 8 (see the footnote to page 226).

A wave packet of the circularly polarized plane waves carrying a finite total angular momentum cannot be formed simply by superposing a continuum of wave-number magnitudes, all that there was the freedom to do in (9.96), for the

† Represented in the expression (9.100) by the necessity of choosing $|c(k \to 0)|^2 \sim k^{n>3} \to 0$ to get a finite result, thereby restricting the relative intensities invested in such very long wavelengths as may not average to zero (interfere destructively) over large distances.

spherical waves of definite j, m. Here that would produce a wave train of finite length but still infinitely wide; its codirectionally propagated constituents all have linear momenta parallel to the propagation direction and uniformly distributed about it, yielding a zero total angular momentum, as each ideal plane wave does. What is necessary is to consider a *beam* of finite *lateral* extent, like the ray mentioned in connection with the complementary relations (7.22). The lateral restrictions to Δx, $\Delta y < \infty$ introduce momentum components $p_{x,y} \sim k_{x,y}$ *transverse* to the beam direction ($k_z \sim p_z$), as in (7.19), and it is those transverse components which can yield moments about the beam direction. The finitely wide beam will also contain such longitudinal *field* (z) components as are found in the spherical waves (because of the transversality to $k_{x,y}$ of the corresponding free-space field components). The formulation of a beam with all those properties and showing that it does in fact carry the expected angular momentum are left to Exercises 9.15 and 9.16, since a less special and more significant approach, presented in the next subsection,† has been found.

Photon Spin

An approach which allows evaluating the angular momenta in plane waves without a detailed construction of a beam was found by expressing the angular-momentum density in terms of a solenoidal (divergenceless) vector potential **A** in place of the fields **E**, **B**. In this approach, the finite spatial extent actually needed for identifying a conserved angular momentum is enforced by imposing the general boundary conditions characteristic of finitely extending fields.

It will initially be sufficient to use only the expression $\mathbf{B} = \nabla \times \mathbf{A}$, so that the momentum density becomes

$$\mathbf{g} = \frac{\mathbf{E} \times (\nabla \times \mathbf{A})}{4\pi c} = \frac{1}{4\pi c} \sum_j (\mathsf{E}_j \nabla A_j - \mathsf{E}_j \nabla_j \mathbf{A}) \tag{9.136}$$

It will be instructive first to show that the final terms here play no role in determining the *resultant* field momentum in a finitely extending, isolated field. They are proportional to

$$(\mathbf{E} \cdot \nabla)\mathbf{A} = \nabla \cdot (\mathbf{E}\mathbf{A}) - \mathbf{A}(\nabla \cdot \mathbf{E}) = \nabla \cdot (\mathbf{E}\mathbf{A})$$

since $\nabla \cdot \mathbf{E} = 0$ in an isolated, free field, and this is just a divergence of a field dyadic which will as usual vanish upon integration over the entire field, enclosable by a surface on which $\mathbf{E} \to 0$. Thus the resultant field momentum will receive an average contribution

$$\langle \mathbf{g} \rangle = \frac{\frac{1}{2} \operatorname{Re} \left(\sum_j \mathsf{E}_j^* \nabla \mathbf{A}_j \right)}{4\pi c} \tag{9.137}$$

† A modification and extension of the one in Ref. 4, vol. 2, page 320.

from unit volumes of the field. For a field representable by an idealized plane wave,† $\mathbf{A} = \mathsf{E}/ik$ and $\nabla\mathbf{A} = ik\mathbf{A} = \hat{\mathbf{k}}\mathsf{E}$, with the usual result, $\langle \mathbf{g} \rangle = \hat{\mathbf{k}}|\mathsf{E}|^2/8\pi c = \hat{\mathbf{k}}\langle w \rangle/c$.

Whereas the final terms in (9.136), those with the negative signature, were found not to contribute to the resultant *linear* field momentum, they are just the ones that will now be found to be responsible for an intrinsic-spin *angular* momentum of the field. When the moment $\mathbf{r} \times \mathbf{g}$ is formed, the terms in question contribute

$$-(4\pi c)^{-1} \sum_j \mathsf{E}_j \mathbf{r} \times (\nabla_j \mathbf{A}) = -(4\pi c)^{-1} \nabla \cdot [\mathsf{E}(\mathbf{r} \times \mathbf{A})]$$
$$+ (4\pi c)^{-1}[(\mathbf{r} \times \mathbf{A})\nabla \cdot \mathsf{E} + \mathsf{E} \times \mathbf{A}]$$

with the lever arm $\mathbf{r}$ eliminated from the last term as a result of the derivative operations $\nabla_j x_i = \delta_{ij}$. The other term in the last line vanishes because $\nabla \cdot \mathsf{E} = 0$. The remaining term, shown in the first line, is again a divergence of a field dyadic that vanishes upon integration over the entire field. Thus, the terms in question make the contribution

$$\boldsymbol{\Sigma} \equiv \oint dV(\mathbf{r}) \frac{\mathsf{E} \times \mathbf{A}}{4\pi c} \tag{9.138}$$

to the resultant field angular momentum. This defines what is called an *intrinsic-spin contribution* because it is completely independent of the momental center ($\mathbf{r} = 0$) that might be used in gauging the field angular momenta.

The total angular momentum that follows from the expression (9.136) of the linear momentum density can now be written as

$$\mathbf{J} = \oint dV(\mathbf{r})\mathbf{r} \times \mathbf{g} = \boldsymbol{\Lambda} + \boldsymbol{\Sigma} \tag{9.139}$$

if $\boldsymbol{\Lambda}$ is defined by

$$\boldsymbol{\Lambda} \equiv \oint dV \sum_j \frac{\mathsf{E}_j \mathbf{r} \times \nabla A_j}{4\pi c} \tag{9.140}$$

Because this part does depend on the lever arms $\mathbf{r}$ of the momentum distribution, it is called the *orbital-momentum contribution.*

A point of some significance‡ is that the infinitesimal rotation operator $\mathsf{L} = -i\mathbf{r} \times \nabla$ of (9.44) occurs in the integral for the orbital momentum $\boldsymbol{\Lambda}$. A

† With the vector potential first noted in the footnote to (7.14).

‡ From the point of view that the Maxwell equations, or the equivalent equations for potentials, constitute a wave mechanics of photons comparable to the Schrödinger equation for mass particles or, better, to the relativistic Dirac equations for electrons. In all these cases, second quantization, which makes each particle a quantum of some field, is not explicitly recognized. The use of $\mathbf{A}$ as the photon wave function, as in the matrix elements (9.141) and (9.142), conforms to a necessity of second quantization.

consequence is that for a monochromatic representative field, when $\mathbf{E} = ik\mathbf{A}(\mathbf{r})$ is an eigenfunction of frequency,

$$\mathbf{\Lambda} = \tfrac{1}{2}\,\mathrm{Re}\left(\frac{k}{4\pi c}\oint dV \sum_j A_j^* \mathbf{L} A_j\right) \tag{9.141}$$

the real part of a type of integral called a matrix element of the operator $\mathbf{L}$. Correspondingly, the intrinsic spin of the field can be expressed in terms of matrix elements of the vector rotation operation $\mathbf{S}_j \mathbf{E} = i\mathbf{e}_j \times \mathbf{E}$ of (9.59*b*):

$$\mathbf{\Sigma} = -\tfrac{1}{2}\,\mathrm{Re}\left(\frac{k}{4\pi c}\oint dV \sum_j A_j^* \mathbf{S}_j \mathbf{A}\right) \tag{9.142a}$$

The vector character of the integral is attributed to the operator by itself in the matrix notation of (9.60) and (9.61)

$$\mathbf{\Sigma} = +\tfrac{1}{2}\,\mathrm{Re}\left(\frac{k}{4\pi c}\oint dV\, A^\dagger \mathbf{S} A\right) \tag{9.142b}$$

where $A^\dagger$ is the complex conjugate of a column vector transposed into a row.†

In the circularly polarized plane waves described as in (9.132) or (9.134) or (9.135), the orbital momentum $\mathbf{\Lambda}$ vanishes. This can be found by putting any one of these forms into either one of the integrals (9.140) or (9.141). It will perhaps be most significant to use the form (9.134) for $\mathbf{E} = ik\mathbf{A}$ in the integral (9.141). The orbital-momentum operator $\mathbf{L}$, constituted of differentiations with respect to angles as in (9.47) and (9.48), acts only on the factors $Y_{l0}(\vartheta, \varphi)$ in the l-wave decomposition (9.134). According to (9.50), the components $\mathbf{L}_\pm \equiv \mathbf{L}_x \pm i\mathbf{L}_y$ of the operator convert each Y_{l0} into $Y_{l,\pm 1}$, respectively, and both these results are orthogonal to the $Y_{l'0}^*$ in the integration (9.141). The Y_{l0} are eigenfunctions (9.54) of the remaining component $\mathbf{L}_z$ but with a vanishing eigenvalue ($m = 0$): $\mathbf{L}_z Y_{l0} = 0$. Thus, all three components of $\mathbf{\Lambda}$ vanish.

It will be equally significant to use either of the forms (9.132) or (9.134) to furnish the $\mathbf{A} = \mathbf{E}/ik$ for the evaluation of the intrinsic-spin contribution (9.142) because the unit-spin operator $\mathbf{S}$ of (9.59), (9.65), or (9.67) acts only on the vector factor $\mathbf{e}_{\pm 1}$ in each wave. If the vector form of the operator $\mathbf{S}_j \mathbf{A} = i\mathbf{e}_j \times \mathbf{A}$ occurring in (9.142*a*) is used, and the basic vector properties $\mathbf{e}_{\pm 1}^* \times \mathbf{e}_{\pm 1} = \pm i\mathbf{e}_z$ and $\mathbf{e}_z \times \mathbf{e}_{\pm 1} = \mp i\mathbf{e}_{\pm 1}$ that follow from the definitions (9.64), each unit volume of the integral for $\mathbf{\Sigma}$ contributes

$$-\frac{ik}{8\pi c}\mathbf{A}_{\pm 1}^* \times \mathbf{A}_{\pm 1} = -\frac{|\mathbf{E}_{\pm 1}|^2}{8\pi\omega} i\mathbf{e}_{\pm 1}^* \times \mathbf{e}_{\pm 1} = \pm\mathbf{e}_z \frac{\langle w\rangle}{\omega} \tag{9.143}$$

These are just the amounts, per unit energy, that were expected from the nature of the spherical-wave constituents of the respectively circularly polarized plane waves (9.135). The outcome that $\mathbf{\Lambda} = 0$ and $\mathbf{\Sigma} = \mathbf{J} \neq 0$ in each circularly polarized wave shows that each carries angular momentum solely in the form of

† A so-called *hermitean conjugate*, see Ref. 17, eq. (10.81), for example.

instrinsic spin. It is said that plane waves of positive helicity consist of photons spinning parallel to their propagation direction and that negative-helicity photons have antiparallel spins.

It should still be emphasized that there is no *observable* difference between the spin and the orbital angular momenta of fields,† fundamentally because only their vector resultant $\mathbf{J} = \boldsymbol{\Lambda} + \boldsymbol{\Sigma}$ ever needs to be conserved at a definite value. The difference is a conceptual one that makes formulations like (9.141) and (9.142) possible, forms that are more essential to quantum electrodynamics. In the general case, the splitting of the observable $\mathbf{J}$ between $\boldsymbol{\Lambda}$ and $\boldsymbol{\Sigma}$ depends on the axis chosen‡ relative to which the observable effects are to be analyzed.

EXERCISES

9.1 (*a*) Show that a factorization of the operator (9.5), alternative to (9.8), is $-(d/d\zeta - l/\zeta)(d/d\zeta + l/\zeta)$ and that the descending ladder operation

$$h_{l-1} = +\zeta^{-(l+1)} \frac{d}{d\zeta} \zeta^{(l+1)} h_l$$

follows.

(*b*) Show that a consequence of (*a*) and (9.9) is the recurrence relation

$$h_{l+1} + h_{l-1} = \frac{2l+1}{\zeta} h_l$$

(c) Show that another consequence is

$$\frac{dh_l}{d\zeta} = \frac{l h_{l-1} - (l+1) h_{l+1}}{2l+1}$$

9.2 Derive the commutation properties (9.52), (9.53), and (9.62) of the rotation operators, as applied to any vector function $\mathbf{f}(\mathbf{r})$. Confirm assertions (9.65) and (9.67). (These results are also useful in quantum mechanics.)

9.3 You may more readily accept the results (9.80), (9.81), (9.88), and (9.90) without seeing their derivation if you confirm them in a sufficient number of special cases ($l \leq 2$, $|m| \leq 1$) to check every term, using the explicit forms (9.11) and (9.77). You can also show quite generally how each of the two curl operations (9.90) follows from the other, as a consequence of the Helmholtz equation being satisfied by the divergenceless fields.

9.4 In applying (9.116) to the case of the linear oscillator, it is simplest to begin by forming a $\mathbf{D}(t_s)$, as in (8.45), *for a given k*. Use

$$j_1(kr_q) \approx \frac{kr_q}{3} \left[1 - \frac{(kr_q)^2}{10} + \cdots \right]$$

† Such situations are well known in the relativistic quantum theory of spinning particles in general. The Dirac electron has continual spin flips, generating exchanges with orbital angular momentum relative to *any* chosen momental center, even during free motion. (See, for example, Ref. 16, page 77.)

‡ See the third footnote to page 261.

which follows from (9.22) for $kr_q \ll 1$, to show that

$$\mathbf{D}(\omega_o) \approx q\mathbf{a}\left[1 - \frac{(k_o a)^2}{40} + \cdots\right]$$

for radiation of frequency $\omega_o = ck_o$. (Such corrections are needed for evaluating the oscillator sum rules of quantum radiation theory, in order to obtain agreement with accurate measurements. The reduction from $q\mathbf{a}$ also accounts exactly for a discrepancy with relativistic corrections, to be discussed on page 292 and in Exercise 10.7. It will have been found that there are additional corrections—but to radiation of frequency $3\omega_o$.)

9.5 Some sources are best described by oscillating magnetization distributions $\mathbf{M}(\mathbf{r}, t)$ equivalent to the microscopic current densities $\mathbf{j}_M = c\nabla \times \mathbf{M}$ of (E.3) that can replace $\mathbf{j}$ of (9.120*a*).

(*a*) A hermiteanship like (4.51) can be used to transfer the curl operation on $\mathbf{M}$ to the remainder of the integrand in (9.120*a*). Show that

$$\mu_{jm} = (2j+1)!!\,k^{1-j}\left(\frac{j}{j+1}\right)^{1/2} \oint dV\,\mathbf{M}(\mathbf{r}) \cdot \mathbf{U}^*_{kjm}(\mathbf{r})$$

where $\mathbf{M}(\mathbf{r})$ is a monochromatic representative of $\mathbf{M}(\mathbf{r}, t)$.

(*b*) Show by using (9.113) to replace the operation in the form (9.91) of $\mathbf{U}_{kjm}$ that

$$\mu_{jm} = -\frac{(2j+1)!!}{(j+1)k^j} \oint dV\,Y^*_{jm}\left[\left(\frac{\partial}{\partial r} rj_j\right)(\nabla \cdot \mathbf{M}) - k^2 j_j(\mathbf{M} \cdot \mathbf{r})\right]$$

which is the conventional form.

(*c*) Show that for small sources ($ka \ll 1$)

$$\mu_{jm} \approx \oint dV \rho_M r^j Y^*_{jm}$$

which is like q_{jm} of (3.101) except that the charge density is replaced by the pole-strength distribution $\rho_M = -(\nabla \cdot \mathbf{M})$ of (E.7). [The moments here are additive to (9.120) when the *total* current density is $\mathbf{j} + \mathbf{j}_M$.]

9.6 The varying magnetization introduced in Exercise 9.5 can also contribute to *electric* multipole radiations.

(*a*) Show why $\mathbf{j}_M = c\nabla \times \mathbf{M}$ cannot contribute to Q_{jm} of (9.115) but when it replaces $\mathbf{j}$ in Q'_{jm}, an immediate result is

$$Q'_{jm} = -\frac{(2j+1)!!}{(j+1)k^{j-1}} \oint dV (j_j Y^*_{jm})\mathbf{L} \cdot \mathbf{M}$$

More conventionally, $\mathbf{L} \cdot \mathbf{M}$ is replaced by its equivalent, $i\nabla \cdot (\mathbf{r} \times \mathbf{M})$.

(*b*) Make use of the hermiteanship of the operator $\mathbf{L}$ to show that Q'_{jm} can be put into the same form as μ_{jm} of Exercise 9.5(*a*) except that $j_j(\mathbf{T}^j_{jm})^*$ replaces $\mathbf{U}^*_{kjm}$. [Such forms take advantage of the explicit results (9.69). For small sources, Q'_{jm} is usually smaller by a factor ka than μ_{jm} of Exercise 9.5(*c*).]

9.7 In the small-source approximation ($\omega_o a \ll c$), the linear oscillator emits electric dipole ($j = 1$) radiation of frequency ω_o and electric quadrupole ($j = 2$) radiation of frequency $2\omega_o$. Show that:

(*a*) A small-source (Ej) moment with $j > 2$ yields radiation of frequencies

$$j\omega_o,\ (j-2)\omega_o,\ \ldots,\ 3\omega_o,\ \omega_o \qquad \text{if } j \text{ is odd}$$

$$j\omega_o,\ (j-2)\omega_o,\ \ldots,\ 4\omega_o,\ 2\omega_o \qquad \text{if } j \text{ is even}$$

(*b*) When $\omega = j\omega_o$ is being observed, the moments of higher order than j contribute negligibly if only $\omega_o \ll c/a$.

9.8 Quantum-mechanical effects generally restrict sources producing radiation of a given frequency to one of two parities: having $\rho(-\mathbf{r}) = +\rho(\mathbf{r})$ [even or (+) parity] or $\rho(-\mathbf{r}) = -\rho(\mathbf{r})$ [odd or (−) parity].

(*a*) Show that $\mathbf{j}(-\mathbf{r}) = \mp\mathbf{j}(\mathbf{r})$, respectively, for even- and odd-parity sources.

(*b*) Show that if $\mathbf{r}$ has the spherical coordinates r, ϑ, and φ, then $\mathbf{r}' = -\mathbf{r}$ has $r, \pi - \vartheta$, and $\varphi + \pi$.

(*c*) Show that $Y_{lm}(-\hat{\mathbf{r}}) = (-)^l Y_{lm}(\hat{\mathbf{r}})$. [Descriptions by $Y_{lm}(\hat{\mathbf{r}})$ are then said to have the parity $(-)^l$.]

(*d*) Find the parities of descriptions by $\mathbf{T}^l_{jm}(\hat{\mathbf{r}})$, $\mathbf{U}_{kjm}(\mathbf{r})$, and $\mathbf{W}_{kjm}(\mathbf{r})$. (Note that $\mathbf{e}_\mu$ describes a frame unchanged in the active space inversion $\mathbf{r} \to -\mathbf{r}$.)

(*e*) Show that $a_m(Ej) = 0$ for even sources if j is odd and for odd sources if j is even. Investigate $a_m(Mj)$ similarly.

(*f*) The (Ej) or (Mj) radiation is assigned a parity equal to the parity of the source that produces it, so that it can be said that parity is conserved. Show that (Ej) radiation must be assigned the parity $(-)^j$ while (Mj) radiation has the parity $(-)^{j+1}$.

(*g*) Show that the parity of a multipole field is the same as the parity of its *magnetic* field description, while the corresponding electric field has a description of the opposite parity (to that of the radiation it describes). [The results are selection rules for the radiation field that will be produced, following from source symmetries in space inversion ($\mathbf{r} \to -\mathbf{r}$), superposed on the selection of specific j values for the field arising from source symmetries in proper rotations. The latter are required by angular-momentum conservation and the former by parity conservation.]

9.9 Any one of the multipole fields (9.94) or (9.95) can be excited inside a spherical cavity in a conductor. Assume the conductor perfect and a cavity radius R.

(*a*) Find orientation-independent transcendental equations from which the resonant frequencies can be determined. [See (9.93) and Exercise 9.1.]

(*b*) Show that $\tan\zeta$, with $\zeta = \omega_{nj}R/c$, must equal ζ and $\zeta/(1-\zeta^2)$, respectively, for the $(M1)$ and $(E1)$ resonances.

(*c*) By sketching graphs of the equalities in (*b*), show that the lowest $(E1)$ resonance has $\pi c/2R < \omega_{11} < \pi c/R$, whereas $(M1)$ has $\pi c/R < \omega_{11} < 3\pi c/2R$.

(*d*) Show that especially the higher frequencies of the full spectrum can be approximated by

$$\omega_{nj} \approx \begin{cases} \dfrac{(2n+j-1)\pi c}{2R} & (Ej) \\[2ex] \dfrac{(2n+j)\pi c}{2R} & (Mj) \end{cases} \qquad n, j = 1, 2, 3, \ldots$$

(See Exercise 9.8 for the parities of these resonance levels.)

9.10 (*a*) Show that any point in a plane wave that is chosen as the $r = 0$ of the analysis (9.135) has an incoming spherical wave converging on it and, superposed, an outgoing part of equal absolute amplitude.

(*b*) Now suppose that $r = 0$ is a point having centered on it some finite obstacle to an incident plane wave like either one of (9.135). Show that the resultant field outside the obstacle must be representable by the original plane wave plus (superposed on it) a purely outgoing wave like

$$\mathbf{E}^o = \sum_{j,m} (A_{jm} h_j \mathbf{T}^j_{jm} + B_{jm}\mathbf{U}^o_{kjm}) \qquad \text{and} \qquad \mathbf{B}^o = \sum_{j,m} (A_{jm}\mathbf{U}^o_{kjm} - B_{jm} h_j \mathbf{T}^j_{jm})$$

with coefficients that depend on the nature of the obstacle. The total fields $\mathbf{E}_{\pm 1} + \mathbf{E}^o$ and $\mathbf{B}_{\pm 1} + \mathbf{B}^o$ must be regarded as fields established after a sufficiently long time of incidence, since they are being treated as monochromatic.

9.11 Suppose that the obstacle of Exercise 9.10 is a perfectly reflecting sphere of radius $r = a$. It will be found economical to express the results in terms of the definitions

$$-\frac{h_j^*(ka)}{h_j(ka)} \equiv e^{2i\delta_j} \qquad -\left[\frac{d(\zeta h_j^*)/d\zeta}{d(\zeta h_j)/d\zeta}\right]_{\zeta = ka} \equiv e^{2i\delta_j'}$$

each being a complex ratio of unit absolute magnitude. The phase angles δ_j, δ_j' are then called *scattering phase shifts* due to the perfectly refecting sphere. (Reference 17, sec. 15.3, for example, explains why the name is appropriate.)

(*a*) Show that the boundary condition $\hat{\mathbf{r}} \cdot \mathbf{B} = 0$ leads to

$$A_{jm} = \mp 4\pi\delta_{m,\pm 1}\sqrt{2j+1}\, i^{j+1} e^{i\delta_j} \sin \delta_j$$

(*b*) Show that $\hat{\mathbf{r}} \times \mathbf{E} = 0$ at $r = a$ yields

$$B_{jm} = +4\pi\delta_{m,\pm 1}\sqrt{2j+1}\, i^{j} e^{i\delta'_j} \sin \delta'_j$$

The resulting outgoing wave, represented as in Exercise 9.10(*b*), is then said to describe the radiation scattered out of the incident beam (which need only be wide enough, relative to *a*, for beam-edge effects to be negligible).

9.12 (*a*) Show that the outgoing wave of Exercise 9.10(*b*) yields a radiation rate given by (9.127) and (9.129), after the replacements $a_m(M_j) \to A_{jm}$ and $a_m(E_j) \to B_{jm}$.

(*b*) Refer to (10.25) for the definition of cross sections and show that it leads to $d\sigma/d\Omega = d\mathscr{P}/(c\, d\Omega)$ and $\sigma = \mathscr{P}/c$, where $\mathscr{P}$ is a rate initiated by the incidence of plane waves normalized to unit energy density, like either of (9.135).

(*c*) For the perfectly reflecting sphere of Exercise 9.11, show that

$$\frac{d\sigma}{d\Omega} = \frac{2\pi}{k^2} \left|\sum_{j=1}^{\infty} \sqrt{2j+1}\, [e^{i\delta_j} \sin \delta_j \mathbf{T}^j_{j,\pm 1} \pm i e^{i\delta'_j} \sin \delta'_j (\hat{\mathbf{r}} \times \mathbf{T}^j_{j,\pm 1})]\right|^2$$

with the amplitude that is squared here having the directions of the polarizations of the waves scattered in various directions, from incident waves of positive and negative helicity, respectively.

(*d*) Show that the integrated cross section corresponding to (*c*) is

$$\sigma = \frac{2\pi}{k^2} \sum_{j=1}^{\infty} (2j+1)(\sin^2 \delta_j + \sin^2 \delta'_j)$$

independent of the incident polarizations and an incoherent sum of magnetic and electric multipole contributions.

9.13 (*a*) Show how $\tan \delta_j = j_j(ka)/n_j(ka)$ follows from the definition in Exercise 9.11 and find the corresponding real expression for $\tan \delta'_j$.

(*b*) Reflections from a sphere will differ most extremely from the reflections by a plane conductor (Exercise 7.6) when the sphere is small compared with one wavelength. Show that for $ka \ll 1$, $\delta_{j>1}$ is negligible relative to $\delta_1 \approx -(ka)^3/3$ and that $\delta'_j \approx -\delta_j(1 + j^{-1})$.

(*c*) Show that the total scattering by the small sphere is given by

$$\sigma \approx \frac{10\pi a^2}{3}\left(\frac{\omega a}{c}\right)^4$$

The proportionality to ω^4 can be understood to arise from dipoles induced in the sphere [compare (8.46), (8.50), and (8.66)].

(*d*) Show that the scattering is mostly backward ($\vartheta > \frac{1}{2}\pi$) in proportion to $5(1 + \cos^2 \vartheta) - 8 \cos \vartheta$ and that this does yield a small forward ($\vartheta = 0$) peak relative to immediately adjacent angles. (Notice that the angular distribution here is independent of thc incident polarization.)

9.14 Show how it follows from (9.137) that the total linear momentum in a frequency eigenfield $\mathbf{E} = ik\mathbf{A}$ can be expressed as

$$\mathbf{P} = \tfrac{1}{2} \operatorname{Re}\left[\frac{k}{4\pi c} \oint dV \sum_j A_j^*(-i\nabla)A_j\right]$$

a matrix element of the infinitesimal translation operator in (9.42) and in the footnote to page 246. Compare (9.141).

9.15 A circularly polarized beam is representable [see (9.132)] as

$$\mathbf{E}_{\pm 1} = [\mathbf{e}_{\pm 1} f_0(x, y) + \mathbf{e}_z f_{\pm 1}(x, y)]e^{ikz} \qquad \mathbf{B}_{\pm 1} \approx \pm i\mathbf{E}_{\pm 1}$$

with f_0 real. Both f_0 and $f_{\pm 1}$ approach zero outside a cross section having $\Delta x\, \Delta y \gg k^{-2} = (\lambda/2\pi)^2$ if the beam is to continue reasonably well defined. Moreover, $|f_{\pm 1}|^2$ must be negligible compared with f_0^2 if it is to be experimentally identifiable as circularly polarized.

(*a*) Show that the longitudinal amplitude must be given by

$$f_{\pm 1} = \pm \frac{1}{\sqrt{2}\, ik}\left(\frac{\partial}{\partial x} \pm i\frac{\partial}{\partial y}\right) f_0$$

to satisfy $\nabla \cdot \mathbf{E} = 0$.

(*b*) Show that $\langle w(x, y)\rangle = (f_0^2 + |f_{\pm 1}|^2)/8\pi \approx f_0^2/8\pi$, whereas

$$\langle \mathbf{g} \rangle = \mathbf{e}_z \frac{f_0^2}{8\pi c} - \frac{f_0}{4\pi c}\,\mathrm{Re}\,(\mathbf{e}_{\pm 1}^* f_{\pm 1})$$

if terms of first order in the small $|f_{\pm 1}|$ are retained.

9.16 (*a*) Show that in the beam of Exercise 9.15

$$(\mathbf{r}\mathrm{x}\langle \mathbf{g} \rangle)_z = \mp \frac{1}{16\pi\omega}\left(\frac{\partial}{\partial x} x f_0^2 + \frac{\partial}{\partial y} y f_0^2\right) \pm \frac{f_0^2}{8\pi\omega}$$

and that after integration over the beam cross section, only the last term contributes to the resultant angular momentum in the beam.

(*b*) Show that $\pm f_0^2/8\pi\omega = \langle w \rangle(\pm 1/\omega)$. (Compare this to the expectations discussed on page 274.)

CHAPTER

TEN

FIELDS OF A MOVING POINT CHARGE

The Lienard-Wiechert potentials · Point-charge radiation rate and pattern formulas · The nonrelativistic Larmor radiation formulas as an electric dipole approximation · Thomson scattering · The collimation of radiation by high source speeds · Continuous spectra · Synchrotron radiation and bremsstrahlung · The moving Coulomb field and its Lorentz contraction

All the sources of radiation treated so far were confined within some finite radius of a *static* center, relative to which the radiation was to be observed. The currents in such sources arise from charges maintained in bound states of motion, and their continued enclosure tends to develop periodicities that are analyzable into discrete frequencies [as in the simple examples leading to (8.67) or (8.77)]. It was therefore appropriate to consider the linearly independent, and discretely enumerable, monochromatic representatives of the fields. If the source had any *net* charge q, it formed a *static* monopole (8.15), and only motions in the charge *distribution* relative to the static center led to any radiation.

Suppose now that the source center is not static—one that it is appropriate to consider when the entire source moves bodily along some extended trajectory in an unbound state of motion. Now even if the distributional oscillations within the source are neglected, the motion of the net charge q may itself yield radiation. The body motion may have no periodicity within extended times of observation, and *immediate* classification by frequency is no longer particularly appropriate. The essentially new aspects of radiative processes thus suggested can already be exhibited by the motion of a simple point charge q, and this will now receive attention.

10.1 THE LIENARD-WIECHERT POTENTIALS

A simple point charge q following some given trajectory $\mathbf{r}_q(t_s)$ is now to be considered. It will generate a field that can be described by the retarded potentials (8.6), or (8.7), and (8.11), with $\rho(\mathbf{r}_s, t_s) = q\delta[\mathbf{r}_s - \mathbf{r}_q(t_s)]$ and $\mathbf{j} = \rho\mathbf{u}$, where $\mathbf{u}(t_s) \equiv \dot{\mathbf{r}}_q(t_s)$. Integrations can then be carried out by making use of certain formal properties of δ distributions (as in Exercise 10.1), but greater insight into the results can follow from an approach used long before δ functions were defined.

The result for the scalar potential at a time t is *not* merely $q/R(t - R/c)$, where the function $R(t_q)$ is the distance from charge to field point at the retarded

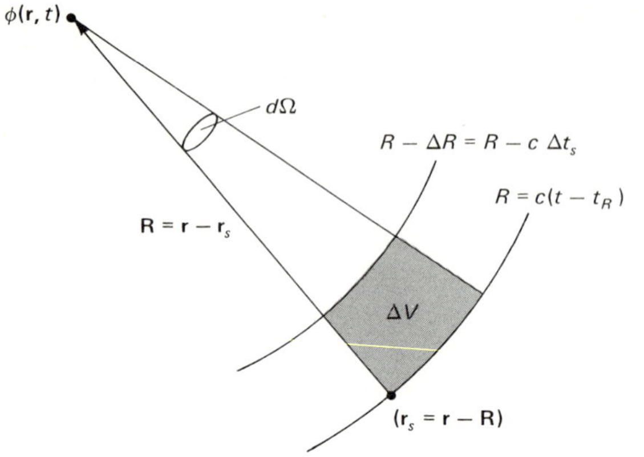

Figure 10.1

time $t_q = t - R/c$, because of a fact alluded to in the discussion (on page 210) of the retarded potential, namely that $\oint dV(\mathbf{r}_s)\rho(\mathbf{r}_s, t_s)$, with $t_s = t - R/c$ variable in the integration, is not generally equal to the charge present at any given moment. A distinction must generally be maintained between any contributing element $\rho(\mathbf{r}_s, t_s)\, dV$ and whatever part dq of the total charge supplies the values for that contribution.

Consider the fixed volume element ΔV indicated in Fig. 10.1, extending between distances $R - \Delta R$ and R from some fixed field point $\mathbf{r}$. For its contribution to the field at the given $\mathbf{r}$, at some fixed time t, $\rho(\mathbf{r}_s, t_s)$ must be evaluated with t_s ranging from $t_R \equiv t - R/c$ on the outer surface of the element to $t - (R - \Delta R)/c = t_R + \Delta R/c$ on the inner surface. It is as if a spherical signal-collecting wavefront† were converging on $\mathbf{r}$ with velocity c, timed to arrive at the moment t, and passing each source point $\mathbf{r}_s$ at just the moment t_s at which $\rho(\mathbf{r}_s, t_s)$ is to be evaluated. Such a picture makes it simple to see the connection between the contribution to $\rho(\mathbf{r}_s, t_s)\, \Delta V$ of the volume element ΔV and the charge Δq responsible for this contribution; it is simply the charge encountered by the collector wavefront as it sweeps across the volume element.

The charge encountered is not merely

$$\rho(\mathbf{r} - \mathbf{R}, t_R)\, \Delta V \equiv \Delta q(t_R)$$

which is the charge present in ΔV at the moment $t_R = t - R/c$ when the sweep begins. During the duration $\Delta R/c$ of the sweep, there may be an influx ($\gtrless 0$) of charge into ΔV through its *inner* surface, which will also be encountered in ΔV. Any influx through the *outer* surface, after the sweep begins, will not add to the

† Such a collector wave was alluded to in the discussion following (8.9) and (8.10); it may be regarded as a *resultant*, converging on the particular field point $\mathbf{r}$, of such signal waves (Huyghens' wavelets) from all the possible source points, as were introduced in the paragraph preceding (8.9).

charge encountered if it is presumed that the speeds u of any charge must be less† than c and hence no charge can catch up with the collector wavefront in its sweep. The influx through the inner surface‡ $(R - \Delta R)^2\, d\Omega$ can be evaluated with the help of the current density $\rho\mathbf{u}$ there by using its projection on the direction of $-\mathbf{R}$ (if $\mathbf{R} \equiv \mathbf{r} - \mathbf{r}_s$ as usual) normal to the surface. Thus $\rho\mathbf{u} \cdot (-\hat{\mathbf{R}}) \times (R - \Delta R)^2\, d\Omega$ is the net charge entering ΔV through its inner surface per unit time, and the expression must still be multiplied by the duration time $\Delta R/c$ of the sweep to get the addition ($\gtrless 0$) to the charge $\Delta q(t_R)$ above that will be encountered in the sweep over ΔV. As $\Delta R \to dR \to 0$, the volume element ΔV approaches $dV = R^2\, dR\, d\Omega$ and the product $(R - \Delta R)^2\, d\Omega(\Delta R/c)$ occurring in the charge supplementation approaches dV/c. Thus the connection to $dq(t_R) \equiv \rho(\mathbf{r} - \mathbf{R}, t_R)\, dV$ of the charge dq *encountered* is

$$dq = \left(1 - \frac{\mathbf{u}}{c} \cdot \hat{\mathbf{R}}\right) \rho(\mathbf{r} - \mathbf{R}, t_R)\, dV$$

This can be put into the integral for the potential, expressed in terms of $\mathbf{R}$ as the integration variable,

$$\phi(\mathbf{r}, t) = \oint \frac{dV(\mathbf{R})\rho(\mathbf{r} - \mathbf{R}, t_R)}{R}$$

to yield

$$\phi(\mathbf{r}, t) = \oint \frac{dq}{R(1 - \mathbf{u} \cdot \hat{\mathbf{R}}/c)}$$

where $\mathbf{R} \equiv \mathbf{r} - \mathbf{r}_s$ and $\mathbf{u}(\mathbf{r}_s, t_s)$ must still be evaluated at the times $t_s = (t_R \equiv)t - R/c$.

For the point charge q, this is all the charge ($\oint dq = q$) encountered, and at any moment its point position relative to the field point gives unique values to $t_s \equiv t_q = t - R/c$, $\mathbf{R} = \mathbf{r} - \mathbf{r}_q(t_q)$, and $\mathbf{u}(t_q)$. Then

$$\phi(\mathbf{r}, t) = \left(\frac{q}{R - \boldsymbol{\beta} \cdot \mathbf{R}}\right)_{t_q = t - R/c} \qquad \boldsymbol{\beta} \equiv \frac{\mathbf{u}}{c} \tag{10.1}$$

is the scalar potential, with $\boldsymbol{\beta}$ the velocity in units of c. The vector potential requires integration over $\mathbf{j}/c = \rho\mathbf{u}/c$ in place of ρ, and so

$$\mathbf{A}(\mathbf{r}, t) = (\boldsymbol{\beta}\phi)_{t_q = t - R/c} \tag{10.2}$$

merely. These are called the *Lienard-Wiechert potentials.*

† This is a significant restriction that is not as easily interpretable in the δ-function formalism, mentioned above as giving the same final results.

‡ Flux through the lateral sides can be ignored since it is contributions per unit solid angle of a complete spherical shell that are eventually wanted.

Requisite Derivative Operations

To check that the potentials found here satisfy the Lorentz condition (see Exercise 10.3) and to derive the electric and magnetic vectors from them, it will be necessary to apply field-gradient ∇ and time-derivative operations to $\mathbf{R} = \mathbf{r} - \mathbf{r}_q(t_q)$, to $t_q = t - R/c$, and to $\boldsymbol{\beta} = \mathbf{u}(t_q)/c$.

Notice first that all these quantities have variations with $\mathbf{r}$ and t arising from their dependence on $t_q = t - |\mathbf{r} - \mathbf{r}_q|/c$. Then, for example,

$$\frac{\partial R}{c\,\partial t} = \frac{1}{2R}\frac{\partial}{c\,\partial t}\mathbf{R}\cdot\mathbf{R} = \hat{\mathbf{R}}\cdot\frac{\partial \mathbf{R}}{c\,\partial t} = \hat{\mathbf{R}}\cdot\left[-\frac{\partial \mathbf{r}_q(t_q)}{c\,\partial t}\right] = -\hat{\mathbf{R}}\cdot\dot{\mathbf{r}}_q\frac{\partial t_q}{c\,\partial t}$$

Since $\dot{\mathbf{r}}_q(t_q) \equiv c\boldsymbol{\beta}(t_q)$, the particle's velocity, and $\partial t_q/\partial t = 1 - \partial R/(c\,\partial t)$,

$$\frac{\partial R}{c\,\partial t} = -\hat{\mathbf{R}}\cdot\boldsymbol{\beta}\left(1 - \frac{\partial R}{c\,\partial t}\right) = -\frac{\boldsymbol{\beta}\cdot\hat{\mathbf{R}}}{1 - \boldsymbol{\beta}\cdot\hat{\mathbf{R}}} \tag{10.3}$$

and

$$\frac{\partial t_q}{\partial t} = \frac{1}{1 - \boldsymbol{\beta}\cdot\hat{\mathbf{R}}} \tag{10.4}$$

The latter relation shows that (except for $\beta \ll 1$) there is a difference of scale between time intervals Δt_q of the source motion and the time intervals Δt during which the corresponding changes occur in the field at any given point. The relation found can be understood in the terms of Fig. 10.2. The length $R = c(t - t_q)$ is the distance traversed by the signal wave from the charge's position at the time t_q to a field point reached at the moment t. The signal wave that originates at the time $t_q + \Delta t_q$ reaches the same field point at a later time $t + \Delta t$ after traversing a radius $R - \mathbf{u}\cdot\hat{\mathbf{R}}\,\Delta t_q$. This equals $c(t + \Delta t) - c(t_q + \Delta t_q) = R + c(\Delta t - \Delta t_q)$, and so†

$$\Delta t = \Delta t_q\left(1 - \frac{\mathbf{u}}{c}\cdot\hat{\mathbf{R}}\right) \tag{10.5}$$

in agreement with $\Delta t_q = (\partial t_q/\partial t)_\mathbf{r}\,\Delta t$ of (10.4).

Relations (10.3) and (10.4) give a start toward establishing (Exercise 10.2) the table of derivatives

$$\begin{aligned} \nabla_i R_j &= \delta_{ij} + \frac{\hat{R}_i\beta_j}{1 - \boldsymbol{\beta}\cdot\hat{\mathbf{R}}} & \qquad \frac{\partial}{c\,\partial t}\mathbf{R} &= -\frac{\boldsymbol{\beta}}{1 - \boldsymbol{\beta}\cdot\hat{\mathbf{R}}} \\ \nabla_i \beta_j &= -\frac{\hat{R}_i\dot{\beta}_j/c}{1 - \boldsymbol{\beta}\cdot\hat{\mathbf{R}}} & \qquad \frac{\partial}{c\,\partial t}\boldsymbol{\beta} &= \frac{\dot{\boldsymbol{\beta}}/c}{1 - \boldsymbol{\beta}\cdot\hat{\mathbf{R}}} \end{aligned} \tag{10.6}$$

where $\nabla\mathbf{R}$ and $\nabla\boldsymbol{\beta}$ are dyadic tensor expressions from which operations like $\nabla\cdot\mathbf{R}$ or $\nabla\times\boldsymbol{\beta}$ are simple to deduce and $c\dot{\boldsymbol{\beta}} = \dot{\mathbf{u}}(t_q)$ is the intrinsic acceleration of the particle on its trajectory $\mathbf{r}_q(t_q)$.

† A correspondence to the Doppler effect will become clear in Sec. 11.1.

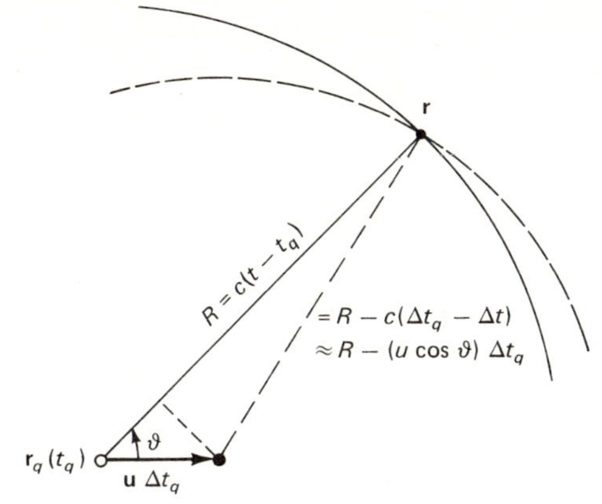

Figure 10.2

As an example, the gradient of the potential (10.1) is

$$\nabla\phi = -\frac{q}{(R - \boldsymbol{\beta}\cdot\mathbf{R})^2}\left[\nabla R - \sum_j (\beta_j \nabla R_j + R_j \nabla\beta_j)\right]$$

where $$\nabla_i R = \sum_j \hat{R}_j \nabla_i R_j = \sum_j \hat{R}_j\left(\delta_{ij} + \frac{\hat{R}_i\beta_j}{1 - \boldsymbol{\beta}\cdot\hat{\mathbf{R}}}\right) = \frac{\hat{R}_i}{1 - \boldsymbol{\beta}\cdot\hat{\mathbf{R}}} \tag{10.7}$$

and the result of such uses of the table (10.6) is

$$\nabla\phi = -\frac{q}{R^2(1 - \boldsymbol{\beta}\cdot\hat{\mathbf{R}})^3}\left[\hat{\mathbf{R}}\left(1 - \beta^2 + \dot{\boldsymbol{\beta}}\cdot\frac{\mathbf{R}}{c}\right) - \boldsymbol{\beta}(1 - \boldsymbol{\beta}\cdot\hat{\mathbf{R}})\right] \tag{10.8}$$

a step essential to obtaining the vector fields **E** and **B**.

The Velocity and Acceleration Fields

The redescription of the field $\phi = q/(R - \boldsymbol{\beta}\cdot\mathbf{R})$, $\mathbf{A} = \boldsymbol{\beta}\phi$ by $\mathbf{E} = -\nabla\phi - \partial\mathbf{A}/(c\,\partial t)$ and $\mathbf{B} = \nabla \times \mathbf{A}$ falls into two parts, as in

$$\mathbf{E} = \mathbf{E}_v + \mathbf{E}_a \qquad \text{and} \qquad \mathbf{B} = \mathbf{B}_v + \mathbf{B}_a$$

that are defined when the acceleration (or radiation) field $\mathbf{E}_a$, $\mathbf{B}_a$ is identified as a part that vanishes when there is no acceleration ($\dot{\boldsymbol{\beta}} = 0$). The remainders $\mathbf{E}_v$, $\mathbf{B}_v$ are called *velocity* or *attached fields*. Such operations as were discussed in the preceding subsection yield (Exercise 10.4)

$$\mathbf{E}_v(\mathbf{r}, t) = \frac{q}{R^2}\frac{1 - \beta^2}{(1 - \boldsymbol{\beta}\cdot\hat{\mathbf{R}})^3}(\hat{\mathbf{R}} - \boldsymbol{\beta}) \qquad \mathbf{B}_v = \boldsymbol{\beta}\times\mathbf{E}_v \tag{10.9}$$

$$\mathbf{E}_a(\mathbf{r}, t) = \frac{q}{cR}\frac{\hat{\mathbf{R}}\times[(\hat{\mathbf{R}} - \boldsymbol{\beta})\times\dot{\boldsymbol{\beta}}]}{(1 - \boldsymbol{\beta}\cdot\hat{\mathbf{R}})^3} \qquad \mathbf{B}_a = \hat{\mathbf{R}}\times\mathbf{E}_a \tag{10.10}$$

Clearly, if $\mathbf{B}_v = \boldsymbol{\beta} \times \mathbf{E}_v$, then $\mathbf{B}_v = \hat{\mathbf{R}} \times \mathbf{E}_v$ also. Thus the entire magnetic field that results from the operation $\mathbf{B} = \nabla \times \boldsymbol{\beta}\phi$ can be written as

$$\mathbf{B} = \hat{\mathbf{R}} \times \mathbf{E}_v + \hat{\mathbf{R}} \times \mathbf{E}_a = \hat{\mathbf{R}} \times \mathbf{E} \tag{10.11}$$

being transverse to the entire electric field and to the field-point direction $\hat{\mathbf{R}}$. Since $\mathbf{E}_a$ is also transverse to $\hat{\mathbf{R}}$, $|\mathbf{B}_a| = |\mathbf{E}_a|$ in magnitude, but $|\mathbf{B}_v|$ equals only the transverse component of $\mathbf{E}_v$.

It can be seen that the velocity fields (10.9) properly reduce to electro- and magneto*static* results as $\boldsymbol{\beta} \to 0$:

$$\mathbf{E}_v \to \hat{\mathbf{R}} \frac{q}{R^2} \qquad \text{and} \qquad \mathbf{B}_v \to \frac{q\mathbf{u} \times \hat{\mathbf{R}}}{cR^2} \tag{10.12}$$

The result for the magnetic field was anticipated in (4.26) on the basis of the magnetostatic Biot-Savart law, applicable only as a quasi-static approximation. Now it is properly justified for the large range of velocities $u \ll c$, as a result of the full, nonstatic theory. The complete results for the velocity field are left to later discussion (Sec. 10.6).

Another characteristic of the velocity field has interest at this point. The magnitudes of $\mathbf{E}_v$ and $\mathbf{B}_v$ tend to vanish for large distances R from charge to field point, in proportion to $1/R^2$. This means that the corresponding energy flux density $\mathbf{N}_v = c\mathbf{E}_v \times \mathbf{B}_v/4\pi$ tends to vanish as $1/R^4$ and, even after multiplication by the surface area $4\pi R^2$ of any sphere centered on the charge, the result for the outward energy transmission vanishes as $1/R^2 \to 0$. Thus none of this energy is ever dissociated from the vicinity of the source, as required for permanent radiation. This means that while the charge moves *uniformly*, so that $\dot{\boldsymbol{\beta}} = 0$ and the velocity field is the entire one, no radiation is freed. A charge must be *accelerated* ($\gtrless 0$) in order to radiate.

The attached velocity field also plays no role in determining the radiated energy when the charge is accelerated. The complete energy flux density $\mathbf{N} = c(\mathbf{E}_v + \mathbf{E}_a) \times (\mathbf{B}_v + \mathbf{B}_a)/4\pi$ contains interferences proportional to $\mathbf{E}_v \times \mathbf{B}_a$ and $\mathbf{E}_a \times \mathbf{B}_v$, but these interferences vanish as $1/R^3 \to 0$, and so the corresponding energy flows are also restricted to the vicinity of the charge. Radiation of energy by a point charge can be evaluated while ignoring the velocity field altogether,† as done in the next sections.

10.2 RADIATION BY A POINT CHARGE

The acceleration fields $\mathbf{E}_a$ and $\mathbf{B}_a = \hat{\mathbf{R}} \times \mathbf{E}_a$ are each transverse to $\hat{\mathbf{R}}$ and to each other and have equal magnitudes, as is characteristic of plane waves in the direction $\hat{\mathbf{R}}$. The corresponding energy-flux density is therefore

$$\mathbf{N}_a(\mathbf{r}, t) = \frac{c}{4\pi} \mathbf{E}_a \times \mathbf{B}_a = \hat{\mathbf{R}} \frac{c\,|\mathbf{E}_a(\mathbf{r}, t)|^2}{4\pi} = \hat{\mathbf{R}} \frac{c\mathbf{B}_a^2}{4\pi} \tag{10.13a}$$

† Just as the near-zone fields of the sources treated in Chaps. 8 and 9 could be ignored, and only the wave-zone fields considered, for the radiation rates (8.31), (8.44), etc.

having the direction of $\mathbf{R} = \mathbf{r} - \mathbf{r}_q(t_q)$ and a magnitude

$$N_a = \hat{\mathbf{R}} \cdot \mathbf{N}_a = \frac{q^2}{4\pi c R^2} \frac{|\hat{\mathbf{R}} \times [(\hat{\mathbf{R}} - \boldsymbol{\beta}) \times \dot{\boldsymbol{\beta}}]|^2}{(1 - \boldsymbol{\beta} \cdot \hat{\mathbf{R}})^6} \tag{10.13b}$$

according to the field expressions (10.10). When the value at a moment t, at some point $\mathbf{r}$ out in the field, is desired, $\hat{\mathbf{R}}$ is directed from the position $\mathbf{r}_q(t_q)$ of the charge at the earlier time $t_q = t - |\mathbf{r} - \mathbf{r}_q|/c$ and the values of the velocity $c\boldsymbol{\beta}(t_q)$ and acceleration $c\dot{\boldsymbol{\beta}}(t_q)$ at that earlier time help determine the result.

For determining the energy per unit time $d\mathscr{P}$ emanating into a solid-angle element $d\Omega$ *at the charge*, N_a can be multiplied by a factor $R^2\, d\Omega$ with any radius, since the product becomes independent of the distances R to which that energy may flow; it becomes dissociated from the charge and constitutes radiation formed at the instant t_q of the evaluation. The complete result for $d\mathscr{P}(t_q)$ must be such that multiplying it by a time interval $\Delta t_q \to 0$ yields the energy lost to radiation while the charge undergoes a displacement by $\Delta \mathbf{r}_q(t_q) = \mathbf{u}\, \Delta t_q$. During this radiation, the energy measured by $N_a(\mathbf{r}, t)$ takes a time $\Delta t \neq \Delta t_q$ to escape through a sphere of radius $R = |\mathbf{r} - \mathbf{r}_q(t_q)|$. [Since $R = c(t - t_q)$, the time at which the escape takes place through a sphere with $R \to \infty$ is a $t \to \infty$!] The conclusion is that $d\mathscr{P}(t_q) = R^2\, d\Omega N_a(\Delta t/\Delta t_q)$, proportional to the ratio of field to source time intervals as given by (10.5). Thus the power loss per unit solid angle is to be calculated as

$$\frac{d\mathscr{P}(t_q)}{d\Omega} = \frac{q^2}{4\pi c} \frac{|\hat{\mathbf{R}} \times [(\hat{\mathbf{R}} - \boldsymbol{\beta}) \times \dot{\boldsymbol{\beta}}]|^2}{(1 - \boldsymbol{\beta} \cdot \hat{\mathbf{R}})^5} \tag{10.14}$$

for the instant t_q of the charge's motion.

The cogency of that result can perhaps be better appreciated if the distribution of the energy

$$w_a(\mathbf{r}, t) = \frac{\mathsf{E}_a^2 + \mathsf{B}_a^2}{8\pi} = \frac{\mathsf{E}_a^2}{4\pi} = \frac{N_a}{c}$$

in the field at some given moment of observation t is considered. Any that was radiated up to a phase at t_q of the point-charge motion is distributed outside a sphere of radius $R = c(t - t_q)$, indicated in Fig. 10.3, which differs from Fig. 10.2 by referring to a fixed moment t of the field distribution rather than to different times at a given field point. Any energy radiated after the moment $t_q + \Delta t_q$ is confined to the smaller of the two spheres indicated, centered on the displaced position $\mathbf{r}_q(t_q) + \mathbf{u}\, \Delta t_q$ and having the radius $c[t - (t_q + \Delta t_q)]$. Then the energy radiated *during* the time interval Δt_q lies within the shell of variable thickness $\Delta R = c(1 - \boldsymbol{\beta} \cdot \hat{\mathbf{R}})\, \Delta t_q$. As $\Delta t_q \to dt_q \to 0$, the part within the solid angle $d\Omega$ amounts to $w_a R^2\, dR\, d\Omega = (N_a/c)R^2\, d\Omega c(1 - \boldsymbol{\beta} \cdot \hat{\mathbf{R}})\, dt_q$, and for a unit of dt_q this is just $d\mathscr{P}(t_q)$, as calculated above.

The energy $\mathscr{P}(t_q)$ being radiated into *all* directions can be obtained (as in Exercise 10.5) by integrating (10.14) over the complete solid angle $\oint d\Omega = 4\pi$.

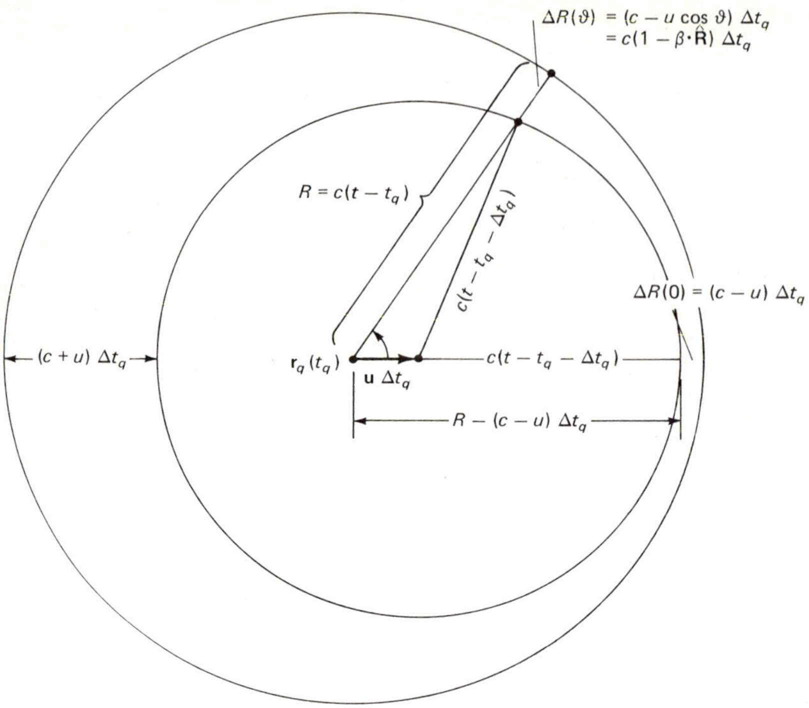

Figure 10.3

However, much more powerful methods will be exhibited, in Exercise 11.9 and Sec. 14.2, and so the result will only be quoted for use here:

$$\mathscr{P}(t_q) = \frac{2}{3}\frac{q^2}{c}\frac{\dot{\beta}^2 - (\boldsymbol{\beta} \times \dot{\boldsymbol{\beta}})^2}{(1 - \beta^2)^3} \tag{10.15}$$

This positive definite emission cannot vanish for any $c\beta = u < c$ whenever $\dot{\boldsymbol{\beta}}(t_q) \neq 0$.

10.3 LOW-VELOCITY RADIATION

The form to which (10.15) reduces for accelerations from rest ($\beta = 0$) and continues as a very good approximation throughout the large range of velocities $u = c\beta \ll c$, is much simpler to obtain.

Start with the forms taken by the radiation field (10.10) for $\beta \to 0$,

$$\mathbf{E}_a \to \frac{q}{c^2 R}\hat{\mathbf{R}} \times (\hat{\mathbf{R}} \times \dot{\mathbf{u}}) \qquad \mathbf{B}_a \to \frac{q}{c^2 R}\dot{\mathbf{u}} \times \hat{\mathbf{R}} \tag{10.16}$$

These have interest in themselves, since they are always valid relative to any instantaneous rest frame of the point charge at a moment when it is undergoing an acceleration $\dot{\mathbf{u}}$ relative to that frame. The radiation pattern follows immediately, as $R^2\hat{\mathbf{R}} \cdot \mathbf{N}_a$ of (10.13),

$$\frac{d\mathscr{P}}{d\Omega} = \frac{q^2}{4\pi c^3} |\hat{\mathbf{R}} \times \dot{\mathbf{u}}|^2 = \frac{q^2\dot{u}^2}{4\pi c^3} \sin^2 \vartheta \qquad (10.17)$$

agreeing with (10.14) for $\beta \to 0$. Here, ϑ is the angle between the acceleration direction of $\dot{\mathbf{u}}$ and the direction $\hat{\mathbf{R}}$ into the solid-angle element $d\Omega(\hat{\mathbf{R}})$. The result shows that the radiation intensity is greatest sideways to the acceleration ($\vartheta = \frac{1}{2}\pi$) and has nodes on its line.

The occurrence of just the one angle in (10.17) makes its integration over all directions, $\oint d\Omega = 2\pi \int_{-1}^{1} d(\cos \vartheta)$, elementary, with the result

$$\mathscr{P} = \frac{2}{3}\frac{q^2\dot{u}^2}{c^3} \qquad u \ll c \qquad (10.18)$$

agreeing with (10.15) for $\beta \to 0$. This is the famous Larmor radiation formula, by far the most used because of the prevalence of situations in which $u \ll c$ (true even for most of the electrons in atoms). Moreover, it is exact relative to every instantaneous rest frame of the charge, no matter how great the speeds relative to frames of observation. [See (14.14).]

The Larmor Formulas as Dipole Approximations

In the special case of the monochromatic radiation from a linearly oscillating point charge, with $\mathbf{r}_q(t_q) = \mathbf{a} \cos \omega_o t_q$, the acceleration is $\dot{\mathbf{u}} = -\omega_o^2 \mathbf{r}_q$, and the Larmor formulas yield

$$\mathscr{P} = \frac{2}{3}\frac{q^2a^2\omega_o^4}{c^3} \cos^2 \omega_o t_q \qquad \left\langle \frac{d\mathscr{P}}{d\Omega} \right\rangle = \frac{q^2}{8\pi c^3} a^2\omega_o^4 \sin^2 \vartheta \qquad (10.19)$$

the latter being time-averaged, with $\langle \cos^2 \omega_o t_q \rangle = \frac{1}{2}$. The time-averaged results coincide with those (8.46) obtained in the dipole approximation relative to the fixed center of oscillation.

The last fact has implications for the so-called *relativistic* corrections to the Larmor formulas arising when the exact result (10.15) is used to take into approximate account the finite velocity $\mathbf{u} = -\omega_o \mathbf{a} \sin \omega_o t_q$ reached in the course of the oscillations. Since this velocity is parallel to the acceleration here, the first-order correction arises entirely from the factor $(1 - \beta^2)^{-3} \approx 1 + 3\beta^2$ in (10.15) when $u \ll c$ continues to hold. The corresponding supplementation $\Delta\mathscr{P}$ of the Larmor rate (10.18) is

$$\Delta\mathscr{P} = \frac{2}{3}\frac{q^2}{c^3} \dot{u}^2 \left(3\frac{u^2}{c^2}\right) = \frac{2q^2a^2\omega_o^4}{c^3} \cos^2 \omega_o t_q \left(\frac{\omega_o a}{c}\right)^2 \sin^2 \omega_o t_q \qquad (10.20)$$

This shows that relativistic correction through expansion in powers of $\beta = u/c \ll 1$ is equivalent to expansion in powers of $\beta_o = \omega_o a/c = k_o a = a/\lambdabar_o \ll 1$, as done to obtain the successive multipole orders of radiation in Sec. 8.3 and Chap. 9.

The time average of the result (10.20), $\langle \Delta\mathscr{P} \rangle = \frac{1}{4}q^2a^4\omega_o^6/c^5$, can be compared to the quadrupole correction, $\mathscr{P}_Q = \frac{4}{15}q^2a^4\omega_o^6/c^5$ (Exercise 8.3). The relativistic correction is different by the factor $\frac{15}{16}$ only because it also incorporates a small diminution of the dipole moment traceable to the modification (9.116) (Exercise 10.7).

The Larmor formulas quite generally yield the electric dipole approximation for any given frequency. Conversely, the dipole approximations correspond to treating the velocities of net charge motions as small: $u \ll c$. Thus, the treatment of the radiation by a circling point charge† in connection with Fig. 8.3, on the basis of source radius a small in comparison to the wavelength $\lambda\!\!\!^-_o = c/\omega_o$, was equivalent to a nonrelativistic approximation, on the basis of a velocity in orbit, $u = \omega_o a$, small in comparison to c: $a/\lambda\!\!\!^-_o \equiv \omega_o a/c \ll 1$. Adding the quadrupole radiation evaluated in (8.79) corresponded to a first relativistic correction of the Larmor ≡ dipole approximation. Results valid for a point charge orbiting with high velocities, $\omega_o a \to c$, will be obtained as synchrotron radiation in Sec. 10.5.

The Thomson Scattering of Radiation

The Larmor approximation is used in obtaining what is called the *Thomson formula* for the elastic scattering of radiation by a free point charge. Any electromagnetic scattering or reflection process can be understood as the result of emissions induced by the incidence of some initial radiation. Such combinations of excitations and reemissions are alluded to in the discussion of Fig. F.2, to help understand reflections and transmissions in macroscopically described bulk matter. Here, a microscopic basis for the behavior will be discussed, namely reflection from a single electron in matter. The electrons are the most important charges to consider; they are the most numerous and lightest constituents, most readily accelerated (by any incident electromagnetic fields) and induced to radiate. Thomson scattering by electrons treated as free is one of the most important processes attending the passage of light or other electromagnetic radiations through matter.

Consider a monochromatic component [like the one in (7.14)]

$$\mathbf{E}(\mathbf{r}, t) = \mathbf{a}e^{i(\mathbf{k}_o \cdot \mathbf{r} - \omega_o t)} \qquad \mathbf{B} = \hat{\mathbf{k}}_o \times \mathbf{E} \tag{10.21}$$

of incident radiation. This will accelerate an electron it meets in accordance with the equation of motion (5.1), but the magnetic part of the Lorentz force $q(\mathbf{E} + \boldsymbol{\beta} \times \mathbf{B})$ in it should be neglected since $\beta \ll 1$ for electrons in matter and the interest is in the instantaneous reradiations, occurring before any appreciable velocity can develop from the acceleration. Thus, the field representative (10.21) will generate an acceleration representative

$$\dot{\mathbf{u}}(t_q) = \frac{q}{m}\mathbf{E}(\mathbf{r}_q, t_q) = \frac{q\mathbf{a}}{m} e^{i(\mathbf{k}_o \cdot \mathbf{r}_q - \omega_o t_q)} \tag{10.22}$$

† Significant here is the fact that, as sources, orbiting charges are equivalent to superpositions of linear oscillations orthogonal to each other.

and this yields a radiation field (10.16) with

$$\mathbf{E}_a(\mathbf{r}, t) \approx \frac{q^2}{mc^2 R}[\hat{\mathbf{k}} \times (\hat{\mathbf{k}} \times \mathbf{a})]e^{i(\mathbf{k}_o \cdot \mathbf{r}_q - \omega_o t_q)} \tag{10.23}$$

The change of notation $\hat{\mathbf{R}} \to \hat{\mathbf{k}}$, for the direction from radiating charge to a field point, has now become appropriate. With $\beta \ll 1$, the two time scales of (10.5) become linearly proportional, and $\mathbf{E}_a(\mathbf{r}, t) \sim \exp(-i\omega_o t)$ in (10.23). Thus the reradiated frequency is the same, ω_o, as in the incident field $\mathbf{E}(\mathbf{r}_q, t_q)$ of (10.21) that has excited it, a result typical of ($\beta \ll 1$, nonrelativistic) *elastic* scattering. This has entailed neglecting *net* recoils of the comparatively massive electrons from the radiation pressure.

As a representative of a monochromatic field, (10.23) yields a time-averaged energy efflux $\langle N_a \rangle = c|\mathbf{E}_a|^2/8\pi$ and the radiation rate per unit solid angle $R^2\langle N_a \rangle$, equal to

$$\left\langle \frac{d\mathscr{P}}{d\Omega} \right\rangle = \frac{c}{8\pi}\left(\frac{e^2}{mc^2}\right)^2 |\hat{\mathbf{k}} \times \mathbf{a}|^2 \tag{10.24}$$

from an electron $q = -e$. This is just the Larmor formula (10.17) after substitution of the time average $\langle \dot{u}^2 \rangle = \frac{1}{2}\dot{\mathbf{u}}^* \cdot \dot{\mathbf{u}}$ obtained from the representative (10.22). The monochromatic rate formula (8.31) likewise yields the same result when $\mathbf{j}(\mathbf{r}_s)e^{-i\omega_o t_q} = -e\delta(\mathbf{r}_s - \mathbf{r}_q)\dot{\mathbf{u}}/(-i\omega_o)$ is put into it.

Measured rates of processes initiated by the incidence of any form of energy on a particle are usually expressed as cross sections $d\sigma$ per target particle. Here,†

$$d\sigma \equiv \frac{\text{energy radiated into } d\Omega/\text{unit time}}{\text{incident energy/unit area and unit time}} = d\Omega \frac{\langle d\mathscr{P}/d\Omega \rangle}{c|a|^2/8\pi} \tag{10.25}$$

since the incident field represented by (10.21) brings in the time-averaged energy-flux density $c|a|^2/8\pi$. Then, in terms of a unit vector $\boldsymbol{\epsilon} \equiv \mathbf{a}/|\mathbf{a}|$ giving the polarization direction of the incident field, (10.24) yields the result

$$\frac{d\sigma}{d\Omega} = \left(\frac{e^2}{mc^2}\right)^2 |\hat{\mathbf{k}} \times \boldsymbol{\epsilon}|^2 \tag{10.26}$$

for the scattering of polarized radiation.

Observing the angular distribution implicit in (10.26) must be done relative to a given incidence direction $\hat{\mathbf{k}}_o$. Suppose first that the incident radiation is *linearly* polarized, so that $\boldsymbol{\epsilon}$ is a *real* vector with a steady direction transverse to $\hat{\mathbf{k}}_o$. Then $\hat{\mathbf{k}}_o$, $\boldsymbol{\epsilon}$, and $\hat{\mathbf{k}}_o \times \boldsymbol{\epsilon}$ form an orthogonal frame on which the scattering direction $\hat{\mathbf{k}}$ can be decomposed, as indicated in Fig. 10.4. The azimuthal angle is called $\varphi - \psi$, so that variations in the orientation of the plane of polarization, relative to the viewing direction $\hat{\mathbf{k}}(\vartheta, \varphi)$, can later be allowed simply by varying ψ. In terms of the angles so defined, the scattering pattern (10.26) becomes

$$\left(\frac{d\sigma}{d\Omega}\right)_\psi = \left(\frac{e^2}{mc^2}\right)^2 [1 - \sin^2\vartheta \cos^2(\varphi - \psi)] \tag{10.27}$$

† Compare the wave-scattering cross section of Ref. 17, eq. (15.55).

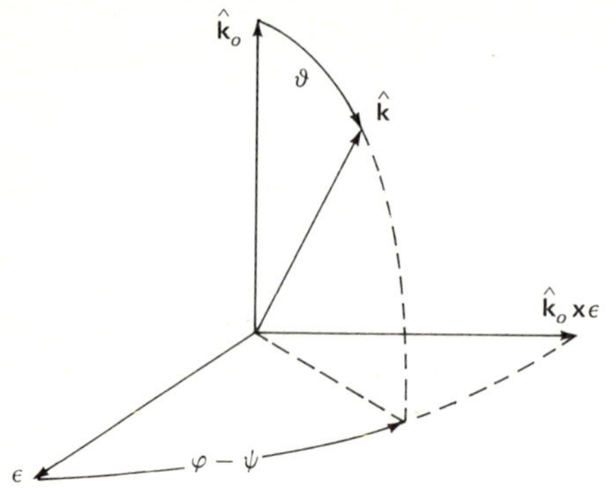

Figure 10.4

In the plane $\varphi - \psi = \frac{1}{2}\pi$, perpendicular to the direction $\boldsymbol{\epsilon} \equiv \mathbf{a}/|\mathbf{a}|$ of the incident and induced oscillations (10.21) and (10.22), the reradiation is isotropic, but within the plane of incident polarization $\varphi = \psi$, the distribution is proportional to $\cos^2 \vartheta$, with no intensity along the direction of the oscillations, as averred in the discussion of Fig. F.2 (where it is the refracted beam that is undergoing the scattering†).

Ordinary incident light is unpolarized, with the orientations of the polarization plane as described by ψ distributed at random relative to any given viewing direction (ϑ, φ), so that $\cos^2 (\varphi - \psi)$ averages to $\frac{1}{2}$. Then

$$\frac{d\sigma}{d\Omega} = \left(\frac{e^2}{mc^2}\right)^2 \tfrac{1}{2}(1 + \cos^2 \vartheta) \tag{10.28}$$

forms a pattern like that of (8.50), characteristic of the radiation from a circling charge, as might be expected from an isotropic distribution of the oscillations (10.22) in the plane normal to the incidence direction. Integration over all directions gives the total *Thomson cross section*

$$\sigma = \frac{8\pi}{3}\left(\frac{e^2}{mc^2}\right)^2 \tag{10.29}$$

If this is interpreted as the area‡ $4\pi r_e^2$ of a reflecting sphere that is bathed by long waves of light and scatters them out of the incident beam, it gives $r_e = (\frac{2}{3})^{1/2}(e^2/mc^2)$ for the radius of the electron as seen in that light.

† This refers to the absence of reflections from a medium at the Brewster angle $R(i_B) = 0$. Of course, there are reflections into other directions, whatever the polarization, from each individual electron. The reradiated fields from the many electrons interfere with each other, with resultant fields as given by the macroscopic treatment in Chap. F. This accounts for the impossibility of resultant reflections except into an angle equal to the angle of incidence.

‡ See, for example, the long-wave scattering cross section [Ref. 17, eq. (15.64)]. See page 430 for limitations on the interpretation here.

When the energy intercepted by any one electron is large enough, the immediate acceleration (10.22) may be great enough for the above restrictions to $\beta \ll 1$ to lose validity. A treatment valid for relativistic velocities $u \to c$ is then called for, but there is limited profit in such a classical consideration because, actually, quantum-mechanical effects (on the *electron's* motion!) intervene seriously. The latter become observable when an energy unit $\hbar\omega$ is intercepted, and this approaches mc^2 in magnitude. Thus, the Thomson results are valid only if $\hbar\omega \ll mc^2$. When $\hbar\omega \to mc^2$, as happens for x-rays scattered by an electron, the process becomes what is called a Compton effect, to be paid some attention in Sec. 11.4. The range of validity of the classical Thomson cross section is greater for scattering from protons, which have $M \approx 1836$ m, and so the approach $\hbar\omega \to Mc^2$ takes very energetic γ-rays. [On the other hand, the Thomson cross section (10.29) is reduced by the factor $(m/M)^2$.]

The Thomson cross section is derived on the basis that the target electron is free, so that no forces aside from those presented by the incident fields need be considered for the acceleration (10.22). Plainly, taking into account such forces as bind a target electron to some atom will allow work to be done on the atomic structure, exciting it to subsequent emissions of frequencies characteristic of an excited atom; inelastic scattering thus ensues. However, when the incident frequency is not in resonance with any of the discrete frequencies of the bound system, the elastic Thomson scattering remains separately discernible. There are still interferences between simultaneous radiations from several atomic electrons that have effect on some observations when the wavelength is long enough to embrace positions of more than one electron (Exercise 10.8). However, the Thomson formula has described the most basic elements of the processes.

10.4 RADIATION AT HIGH SPEEDS

Radiation rates that may be moderate when a low-velocity particle is accelerated become enormous when the same acceleration is given to the charge while it is traveling at a high speed $u \to c$. It is customary to express such so-called *extreme relativistic* results in terms of the quantity

$$\gamma \equiv \frac{1}{\sqrt{1-\beta^2}} \qquad (\to \infty \text{ as } u \to c) \tag{10.30}$$

because this will turn out to be proportional to the energy of the particle (Chap. 11). Then, for an instant at which the direction of the acceleration makes an angle χ with the velocity vector $\mathbf{u} = c\boldsymbol{\beta}$, the rate (10.15) can be expressed as

$$\mathscr{P}(t_q) = \frac{2}{3}\frac{q^2}{c^3}\dot{u}^2\gamma^4(\gamma^2\cos^2\chi + \sin^2\chi) \tag{10.31}$$

It tends to an infinite rate in proportion to $\gamma^6 \to \infty$ for acceleration ($\gtrless 0$) parallel to the velocity and in proportion to $\gamma^4 \to \infty$ for $\chi = \frac{1}{2}\pi$. This makes it impossible to accelerate any charge to the speed of light; it also makes it costly in energy just to try bending a charge from a straight-line path while it has a speed close to the velocity of light.

The angular distribution of the radiation as given by (10.14) can become quite complicated at any instant since it depends on the directions of both the velocity and the acceleration. It is therefore most useful to consider separately the patterns to be expected for acceleration components parallel and perpendicular to the velocity.

$\dot{\boldsymbol{\beta}}$ Parallel to $\pm\boldsymbol{\beta}$

The acceleration of charged particles, like electrons or protons, parallel to their directions of motion is characteristic of devices called *linear accelerators.*

With $\boldsymbol{\beta} \times \dot{\boldsymbol{\beta}} = 0$, the angular distribution (10.14) becomes

$$\frac{d\mathscr{P}}{d\Omega} = \frac{q^2\dot{u}^2}{4\pi c^3}\frac{\sin^2\vartheta}{(1-\beta\cos\vartheta)^5} \tag{10.32}$$

in terms of the angle ϑ from the velocity and acceleration direction to the direction $\hat{\mathbf{R}}$ of radiation. As might be expected, the pattern has nodes fore and aft ($\vartheta = 0$ and π). An angle ϑ_m at which the intensity reaches a maximum can be found by setting equal to zero the derivative $d/d(\cos\vartheta)$ of the expression (10.32); the result is quadratic in $\cos\vartheta_m$ but has only one root for which ϑ_m is real ($|\cos\vartheta_m| \le 1$):

$$\cos\vartheta_m = \frac{1}{3\beta}(\sqrt{1+15\beta^2}-1) = \frac{4(1-15/16\gamma^2)^{1/2}-1}{3(1-1/\gamma^2)^{1/2}} \tag{10.33}$$

This is consistent with the expectation from the Larmor formula in (10.17) since $\cos\vartheta_m \to 0$ for $\beta \to 0$, and the radiation maximum is sideways ($\vartheta_m \to \frac{1}{2}\pi$) to the acceleration direction. For any $\beta \ne 0$, the intensity peak is tipped forward ($\cos\vartheta_m > 0$ and $\vartheta_m < \frac{1}{2}\pi$). For extreme relativistic velocities $\beta \to 1$, expression (10.33) can be expanded in powers of $1/\gamma^2 \ll 1$ to give $1 - 1/8\gamma^2$; then $\cos\vartheta_m$ is close to unity and can be approximated as $1 - \frac{1}{2}\vartheta_m^2$. Thus

$$\vartheta_m \approx \frac{1}{2\gamma} \ll 1 \tag{10.34}$$

at the high velocities.

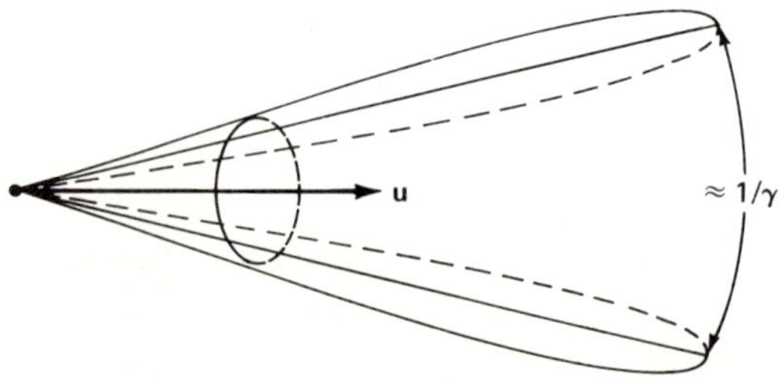

Figure 10.5

The pattern for $\beta = 0.9$ is shown in Fig. 10.5. Here $\gamma \approx 2.29$ only, yet $\vartheta_m \approx 0.224$, as against $1/2\gamma \approx 0.218$. The peak intensity is approximately 1700 times as great as the intensity sideways ($\vartheta = \frac{1}{2}\pi$), which equals the *peak* Larmor ($\beta \to 0$) intensity for the same acceleration. Most of the radiation forms a cone of opening angle $1/\gamma$ about the forward direction, along which a hollow in the cone of radiation remains.

$\dot{\boldsymbol{\beta}}$ Perpendicular to $\boldsymbol{\beta}$

Accelerations of a particle perpendicular to its velocity are characteristic of deflections by an imposed magnetic field, as seen in Sec. 5.1. These occur, for example, in the synchrotron devices for accelerating particles, and emissions arising from deflections of high velocities by any magnetic field in any circumstances are customarily referred to as synchrotron radiation. (Sometimes cyclotron radiation is also spoken of, referring to emissions arising from the spiraling of $\beta \ll 1$ particles about magnetic field lines, treated like the radiation from a circling charge in connection with Fig. 8.3. See Exercise 8.2.)

With $\dot{\boldsymbol{\beta}} \cdot \boldsymbol{\beta} = 0$ here, the square of the vector

$$\hat{\mathbf{R}} \times [(\hat{\mathbf{R}} - \boldsymbol{\beta}) \times \dot{\boldsymbol{\beta}}] = (\hat{\mathbf{R}} - \boldsymbol{\beta})(\dot{\boldsymbol{\beta}} \cdot \hat{\mathbf{R}}) - \dot{\boldsymbol{\beta}}(1 - \boldsymbol{\beta} \cdot \hat{\mathbf{R}})$$

needed for the radiation pattern (10.14), is

$$\dot{\beta}^2(1 - \boldsymbol{\beta} \cdot \hat{\mathbf{R}})^2 - (1 - \beta^2)(\dot{\boldsymbol{\beta}} \cdot \hat{\mathbf{R}})^2$$

Then, in terms of the angles ϑ, φ indicated in Fig. 10.6, having the instantaneous velocity as polar axis and with φ measured from the orbital plane (of $\boldsymbol{\beta}$ and $\dot{\boldsymbol{\beta}}$)

$$\frac{d\mathscr{P}}{d\Omega} = \frac{q^2\dot{u}^2}{4\pi c^3}\frac{1}{(1 - \beta \cos \vartheta)^3}\left[1 - \frac{1 - \beta^2}{(1 - \beta \cos \vartheta)^2} \sin^2 \vartheta \cos^2 \varphi\right] \tag{10.35}$$

This properly reduces to the Larmor result (10.17) for $\beta = 0$, since the angle used in that expression has a cosine equal to $\sin \vartheta \cos \varphi$ in the present terms. For $\beta \neq 0$, the radiation does not actually vanish along the line of acceleration ($\vartheta = \frac{1}{2}\pi$, $\varphi = 0$ or π), as might have been expected; this is fundamentally because that line does not have a steady direction but one that varies within $d\Omega(\hat{\mathbf{R}})$. The intensity in question does remain small, having the ratio $\beta^2(1 - \beta)^3$ ($\to 0.81 \times 10^{-3}$ for $\beta = 0.9$) to the peak intensity.

Also indicated in Fig. 10.6 is the intensity pattern in the half plane $\varphi = \pi$ for the case $\beta = 0.9$. The complete pattern in the orbital plane (which includes the half plane $\varphi = 0$) is symmetrical with respect to $\boldsymbol{\beta}$. The orbital plane was chosen for the display because it is the only one (through $\boldsymbol{\beta}$) in which nodes occur, at $\vartheta_o = \cos^{-1} \beta$ ($\approx 1/\gamma$ as $\beta \to 1$). Most of the radiation is confined within this angle of the velocity direction, where the intensity peaks. There does exist a subsidiary maximum [at $\cos \vartheta = (5\beta^2 - 2)/3\beta$], but it had to be multiplied by the factor 10 to be as visible relative to the main peak as indicated (for the case $\beta = 0.9$).

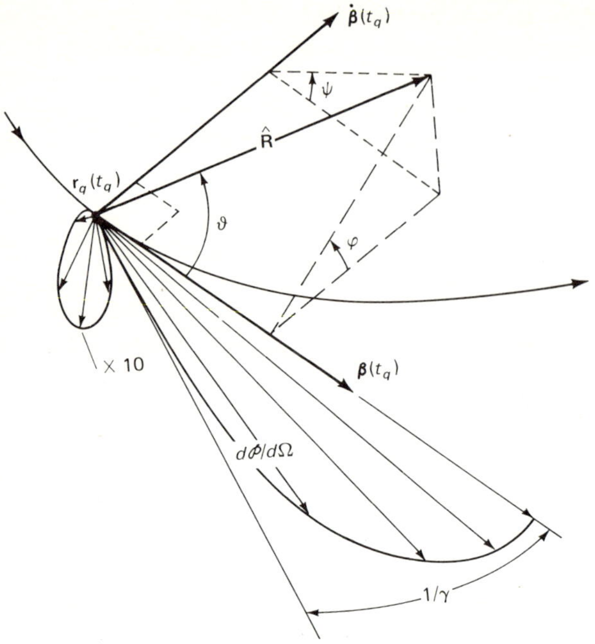

Figure 10.6

Although the patterns in other planes through the velocity vector ($\varphi \neq 0, \pi$) are nodeless, they are all sharply peaked along the velocity direction for $\beta \to 1$. It is simple to work out (Exercise 10.9) that the mean emission angle, obtained by averaging the intensity over all directions, is $\langle \vartheta^2 \rangle^{1/2} \approx 1/\gamma \ll 1$ for $\beta \to 1$.

The studies in this section have shown that radiation at very high velocities is mostly confined within a cone about the forward direction having an opening angle ($\sim 2/\gamma \ll 1$) so small that the radiation is almost raylike. Moreover, the radiation can reach an enormous intensity compared with that generated by the same acceleration at nonrelativistic velocities ($\beta \ll 1$).

10.5 CONTINUOUS SPECTRA

Among the characterizing observables in any radiation is its *spectrum*, i.e., its intensity as a function of frequency per unit range of frequencies. A visible portion of any such spectrum gives the fraction of the radiated energy that is invested in each color. A means of calculating the radiation with any one, discrete frequency has already been presented in the result (8.31) but in a way best suited to confined sources emitting *discrete* spectra of frequencies. The more essentially aperiodic charge motions of interest now can also have their fields decomposed into frequencies, in correspondence with the spectral resolutions afforded by frequency-sensitive detectors, but these decompositions can be expected to furnish *continua* of emitted frequencies, and adaptations to such continua are now desirable.

The radiated energy is expressible in terms of just the electric vector field, as in the flux-density expressions (10.13), and so consideration of the Fourier analysis [see (6.50)]

$$\mathbf{E}(\mathbf{r}, t) = \int_{-\infty}^{\infty} d\omega \mathbf{E}_{\omega}(\mathbf{r}) e^{-i\omega t}$$
$$\mathbf{E}_{\omega}(\mathbf{r}) = \int_{-\infty}^{\infty} \frac{dt}{2\pi} \mathbf{E}(\mathbf{r}, t) e^{+i\omega t} = \mathbf{E}^{*}_{-\omega} \tag{10.36}$$

will suffice. Note that it is the time scale of t, in the field at points of observation, relative to which the periods $2\pi/|\omega|$ and frequencies ω are defined, rather than the scale of the emission times t_q. If $\mathbf{E}(\mathbf{r}, t)$ is the field in a radiation zone, the flux density there (10.13a) can be expressed as

$$N(\mathbf{r}, t) = \frac{c}{4\pi} \mathsf{E}^2 = \frac{c}{4\pi} \iint_{-\infty}^{\infty} d\omega\, d\omega' \mathbf{E}^{*}_{\omega} \cdot \mathbf{E}_{\omega'} e^{i(\omega - \omega')t}$$

The most sharply defined contribution to each frequency interval $d\omega \to 0$ is obtainable for the energy radiated over all time, since $\int_{-\infty}^{\infty} dt \exp i(\omega - \omega')t = 2\pi\delta(\omega - \omega')$ according to the Fourier-integral representation (3.48) of a δ distribution. Then the *total* energy radiated, through an $r \to \infty$ sphere centered on the origin chosen for the position vectors $\mathbf{r}$ of the radiation-zone field points, is

$$\int_{-\infty}^{\infty} dt \oint d\Omega r^2 N = \tfrac{1}{2}c \oint d\Omega r^2 \int_{-\infty}^{\infty} d\omega |\mathbf{E}_{\omega}|^2_{r\to\infty}$$
$$= \oint d\Omega \int_{0}^{\infty} d\omega c r^2 |\mathbf{E}_{\omega}|^2 \equiv \int_{0}^{\infty} d\omega \oint d\Omega \frac{dI(\omega)}{d\Omega} \tag{10.37}$$

where it has been recognized that the negative frequencies correspond to observable periodicities indistinguishable from their positive counterparts. A new quantity $I(\omega)$, to be called the *intensity per unit frequency range*, is defined in (10.37) and

$$\frac{dI(\omega)}{d\Omega} \equiv cr^2 |\mathbf{E}_{\omega}|^2_{r\to\infty} \tag{10.38}$$

is such an intensity per unit of the solid-angle element $d\Omega$.

A time-integrated energy like that considered here corresponds to what must be admitted by a detector that is to discriminate between frequencies. As is also indicated formally by the infinite range of the time integration for $\mathbf{E}_{\omega}$ in (10.36), observation over a time large compared with a period $2\pi/\omega$ is needed to establish the existence of a given, well-defined frequency ω of variation in time. Of course, only finite observation times $\Delta t = \int dt < \infty$ are actually possible, but the flux of radiant energy being furnished for the integrands in (10.37) is also generated over some finite time interval Δt_q, and so the integrands can be taken to vanish outside some finite Δt; then there is no error in the formal extensions to

$\int_{-\infty}^{\infty} dt$. It is desirable to thus keep the integration span unlimited because the effectively contributing emission intervals Δt_q are usually ill defined and vary with the frequency considered.

The connection between the spectral intensity $dI/d\Omega$ and the instantaneous radiation rates $d\mathscr{P}(t_q)/d\Omega$, found in the preceding subsections, follows from the fact that, in terms of $\mathscr{P}(t_q)$, the total radiated energy (10.37) must be equal to $\int_{-\infty}^{\infty} dt_q \mathscr{P}(t_q)$. Then

$$\int_0^\infty d\omega \frac{dI}{d\Omega} = \int_{-\infty}^{\infty} dt r^2 N = \int_{-\infty}^{\infty} dt_q \frac{r^2 N}{\partial t_q/\partial t} = \int_{-\infty}^{\infty} dt_q \frac{d\mathscr{P}(t_q)}{d\Omega} \tag{10.39}$$

The intermediate expressions here show the consistency with the way $d\mathscr{P}(t_q)$ was evaluated from $N(\mathbf{r}, t)$ when (10.14) was obtained.

A detector of radiation into some direction from an accelerated point charge will be focused on some finite sector of any extended trajectory the charge may be following. The contributing sector may also have natural limits, with $\dot{\boldsymbol{\beta}} \approx 0$ outside them, as when the only accelerations are provided during passage through a force field of limited effective range (characteristic of charged particles in collision with atoms). In any case, it is possible to speak of a wave zone remote from the region in which the detected radiation originates and to consider the detector to be placed there, as indicated in Fig. 10.7. The field at the remote detector can be obtained from $\mathbf{E}_a(\mathbf{r}, t)$ of (10.10) by putting $R = |\mathbf{r} - \mathbf{r}_q(t_q)| \approx r$ and $\hat{\mathbf{R}} \to \hat{\mathbf{r}} \equiv \hat{\mathbf{k}}$ (as in Sec. 8.3) into that expression, with the result

$$\mathbf{E}(\mathbf{r}, t) \approx \frac{q}{cr} \frac{\hat{\mathbf{k}} \times [(\hat{\mathbf{k}} - \boldsymbol{\beta}) \times \dot{\boldsymbol{\beta}}]}{(1 - \hat{\mathbf{k}} \cdot \boldsymbol{\beta})^3} \tag{10.40}$$

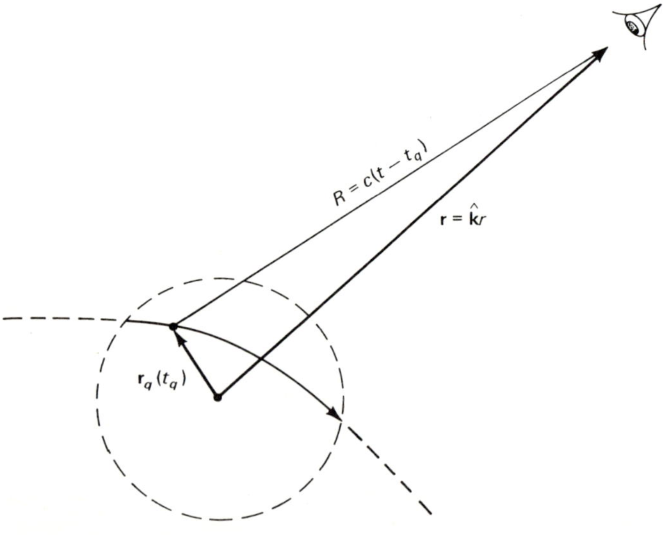

Figure 10.7

This is entirely derivable, as $\mathbf{E} = -\partial \mathbf{A}_\tau/(c\,\partial t)$, from a transverse vector potential† that may be easiest to find here after noticing that (10.40) can be written

$$\mathbf{E}(\mathbf{r}, t) = \frac{1}{1 - \hat{\mathbf{k}} \cdot \boldsymbol{\beta}} \left(\frac{\partial}{c\,\partial t_q} \right)_{\mathbf{r}} \left[\frac{q}{r} \frac{\hat{\mathbf{k}} \times (\hat{\mathbf{k}} \times \boldsymbol{\beta})}{1 - \hat{\mathbf{k}} \cdot \boldsymbol{\beta}} \right] \equiv -\frac{\partial}{c\,\partial t} \mathbf{A}_\tau(\mathbf{r}, t) \tag{10.41}$$

as is readily confirmed by performing the indicated partial derivative with respect to t_q. Since the factor $(1 - \hat{\mathbf{k}} \cdot \boldsymbol{\beta})^{-1}$ is just a wave-zone value of $\partial t_q/\partial t$ in (10.4), the vector potential introduced thus is

$$\mathbf{A}_\tau(\mathbf{r}, t) = -\frac{q}{r} \frac{\hat{\mathbf{k}} \times (\hat{\mathbf{k}} \times \boldsymbol{\beta})}{1 - \mathbf{k} \cdot \boldsymbol{\beta}} \approx -\hat{\mathbf{k}} \times (\hat{\mathbf{k}} \times \mathbf{A}) = \mathbf{A} - \hat{\mathbf{k}}(\hat{\mathbf{k}} \cdot \mathbf{A}) \tag{10.42}$$

being just the transverse component of the Lienard-Wiechert potential $\mathbf{A} = \boldsymbol{\beta}\phi \approx \boldsymbol{\beta} q/(r - \boldsymbol{\beta} \cdot \mathbf{r})$ in the wave zone.

To obtain the spectral distribution (10.38), it is necessary to find the monochromatic components $\mathbf{E}_\omega(\mathbf{r})$ of the field by way of the time integration in (10.36). When $\mathbf{E} = -\partial \mathbf{A}_\tau/(c\,\partial t)$ is substituted into that integral, a partial integration to

$$\mathbf{E}_\omega(\mathbf{r}) = -\frac{1}{2\pi c} [\mathbf{A}_\tau e^{i\omega t}]_{-\infty}^{\infty} + \frac{i\omega}{2\pi c} \int_{-\infty}^{\infty} dt \mathbf{A}_\tau e^{i\omega t} \tag{10.43}$$

becomes possible. The terms to be evaluated for $t \to \pm\infty$ can be taken to vanish as functions of ω; over any interval $d\omega$ of ω, the exponentials $e^{\pm i\omega t}$ perform oscillations of zero average and the wavelength $2\pi/|t| \to 0$. The remainder in (10.43) yields

$$\mathbf{E}_\omega(\mathbf{r}) = -\frac{i\omega q}{2\pi c r} \int_{-\infty}^{\infty} dt \frac{\hat{\mathbf{k}} \times (\hat{\mathbf{k}} \times \boldsymbol{\beta})}{1 - \hat{\mathbf{k}} \cdot \boldsymbol{\beta}} e^{i\omega t} \tag{10.44}$$

after a substitution of (10.42) for $\mathbf{A}_\tau$. The integrand here (or $\mathbf{A}_\tau$) need not vanish for a uniform ($\dot{\boldsymbol{\beta}} = 0$), and hence nonradiating, motion. However, the integrated result does vanish in that case ($\boldsymbol{\beta}$ constant); it becomes identical with the negative of the first term given in (10.43) and vanishes, again individually, for the same reason.

Since $dt/(1 - \hat{\mathbf{k}} \cdot \boldsymbol{\beta}) = dt_q$ in the wave-zone expression (10.44), and since $\boldsymbol{\beta}(t_q)$ is most immediately given as a function of the retarded time $t_q = t - R(t_q)/c$, it is

† Such a Lorentz potential field, with $\phi_\tau \equiv 0$, is always possible to find for radiation, or wave-zone, fields, as also seen in (9.110). Such fields have the transverse-plane-wave characteristics; that the latter allow a transverse ≡ solenoidal Lorentz gauge was shown in (7.13).

The $\mathbf{A}_\tau$ of (10.42) could also be found in an alternative way, by introducing the transverse component $\mathbf{A}_\tau = \mathbf{A} - \hat{\mathbf{k}}(\hat{\mathbf{k}} \cdot \mathbf{A})$ into $\mathbf{E} = -\nabla\phi - \partial\mathbf{A}/(c\,\partial t)$ and showing that $\hat{\mathbf{k}}[\hat{\mathbf{k}} \cdot \partial\mathbf{A}/(c\,\partial t)]$ just cancels $\nabla\phi$ in the wave zone. The Lorentz condition $\partial\phi_\tau/(c\,\partial t) = -\nabla \cdot \mathbf{A}_\tau = 0$ is satisfied by $\phi_\tau \equiv 0$ in the wave zone.

convenient to change from t to t_q as the integration variable. In doing so, the t occurring explicitly in the exponential must be replaced, as in

$$\omega t = \omega\left(t_q + \frac{|\mathbf{r} - \mathbf{r}_q(t_q)|}{c}\right) \approx \omega t_q + kr - \mathbf{k} \cdot \mathbf{r}_q(t_q) \tag{10.45}$$

on the same basis as that discussed in connection with (8.23). Here $\mathbf{k} \equiv \hat{\mathbf{r}}\omega/c$, as usual. Thus

$$\mathbf{E}_\omega(\mathbf{r}) = -\frac{ik}{2\pi}\frac{qe^{ikr}}{r}\left[\hat{\mathbf{k}} \times \left(\hat{\mathbf{k}} \times \int_{-\infty}^{\infty} dt_q\, \boldsymbol{\beta} e^{i\omega t_q - i\mathbf{k}\cdot\mathbf{r}_q}\right)\right] \tag{10.46}$$

for the wave-zone field, a superposition of such outgoing waves, formed out of plane-wave sectors, as were found for the monochromatic representatives of the radiation fields in Chaps. 8 and 9. The outcome for the spectrum (10.38) is

$$\frac{dI}{d\Omega} = \frac{q^2\omega^2}{c}\,|\hat{\mathbf{k}} \times (\hat{\mathbf{k}} \times \langle\boldsymbol{\beta}\rangle_\omega)|^2$$

with

$$\langle\boldsymbol{\beta}\rangle_\omega \equiv (2\pi)^{-1}\int_{-\infty}^{\infty} dt_q\, \boldsymbol{\beta}(t_q) e^{i\omega t_q - i\mathbf{k}\cdot\mathbf{r}_q(t_q)} \tag{10.47}$$

which differs from a Fourier transform of the velocity $\boldsymbol{\beta}(t_q)$, as a function of time, only in that it includes a retardation factor like the one first seen in (8.19), owing to the fact that ω is a frequency on the scale of field times t Doppler-shifted from the scale of the source times t_q. The squared magnitude of the amplitude $\hat{\mathbf{k}} \times (\hat{\mathbf{k}} \times \langle\boldsymbol{\beta}\rangle_\omega)$ would not be changed if this were replaced by $\hat{\mathbf{k}} \times \langle\boldsymbol{\beta}\rangle_\omega$, but the full expression is valuable to retain for discriminating between contributions of different polarization. The polarizations are given by component directions of $\mathbf{E}_\omega(\mathbf{r})$ in (10.46), since they are well-defined observables only for monochromatic plane-wave sectors of the wave-zone field.

The result (10.47) might be compared to the monochromatic radiation rate in (8.31) by replacing the monochromatic representative of the current density in that by

$$\mathbf{j}_\omega(\mathbf{r}_s) = \int_{-\infty}^{\infty} \frac{dt_q}{2\pi}\,\mathbf{j}(\mathbf{r}_s, t_q)e^{-i\omega t_q}$$

with

$$\mathbf{j}(\mathbf{r}_s, t_q) = q\mathbf{u}(t_q)\delta[\mathbf{r}_s - \mathbf{r}_q(t_q)]$$

Then

$$\oint dV(\mathbf{r}_s)\mathbf{j}_\omega(\mathbf{r}_s)e^{-i\mathbf{k}\cdot\mathbf{r}_s} = \frac{qc}{2\pi}\int_{-\infty}^{\infty} dt_q\, \boldsymbol{\beta} e^{i\omega t_q - i\mathbf{k}\cdot\mathbf{r}_q} \tag{10.48}$$

and the squared amplitude in $d\mathscr{P}/d\Omega$ of (8.31) becomes proportional to the one in (10.47), as should be expected from the connection (10.39) between radiation rates and time-integrated intensities. The rates given in (8.31) were *steady* ones, based on time-averaged fluxes, suitable to use only for the steadily maintained, discrete frequencies presumed for (8.31).

The result (10.47) may include contributions by time intervals dt_q, and hence stretches of trajectory $d\mathbf{r}_q = \mathbf{u}\, dt_q$ on which $\dot{\boldsymbol{\beta}}(t_q) = 0$, despite the fact that no radiation is supposed to arise while there is no acceleration. The stretches of uniform motion are important for determining the relative phase shifts between moments of emission when the results are calculated as in (10.47). It is possible to reformulate the result in terms of only such intervals dt_q as do have $\dot{\boldsymbol{\beta}}(t_q) \neq 0$. It is only necessary to forgo the partial integration (10.43) and insert the form (10.40) directly into the time integral for $\mathbf{E}_\omega(\mathbf{r})$ of (10.36). Then, after a change of variable from t to t_q,

$$\mathbf{E}_\omega(\mathbf{r}) = \frac{qe^{ikr}}{2\pi cr} \int_{-\infty}^{\infty} dt_q \frac{\hat{\mathbf{k}} \times [(\hat{\mathbf{k}} - \boldsymbol{\beta}) \times \dot{\boldsymbol{\beta}}]}{(1 - \hat{\mathbf{k}} \cdot \boldsymbol{\beta})^2} e^{-i(\mathbf{k} \cdot \mathbf{r}_q - \omega t_q)} \tag{10.49}$$

and the spectral distribution (10.38) becomes

$$\frac{dI(\omega)}{d\Omega} = \frac{q^2}{4\pi^2 c} \left| \int_{-\infty}^{\infty} dt_q \frac{\hat{\mathbf{k}} \times [(\hat{\mathbf{k}} - \boldsymbol{\beta}) \times \dot{\boldsymbol{\beta}}]}{(1 - \hat{\mathbf{k}} \cdot \boldsymbol{\beta})^2} e^{-i(\mathbf{k} \cdot \mathbf{r}_q - \omega t_q)} \right|^2 \tag{10.50}$$

This form plainly includes contributions only from moments of emission, each already in the proper phase for superposition. In the Larmor ($\beta \ll 1$) approximation, it differs from (10.47) only in the replacement of $-i\omega\boldsymbol{\beta}$ by $\dot{\boldsymbol{\beta}}$, as for any monochromatic representative of frequency ω. The form (10.50) is much more complicated to evaluate than (10.47) but is advantageous for certain types of approximation.

Synchrotron Radiation

The result (10.47) can be applied to finding the spectrum for the important case discussed in connection with Fig. 10.6, radiation from a point charge traversing a sector of circular path. A case of low-velocity, $\beta \ll 1$ (the Larmor, or dipole, approximation), circular motion has already been investigated in connection with Fig. 8.3, and so the attention here will be focused on the opposite extreme: $\beta \to 1$. It is such radiation, at high velocities, that becomes substantial enough to have interest when an electron follows a path with *macroscopic* radii of curvature, as in a synchrotron or in the galaxies, rather than with a radius of atomic dimensions, for which the discussion of Fig. 8.3 was adequate. In the macroscopic situations, a detector can be focused on a small sector of the trajectory, traversed in some brief time Δt_q, which need not even complete a circle. The restriction to times Δt_q will permit conclusions only about frequencies $\omega \gg 2\pi/\Delta t_q$, according to the discussion on page 299, but it will turn out that this excludes only a negligible low-frequency portion of the spectrum when the velocity is comparable to c.

It is already known, from the consideration of the *instantaneous* radiation rates in connection with Fig. 10.6, that the radiation at each instant is strongly peaked into the direction of the velocity. Thus, to receive most of the radiation, a detector of it should be set in a direction tangential to the particle trajectory, as indicated in Fig. 10.8. Actually, to discriminate frequencies, the detector must

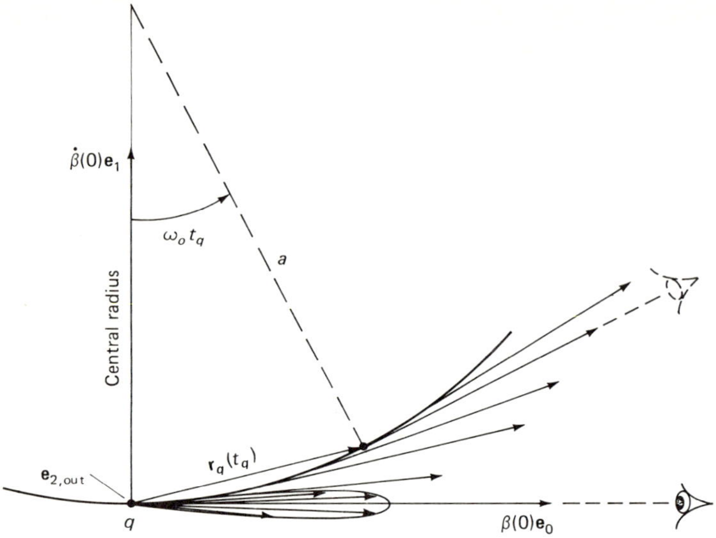

Figure 10.8

accept a *time-integrated* intensity distribution, like (10.47), from a finite sector of trajectory. It is then plain that the effect of the time integration is to broaden the peaking toward isotropy in the orbital plane. Changing the angle of observation *within the orbital plane*, as indicated in Fig. 10.8, corresponds to observing like radiation from neighboring sectors, thus finding a featureless uniform distribution rather than discriminating features in the pattern from any one sector. A like situation prevails in each plane through the central radius of the sector to be focused on, i.e., in every plane containing the acceleration vector marked $\dot{\beta}(0)\mathbf{e}_1$ in the diagram. On the other hand, the intensity must be expected to diminish rapidly away from the orbital plane and to depend sensitively on an azimuthal angle about the central radius. Such an angle, called ψ here, has already been indicated in Fig. 10.6 [and is to be distinguished from the azimuth φ about the *velocity* vector, used for the expression (10.35)]. The azimuth ψ is to start from a zero value at the orbital plane, to increase in a direction which is initially perpendicular to it, and to give the positions of various planes through the central radius. The distribution in ψ is obtained by considering the viewing directions

$$\hat{\mathbf{k}} = \mathbf{e}_0 \cos\psi + \mathbf{e}_2 \sin\psi \qquad \text{with } \mathbf{e}_2 = \mathbf{e}_0 \times \mathbf{e}_1 \tag{10.51}$$

in planes perpendicular to the orbital plane and parallel to the velocity $\beta(0)\mathbf{e}_0$ at the central position of the sector. Only these need be given attention because the isotropy to be expected in each plane of given ψ should make the results relatively† independent of other components of $\hat{\mathbf{k}}$.

† A more detailed investigation can be found in a definitive paper by Schwinger (Ref. 28). His angle ψ is not quite the same as the one defined here.

The velocity at all points of the trajectory is naturally confined to the orbital plane, and so that velocity matrix element $\langle \boldsymbol{\beta} \rangle_\omega$ in (10.47) is resolvable into just the two components in

$$\langle \boldsymbol{\beta} \rangle_\omega = \mathbf{e}_0 \langle \beta_0 \rangle_\omega + \mathbf{e}_1 \langle \beta_1 \rangle_\omega$$

With (10.51), this leads to

$$\hat{\mathbf{k}} \times \langle \boldsymbol{\beta} \rangle_\omega = \mathbf{e}_1 \sin \psi \langle \beta_0 \rangle_\omega + (\hat{\mathbf{k}} \times \mathbf{e}_1) \langle \beta_1 \rangle_\omega \tag{10.52}$$

The square of this has the same magnitude as the square of the amplitude shown in (10.47), and since the two terms of (10.52) are orthogonal to each other, the result is

$$\frac{dI}{d\Omega} = \frac{q^2 \omega^2}{c} \left(|\langle \beta_1 \rangle_\omega|^2 + |\langle \beta_0 \rangle_\omega|^2 \sin^2 \psi \right) \tag{10.53}$$

a superposition of two uninterfering intensity components. The two parts have linear polarizations that are perpendicular to each other. That can be seen from the fact that the electric vector (10.46) has the direction of

$$\hat{\mathbf{k}} \times (\hat{\mathbf{k}} \times \langle \boldsymbol{\beta} \rangle_\omega) = (\hat{\mathbf{k}} \times \mathbf{e}_1) \sin \psi \langle \beta_0 \rangle_\omega - \mathbf{e}_1 \langle \beta_1 \rangle_\omega$$

Thus, the intensity measured by $|\langle \beta_1 \rangle_\omega|^2$ is linearly polarized parallel to the acceleration-vector direction $\mathbf{e}_1$, in the orbital plane, whereas the remainder is polarized parallel to $\hat{\mathbf{k}} \times \mathbf{e}_1$, which lies in planes perpendicular to the orbital plane. The latter intensity component is proportional to $|\langle \beta_0 \rangle_\omega|^2 \sin^2 \psi$ and hence vanishes in the $\psi = 0$ orbital plane. As a result, the maximum of the radiation, which emerges in the orbital plane, has the polarization parallel to it. Such a polarization of the radiation into the orbital plane was also found in the low-velocity case of Fig. 8.3; however, in that case there was more radiation emerging normally to the orbital plane, with its different polarization.

To evaluate the amplitudes $\langle \beta_{0,\,1} \rangle$ in (10.53), an orbit description $\mathbf{r}_q(t_q)$ must be constructed, as reference to (10.47) shows. That must be done relative to a fixed origin close to the radiating sector, as in Fig. 10.7, and so an origin right on the trajectory itself, a point through which the charge passes at a moment taken as $t_q = 0$, is chosen. Then Fig. 10.8 makes it simple to see that

$$\begin{aligned} \mathbf{r}_q(t) &= a[\mathbf{e}_0 \sin \omega_o t + \mathbf{e}_1 (1 - \cos \omega_o t)] \\ \boldsymbol{\beta}(t) &= \beta[\mathbf{e}_0 \cos \omega_o t + \mathbf{e}_1 \sin \omega_o t] \end{aligned} \tag{10.54a}$$

with $\beta = a\omega_o/c$ resulting from $\boldsymbol{\beta} = \dot{\mathbf{r}}_q/c$. The subscript on t_q has been dropped here, since this is the only time variable that will appear in the present considerations. The interest is in conclusions valid for a small segment of arc, $\omega_o t \ll 1$, and so the approximations

$$\begin{aligned} \mathbf{r}_q(t) &\approx a\mathbf{e}_0 \omega_o t(1 - \tfrac{1}{6}\omega_o^2 t^2) + a\mathbf{e}_1 \tfrac{1}{2}\omega_o^2 t^2 \\ \boldsymbol{\beta}(t) &\approx \beta(\mathbf{e}_0 + \mathbf{e}_1 \omega_o t) \end{aligned} \tag{10.54b}$$

will be adopted. They give $\beta_0 \approx \beta$ and $\beta_1 \approx \beta\omega_o t$ for the components needed in (10.53). The expression for $\mathbf{r}_q(t)$ itself is needed for the phases $\omega t - \mathbf{k} \cdot \mathbf{r}_q$ of the contributions to the integral in (10.47), and those become

$$\omega t - \mathbf{k} \cdot \mathbf{r}_q \approx \omega t(1 - \beta \cos \psi) + \tfrac{1}{6}\omega\omega_o^2 t^3 \beta \cos \psi \tag{10.55}$$

after use of (10.51) for $\hat{\mathbf{k}} = c\mathbf{k}/\omega$ and of $a = c\beta/\omega_o$. Thus the integrations are to be carried out in the approximate forms

$$\left.\begin{matrix}\langle\beta_0\rangle_\omega \\ \langle\beta_1\rangle_\omega\end{matrix}\right\} \approx \frac{\beta}{2\pi}\int_{-\infty}^{\infty} dt \left\{\begin{matrix}1 \\ \omega_o t\end{matrix}\right\} e^{i\omega t(1-\beta\cos\psi)+(1/6)i\omega\omega_o{}^2 t^3 \beta \cos\psi} \tag{10.56}$$

It is essential for these to have retained the term of order t^3 in the approximation of the trajectory. Without it, the integrands would merely have simple harmonic variations of frequency $\omega(1 - \beta \cos \psi)$ or this $\pm\omega_o$ (the factor $\omega_o t$ cannot replace the original $\sin \omega_o t$ when the term of order t^3 is not retained); then the integrals would vanish on the same grounds as those discussed in connection with (10.44). The term of order t^3 represents effects of an acceleration that is symmetric about the central radius of the sector and is essential for a net radiation; leaving the term out corresponds to using a uniform, nonradiating motion of the charge for the evaluation of the phases (10.55).

The presence of the term of order t^3 also permits the formal extension of the integrations to $t \to \pm\infty$ despite the use of the approximations valid only for $\omega_o t \ll 1$. This is because, for each radiated frequency $\omega \gg \omega_o$, only a limited sector $(\Delta t)_\omega$ actually contributes effectively regardless of the extension of the integration range. The simple harmonic variations mentioned in the preceding paragraph merely modulate oscillations having an effective frequency $\approx \frac{1}{6}\beta\omega(\omega_o t)^2 \cos \psi$, proportional to ω, that increases with $|t|$ at an accelerated rate and becomes much larger than the modulating frequency outside some period $(\Delta t)_\omega \sim 1/\omega$, inversely proportional to the radiated frequency. Outside that roughly defined interval the contributions are averaged to zero by the accelerated oscillations. The conclusion is that the effectively contributing span $\omega_o(\Delta t)_\omega \sim \omega_o/\omega$ can be as small as has been presumed for the approximations if the radiation frequencies to be considered are large enough multiples of the angular velocity ω_o. It should also be noted that the accelerated frequency above is proportional to $\cos \psi$ and so tends to vanish at observation angles too far from the orbital plane ($\psi = 0$). This means that the procedure loses validity for large angles ψ, not an important restriction since most of the radiation is confined close to the orbital plane.

Formally, the integration forms in (10.56) are proportional to what are called *Airy integrals*, first met in an optical problem having to do with caustics. In forms presented by Schwinger (Ref. 28) they are

$$\begin{aligned}\int_0^\infty dx \cos\left[\tfrac{3}{2}\xi\left(x + \tfrac{1}{3}x^3\right)\right] &= 3^{-1/2} K_{1/3}(\xi) \\ \int_0^\infty dx\,x \sin\left[\tfrac{3}{2}\xi\left(x + \tfrac{1}{3}x^3\right)\right] &= 3^{-1/2} K_{2/3}(\xi)\end{aligned} \tag{10.57}$$

The functions $K_m(\xi)$ representing the integrated results are just modified Bessel functions, of the family defined in (C.34). Their relations to $\langle\beta_{0,1}\rangle_\omega$ of (10.56) begin to be apparent when it is recognized that the exponentials there have the form $e^{if(t)} = \cos f(t) + i \sin f(t)$ with an argument $f(t) = -f(-t)$ that is odd in t. Since terms of the entire integrands that are odd in t must vanish upon integration over the symmetrical interval $-\infty < t < \infty$, only the $\cos f(t)$ will survive in $\langle\beta_0\rangle_\omega$ and only the $i \sin f(t)$ in $\langle\beta_1\rangle_\omega$. The argument $f(t)$ can be given the form $\frac{3}{2}\xi(x + \frac{1}{3}x^3)$ by trying an arbitrary linear substitution $t \sim x$, which will make it apparent that it is the substitutions

$$x = \frac{\omega_o t}{2}\left(\frac{2\beta \cos \psi}{1 - \beta \cos \psi}\right)^{1/2} \quad \text{and} \quad \xi \equiv \frac{4}{3}\frac{\omega}{\omega_o}\left[\frac{(1 - \beta \cos \psi)^3}{2\beta \cos \psi}\right]^{1/2} \tag{10.58}$$

that are needed. It then becomes simple to conclude that

$$\langle\beta_0\rangle_\omega = \frac{2\beta}{\pi\omega_o}\left(\frac{1 - \beta \cos \psi}{6\beta \cos \psi}\right)^{1/2} K_{1/3}(\xi)$$

$$\langle\beta_1\rangle_\omega = \frac{2i}{\pi\omega_o}\frac{1 - \beta \cos \psi}{\cos \psi} 3^{-1/2} K_{2/3}(\xi) \tag{10.59}$$

These representations will make it possible to draw upon the extensively recorded properties of the Bessel functions for interpreting results.

Substituting the results for $\langle\beta_{0,1}\rangle_\omega$ into (10.53) yields

$$\frac{dI}{d\Omega} = \frac{4q^2}{3\pi^2 c}\left(\frac{\omega}{\omega_o}\right)^2\left(\frac{1 - \beta \cos \psi}{\cos \psi}\right)^2\left[K_{2/3}^2(\xi) + \frac{1}{2}\frac{\beta \cos \psi \sin^2 \psi}{1 - \beta \cos \psi} K_{1/3}^2(\xi)\right] \tag{10.60}$$

for the angular distribution of the intensity with a given frequency. The most important dependences on the emission angle ψ are hidden in the argument ξ [Eq. (10.58)] of the Bessel functions. This is easiest to appreciate after an integration over all frequencies, with the help of the Bessel-function properties (C.38)

$$\int_0^\infty d\xi \xi^2 K_{1/3}^2 = \frac{5\pi^2}{144} \quad \text{and} \quad \int_0^\infty d\xi \xi^2 K_{2/3}^2 = \frac{7\pi^2}{144}$$

Then using the ratio $\omega^2\, d\omega/(\xi^2\, d\xi)$ that follows from (10.58) gives

$$\int_0^\infty d\omega \frac{dI}{d\Omega} = \frac{7q^2\omega_o}{128c}\left(\frac{2\beta^3}{\cos \psi}\right)^{1/2}(1 - \beta \cos \psi)^{-5/2}\left(1 + \frac{5}{14}\frac{\beta \cos \psi \sin^2 \psi}{1 - \beta \cos \psi}\right) \tag{10.61}$$

This exhibits proportionalities to substantial inverse powers of $1 - \beta \cos \psi$ like those found responsible for the strong peaking discussed in connection with the instantaneous radiation (10.35). With most of the intensity thus confined to small angles ψ, the fact that the results cannot be valid for $\cos \psi \to 0$ [pointed out in the discussion following (10.56)] is of no importance. They approach precision for the extreme relativistic velocities $\beta \approx 1 - 1/2\gamma^2 \to 1$ when the intensity is confined to such small angles that $\cos\psi \approx 1 - \frac{1}{2}\psi^2$ is a good approximation and

$$1 - \beta\cos\psi \approx 1 - \left(1 - \frac{1}{2\gamma^2}\right)\left(1 - \frac{\psi^2}{2}\right) \approx \frac{1}{2}\left(\frac{1}{\gamma^2} + \psi^2\right)$$

In other factors than $1 - \beta\cos\psi$, the approximations $\beta = a\omega_o/c \approx 1$, $\cos\psi \approx 1$, and $\sin\psi \approx \psi$ are sufficient for presenting the extreme relativistic versions of (10.60) and (10.61):

$$\frac{dI}{d\Omega} \approx \frac{q^2}{3\pi^2 c}\left(\frac{\omega}{\omega_o}\right)^2\left(\frac{1}{\gamma^2} + \psi^2\right)^2\left[K_{2/3}^2(\xi) + \frac{\psi^2}{\gamma^{-2} + \psi^2}K_{1/3}^2(\xi)\right] \tag{10.62}$$

where

$$\xi \approx \frac{\omega}{2\omega_c(\psi)} \qquad \text{with } \omega_c(\psi) \equiv \frac{3\omega_o}{2(\gamma^{-2} + \psi^2)^{3/2}} \tag{10.63}$$

follows from (10.58), and

$$\int_0^\infty d\omega \frac{dI}{d\Omega} \approx \frac{7}{16}\frac{q^2\omega_o}{c}\frac{1}{(\gamma^{-2} + \psi^2)^{5/2}}\left(1 + \frac{5}{7}\frac{\psi^2}{\gamma^{-2} + \psi^2}\right) \tag{10.64}$$

Each power of $1/(\gamma^{-2} + \psi^2)$ reduces the corresponding intensity contribution by a factor 2 as $\psi \to 1/\gamma$, the characteristic angle found in the instantaneous radiation pattern (10.35). The last part of the intensity, proportional to $\sin^2\psi \approx \psi^2$, represents the component that is polarized perpendicularly to the orbital plane, as found in the discussion of (10.53); it vanishes in the orbital plane ($\psi = 0$) and still constitutes only the fraction $\frac{5}{19} \approx 25$ percent of the weaker total radiation at the extreme angle $\psi = 1/\gamma \ll 1$.

To gain a more explicit idea of the *spectrum* shape at any one angle, as given by (10.60) or (10.62), it is necessary to draw upon the recorded properties of the modified Bessel functions. Reference to (C.37) and (C.35) is sufficient for seeing the proportionalities

$$\omega^2 K_{2/3}^2\left(\frac{\omega}{2\omega_c} \ll 1\right) \sim \omega^{2/3} \qquad \omega^2 K_{1/3}^2\left(\frac{\omega}{2\omega_c} \ll 1\right) \sim \omega^{4/3}$$
$$\omega^2 K_m^2\left(\frac{\omega}{2\omega_c} \gg 1\right) \sim \omega e^{-\omega/\omega_c} \qquad \text{for any order } m \tag{10.65}$$

Thus the spectrum shapes (intensities as functions of the radiated frequency ω) begin by rising at the smallest frequencies but then must go through a maximum since they eventually fall off exponentially beyond the main frequency

$$\omega_c(0) = \tfrac{3}{2}\gamma^3\omega_o \tag{10.66}$$

which is the greatest value (the one in the orbital, $\psi = 0$, plane) that is given by (10.63). A sample plot is shown in Fig. 10.9.

To be appreciated is the fact that at high velocities, when $\gamma \equiv (1 - \beta^2)^{-1/2} \gg 1$, the radiated frequencies over practically their entire range are much greater than the angular frequency ω_o. The latter, on a radius $a \approx c/\omega_o$

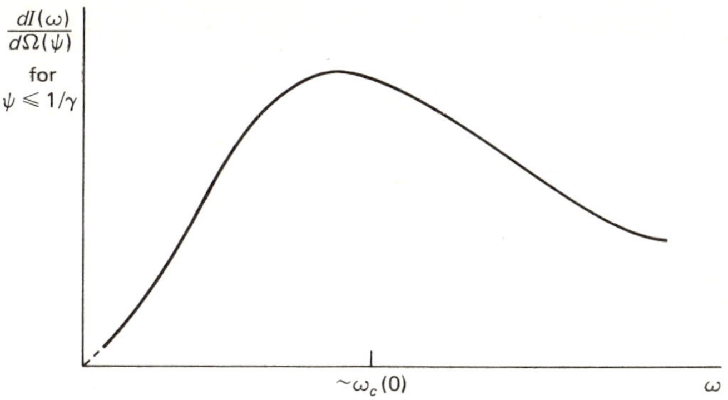

Figure 10.9

of a meter or so, may be only of the order $\omega_o \approx 3 \times 10^8 \text{ s}^{-1}$, whereas 100-Mev electrons have $\gamma \approx 200$, making the main frequency about $\omega_c(0) \approx 4 \times 10^{15} \text{ s}^{-1}$, corresponding to a wavelength $\lambda_c \approx 5 \times 10^{-5}$ cm = 5000 Å in the visible range. With $\omega_o/\omega_c(0) \approx 10^{-7}$, the radiated wavelengths are so tiny a fraction of even a small sector of the circular path that the approximations used here approach precision for all but a vanishingly small part (with $\omega \to \omega_o$) of the total spectrum.

Visible radiation was also found possible from the circular motion of low velocity, $\beta = \omega_o a/c \ll 1$, considered in connection with Fig. 8.3. That occurs when the orbit is only of atomic dimensions, so that even a velocity $\beta \ll 1$ makes the charge traverse the whole of the tiny orbit with a high enough frequency $\omega_o = c\beta/a$ to be in the visible-frequency range.

It should still be understood just why the low velocities considered in connection with Fig. 8.3 led to dipole, quadrupole, ... radiations with the *discrete* frequencies $\omega_o, 2\omega_o, \ldots$, whereas the high velocities as treated here yield *continuous* spectra like that in Fig. 10.9. The low-velocity treatment presumed that a steadily maintained orbit was completed many times in the course of the radiation observed. (A generation time Δt of many periods $T = 2\pi/\omega_o$ is necessary to define a radiated frequency $\omega = \omega_o$ with a $\Delta\omega = 2\pi/\Delta t \ll \omega_o$.) The high velocities would also have yielded a discrete spectrum if it had been presumed that the simple harmonic orbit (10.54*a*) was followed precisely without approximating it by (10.54*b*). Then the radiation field (10.40) would have had precisely repeated values in every period $T = 2\pi/\omega_o$, and its Fourier analysis would have consisted of discrete radiated frequencies $\omega = n(2\pi/T) = n\omega_o$ that are integer multiples of the fundamental ω_o. Any function defined on a *finite* interval T and merely repeated outside it leads to such a discrete Fourier series, as was learned in Sec. 3.2.

The conclusion is that a charge following a perfectly circular orbit with *any* velocity $\beta = a\omega_o/c$ will radiate a discrete spectrum of frequencies $\omega = n\omega_o$ that is a continuation of the spectrum $\omega_o, 2\omega_o, \ldots$ found for the successive low-order multipole radiations. The intensity radiated by each successive multipole was also found to increase with a higher power of the radiated frequency, and so it is to be expected that at high velocities most of the intensity will become invested in high frequencies, like the $\omega_c = \frac{3}{2}\gamma^3\omega_o \gg \omega_o$ of (10.66). These correspond to high-order harmonics, with $n \approx \omega_c/\omega_o \approx \frac{3}{2}\gamma^3 \gg 1$, that are very closely spaced, with $\Delta n/n = 1/n \ll 1$, on the scale of ω_c. The continuous-spectrum curve of Fig. 10.9 is then replaced by, and becomes only an average of, a series of closely spaced spectral lines if the orbit is sufficiently well defined. However, conditions sufficiently stable to make the closely spaced lines resolvable would be difficult to achieve for high-velocity particles, particularly when the observations must be made on beams of particles, each with a slightly different trajectory and period. The continuous-spectrum results are far the more general in scope. They apply in some approximation to any smooth curve followed with a high enough velocity for small sectors of it to be able to define most of the high frequencies mainly emitted. A radius of curvature is definable at every point of a continuous trajectory, and each sector is representable as a circular arc in the approximation of (10.54*b*).

In point of fact, no spectrum is absolutely discrete in physically realized situations. Each of the so-called discrete frequencies emitted by atoms, for instance, is broadened in accordance with $\Delta\omega = 2\pi/\Delta t$ by the finite durations Δt of the states of motion that supply the radiated energy. There is at least a so-called *natural broadening* just because the losses to radiation must change the energy for the motion. (Some estimates will be presented in Secs. 11.4 and 14.7.) For like reasons, electrons in a synchrotron tend to spiral out of their orbits, and this may broaden each of the harmonics $\omega = n\omega_o$ sufficiently to result in a continuous spectrum like that of Fig. 10.9.

Bremsstrahlung Spectra

Continuous spectra can also be generated at nonrelativistic velocities, low enough for the Larmor formulas (10.17) and (10.19) to be applicable. Important examples are provided by charged particles penetrating through matter and undergoing collisions with the atoms in it. The accompanying radiation is usually called *bremsstrahlung* (German for "braking radiation"). X-rays are often produced by letting a beam of electrons slam into a massive target, as in a Coolidge tube. To obtain a high proportion of wavelengths in the x-ray range, quite energetic electrons—10 kev, say—may be used, but that still corresponds to a $\beta \approx 0.2$ (and $\gamma \approx 1.02 \approx 1$), which can be treated as nonrelativistic; an important point is that this remains large in comparison to the $\beta \approx 0.01$ typical of electrons bound to atoms (which then cannot change significantly the momentum of the particles traversing clouds of atomic electrons).

In each of the collisions, the incident particle approaches a target atom with a nearly uniform velocity, undergoes a momentum change $mc\,\Delta\boldsymbol{\beta}$ over a brief interval Δt_q of collision interaction, then goes on with a resultant uniform velocity. This makes it advantageous to use the expression (10.50) for the spectrum, since this form includes contributions only from the interval Δt_q in which the velocity is changing, and for $\beta \ll 1$ it reduces to

$$\frac{dI}{d\Omega} \approx \frac{q^2}{c} |\hat{\mathbf{k}} \times (\hat{\mathbf{k}} \times \langle\dot{\boldsymbol{\beta}}\rangle_\omega)|^2 = \frac{q^2}{c} |\hat{\mathbf{k}} \times \langle\dot{\boldsymbol{\beta}}\rangle_\omega|^2 \tag{10.67a}$$

where
$$\langle\dot{\boldsymbol{\beta}}\rangle_\omega \equiv \int_{-\infty}^{\infty} \frac{dt_q}{2\pi} \dot{\boldsymbol{\beta}}(t_q) e^{i(\omega t_q - \mathbf{k}\cdot\mathbf{r}_q)} = \int_{-\infty}^{\infty} \frac{dt}{2\pi} \dot{\boldsymbol{\beta}} e^{i\omega t} \tag{10.67b}$$

since $\partial t_q/\partial t = (1 - \hat{\mathbf{k}} \cdot \boldsymbol{\beta})^{-1} \approx 1$ for $\beta \ll 1$. With equal scales for t and t_q, $\langle\dot{\boldsymbol{\beta}}\rangle_\omega$ becomes just a Fourier transform of $\dot{\boldsymbol{\beta}}$, and (10.67a) might have been obtained directly from the Larmor formula (10.17) by a simple Fourier analysis of its amplitude, in such steps as those from (10.36) to (10.38). Neglecting the distinction between $\dot{\boldsymbol{\beta}}(t_q)$ and $\dot{\boldsymbol{\beta}}(t)$ and using the last form in (10.67b) corresponds to setting equal to unity the retardation factor distinguished in the intermediate form, as is generally done in dipole ($\equiv$ Larmor) approximations [see the conversion of (8.26) to (8.37) and (8.39), based on $k|\mathbf{r}_q| = |\mathbf{r}_q|/\lambda\!\!\!^- \ll 1$].

Immediate qualitative conclusions are possible about the bremsstrahlung spectrum from any collision. With $\dot{\boldsymbol{\beta}} \approx 0$ outside a collision period Δt_q, the limits on the last integral in (10.67b) can be taken to be $\pm\frac{1}{2}\Delta t \approx \pm\frac{1}{2}\Delta t_q$. Then consider frequencies $\omega < 1/\Delta t$ that might be radiated. For frequencies so low, the variation from unity of the exponential $e^{i\omega t}$ within the integration span can be regarded as negligible, and $2\pi\langle\dot{\boldsymbol{\beta}}\rangle_\omega \approx \Delta\boldsymbol{\beta}$, the velocity change produced by the collision. Higher frequencies $\omega > 2\pi/\Delta t$ will make the integrand oscillate about a zero average within the integration span, and the result for $\langle\dot{\boldsymbol{\beta}}\rangle_\omega$ can be considered negligible. The outcome for the spectrum (10.67a) is

$$\frac{dI}{d\Omega} \approx \frac{q^2}{4\pi^2 c} |\Delta\boldsymbol{\beta}|^2 \sin^2\chi \qquad \omega < \omega_c \tag{10.68a}$$

where χ is the angle between $\Delta\boldsymbol{\beta}$ and the radiation direction $\hat{\mathbf{k}}$ and ω_c is some critical frequency of the order $1/\Delta t$; for $\omega > \omega_c$, $dI/d\Omega \approx 0$. Integration over all directions yields

$$I(\omega < \omega_c) \approx \frac{2q^2}{3\pi c} |\Delta\boldsymbol{\beta}|^2 \qquad I(\omega > \omega_c) \approx 0 \tag{10.68b}$$

as pictured in Fig. 10.10a. The conclusions here are least trustworthy in the vicinity of $\omega \approx \omega_c$, and a possible departure in the ω independence of $I(\omega < \omega_c)$, and from $I(\omega > \omega_c) = 0$, that might occur in some particular case of collision is represented by the curve. Figure 10.10b shows the corresponding intensity per unit of wavelength range, as follows from $d\omega/d\lambda = d(2\pi c/\lambda)/d\lambda = -2\pi c/\lambda^2$.

Better evaluations require looking into some of the details of collision processes. A charged particle penetrating through matter passes through clouds of light electrons surrounding the atomic nuclei, but only rarely does it meet an electron head on. (This is borne out by the relative rarity of the resulting knock-on electrons in cloud-chamber tracks. Most collision energy losses occur in very

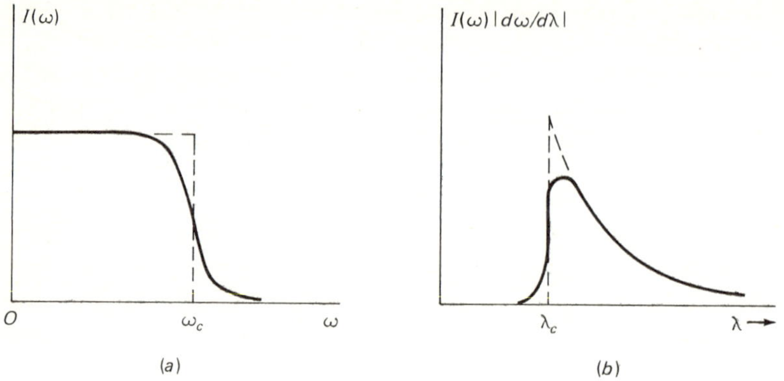

Figure 10.10

small steps as fractional energy transfers, including ones sufficient to produce ionization.) The consequence for the energetic electrons of concern here is that each suffers negligible momentum deflections from its passage through just the light atomic electrons around a nucleus; the principal velocity changes come from the interaction with higher individual charge and much greater mass of the atomic nucleus itself. It is the velocity changes that generate radiation, and x-ray production, for example, is well understood as chiefly the result of encounters of the incident electrons with practically immovable nuclei.

A trajectory of an electron with $q = -e$ past an atomic nucleus of charge $+Ze$ is indicated in Fig. 10.11. The electron is deflected from a straight-line path by a Coulomb attraction of magnitude Ze^2/r^2, and what the eventual deflection angle Θ will be plainly depends on the distance b, called the *impact parameter*, by which the electron fails to approach head on ($b = 0$). The relative probability that an electron will approach with b in a range db is measured by the ring area $2\pi b\, db$ through which it passes, and so a prevalence of larger impact parameters, and consequent small deflections by a force passing through a maximum $\approx Ze^2/b^2$, is to be expected (hence also the rarity of head-on collisions on *any* point of charge). With the nucleus practically immovable, the electron's energy is nearly conserved by itself, and it will undergo an eventual momentum change in direction rather than magnitude. Angular-momentum conservation will require

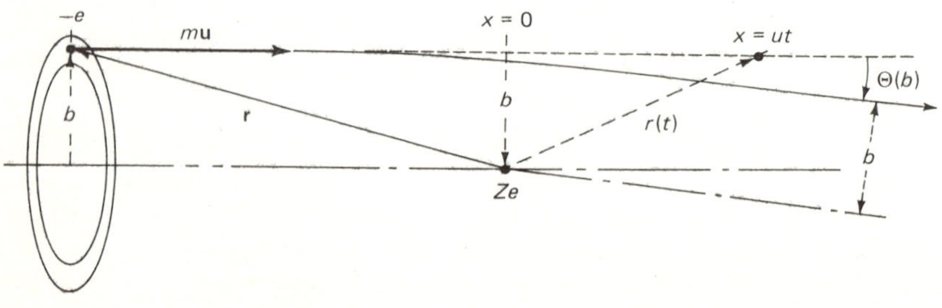

Figure 10.11

an eventual restoration of the lever arm b, as indicated in the diagram. Consistency with linear-momentum convervation comes from the fact that the heavy nucleus can absorb any amount p of order mu without taking appreciable energy ($p^2/2M \to 0$ for $M \gg m$). All the foregoing suggests that it will be a fair approximation to evaluate Ze^2/r^2 with $r^2 = b^2 + u^2t^2$ (see the diagram), as if the electron were continuing along its initial line, without missing essentials about the most prevalent collisions.

The momentum transfer is practically all transverse† to the initial velocity direction, in the prevalent small-angle collisions, and the rate of transverse momentum change is given by the force component $(Ze^2/r^2)(b/r)$ ($\approx m\dot{u} = mc\dot{\beta}$). Accordingly, the magnitude of the integral $\langle\dot{\boldsymbol{\beta}}\rangle_\omega$ of (10.67b), needed for evaluating the radiation, will be calculated as

$$2\pi\langle\dot{\beta}\rangle_\omega = \frac{1}{mc}\int_{-\infty}^{\infty} dt\,\frac{Ze^2b}{(b^2+u^2t^2)^{3/2}}\,e^{i\omega t} = \frac{2Ze^2b}{mc}\int_0^{\infty} dt\,\frac{\cos\omega t}{(b^2+u^2t^2)^{3/2}} \tag{10.69}$$

with the last equivalence following from the evenness in t of the factor multiplying $e^{i\omega t}$. The integral in (10.69) is proportional to a special case of a *Bassett integral* (Ref. 30, p. 172), one of the many integral representations of the modified Bessel functions (C.34)

$$K_m(\xi) = \frac{(m-\frac{1}{2})!}{(-\frac{1}{2})!}(2\xi)^m\int_0^{\infty} dx\,\frac{\cos x}{(x^2+\xi^2)^{m+1/2}} \tag{10.70}$$

It is easy to see that, in such terms, (10.69) becomes

$$2\pi\langle\dot{\beta}\rangle_\omega = \frac{2Ze^2}{mu^2}\frac{\omega}{c}K_1\left(\frac{b\omega}{u}\right) \tag{10.71}$$

which is properly dimensionless (remember that $\langle\dot{\beta}\rangle_\omega$ is a time integral with the same lack of dimensions as $\Delta\beta$). Now the result (10.68b) for the directionally integrated spectrum is to be replaced by

$$I(\omega) = \frac{2e^2}{3\pi c}\,|2\pi\langle\dot{\beta}\rangle_\omega|^2 = \frac{2e^2}{3\pi c^3}\left(\frac{2Ze^2}{mu^2}\right)^2\omega^2K_1^2\left(\frac{b\omega}{u}\right) \tag{10.72}$$

This intensity distribution in frequency, from collisions characterized by any (not too small) impact parameter b, starts out at low frequencies with a constant value as in Fig. 10.10a, since reference to (C.37) shows that $\omega K_1(b\omega/u \ll 1) \approx u/b$ is independent of ω. At frequencies $\omega \gg u/b$ it falls off exponentially (C.35), as

$$\omega^2K_1^2\left(\frac{b\omega}{u} \gg 1\right) \approx \frac{\pi u\omega}{2b}e^{-2b\omega/u} \tag{10.73}$$

† Not at all like the slowing down of the particle along its line that might have been suggested by the words "braking radiation."

The total energy in the spectrum (10.72) can be calculated with the help of $\int_0^\infty d\omega\omega^2 K_1^2 = (u/b)^3 3\pi^2/32$ (C.38) to be

$$\int_0^\infty d\omega I(\omega) = \frac{\pi e^2}{4c}\left(\frac{Ze^2/b}{mcu}\right)^2 \frac{u}{b} = I(0)\frac{3\pi^2}{32}\frac{u}{b} \tag{10.74}$$

equivalent to the energy in a spectrum that is uniform in frequency, with $I(\omega < \omega_c) = I(0)$, up to a critical frequency $\omega_c = (3\pi^2/32)u/b \approx 0.93u/b \approx u/b$.

The result (10.72) is supposed to give a fair approximation to the energy radiated per unit frequency range from a single electron-atom encounter with a specific, not too small, impact parameter b. What can actually be observed is only the results of many encounters with a whole range of b values. Measurements are possible on the energy per unit frequency range that is emitted while a known number of electrons is passing into a target containing a known number of atoms per unit volume. A standard way to express the measurements is per unit incident beam, consisting of one electron incident per unit time and per unit of beam cross section, and per atom in the path of the beam. Then the number of encounters, per unit time, with a given b value is plainly just $2\pi b\, db$ per atom. Moreover, it is nowadays customary to express radiated energies at a given frequency in units $\hbar\omega$, each called a quantum of energy or photon, so that $I(\omega)\, d\omega/\hbar\omega$ is the number of the units emitted in the frequency range $d\omega$ from one of the electron-atom encounters. Thus, the number of the energy units that will be emitted per atom, during the incidence of a unit beam, is

$$d\sigma \equiv d\omega \int \frac{I(\omega)}{\hbar\omega} 2\pi b\, db \tag{10.75}$$

a measurable quantity called a *differential cross section*; it has the dimensions of an area as a result of the standardization to the unit incident beam.

For the theoretical expectations about the cross section (10.75), it will be assumed sufficient to approximate $I(\omega)$ as in (10.68*b*), that is, to take $I(\omega < u/b) = I(0)$ and $I(\omega > u/b) = 0$, with the value $\omega_c \approx u/b$ pointed out just below (10.74) for the critical frequency. Since $\omega^2 K_1^2 = u^2/b^2$ for $\omega \to 0$, (10.72) yields

$$I\left(\omega < \frac{u}{b}\right) \approx \frac{8e^2}{3\pi c}\left(\frac{Ze^2/b}{mcu}\right)^2 \qquad I\left(\omega > \frac{u}{b}\right) \approx 0 \tag{10.76}$$

Where it exists, this is proportional to $1/b^2$, and so the integration (10.75) will make the result proportional to $\ln b| = \ln (b_{max}/b_{min})$, where b_{max} and b_{min} are upper and lower limits on the b values that can contribute significantly (plainly $b_{max} < \infty$ and $b_{min} > 0$ are needed for a finite result). In these terms,

$$\frac{d\sigma}{d\omega} \approx \frac{16}{3}\left(\frac{e^2}{\hbar c}\right)\left(\frac{Ze^2}{mc^2}\right)^2\left(\frac{c}{u}\right)^2 \frac{\ln (b_{max}/b_{min})}{\omega} \tag{10.77}$$

is the result for the bremsstrahlung cross section. Actually, the assumptions (10.76) set an effective upper limit b_{max} at each frequency in that the contributions from $b > u/\omega$ are supposed to be negligible and so $b_{max} \approx u/\omega$ might be taken; however, a stricter upper limit can be set for the low frequencies at which u/ω is large by the fact that when the electron's passage by the nucleus is more distant† than about 10^{-8} cm, enough atomic electrons may intervene to *screen* the incident particle from the coulombic force of the nucleus. A lower limit b_{min} can be considered set by the fact that the above development has some validity only for b's that are "not too small," but this by itself does not preclude significant contributions from $b \to 0$. A treatment valid for close (small b) collisions, in which the great accelerations close to the nucleus can generate velocities approaching c is called for. There would be little profit from a classical treatment of these, because of quantum-mechanical effects. A point significant in this connection is that an electron incident with a finite energy $\frac{1}{2}mu^2$ may not have enough to radiate an entire unit $\hbar\omega$ when $\omega \to u/b_{min}$. Sometimes this fact is used to put $b_{min} \approx 2\hbar/mu$, as follows from $\hbar u/b < \frac{1}{2}mu^2$. That, together with the $b_{max} \approx u/\omega$ quoted above, yields $b_{max}/b_{min} \approx \frac{1}{2}mu^2/\hbar\omega$ for the argument of the logarithm, making $d\sigma/d\omega \to 0$ for $\hbar\omega \to \frac{1}{2}mu^2$. The ambiguities discussed here are not crucial since a logarithmic function is less sensitive to its argument than any power.

The form (10.77) for the bremsstrahlung cross section describes fairly well most of the features that are actually observed, e.g., the proportionality to the square of the nuclear charge responsible for the accelerations and the inverse proportionality to the squared mass of the incident particle. The latter fact is responsible for the lesser importance of direct bremsstrahlung during the passage of heavy particles, like protons or alpha particles, through matter; more is generated by electrons, including those atomic electrons which are accelerated by collisions with heavy incident particles. The inverse proportionality of (10.77) to $(mu)^2$ comes from the fact that high momenta are more difficult to deflect. The shape of the described spectrum is determined principally by the inverse proportionality to the frequency; there are divergently many soft photons, with $\hbar\omega \to 0$, a feature sometimes referred to as an *infrared catastrophe*. Actually, the property $d\sigma/d\omega \to \infty$ as $\omega \to 0$ is a result of overidealization, of treating the Coulomb field as extending uninterrupted to all distances and hence causing slight deflections even at $r \to \infty$. At least the screening effects mentioned above must interrupt this behavior at the lowest frequencies.

10.6 THE FIELD OF A UNIFORMLY MOVING POINT CHARGE

The entire acceleration field of a charged particle, reviewed in the preceding sections, was found to constitute energy lost from the particle permanently, i.e., radiated away. Only its attached velocity field, as identified in (10.9), continues

† There remains a comparatively great range for b, between 5×10^{-13} and 10^{-8} cm, from a typical nuclear radius to an atomic radius enclosing the neutralizing electron cloud around the nucleus.

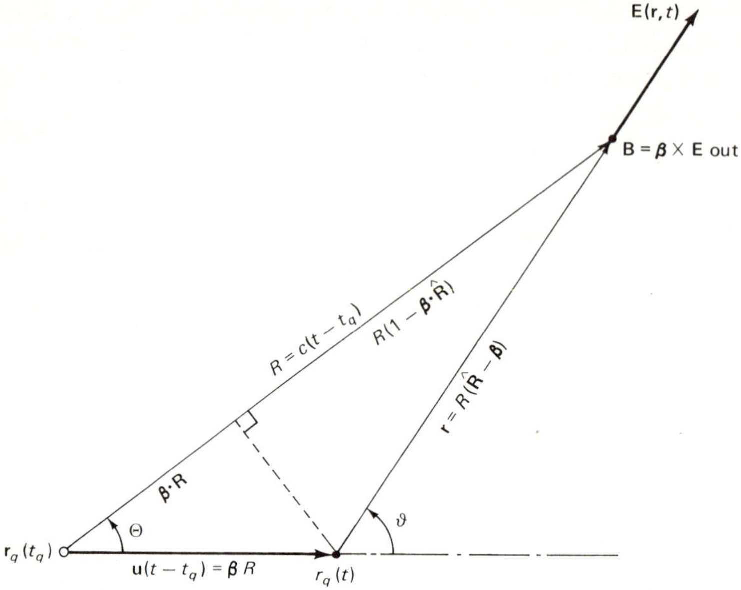

Figure 10.12

to be associated with the particle. It is the *entire* field of a uniformly moving point charge, when $\boldsymbol{\beta} = \mathbf{u}/c$ is a given constant, and will now be discussed as such.

According to (10.9), the *electric* vector at time t has the direction of $\hat{\mathbf{R}} - \boldsymbol{\beta}$ if $\mathbf{R} = \mathbf{r} - \mathbf{r}_q(t_q)$ is the position vector of the field point relative to the charge's position at the retarded time $t_q = t - R/c$, as indicated in Fig. 10.12. The diagram should help make it clear that an $\mathbf{r} \equiv R(\hat{\mathbf{R}} - \boldsymbol{\beta}) = \mathbf{R} - \boldsymbol{\beta}R$ is just the position vector of the same field point relative to the charge's position $\mathbf{r}_q(t)$, reached by the uniformly moving charge while the field $\mathbf{E}(\mathbf{r}, t)$ was being propagated to $\mathbf{r}$. Thus the entire field at any given moment t is formed of radial lines emanating from the position of the charge at the same moment, despite the fact that parts of the field away from the charge were propagated from earlier positions of the charge. The instantaneous field is an anisotropic, radial (longitudinal) field $\mathbf{E} = \mathbf{E}_r$ that can be described as a function of the radii r and the angles ϑ these make with the velocity direction. Multiplying numerator and denominator of the expression in (10.9) by the magnitude R yields

$$\mathbf{E}(\mathbf{r}, t) = \frac{\mathbf{r}q(1 - \beta^2)}{(\mathbf{r} \cdot \hat{\mathbf{R}})^3}$$

and

$$(\mathbf{r} \cdot \hat{\mathbf{R}})^2 = r^2 - \beta^2 R^2 \sin^2 \Theta = r^2 - \beta^2 r^2 \sin^2 \vartheta \tag{10.78}$$

in the terms of Fig. 10.12. Thus

$$\mathbf{E}(r, \vartheta, t) = \hat{\mathbf{r}}\,\frac{q}{r^2}\,\frac{1 - \beta^2}{(1 - \beta^2 \sin^2 \vartheta)^{3/2}} \tag{10.79}$$

is the desired expression. For $\beta^2 \ll 1$, it describes the *isotropic*, Coulomb field of a point charge at rest, but at high velocities the electric flux is concentrated sideways to the velocity, as can be seen from the ratio

$$\frac{\mathbf{E}(r, \frac{1}{2}\pi, t)}{\mathbf{E}(r, 0, t)} = \frac{1}{(1-\beta^2)^{3/2}} \equiv \gamma^3 \qquad (\to \infty \text{ as } \beta \to 1) \tag{10.80}$$

of the transverse to longitudinal flux densities on a given radius r. It can readily be checked that the total flux through every static sphere is just $\oint d\mathbf{S} \cdot \mathbf{E} = 4\pi q$, in conformity with the Gauss law. The field is not static, since the center $\mathbf{r} = 0$ moves with velocity $c\boldsymbol{\beta}$, but the point charge provides the only singularity in it at every moment.

The expression (10.79) for the field, in coordinates based on the charge as origin, has no explicit dependence on time, and this suggests that the field is not static *only* because it must move with the charge. If that is so, the field must have the property

$$\mathbf{E}(\mathbf{r}, t + dt) = \mathbf{E}(\mathbf{r} - \mathbf{u}\, dt, t) \tag{10.81a}$$

i.e., the field vector at a point $\mathbf{r}$ and the instant $t + dt$ must be the same as the instantaneous field at t as evaluated at a point farther back by $\mathbf{u}\, dt$ along the velocity direction. The entire velocity field does indeed have that property, as is most readily confirmed in the form following from the subtraction of $\mathbf{E}(\mathbf{r}, t)$ from both sides of (10.81a); that yields $dt(\partial\mathbf{E}/\partial t) = -(\mathbf{u}\, dt \cdot \nabla)\mathbf{E}$, and hence

$$\frac{\partial \mathbf{E}}{c\, \partial t} = -(\boldsymbol{\beta} \cdot \nabla)\mathbf{E} \tag{10.81b}$$

This can be checked most directly† for $\mathbf{E}$ in the original form (10.9) through the use of the operations in (10.3), (10.5), and (10.6). However, it will prove more instructive to rederive $\mathbf{E}$ in the form (10.79) with the requirement (10.81) as a condition on the solution, since a significant new connection will then be suggested between the velocity field and the simple, isotropic Coulomb field of a static point charge.

All *static* fields ϕ, and $\mathbf{A}$, and hence their derivatives $\mathbf{E} = -\nabla\phi$ and $\mathbf{B} = \nabla \times \mathbf{A}$, must satisfy the Laplace equation $\nabla^2 = 0$ outside the singularities, the charges, in the field. Similarly, all nonstatic fields must be solutions of the wave equation $\nabla^2 - \partial^2/(c^2\, \partial t^2) = 0$. Of the many solutions of the wave equation, the ones that satisfy the condition (10.81) will have

$$\frac{\partial \mathring{\mathbf{E}}}{c^2\, \partial t} = -\frac{(\boldsymbol{\beta} \cdot \nabla)\mathring{\mathbf{E}}}{c} \equiv \frac{\partial^2 \mathbf{E}}{c^2\, \partial t^2} = +(\boldsymbol{\beta} \cdot \nabla)^2 \mathbf{E}$$

† The reader may prefer granting that the velocity field (10.9) satisfies the Maxwell equations $\partial\mathbf{E}/(c\,\partial t) = \nabla \times \mathbf{B}$ and $\nabla \cdot \mathbf{E} = 0$ outside q and that $\mathbf{B} = \boldsymbol{\beta} \times \mathbf{E}$ follows from them. Then

$$\frac{\partial \mathbf{E}}{c\, \partial t} = \nabla \times (\boldsymbol{\beta} \times \mathbf{E}) = \boldsymbol{\beta}(\nabla \cdot \mathbf{E}) - (\boldsymbol{\beta} \cdot \nabla)\mathbf{E} = -(\boldsymbol{\beta} \cdot \nabla)\mathbf{E}$$

is sufficient to show that $\mathbf{E}$ does have the property (10.81b).

since the same statements can be made about the field $\dot{\mathbf{E}}$ as were made about $\mathbf{E}$ in that connection. The last equality simplifies to $\ddot{\mathbf{E}}/c^2 = \beta^2\, \partial^2\mathbf{E}/\partial z^2$ if the z axis is chosen to lie along the velocity direction, and so the wave equation for $\mathbf{E}$ becomes equivalent to

$$\left[\frac{\partial^2}{\partial x^2} + \frac{\partial^2}{\partial y^2} + (1 - \beta^2)\frac{\partial^2}{\partial z^2}\right]\mathbf{E}(\mathbf{r}, t) = 0 \tag{10.82}$$

at any one instant t. This is just a Laplace equation in a space distorted by rescaling distances parallel to the velocity ($u = u_z$) to $z_o \equiv z/(1 - \beta^2)^{1/2} \equiv \gamma z$. Then if

$$r_o^2 \equiv x^2 + y^2 + z_o^2 = x^2 + y^2 + \gamma^2 z^2 \tag{10.83}$$

the particular solution with the right total flux $4\pi q$ from the origin $\mathbf{r} = 0$ is the radial one of magnitude $\mathsf{E}_o = q/r_o^2$, a simple Coulomb field in the distorted space, isotropically distributed on every charge-centered equipotential sphere in it, as indicated in Fig. 10.13*a*. The transformation to undistorted space requires a redistribution of the radial field lines to give equal fluxes into corresponding solid angles of the two spaces

$$\mathsf{E}_o r_o^2\, d(\cos\vartheta_o)\, d\varphi = \mathsf{E} r^2\, d(\cos\vartheta)\, d\varphi$$

Since $\mathsf{E}_o = q/r_o^2$, the flux density in ordinary space must be

$$\mathsf{E} = \frac{q}{r^2}\frac{d(\cos\vartheta_o)}{d(\cos\vartheta)} \tag{10.84}$$

Now

$$\cos\vartheta_o = \frac{z_o}{r_o} = \frac{\gamma\cos\vartheta}{(\sin^2\vartheta + \gamma^2\cos^2\vartheta)^{1/2}} = \frac{\cos\vartheta}{(1 - \beta^2\sin^2\vartheta)^{1/2}} \tag{10.85a}$$

so that

$$\frac{d(\cos\vartheta_o)}{d(\cos\vartheta)} = \frac{1 - \beta^2}{(1 - \beta^2\sin^2\vartheta)^{3/2}} \tag{10.85b}$$

is just the factor needed for agreement between (10.84) and the expression (10.79) derived earlier.

The result for the electric field distribution with $\beta = 0.8$, when $\gamma = \frac{5}{3}$, is indicated in Fig. 10.13*b*. The inner ellipsoid is just the equipotential sphere r_o with its longitudinal dimensions (parallel to the velocity) contracted in accordance with $z = z_o/\gamma = (1 - \beta^2)^{1/2} z_o$, so that on the ellipsoid†

$$r(\vartheta) = \frac{r_o}{\gamma(1 - \beta^2\sin^2\vartheta)^{1/2}} \tag{10.86}$$

† That this describes an ellipsoid can be seen, for example, in Ref. 17, eq. (4.6).

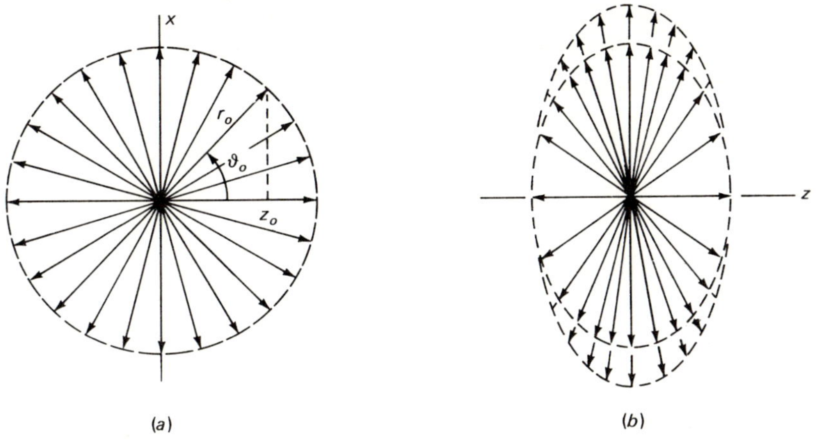

Figure 10.13 (*a*) $\beta_o = 0$; (*b*) $\beta = 0.8$.

as follows from (10.83) with $x^2 + y^2 = r^2 \sin^2 \vartheta$ and $z = r \cos \vartheta$. This is also an equipotential surface, since (10.1) can be written as

$$\phi = \frac{q}{R - \boldsymbol{\beta} \cdot \mathbf{R}} = \frac{q}{\mathbf{r} \cdot \hat{\mathbf{R}}} = \frac{q/r}{(1 - \beta^2 \sin^2 \vartheta)^{1/2}} \tag{10.87}$$

and so $\phi = \gamma q / r_o$ everywhere on the ellipsoid. This equipotential ($\mathbf{A} = \boldsymbol{\beta}\phi$ likewise has a constant magnitude on it!) is distinguished by the fact that the field strength $\mathsf{E}_o = q/r_o^2$ is retained at its "front," $\vartheta = 0$. It is then said that the static field suffers a *Lorentz contraction* in the direction of the velocity by the factor $1/\gamma = \sqrt{1 - \beta^2}$ when the charge is given a velocity $c\boldsymbol{\beta}$. For some purposes, it is more significant to consider a surface on which a given field strength q/r_o^2 is retained everywhere, the outer surface of revolution indicated in Fig. 10.13*b*. On this, as equating (10.79) to q/r_o^2 shows,

$$r(\vartheta) = \frac{r_o}{\gamma(1 - \beta^2 \sin^2 \vartheta)^{3/4}} \tag{10.88}$$

It can then be said that while the longitudinal range of the electric force is Lorentz-contracted, its transverse range is distended by the factor $r(\frac{1}{2}\pi)/r_o = \gamma^{1/2}$. This can be a very long range indeed for $\beta \to 1$ and $\gamma \to \infty$. *In toto*, the radial field assumes a disklike shape, with the disk facing into the direction of motion, contrasting with the narrow, forwardly directed cone formed by the transverse radiation field that is superposed at the high velocities whenever the velocity changes.

While the point charge's Coulomb field suffers the longitudinal Lorentz contraction during its translation, a magnetic field $\mathbf{B} = \boldsymbol{\beta} \times \mathbf{E}$ [Eq. (10.9)] arises. This can be traced to a Maxwell induction of mmf by the electric-flux changes through every circuit in the field, in accordance with (1.39). Consider the electric

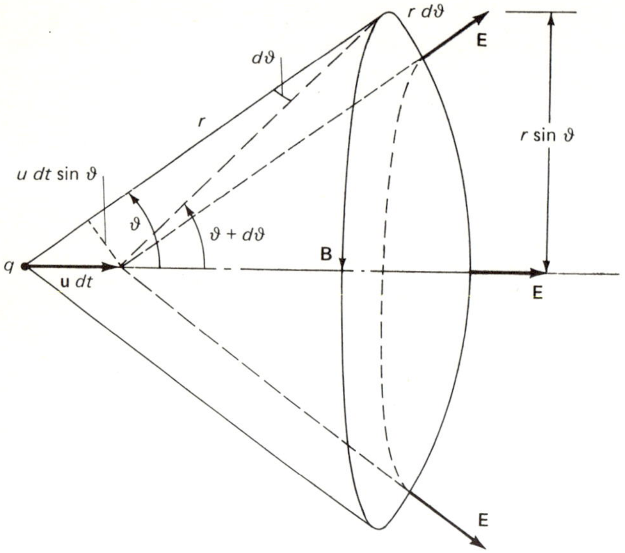

Figure 10.14

flux through a ring of radius $r \sin \vartheta$, having the velocity direction as axis, indicated in Fig. 10.14. A cone of the radial electric field lines making angles up to ϑ passes through the ring initially, but after a displacement $\mathbf{u}\,dt$ of the charge, the same flux passes through a smaller central part of the spherical surface area enclosed by the ring, the remainder of the enclosed area forming an annulus of breadth $r\,d\vartheta = u\,dt \sin \vartheta$, as reference to the figure shows. Now passing through the ring are lines of flux up to the angle $\vartheta + d\vartheta$. There has thus been an increase of electric flux by $2\pi r \sin \vartheta (r\,d\vartheta)\mathsf{E} = 2\pi r \mathsf{E} u\,dt \sin^2 \vartheta$. The *rate* of the flux change, divided by c, has been $(2\pi r u \mathsf{E} \sin^2 \vartheta)/c$, and this is to be equated to the mmf in the circumference $2\pi r \sin \vartheta$ of the ring. The ring itself must form one of the magnetic field lines since these are being induced by an azimuthally symmetric flux (or field current), and so the mmf is just $2\pi r \sin \vartheta\, \mathsf{B}$. The field B must then have the magnitude $\mathsf{B} = (u\mathsf{E} \sin \vartheta)/c = |\boldsymbol{\beta} \times \mathbf{E}|$, as anticipated.

The principal point learned from the present rederivations of the velocity fields $\mathbf{E}, \mathbf{B}$ is a connection between the fields $\mathbf{E}_o = \hat{\mathbf{r}}_o q/r_o^2$ and $\mathbf{B}_o = 0$, as observed in a frame relative to which the particle is static, i.e., its proper rest frame that moves along with it, and the frame of $\mathbf{E}, \mathbf{B}$ in which that rest frame has a uniform velocity $c\boldsymbol{\beta}$. Compare the results thus found with the expectations from galilean relativity, as expressed in (4.68) and (4.69) only for low velocities, $v \ll c$, of two frames in uniform relative motion. For the comparison $\mathbf{v} = c\boldsymbol{\beta}$ should be considered, with $\mathbf{E}' \equiv \mathbf{E}_o$, $\mathbf{B}' \equiv \mathbf{B}_o = 0$. Then (4.69) becomes $0 = \mathbf{B} - \boldsymbol{\beta} \times \mathbf{E}$, and the galilean expectation is $\mathbf{B} = \boldsymbol{\beta} \times \mathbf{E}$ for the low velocities. The finding in this chapter is that this relation continues to hold at high velocities, when $\mathbf{E}$ is evaluated for them, and now the galilean low-velocity expectation is better justified. The expectation from the galilean connection reciprocal to (4.68) is that

$$\mathbf{E} \approx \mathbf{E}_o - \boldsymbol{\beta} \times \mathbf{B}_o = \mathbf{E}_o$$

if only $\beta^2 \ll 1$. This also remains correct *as far as it goes* ($\beta^2 \ll 1$). However, the galilean ideas could not yield consistent results for high velocities, as was seen in the discussion of (4.69). This should perhaps not be entirely surprising since the galilean ideas were formed on the basis of experience with velocities $v \ll c$ only. More surprising, at least initially, is that the "experience" with high velocities, as represented by the connection of the rest-frame field in Fig. 10.13*a* and the same field as observed in a frame relative to which the charge and its field has a high velocity, should require treating spatial distances parallel to the velocity as being Lorentz-contracted. A basis on which such phenomena are only to be expected was formulated by Einstein in his special theory of relativity, which will be reviewed next.

EXERCISES

10.1 (*a*) Show that a substitution of $\rho = q\delta[\mathbf{r}_s - \mathbf{r}_q(t_s)]$ into (8.7) permits immediate integration over the source volume, to produce

$$\phi = q \int \frac{dt_s \delta(\zeta - t)}{R(t_s)}$$

with $\zeta(t_s) \equiv t_s + R(t_s)/c$ and $\mathbf{R} = \mathbf{r} - \mathbf{r}_q(t_s)$.

(*b*) The integration variable can be changed to ζ through $d\zeta(t_s) = (\partial\zeta/\partial t_s)\, dt_s$, and this makes the remaining integration simple. Show that

$$\frac{\partial \zeta}{\partial t_s} = 1 - \frac{\hat{\mathbf{R}} \cdot \dot{\mathbf{r}}_q}{c}$$

and hence that the result for $\phi(\mathbf{r}, t)$ is just (10.1).

10.2 Using techniques illustrated in obtaining (10.4), derive the table of derivatives (10.6).

10.3 Show that the Lienard-Wiechert potentials properly satisfy the Lorentz gauge condition.

10.4 Derive the velocity and acceleration fields (10.9) and (10.10).

10.5 Show that the solid-angle integration of (10.14) for an arbitrary fixed angle between $\dot{\boldsymbol{\beta}}$ and $\boldsymbol{\beta}$ can be reduced to a linear combination of the simple integrals

$$\int_{1-\beta}^{1+\beta} \frac{ds}{s^n} \qquad \text{with } n = 3, 4, 5$$

leading to the result (10.15).

10.6 Extend the calculations of Exercise 10.5 to finding the emission rate of *field momentum* that accompanies the field energy of $\mathscr{P}$ (10.15). Show that the result can be expressed as

$$\oint \frac{d\mathscr{P}}{c} \hat{\mathbf{R}} = \frac{\mathscr{P}}{c^2} \mathbf{u}$$

10.7 Show that the correction to the dipole moment $q\mathbf{a}$ found in Exercise 9.4 accounts exactly for the discrepancy between the relativistic correction and the quadrupole correction noted in the discussion of (10.20).

10.8 Suppose that the wavelength of the incident radiation (10.21) is long enough to encompass the instantaneous positions $\mathbf{r}_j$, with $j = 1, \ldots, Z$, of Z electrons contained within a mean radius s, so that $|\mathbf{r}_{j,\,\max}| < s$ and $k_0 s \lesssim 1$. The reradiated field is then a sum of terms like (10.23), different from each other only in that the jth term has $\mathbf{R} \to \mathbf{R}_j = \mathbf{r} - \mathbf{r}_j$, $\mathbf{r}_q \to \mathbf{r}_j$ and $t_q \to t - R_j/c$.

(*a*) Show that the usual radiation-zone approximation (8.23) for $\omega_0 R_j/c$ leads to multiplying the radiation rate (10.24) by

$$\left\langle \left| \sum_{j=1}^{Z} e^{-i\mathbf{q}\cdot\mathbf{r}_j} \right|^2 \right\rangle$$

where $\mathbf{q} \equiv \mathbf{k} - \mathbf{k}_o$, so that $q = 2k_o \sin \frac{1}{2}\vartheta$ (the quantity $\hbar\mathbf{q}$ is called a *momentum transfer*; see Fig. 11.7).

(*b*) For large enough scattering angles ϑ and wavelengths $2\pi/k_o$ that are not too long, $qs \gg 1$ is possible and the interference (cross) terms of the average introduced in (*a*) oscillate to a zero average; but, for $qs \ll 1$, each of the Z terms forming the amplitude that is squared in (*a*) can be approximated by unity. Show that the consequences for the scattering of unpolarized light are that for any angles having $2 \sin \frac{1}{2}\vartheta > 1/k_o s$, the differential cross section approaches (10.28) multiplied merely by Z, as to be expected at all angles when $k_o s \gg 1$, but for $\vartheta \ll 1/k_o s$, the scattered intensities can be much larger, approaching (10.28) multiplied by Z^2. (The latter characterizes a coherence in the effects of the individual particles.)

(*c*) Why should just the result (10.29) with e replaced by Ze be *expected* (without calculation) for incident wavelengths that are very long in comparison to the region occupied by the Z electrons?

10.9 Because of the sharp forward peaking of the angular distribution (10.35) for $\beta \to 1$, approximations $\sin^2 \vartheta \approx \vartheta^2$ and $\cos \vartheta \approx 1 - \frac{1}{2}\vartheta^2$ should be adequate in it for most of the intensity it represents.

(*a*) Show that the corresponding approximation of $1 - \beta \cos \vartheta$ should be taken to be $\frac{1}{2}(\vartheta^2 + \gamma^{-2})$.

(*b*) In view of the above, $\int_0^\pi d\vartheta \sin \vartheta$ should be approximately replaceable by $\int_0^\infty d\vartheta\vartheta$ when integrating (10.35) over all directions. Show that the approximation procedure suggested here yields the exact results (10.31) for $\chi = \frac{1}{2}\pi$.

(*c*) Show that the same approximation procedure yields $\langle \vartheta^2 \rangle = 1/\gamma^2$ for the indicated average over the angular distribution.

10.10 Losses to radiation must be offset when attempting to accelerate a charged particle in a synchrotron, a device in which the particle follows a circular track of some radius a along which there is a uniform magnetic field.

(*a*) Find an expression for the energy loss per cycle, of period $2\pi a/c\beta$, in which the particle has the speed $c\beta$.

(*b*) Show that the loss per cycle is inversely proportional to the radius a at a given speed. [That is why the highest-energy electron accelerators use a long linear track ($a \to \infty$).]

10.11 Spectral resolutions can also be provided for monodirectional ($\|\hat{\mathbf{k}}$) but nonmonochromatic beams of radiation that are described by some $\mathbf{E}(\mathbf{r}, t)$ with its accompanying $\mathbf{B} = \hat{\mathbf{k}} \times \mathbf{E}$ in a way paralleling the definition (10.38). Now let $\mathscr{I}(\mathbf{r}, \omega)$ represent the time-integrated energy per unit of frequency range $d\omega$ *and* per unit of beam cross section, a *flux* passing a given point $\mathbf{r}$ in the beam. Show that

$$\mathscr{I}(\mathbf{r}, \omega) = c\,|\,\mathbf{E}_\omega(\mathbf{r})\,|^2$$

10.12 A point charge q having a high uniform velocity $\mathbf{u} = c\boldsymbol{\beta}$ passes a fixed observation point that lies at a (least) distance b from the line of motion.

(*a*) Find the transverse (perpendicular to $\boldsymbol{\beta}$) and longitudinal (parallel to $\boldsymbol{\beta}$) electric field components at the observation point as functions of time, taking $t = 0$ to be the moment of closest approach. Make plots of each component vs. t.

(*b*) Find the time interval during which the transverse component exceeds half its maximum value.

(*c*) Find the time interval between the maxima in the magnitude of the longitudinal field.

(*d*) Construct an expression $\mathbf{N}(t)$ for the field-energy flux passing through the observation point.

Show that its longitudinal component depends only on the transverse field components and has a finite time-integrated value whereas the transverse component reverses direction at the moment of closest approach and has a vanishing time integral.

10.13 The transverse electric field $\mathbf{E}_\perp(t)$ of Exercise 10.12 is accompanied by $\mathbf{B} = \boldsymbol{\beta} \times \mathbf{E} = \boldsymbol{\beta} \times \mathbf{E}_\perp$, also transverse to the velocity $c\boldsymbol{\beta}$ of the point charge. For extreme relativistic particles ($\beta \to 1$), $\mathbf{B} \approx \mathbf{E}_\perp$, and so effects of the point-charge field approach equivalence to those of a longitudinally directed pulse of plane waves (said to be constituted of virtual photons accompanying the charge) passing through the observation point. Use (10.70) to show that such a pulse has a flux spectrum, as defined in Exercise 10.11, given by

$$\mathscr{I}_{\parallel}(b, \omega) = \frac{q^2}{\pi^2 c^3} \frac{\omega^2}{\beta^4 \gamma^2} K_1^2\left(\frac{b\omega}{c\beta\gamma}\right)$$

having the same shape (in ω) as the bremsstrahlung (10.72) for a given b. (The effect of the longitudinal electric field can be similarly represented but is comparatively negligible, especially for $\gamma \gg 1$, as reference to Exercise 10.12 shows. The procedure suggested here is known as the *Weizsacker-Williams method.*)

CHAPTER

ELEVEN

EINSTEIN'S SPECIAL THEORY OF RELATIVITY

The Lorentz transformation · Length contraction and time dilation · The relativity of simultaneity · c as the limit on all physically effective velocities · Relativistic Doppler effects · Velocity and momentum transformations · The equivalence of mass and energy · The event and energy-momentum 4-vectors · Effects of energy-momentum conservation on elastic and inelastic processes · Mandelstam variables · Reaction thresholds · Compton scattering and the identification of photon momentum · Recoils from radiation · Mössbauer effect and the gravitation of electromagnetic mass · Annihilation radiation

The classical descriptions of motion possess a type of special relativistic invariance, embodying a version of the idea that only *relative* motion can have meaning. On the basis of the apparently obvious galilean connection $\mathbf{u}'(t) = \mathbf{u}(t) + \mathbf{v}$ [Eq. (4.67)] between velocities with respect to reference frames in uniform relative motion, the same equations of motion, $\mathbf{F} = m\dot{\mathbf{u}} = m\dot{\mathbf{u}}'$, are equally applicable in either frame, to the extent that the same forces can be held responsible for any velocity changes in either. The classical theory of mechanics was thus developed to agree with the experience that, whenever two reference systems are in really uniform relative motion (say† a vessel afloat on a smoothly flowing river and a bank nearby), it is impossible to decide which is "actually" moving. The difference between rest and uniform motion becomes merely a matter of viewpoint.‡ Because of the way the idea is embodied in the classical theory of mechanics, this is said to conform to a galilean principle of relativity.

The beautiful equations that help describe the electromagnetic fields do *not* conform to the galilean principle, except in an approximation for low relative velocities, $v \ll c$, of reference frames. This was seen in the attempt, based on the galilean ideas, to find the connection between field descriptions in relatively moving frames that led to (4.68) and (4.69). Those results are actually inconsistent with each other unless all terms of order $v^2/c^2 \ll 1$ are systematically dropped. A connection putatively valid for the full range $0 < v < c$ was found in the last pages of the preceding chapter for the basic case of a point-charge field. The new connection involves a Lorentz contraction of space that is alien to the galilean principle. Both approaches tacitly assumed that the same constant c

† Another example: passengers aboard an airplane in smooth level flight find that they can trust coffee to pour exactly as it does at home.

‡ How postulating this effectively displaces Newton's first law in the starting basis of mechanical description is emphasized in Ref. 17, sec. 2.1.

should be used in any reference frame, and it is to this treatment that the difficulty of applying the galilean ideas can be traced.

The treatment of c as an invariant constant stems from its first introduction as a mere conversion factor between systems of units, discussed in connection with (4.56). On the other hand, c has also acquired interpretation as the speed of light. Then the galilean velocity connection requires that if $\mathbf{c}$ is the directed velocity of a light flash in one frame, it should be $\mathbf{c}' = \mathbf{c} + \mathbf{v}$ in a frame with respect to which the first frame has the velocity $\mathbf{v}$; not only are the magnitudes of $\mathbf{c}'$ and $\mathbf{c}$ expected to be different, but in the space of at least one of the frames the speed of light is expected to become anisotropic. If the Maxwell equations as they stand are found valid in one frame, they must be expected to take on more complicated forms, yielding anisotropic propagation speeds, in frames moving relative to the first. The frames in which they have their simplest form become *special* ones, and velocities with respect to them gain an absolute significance, quite as if some ethereal medium pervaded all space† and were providing a reference frame at absolute rest.

The questions that thus arose called for direct measurements of effects ascribable to frame motion on the speed of light. A difficulty is that for the purpose measuring apparatus must be given velocities sufficiently comparable to the enormous speed of light to make a detectable difference. The famous Michelson-Morley experiment took advantage of the comparatively high (and semiannually reversed) velocity $v \approx 10^{-4}c$ of the earth in its orbit around the sun. The speed of light parallel to the earth's motion was compared with its speed in transverse directions, and *no* difference could be found, at any time of the year. It was concluded that the speed of light is independent of reference frame, that it is a universal constant.

Much effort was expended on trying to escape that conclusion, simply because the galilean connections were regarded as inescapable. None of the many ingenious alternative explanations could avoid contradicting other evidence. Meanwhile, any necessity for avoiding the Michelson-Morley conclusion was removed by Einstein. Moreover, his reformulations have led to so many consequences—confirmed in so much detail by an enormous variety of applications, including nonelectromagnetic ones—that it would be misleading to leave the impression that they any longer depend solely on any one, or any few, crucial experiments.

By his own account, Einstein paid little attention to the direct tests of the constancy of the speed of light. He presumed that the beautiful simplicity‡ of the Maxwell formulation with a constant c was sufficient ground for expecting it to hold with the same light velocity in every equivalent reference frame. He also assumed that all frames in uniform relative motion should continue to be held

† Compare similar considerations about the galilean relativity of friction in classical mechanics (Ref. 17, pages 40–41).

‡ Not an argument to sway "hard-nosed" scientists unless they recognize the part they play in picking out of experience what is most meaningful to them.

equivalent, as in classical mechanics. It was the use of the galilean connections that he questioned. He recognized that the expectations they stand for were formed from experiences limited to velocities much smaller than the speed of light; it should not be surprising that they might require modification at the enormous velocities comparable to the speed of light.

Like its galilean approximation for low velocities, Einstein's theory is given the qualification "special" because it deals only with reference systems in *uniform* relative motion. It provides a succinct deductive basis for a vast field of verifiable physical expectations by starting from the following two postulates:

Postulate 1 A principle of relativity, reaffirming that *valid physical laws should apply equally well for both observers of any pair in uniform relative motion, each being equally entitled to consider himself at rest, and neither being able to detect within his own system of reference any effects of the motion ascribed to it by the other.*

Postulate 2 An invariance of the speed of light, asserting that *all observers, even when in uniform relative motion, will find the same value c for the speed of light in empty space.*

The *formal* connections between frames that follow from these postulates are considered next.

11.1 LORENTZ TRANSFORMATIONS

To be considered now is how the galilean connections between a pair of reference frames in uniform relative motion must be modified to allow using the same speed c for light relative to either frame.

The connections can be formulated by considering a pair of rectangular frames that coincide at a moment taken to be the zero of time. (Other frames might differ from these by merely *static* displacements, of their origins and/or their orientations.) Also take the z axis parallel to the uniform relative velocity $\mathbf{v}$ of the two reference systems, all as indicated in Fig. 11.1. The relations between the two frames can be discussed in terms of conclusions reached by two *observers, O* and *O'*, moving with respect to each other and at rest in the respective frames, each using his own rest frame for reference. Each of these observers may consider himself to be at rest and the other to be steadily approaching ($t < 0$) and then receding from ($t > 0$) him.

The experience with velocities small in comparison to the speed of light made it seem obvious that the connections

$$x' = x \qquad y' = y \qquad z' = z - vt \qquad \text{for } v \ll c \tag{11.1}$$

are to be expected. These constitute what is called a galilean frame transformation. Implicit in it is a velocity connection $dz'/dt = dz/dt - v$ that can no longer

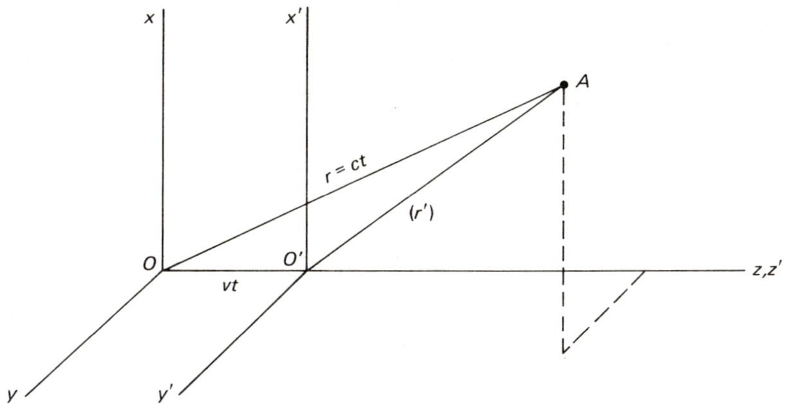

Figure 11.1

be accepted for high velocities, as when $dz/dt = c$ and dz, dz' refer to an interval traversed by light traveling in the z direction.

To introduce the requirement of an invariant light speed, suppose a flash of light originates at $t = 0$ and at the then common origin of the two frames. Consider what is observed when the flash arrives at some point A of space (where there may be a detector, with an observable response, at the moment of arrival). The observer O will assign coordinates x, y, z to the arrival event A and will report $t = r/c$, with $r = (x^2 + y^2 + z^2)^{1/2}$, as the arrival time. The observer O' will report a $t' = r'/c$ with $r' = [(x')^2 + (y')^2 + (z')^2]^{1/2}$. An inequality $r' \neq r$ must certainly be allowed, since the two origins are supposed to have separated during the propagation time of the flash, and so the possibility that $t' \neq t$ must also be considered. Admitting this even as a possibility was difficult to accept on the basis of the experience limited to low velocities. That had led to a conviction that the flow of time is independent of all circumstances, so habitual that it did not ordinarily occur to anyone to make $t' = t$ explicit, as part of the galilean transformation (11.1). Admitting now that different time scales, $t' \neq t$, may be experienced by observers in relative motion amounts to supposing that clocks, performing identically when relatively at rest, will run at different rates when in relative motion.

The above considerations make the problem one of finding a connection between x, y, z, t and x', y', z', t' that makes $r' = ct'$ whenever $r = ct$. It will prove most significant to put this criterion in another way, in terms of the quadratic expressions $r^2 - c^2t^2$ and $(r')^2 - c^2(t')^2$ that can be formed by the respective observers for *any* single point event observed by both. Then the criterion is that both quadratic expressions must vanish whenever and wherever the point event is an arrival of a light flash originating at $\mathbf{r} = \mathbf{r}' = 0$ and $t = t' = 0$. They must consequently *vary* together, for various events in a homogeneous space, occurring at times assigned by identical procedures, so that the equality

$$r^2 - c^2t^2 \doteq (r')^2 - c^2(t')^2 \tag{11.2}$$

should be expected to hold everywhere and at all times for events in which neither side vanishes. The expression $r^2 - c^2t^2$, formed for any one event, should be *invariant*, having the same value whatever the frame to which it refers (and an invariantly zero value for traversals by light).

Some presumptions are discernible in coming to the conclusion (11.2). It is being presumed that the relationships between the event coordinates themselves, i.e., between the values assigned to $\mathbf{r}$, t and $\mathbf{r}'$, t' by the respective observers, are *linear* ones. This is necessary for a uniqueness† in *reciprocal* correspondences between conclusions of equivalent observers. Moreover, changes of scale between the respective values of the coordinates are still to be allowed, and so an important presumption was the absence of an *overall* scale factor between the two sides of (11.2), one that would be common to every dimension.

That a scale change common to every dimension, so that even lengths perpendicular to the direction of the relative motion differ in scale, cannot be admitted becomes evident when observations are made on two identical measuring sticks respectively laid out along the x axes of the two frames indicated in Fig. 11.1. Place each stick symmetrically about the origin of the frame in which it is at rest so that the sticks can coincide at the moment of passage, $t = t' = 0$. Both the observers see a set of *simultaneous*, momentary matchings of graduation marks all along each stick; that is what is *meant* by an orthogonality of the sticks to their relative direction of motion. If there were to be the overall scale change, one of the observers would have to report that a stick moving past him perpendicularly to itself is shrunk by its motion, and then the other observer, seeing the same matchings (or mismatchings!), would have to conclude that a stick with such a motion is lengthened by it. Two equivalent observers would be reaching different conclusions about a phenomenon that ought to be the same for both, according to the principle of relativity. It must therefore be expected that, actually, length scales perpendicular to the direction of motion neither shrink nor lengthen, that the part of the galilean transformation (11.1) that is represented by $x' = x$, $y' = y$ must continue to be valid to high velocities. This also precludes introducing a nonunit ratio of the two sides of the relation (11.2) between coordinates of any event.

The problem has now been reduced to finding a frame transformation with the invariance property (11.2). This is a significant approach to adopt because it constitutes a simple generalization of the problem of finding a connection between frames at some given moment ($t' = t = 0$) and having a common origin. Such frames can differ only in orientation, i.e., be at most relatively rotated, so that $(r')^2 = r^2$ is a necessary invariance. It is quite plain that all relatively rotated orthogonal frames have connections like $x_i' = \sum_j a_{ij} x_j$, where a_{ij} is the cosine of the angle between the axes of x_i' and x_j. The generalization to $(r')^2 - c^2(t')^2 = r^2 - c^2t^2$ can be expressed similarly by introducing, besides $x_1^{(\prime)} \equiv x^{(\prime)}$,

† More evident after (11.3).

$x_2^{(\prime)} \equiv y^{(\prime)}$, and $x_3^{(\prime)} \equiv z^{(\prime)}$, a *fourth* coordinate $x_4^{(\prime)} \equiv ict^{(\prime)}$ into

$$x'_\mu = \sum_{\nu=1}^{4} a_{\mu\nu} x_\nu \qquad \text{such that } \sum_{\mu=1}^{4} (x'_\mu)^2 = \sum_{\nu=1}^{4} x_\nu^2 \tag{11.3}$$

The last equation is just a transcription of $(r')^2 - c^2(t')^2 = r^2 - c^2t^2$, using $i^2 = -1$. The form of connection thus being tried allows only for a most general *linear* relation between the coordinates; a nonlinear one cannot be admitted, for that would lead to nonunique reciprocal correspondences† between equivalent observers.

The definition of a fourth coordinate here amounts to introducing a four-dimensional *space-time* (called a *Minkowski space*) on which each observer can map his experiences. The homogeneous linear connections between alternative frames in the four dimensions, when required to retain the four-dimensional squared lengths in (11.3) unchanged, describe rotations in space-time, and time becomes replaceable by space traversed at light speed, in a certain way. All this serves to emphasize that the concepts of space and time can be regarded as mental constructs adopted to provide terms and frames of reference.‡ Of course, physicists find it desirable to adopt space-time *coordinates* that will have linearly proportional correspondences with what they classify as their "length" and "time" *measurements*.

The problem has now become a matter of finding each of the 16 elements $a_{\mu\nu}$ constituting the matrix of coefficients in the linear transformation (11.3). With the invariance

$$\sum_\mu (x'_\mu)^2 = \sum_{\nu,\rho} x_\nu x_\rho \sum_\mu a_{\mu\nu} a_{\mu\rho} = \sum_\nu x_\nu^2$$

to be required, the 16 coefficients must satisfy

$$\sum_{\mu=1}^{4} a_{\mu\nu} a_{\mu\rho} = \delta_{\nu\rho} \qquad \nu, \rho = 1, 2, 3, 4 \tag{11.4}$$

known as an *orthonormality condition*. This provides only 10 independent equations for the 16 coefficients, since (11.4) is symmetric in the interchange $\nu \leftrightarrow \rho$ and so, besides the four equations with $\delta_{\nu\nu} = 1$, there are here only six independent off-diagonal equations with $\delta_{\nu \neq \rho} = \delta_{\rho \neq \nu} = 0$. The arbitrariness that is left, with its six degrees of freedom, corresponds to the arbitrary choices that must exist for the three components of the relative frame velocity **v** and the three (Euler) angles that can specify the possible relative orientations of the frames in ordinary three-dimensional space.

If $a_{i \neq j} \neq 0$ for $i, j = 1, 2, 3$, the primed and unprimed spatial axes have projections on each other, and so, to obtain connections between such *parallel* three-dimensional frames as the ones indicated in Fig. 11.1, the six conditions

† A primitive example is $x' = x^2 \rightarrow x = +\sqrt{x'}$ or $-\sqrt{x'}$.

‡ Compare remarks on page 31 of Ref. 17.

$a_{i \neq j} = 0$ will be added. This eliminates all but 10 of the 16 coefficients to be determined from the 10 independent orthonormality equations. Of the latter, the three off-diagonal equations with $ij = 12$, 23, 31 immediately reduce to $\sum_\mu a_{\mu i} a_{\mu j} = a_{4i} a_{4j} = 0$, and so at least two of the three coefficients a_{41}, a_{42}, a_{43} must also vanish. The remaining three off-diagonal equations ($\rho = 4$, $\nu = i = 1$, 2, 3) yield

$$\sum_{\mu=1}^{4} a_{\mu i} a_{\mu 4} = a_{ii} a_{i4} + a_{4i} a_{44} = 0$$

Here, none of the four diagonal coefficients $a_{\nu\nu}$ can be allowed to vanish, nor can a_{34}, if there is to be agreement with (11.1) for low velocities. Thus satisfying all the off-diagonal equations requires

$$a_{43} = -a_{34} \frac{a_{33}}{a_{44}} \neq 0 \qquad \text{and} \qquad a_{41} = a_{42} = a_{14} = a_{24} = 0$$

The four diagonal conditions, $\sum_\mu a_{\mu\nu}^2 = 1$, have been reduced to

$$a_{11}^2 = a_{22}^2 = 1 \qquad a_{33}^2 + a_{43}^2 = 1 \qquad a_{34}^2 + a_{44}^2 = 1$$

For the four coefficients a_{33}, a_{44}, a_{34}, a_{43} there remain only three equations to be satisfied, and their solution can be expressed in terms of a_{33}, which will now be denoted $a_{33} \equiv \gamma(v)$, some function of the relative frame velocity v. When signs are adopted (all $a_{\nu\nu} > 0$) to give like senses to variables increasing in parallel, the results for the diagonal coefficients can be written

$$a_{11} = a_{22} = +1 \qquad a_{33} = a_{44} = \gamma > 0$$

Thus $x' = x$ and $y' = y$, as anticipated on page 328.

The only surviving off-diagonal coefficients are related as

$$a_{43} = -a_{34} \qquad \text{and} \qquad a_{34}^2 = a_{43}^2 = 1 - \gamma^2 \tag{11.5}$$

Caution is needed at this point because imaginary scales have been used for $x_4 = ict$ and $x_4' = ict'$. It will now be simplest to reexpress the linear connections in terms of real variables,

$$x' = x \qquad y' = y \qquad z' = \gamma z + ica_{34} t \qquad t' = \gamma t + \frac{ia_{34} z}{c}$$

the conclusions from just the orthonormality conditions about frames moving parallel to each other. The relative velocity of the frames is still to be introduced, and this is most simply done by noting that the observer O attributes the velocity $v_z = v$, and hence $z = vt$, to the origin O', where $z' = 0$. It follows directly that $a_{34} = i\beta\gamma$, with $\beta \equiv v/c$ and $\gamma^2 = 1 - a_{34}^2 = 1 + \beta^2\gamma^2 = 1/(1 - \beta^2)$.

Thus derived is a so-called *Lorentz transformation*

$$\begin{aligned} x' &= x & z' &= \gamma(z - vt) \\ y' &= y & t' &= \gamma\left(t - \frac{vz}{c^2}\right) & \gamma &\equiv \frac{1}{\sqrt{1 - v^2/c^2}} \end{aligned} \tag{11.6}$$

which reduces to the galilean one (11.1), with $t' = t$, for $v \ll c$ (as if $c \to \infty$!). It constitutes the formal basis of Einstein's theory. The name of Lorentz is attached to it because this great investigator had already formulated it as a means of transforming the equations of electromagnetism for application in frames he regarded as having absolute motion, in an ether, relative to which alone was the light speed to be c. It was Einstein who recognized that the conception of the elusive ether was unnecessary, that the existence of absolute motions need not be accepted, and that the same connections were the proper generalizations of the galilean ones to velocities approaching the speed of light. Thus, to Einstein, the Lorentz transformation is not restricted to electromagnetism but helps describe properties of space and time measurements in general. The role of electromagnetism was merely to provide physicists with their first well-defined experiences of very high velocities, like that of light.

The Reciprocal Transformation

It is essential for the applicability of Lorentz transformations to any pair of frames in uniform relative motion that solving for x, y, z, t in Eqs. (11.6) yield, as it does,

$$x = x' \qquad y = y' \qquad z = \gamma(z' + vt') \qquad t = \gamma\left(t' + \frac{vz'}{c^2}\right) \tag{11.7}$$

since the observer O' must attribute the velocity $v_z' = -v$ to the frame of O if v is to be their relative velocity. This will be referred to as the transformation *reciprocal* to (11.6), although the term "inverse" is perhaps more usual.

It will be useful to make explicit the matrices of coefficients in both the transformations (11.6) and (11.7) when they are given the forms

$$x_\mu' = \sum_{\nu=1}^{4} a_{\mu\nu} x_\nu \qquad \text{and} \qquad x_\nu = \sum_{\mu=1}^{4} (a^{-1})_{\nu\mu} x_\mu' \tag{11.8}$$

The coefficients $a_{\mu\nu}$ needed for (11.6) can be presented in the array

$$a \equiv \begin{bmatrix} a_{11} & a_{12} & a_{13} & a_{14} \\ a_{21} & a_{22} & a_{23} & a_{24} \\ a_{31} & a_{32} & a_{33} & a_{34} \\ a_{41} & a_{42} & a_{43} & a_{44} \end{bmatrix} = \begin{bmatrix} 1 & 0 & 0 & 0 \\ 0 & 1 & 0 & 0 \\ 0 & 0 & \gamma & i\beta\gamma \\ 0 & 0 & -i\beta\gamma & \gamma \end{bmatrix} \tag{11.9}$$

and then
$$(a^{-1})_{\nu\mu} = a_{\mu\nu}(\beta) = a_{\nu\mu}(-\beta) \tag{11.10}$$

are obvious from (11.7). The coefficients in the reciprocal transformation are represented as elements of a matrix denoted a^{-1} because of the property

$$\sum_\mu (a^{-1})_{\nu\mu} a_{\mu\rho} = \sum_\mu a_{\nu\mu} (a^{-1})_{\mu\rho} = \delta_{\nu\rho} \tag{11.11}$$

which is written $a^{-1}a = aa^{-1} = 1$ in the language of matrices. The property (11.10) is expressed as $a^{-1} = \tilde{a}$, and is common to all orthonormal frame transformations, the rotations in any number of dimensions.† Other choices than (11.9) for the transformation coefficients can describe connections between nonparallel frames.

The Lorentz-FitzGerald Contraction

Among the many attempted explanations of the Michelson-Morley result was one offered by FitzGerald, who pointed out that the failure to detect a difference between light speeds parallel and perpendicular to the earth's motion could be accounted for by supposing that dimensions parallel to the earth's velocity, of any measuring apparatus, contract by just the factor $(1-\beta^2)^{1/2}$ relative to the perpendicular dimensions. Lorentz expected just such a phenomenon as a result of the contraction in longitudinal range of all electromagnetic forces, as reviewed in connection with Fig. 10.13, since it is equilibria between such forces that can be held primarily responsible for the dimensions of any bodies. However, those explanations presumed the effect to be due to motion relative to an ether. In Einstein's theory, such an effect follows from properties to be attributed to a space spanned by any length measurements. A Lorentz contraction is implicit in the transformation relations (11.6).

Consider a rigid stick of length $L_o = z_2' - z_1'$ lying at rest along the z' axis in the frame of O' indicated in Fig. 11.1. The observer O has that stick passing by him and, at some moment $t = t_1 = t_2$, ascribes coordinates z_1, z_2 to its ends, related to the coordinates $z'_{1,2}$ by (11.6), as $z'_{1,2} = \gamma(z_{1,2} - vt)$. He thus attributes the contracted length

$$L = z_2 - z_1 = \frac{z_2' - z_1'}{\gamma} = \sqrt{1-\beta^2}\,L_o < L_o \tag{11.12}$$

to the stick, moving past him with a velocity $\mathbf{v} = c\boldsymbol{\beta}$ parallel to itself. This, despite the expectation that if the same stick were in O's possession, at rest in his frame, he would also ascribe a length $L_o(= z_2^o - z_1^o)$ to it. It would then have the velocity $v_z' = -v$ relative to observer O', and the latter would now ascribe to it a length given by

$$L' = (z_2^o)' - (z_1^o)' = \gamma[(z_2^o - vt_2^o) - (z_1^o - vt_1^o)]$$

where $t_{1,2}^o$ are the different times that correspond to some one moment t' at which O' observes the end positions $(z_{1,2}^o)'$:

$$t' = \gamma\left(t_1^o - \frac{vz_1^o}{c^2}\right) = \gamma\left(t_2^o - \frac{vz_2^o}{c^2}\right)$$

† Readers unfamiliar with such matters may want to review sec. 9.3 of Ref. 17 or any equivalent account of linear vector spaces. They can also learn that transformations between nonparallel frames, like a general rotation itself, are best performed by *separate* operations, a rotation in 3-space which makes the z axis coincide with the relative velocity direction and then any desired 3-space rotation after the parallel-frame Lorentz transformation. Thus, considering only $a_{11} = a_{22} = 1$, $a_{33} = a_{44} = \gamma$, $a_{34} = -a_{43} = i\beta\gamma$, with all other coefficients zero, is sufficient.

so that $t_2^o - t_1^o = (z_2^o - z_1^o)v/c^2 = L_o v/c^2$ and

$$L' = \gamma L_o\left(1 - \frac{v^2}{c^2}\right) = \sqrt{1-\beta^2}\, L_o < L_o \tag{11.13}$$

Both observers come to the same conclusions about the effect on a stick of motion parallel to it, as required by the principle of relativity, in contrast to the contradiction of the principle by any contractions that might have resulted from motion perpendicular to the stick, discussed following (11.2). The result (11.13) could have been obtained just as briefly as (11.12) was, without resort to the time transformation, if the reciprocal relations (11.7) had been used.

The discussion here has avoided referring to what an observer actually sees, to the visual appearance of a moving body, as seen by an observer of the motion, as against conclusions he must come to about its dimensions while it is in motion. For correct conclusions, the observer must take into account the fact that different parts of the body are at different distances from him and that light reflected from them requires various times to arrive. The consequences for the visual appearances of Lorentz-contracted bodies were pointed out relatively recently by Terrell (Ref. 29), who finds, for example, that a sphere in motion will continue to appear spherical and that a moving cube will appear merely to have had its orientation rotated (with light from the back face of the cube coming into view). Corrections for the effects of ordinary perspectives are habitual to people, but at high velocities more careful analysis is needed to come to correct conclusions from a visual appearance.

Lorentz contractions of dimensions parallel to motion are quite real. Designs of high-energy linear accelerators depend quite indispensably on their existence. The Stanford accelerator is nearly 2 miles long, and that could present insurmountable problems of aiming and focusing particles it accelerates down its narrow tube; even a slight angular spread at the starting point would be enormously magnified for slow particles traveling 2 miles without refocusing of some kind. However, a 10-Gev electron has $\gamma \approx 2 \times 10^4$ and, for it, the 2 miles are contracted to little more than 6 inches, a fact that makes the focusing problem much less formidable.

Time Dilatation, Simultaneity, and Causality

It was apparent in the discussion of the Lorentz contraction, of lengths parallel to their motion, that it is accompanied by relative distortions of time intervals. Consider a succession of two events that, according to observer O' of Fig. 11.1, both take place at the same point in space, with $z_1' = z_2' = z'$. If he finds that a time $\Delta t' \equiv t_2' - t_1'$ has elapsed between the two events, observer O finds them occurring at different places, having

$$z_1 = \gamma(z' + vt_1') \neq z_2 = \gamma(z' + vt_2')$$

as O' passes by, and judges the time elapsed to be

$$\Delta t \equiv t_2 - t_1 = \gamma\left[t_2' - t_1' + v\frac{(z_2' - z_1')}{c^2}\right] = \frac{\Delta t'}{\sqrt{1-\beta^2}} \tag{11.14}$$

Thus an observer (here, O) finds a *dilatation* ($\Delta t > \Delta t'$) of the time between two events taking place at a point moving by him, relative to the time interval reported from the rest frame of that point.

The phenomenon has been confirmed many times and in many ways since the expectation first arose. One of the most direct types of confirmation comes from observing the decays of unstable particles. Each species has a definite mean life τ when it decays at (or near) rest. When it is observed decaying in flight with a high velocity v, it is found to do so more slowly, to have just the mean life $\tau/\sqrt{1 - v^2/c^2}$. Muons generated by cosmic rays at the top of the atmosphere manage to survive the long distance they must travel to the earth's surface, despite a mean life (at rest) of only 2.2 μs, because of their high speeds. As soon as they are slowed down (in solid matter) upon reaching the earth, they vanish within the shorter distances they manage to travel at low velocities in just 2.2 μs. Recently, a telegraphic system based on modulating the intensity of fast muon beams has been devised, relying indispensably on the longer life of muons traveling at a high speed.

The time dilatation implies that a given observer will also find that clocks passing by at a high speed run *slow* in comparison to his own clocks (at rest in his frame). Consider, as an idealized prototype clock, a rod with a mirror at each end that has a pulse of light bouncing back and forth between the mirrors,† being ideally reflected and re-reflected. When this clock is at rest, what can be taken as its period, i.e., the time between successive ticks, is $T_o = 2L/c$ if L is the length of the rod. Now suppose that this clock is in motion, with the rod perpendicular to the direction of the velocity so that no change of its length is observed. An observer of this motion finds that, between ticks, the bouncing light pulse travels some distance vT perpendicular to the rod as well as the distance $2L$ parallel to the rod, the resultant distance being $[(2L)^2 + (vT)^2]^{1/2}$. The light is traversing this distance with a velocity c, as in any such circumstances, and so

$$T = \frac{[(2L)^2 + (vT)^2]^{1/2}}{c}$$

Solving for the tick period of the *moving* clock here gives

$$T = \frac{T_o}{\sqrt{1 - v^2/c^2}} \tag{11.15}$$

in terms of $T_o = 2L/c$, and so a dilated period for the clock in motion. This shows how primitively the effect follows from the constancy of the light velocity. It should be emphasized that while the observer O is finding the clocks in a frame of O' running slow compared with O clocks at rest in his own frame, the observer in O' is finding the O clocks running slow relative to the O' clocks, quite

† Any unidealized clock also has periodic motions, and its various parts communicate changes of their positions to each other via forces propagated with the signal velocity c.

in keeping with the principle of relativity, which entitles each observer to consider himself at rest.

Further implications of the relative time distortions become evident when connections are considered between two events that take place at different points of space for both observers. Suppose O of Fig. 11.1 finds them occurring at z_1, t_1 and z_2, t_2, respectively. O' will find the same events separated by the time interval

$$t'_2 - t'_1 = \gamma\left[(t_2 - t_1) - \frac{v(z_2 - z_1)}{c^2}\right] \tag{11.16}$$

A first conclusion from this expression is that two events that are *simultaneous* for O $(t_2 = t_1)$ are *not* simultaneous for O', in general. The event farther away along the direction of motion (say $z_2 > z_1$) appears to O' to have occurred earlier and at a greater distance from the other, since $z'_2 - z'_1 = \gamma(z_2 - z_1)$ with $\gamma > 1$ for events simultaneous to O. Simultaneity no longer has an absolute meaning but depends on the observer reporting it. Of course when two events are not only simultaneous but taking place at the same point of space for either observer, this is also so for the other, since the events are then virtually indistinguishable from a single event. Notice, moreover, that when two events are apart only in directions perpendicular to the relative motion $(z_1 = z_2,\ x_1 = x'_1 \neq x_2 = x'_2)$, if they are simultaneous for O, they are also simultaneous for O'. That was the situation for the measuring sticks perpendicular to their relative motion, as discussed following (11.2).

It is also possible for the *order* in time of two separated events to be reversed for the two observers, as when $z_2 - z_1 > (t_2 - t_1)c^2/v$ in (11.16). This poses a problem when one of the events, say (z_1, t_1), is held to be a *cause* of the other, (z_2, t_2). However, notice that the interaction responsible for the cause-effect relationship would have to be propagated from z_1 to z_2 with a velocity

$$\frac{z_2 - z_1}{t_2 - t_1} > \frac{c^2}{v} > c \qquad \text{for } v < c \tag{11.17}$$

in order that the reversal in time order occur. It must simply be held that no such causal effects can be propagated with greater than light velocity, just as no $v > c$ can be contemplated in applying the transormation (11.6). This is a highly significant outcome. It promotes the velocity c from its status as just the velocity of light to a universal limit on all effective velocities.

The Relativistic Doppler Effect

The time-dilatation phenomena will also influence what period T or T', and hence frequency $\omega = 2\pi/T$ or $\omega' = 2\pi/T'$, will be ascribed to any harmonic wave by observers in relative motion.

Consider first ascriptions by an observer O, whom a specific wave crest passes at a moment t_1, as indicated in Fig. 11.2*a*. Suppose that the wave is directed toward O' receding from O and at some distance $z(t_1)$ from O at the

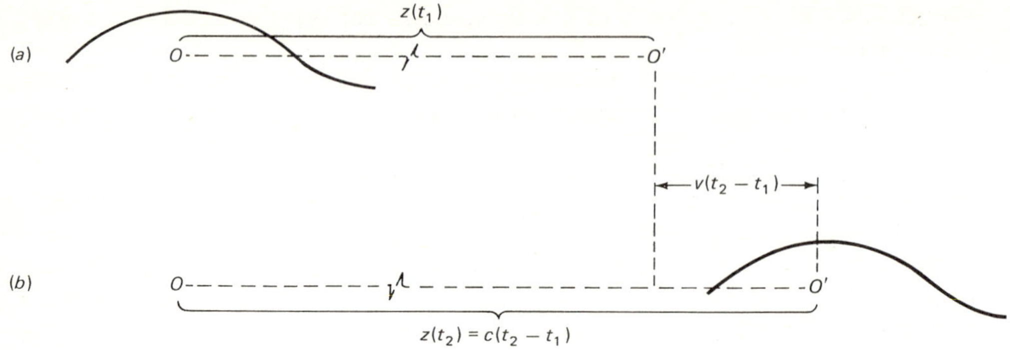

Figure 11.2 Observations by O: (*a*) at $t = t_1$; (*b*) at $t = t_2$.

moment the crest passes O. Let t_2 be the instant, again by O's clock, at which the *same* crest passes O', now at the distance $z(t_2) = z(t_1) + v(t_2 - t_1)$ from O. That length $z(t_2)$ is the entire distance traversed by the wave crest during $t_2 - t_1$, and hence $z(t_2) = c(t_2 - t_1)$ if c is the phase velocity of the wave. A relation

$$(c - v)(t_2 - t_1) = z(t_1) \tag{11.18a}$$

has thus been established. If the wave has a period $\Delta t_1 \equiv T = 2\pi/\omega$ by O's clock, the *next* crest will leave O at his time $t_1 + T$, when O' is at distance $z(t_1 + T) = z(t_1) + vT$, and will arrive at O' at a moment $t_2 + \Delta t_2$ as given by (11.18*a*) for the shifted times,

$$(c - v)[(t_2 + \Delta t_2) - (t_1 + T)] = z(t_1) + vT \tag{11.18b}$$

The difference of this from (11.18*a*) is $(c - v)(\Delta t_2 - T) = vT$, and so

$$\Delta t_2 = \frac{T}{1 - v/c} \tag{11.19}$$

is the time interval (still by O's clock!) between the arrivals of successive crests at O'. This would be the nonrelativistic period $2\pi/\omega_{NR}$ ascribed to the wave by O' if *his* velocity relative to O were small in comparison to the velocity of *light* so that no time dilatation would be detectable and $\Delta t_2' \approx \Delta t_2$. Such a result, $\omega_{NR} \approx (1 - v/c_s)\omega$, is just the Doppler effect on *sound* if the c_s in this result is the velocity of sound (*much* smaller than that of light) for a hearer moving with subsonic velocity through still air from a source of sound at rest in the air.

For the Doppler effect on electromagnetic waves, the c in the result (11.19) refers to the speed of light, and relative velocities v up to this enormous speed can be contemplated. Then an observer moving with O' reports time intervals $\Delta t_2'$ between crests that differ from Δt_2. This is a time interval between events at rest in O' (passages of successive crests *at* O') and so $\Delta t_2 = \gamma\, \Delta t_2'$ is dilated relative to $\Delta t_2' = T' = 2\pi/\omega'$, with T' and ω' the period and frequency as observed by O', gauged by his clocks. Now equating (11.19) to $\gamma\ \Delta t'_2 = 2\pi\gamma/\omega'$ and an inversion lead to

$$\omega' = \gamma\left(1 \mp \frac{v}{c}\right)\omega \tag{11.20}$$

for the Doppler effects as detected by observers respectively moving away from or toward a source of frequency ω. These are only the collinear Doppler effects, with the relative velocity parallel or antiparallel to the direction of the wave. Generalization to other relative directions merely requires replacing v by its component on the direction of the light, as will be seen in a more formal derivation of the entire phenomenon in Sec. 12.1.

Since $\gamma = (1 - v^2/c^2)^{-1/2}$, the shift to the lower frequency (red shift) of the electromagnetic waves can also be written

$$\omega' = \gamma\left(1 - \frac{v}{c}\right)\omega = \left(\frac{1 - v/c}{1 + v/c}\right)^{1/2}\omega = \frac{\omega}{\gamma(1 + v/c)} < \omega \tag{11.21}$$

When the (relative) velocity of source and detector is reversed,

$$\omega' = \gamma\left(1 + \frac{v}{c}\right)\omega = \left(\frac{1 + v/c}{1 - v/c}\right)^{1/2}\omega = \frac{\omega}{\gamma(1 - v/c)} > \omega \tag{11.22}$$

higher than the source frequency ω. These results can be compared to the Doppler effects on sound: $\omega_{NR} = (1 \mp v/c_s)\omega$ for the shifts to lower (higher) frequencies when the source is stationary (in air) and the hearer is moving, while $\omega_{NR} = \omega/(1 \pm v/c_s)$ when the source moves and the listener is stationary in the air. Thus, for sound, velocities relative to the air also matter. For electromagnetic waves, there is no ethereal medium to help define a word like "stationary," and the formal switchovers from $1 \mp v/c$ to $1/(1 \pm v/c)$ are replaced by the equivalences $\gamma(1 \mp v/c) = 1/\gamma(1 \pm v/c)$.

The Relativistic Addition of Velocities

The Lorentz transformation of coordinates implies relations between velocities as they are observed with respect to frames in uniform relative motion. Suppose that observer O of Fig. 11.1 finds that a point particle undergoes a displacement $d\mathbf{r}(dx, dy, dz)$ during a time interval dt. Observer O' will gauge the same displacement as $d\mathbf{r}'(dx', dy', dz')$ and its duration as dt'. According to the transformation (11.6), the relations between them must be such that

$$\frac{dx'}{dt'} = \frac{dx}{\gamma(dt - v\ dz/c^2)} \qquad \frac{dy'}{dt'} = \frac{dy}{\gamma(dt - v\ dz/c^2)} \qquad \frac{dz'}{dt'} = \frac{dz - v\ dt}{dt - v\ dz/c^2}$$

Expressed in terms of the velocities $\mathbf{u} = d\mathbf{r}/dt$ and $\mathbf{u}' = d\mathbf{r}'/dt'$ relative to the respective frames, they are

$$u'_{x,\,y} = \frac{u_{x,\,y}}{\gamma(1 - vu_z/c^2)} \qquad u'_z = \frac{u_z - v}{1 - vu_z/c^2} \tag{11.23}$$

connections that reduce to the galilean expectations, $u'_{x,y} = u_{x,y}$ and $u'_z = u_z - v$ for $v \ll c$, when $\gamma \to 1$. The reciprocal relations

$$u_{x,y} = \frac{u'_{x,y}}{\gamma(1 + vu'_z/c^2)} \qquad u_z = \frac{u'_z + v}{1 + vu'_z/c^2} \tag{11.24}$$

can be obtained either by inverting (11.23) or from the reciprocal transformation (11.7). The last of Eqs. (11.24) shows that when the velocities v and u'_z are appreciable fractions of c, adding them relativistically does not give $v + u'_z$, the galilean expectation, but a smaller result instead, *always less than* c for $v, u'_z < c$, even when $v + u'_z > c$ (see Exercise 11.8).

Notice that when the velocity $\mathbf{u}$ observed by O is that of a light pulse propagated in the z direction, so that $u = u_z = c$, even if the observer O' chases after it with a velocity $v = 0.999c$, it still runs away from him at the full speed

$$u' = u'_z = \frac{c - v}{1 - v/c} = c$$

Conversely, trying to add to the velocity of light by giving its source a high velocity, as described by $u' = u'_z = c$ and $v \to c$, still leaves it with the speed

$$u = u_z = \frac{c + v}{1 + v/c} = c$$

according to the relation in (11.24). Of course just such results were arranged for in constructing the Lorentz transformation.

The consistencies with an invariant light velocity naturally extend to rays not directed parallel to the relative motion of the frames. If, in O', the ray makes polar angles ϑ', φ' with respect to the direction of relative motion as z axis,

$$u'_x = c \sin \vartheta' \cos \varphi' \qquad u'_y = c \sin \vartheta' \sin \varphi' \qquad u'_z = c \cos \vartheta' \tag{11.25}$$

Then the transformation (11.24) gives as components of the light velocity

$$u_x = \frac{c \sin \vartheta' \cos \varphi'}{\gamma(1 + \beta \cos \vartheta')} \qquad u_y = \frac{c \sin \vartheta' \sin \varphi'}{\gamma(1 + \beta \cos \vartheta')} \qquad u_z = \frac{c(\cos \vartheta' + \beta)}{1 + \beta \cos \vartheta'} \tag{11.26}$$

when $\beta \equiv v/c$. It is easy to confirm that $u_x^2 + u_y^2 + u_z^2 = c^2$. The angles ϑ, φ as observed in O are given by

$$\tan \varphi = \frac{u_y}{u_x} = \tan \varphi' \qquad \tan \vartheta = \frac{u_\perp}{u_z} = \frac{1}{\gamma} \frac{\sin \vartheta'}{\cos \vartheta' + \beta} \tag{11.27}$$

where $u_\perp = \sqrt{u_x^2 + u_y^2}$. The result $\varphi = \varphi'$ is due to the fact that the Lorentz transformation is azimuthally symmetric about the direction of the relative frame velocity.

Attention ought still be paid to the consequences of the velocity transformations for the important factors by which lengths are contracted and time intervals dilated. When they are given by $\gamma(u) = (1 - u^2/c^2)^{-1/2}$ for one observer, another will have in its place $\gamma(u') = [1 - (u')^2/c^2]^{-1/2}$, where u' is related to u through the relative velocity v of the two observers, as in (11.23). Starting from

$$(u')^2 = (u'_x)^2 + (u'_y)^2 + (u'_z)^2$$
$$= \left(1 - \frac{vu_z}{c^2}\right)^{-2} \left[u^2 - 2u_z v + v^2 - \left(\frac{v}{c}\right)^2 (u^2 - u_z^2)\right] \tag{11.28}$$

it is easy to find that

$$\gamma(u') = \gamma(u)\gamma(v)\left(1 - \frac{\mathbf{v} \cdot \mathbf{u}}{c^2}\right) \tag{11.29}$$

where $\mathbf{v} \cdot \mathbf{u} \equiv vu_z$ for the frames in relative motion along the z direction.

The relationships between positions and times, as measured not only by velocities but by accelerations (see Exercise 11.9), are the central concern of the classical *kinematics* of point particles. The development of a relativistic kinematics here is more useful to carry on for particles that have been endowed with *mass*, as in the next section.

11.2 THE RELATIVISTIC MASS PARTICLE

Attributing *mass* to a particle† makes it possible to represent its inertia to acceleration, as by the impact on it of another particle. If a particle initially at rest is accelerated to some velocity u by an impinging particle which concurrently undergoes a velocity change Δu_1, then $\Delta u_1/u$ is taken to be the ratio m/m_1 of their masses. Thus, a particle's mass m has always been taken to amount to a scalar coefficient by which its velocity u should be multiplied in order to represent its *momentum* mu, as this engages in processes conforming to the universal conservation of momentum. The interest here is in how such attributions of mass might be influenced by the different conclusions about high velocities now drawn by observers in uniform relative motion. The possibility that differently moving observers will consequently attribute different masses to the same particle should be investigated.

Such a possibility implies that the mass attributed to a particle may vary with the velocity it is given, since to relatively moving observers it differs primarily in velocity. It is the possibility that a particle's mass should be made some function $M(u) \neq M(0) \equiv m$ of its speed‡ that is to be considered. To test the point, it will be simplest to compare observations on a collision of two identical, elastically interacting particles. The equality of the two masses at equal speeds

† See Ref. 17, eq. (2.5) and discussions leading to and following it for a review of the way mass is introduced into classical descriptions of motion.

‡ The mass should not be expected to depend on the velocity *direction*, or on its position, in a space that must be supposed isotropic and homogeneous to the motions of an isolated particle, like *each* of a pair of colliding particles before and after collision.

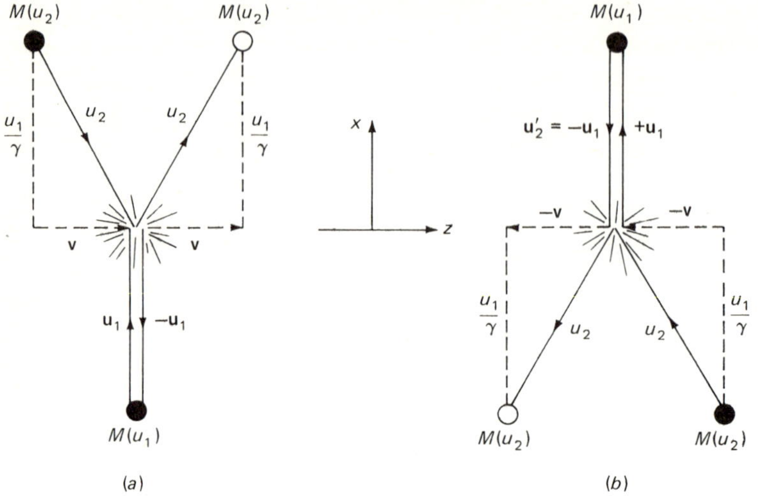

Figure 11.3 (*a*) As observed by *O*; (*b*) as observed by *O'*.

can be established by any observer's simply giving the particles equal and opposite velocities for a head-on collision when a simple reversal of the velocities must be expected of equal masses. Subsequently, a comparison of the same particles at unequal speeds is wanted.

Since the conclusions about the masses at *unequal* speeds are to come from comparing observations in relatively moving reference frames, it is desirable to arrange for a collision that is symmetrical between two observers in relative motion, like O and O' of Fig. 11.1. That can be done† as indicated in Fig. 11.3. Each observer takes charge of one of the like particles and gives it a velocity along his x axis equal and opposite to the velocity given by the other observer to his particle. Thus the observer O' of Fig. 11.2*b* gives his particle a velocity $u'_2 = -u'_{2x} = u_1$ when O gives his particle $u_1 = +u_{1x}$; the observer O then ascribes the velocity components $u_{2x} = -u_1/\gamma$, $u_{2z} = v$ to the O' particle, according to the transformation (11.24). The incidence to collision can be arranged so that each observer has the velocity of his particle simply reversed, with the results indicated in the figure. Then, an application of the momentum-conservation principle by *either* observer leads to the equality of momentum transfers, $2M(u_1)u_1 = 2M(u_2)u_1/\gamma$, and hence

$$M(u_2) = \gamma(v)M(u_1) = \frac{M(u_1)}{\sqrt{1 - v^2/c^2}} \tag{11.30}$$

whenever $\mathbf{u}_1$ is perpendicular to $\mathbf{v}$.

† As suggested by Lewis and Tolman (Ref. 19).

The finding (11.30) might be taken to imply that when $M(u_1) \to M(0) = m$, a mass at rest (or the result of a galilean, low-velocity measurement) so that $u_2 \equiv v$, then $M(u_2) = \gamma(u_2)m$, and so also $M(u_1) = \gamma(u_1)m$. Actually, the finding (11.30) was restricted to a case of $\mathbf{u}_1$ perpendicular to the relative frame velocity $\mathbf{v}$, when the connection (11.29) reduces to $\gamma(u_2) = \gamma(u_1)\gamma(v)$. This transforms (11.30) into an expression of an *invariance*, to the speed of a given mass, of ratios that are independent of the relative frame velocity v

$$\frac{M(u_2)}{\gamma(u_2)} = \frac{M(u_1)}{\gamma(u_1)} = \frac{M(0)}{\gamma(0)} = m$$

since $M(0) \equiv m$ and $\gamma(0) = 1$, confirming the tentative conclusions just above. Thus it is to be expected that whenever a particle of rest mass m is given some velocity u, the mass

$$M(u) = \frac{m}{\sqrt{1 - u^2/c^2}} \tag{11.31}$$

will be attributed to the particle, as the coefficient of u in the expression of its conservable momentum $\mathbf{p} = M\mathbf{u} = \gamma(u)m\mathbf{u}$. The kinetic mass $M(u)$ of a particle in motion is always greater than its rest mass m.

The preliminary experiment suggested above for testing the equality of two masses now has a further implication that becomes most apparent if it is done with two macroscopic elastic spheres with equal rest masses $m \equiv M(0)$. When they are allowed to collide with equal and opposite velocities of magnitude u, the total momentum of the system, $M(u)u - M(u)u = 0$, is conserved at its zero initial value by a final outcome with a simple reversal of the individual velocities. Now consider the fact that there is an intermediate stage at which the zero total momentum is conserved through both particles bringing each other to *rest* with the spheres elastically deformed and all the kinetic energy of the collision absorbed into a potential energy. At this stage of rest, the sum of the masses is $2m < 2M(u)$ and "mass" is not being conserved! What this indicates is that the kinetic masses needed for gauging the momenta of particles in motion differ from their rest masses by an *energy*, in some units, that can be converted into potential energy by such an interaction as that considered here. Such equivalences of mass and energy will be explored further after the next considerations.

The Energy-Momentum 4-Vector

The quantity $M(u) = \gamma(u)m$ is called the *kinetic mass* of a particle having a mass m at rest because it is the coefficient by which its velocity must be multiplied in expressing its momentum

$$\mathbf{p} = M\mathbf{u} = \gamma(u)m\mathbf{u} \equiv \frac{m\mathbf{u}}{\sqrt{1 - u^2/c^2}} \tag{11.32}$$

as this participates in momentum conservation. Now suppose that this refers to the frame O of Fig. 11.1 and that the corresponding momentum in the relatively moving frame O' is denoted

$$\mathbf{p}' = \gamma(u')m\mathbf{u}'$$

Then the relation of $\mathbf{p}'$ to $\mathbf{p}$ is obtainable from the transformation relations for the velocities, (11.23), and for the mass-dilatation factor (11.29):

$$\begin{aligned} p'_{x,\,y} &= \gamma(u)mu_{x,\,y} = p_{x,\,y} \\ p'_z &= \gamma(u)\gamma(v)m(u_z - v) = \gamma(v)[p_z - \gamma(u)mv] \end{aligned} \tag{11.33}$$

To be noticed here is that the momentum components transform exactly like the coordinates themselves, in the Lorentz transformation (11.8), i.e., with the same matrix of coefficients, if a *fourth* component,

$$p_4 \equiv i\gamma(u)mc = iM(u)c \tag{11.34}$$

is defined. Then, also, the transformation relations (11.33) are supplemented by the fourth one:

$$p'_4 \equiv i\gamma(u')mc = \gamma(v)\left(p_4 - \frac{ivp_z}{c}\right) \qquad \text{or} \qquad \gamma(u') = \gamma(v)\gamma(u)\left(1 - \frac{vu_z}{c^2}\right) \tag{11.35}$$

just a reiteration of (11.29), already found valid.

This is an example of a space-time 4-vector, $p_\mu(p_x, p_y, p_z, p_4) \equiv p_\mu(\mathbf{p}, iMc)$, a name given to any four-component quantity that transforms like the event 4-vector $x_\mu(\mathbf{r}, ict)$. The property that, whenever

$$x'_\mu = \sum_\nu a_{\mu\nu} x_\nu \qquad \text{then} \qquad p'_\mu = \sum_\nu a_{\mu\nu} p_\nu \tag{11.36}$$

is said to describe the *covariance* of 4-vectors with frame transformations. Just as the orthonormality property (11.4) of the coefficients ensures that $\sum x_\nu^2 = r^2 - c^2t^2$ has the same invariant value in every frame, so

$$\sum p_\nu^2 = \mathbf{p}^2 - M^2c^2 = \gamma^2(u)m^2(u^2 - c^2) = -m^2c^2 \tag{11.37}$$

has the same value in every frame for a particle of a given mass; the value $-m^2c^2$ of the invariant also follows simply from letting $u \to 0$ on the left side of the last equality, that being characteristic of the particular (proper) frame of the particle at rest.

The quantity $-icp_4 = Mc^2$ is also called the *energy* of the free particle while it has the velocity that determines $M(u)$ and is denoted

$$E \equiv Mc^2 = \frac{mc^2}{\sqrt{1 - u^2/c^2}} = \gamma mc^2 \tag{11.38}$$

This is considered appropriate particularly because an expansion in powers of u^2/c^2 yields

$$E \approx mc^2\left(1 + \frac{u^2}{2c^2} + \frac{3}{8}\frac{u^4}{c^4} + \cdots\right) \tag{11.39}$$

and so $E - mc^2 \approx \frac{1}{2}mu^2$ for $u \ll c$, just the c-independent galilean kinetic energy of a particle with a rest mass m and a velocity u. In general,

$$T = E - mc^2 = mc^2\left(\frac{1}{\sqrt{1 - u^2/c^2}} - 1\right) = (\gamma - 1)mc^2 \tag{11.40}$$

is called the *relativistic kinetic energy* of the particle. Its total energy at velocity u

$$E = T + mc^2 = Mc^2 = \gamma mc^2 \tag{11.41}$$

is said to include, besides its relativistic kinetic energy, a rest energy

$$E_o \equiv M(0)c^2 = mc^2 \tag{11.42}$$

Such an equivalence between energy and mass has already been indicated by the example discussed in the last paragraph of the preceding subsection, but only for $[M(u) - m]c^2 = T$. However, the convertability of even the rest masses into kinetic energies (usually of particles with smaller rest masses, into which the initial mass divides) has been frequently confirmed (and much publicized). A consequence is that the separate conservation of mass and of energy, as known in classical physics, is replaced by the joint conservation of a total mass-energy, $\sum M_n c^2 = \sum (T_n + m_n c^2)$.

A significant implication of the mass-energy relation (11.38) is that it would cost an infinite amount of energy to accelerate a rest mass m to the speed of light, and hence $u < c$ always for particles with rest mass. This is quite aside from the infinite energy needed to accelerate a *charged*, and hence radiating, particle to the light velocity, as reviewed in connection with the radiation rate (10.31).

The equivalence $E = Mc^2$ has led to expressing many of the relations reviewed above in terms of E in place of $M(u)$. Thus the connection between velocity and momentum, $\mathbf{u} = \mathbf{p}/M$, is frequently to be seen as

$$\mathbf{u} = \frac{c^2\mathbf{p}}{E} \tag{11.43}$$

The 4-vector momentum p_μ of (11.36) can now be called the energy-momentum 4-vector of a particle, since it is decomposable as indicated by

$$p_\mu\left(\mathbf{p}, \frac{iE}{c}\right) \tag{11.44}$$

The invariant sum (11.37) of the squared components can now be written as

$$\sum_\mu p_\mu^2 = \mathbf{p}^2 - \frac{E^2}{c^2} = -m^2c^2 \qquad \left(\equiv -\frac{E_o^2}{c^2}\right) \tag{11.45}$$

Thus the energy-momentum relation

$$E(\mathbf{p}) = \sqrt{c^2\mathbf{p}^2 + m^2c^4} \tag{11.46}$$

replaces the $T = p^2/2m$ ($\approx E - mc^2$) of low-velocity physics. It also becomes appropriate to express the 4-momentum transformation relations (11.33) and (11.35) as

$$p'_x = p_x \qquad p'_y = p_y \qquad p'_z = \gamma(v)\left(p_z - \frac{Ev}{c^2}\right) \qquad E' = \gamma(v)(E - vp_z) \quad (11.47)$$

It is easy to recheck their consistency with the invariance (11.45) by forming $(p')^2 - (E')^2/c^2$. The reciprocal of (11.47) can, as usual, be formed merely by interchanging the primed and unprimed components and reversing the sign of the relative frame velocity v. If the primed frame is the rest frame of the particle, so that in it the momentum 4-vector has the components indicated by $p'_\mu(\mathbf{p}' \equiv 0, iE_o/c = imc)$, the (reciprocal) transformation to the unprimed frame yields

$$p_{x,y} = 0 \qquad p_z = \gamma(v)mv = M(v)v \qquad E = \gamma(v)mc^2 = M(v)c^2 \quad (11.48)$$

as it should, since the particle has the velocity $v = v_z$ in this frame. The particle is said to have received a Lorentz *boost* by referring it to a frame in which it has velocity.

The relations developed here take on particularly simple forms for particles that have been found to possess *no rest mass*, the photons and neutrinos. These particles can have finite energies E in accordance with the relation (11.38), despite $m = 0$ for them, if, and only if, they always travel with the limiting velocity c relative to every possible reference frame. Indeed, no way has ever been found to slow down a photon or a neutrino without destroying it (as by an absorption process). The conclusions that $m = 0$ came most directly from collision processes in which particles of energy E are found (as reviewed in Sec. 11.4) to participate in momentum conservation with magnitudes

$$p_o = \frac{E}{c} \qquad \text{for } m = 0 \quad (11.49)$$

of their momenta, as to be expected from the velocity-momentum relation (11.43) for $u = c$, or a zero value of the invariant (11.45). The 4-vector momenta (11.44) of such particles can be decomposed as in $p_\mu(\mathbf{p}, ip_o)$ and the relations (11.47) between their momentum components in relatively moving frames reduce to

$$\begin{aligned} p'_x &= p_x \qquad & p'_z &= \gamma(p_z - \beta p_o) \\ p'_y &= p_y \qquad & p'_o &= \gamma(p_o - \beta p_z) \end{aligned} \qquad m = 0 \quad (11.50)$$

with $c\beta$ the relative frame velocity and $\gamma \equiv 1/\sqrt{1 - \beta^2}$. No frame exists in which $\mathbf{p}' = 0$ unless $p'_o = E'/c = 0$ also, and then the particle does not exist! The invariant velocity c of the particles cannot be transformed away, as any speed $u < c$ of a particle with rest mass can, simply by referring it to its rest frame. Photons and neutrinos have no rest frames.

Rest-mass-less particles fit into a review of mass particles in that kinetic masses $M(c) = p/c = E/c^2$ may still be ascribed to them. There is point in doing so because photons passing close by the sun have been found subject to its gravitational attraction† in proportion to their energies $E = M(c)c^2$. $M(u)$ is not only an inertial ($\equiv p/u$) but also a gravitational mass of a particle that has velocity u.

It is perhaps also significant that no *charged* particle has ever been found without rest mass. If such a particle had a velocity c, it would presumably radiate at an infinite rate, as suggested by the rate formula (10.31), simply by being looked at from a frame that is accelerated, *no matter how slightly*. A rest-mass-less charge would then have to destroy itself, in violation of the conservation of charge!

Energy-Momentum Conservation

Any number of particles with individual momenta $\mathbf{p}_1, \mathbf{p}_2, \ldots, \mathbf{p}_n, \ldots$ might be brought together to form a composite system with the resultant momentum $\mathbf{P} = \sum_n \mathbf{p}_n$. The individual momenta in this sum may change with time if the constituents of the composite undergo mutual interactions and hence momentum exchanges, but the total $\mathbf{P}$ must remain conserved as long as the entire system is isolated from any external influence. That is just what is meant by the *isolation* of a system. After a time of the mutual interactions, the total momentum might again become decomposable into constituent free-particle momenta, generally different from the initial ones, and

$$\mathbf{P}^f \equiv \sum_m \mathbf{p}_m^f = \mathbf{P}^i \equiv \sum_n \mathbf{p}_n^i \tag{11.51}$$

or $\Delta\mathbf{P} = \mathbf{P}^f - \mathbf{P}^i = 0$ is to be expected. The superscripts i and f refer to initial and final phases in the interaction process. Such momentum-conservation relations as (11.51) are helpful in determining individual momenta $\mathbf{p}_m^f$ that might be freed as a result of a process initiated with given momenta $\mathbf{p}_n^i$.

Fourth components like iE_n/c of the individual energy-momentum 4-vectors can also be added together to constitute a component $P_4 = iE/c$ with

$$E = \sum_n E_n \tag{11.52}$$

representing a total energy of the composite system. The identifiability of a conserved total energy has always been found to be an additional characteristic of isolated systems, but now, in the new relativistic context, a conservation $\Delta E = E^f - E^i = 0$ of such energies as (11.52) *follows* from just the momentum conservation $\Delta\mathbf{P} = 0$ (and vice versa). This stems from the expectation that the conservation principles will hold whatever reference frame might be used in applying them (the relativity principle). With the constituent-particle energy-momentum 4-vectors, as viewed from relatively moving frames, connected by the transformation relations (11.47), their sums $P_\mu(\mathbf{P}, iE/c)$ and $P'_\mu(\mathbf{P}', iE'/c)$ will be

† Not until after Einstein predicted that this would be found, on the basis of a *principle of equivalence* embodied in his *general* theory of relativity.

related in exactly the same way, with the same matrix of coefficients. This is also true for $\Delta\mathbf{P} = \mathbf{P}^f - \mathbf{P}^i$ and $\Delta E = E^f - E^i$, so that

$$\Delta P'_{x,y} = \Delta P_{x,y} \qquad \Delta P'_z = \gamma\left(\Delta P_z - \frac{v\,\Delta E}{c^2}\right) \qquad \Delta E' = \gamma(\Delta E - v\,\Delta P_z)$$

It is then evident that $\Delta\mathbf{P} = 0$ and $\Delta\mathbf{P}' = 0$ together lead to the expectations $\Delta E = \Delta E' = 0$. Conversely, $\Delta E = 0$ and $\Delta E' = 0$ immediately yield $\Delta P_z = \Delta P'_z = 0$ and also $\Delta\mathbf{P} = \Delta\mathbf{P}' = 0$ for all components, since the relative frame velocity could be chosen to have any direction. The basic reason that the momentum and an energy must be thus conserved together is that they transform into each other merely by being viewed from different reference frames.

The result of the considerations here is that the momentum-conservation relation (11.51) is generalized to the four-component relation

$$P^f_\mu \equiv \sum_m p^f_{m\mu} = P^i_\mu \equiv \sum_n p^i_{n\mu} \tag{11.53}$$

or $\Delta P_\mu \equiv P^f_\mu - P^i_\mu = 0$. Each of the resultant 4-vectors transforms like the individual 4-momenta in (11.36) upon reference to relatively moving frames.

The transformation properties of the 4-momenta, as extended to the composites like $P_\mu = \sum_n p_{n\mu}$, lead to an invariance like (11.45)

$$\sum_\mu P^2_\mu = \mathbf{P}^2 - \frac{E^2}{c^2} = -\frac{E^2_o}{c^2} \tag{11.54}$$

where E_o is the energy of the composite in the special frame relative to which $\mathbf{P}_o = \sum_n \mathbf{p}^o_n = 0$. This special frame is the *center-of-mass* (CM) frame of the system, defined by the vanishing of the total 3-momentum relative to it, exactly as in the procedures of classical mechanics.

The velocity of the CM frame with respect to the initial frame (of $\mathbf{P} \neq 0$) can be found from the transformation relations (11.47) if, for the purpose, the z axis is chosen parallel to $\mathbf{P}$, so that $P_{x,y} = 0$ and $P_z = P$. Then

$$P_o = \gamma(v)\left(P - \frac{Ev}{c^2}\right) = 0$$

and the CM-frame velocity is

$$\mathbf{v} = \frac{c^2\mathbf{P}}{E} \tag{11.55}$$

as for the velocity (11.43) of an individual particle except that here $\mathbf{P}$ and E/c^2 are the total momentum and kinetic mass of a composite system. [The remaining one of the transformation relations (11.47) now merely reaffirms the connection (11.54) between E_o and E.]

At a phase in the system motion at which E is decomposable as in (11.52),

$$\mathbf{v} = c^2\frac{\sum_n \mathbf{p}_n}{\sum_n E_n} = \frac{\sum_n M_n(u_n)\mathbf{u}_n}{\sum_n M_n(u_n)} \tag{11.56}$$

where $\mathbf{u}_n$ is a velocity of the nth constituent and $M(u_n) = \gamma(u_n)m_n$ is its kinetic mass if m_n is its rest mass. If the constituent is one of the rest-mass-less particles, $M_n \equiv E_n/c^2$ and $u_n = c$ in magnitude. The expression (11.56) reduces to the classical definition of a CM velocity $\mathbf{v} \to \sum m_n \mathbf{u}_n / \sum m_n$ when all the constituents have rest mass and $u_n \ll c$ for each. The generalization (11.56) to relativistic particles merely requires weighting by the kinetic masses in place of the rest masses that are used in the classical mechanics.

It is a remarkable fact that a CM frame with $v < c$ can exist even for a system composed *entirely* of rest-mass-less particles with individual, directed velocities $\mathbf{u}_n \equiv \mathbf{c}_n$ each of magnitude c. Then the resultant CM velocity becomes c in magnitude only if the individual velocities $\mathbf{c}_n$ are all parallel to each other.

The fact that an energy-momentum 4-vector can be ascribed to a composite system implies that this can thus participate "as a body" in the conservation of energies and momenta during interactions with other systems external to the composite in question. The composite body behaves in this respect like a *single* particle, of rest mass

$$m^* = \frac{E_o}{c^2} \tag{11.57}$$

since E_o of (11.54) is the energy of the composite in the frame relative to which the CM is at rest. Each of the individual component momenta discriminated in such resolutions as (11.53) might itself belong to some composite of further individualized particles.

Particle Intertransformations

The two elastic spheres engaged in collision interaction with equal and opposite velocities u, discussed on page 341, can be regarded as forming a composite system. The frame of those velocities is itself the CM frame of the composite, since $\mathbf{v} = [M(u)\mathbf{u} - M(u)\mathbf{u}]/2M(u) = 0$ in it, and so the total energy $2M(u)c^2 = E_o$ of the two spheres forms the total rest mass $m^* = 2M(u)$ of the composite. However, the elastic constituents of this rest mass are at *relative* rest only in the instantaneous phase in which their kinetic energies are entirely converted into the elastic potential energy.

There is perhaps more point in identifying such a composite rest mass when, instead of being elastic, the two spheres are made up of some puttylike substance that causes them to remain stuck together at rest after the collision. The initial kinetic energy $2[M(u) - m]c^2 = (m^* - 2m)c^2$ is now converted into internal thermal motions, a hot (excited) particle being the result. The "same" particle may be regarded as being formed, but with a rest mass $m^o = 2M(0) = 2m$, when the two spheres are simply pressed together with negligible relative velocity. Then the above collision process is said to have formed the same particle in an excited state, with the excitation energy $(m^* - m^o)c^2$.

Suppose that this excitation is great enough, i.e., the conglomerate hot enough, to boil off a fragment of considerable mass. This fragment frees itself with some momentum, and the residual mass will recoil with an equal and opposite momentum. The outcome is a coming apart into generally unequal, new masses replacing the initial spheres. Such a process, in which new particles are formed, is called an *inelastic collision*, contrasting with the elastic collision in which the equal initial spheres sprang apart again.

Instead of boiling off fragments with some velocity less than c, the conglomerate may have become red-hot, ridding itself of the excitation energy $(m^* - m^o)c^2$ in the form of radiation, the final cold residuum having the rest mass m^o. If the heating is somehow nonuniform, so that the radiation comes off one-sidedly, it carries off a *net* field momentum like that first discussed in Chap. 6. The residual mass recoils from this just as from the boiling off of massive particles, and part of the excitation energy appears as a recoil kinetic energy, in place of radiation energy.

Attention was paid to these rather primitive examples only because they provide simple counterparts of processes important in molecular, atomic, nuclear, and elementary-particle physics and also in the interactions of particles with radiation. Besides the simple elastic scattering of the various types of particles, there are numerous processes in which new particles are formed, with particle intertransformations taking place. Interatomic collisions may result in molecular formation, and/or the excitation of radiation, or in ionization, with electrons ejected and charged ions formed. Transformations into new species can occur in collisions of atomic nuclei. Some of the new nuclei may be radioactive, emitting β electrons and neutrinos in place of electromagnetic radiation. The high-energy collisions of elementary particles have resulted in the formation of a variety of "strange particles" that decay into stabler fragments in short times.

Expectations about all such processes can be derived from the conservation relations (11.53). The possibility of particle intertransformations means that the final resolution $\sum_m p^f_{m\mu}$ may refer to an entirely different set of particles from the initial one $\sum_n p^i_{n\mu}$; even the total number may be different at the two stages (and hence the use of different summation indices, m and n). Some instructive examples will be treated in the following sections.

11.3 ENERGY-MOMENTUM CONSERVATION IN PARTICLE REACTIONS

More explicit conclusions from energy-momentum conservation about processes involving particles, simple or composite, will now be expressed for initial and final resolutions into two-particle states

$$P_\mu = p_{1\mu} + p_{2\mu} = p_{3\mu} + p_{4\mu} \tag{11.58}$$

where each $p_{n\mu}$ is the 4-momentum in some frame of a particle with rest mass m_n (and some $m_n = 0$ *not* excluded). The two-particle resolutions are particularly apt for *elastic* collisions, when $m_3 \equiv m_1$ (but $p_{3\mu} \neq p_{1\mu}$) and $m_4 \equiv m_2$. Allowing for the possibility that $m_{3,4} \neq m_{1,2}$ will make it possible to draw conclusions about

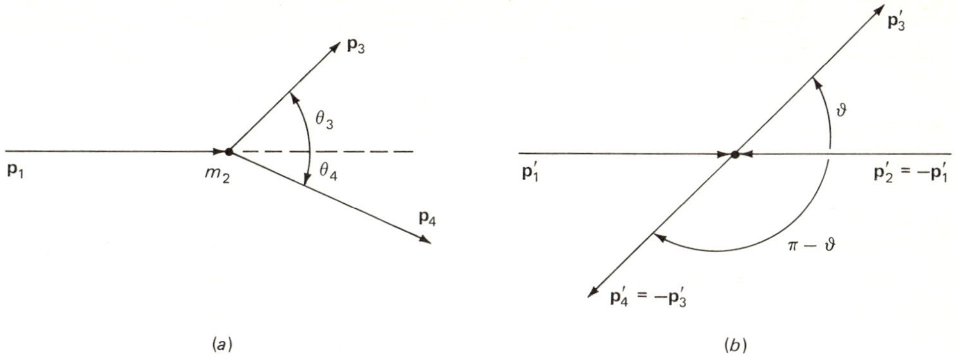

Figure 11.4 (*a*) In the laboratory frame; (*b*) in the CM frame.

inelastic processes. Moreover, some of the results will be applicable to the spontaneous disintegration of an isolated particle, m_1 say, through putting $p_{2\mu} \equiv 0$. Conclusions about other than two-particle final states can also be rendered obtainable, through putting $p_{4\mu} \equiv 0$ or by letting $p_{3\mu}$ and/or $p_{4\mu}$ stand for some sum of further individualizable 4-momenta.

The two-particle resolution $P_\mu = p_{1\mu} + p_{2\mu}$ of an *initial* state is particularly important to consider because, in the laboratory, it is not practical to arrange for a significantly close encounter of more than two particles at a time. If either of these particles is a composite system, it must be a sufficiently stable one, with its constituents interacting so strongly that its momentum cannot be expressed as any sum of such constituent momenta as characterize particles. The standard way processes encompassed by (11.58) are initiated in the laboratory is indicated in Fig. 11.4*a*. Some particle of rest mass m_1 is given a momentum $\mathbf{p}_1$ toward a particle m_2 that is practically at rest in the laboratory, so that $\mathbf{p}_2 = 0$, $E_2 = m_2 c^2$ is taken to define a laboratory frame. The product state is analyzed into two momenta $\mathbf{p}_{3,4}$ that may or may not be composites of several free-particle momenta. Thus, the consequences of energy-momentum conservation will be sought for a final state description by $E_{3,4} = (p_{3,4}^2 + m_{3,4}^2)^{1/2}$ and emergence angles $\theta_{3,4}$. As indicated by these expressions for $E_{3,4}$, *such units will hereafter be used that* $c = 1$ *in them.*

Energy-momentum conservation, (11.58) as applied in the laboratory frame, provides just the *three* relations

$$
\begin{aligned}
E_1 + m_2 &= E_3 + E_4 \\
p_1 &= p_3 \cos\theta_3 + p_4 \cos\theta_4 \qquad (11.59)\\
0 &= p_3 \sin\theta_3 - p_4 \sin\theta_4
\end{aligned}
$$

to help determine the *four* quantities $E_{3,4}$, $\theta_{3,4}$ in terms of a given $E_1 = (p_1^2 + m_1^2)^{1/2}$ (and $m_{1,2,3,4}$). Thus the conservation relations are not by themselves sufficient to determine the outcome, and any results for $E_{3,4}$, $\theta_{3,4}$

should be expressed in terms of some one additional parameter that is unrestricted by the conservation conditions.

The situation becomes plainer when it is viewed from the CM frame defined by $\mathbf{p}'_1 + \mathbf{p}'_2 = 0$ and indicated in Fig. 11.4*b*. Relative to this frame, the total 4-momentum has the components indicated in

$$P'_\mu[(\mathbf{p}'_1 + \mathbf{p}'_2), i(E'_1 + E'_2)] = P'_\mu[(\mathbf{p}'_3 + \mathbf{p}'_4), i(E'_3 + E'_4)] = P'_\mu(0, im^*) \quad (11.60)$$

where $m^* \equiv E_o$ of (11.54) and (11.57) represents the rest energy of the entire system as a composite. It is determined by the laboratory energy E_1 in accordance with the invariance (11.54) when it is recognized that $\mathbf{p}_1$ is the total laboratory momentum of the system and $E_1 + m_2$ is the total laboratory energy, so that $P_\mu[\mathbf{p}_1, i(E_1 + m_2)]$. Then

$$(m^*)^2 = -\sum_\mu (P'_\mu)^2 = -\sum_\mu P_\mu^2 = (E_1 + m_2)^2 - \mathbf{p}_1^2 = m_1^2 + m_2^2 + 2E_1 m_2 \quad (11.61)$$

It can thus be said that bringing together the particles $m_{1,2}$ with a given energy $E_1 + m_2$ amounts to forming a rest mass m^* which can subsequently break up in various ways. Expression (11.60) gives the results for $m^* = E'_3 + E'_4$ with $\mathbf{p}'_4 = -\mathbf{p}'_3$, and these represent the entire effect of the conservation conditions. The *orientation* of the direction along which the product momenta $\mathbf{p}'_3 = -\mathbf{p}_4$ come apart, measured by the CM angle $0 < \vartheta < \pi$ in the figure, is completely unrestricted† by the energy-momentum conservation. Thus the CM angle ϑ can serve as the additional parameter in terms of which the results for the laboratory observables $E_{3,4}$, $\theta_{3,4}$ can be expressed.

Descriptions in the CM Frame

The connections between the laboratory and CM-frame descriptions are obtainable by Lorentz transformations with a relative frame velocity given by the CM velocity expression (11.55). Here

$$\boldsymbol{\beta} = \frac{\mathbf{p}_1}{E_1 + m_2} \qquad \text{and} \qquad \gamma = \frac{E_1 + m_2}{m^*} \quad (11.62)$$

for $\gamma = (1 - \beta^2)^{-1/2}$, in which $m^*(E_1)$ of (11.61) occurs. The relation of $\boldsymbol{\beta}$ to the total laboratory momentum here, $\mathbf{p}_1 = (\gamma m^*)\boldsymbol{\beta}$, is to be expected because γm^* is the total mass-energy of the system in the laboratory frame when its rest mass is m^* and its CM velocity is $\boldsymbol{\beta}$.

† It is instead determined by the *dynamics* of the interaction, i.e., the particular internal forces responsible for it and for the character of the products, as a function of the impact parameter (a distance by which the initial particles miss head-on collision) and of any definable orientations that each particle might have. Simple examples of the role played by forces are treated in Ref. 17, chap. 5, but only for low-velocity (nonrelativistic) collisions, when a representation by newtonian forces can be sufficient. The CM orientation angle might instead be completely random, as for so-called δ-function interactions, i.e., isotropic ones between spinless point-particles.

Now the *initial* state description in the CM frame follows from appropriate adaptations of transformation (11.47)

$$\begin{aligned} p'_1 &= \gamma(p_1 - \beta E_1) \qquad & E'_1 &= \gamma(E_1 - \beta p_1) \\ -p'_2 &= -\gamma\beta m_2 & E'_2 &= \gamma m_2 \end{aligned} \tag{11.63}$$

since $\mathbf{p}_2 = 0$ and $E_2 = m_2$ in the laboratory. Because m_2 is at rest in the laboratory, all its CM kinetic energy $T'_2 = E'_2 - m_2 = (\gamma - 1)m_2$ comes from the frame motion. With (11.62), the results are

$$p'_1 = p'_2 = \frac{m_2}{m^*} p_1 \tag{11.64}$$

$$E'_1 = \frac{m_1^2 + E_1 m_2}{m^*} \qquad E'_2 = \frac{m_2(E_1 + m_2)}{m^*} \tag{11.65}$$

The equality of the individual momentum magnitudes must of course be expected in the CM frame. The CM energies add up to $E'_1 + E'_2 = m^*$ properly.

The results (11.64) and (11.65) reduce to familiar† nonrelativistic (galilean) ones when the initial laboratory velocity is $u_1 = p_1/m_1 \ll c$ and $T_1 = E_1 - m_1 c^2 \approx \frac{1}{2}m_1 u_1^2 \ll m_1 c^2$ [Eq. (11.39)]. Then

$$m^* = [(m_1 + m_2)^2 + 2m_2 T_1]^{1/2} \rightarrow M_o c^2 + \tfrac{1}{2}\mu u_1^2 \tag{11.66}$$

where $M_o \equiv m_1 + m_2$ is the total initial rest mass and $\mu \equiv m_1 m_2/M_o$ is the reduced mass. The individual CM momentum values (11.64) reduce to

$$p'_1 = p'_2 \approx \mu u_1 \tag{11.67}$$

the familiar nonrelativistic results. The individual particle velocities relative to their CM become $\mathbf{u}'_1 = \mathbf{p}'_1/m_1 = (m_2/M_o)\mathbf{u}_1$ and $\mathbf{u}'_2 = \mathbf{p}'_2/m_2 = -\mathbf{p}'_1/m_2 = -(m_1/M_o)\mathbf{u}_1$, the appropriate fractions of the relative velocity $\mathbf{u}_1$, this being a *galilean* invariant $\mathbf{u}'_1 - \mathbf{u}'_2 = \mathbf{u}_1 - 0$. The CM energies (11.65) reduce to

$$E'_1 \approx m_1 c^2 + \tfrac{1}{2}m_1(u'_1)^2 \qquad E'_2 \approx m_2 c^2 + \tfrac{1}{2}m_2(u'_2)^2 \tag{11.68}$$

yielding a total CM kinetic energy

$$\tfrac{1}{2}m_1(u'_1)^2 + \tfrac{1}{2}m_2(u'_2)^2 = \tfrac{1}{2}\mu u_1^2 \tag{11.69}$$

properly equal to $(m^* - M_o)c^2$ of (11.66).

The *final* state descriptions, for particles $m_{3,4}$, are simplified in the CM frame by the fact that momentum conservation requires $\mathbf{p}'_4 = -\mathbf{p}'_3$, just as it does $\mathbf{p}'_2 = -\mathbf{p}'_1$. Then the energy-conservation condition $E'_1 + E'_2 = m^* = E'_3 + E'_4$ of (11.60) can be expressed as

$$[m_1^2 + (p'_1)^2]^{1/2} + [m_2^2 + (p'_1)^2]^{1/2} = [m_3^2 + (p'_3)^2]^{1/2} + [m_4^2 + (p'_3)^2]^{1/2} \tag{11.70}$$

† See Ref. 17, chap. 5, for example.

The chief value of this expression is that, for the special case of *elastic* collisions ($m_3 \equiv m_1$ and $m_4 \equiv m_2$), it demonstrates most directly that *all* the individual particle momenta relative to the CM have equal magnitudes $p'_3 = p'_4 = p'_1 = p'_2$. There is merely a reorientation by some ϑ, in the CM frame, during an elastic scattering.

For the general case, with any $m_{3,4}$, it is more significant to replace the left side of (11.70) with its equivalent, m^*, before solving for $p'_3 = p'_4$. The results are simplest to express for the corresponding CM energies

$$E'_{3,4} = \frac{(m^*)^2 \pm (m_3^2 - m_4^2)}{2m^*} \tag{11.71}$$

This completes the descriptions relative to the CM frame for any initial laboratory energy E_1 that may be used in forming m^* of (11.61).

The Final State in the Laboratory Frame

The preceding discussion has shown how the results of energy-momentum conservation can be determined for the description of the final state relative to its CM frame by $E'_{3,4}$ and ϑ. The corresponding laboratory observables $E_{3,4}$ and $\theta_{3,4}$ can be obtained by transforming back to the laboratory frame. This roundabout alternative to (11.59) for finding the effects of the energy-momentum conservation has the advantage of being able to yield explicit results, in terms of the fundamental parameters $m^*(E_1)$ and ϑ. The transformation reciprocal to (11.47) yields

$$\begin{aligned} p_{3,4} \cos \theta_{3,4} &= \gamma(\pm p'_3 \cos \vartheta + \beta E'_{3,4}) \\ p_{3,4} \sin \theta_{3,4} &= p'_3 \sin \vartheta \\ E_{3,4} &= \gamma(E'_{3,4} \pm \beta p'_3 \cos \vartheta) \end{aligned} \tag{11.72}$$

with β, γ given by (11.62) again. The terms $\pm p'_3 \cos \vartheta$ are respectively the components of $\mathbf{p}'_3$ and $\mathbf{p}'_4 = -\mathbf{p}'_3$ on the relative frame velocity (parallel to $\mathbf{p}_1$ and to $\mathbf{p}'_1$), as reference to Fig. 11.4*b* shows.

The last line of (11.72) yields $E_{3,4}$ in terms of $m^*(E_1)$ and ϑ after substitutions for β, γ from (11.62) and for $E'_{3,4}(\mathbf{p}'_{3,4})$ from (11.71). The explicit results are not presented here because they have been superseded, in late practice, by expressions [see (11.87)] in terms of frame invariants that can be associated with the process, to be introduced in the next subsection. However, the transformations (11.72) still yield the best physical insight into the connections of the emergence angles $\theta_{3,4}$ to the CM orientation ϑ.

A ratio of the first two lines of (11.72) yields expressions for $\tan \theta_{3,4}$ independent of $p_{3,4}$. Each depends on the CM velocity $v = c\beta$ and also on a particle velocity $u'_{3,4} = c^2 p'_{3,4}/E'_{3,4}$ relative to the mass center

$$\tan \theta_{3,4} = \frac{u'_{3,4} \sin \vartheta}{\gamma(v \pm u'_{3,4} \cos \vartheta)} \tag{11.73}$$

These are generalizations to $m_{3,4} \neq 0$ of the relation (11.27), which was obtained for $u'_3 = c$. The labels 3, 4 will be dropped for the ensuing discussion because it is a matter of choice which of the products is labeled m_3 and which m_4. The emergence angle of either may be called ϑ if the other's is called $\pi - \vartheta$. Thus, the case of the plus sign in the denominator of (11.73) will be sufficient to consider.

The form of (11.73) to be discussed can be written

$$\tan\theta = \frac{1}{\gamma}\tan\theta^o \qquad \text{if} \qquad \tan\theta^o \equiv \frac{\sin\vartheta}{v/u' + \cos\vartheta} \tag{11.74}$$

Plainly, $\theta \to \theta^o$ for nonrelativistic CM velocities $v \ll c$ when $\gamma \to 1$. For higher velocities, θ will differ from θ^o only through what is called the collimating effect of the factor $1/\gamma$, to be left to later discussion (it has already been met in connection with the high-velocity radiation patterns of Figs. 10.5 and 10.6). Since $1/\gamma$ varies monotonically with v, it leaves common to both $\tan\theta$ and $\tan\theta^o$ features arising from their common zeros ($\theta, \theta^o = 0, \pi$) and their infinities ($\theta, \theta^o = \frac{1}{2}\pi$). The explicit results are simpler to express for $\tan\theta^o$ because it is a function of only one parameter, the velocity ratio v/u', besides ϑ. Because the velocities occur only in a ratio, for a given value of this ratio the individual velocities may be either relativistic or nonrelativistic without changing the conclusions from θ^o.

Plots† of $\theta^o(\vartheta)$ for some ratios u'/v are shown in Fig. 11.5. Both the angles θ and θ^o have the values 0, $\frac{1}{2}\pi$, and π for the same values of ϑ. The switchovers

† Similar plots are shown in fig. 5.4 of Ref. 17, labeled appropriately for nonrelativistic velocities only.

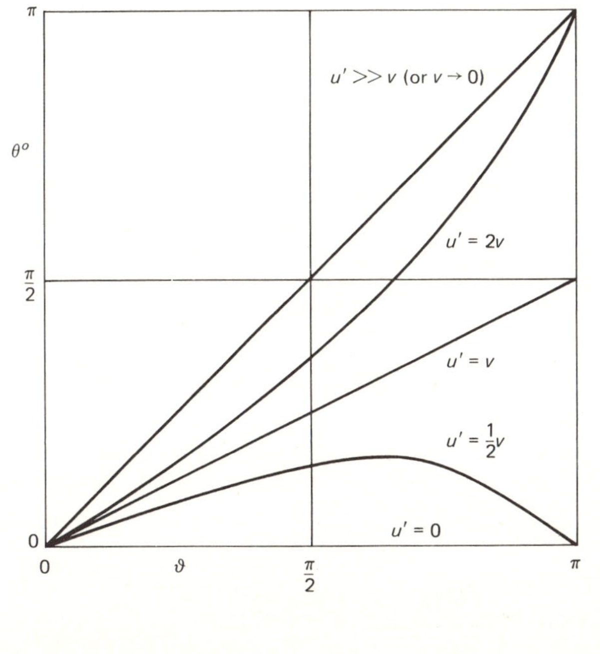

Figure 11.5

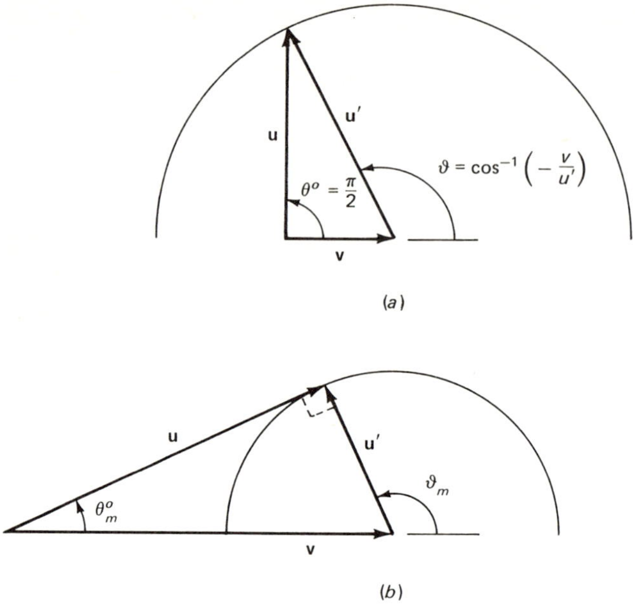

Figure 11.6

from the forward to backward cones ($\theta, \theta^o < \frac{1}{2}\pi$ to $\theta, \theta^o > \frac{1}{2}\pi$) occur only for $u' > v$ and at $\cos \vartheta = -v/u'$ (where $\tan \theta^o$ and $\tan \theta \to \infty$). This can be understood from the vector diagram of the velocities in Fig. 11.6*a*, which can be strictly valid only for nonrelativistic values, when the galilean velocity addition is valid. However, as already remarked, conclusions from it about the velocity *ratio* are also valid relativistically. Notice that backward emission in the laboratory can occur only for $\vartheta > \frac{1}{2}\pi$ but emission remains in the forward cone for those $\vartheta > \frac{1}{2}\pi$ having $u'|\cos \vartheta| < v$.

For each $u' < v$ there is a maximum emission angle given by

$$\tan \theta_m = \frac{1}{\gamma} \frac{u'}{[v^2 - (u')^2]^{1/2}} \tag{11.75a}$$

and occuring at a CM angle $\vartheta_m > \frac{1}{2}\pi$, with

$$\cos \vartheta_m = -\frac{u'}{v} \tag{11.75b}$$

that is also independent of γ. This can be understood from the galilean diagram in Fig. 11.6*b*, which makes it plain why $\theta < \theta_m$ for both $\vartheta \gtrless \vartheta_m$.

For high initial energies $E_1 \gg m_{1,2}c^2$, when (11.62) gives

$$\frac{1}{\gamma} = \frac{(m_1^2 + m_2^2 + 2E_1 m_2)^{1/2}}{E_1 + m_2} \approx \left(\frac{2m_2 c^2}{E_1}\right)^{1/2} \ll 1 \tag{11.76}$$

for the ratio $(\tan \theta)/(\tan \theta^o)$, forward angles $\theta < \frac{1}{2}\pi$ are substantially reduced, and backward angles become more backward (closer to π) than those indicated for θ^o in Fig. 11.5. The backward cone can exist only for a range of velocities $v < u' < c$, which becomes very narrow for $v \to c$. The striking effect is the collimation of the forward cone to a narrow sheaf. For $v = u' = 0.98c$ ($\gamma \approx 5$), the angle θ is narrowed to approximately 11° from the $\theta^o = 45°$ given by the figure. The value given for the maximum emission angle by (11.75*a*), in the case $u' = \frac{1}{2}v = \frac{1}{2}(0.98c)$, is $\theta_m \lesssim 4°$, as against $\theta^o = 30°$, but this still occurs at the angle $\vartheta_m = 120°$ in the CM frame, as for nonrelativistic velocities.

The Invariants *s*, *t*, and *u*

Instead of expressions in terms of a given laboratory energy E_1 and CM orientation ϑ, customary in nonrelativistic descriptions, there has come into widespread use a parametrization by three related invariants (often called *Mandelstam variables*):

$$\begin{aligned} s &\equiv -(p_{1\mu} + p_{2\mu})^2 = -(p_{3\mu} + p_{4\mu})^2 \\ t &\equiv -(p_{1\mu} - p_{3\mu})^2 = -(p_{4\mu} - p_{2\mu})^2 \\ u &\equiv -(p_{1\mu} - p_{4\mu})^2 = -(p_{3\mu} - p_{2\mu})^2 \end{aligned} \tag{11.77}$$

with the equalities following from the 4-momentum conservation (11.58). In each of these expressions *summation* over $\mu = 1, 2, 3, 4$ is to be understood,† and it is this that gives the quantities the same values whatever the frame used in evaluating them, a consequence of the 4-vector character of every composite momentum. Thus, s, t, u have the same values whether the momentum components refer to the laboratory or CM frames of Fig. 11.4, quite unlike such parameters as E_1 and ϑ, each of which refers to a different one of the special frames.

The invariant s is plainly identical with the square of the composite rest mass in (11.61): $s = -P_\mu^2 = (m^*)^2$. The invariants t and u similarly amount to squared invariant masses in processes initiated by bringing together the 4-momenta $p_{1\mu}$ and $-p_{3\mu}$, or $p_{2\mu}$ and $-p_{4\mu}$, or $p_{1\mu}$ and $-p_{4\mu}$, or $p_{3\mu}$ and $-p_{2\mu}$; these alternatives to the interaction of $p_{1\mu}$ with $p_{2\mu}$ are called its *cross-channel processes*. When it is $m^* = \sqrt{s}$ that is formed, t and u are referred to as the 4-momentum *transfers* (squared) in the process.

The connections of the parameters s, t, u to the laboratory observables E_1, $E_2 = m_2$, E_3, and E_4 can be found from the expressions in (11.77) containing $p_{2\mu}(0, im_2)$, which defines the laboratory frame as having $\mathbf{p}_2 = 0$. Using the invariants $p_{n\mu}^2 = -m_n^2$ for the individual 4-momenta yields

$$\begin{aligned} s &= m_1^2 + m_2^2 - 2p_{1\mu}p_{2\mu} = m_1^2 + m_2^2 + 2E_1 m_2 \\ t &= m_2^2 + m_4^2 + 2p_{2\mu}p_{4\mu} = m_2^2 + m_4^2 - 2m_2 E_4 \\ u &= m_2^2 + m_3^2 + 2p_{2\mu}p_{3\mu} = m_2^2 + m_3^2 - 2m_2 E_3 \end{aligned} \tag{11.78}$$

† In accord with a summation convention, by which, e.g., in $\sum_\mu P_\mu^2 \equiv (\sum_\mu) P_\mu P_\mu \equiv P_\mu^2$, the occurrence of a repeated index in a product of two covariants implies summation over the range of the index.

Adding these together gives

$$s + t + u = m_1^2 + m_2^2 + m_3^2 + m_4^2 \tag{11.79}$$

since the remaining terms add up to

$$2m_2[(E_1 + m_2) - (E_3 + E_4)] = 0$$

because of the total energy conservation expressed as in (11.59). The existence of the relation (11.79) shows that only two of the three invariants can be treated as independent parameters, just as only the two parameters E_1 and ϑ were left independent by the energy-momentum conservation.

The CM energies and momenta are all parametrized by $s \equiv (m^*)^2$ alone, because they are independent of the CM orientation ϑ, as discussed in connection with Fig. 11.4*b*. The relations can be found by evaluating the expressions for s in (11.77) in terms of the CM 4-momenta, characterized by $\mathbf{p}_1' + \mathbf{p}_2' = \mathbf{p}_3' + \mathbf{p}_4' = 0$. Thus $s = -(p_{1\mu}' + p_{2\mu}')^2$ becomes

$$s = (E_1' + E_2')^2 = m_1^2 + m_2^2 + 2[E_1' E_2' + (p_1')^2] \tag{11.80}$$

The connection of s to the laboratory energy E_1 in (11.78) shows that

$$E_1 m_2 = E_1' E_2' + (p_1')^2 \tag{11.81}$$

equivalent to the useful invariance $p_{1\mu} p_{2\mu} = p_{1\mu}' p_{2\mu}'$ (summation over μ understood). Solving it for $(p_1')^2 = (E_{1,2}')^2 - m_{1,2}^2$ yields $p_1' = m_2 p_1 / \sqrt{s}$ of (11.64) directly. The expression of p_1' entirely in terms of the invariant s involves a frequently occurring type of combination of invariant (squared) masses that can be defined by

$$\Delta_{ab}^2(c) \equiv a^2 + b^2 + c^2 - 2ab - 2bc - 2ca \tag{11.82}$$

This will be denoted $\Delta_{12}^2(s)$ when $a = m_1^2$, $b = m_2^2$, and $c = (m^*)^2 = s$ as needed for the

$$p_1' = p_2' = \frac{\Delta_{12}(s)}{2\sqrt{s}} \tag{11.83a}$$

that follows from (11.80). When $E_{1,2}' = [(p_1')^2 + m_{1,2}^2]^{1/2}$ is used, this leads to

$$E_{1,2}' = \frac{s \pm (m_1^2 - m_2^2)}{2\sqrt{s}} \tag{11.83b}$$

which are also derivable from (11.65).

The final state CM energies and momenta follow similarly from $s = -(p_{3\mu}' + p_{4\mu}')^2$, with the results

$$p_3' = p_4' = \frac{\Delta_{34}(s)}{2\sqrt{s}} \tag{11.84a}$$

$$E_{3,4}' = \frac{s \pm (m_3^2 - m_4^2)}{2\sqrt{s}} \tag{11.84b}$$

These could have been written down immediately from (11.83), since $m_{3,4}$ and $m_{1,2}$ are merely alternative pairs into which $m^* = \sqrt{s}$ can break up. The expression (11.84*b*) has already been found in (11.71). Notice that the results (11.83*b*) and (11.84*b*) yield

$$E'_1 + E'_2 = E'_3 + E'_4 = \sqrt{s} = m^*$$

as they should.

The invariant $\Delta_{12}(s)$ just equals the initial *laboratory* momentum multiplied by $2m_2$

$$p_1 = \frac{\Delta_{12}(s)}{2m_2} \tag{11.85a}$$

This follows from the expression (11.83*a*) for p'_1 and the connection $p'_1 = m_2 p_1/\sqrt{s}$ last noted just below (11.81). The corresponding expression for the laboratory energy $E_1 = (p_1^2 + m_1^2)^{1/2}$ is

$$E_1 = \frac{s - (m_1^2 + m_2^2)}{2m_2} \tag{11.85b}$$

which also follows from (11.78) or (11.61).

The *final* state laboratory energies and momenta require parametrization not only by s but also by t (or u), because they depend on the CM angle ϑ in addition to the initial laboratory energy E_1. The connection of t to ϑ can be found by evaluating $t = -(p'_{1\mu} - p'_{3\mu})^2$ of (11.77) in the CM frame, where $\mathbf{p}'_1 \cdot \mathbf{p}'_3 = p'_1 p'_3 \cos\vartheta$ (Fig. 11.4*b*). Thus

$$t = m_1^2 + m_3^2 + 2p'_{1\mu}p'_{3\mu} = m_1^2 + m_3^2 - 2(E'_1 E'_3 - p'_1 p'_3 \cos\vartheta)$$

Using (11.83) and (11.84) to put the CM energies and momenta in terms of the parameter s yields

$$t = (4s)^{-1}[(m_1^2 - m_2^2 - m_3^2 + m_4^2)^2 - (\Delta_{12}^2 + \Delta_{34}^2 - 2\Delta_{12}\Delta_{34}\cos\vartheta)] \tag{11.86a}$$

Conversely,

$$\cos\vartheta = \frac{s^2 + s[2t - (m_1^2 + m_2^2 + m_3^2 + m_4^2)] + (m_1^2 - m_2^2)(m_3^2 - m_4^2)}{\Delta_{12}(s)\Delta_{34}(s)} \tag{11.86b}$$

There is thus a *linear* connection between t and $\cos\vartheta$ for each value of the energy $\sqrt{s}$. A similar linear connection can be found between $\cos\vartheta$ and u; for it, $t \to u$, $\cos\vartheta \to -\cos\vartheta$, and $m_3 \leftrightarrow m_4$ [which has no effect on $\Delta_{34}(s)$].

Expressions for $E_{3,4}$ in terms of the invariants s, $t(u)$ follow directly from (11.78) and (11.79)

$$E_3 = \frac{(m_2^2 + m_3^2) - u}{2m_2} = \frac{s + t - (m_1^2 + m_4^2)}{2m_2} \tag{11.87a}$$

$$E_4 = \frac{(m_2^2 + m_4^2) - t}{2m_2} \tag{11.87b}$$

From these, the laboratory momentum expressions

$$p_3 = \frac{\Delta_{23}(u)}{2m_2} \qquad \text{and} \qquad p_4 = \frac{\Delta_{24}(t)}{2m_2} \tag{11.87c}$$

follow, as does p_1 (11.85*a*) from E_1 (11.85*b*). Such expressions of the consequences of energy-momentum conservation for the final state observables in terms of frame invariants are more significant than expressions in terms of E_1, ϑ that would follow directly from the transformation (11.72).

Completing the final state description still requires finding similar parametrizations of $\theta_{3,4}$. The expression for the laboratory angle θ_3, as defined by $\mathbf{p}_1 \cdot \mathbf{p}_3 = p_1 p_3 \cos\theta_3$ (Fig. 11.4*a*), can be found from evaluating $t = -(p_{1\mu} - p_{3\mu})^2$ as for (11.86), but now in the laboratory rather than CM frame. The result is

$$\cos\theta_3 = \frac{[s - (m_1^2 + m_2^2)][(m_2^2 + m_3^2) - u] + 2m_2^2[t - (m^2 + m_3^2)]}{\Delta_{12}(s)\Delta_{23}(u)} \tag{11.87d}$$

The expression for $\cos\theta_4$, to be found from $u = -(p_{1\mu} - p_{4\mu})^2$, is just like this except for the interchanges $t \leftrightarrow u$ and $m_3 \leftrightarrow m_4$. These results are actually far too general to be convenient to use in the simple cases to be discussed in the following pages, for which the simpler expressions (11.73) will be adequate.

Energy-Momentum Conservation in Elastic Collisions

In the important special case of elastic scattering, with $m_3 \equiv m_1$ and $m_4 \equiv m_2$, the individual CM momenta before and after collision are all equal,

$$p'_1 = p'_2 = p'_3 = p'_4 = \frac{m_2}{m^*} p_1 \tag{11.88a}$$

as already noted in connection with (11.70), and the individual CM energies are

$$E'_1 = E'_3 = \frac{m_1^2 + E_1 m_2}{m^*} \qquad E'_2 = E'_4 = \frac{m_2(E_1 + m_2)}{m^*} \tag{11.88b}$$

The connections to the initiating laboratory momentum $p_1 = (E_1^2 - m_1^2)^{1/2}$ that are specified here all follow from (11.64) and (11.65). The same results also follow from specializing the s, t, u parametrizations (11.83), (11.84), and (11.85) to $m_3 = m_1$ and $m_4 = m_2$, which makes $\Delta_{34}(s) = \Delta_{12}(s) = 2m_2 p_1$.

The parameter $-t$ becomes equal to just the square of the CM *3-momentum* transfer in the elastic collisions because $E'_3 = E'_1$ makes $t = -(p'_{1\mu} - p'_{3\mu})^2 = -(\mathbf{p}'_1 - \mathbf{p}'_3)^2$. A special symbol

$$\mathbf{q} \equiv \mathbf{p}'_3 - \mathbf{p}'_1 = \mathbf{p}'_2 - \mathbf{p}'_4 \tag{11.89a}$$

is frequently used for this vector change in the CM momenta resulting from the collision; there is only an orientational change, by ϑ, since $p'_3 = p'_1$ and $p'_4 = p'_2$

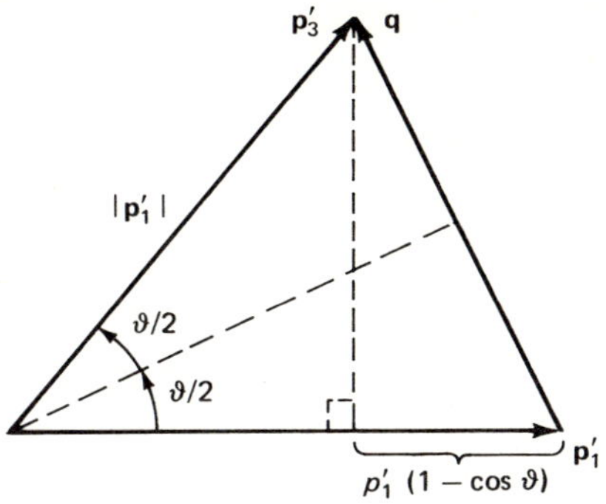

Figure 11.7

in magnitude. Figure 11.7, adapted from the CM picture in Fig. 11.4*b*, shows that

$$\sqrt{-t} \equiv q = 2p'_1 \sin \tfrac{1}{2}\vartheta = p'_1[2(1 - \cos \vartheta)]^{1/2} \tag{11.89b}$$

The result

$$t = -2(p'_1)^2(1 - \cos \vartheta) = -2\left(\frac{m_2}{m^*}\right)^2 p_1^2(1 - \cos \vartheta) \tag{11.90}$$

also follows from the expression (11.86*a*), since $\Delta_{12} = \Delta_{34} = 2m_2 p_1$.

The final laboratory energies as given by (11.87) reduce to

$$E_3 = E_1 + \frac{t}{2m_2} \equiv E_1 - \frac{q^2}{2m_2} \qquad E_4 = m_2 - \frac{t}{2m_2} \equiv m_2 + \frac{q^2}{2m_2} \tag{11.91}$$

Thus, the elastic processes lead to the relativistic kinetic-energy transfers

$$\Delta T = E_4 - m_2 = E_1 - E_3 = -\frac{t}{2m_2} = \frac{q^2}{2m_2} \tag{11.92}$$

Using the form (11.90) for t yields

$$\Delta T = \frac{p_1}{m^*} p'_1(1 - \cos \vartheta) = -\gamma \boldsymbol{\beta} \cdot \mathbf{q} \tag{11.93}$$

where $\boldsymbol{\beta} = \mathbf{p}_1/\gamma m^*$ is the CM velocity of (11.62). That $-p'_1(1 - \cos \vartheta)$ is just the projection of the CM momentum transfer $\mathbf{q}$ on the direction of $\boldsymbol{\beta} \| \mathbf{p}_1 \| \mathbf{p}'_1$ is clear from Fig. 11.7. The last equality in (11.93) is the relativistic generalization of a well-known† classical result $\Delta T \approx -\mathbf{v} \cdot \mathbf{q}$.

There remains consideration of the laboratory angle θ_3 by which an elastic collision will deflect the incident particle m_1 and of the recoil angle θ_4 at which the target particle m_2 will emerge. These can be obtained from (11.73), which is

† See, for example, Ref. 17, eq. (5.70).

parametrized by the velocities relative to the mass center $u'_3 = p'_3/E'_3$ and $u'_4 = p'_4/E'_4$. According to (11.88),

$$\begin{aligned} u'_3 &= \frac{p'_1}{E'_1} = u'_1 = \frac{m_2 p_1}{m_1^2 + E_1 m_2} = \frac{m_2(E_1 + m_2)}{m_1^2 + E_1 m_2} v = \frac{E'_2}{E'_1} v \\ u'_4 &= \frac{p'_2}{E'_2} = u'_2 = \frac{p_1}{E_1 + m_2} = v \end{aligned} \tag{11.94}$$

where $v \equiv \beta$ is the CM velocity of (11.62). The outcome $u'_4 = u'_2 = v$ is to be expected since the target particle is at rest in the laboratory frame and must have the velocity $\mathbf{u}'_2 = -\mathbf{v}$ relative to a CM frame moving with the velocity $\mathbf{v}$. With (11.94), expressions (11.73) yield

$$\tan \theta_3 = \frac{m^* m_2 \sin \vartheta}{(m_1^2 + E_1 m_2) + m_2(E_1 + m_2) \cos \vartheta} \tag{11.95a}$$

and

$$\tan \theta_4 = \frac{m^*}{E_1 + m_2} \cot \tfrac{1}{2}\vartheta \tag{11.95b}$$

for the recoil angle, which is then properly restricted to the forward cone $\theta_4 < \frac{1}{2}\pi$. These correspond to well-known† nonrelativistic results,

$$\tan \theta_3 \approx \frac{m_2 \sin \vartheta}{m_1 + m_2 \cos \vartheta} \qquad \text{and} \qquad \theta_4 \approx \tfrac{1}{2}(\pi - \vartheta)$$

following from (11.95) for $E_1 \approx m_1$ and $m^* \approx m_1 + m_2$.

Reaction Thresholds

More particulars about the *inelastic* processes, ones in which $m_3 \neq m_1$ and/or $m_4 \neq m_2$, also have general interest.

A final state of CM energy $E'_3 + E'_4 = m^*$ is to be formed, and since the individual energies $E'_{3,4} = [m_{3,4}^2 + (p'_3)^2]^{1/2}$ cannot be smaller than $m_{3,4}$, respectively, the process can occur at all only if

$$m^* = [(m_1 + m_2)^2 + 2m_2 T_1]^{1/2} \geq m_3 + m_4 \tag{11.96}$$

The rest masses $m_{3,4}$ must be created out of the energy m^* made available in the CM frame, which includes the initial masses, $m_1 + m_2$. It is more than sufficient when the total mass $m_3 + m_4$ to be produced is smaller than the total initial rest mass $m_1 + m_2$, but when the mass increment

$$\Delta M = (m_3 + m_4) - (m_1 + m_2) \tag{11.97}$$

is to be positive, this must be "created" out of kinetic energy T_1 given to the beam of particles m_1 that bombard m_2. According to (11.96), it is not sufficient

† See Ref. 17, eqs. (5.65) and (5.66).

to supply a T_1 equal to ΔM; rather

$$T_1 \geq T_0 = \frac{(m_3 + m_4)^2 - (m_1 + m_2)^2}{2m_2} = \frac{\Delta M[(m_1 + m_2) + (m_3 + m_4)]}{2m_2}$$
$$= \Delta M \left(1 + \frac{m_1}{m_2} + \frac{\Delta M}{2m_2}\right) \tag{11.98}$$

is required if $\Delta M > 0$. The minimum requirement T_0 is then called the *threshold kinetic energy* for the process.

The requisite kinetic energy T_0 is greater than ΔM because in a bombardment of target particles m_2 that are at rest in the laboratory† some of the energy must be "wasted" in first giving the whole system a translational CM energy. This can be evaluated as the kinetic energy of the system's rest mass m^*

$$(\gamma - 1)m^* = E_1 + m_2 - m^* = T_1 - [m^* - (m_1 + m_2)] \tag{11.99}$$

since the entire system as a body has the CM velocity, given by (11.62). It can be seen that at threshold $T_1 \to T_0$ this translational energy is just the extra amount $T_0 - \Delta M$ in (11.98).

Quite generally in any process initiated by bombarding particles m_2 at rest with particles of energy $E_1 = m_1 + T_1$ only part of the kinetic energy T_1 minus the CM translation energy

$$T_1 - (\gamma - 1)m^* = m^* - (m_1 + m_2)$$
$$= [(m_1 + m_2)^2 + 2m_2 T_1]^{1/2} - (m_1 + m_2) \tag{11.100}$$

is available for interaction in the CM frame. In low-velocity nonrelativistic collisions, when $T_1 \approx \frac{1}{2}m_1 u_1^2$ and the CM velocity is $v \approx m_1 u_1/(m_1 + m_2)$, the result (11.100) reduces to a well-known nonrelativistic expression for the available energy

$$T_1 - \tfrac{1}{2}(m_1 + m_2)v^2 = \frac{m_2}{m_1 + m_2} T_1 \tag{11.101}$$

When all this is used to produce a (sufficiently small) mass increment ΔM, it is to be equated to ΔM to give a $T_1 \to T_0$,

$$T_0 \approx \frac{m_1 + m_2}{m_2} \Delta M \tag{11.102}$$

Thus, the added fraction $\Delta M/2m_2$ in the last line of (11.98) represents a relativistic correction.

† Some particle accelerators are equipped with storage rings in which high-speed beams of the particles are accumulated to serve as targets for the primary beam. A head-on clash, in the laboratory, of equally massive particles has a CM frame that coincides with the laboratory frame, in which case all the kinetic energy $2T_1 = m^* - 2m_1$ becomes available for creating the product rest masses.

Two-Particle Decays

There are particles that have finite mean lives because they disintegrate *spontaneously*, without encountering a second particle, subject to the energy-momentum conservation (11.58) with $p_{2\mu} \equiv 0$. An example of a disintegration into two products is provided by α decay, as of the nucleus ^{210}Po into ^{206}Pb and a helium nucleus (^{4}He, the α particle). In pion physics there are decays of pions into muons plus neutrinos, $\pi \to \mu + \nu$, of Λ hyperons into nucleons and pions, $\Lambda \to N + \pi$, and many other such processes to consider.

In the rest frame of an unstable parent m_1, energy conservation requires

$$m_1 = E'_3 + E'_4 = (m_3 + m_4) + T'_3 + T'_4 > m_3 + m_4 \tag{11.103}$$

Primed symbols are used for the energies here because the rest frame of m_1 is also the CM frame of the process (and its products); it coincides with the laboratory frame (and then $E'_{3,4} \equiv E_{3,4}$) for parents that decay while at or near rest in the laboratory. The two-particle decays of particles at rest are observed as emergences of products m_3 and m_4 with equal and opposite momenta $\mathbf{p}'_3 = -\mathbf{p}'_4$. Putting $p'_3 = p'_4$ into (11.103) allows finding $E'_{3,4} = [m_{3,4}^2 + (p'_3)^2]^{1/2}$ from it, but the results have already been given in terms of $m^* \equiv m_1$ in (11.71),

$$E'_{3,4} = \frac{m_1^2 \pm (m_3^2 - m_4^2)}{2m_1} \tag{11.104}$$

Expressed in terms of the rest-mass decrement $|\Delta M| = m_1 - (m_3 + m_4)$, the kinetic energies that follow from (11.104) are

$$T'_3 = |\Delta M| \frac{m_1 - m_3 + m_4}{2m_1} \qquad T'_4 = |\Delta M| \frac{m_1 + m_3 - m_4}{2m_1} \tag{11.105}$$

In many cases one of the products is much lighter than the other, say $m_4 \ll m_3$, and then it gets most of the kinetic energy,

$$T'_4 \approx \frac{m_1 + m_3}{m_1 - m_3 + m_4} T'_3 \gg T'_3$$

because the momenta must match ($p'_4 = p'_3$).

A laboratory frame different from the CM frame must be used for gauging results from decays in flight, when the parent has some velocity $u_1 = p_1/E_1 = v$ that determines the CM velocity v. This can be pictured as in Fig. 11.4*a*, but with $m_2 \equiv 0$ and the position marked by it only a point at which the spontaneous disintegration happens to take place. The transformation (11.72) can be used to find the product energies that will be observed in the laboratory

$$E_{3,4} = \frac{[m_1^2 \pm (m_3^2 - m_4^2)]E_1 \pm \Delta_{34}(m_1^2)p_1 \cos \vartheta}{2m_1^2} \tag{11.106}$$

since $\gamma = E_1/m_1$ when the CM velocity is $v = p_1/E_1$. The convenient invariant† $\Delta_{34}(m_1^2) = 2m_1 p_3' = 2m_1 p_4'$ of (11.84) is here used to replace $p_{3,4}'$ by their explicit dependence on the masses, which also follows from (11.104).

The emission directions $\theta_{3,4}$ can be obtained from (11.73), which is valid however $m_{3,4}$ are produced. Those expressions are parametrized by the velocities of the products relative to their CM frame

$$u_{3,4}' = \frac{p_{3,4}'}{E_{3,4}'} = \left\{1 - \frac{4m_1^2 m_{3,4}^2}{[m_1^2 \pm (m_3^2 - m_4^2)]^2}\right\}^{1/2} \tag{11.107}$$

as follows from $p_{3,4}' = [(E_{3,4}')^2 - m_{3,4}^2]^{1/2}$ and (11.104). (See also Exercise 11.16.)

11.4 PARTICLE-PHOTON INTERACTIONS

Energy-momentum conservation also plays a role in the interactions of charged particles with electromagnetic wave fields. A part of the interaction can always be considered in which a nearly monochromatic and well-directed component of the wave field participates. This is representable, at least in the vicinity of the particle, as a *plane* wave with some definite wave vector **k**. Such a plane wave has a distributed field energy of some density w and a field *momentum* (7.17) in the density $\hat{\mathbf{k}}w/c$. The *ratio* of this momentum to the energy is just like that, $p_o/E = 1/c$ [Eq. (11.49)], characteristic of particles with zero rest mass. Indeed, the rest-mass-less particles called photons in connection with (11.49) have been identified as each possessing for its energy a quantum of electromagnetic field energy. This is what makes processes involving photons of special interest to electromagnetic theory.

Compton Scattering

The type of process through which photons were first proved to be carriers of electromagnetic field *momentum* is a modification to higher (x-ray) frequencies of the process that was discussed as elastic Thomson scattering of a plane electromagnetic wave (10.21) by interaction with a free electron at rest. That treatment was nonrelativistic, with the result that the reradiated scattered wave had the same frequency as the incident one, and no account was taken of any *net* recoil that might be imparted to the (comparatively massive) electron by the radiation pressure. The electron's mass value did enter the results, but only through having helped determine the periodic accelerations of zero average that generated the reradiation.

† The general s, t, u parametrization becomes useless for obtaining $E_{3,4}$, as in (11.87), because s, t, u of (11.78) become independent of $E_{3,4}$ for $m_2 \equiv 0$.

Effects on scattered x-radiation of a net recoil momentum imparted to the electron were first identified by Compton. He interpreted his observations in terms of such conservation relations as (11.59), adapted to an elastic process with $m_1 = m_3 = 0$, for the incident and scattered radiation, and $m_2 = m_4 = m$ the electron's rest mass. Then, with $p_{1,3} = E_{1,3}/c$,

$$\begin{aligned} E_1 + mc^2 &= E_3 + E_4 \\ E_1 &= E_3 \cos\theta_3 + cp_4 \cos\theta_4 \\ 0 &= E_3 \sin\theta_3 - cp_4 \sin\theta_4 \end{aligned} \tag{11.108}$$

There is no need here to consider the CM-frame description, as in the preceding section, because Compton found it simplest to observe the radiation energies E_3 as a function of their *laboratory* emission angle θ_3. For this, the recoil electron's momentum p_4 and recoil angle θ_4 are eliminated between the momentum-conservation relations by forming

$$(E_1 - E_3 \cos\theta_3)^2 + E_3^2 \sin^2\theta_3 = c^2 p_4^2 = E_4^2 - m^2c^4$$

and putting $E_4 = (E_1 - E_3) + mc^2$, as follows from the energy conservation. The result can be expressed as

$$\frac{1}{E_3} - \frac{1}{E_1} = \frac{1 - \cos\theta_3}{mc^2} \tag{11.109}$$

Its development has presumed that each elementary instance of the process involved some definite amount E_1 of the x-radiation field energy, which is then changed to E_3 because of recoil energy imparted to the electron.

Compton supposed that the field energy E_1 participating in each process is determined by the *frequency* ω_1 of the incident field, as $E_1 = \hbar\omega_1$, because of earlier findings by Planck and Einstein. He also expected that $E_3 = \hbar\omega_3$ for the scattered radiation and hence a scattered frequency ω_3 that is different from the incident one. Then the inverse energy difference (11.109) becomes

$$\frac{1}{\hbar\omega_3} - \frac{1}{\hbar\omega_1} = \frac{\Delta\lambda}{2\pi\hbar c} = \frac{1 - \cos\theta_3}{mc^2} \tag{11.110}$$

where $\Delta\lambda = \lambda_3 - \lambda_1$ is an expected Compton shift of the x-ray wavelength by the scattering. It was through observing exactly this, with variations depending on the scattering angle as required by (11.110), that all the expectations were confirmed. The quantity amounting to the Compton shift at right angles $\Delta\lambda = 2\pi\hbar/mc$ has ever since been used as a significant unit called the *Compton wavelength of the electron*, $\lambdabar_c \equiv \hbar/mc \approx 3.86 \times 10^{-11}$ cm.

A consequence of the findings is that the energies and momenta of photons are expressed as

$$E = \hbar\omega \qquad \mathbf{p} = \frac{\hat{\mathbf{k}}\hbar\omega}{c} = \hbar\mathbf{k} \tag{11.111}$$

The energy-momentum 4-vector becomes replaceable by the *wave* 4-vector

$$k_\mu\left(\mathbf{k}, \frac{i\omega}{c}\right) = \hbar^{-1} p_\mu\left(\mathbf{p}, \frac{iE}{c}\right) \tag{11.112}$$

and the vanishing of the invariant $p^2 - E^2/c^2 = \hbar^2(k^2 - \omega^2/c^2) = 0$, characteristic of rest-mass-less particles, becomes equivalent to the earlier definition $k \equiv \omega/c$.

In the case of the nonrelativistic Thomson scattering, it was also possible to find valid expressions for the scattered intensities, as represented by the cross-sectional formulas (10.28) and (10.29). That was simple only because any net energy transfer to the electron is negligible when $\hbar\omega_1 \ll mc^2$. At relativistic energies, the dynamics of the electron's motions becomes important, and quantum-mechanical strictures on it prevent the classical treatments from giving significant results.†

Recoils from Radiation

Molecular, atomic, and nuclear systems enclose steadily periodic internal motions that lead to the expectation of discrete values for the frequencies radiated by them. If a process of radiating a given frequency ω requires formation of a quantum of field energy $\hbar\omega$, the system's internal energy must be reduced by the same discrete amount in the process. This led Bohr to suppose that closed systems can exist only in states of internal motion characterized by discrete values for their energies. Each of these can be represented by a rest energy for the system as a body, $m_o c^2$ in the ground state of the system and $m^* c^2$ while it is in some one of the discrete excited states. Then

$$(m^* - m_o)c^2 \equiv \hbar\omega_o \tag{11.113}$$

defines an energy that must be lost by the internal motions during a radiation process that deexcites the system. It will next be seen that the radiated frequency ω must be expected to differ from ω_o because of recoil effects.

Whenever radiation of definitely directed momentum $\hbar\mathbf{k}$ is detected, it must be a product of a radiating system that suffered a recoil $\mathbf{p} = -\hbar\mathbf{k}$ in producing it. Thus

$$\frac{\hbar\omega}{c} = p = c^{-1}(E^2 - m_o^2 c^4)^{1/2}$$

if m_o is the rest mass of the system after the separation of the quantum of radiation from it and $E - m_o c^2$ is the recoil kinetic energy. Substituting the result of energy conservation

$$E = m^* c^2 - \hbar\omega = \hbar(\omega_o - \omega) + m_o c^2$$

† The relativistic quantum-mechanical generalization of the Thomson formula is known as the *Klein-Nishina formula*; see Ref. 12, for example. The Klein-Nishina formula agrees with the Thomson formula for $\hbar\omega_1 \ll mc^2$ but deviates substantially from any classical-mechanical expectations for $\hbar\omega_1 \gtrsim mc^2$.

into the momentum relation yields†

$$\hbar\omega = \frac{(m^*)^2 - m_o^2}{2m^*}c^2 = \hbar\omega_o\left(1 - \frac{\hbar\omega_o}{2m^*c^2}\right) \tag{11.114}$$

Frequently the radiating system is much more massive than the radiated energy, so that $m^*c^2 = m_o c^2 + \hbar\omega_o \approx m_o c^2$ and the fractional difference $(\omega_o - \omega)/\omega_o$ becomes $\approx \hbar\omega_o/2m_o c^2$.

The frequency shift arising from the recoil of the source has importance for the observability of *resonance scattering*, or fluorescence. This refers to a process in which radiation from an excited system is allowed to fall on samples of the same system in their ground states, thus exciting them to the same level as the source. Without the intervention of recoils, the incident frequency would be in resonance with the frequency ω_o needed to excite the absorbers, resulting in scattered reradiations of particularly great intensity. However, exact resonance is prevented not only by the fact that the source frequency ω will actually be smaller than ω_o because of the energy loss to the *source* recoil but also because a greater frequency than ω_o is required for the absorption step in the process, to compensate for the recoil of the *absorbing system* from the radiation falling on it.

The greater frequency ω_a needed for an excitation by $\hbar\omega_o \equiv (m^* - m_o)c^2$ of an absorber is as easy to find as the frequency shift in (11.114) was. The result [Exercise 11.21 or (11.85*b*) with $m_1 \equiv 0$, $m_2 \equiv m_o$] is

$$\hbar\omega_a = \frac{(m^*)^2 - m_o^2}{2m_o}c^2 = \hbar\omega_o\left(1 + \frac{\hbar\omega_o}{2m_o c^2}\right) \tag{11.115}$$

The consequent fractional difference between ω_a and the frequency ω of (11.114), made available by the recoiling source, is

$$\frac{\omega_a - \omega}{\omega_o} = \frac{\hbar\omega_o}{2m_o c^2} + \frac{\hbar\omega_o}{2m^*c^2} \approx 2\frac{\hbar\omega_o}{2m_o c^2} \tag{11.116}$$

with the last approximation valid for the usual case of $m_o c^2 \gg \hbar\omega_o$.

The discrepancy (11.116) is not serious for the resonance fluorescence of *atomic* radiations. The frequencies emitted and absorbable by any system are never ideally discrete because of the finite durations of the states of motion involved, consequences of just the excitation and deexcitation processes changing the states of motion, among other effects. If Δt is a duration time of an otherwise periodic motion, its frequency is replaced by a range of frequencies with $\Delta\omega \approx 2\pi/\Delta t$ centered on the main one, a phenomenon discussed‡ in connection with (7.23). What is called a *natural broadening of a line* (a frequency) arises from the finite time it takes to form a quantum approximately equal to

† A case of $E_3' \equiv \hbar\omega$, $m_3 = 0$ in (11.104), with $m_1 \equiv m^*$ and $m_4 \equiv m_o$.

‡ A more detailed example is treated in connection with Fig. 7.6 for a system entirely constituted of an electromagnetic wave field confined within a cavity resonator and having a finite duration because of Joule losses in the cavity walls.

$\hbar\omega_o$ of radiation, estimated as $\Delta t \approx \hbar\omega_o/\mathscr{P}$ if $\mathscr{P}$ is the radiation rate. It was found in the discussion of (8.79) that atoms emit mostly electric dipole radiation because of the small ratio $a/\lambda\!\!\!^{-} \approx 10^{-3}$ of an atomic radius $a \approx 10^{-8}$ cm to the wavelength $\lambda\!\!\!^{-} = c/\omega_o \approx 10^{-5}$ cm arising from the energies $\hbar\omega_o$ made available by the internal motions in atoms. Thus the dipole rate $\mathscr{P} \approx e^2a^2\omega_o^4/c^3$ of (8.50) can be used for the estimate

$$\frac{\Delta\omega}{\omega_o} = \frac{2\pi}{\omega_o\,\Delta t} \approx \frac{2\pi\mathscr{P}}{\hbar\omega_o^2} \approx \frac{2\pi e^2}{\hbar c}\left(\frac{a}{\lambda\!\!\!^{-}}\right)^2 \approx 5 \times 10^{-8}$$

as typical of the natural broadening in atomic spectra. This is far greater than the frequency shifts (11.116) arising from the correspondent recoils, since

$$\frac{\hbar\omega_o}{m_o c^2} \approx \frac{\hbar/Mc}{A\lambda\!\!\!^{-}} \approx \frac{2 \times 10^{-9}}{A}$$

for an atomic mass of approximately AM, where M is a proton mass and $\hbar/Mc \approx 2.1 \times 10^{-14}$ cm (the *proton Compton wavelength*). In the much investigated case of mercury fluorescence, $A \approx 200$, and the recoil shift is less than a fraction 10^{-3} of the natural width; there is then no appreciable displacement from resonance.

The situation is otherwise for nuclear gamma radiations, principally because the radiated energies, and hence the recoils, are of the order of 10^5 times as great, yet the fractional broadening of the resonances is only of the same order as for the atomic radiations. Measured radiation widths $\hbar\,\Delta\omega = h/\Delta t < 10^{-2}$ eV are typical, and even for a relatively low-energy gamma ray, $\hbar\omega_o \approx 10^5$ eV, that makes $\Delta\omega/\omega_o < 10^{-7}$. Such a line is sharp in comparison to the recoil effect

$$\frac{\hbar\omega_o}{m_o c^2} \approx \frac{0.1\text{ MeV}}{A(931\text{ MeV})} \approx \frac{10^{-4}}{A}$$

even for nuclei of mass number $A \approx 200$. The recoils throw the radiation off resonance by several line breadths even in the most favorable cases.

The situation makes the resonance scattering of nuclear gamma rays difficult to observe, but it has been managed in a few cases by compensating for the recoil effects through Doppler shifts of the radiated frequencies. The source atoms are given a velocity relative to the absorber, either by heating them to greater thermal velocities or, more controllably, by mounting the source on the periphery of a rotator (Ref. 20). The radiated frequency ω can be raised to $\omega' = \gamma\omega(1 + v/c)$ by a velocity v, according to (11.20), and the Doppler shift $\Delta\omega'/\omega_o \approx v/c$ can compensate for the discrepancy $\hbar\omega_o/m_o c^2$ of (11.116) by using $v \approx (\hbar\omega_o/m_o c^2)c$. Moon and Davey (Ref. 20) did this for the 0.411-MeV gamma ray of ^{198}Hg and found a sharp enhancement of scattering by mercury as the velocity was passed through the value $v \approx 2.3 \times 10^{-6}c \approx 7 \times 10^4$ cm/s, a quite attainable magnitude for a rotator.

Resonance absorption attained great importance after the discovery (Ref. 21) of what is now called the Mössbauer effect, i.e., a finding that the attachment of a source nucleus to a metal lattice prevents appreciable recoil energy (and frequency) loss by the radiation in certain circumstances. Ideally, fixing an atom in a crystal structure results in imparting the recoil momentum from the radiation to the entire macroscopic mass of the structure, which absorbs the momentum without taking appreciable energy† ($T = p^2/2M = (\hbar\omega)^2/2Mc^2 \to 0$ as $M \to \infty$). The sharpness of resonance with the absorption mechanism of a like atom similarly fixed is then not appreciably spoiled by recoil effects. The comparatively low-energy ($\hbar\omega_o = 14.4$ keV) radiation from ^{57}Fe has been found particularly suitable; its momentum is low, for a nuclear radiation, and the natural breadth is also small because of the comparatively long life ($\Delta t \approx 1.4 \times 10^{-7}$ s, making $\Delta\omega/\omega_o \approx 2 \times 10^{-12}$). There is thus made available a resonance breadth of uncommon sharpness (monochromaticity), making it a sensitive tool for detecting any processes that might cause extremely small shifts from resonance.

The detectability of extremely small frequency shifts made possible a terrestrial confirmation of the equivalence of photon energy and gravitational mass, alluded to on page 345. Pound and Rebka‡ allowed photons from ^{57}Fe embedded in an iron lattice to fall through a height h of some meters, with an expected increase of the photon energy by

$$[\Delta(\hbar\omega)]_g = \frac{\hbar\omega_o}{c^2} gh$$

because of a gravitational acceleration of the mass $\hbar\omega_o/c^2$ by $g \approx 980$ cm/s^2. The fractional frequency shift here is only $(\Delta\omega/\omega_o)_g = gh/c^2 \approx 2.5 \times 10^{-15}$ for the $h \approx 23$ m to which Pound and Rebka were limited by the inverse-square law of intensity diminution. The shift is small in comparison to the natural width $\Delta\omega/\omega_o \approx 10^{-12}$ characteristic of the resonance absorption in iron, but it nevertheless proved detectable through effects of superposing easily controlled Doppler shifts, induced by vibrating the source. A confirmation of the gravitational effect on photons to an accuracy of 5 percent was thus achieved.

Annihilation Processes

Further illustrations of energy-momentum conservation in electromagnetic radiations are provided by processes in which *entire* rest masses are converted into radiation, symbolized as $e^+ + e^- \to 2\gamma$ and $\pi^0 \to 2\gamma$. The first of these is the annihilation of a positron e^+ by an electron e^-, and the second represents spontaneous decay of a neutral pion (which may be viewed as self-annihilation, or suicide). In both cases, just two photons may appear, each symbolized by γ (gamma ray). The necessity for *at least* two photons to be produced in the

† For energy transfers between an atom and a lattice, vibrations of the latter (*phonons*) must be excited or deexcited. The fact that the energy transfers are restricted to multiples of $\hbar\omega_p$, where ω_p is a lattice vibration frequency, helps damp transfers at low temperatures (when most $\hbar\omega_p > kT$). The experiments are therefore best performed at low temperatures.

‡ Pound and Rebka, Ref. 24. Crenshaw et al. (Ref. 7) obtained somewhat less definitive results at about the same time.

annihilation of an isolated system becomes obvious when it is viewed in its CM frame. The total CM energy can have associated with it only a vanishing *resultant* of CM momenta, and such can be provided by two oppositely directed photons. Sharing a given initial energy between more than just two photons is usually much rarer, as might be expected from the fact that the rate of radiation decreases rapidly with diminutions of the energy (frequencies) radiated.

In the pair annihilation $e^+ + e^- \to 2\gamma$ the oppositely charged initial particles have equal rest masses $m_1 = m_2 = m$, each the characteristic rest mass of an electron ($m = 0.511$ MeV$/c^2$). Most frequently observed is the incidence of a positron with some momentum $\mathbf{p}_1$ onto an electron that is practically at rest in the laboratory, the situation presented in Fig. 11.4*a*. Application of the energy-momentum conservation conditions (11.58) here provides limits on the photon energies $E_{3,4}$, as functions of their emission angles $\theta_{3,4}$, that will be observed for any given initial momentum $\mathbf{p}_1$.

No limit smaller than π can immediately be expected only for the CM angle at which the oppositely directed photons of equal CM energy will come off, relative to the CM velocity $\boldsymbol{\beta} = \mathbf{p}_1/(E_1 + m)$. Since the final particles are photons with velocities $u'_{3,4} = c$, relations (11.73) become

$$\tan \theta_{3,4} = \frac{\sin \vartheta}{\gamma(\beta \pm \cos \vartheta)} \tag{11.117}$$

results that are also obtainable from adaptations of (11.27). As $\vartheta \to \pi$, 0 and $\cos \vartheta \to \mp 1$ here, $\tan \theta_{3,4}$ can take on negative values. Each emission angle can therefore have the full range $0 < \theta_{3,4} < \pi$ for any initial momentum $\mathbf{p}_1 = \boldsymbol{\beta}(E_1 + m)$, in correspondence to curves in Fig. 11.5 with $u'_3 = c > v = c\beta$. The photon energies $E_{3,4}(\theta_{3,4})$ at each emission angle follow most directly from the conservation relations (11.59), which readily yield

$$E_{3,4} = \frac{mc^2}{1 - \beta \cos \theta_{3,4}} \tag{11.118}$$

With $\theta_{3,4}$ unrestricted in range, any photon energies between the limits given by

$$\frac{mc^2}{1 + \beta} < E_{3,4} < \frac{mc^2}{1 - \beta} \tag{11.119}$$

can be expected. When one of the energies has one of the limit values, the other is at the opposite limit, in agreement with energy conservation $E_3 + E_4 = E_1 + m$. Another point made evident by (11.118) is that whenever one of the photons comes off at right angles (say $\theta_3 = \frac{1}{2}\pi$) to the incident positron's direction, it receives just the energy $E_3 = mc^2$ of the rest electron, while the other photon acquires all the energy of the positron, $E_4 = E_1$. (See Exercise 11.22 for the angle θ_4 at which the more energetic photon will be found.)

Each instance of the pion decay process $\pi^0 \to 2\gamma$ takes place while the initial particle is in flight† with some velocity $\boldsymbol{\beta} = \mathbf{p}_1/E_1$, and this becomes the CM

† Because of its short life and the paucity of processes by which neutral particles can be slowed down.

velocity of the product photons. Relative to the CM frame, the equal and opposite momenta $\mathbf{p}'_3 = -\mathbf{p}'_4$ of the photons may take any direction, and this will lead to a variety of directions relative to the laboratory. It is the opening angle $\theta \equiv \theta_3 + \theta_4$ between the photon directions that is the more directly observable than the individual angles $\theta_{3,4}$ relative to $\boldsymbol{\beta}$, because the initial particle is uncharged, making its direction of motion before decay difficult to determine.

The opening angle θ enters into the conservation relations [(11.58) with $p_{2\mu} \equiv 0$] through the projections $\mathbf{p}_3 \cdot \mathbf{p}_4 = E_3 E_4 \cos\theta$ of the product laboratory momenta on each other, as these occur in the invariant (11.77),

$$s = -p_{1\mu}^2 = -(p_{3\mu} + p_{4\mu})^2 \equiv m_1^2 = 2(E_3 E_4 - \mathbf{p}_3 \cdot \mathbf{p}_4) \tag{11.120a}$$

after use of $p_{3\mu}^2 = p_{4\mu}^2 = 0$ for the rest-mass-less photons. Thus

$$m_1^2 = 2E_3 E_4(1 - \cos\theta) = 4E_3 E_4 \sin^2 \frac{\theta}{2} \tag{11.120b}$$

and

$$\sin \frac{\theta}{2} = \frac{m_1}{2(E_3 E_4)^{1/2}} \tag{11.121}$$

is a relation of the opening angle to the laboratory energies of the product photons. The connection to the initial flight velocity $\boldsymbol{\beta}$ and to the emission angle ϑ that occurs in the CM frame relative to that velocity can be found by substituting for $E_{3,4}$ the results (11.106), adapted to $m_3 = m_4 = 0$ and so having $\Delta_{34}(m_1^2) = m_1^2$,

$$E_{3,4} = \tfrac{1}{2}E_1(1 \pm \beta \cos\vartheta) \tag{11.122}$$

Here $E_1 = m_1/(1 - \beta^2)^{1/2} = \gamma m_1$ is the total energy of the pion in flight and $E_3 + E_4 = E_1$ properly. Now relation (11.121) can be expressed as

$$\sin \frac{\theta}{2} = \frac{m_1/E_1}{(1 - \beta^2 \cos^2\vartheta)^{1/2}} = \left(\frac{1 - \beta^2}{1 - \beta^2 \cos^2\vartheta}\right)^{1/2} \tag{11.123}$$

This immediately shows that the opening angle π in the rest frame of the pion cannot be collimated to less than $\theta_{\min} = 2 \sin^{-1}(1/\gamma)$ by the frame motion and that the minimum laboratory angle occurs when the photons are emitted at right angles ($\vartheta = \frac{1}{2}\pi$) relative to that frame.

The result (11.123) also shows that the opposite emissions described by $\theta \to \pi$ and characteristic of decays at rest still occur at high velocities whenever one of the two photons happens to be emitted parallel to the flight velocity ($\vartheta = 0$). In such cases, each photon acquires one of the two extreme energy values of the range

$$\tfrac{1}{2}E_1(1 - \beta) \le E_{3,4} \le \tfrac{1}{2}E_1(1 + \beta) = \tfrac{1}{2}(E_1 + p_1) \tag{11.124}$$

afforded by (11.122). Thus one of the photons can be quite "soft" (E_3 or $E_4 \to 0$) when it is emitted against the direction of a high ($\beta \to 1$) flight velocity; of course a photon velocity c cannot actually be reversed, as the smaller velocity of a particle with rest mass could be. The difference of the extremes in (11.124) is just

$E_1 \beta = p_1$, and this approaches zero as $\beta \to 0$ and $E_1 \to m_1$, when the equal values $E_3 = E_4 = m_1/2$ should be expected.

The relative frequency with which various of the possible opening angles $\theta_{\min} < \theta < \pi$ will occur can actually be predicted because there can be no preferred emission direction in the pion's rest frame (pions are spinless, and no orientation of one can be defined). Every CM angle of the range $0 < \vartheta < \frac{1}{2}\pi$ is equally likely, this being a sufficient range to consider here because taking larger values amounts to transferring attention to the direction of a partner photon. Then the isotropy of emission in the CM frame can be represented by an expression for the fraction $d^2\Gamma/\Gamma$ of the total decay rate Γ that is devoted to producing a direction ϑ ($< \frac{1}{2}\pi$), within a solid-angle element $d\Omega(\vartheta, \varphi) = d(\cos\vartheta)\, d\varphi$, given by

$$\frac{d^2\Gamma(\vartheta, \varphi)}{\Gamma} = \frac{d\Omega(\vartheta, \varphi)}{2\pi} = \frac{1}{2\pi}\, d(\cos\vartheta)\Big|_0^1 \, d\varphi\Big|_0^{2\pi} \tag{11.125}$$

In the *ratio* on the left, it does not matter whether Γ refers to the number of processes per unit of time in the proper (rest) frame of the pion or to the laboratory frame. Individually, $d^2\Gamma$ and Γ *are* different in the two frames because of the time-dilatation effects discussed on page 334 (evaluated per pion, the rates Γ are just inverses of the mean lives τ discussed there). However, the ratio in (11.125) refers to an invariant number of processes, a count delimited by a given magnitude of CM solid-angle element in every CM direction and independent of the frame from which it might be viewed. When this element is redescribed in terms of corresponding laboratory angles, it will generally become nonisotropic as a function of them.

To obtain a relative frequency $d\Gamma(\theta)/\Gamma$ for the laboratory opening angle θ to have a cosine in the range $d(\cos\theta)$, the second-order differential in (11.125) is integrated over the azimuth, to yield

$$\frac{d\Gamma}{\Gamma} = d(\cos\vartheta)\Big|_0^1 = -\frac{d(\cos\vartheta)}{d(\cos\theta)}\, d(\cos\theta)\Big|_{-1}^{\cos\theta_{\min}} \tag{11.126}$$

The trivial minus sign stems from the interchange of the indicated integration limits. The differential quotient in (11.126) can be evaluated from the inverse of expression (11.123)

$$\cos\vartheta = \frac{1}{\beta}\left[1 - \frac{1-\beta^2}{\sin^2(\theta/2)}\right]^{1/2} = \frac{1}{\beta}\left[1 - \frac{2(1-\beta^2)}{1-\cos\theta}\right]^{1/2}$$

to produce

$$\frac{d\Gamma}{\Gamma\, d(\cos\theta)} = \frac{1-\beta^2}{4\beta}\, \frac{1}{\sin^3(\theta/2)}\, \frac{1}{[\sin^2(\theta/2) - (1-\beta^2)]^{1/2}} \tag{11.127}$$

defined only for $\theta > \theta_{\min} = 2\sin^{-1}\sqrt{1-\beta^2}$. It can be checked that integrating this over $\theta_{\min} < \theta < \pi$ yields unity, as it should.

The result (11.127) indicates that of all the opening angles possible in the range afforded by a given flight velocity, the smallest will appear with the greatest frequency, when the "bins" $d(\cos\theta)$, in which counts are made, are chosen uniformly on the scale of $\cos\theta$. The full range of θ variations becomes very narrow, with $\theta \to \pi$ for $\beta \to 0$, but $\theta \to 0$ becomes possible as $\beta \to 1$, and then the frequency of small angles is relatively greater than ever. The smallest angles in the possible ranges are those for which the geometric mean $(E_3 E_4)^{1/2}$ of the photon energies is greatest, as (11.121) shows most directly, and corresponds to the most equal possible sharing $E_3 + E_4 = E_1$ of the available energy. However, the relative frequency of a given photon energy, say E_3 accompanied by $E_4 = E_1 - E_3$, should properly be judged by comparing counts in bins dE_3 of equal size, in the range (11.124) of these energies.

The relative frequency of energies E_3 in equal intervals dE_3 is most readily found by transforming the isotropic distribution in CM angles ϑ,

$$\frac{d\Gamma}{\Gamma} = \frac{1}{2}\, d(\cos\vartheta)\Big|_{-1}^{1}$$

while allowing $0 < \vartheta < \pi$ in order to produce the full range of energies in (11.124). It then follows from the relation (11.122) of $\cos\vartheta$ to E_3 that

$$\frac{d\Gamma}{\Gamma\, dE_3} = \frac{1}{2}\frac{d(\cos\vartheta)}{dE_3} = \frac{1}{\beta E_1} = \frac{1}{cp_1} \tag{11.128}$$

This is independent of E_3 in the range (11.124), showing that the distribution in energy is *flat*, with every possible photon energy equally likely, i.e., equal numbers of the processes falling into any one of the equal bins dE_3. Such a flat energy distribution is characteristic of every isotropic two-particle decay, even when the two products have unequal, nonzero rest masses. (See Exercise 11.23.)

EXERCISES

11.1 (*a*) Show that the Lorentz transformation of the space coordinates in (11.6) can be reexpressed as

$$\mathbf{r}' = \mathbf{r} + (\gamma - 1)(\mathbf{r}\cdot\hat{\mathbf{v}})\hat{\mathbf{v}} - \gamma\mathbf{v}t$$

This can now be decomposed on *any* three orthogonal directions to give a connection between parallel frames having an *arbitrarily* directed relative velocity $\mathbf{v} = \hat{\mathbf{v}}v$ of their origins.

(*b*) Find the correspondingly frame-independent expression for the connection of t' to t.

(*c*) Similarly, show that the velocity transformation (11.23) can be generalized to

$$\mathbf{u}' = \frac{\gamma(\mathbf{u} - \mathbf{v}) + (\gamma - 1)\hat{\mathbf{v}} \times (\hat{\mathbf{v}} \times \mathbf{u})}{\gamma(1 - \mathbf{u}\cdot\mathbf{v}/c^2)}$$

11.2 The origin O' of a frame has a velocity $\mathbf{v}$ with respect to a frame O in which $v_z = v\cos\alpha$, $v_x = v\sin\alpha$, $v_y = 0$.

(*a*) Construct a matrix of coefficients which will rotate frame O about its y axis so that its new z axis has the direction of $\mathbf{v}$.

(*b*) Show that a suitable product of the matrix in (*a*) with (11.9) will yield the connection to the *initial* frame O of an O' frame having its z' axis along the direction of $\mathbf{v}$. [See (11.11) for an example of a matrix resultant of a matrix product.]

(*c*) Multiply the matrix product found in (*b*) with a third matrix which will produce a frame O' entirely parallel to the initial O frame and check your results against Exercise 11.1.

11.3 Suppose that two frames $O'(\mathbf{v})$ and $O'(\mathbf{v} + \Delta\mathbf{v}) \equiv O_1$ both have space axes parallel to those of O but differ in velocity with respect to O by a small amount $\Delta\mathbf{v} \to 0$.

(*a*) Find expressions for the coordinates $\mathbf{r}_1$, t_1 in O_1 in terms of $\mathbf{r}$, t, $\mathbf{v}$, $\gamma(v)$, and $\Delta\mathbf{v}$ to an approximation linear in $\Delta\mathbf{v}$.

(*b*) Find what the result of Exercise 11.1(*c*) gives for the velocity $\Delta\mathbf{v}'$ of O_1 relative to $O'(\mathbf{v})$.

(*c*) By using the results of (*a*) together with the reciprocal of the transformation given in Exercise 11.1(*a*) show that the connection of $\mathbf{r}_1$ to the coordinates $\mathbf{r}'$, t' in $O'(\mathbf{v})$ can be expressed as

$$\mathbf{r}_1 = \mathbf{r}' + \Delta\boldsymbol{\theta} \times \mathbf{r}' - \Delta\mathbf{v}'t' \qquad \text{with } \Delta\boldsymbol{\theta} \equiv \frac{(\gamma - 1)(\Delta\mathbf{v} \times \mathbf{v})}{v^2}$$

[Thus, for both $O'(\mathbf{v})$ and $O'(\mathbf{v} + \Delta\mathbf{v})$ to have axes parallel to those of O, they must be *rotated* (when $\Delta\boldsymbol{\theta} \neq 0$) relative to *each other*.]

11.4 Sometimes forces applied to a particle are first more readily formulated with respect to its *instantaneous* rest frames, each a frame $O'(t)$ with a translational velocity taken equal to the velocity $\mathbf{v}(t)$ that the particle itself has at the instant (t) relative to the laboratory frame O. To avoid an appearance in $O'(t)$ of undetermined inertial (centrifugal and Coriolis) forces, extra to the applied ones, it must be chosen as a $\Delta\mathbf{v} \equiv \dot{\mathbf{v}}\,\Delta t \to 0$ limit of an $O'(t + \Delta t) \equiv O_2$ that has only a uniform translational motion $\Delta\mathbf{v}'$ relative to $O'(t)$ *without* the rotation (here during $\Delta t \to 0$) found for O_1 of Exercise 11.3.

(*a*) Show that the parallel-, i.e., unrotated-, axis transformations of Exercise 11.1 here yield $t_2 \equiv t_1$ but $\mathbf{r}_2 = \mathbf{r}' - \Delta\mathbf{v}'t'$ in place of the $\mathbf{r}_1$ in O_1.

(*b*) Using the fact that it is the O_1 axes that are parallel to those of the laboratory frame O, show that the properly chosen particle frame $O'(t) = \lim O_2$, with axes only *instantaneously* parallel to the O axes, will have not only the translation $\mathbf{v}(t)$ but also an instantaneous angular velocity with magnitude and *direction* given by

$$\boldsymbol{\omega}_T = (\gamma - 1)\frac{\dot{\mathbf{v}} \times \mathbf{v}}{c^2}$$

[This is the famous *Thomas precession*, first discovered in a problem of atomic physics (Exercise 12.6).]

11.5 A long rod moving parallel to its length passes close by an observer O, who attributes the speed $c\beta$ to it. At a moment at which the back end of the rod passes through the position of O, he *sees* the front end at a distance s from himself (perhaps that end activates a light signal as it passes, one known by O to be at the distance s). Find the *rest* length L_o of the rod in terms of s and β.

11.6 A solid rod of rest length L_o is traveling with a speed $u = 4c/5$, and parallel to its length when its forward end impinges normally on an immovable wall. Suppose that the rod comes to rest intact and without change of orientation, with its forward end remaining stuck to the wall thereafter.

(*a*) How far from the wall (and in its rest frame) was the far end of the rod at the moment of impact and after it has come to rest?

(*b*) Find a minimum time interval Δt after first impact that is needed for the rod to come entirely to rest. (Think of shock fronts that are reflected only at discontinuities of the rod medium. Then $\Delta t \geq 5L_o/3c$ when shock-front velocities c are taken to characterize the most perfect rigidity consistent with relativity.)

11.7 An unmanned rocket is receding from the earth, and a measurement of its speed is wanted.

(*a*) An earth station sends electromagnetic laser pulses to a mirror on the rocket. If the pulses are sent at periodic time intervals τ, at what intervals must the reflected pulses be returning to warrant a conclusion that the rocket's speed is $u = 0.6c$?

(*b*) As a check, the earth station also sends a continuous radio beam to a receiver-transmitter on the rocket pretuned before takeoff to respond to or with a fixed frequency ν_o. What frequency must the earth send to get a response if the rocket speed is indeed that found in (*a*)?

(*c*) To what frequency should an earth receiver be tuned in order to detect the rocket's return signal?

11.8 Prove that (11.24) always yields $u_z < c$ if only v, $u'_z < c$.

11.9 (*a*) Find the acceleration transformation $\dot{\mathbf{u}} \to \dot{\mathbf{u}}'$ corresponding to the velocity transformation (11.23).

(*b*) Use the result in (*a*) to derive (10.15) from the Larmor ($\mathbf{u} = 0$) radiation rate (10.18). (Include an argument that $\mathscr{P}$, an energy increment divided by a time increment, must be the same in relatively moving frames.)

11.10 Find what the velocity transformation as given in Exercise 11.1(*c*) yields for the momentum transformation of a rest mass m having that velocity.

11.11 A mass M at rest undergoes spontaneous fission into two like fragments of rest mass $m = 0.3M$ each.

(*a*) What will the *relative* velocity of the two fragments be?

(*b*) How rapidly, in centimeters per second, will their distance apart be *observed* to increase in the laboratory frame (of the initial M at rest)?

11.12 It is well known that when a mass $m_1 = m$ with a low speed ($u_1 \ll c$) collides head on with an *equal* mass $m_2 = m$ at rest, m_1 stops dead and m_2 gets all its kinetic energy (in an elastic collision).

(*a*) Show that if $m_1 \gg m_2$, instead (and $u_1 \ll c$ still), m_1 is hardly slowed down while m_2 acquires a velocity approximately equal to $2u_1$.

(*b*) Now find what happens if the rest masses are equal ($m_1 = m_2 = m$) but $u_1 \to c$, so that $M_1(u_1) \gg m_2 = m$.

11.13 In so-called *missing-mass* experiments, a product m_4 of a process like that indicated in Fig. 11.4*a* is unobservable (perhaps because it is uncharged). Develop a formula for calculating the mass m_4 from $m_{1,2,3}$ and observations on p_1, p_3, and θ_3.

11.14 Even a high-energy neutron encountering a proton at rest is sometimes captured (forming a deuteron) with the emission of a single photon. For a high enough neutron kinetic energy T, the neutron and proton rest masses can be treated as equal (M) and the deuteron rest mass as just twice as great ($2M$). Find the greatest angle from the incidence direction at which the deuteron can be expected to emerge and the highest photon energy that should be observable in terms of T and M. [For $T = 2Mc^2$, $\sin^{-1}(1/2\sqrt{2})$ and $(1 + \frac{1}{2}\sqrt{2})Mc^2$.]

11.15 A pion π can be produced in the collision of nucleons, via $N + N \to N + N + \pi$. In units of the nucleon rest mass M, the pion rest mass is about $0.15M$.

(*a*) The IUCF cyclotron gives nucleons (protons) a kinetic energy $T_1 \approx 0.2Mc^2$. Show that this is *not* sufficient to produce a pion through striking a *stationary* proton.

(*b*) Target nucleons that are bound inside atomic nuclei are not stationary but may have kinetic energies up to $T_2 \approx 0.02Mc^2$. Show that the T_1 protons should be able to produce pions at least in head-on collisions with T_2 nucleons.

11.16 Show that the velocities (11.107), equivalent to those of the products of an $m_1 \to m_3 + m_4$ decay at rest, can also be expressed as

$$u'_{3,4} = \frac{\Delta_{34}(m_1^2)}{m_1^2 \pm (m_3^2 - m_4^2)}$$

and check the consistency of this expression with (11.84).

11.17 Preparations are to be made to detect muon (μ) decay products of pions ($\pi \to \mu + \nu$) in a beam which has had to traverse a length of 3 m before arriving at the detection site. In units of the electron mass m, $m_\pi \approx 273m$, $m_\mu = 207m$, $m_\nu = 0$.

(*a*) What is the velocity of most the pions that decay at the detection site, given that their mean life at rest is 2.6×10^{-8} s?

(*b*) What is the velocity, relative to its parent, of the muon?

(*c*) Up to what maximum angle from the pion beam direction should the muon detectors be arrayed?

(*d*) What will be the laboratory kinetic energy of the muons at the maximum angle found in (*c*)? *Ans.* $\approx 10mc^2 \approx 5$ MeV.

11.18 Gamma rays incident on electrons can be converted into $e^{\pm}$ pairs. What is the threshold γ-ray energy needed for this conversion (in units mc^2) when it is incident on an electron at rest?

11.19 Show that, quite apart from the electromagnetic theory of radiation, a *uniformly* moving electron cannot radiate photons, in any number, because that would violate energy-momentum conservation.

11.20 High-energy photons can be produced by a collision transfer of energy from a high-energy electron beam to low-energy (visible) photons. Suppose that an intense laser beam of visible photons is made to intersect the electron beam at a right angle.

(*a*) Derive an expression for the energy of the photons that are propelled into the same direction as that of the initial electron beam, in terms of the initial energies.

(*b*) Show that for extremely relativistic initial electrons, $E_1 \gg mc^2$, the difference of the final photon energies in (*a*) from E_1 becomes proportional to the photon wavelength used.

11.21 Gamma rays with a broad range of energies pass through a sample of atomic nuclei having rest masses m_o. The transmitted intensity shows a peak absorption of the photons with energy $\hbar\omega_a$. Derive an expression for the energy $\hbar\omega_o$ by which each absorbing nucleus is excited, in terms of $\hbar\omega_a$ and $m_o c^2$.

11.22 A beam of positrons is incident on a target of electrons at rest. One annihilation-photon counter is placed in a direction at right angles to the beam. Where should a second such counter be placed, for a given beam momentum p, if it is to register coincidences with the first counter?

11.23 A mass m_1 at rest decays isotropically into products m_3, m_4. Suppose that decays from m_1 in flight, with momentum p_1, are observed. Show that the energy distributions of each product are given by

$$\frac{d\Gamma}{\Gamma\, dE_{3,4}} = \frac{m_1^2}{cp_1 \Delta_{34}(m_1^2)}$$

being uniform in the product energies. What are the lower and upper limits on the product energies? Are they consistent with the normalization of the distribution expression here? Is this also consistent with the special case (11.128)?

CHAPTER

TWELVE

FRAME-INDEPENDENT REPRESENTATIONS

The reformulation of electromagnetic theory in four-dimensional space-time · The Lorentz-potential 4-vector and the electromagnetic field tensor · Transformations of fields and waves · The work-force density 4-vector and the field energy-momentum tensor · The angular-momentum tensor and the CM motion of a free field · The relativistic equation of motion for a charged particle · Minkowski force · The 4-vector velocity · Relativistic motions in static fields

Consistency with relativistic principles requires that the expressions (equations) used as general bases for physical expectations be equally applicable in every equivalent reference frame. This requirement is automatically satisfied by expressions in terms of quantities that may be taken to refer to any one of the frames in question. Thus the newtonian expressions of mechanics, like $m\dot{\mathbf{u}} = \mathbf{F}$, are equally valid in any one of variously *oriented* spatial frames because they involve, besides scalar numbers, only vectors that can be decomposed with equal validity on any one of those frames. Such expressions are then said to possess a *manifest* rotational invariance (although the components of the vectors are *co*variant with the coordinate changes, since a vector has different projections on axes of differently oriented frames).

The newtonian descriptions are also invariant to galilean transformations like (11.1), as discussed in the first paragraphs of Chap. 11 and in connection with the galilean velocity addition (4.67). Just because this is so, the newtonian force invariance must fail at high speeds, since the different, nongalilean velocity connections (11.23) must now be expected to hold. Thus, the usual classical formulations of mechanics are not Lorentz-invariant and must be modified before they can become applicable to high velocities.

On the other hand, the formulations of electromagnetic theory might well be expected to possess the Lorentz invariance without having to be modified, since they were developed in consistency with an invariant velocity for light. However, when the relations are expressed in terms of 3-vectors, as the Maxwell equations in the forms (1.40) are, a Lorentz invariance is not as immediately evident as the invariance to merely static frame rotations. To make both types of invariance equally manifest, as to a general four-dimensional rotation like (11.36), the electromagnetic formulations should be reexpressed in terms of 4-vectors and other four-dimensional covariants. That will be the first objective of this chapter, together with the derivation of various conclusions that follow most readily from the new expressions, now explicitly mapped on four-dimensional space-time.

Any formulations that are to be consistent with einsteinian relativity should be expressible in terms of space-time covariants. This will serve as a guide to reformulating classical mechanics for applicability to high speeds.

12.1 RELATIVISTICALLY COVARIANT FIELD DESCRIPTIONS

The formulations of electromagnetic theory require considering space derivatives represented by the gradient operator ∇ and time derivatives $\partial/\partial t$. In terms of the event 4-vector $x_\mu(x, y, z, ict)$ of (11.36), the four component derivatives $\partial/\partial x_k \equiv \partial_k$ $(k = 1, 2, 3)$ and $\partial_4 \equiv \partial/\partial x_4 \equiv \partial/(ic\, \partial t)$ can be taken to constitute a four-dimensional resolution of a *space-time gradient*

$$\partial_\mu\left(\nabla, \frac{\partial}{ic\, \partial t}\right) \qquad \text{with } \mu = 1, 2, 3, 4 \tag{12.1}$$

This has the covariance, with the Lorentz transformation (11.8), characteristic of a 4-vector like x_μ or p_μ of (11.36), since

$$\partial'_\mu \equiv \frac{\partial}{\partial x'_\mu} = \frac{\partial x_\nu}{\partial x'_\mu}\frac{\partial}{\partial x_\nu} = (a^{-1})_{\nu\mu}\,\partial_\nu = a_{\mu\nu}\,\partial_\nu \tag{12.2}$$

follows from the orthonormality property (11.10) of the transformation coefficients. Here, summations over the repeated index ν are implied, in accordance with the summation convention introduced in (11.77); the index μ chosen for the left side is a free index, not to be summed over.

An immediate consequence of the 4-vector character of the space-time gradient operator is that the d'alembertian

$$\square^2 \equiv \nabla^2 - \frac{\partial^2}{c^2\, \partial t^2} = \partial_\mu^2 \qquad \left(\equiv \sum_\mu \partial_\mu^2\right) \tag{12.3}$$

is *in*variant to the frame changes, like the squared length of any 4-vector, as in (11.3) or (11.37). That is shown formally by the result

$$(\partial'_\mu)^2 = a_{\mu\nu} a_{\mu\rho}\, \partial_\nu\, \partial_\rho = \delta_{\nu\rho}\, \partial_\nu\, \partial_\rho = \partial_\nu^2 \tag{12.4}$$

that follows from the orthonormality (11.4); here, *all* the indices are repeated ones, to be summed over, with an indexless 4-scalar (invariant) $\square^2$ as the result.

The Space-Time Source Density

The basic equations of electromagnetic theory are expressions for field increments that are generated by *conserved* charges, the property represented by the continuity equation (1.14), which can be written

$$\nabla \cdot \mathbf{j} + \frac{\partial(ic\rho)}{ic\, \partial t} = 0$$

This can be reexpressed with the help of the space-time gradient operator as

$$\partial_\mu j_\mu = 0 \tag{12.5}$$

if a source-density 4-vector

$$j_\mu(\mathbf{j}, ic\rho) \tag{12.6}$$

is defined. The characteristic 4-vector transformation $j'_\mu = a_{\mu\nu} j_\nu$ thus implied is essential for the invariant validity of (12.5), as explicitly shown by

$$\partial'_\mu j'_\mu = a_{\mu\nu} a_{\mu\rho}\, \partial_\nu j_\rho = \delta_{\nu\rho}\, \partial_\nu j_\rho = \partial_\nu j_\nu = 0 \tag{12.7}$$

and such an invariance, at the zero value, must hold if the charge is to be conserved for every observer evaluating it. The expression $\partial_\mu j_\mu$ is a 4-scalar space-time divergence of a 4-vector field $j_\mu(x)$, and vanishing space-time divergences are standard expressions of conservation laws for quantities distributed over fields.

The definition (12.6) implies that to charge distributions ρ, formerly treated as scalar quantities in 3-space, must now be attributed the properties expected of fourth components of 4-vectors in space-time, like the dilatation by motion exhibited by time intervals as fourth components of dx_μ/ic and by particle masses proportional to fourth components of energy-momentum 4-vectors (11.36).

Consider a distribution of charge elements that are everywhere at rest relative to each other, so that there is a unique proper (rest) frame of the entire distribution which can be represented by a 4-vector $j^o_\mu(0, ic\rho^o)$. In a frame with respect to which the entire distribution moves bodily, with some uniform velocity $c\beta = u = u_z$, it has the 4-vector components

$$\begin{aligned} j_x &= 0 \qquad j_z = \gamma(0 - i\beta \cdot ic\rho^o) = \gamma\rho^o u \\ j_y &= 0 \qquad \rho = \gamma\rho^o \qquad \gamma = \left(1 - \frac{u^2}{c^2}\right)^{-1/2} \end{aligned} \tag{12.8}$$

as follows from the inverse (11.10) of the matrix (11.9) of transformation coefficients. Thus, there is the expected current density $\mathbf{j} = \rho\mathbf{u}$, and a charge density $\rho = \gamma\rho^o$ that has a magnitude greater than in the proper frame by just the factor $\gamma > 1$ characteristic of time dilatation and of the mass magnification $M(u) = \gamma m$ (11.31). The $j_\mu(\rho\mathbf{u}, ic\rho)$ is said to be the result of boosting $j^o_\mu(0, ic\rho^o)$ by the velocity $\mathbf{u}$.

The greater than rest density of the charge is consistent with charge conservation because it measures charge *per unit volume*, and every volume element undergoes Lorentz contraction relative to a frame in which it moves. Thus

$$dq \equiv \rho\, dx\, dy\, dz = \gamma\rho^o\, dx^o\, dy^o \frac{dz^o}{\gamma} = \rho^o\, dV^o \tag{12.9}$$

is invariant, as every element of charge itself must be. For similar reasons, four-dimensional space-time volume elements, denoted $dx_1\, dx_2\, dx_3\, dx_4 \equiv i\, d^4x$, are also invariant, as in

$$d^4x = dx^o\, dy^o \frac{dz^o}{\gamma}(c\gamma\, dt^o) \equiv d^4x^o \tag{12.10}$$

with the time dilation $dx_4/ic = dt = \gamma\, dt^o$ compensating for the contraction of the 3-space volume elements.

4-Vector Potentials

The charge and current distributions define Poisson equations (6.47) for Lorentz-gauged potentials that can be reexpressed with the help of the manifestly invariant form (12.3) of the d'alembertian operation as

$$\partial_\mu^2 A_\nu = -\frac{4\pi j_\nu}{c} \qquad \nu = 1, 2, 3, 4 \tag{12.11}$$

if only a Lorentz potential 4-vector

$$A_\nu(\mathbf{A}, i\phi) \tag{12.12}$$

is defined. The 4-vector character is imposed on solutions of the four component equations in (12.11) by the behavior of j_ν in frame transformations, since the operation by ∂_μ^2 stays invariant. Formally,

$$(\partial'_\mu)^2 A'_\nu = -\frac{4\pi j'_\nu}{c} \equiv \partial_\mu^2 A'_\nu = -\frac{4\pi a_{\nu\rho} j_\rho}{c}$$

and multiplication of both sides by $a_{\nu\sigma} = (a^{-1})_{\sigma\nu}$ followed by summation over ν yields

$$\partial_\mu^2[(a^{-1})_{\sigma\nu} A'_\nu] = -\frac{4\pi \delta_{\sigma\rho} j_\rho}{c} = -\frac{4\pi j_\sigma}{c}$$

The solutions of this, $A_\sigma \equiv (a^{-1})_{\sigma\nu} A'_\nu$, have thus imposed on them the 4-vector covariance in the reciprocal form shown for the event 4-vector in (11.8).

The potential 4-vectors will satisfy the Lorentz gauge condition (6.46) in every frame if they do so in any one, because that condition can now be given the manifestly invariant form†

$$\partial_\mu A_\mu = 0 \tag{12.13}$$

of a space-time divergence. The gauge invariance (6.45) common to potential descriptions can now be described by the four components of

$$A_\mu \to \bar{A}_\mu = A_\mu + \partial_\mu \chi \tag{12.14}$$

† Becoming the familiar $\nabla \cdot \mathbf{A} = 0$ in static situations.

Any thus regauged 4-vector potential $\bar{A}_\mu$ will satisfy the Lorentz condition (12.13) whenever A_μ does if only the 4-scalar gauge function χ is chosen to be some solution of the invariant wave equation $\partial_\mu^2 \chi = 0$, as in (6.49).

The 4-vector behavior of Lorentz potentials makes it possible to derive the field of a uniformly moving point charge from its static proper-frame distribution $A_\mu^o(0, iq/r_o)$ by transforming it to a laboratory frame in which q has the requisite uniform velocity u ($= u_z = c\beta$), the procedure being called a boost by the velocity u. In the laboratory frame

$$A_{x,y} = 0 \qquad A_z = \gamma(-i\beta)(i\phi^o) = \beta\gamma\phi^o \qquad \phi = \gamma\phi^o \tag{12.15}$$

where $\phi^o \equiv q/r_o$ with $r_o = (x_o^2 + y_o^2 + z_o^2)^{1/2}$, the separation of field point and charge as measured in the proper frame. The laboratory potential should be expressed in terms of the separation distance

$$r = [x^2 + y^2 + (z - ut)^2]^{1/2} \tag{12.16}$$

between the field point and the moving charge, as measured in the laboratory, and the angle defined by $\cos\vartheta = (z - ut)/r$, making $x^2 + y^2 = r^2 \sin^2\vartheta$ (see Fig. 10.12), can also be introduced. The Lorentz transformation (11.6) shows that $z - ut = z_o/\gamma$ is just the Lorentz-contracted component of the proper separation that is parallel to the velocity. Thus

$$\phi = \gamma\phi^o = \frac{\gamma q}{[x^2 + y^2 + \gamma^2(z - ut)^2]^{1/2}} = \frac{q}{r[(1 - \beta^2)\sin^2\vartheta + \cos^2\vartheta]^{1/2}}$$

equivalent to the form shown in (10.87). In addition, (12.15) yields $\mathbf{A} = \boldsymbol{\beta}\phi$, as in (10.2). Reexpression in terms of the propagation distance $R = c(t - t_q)$ to the field point from the charge's position at the retarded time $t_q = t - R/c$ would yield a reversion to the form (10.1). However, the validity of that result for an *arbitrarily* moving point charge cannot be safely concluded from just the connection between frames in *uniform* relative motion. A basis for expecting an acceleration independence of the potential distributions must still be made evident, as done in the earlier development.

The example just considered, of the 4-vector potential arising from a simple point charge, is a basic one in that quite arbitrary physical sources can be analyzed into superpositions of moving point charges. However, the more powerful approach is to describe sources by continuous distributions $j_\nu(x)$. The 4-vector potential arising from any given distribution has the retarded components (8.6) and (8.11), forming

$$A_\nu(\mathbf{r}, t) = \oint \frac{j_\nu(\mathbf{r}_s, t_s)\, dV(\mathbf{r}_s)}{c|\mathbf{r} - \mathbf{r}_s|}\bigg|_{t_s = t - |\mathbf{r} - \mathbf{r}_s|/c} \tag{12.17}$$

The expression (8.10) for the fourth component makes it plain that the entire result can also be given the form

$$A_\nu(x) = \oint \chi(x - x') j_\nu(x') \frac{d^4x'}{c^2} \tag{12.18}$$

where the arguments x and x' stand for dependences on $x_\mu(\mathbf{r}, ict)$ and $x'_\mu(\mathbf{r}_s, ict_s)$, respectively the space-time coordinate 4-vectors of the field and the source points. The propagator, or Green's function,

$$\chi(x - x') = \frac{\delta[(t_s - t) + |\mathbf{r} - \mathbf{r}_s|/c]}{|\mathbf{r} - \mathbf{r}_s|} \tag{12.19}$$

of (8.8), is a solution of (8.9), now written

$$\partial_\mu^2 \chi = -4\pi c \delta(x - x') \tag{12.20}$$

with $\delta(x - x') \equiv \delta(\mathbf{r} - \mathbf{r}_s)\delta(ct - ct_s)$, a 4-scalar since it must behave like the inverse of an invariant space-time volume element $d^4(x - x')$. Thus the solutions χ have a 4-scalar (invariant) character imposed on them, and results of the integration (12.18) will have the 4-*vector* behavior of j_ν itself.

The Electromagnetic Field Tensor

Manifestly invariant reexpressions of the Maxwell equations themselves can be found after deriving the 3-vector fields $\mathbf{E}$ and $\mathbf{B}$ from the covariant potential components. Most immediately evident is that the relation $\mathbf{B} = \nabla \times \mathbf{A}$ can be decomposed into

$$\mathsf{B}_k = \partial_l A_m - \partial_m A_l \tag{12.21}$$

if the indices here are chosen to be $klm = 123$ or 231 or 312, each set in cyclic order. Generalization to the four space-time dimensions suggests defining a second-rank antisymmetric tensor

$$F_{\mu\nu} = \partial_\mu A_\nu - \partial_\nu A_\mu = -F_{\nu\mu} \tag{12.22}$$

with a transformation property

$$F'_{\mu\nu} = \partial'_\mu A'_\nu - \partial'_\nu A'_\mu = a_{\mu\rho} a_{\nu\sigma}(\partial_\rho A_\sigma - \partial_\sigma A_\rho) = a_{\mu\rho} a_{\nu\sigma} F_{\rho\sigma} \tag{12.23}$$

characteristic of second-rank tensors. Like any antisymmetric tensor, this has no diagonal components ($F_{11} = F_{22} = F_{33} = F_{44} = 0$) and altogether just six independent ones, of which three are $\mathsf{B}_1 = F_{23} = -F_{32}$, $\mathsf{B}_2 = F_{31}$, and $\mathsf{B}_3 = F_{12}$ here. The remaining three, $F_{k4} = -F_{4k}$, can be identified from

$$F_{k4} = \nabla_k(i\phi) - \frac{\partial A_k}{ic\,\partial t} = -i\mathsf{E}_k$$

according to (6.42). Thus the field-tensor components constitute the array

$$(F_{\mu\nu}) = \begin{bmatrix} 0 & \mathsf{B}_z & -\mathsf{B}_y & -i\mathsf{E}_x \\ -\mathsf{B}_z & 0 & \mathsf{B}_x & -i\mathsf{E}_y \\ \mathsf{B}_y & -\mathsf{B}_x & 0 & -i\mathsf{E}_z \\ i\mathsf{E}_x & i\mathsf{E}_y & i\mathsf{E}_z & 0 \end{bmatrix} \tag{12.24}$$

The two 3-vector fields $\mathbf{E}$ and $\mathbf{B}$ are hereby knit into a single entity, resolvable on the four dimensions of space-time in ways that convert electric into magnetic fields, and vice versa, merely by viewing the electromagnetic field from relatively moving frames.

The Lorentz field transformation (12.23) implies the connections of fields $\mathbf{E}$, $\mathbf{B}$ as observed in one frame to fields $\mathbf{E}'$, $\mathbf{B}'$ that will be observed in a relatively moving frame. Reference to the matrix of transformation coefficients (11.9) shows that (12.23) yields, for example,

$$\begin{aligned}\mathsf{E}'_x = iF'_{14} = ia_{1\rho}a_{4\sigma}F_{\rho\sigma} &= ia_{1\rho}(-i\beta\gamma F_{\rho 3} + \gamma F_{\rho 4})\\ &= i(-i\beta\gamma F_{13} + \gamma F_{14}) = \gamma(\mathsf{E}_x - \beta \mathsf{B}_y)\end{aligned}$$

when $\beta = \beta_z$. Altogether, the results are

$$\begin{aligned}\mathbf{E}'_{\parallel} &= \mathbf{E}_{\parallel} \qquad & \mathbf{E}'_{\perp} &= \gamma(\mathbf{E}_{\perp} + \boldsymbol{\beta} \times \mathbf{B})\\ \mathbf{B}'_{\parallel} &= \mathbf{B}_{\parallel} \qquad & \mathbf{B}'_{\perp} &= \gamma(\mathbf{B}_{\perp} - \boldsymbol{\beta} \times \mathbf{E})\end{aligned} \tag{12.25}$$

where the subscripts $\parallel$ denote components parallel to the relative frame velocity $c\boldsymbol{\beta}$ ($\mathsf{E}_{\parallel} \equiv \mathsf{E}_z$ when $\beta = \beta_z$) and subscripts $\perp$ refer to transverse (x, y) components.

The results (12.25) agree with the galilean expectations (4.68) and (4.69) for low velocities ($\gamma \to 1$). The new expressions differ only by the factors γ in the transverse components, and this is just sufficient to overcome the inconsistency noted in the galilean expressions, revealed when a boost by a given velocity was followed by its reciprocal. Now this succession of transformations yields

$$\begin{aligned}\mathbf{E}_{\perp} &= \gamma(\mathbf{E}'_{\perp} - \boldsymbol{\beta} \times \mathbf{B}') = \gamma^2[\mathbf{E}_{\perp} + \boldsymbol{\beta} \times \mathbf{B} - \boldsymbol{\beta} \times (\mathbf{B}_{\perp} - \boldsymbol{\beta} \times \mathbf{E}_{\perp})]\\ &= \gamma^2\mathbf{E}_{\perp}(1 - \beta^2) = \mathbf{E}_{\perp}\end{aligned}$$

properly, whereas the absence of the factors γ^2 in the corresponding galilean expressions required dropping terms proportional to β^2 as negligible, for a consistency then restricted to $\beta^2 \ll 1$.

The transformation (12.25) could be used to boost the proper field of a point charge, $\mathbf{E}^o = \hat{\mathbf{r}}_o q/r_o^2$ and $\mathbf{B}^o \equiv 0$, and the results (Exercise 12.1) would be found to agree with expressions (10.79) and $\mathbf{B} = \boldsymbol{\beta} \times \mathbf{E}$. That might be regarded as the more meaningful way to obtain the field of a uniformly moving point charge, but the earlier development also showed its congruence with the velocity field of an *arbitrarily* moving point charge.

It will be useful to point out that just as the sum of the squares of the four components of any 4-vector is an invariant (4-scalar) quantity, so is the sum of the squares of the 16 components of any second-rank 4-tensor. Thus

$$\begin{aligned}(\textstyle\sum_{\mu,\nu})(F'_{\mu\nu})^2 &= (a_{\mu\alpha}a_{\nu\beta}F_{\alpha\beta})(a_{\mu\gamma}a_{\nu\delta}F_{\gamma\delta})\\ &= \delta_{\alpha\gamma}\delta_{\beta\delta}F_{\alpha\beta}F_{\gamma\delta} = F^2_{\alpha\beta}\end{aligned} \tag{12.26}$$

Expressed in terms of $\mathbf{E}$, $\mathbf{B}$, the value of this invariant in the special case of the field tensor (12.24) is

$$F^2_{\alpha\beta} = 2(\mathsf{B}^2 - \mathsf{E}^2) \tag{12.27}$$

a quantity that vanishes in every frame for free-space plane-wave fields (when $B^2 = E^2$). The trace of any tensor $(\sum_\nu)F_{\nu\nu}$ is likewise an invariant, identically zero for antisymmetric tensors like (12.24), this being essential for the antisymmetry to persist in every frame.

The Covariant Maxwell Equations

The Maxwell equations (1.40) can be classified into a source-dependent pair, containing ρ or $\mathbf{j}$, and a source-independent pair (the homogeneous equations). Consider the first component of the Ampère-Maxwell law

$$(\nabla \times \mathbf{B})_1 = \partial_2 B_3 - \partial_3 B_2 = \frac{4\pi j_1}{c} + \partial_4(iE_1)$$

Using (12.24) to put $B_3 = -F_{21}$, $B_2 = F_{31}$, and $iE_1 = F_{41}$ shows that this is just the $\nu = 1$ component of the 4-vector equation

$$\partial_\mu F_{\mu\nu} = -\frac{4\pi j_\nu}{c} \tag{12.28}$$

The remaining two components of the Ampère-Maxwell law are represented by the $\nu = 2$ and $\nu = 3$ components of this equation, while the $\nu = 4$ component is equivalent to

$$\nabla_k(-iE_k) = -i\nabla \cdot \mathbf{E} = -\frac{4\pi(ic\rho)}{c}$$

the Gauss law. The 4-vector relation (12.28) can with equal validity be decomposed on any one of frames in uniform relative motion, and so the consistency of the inhomogeneous Maxwell equations with Einstein's principle of relativity has become manifest.

The equation expressing the nonexistence of magnetic charge can be written

$$\nabla \cdot \mathbf{B} = \partial_1 F_{23} + \partial_2 F_{31} + \partial_3 F_{12} = 0$$

and the first component of the Faraday law becomes

$$(\nabla \times \mathbf{E})_1 + \frac{\partial B_1}{c\,\partial t} = \partial_2 E_3 - \partial_3 E_2 + \partial_4(iB_1)$$

$$= i(\partial_2 F_{34} + \partial_3 F_{42} + \partial_4 F_{23}) = 0$$

All the homogeneous Maxwell equations are encompassed by

$$\partial_\mu F_{\nu\rho} + \partial_\nu F_{\rho\mu} + \partial_\rho F_{\mu\nu} = 0 \tag{12.29}$$

for choices $\mu \neq \nu \neq \rho \neq \mu$ from the four integers 1, 2, 3, 4. The left side transforms like the $4^3 = 64$ components of a *third*-rank four-dimensional tensor in a frame change like (11.3). There is then Lorentz invariance only if every one of the 64 components vanishes in each one of any relatively moving frames. Actually, because of the antisymmetry $F_{\nu\mu} = -F_{\mu\nu}$, 40 components vanish identically. The remaining 24 components constitute six reiterations of the four homogeneous Maxwell equations. Thus the 64-component expression (12.29) has only four *independent* components, as manifestly invariant reexpressions of the four source-independent Maxwell equations. Because Eq. (12.29) has so much redundancy, it is often at once replaced by the requirement $F_{\mu\nu} = \partial_\mu A_\nu - \partial_\nu A_\mu$ with some $A_\nu(x)$, equivalent in that it ensures the automatic satisfaction of (12.29).

In the outcome, the Maxwell equations, and hence their contributions to the multiplicity of problems already considered, are seen to have been consistent with Einstein's relativity principle despite the fact that no conscious effort was made to maintain such a consistency. Indeed, a large proportion of the valid conclusions about electromagnetism had been reached before the principle was known. That was a reward of having paid close attention to observed facts in developing both theories.

Implications for Plane-Wave Fields

Among the most important explicit examples of a field to consider, aside from the point-charge field already given attention above, is a plane wave, since any field outside its sources (in free space) can be decomposed into transverse plane-wave components. That was demonstrated in (7.7) and (7.8) for descriptions by **E** and **B** and in (7.12) for the potential description. How to transcribe those expressions into manifestly Lorentz-invariant forms is not immediately obvious because a different three-dimensional frame was chosen for the definition of each plane-wave amplitude, designed to make the polarization properties evident instead. (Indeed, the manifestly invariant formulations frequently become inconvenient when the interest is only in 3-space distributions relative to some particular laboratory frame, and hence the intentional delay of the explicitly relativistic considerations to this late stage.)

Plane-wave decompositions are results of Fourier-integral analyses of dependences on coordinates, and Lorentz invariance is simplest to maintain by starting with such an analysis in the four dimensions of space-time. It will be most economical to do this first for a description by the four components of a Lorentz potential $A_\nu(x)$ [rather than the 16 components of $F_{\mu\nu}(x)$, to parallel what was done for (7.7) and (7.8)]. The generalization to four dimensions is quite plain from a comparison of the one- and three-dimensional versions, (3.44) and (3.53), of the Fourier-integral theorem

$$A_\nu(x) = \oint \frac{d^4k}{(2\pi)^2}\,\alpha_\nu(k)e^{ik_\rho x_\rho} \qquad \alpha_\nu(k) = \oint \frac{d^4x}{(2\pi)^2}\,A_\nu(x)e^{-ik_\rho x_\rho} \tag{12.30}$$

Here $k_\rho x_\rho$ must be a real, dimensionless number and hence an invariant scalar product, with $x_\rho(\mathbf{r}, ict)$, of a 4-vector $k_\rho(\mathbf{k}, i\omega/c)$, producing just the characteristic plane-wave phase

$$k_\rho x_\rho = \mathbf{k} \cdot \mathbf{r} - \omega t \tag{12.31}$$

The 4-vector k_ρ has already been introduced in (11.112), but now its 4-vector character is not predicated on its identification with a photon 4-momentum (in units of $\hbar$). With d^4x defined as in the discussion leading to (12.10), the complementary integration elements should be defined by $id^4k = dk_1\, dk_2\, dk_3\, dk_4$ and $d^4k = dV(\mathbf{k})\, d\omega/c$. These are invariants, on the same basis as d^4x, and so each amplitude $\alpha_\nu(k)$ must be a 4-vector component if $A_\nu(x)$ is.

To be a Lorentz potential in free space, $A_\nu(x)$ must satisfy the wave equation $\partial_\mu^2 A_\nu = 0$, and this restricts the linearly independent amplitudes to such that $k_\mu^2 \alpha_\nu(k) = 0$. Thus, nonvanishing amplitudes can exist only for $k_\mu^2 = \mathbf{k}^2 - (\omega/c)^2 = 0$, and $\alpha_\nu(k)$ must be expressible as

$$\alpha_\nu(k) = \delta(k_\mu^2) c_\nu^\pm(\pm\mathbf{k}) \tag{12.32}$$

where $\delta(k_\mu^2)$ is a one-dimensional δ function of an invariant argument and has the property

$$\int d\left(\frac{\omega^2}{c^2}\right)\delta\left(\mathbf{k}^2 - \frac{\omega^2}{c^2}\right) = 1 \tag{12.33}$$

For each choice of 3-vector $\mathbf{k}$ two independent choices of the factor in (12.32), $c_\nu^+(\mathbf{k})$ or $c_\nu^-(-\mathbf{k})$, can be made, corresponding to the two values $\omega = \pm c|\mathbf{k}|$ that can be used for the fourth component of k_ρ for a given $\mathbf{k}$. Because an invariant δ function was chosen for the expression (12.32), each $c_\nu^\pm(\mathbf{k})$ has the 4-vector character of $\alpha_\nu(k)$.

The result (12.32) makes it possible to carry out the $\int_{-\infty}^{+\infty} d\omega$ part of the integration $\int d^4k = \int dV(\mathbf{k}) \int d\omega/c$ in (12.30) at once, paying due attention to the fact that the δ function permits contributions from only the two points $\omega = \pm c|\mathbf{k}|$ of the integration range for a given $\mathbf{k}$. Using the equivalences $d\omega = d(\omega^2/c^2)(c^2/2|\omega|)$ at each contributing point and also $\oint dV(-\mathbf{k}) = \oint dV(\mathbf{k})$ at $\omega = -c|\mathbf{k}|$, shows that (12.30) can be written as

$$A_\nu(x) = \oint \frac{dV(\mathbf{k})}{(2\pi)^2(2|\mathbf{k}|)}[c_\nu^+(\mathbf{k})e^{ik_\rho x_\rho} + c_\nu^-(+\mathbf{k})e^{-ik_\rho x_\rho}] \tag{12.34}$$

where $k_\rho x_\rho \equiv \mathbf{k}\cdot\mathbf{r} - c|\mathbf{k}|t$ in both terms. It is then plain that $\mathbf{c}^-(\mathbf{k}) = [\mathbf{c}^+(\mathbf{k})]^*$ is necessary for $\mathbf{A}(x)$ to be real. The corresponding strictures on $c_4^\pm(\mathbf{k})$ follow from the gauge condition $\partial_\nu A_\nu = 0$, which requires that $k_\nu c_\nu^\pm(\mathbf{k}) = 0$. Since $k_4 = i|\mathbf{k}|$ in (12.34), the gauge condition is satisfied by any choices with $c_4^\pm = +i\hat{\mathbf{k}}\cdot\mathbf{c}^\pm$, and so $c_4^- = -(c_4^+)^*$, exactly as needed for a purely imaginary† $A_4 = i\phi$. Meanwhile, the 4-vector characters of $c_\nu^\pm$ in (12.34) are sufficient to demonstrate that $dV(\mathbf{k})/|\mathbf{k}|$ is an invariant.

The form (12.34) of the manifestly invariant expression in (12.30) makes it plain how the earlier formulation (7.12) can be consistent with the Lorentz

† To avoid such formal manipulations in connection with enforcing reality conditions, a language in which 4-vectors have only real components, as denoted in $a_\mu(a_o, \mathbf{a})$ or $b_\nu(b_o, \mathbf{b})$, is sometimes introduced and a metric adopted in which $a_\mu b_\mu \equiv a_o b_o - \mathbf{a}\cdot\mathbf{b}$ becomes the definition of a scalar product. The formal manipulations are thus removed to another stage of the development.

invariance. For the earlier form a special choice of frame in which $c_4^{\pm} \equiv 0$ was made, thus adopting a gauge in which $\hat{\mathbf{k}} \cdot \mathbf{c}^{\pm} = 0$ and $\nabla \cdot \mathbf{A} = 0$, $A_4 \equiv 0$. This shows that the gauge can be made divergenceless, with the potential containing just the transverse plane waves, only in a given frame; viewed from another frame, where $A_4 \neq 0$, the same potential distribution will have such longitudinal components as those discussed following (7.12), ones not contributing to the determination of **E**, **B**, and hence $F_{\mu\nu}$ in free space. That discussion showed that $F_{\mu\nu}$ will contain only the transverse plane-wave components even if $c_4^{\pm} \neq 0$ and hence that the transverse character of the field-tensor plane waves themselves persists in every frame. This also follows from the manifestly covariant expressions for $F_{\mu\nu} = \partial_\mu A_\nu - \partial_\nu A_\mu$, having forms exactly like (12.30) or (12.34) but with $c_\nu^{\pm}$ replaced by $\pm i(k_\mu c_\nu^{\pm} - k_\nu c_\mu^{\pm})$. Choose the propagation direction of some plane-wave component as x axis, so that $|\mathbf{k}| = k_1$ and $c_4^{\pm} = ic_1^{\pm}$ follows from the considerations below (12.34) for that $\mathbf{k}$. Then $\mathsf{E}_i = -iF_{4i}$ contains that component in proportion to

$$k_4 c_i^{\pm} - k_i c_4^{\pm} = i|\mathbf{k}|(c_i^{\pm} - \delta_{i1} c_1^{\pm})$$

This vanishes for just the longitudinal (here $i = 1$) component of **E**, leaving only a transverse wave, despite a longitudinal part (when $c_4^{\pm} \neq 0$) in the potential. Any component can be examined in this way, with the same result.

The principal finding in the present considerations that has more than formal interest is the invariance of the plane-wave phase (12.31), which implies

$$\mathbf{k} \cdot \mathbf{r} - \omega t = \mathbf{k}' \cdot \mathbf{r}' - \omega' t' \tag{12.35}$$

as a connection between frequencies $\omega = c|\mathbf{k}|$ and $\omega' = c|\mathbf{k}'|$ and between propagation directions $\hat{\mathbf{k}}$ and $\hat{\mathbf{k}}'$ as viewed from relatively moving frames. Using the Lorentz transformation (11.7) to express $\mathbf{r}(x, y, z)$ and t in terms of $\mathbf{r}'(x', y', z')$ and t' yields an expression that must be equal to the right side of (12.35) for any x', y', z', t', possible only if

$$k'_x = k_x \qquad k'_y = k_y \qquad k'_z = \gamma\left(k_z - \frac{\omega v}{c^2}\right) \qquad \omega' = \gamma(\omega - k_z v) \tag{12.36}$$

This, of course, is just the expression $k'_\mu = a_{\mu\nu} k_\nu$ of the 4-vector character of k_ν, now seen to follow from the invariance of the plane-wave phase. It is equivalent to (11.50) when the identification $p_\nu = \hbar k_\nu$ with the 4-momentum of a photon is granted.

When the angle ϑ between the propagation direction and the relative frame velocity is introduced by putting $k_z = (\omega/c)\cos\vartheta$, the last expression in (12.36) yields

$$\omega' = \gamma\omega\left(1 - \frac{v}{c}\cos\vartheta\right) \tag{12.37}$$

as a generalization of the longitudinal ($\vartheta = 0$ and π) Doppler effects (11.20). For large relative velocities ($\gamma \neq 1$) of source and observer there exists a Doppler shift to a higher frequency $\omega' = \gamma\omega$ even while the source is passing by normally

($\vartheta = \frac{1}{2}\pi$) to the line of sight, a *transverse* Doppler effect unknown for sound and subsonic relative velocities of source and observer.

When $k'_z = (\omega'/c) \cos \vartheta'$ is put into (12.36), a relation

$$\cos \vartheta' = \frac{\cos \vartheta - v/c}{1 - (v/c) \cos \vartheta} \tag{12.38}$$

is found between angles of propagation as viewed by relatively moving observers. The reciprocal of this relationship has already been seen in (11.26).

12.2 ELECTROMAGNETIC FORCES AND FIELD ENERGY-MOMENTUM

The fields **E** and **B** are *operationally* defined by the force per unit of nonstatic test charge (4.65). This can be reexpressed for the field-tensor components by transforming the force-density expression (4.64) with the help of the correspondences (12.6) and (12.24). For example,

$$f_1 = \rho \mathsf{E}_1 + \frac{j_2 \mathsf{B}_3 - j_3 \mathsf{B}_2}{c} = \frac{j_4}{ic}(-iF_{41}) + \frac{j_2}{c}(-F_{21}) - \frac{j_3}{c} F_{31}$$

Since $F_{11} = 0$, this is just the $\nu = 1$ component of the 4-vector equation

$$f_\nu = -\frac{j_\mu F_{\mu\nu}}{c} \tag{12.39}$$

(with summation over μ understood), for which the fourth component

$$f_4 \equiv -\sum_{k=1}^{3} \frac{j_k F_{k4}}{c} = \frac{i(\mathbf{j} \cdot \mathsf{E})}{c} \tag{12.40}$$

must be defined. Here, $-icf_4 = \mathbf{j} \cdot \mathsf{E}$ is just the work rate, per unit volume of the source distribution, by the electromagnetic field, as first identified in (4.81) and used in evaluating the integrated work rate (6.2). Notice that f_4 has a relationship to the other three components of f_ν that is expressible for $\mathbf{j} = \rho\mathbf{u}$ as

$$f_4 = \frac{i\rho(\mathbf{u} \cdot \mathsf{E})}{c} = \frac{i(\mathbf{f} \cdot \mathbf{u})}{c} \tag{12.41}$$

This relates an energy-changing agency, work, to a momentum-changing one, represented by **f**, in a way comparable to

$$dp_4 = \frac{i\,dE}{c} = ic\mathbf{p} \cdot \frac{d\mathbf{p}}{E} = \frac{i(d\mathbf{p} \cdot \mathbf{u})}{c} \tag{12.42}$$

for the 4-momentum (11.44) of a particle [see (11.46) and (11.43)].

The Stress-Energy-Momentum Tensor

It is through the energy transfers between matter and field as represented by the work rate $\mathbf{j} \cdot \mathbf{E} = -icf_4$ that the field energy and its flux density were identified for their role in energy conservation as represented by the continuity equation (6.7). Field momentum and its flux were identified through the remaining components of f_ν, in leading to the conservation equation (6.26). To obtain those identifications, the source descriptions were eliminated from the component equations of (12.39), in favor of the fields generated, as given by the source-dependent Maxwell equations (12.28). The same steps now yield

$$f_\nu = \frac{F_{\mu\nu}\,\partial_\rho F_{\rho\mu}}{4\pi} \tag{12.43}$$

This can be reexpressed as a conservation equation like (12.5), except that a 4-vector expression is needed here. It is provided by the space-time divergence $\sum_\rho \partial_\rho \ldots$ of a suitably defined *tensor* that can be identified with the help of the field properties described by the remaining source-independent Maxwell equations (12.29). The result will be a 4-vector equation with components equivalent to the continuity equations (6.7) and (6.26) but expressed in a manifestly covariant form.

As a first step in transforming (12.43) into a space-time divergence, it can be written

$$4\pi f_\nu = \partial_\rho(F_{\rho\mu}F_{\mu\nu}) - F_{\rho\mu}\,\partial_\rho F_{\mu\nu} \tag{a}$$

The final term can be reexpressed in three alternative ways.† First,

$$F_{\rho\mu}\,\partial_\rho F_{\mu\nu} = (-F_{\mu\rho})\,\partial_\rho(-F_{\nu\mu}) = F_{\rho\mu}\,\partial_\mu F_{\nu\rho} \tag{b}$$

follows from the antisymmetry of the field tensor, with the last equality obtained by interchanging the names of the two dummy summation variables, $\mu \leftrightarrow \rho$. Next, substitution of the Maxwell equation (12.29) yields

$$F_{\rho\mu}\,\partial_\rho F_{\mu\nu} = -F_{\rho\mu}(\partial_\mu F_{\nu\rho} + \partial_\nu F_{\rho\mu}) \tag{c}$$

Half the sum of the two equivalents (b) and (c) results in a cancellation that leaves

$$F_{\rho\mu}\,\partial_\rho F_{\mu\nu} = -\tfrac{1}{2}F_{\rho\mu}\,\partial_\nu F_{\rho\mu} = -\tfrac{1}{4}\partial_\nu(F_{\rho\mu}^2) \tag{d}$$

a space-time gradient of the invariant sum $F_{\rho\mu}^2 \equiv F_{\alpha\beta}^2$ introduced in (12.26). Thus (a) can be written

$$4\pi f_\nu = \partial_\rho(F_{\rho\mu}F_{\mu\nu}) + \tfrac{1}{4}\,\partial_\nu(F_{\alpha\beta}^2) = \partial_\rho(F_{\rho\mu}F_{\mu\nu} + \tfrac{1}{4}\delta_{\rho\nu}F_{\alpha\beta}^2) \tag{e}$$

just the form sought.

† Thus initiating a type of procedure first illustrated in obtaining (4.33).

The outcome defines an obviously symmetrical second-rank tensor

$$T_{\mu\nu} \equiv \frac{1}{4\pi}(F_{\mu\rho}F_{\rho\nu} + \tfrac{1}{4}\delta_{\mu\nu}F_{\alpha\beta}^2) = T_{\nu\mu} \tag{12.44}$$

such that†

$$f_\nu = \partial_\mu T_{\mu\nu} \tag{12.45}$$

The various components of the tensor can be identified with field properties introduced in Chap. 6 by using the decomposition (12.24) of the field tensor. Thus

$$T_{11} = -\frac{F_{1\rho}^2}{4\pi} + \frac{F_{\alpha\beta}^2}{16\pi} = -\frac{\mathsf{B}_3^2 + \mathsf{B}_2^2 - \mathsf{E}_1^2}{4\pi} + \frac{\mathsf{B}^2 - \mathsf{E}^2}{8\pi}$$
$$= \frac{\mathsf{E}_1^2 + \mathsf{B}_1^2}{4\pi} - \frac{\mathsf{E}^2 + \mathsf{B}^2}{8\pi}$$

is just a diagonal component of the three-dimensional stress tensor (6.24). A sample off-diagonal component is

$$T_{12} = \frac{F_{1\rho}F_{\rho 2}}{4\pi} = \frac{F_{13}F_{32} + F_{14}F_{42}}{4\pi} = \frac{\mathsf{E}_1\,\mathsf{E}_2 + \mathsf{B}_1\,\mathsf{B}_2}{4\pi}$$

When the indices μ and/or ν take on the value 4,

$$T_{44} = -\frac{F_{4\rho}^2}{4\pi} + \frac{\mathsf{B}^2 - \mathsf{E}^2}{8\pi} = \frac{\mathsf{E}^2 + \mathsf{B}^2}{8\pi} = w$$

is just the field-energy density (6.3) and

$$T_{14} = \frac{F_{1k}F_{k4}}{4\pi} = -\frac{i}{4\pi}(\mathsf{B}_3\,\mathsf{E}_2 - \mathsf{B}_2\,\mathsf{E}_3) = -i\frac{(\mathbf{E} \times \mathbf{B})_1}{4\pi}$$

can be recognized as the field-momentum density $g_1 = N_1/c^2$ of (6.14) multiplied by $-ic$. Thus, the tensor (12.44) has the array of components

$$(T_{\mu\nu}) = \begin{bmatrix} T_{xx} & T_{xy} & T_{zx} & -icg_x \\ T_{xy} & T_{yy} & T_{yz} & -icg_y \\ T_{zx} & T_{yz} & T_{zz} & -icg_z \\ -icg_x & -icg_y & -icg_z & w \end{bmatrix} \tag{12.46}$$

representing the field's energy-momentum densities and their flux. The tensor is a traceless one, since

$$T_{\mu\mu} = \sum_{k=1}^{3} T_{kk} + w = 2w - 3w + w = 0$$

follows from (6.24) and even more directly from the definition (12.44).

† The generalization to four dimensions of the equation $\mathbf{f} = \nabla \cdot \mathbf{T}$ used for the static force transmissions in (6.31).

The decomposition (12.46) shows that a space component of the space-time divergence (12.45) is

$$f_k = \nabla_l T_{lk} + \frac{\partial T_{4k}}{ic\,\partial t} = -\frac{\partial g_k}{\partial t} - [\nabla \cdot (-\mathbf{T})]_k \tag{12.47a}$$

just the continuity equation (6.26) that describes the conservation of field momentum in its flow through all points except those at which there is matter to which momentum can be transferred. The fourth component is proportional to

$$icf_4 = -\mathbf{E} \cdot \mathbf{j} = ic\nabla_l T_{l4} + \frac{\partial T_{44}}{\partial t} = \frac{\partial w}{\partial t} + \nabla \cdot \mathbf{N} \tag{12.47b}$$

since $c^2\mathbf{g} = \mathbf{N}$ is the Poynting vector. This is just (6.7), the equation that describes the conservation of field energy except where work on matter is done.

Torque and Field-Angular-Momentum Tensors

A manifestly Lorentz-invariant reformulation of the continuity equation (6.37) that describes angular-momentum conservation can be derived from the 4-momentum conservation equation (12.45) by a procedure paralleling the derivation of (6.37) from (6.26), which is the equation giving just the three spatial components of the 4-vector equation (12.45). As seen in the step from (6.32) to (6.33), the procedure requires forming the density $\mathbf{r} \times \mathbf{f}$ of torques exerted by the field on any charged matter that may be present. For generalization to the four space-time dimensions, it must be recognized that the representability of any quantity by a three-component vector product is due† to the circumstance that in three dimensions any second-rank antisymmetrical tensor has just three independent components. The analogous circumstance does not prevail in four dimensions, where antisymmetrical tensors can have six independent components‡ and cannot be replaced by any single 4-vector. Thus, the more readily generalizable 3-space descriptions represent a vector product like $\mathbf{r} \times \mathbf{f}$ by an antisymmetrical tensor which will here be taken to be formed of the "space-space" components of the four-dimensional tensor defined by

$$\theta_{\mu\nu} \equiv x_\mu f_\nu - x_\nu f_\mu \tag{12.48}$$

and displayable as

$$(\theta_{\mu\nu}) = \begin{bmatrix} 0 & (\mathbf{r}\times\mathbf{f})_z & -(\mathbf{r}\times\mathbf{f})_y & \theta_{14} \\ -(\mathbf{r}\times\mathbf{f})_z & 0 & (\mathbf{r}\times\mathbf{f})_x & \theta_{24} \\ (\mathbf{r}\times\mathbf{f})_y & -(\mathbf{r}\times\mathbf{f})_x & 0 & \theta_{34} \\ \theta_{41} & \theta_{42} & \theta_{43} & 0 \end{bmatrix} \tag{12.49}$$

† See, for example, Ref. 17, eqs. (9.84). It is for this reason that axial vectors like $\mathbf{r} \times \mathbf{f}$ are often called *pseudovectors*.

‡ Sometimes Lorentz 6-vectors, like $\mathbf{B} - i\mathbf{E}$, are defined to represent these components, as the closest analogs of the vector products in three dimensions. Comparison of (12.49) with the antisymmetric field-tensor display (12.24) indicates why the magnetic vector $\mathbf{B}$ has the transformation properties of an *axial* (or pseudo) vector, like those of any vector product of two *polar* ($\mathbf{r}$-like) vectors. The distinction between polar and axial vectors is discussed in Ref. 17, sec. 9.4.

where

$$\theta_{k4} = -\theta_{4k} = i\left[x_k\left(\mathbf{f}\cdot\frac{\mathbf{u}}{c}\right) - ctf_k\right] \tag{12.50}$$

according to the result (12.41) for f_4.

The suggested procedure requires using the expression (12.45) of f_ν to form

$$\theta_{\mu\nu} = x_\mu\,\partial_\rho T_{\rho\nu} - x_\nu\,\partial_\rho T_{\rho\mu} = \partial_\rho(x_\mu T_{\rho\nu} - x_\nu T_{\rho\mu}) \equiv \partial_\rho\Lambda_{\rho\mu\nu} \tag{12.51}$$

which becomes a space-time divergence (a four-dimensional continuity equation where $\theta_{\mu\nu} \equiv 0$) because of the symmetry $T_{\mu\nu} = T_{\nu\mu}$ noted in (12.44). The procedure has defined a third-rank tensor

$$\Lambda_{\rho\mu\nu} \equiv x_\mu T_{\rho\nu} - x_\nu T_{\rho\mu} = -\Lambda_{\rho\nu\mu} \tag{12.52}$$

antisymmetric in its last two indices (to $\mu \leftrightarrow \nu$).

A third-rank tensor normally has $4^3 = 64$ components, but here the 16 components $\Lambda_{\rho\mu\mu}$ vanish identically in every frame, because of the antisymmetry. For the same reason only half the remaining 48 components, $\Lambda_{\rho\nu\neq\mu} = -\Lambda_{\rho\mu\nu}$ are independent numbers. Of the 24 independent components only half, $\Lambda_{\rho kl} = -\Lambda_{\rho lk}$ with $kl = 12, 23, 31$, are concerned with the description of field angular momenta and their flows. Nine of the latter components (those with $\rho = 1, 2, 3$) are identifiable with the components of the second-rank pseudo tensor $\mathbf{M} = \mathbf{T} \times \mathbf{r}$ that was introduced in (6.34) to represent the flux density of field angular momentum. The remaining three, $\Lambda_{4kl} = -\Lambda_{4lk}$, are the components of the vector $-ic\mathbf{r} \times \mathbf{g}$, as reference to the tensor array (12.46) shows, and so represent the field-angular-momentum density itself. Thus the three independent components of Eq. (12.51) with $kl = 12, 23, 31$ are just decompositions of the axial 3-vector equation (6.37) that describes how angular momenta are conserved.

There remain to be discussed just three independent components of Eq. (12.51), those with $\mu\nu = 14, 24, 34$. Their interpretation is best seen for an entire field that is isolated from matter, so that $\theta_{\mu\nu} \equiv 0$ and

$$\partial_\rho\Lambda_{\rho\mu\nu} = 0 \tag{12.53}$$

The components in question contain the 12 remaining components $\Lambda_{\rho k4}$ of the third-rank tensor. The tensor array (12.46) helps show that

$$\begin{aligned}\Lambda_{lk4} &= -ic(x_k g_l + tT_{lk})\\ \Lambda_{4k4} &= x_k w - c^2 t g_k \qquad k, l = 1, 2, 3\end{aligned} \tag{12.54}$$

Then $ic\,\partial_\rho\Lambda_{\rho k4} = 0$ becomes the kth component of the 3-vector equation

$$\frac{\partial}{\partial t}(\mathbf{r}w - c^2 t\mathbf{g}) + c^2\nabla\cdot(\mathbf{g}\mathbf{r} + t\mathbf{T}) = 0 \tag{12.55}$$

At a given moment t relative to some frame, this can be integrated† over all the space $\oint dV(\mathbf{r})$, the type of step by which the total energy $W = \oint dVw$ of an isolated field was found to be a properly conserved constant [in the discussion of (6.4) and (7.11)] and by which the linear field momentum $\mathbf{P} = \oint dV\mathbf{g}$ is conserved [(6.16) when no matter momentum $\mathbf{p}$ is present]. For those conclusions, it was held that all field quantities like $\mathbf{T}$ vanish on the remote surface enclosing all space (when it only contains an isolated field of finite energy) and that also the moments like $\mathbf{rg}$ vanish there (as needed for a field of finite conserved angular momentum). Then the volume integrals of the divergences in (12.55), being equivalent to integrals over the remote surface, can be set equal to zero. The integral $\oint dV\mathbf{r}w$ that also occurs can be used to define a center of mass for the finite isolated field

$$\bar{\mathbf{r}} = \frac{\oint dVw\mathbf{r}}{\oint dVw} = \frac{1}{W}\oint dV\mathbf{r}w \tag{12.56}$$

since W/c^2 is the mass equivalent of the total energy. Thus the result of volume integrating (12.55) becomes

$$\frac{d}{dt}(W\bar{\mathbf{r}} - c^2t\mathbf{P}) = 0 \tag{12.57}$$

This is a statement that a uniform, linear CM motion of the entire field is maintained whenever it has a nonvanishing constant resultant linear momentum $\mathbf{P}$. The CM velocity is

$$\frac{d\bar{\mathbf{r}}}{dt} = \frac{c^2\mathbf{P}}{W} \tag{12.58}$$

exactly as in (11.55), which can apply to systems composed of photons. There is implicit here a mechanical picture of the electromagnetic field, an electrodynamical one, that will be more fully exploited in the next chapter.

The association of the field translations here with the spatial rotations involved in the other components of one tensor $\Lambda_{\rho\mu\nu}$ should not be surprising. The frame rotations in space-time, described by $x'_\mu = a_{\mu\nu}x_\nu$ of (11.8), associate spatial rotations represented by a_{kl} with such translations, represented by a_{k4}, as in Fig. 1.1.

For fields not isolated from charged matter, the space-time divergence (12.55) does not vanish but must be equated to the vector $-\mathbf{r}(\mathbf{f}\cdot\mathbf{u}) + c^2t\mathbf{f}$ having the components $ic\theta_{k4}$ of (12.50). Then W and $\mathbf{P}$ are no longer constants, and defining even an instantaneous mass center is no longer profitable; the

† To carry out such a step in a way that is manifestly invariant at every stage requires integration over a *hypersurface* in space-time that is orthogonal to timelike directions in it, and then each element of the hypersurface in the four dimensions is a three-dimensional volume element. However, there is no harm in adopting some one special frame for interpretations, as actual physical observers do. It avoids excessive abstraction.

interactions of matter and field acquire the complex of possibilities to which all the preceding chapters have been devoted. More field degrees of freedom than just $\bar{\mathbf{r}}$, including internal ones, must be treated, as in the next chapter.

12.3 RELATIVISTIC PARTICLE DYNAMICS

The field energy-momentum described in the preceding section can be transferred to charged matter through the work-force density f_ν in (12.45). The concern now will be with *motions* thus imparted to matter in the form of a mass particle, which is also the primary subject of newtonian formulations of motion. The immediate result will be a relativistic generalization of the equation of motion (5.1) for a point charge in a given external electromagnetic field. It will provide an example of how any force can be treated in consistency with Einstein's relativity principle.

Relativistic Equations of Motion

The rate $d\mathbf{p}/dt$ at which matter in a volume V acquires momentum as a reaction to incident field momentum has already been given in (6.11), the equation which, together with (6.12), provided the basis for identifying field momenta in the first place. It gives components of the matter-momentum increment $d\mathbf{p}$ acquired during a time interval dt as spatial components of the 4-vector expression

$$dp_\mu = dt \int_V dV f_\mu \tag{12.59}$$

Here, $\mathbf{f} = \rho(\mathbf{E} + \mathbf{u} \times \mathbf{B}/c)$ when $\mathbf{u}(\mathbf{r}, t)$ is the instantaneous velocity of the charge in the element $dV(\mathbf{r})$, and $f_4 = i\mathbf{u} \cdot \mathbf{f}/c$ according to (12.41). The fourth component of (12.59) yields an energy increment like $dE = -ic\, dp_4$ of (12.42), acquired by the matter through the work on it by the field, done at the rate $\mathbf{f} \cdot \mathbf{u}$ per unit volume of the matter. This, in the form $\rho\mathbf{u} \cdot \mathbf{E} = \mathbf{j} \cdot \mathbf{E}$, provided the starting point for identifying the *field energy*, through Eq. (6.2). Each of the elements $dt\, dV \equiv d^4x/c$ in (12.59) is a scalar invariant (12.10), and so each element of the integration will properly make a 4-vector contribution to dp_μ.

The expression is now to be specialized to matter in the form of a point particle having some charge q and rest mass m. The requisite volume integration will be least ambiguous to carry out for the spatial components that survive in the *instantaneous* rest frame, the *proper* frame, of the particle. Here the point-charge distribution can be taken to be $\rho^o = q\delta(\mathbf{r}^o - \mathbf{r}_q^o)$, with $\mathbf{r}_q^o$ a *constant* point position and $\mathbf{u}_q(t) \equiv \dot{\mathbf{r}}_q(t) \to 0$. Since $\mathbf{f}^o = \rho^o\mathbf{E}^o$ and $f_4 = i(\mathbf{u}_q \cdot \mathbf{f})/c \to 0$ because $\mathbf{u}_q^o \equiv 0$, the reexpression of (12.59) in the proper frame becomes

$$d\mathbf{p}^o = d\tau \int_{V^o} dV^o \rho^o \mathbf{E}^o = d\tau q \mathbf{E}^o \qquad \frac{dE^o}{d\tau} = 0 \tag{12.60}$$

The proper-time interval $d\tau \equiv dt^o = dt/\gamma$ here is one corresponding, in the frame of (12.59), to a time interval dt during which the particle is displaced by $\mathbf{u}_q(t)\,dt$ with $dt = \gamma\, d\tau$ dilated relative to $d\tau$ because of the proper frame's motion with the instantaneous particle velocity $\mathbf{u}_q(t)$. The particle is instantaneously at rest in the proper frame, but it is undergoing acceleration by the finite force $d\mathbf{p}^o/d\tau = q\mathbf{E}^o$ at the instant. The instantaneous value of $dE^o/d\tau$, $= \mathbf{u}_q^o \cdot d\mathbf{p}^o/d\tau$ by (12.42), vanishes as a first-order effect only because $\mathbf{u}_q^o \equiv 0$ at the instant.

The proper-time interval used in (12.60) is given the special symbol $d\tau$ because it is a frame invariant, a 4-scalar, with a value that can be agreed on by all relatively moving observers; each needs merely to use the instantaneous particle velocity $\mathbf{u}(t)$ with respect to his own frame to evaluate $d\tau = dt/\gamma$. Indeed, $-c^2\, d\tau^2$ is the special case of the general space-time-interval invariant

$$(dx_\mu)^2 = dx^2 + dy^2 + dz^2 - c^2\, dt^2 \tag{12.61}$$

in which $d\mathbf{r} = \mathbf{u}\, dt$, so that†

$$(dx_\mu)^2 = (u^2 - c^2)\, dt^2 = \frac{-c^2\, dt^2}{\gamma^2} = -c^2\, d\tau^2 \tag{12.62}$$

Even more briefly, it is only necessary to recognize that the invariance entails $(dx_\mu)^2 = (dx_\mu^o)^2$ and $(dx_\mu^o)^2 = -c^2(dt^o)^2 \equiv -c^2\, d\tau^2$ when $d\mathbf{r}^o = \mathbf{u}^o\, dt^o \equiv 0$.

The invariance of $d\tau$ makes it possible to retain it unchanged in transforming the results (12.60) so that they refer to a frame in which the particle can have a variable velocity $\mathbf{u}(t)$ as time goes on. When (12.60) refers to an instant at which the particle's proper frame has the velocity $\mathbf{u}(t)$, the components of the force $q\mathbf{E}^o$ parallel and perpendicular to $\mathbf{u}$ respectively become

$$q\mathbf{E}_\parallel^o = q\mathbf{E}_\parallel \qquad \text{and} \qquad q\mathbf{E}_\perp^o = q\gamma\left(\mathbf{E}_\perp + \mathbf{u} \times \frac{\mathbf{B}}{c}\right)$$

according to the field transformation relations (12.25). The momentum transformation relations (11.47) lead to $d\mathbf{p}_\perp^o = d\mathbf{p}_\perp$ and

$$dp_\parallel^o = \gamma\left(dp_\parallel - \frac{u\, dE}{c^2}\right) \qquad dE^o = \gamma(dE - u\, dp_\parallel)$$

With $dE^o/d\tau = 0$ in (12.60), $dE = u\, dp_\parallel = \mathbf{u} \cdot d\mathbf{p}$ during $dt = d\tau/\gamma$, and so $dp_\parallel^o = \gamma\, dp_\parallel(1 - u^2/c^2) = dp_\parallel/\gamma$. Thus the relations (12.60) lead to

$$\frac{d\mathbf{p}}{d\tau} = \gamma q\left(\mathbf{E} + \frac{\mathbf{u}}{c} \times \mathbf{B}\right) \equiv \gamma \mathbf{F}_q \qquad \frac{dE}{d\tau} = \mathbf{u} \cdot \frac{d\mathbf{p}}{d\tau} = \gamma(\mathbf{u} \cdot \mathbf{F}_q) \tag{12.63}$$

† The restriction to a point particle through the definition of a $\mathbf{u}(t) = d\mathbf{r}/dt$ thus engenders an invariant connection between the four components of its space-time displacement dx_μ, much as the 4-momentum components are related through the invariant $p_\mu^2 = -m^2c^2$ for a particle of definite rest mass. Indeed, $dx_\mu = p_\mu\, d\tau/m$ since $\mathbf{p} = \gamma m\mathbf{u}$ and $p_4 = i\gamma mc$. The differential $dx_\mu/d\tau = p_\mu/m \equiv w_\mu(\gamma\mathbf{u}, i\gamma c)$ is sometimes separately defined [see (12.66)] and called the 4-vector *world velocity* of the particle.

where $\mathbf{F}_q$ is just the Lorentz force of (4.66) and (5.1), with $(\mathbf{F}_q)_{\|} = q\mathsf{E}_{\|}$. Comparison to the manifestly invariant general form (12.59) shows that

$$dp_\mu = d\tau K_\mu\left(\gamma\mathbf{F}_q, i\gamma\mathbf{u}\cdot\frac{\mathbf{F}_q}{c}\right) \tag{12.64}$$

must be the 4-vector result of the integrations over a point particle. The new symbol here, K_μ, stands for a 4-vector with the indicated components because $d\tau$ is a 4-scalar (an invariant).

The equation of motion for m, q in a given field **E**, **B**, the relativistic generalization of (5.1), is implicit in the results (12.63). Putting $d\tau = dt/\gamma$ shows that

$$\frac{d\mathbf{p}}{dt} = \frac{d}{dt}\gamma m\mathbf{u} = q\left(\mathsf{E} + \frac{\mathbf{u}}{c}\times\mathsf{B}\right) \tag{12.65a}$$

exactly as in (5.1) except that the momentum must be given its relativistic (4-momentum-component) values $\mathbf{p} = \gamma m\mathbf{u}$. Then $\dot{\mathbf{p}} = \gamma m\dot{\mathbf{u}} + m\dot{\gamma}\mathbf{u}$, and not merely $m\dot{\mathbf{u}}$ as before. The remaining, fourth-component equation in (12.63) gives the rate of kinetic-energy change

$$\frac{dT}{dt} = \frac{d}{dt}(E - mc^2) = q\mathbf{u}\cdot\mathsf{E} \tag{12.65b}$$

that *follows* from (12.65*a*).

It is standard to reexpress the components of K_μ, as given by (12.64), in a manifestly covariant way, with the help of a *4-vector* velocity defined by

$$w_\mu \equiv \frac{p_\mu}{m} = w_\mu(\gamma\mathbf{u}, ic\gamma) \tag{12.66}$$

(see page 394*n*.). A corresponding step has already been taken for the work-force *density* in (12.39), and the proportionality of that expression to $j_\mu(\rho\mathbf{u}, ic\rho) \equiv \rho w_\mu/\gamma$ leads to an expectation that the equation of motion (12.64) in the form

$$\frac{d}{d\tau}mw_\mu = K_\mu = \frac{qF_{\mu\nu}w_\nu}{c} \tag{12.67}$$

should be a result of integrating the density (12.39) over a point charge, a process already carried out in obtaining (12.64). It is simple to check that the form here does yield

$$K_1 = \frac{q}{c}\gamma(F_{12}u_2 + F_{13}u_3 + icF_{14})$$

$$= q\gamma\left[\frac{(\mathbf{u}\times\mathsf{B})_1}{c} + \mathsf{E}_1\right] = \gamma(\mathbf{F}_q)_1$$

and

$$K_4 = \frac{q}{c}\gamma\sum_{k=1}^{3}F_{4k}u_k = \frac{i\gamma q(\mathbf{u}\cdot\mathsf{E})}{c}$$

exactly equivalent to the components shown in (12.64).

Expression (12.64) leads to the manifestly covariant equivalent of (12.65) and (12.67) as the 4-vector equation

$$\frac{dp_\mu}{d\tau} = K_\mu \tag{12.68}$$

This is a form which might also be given to the equation of motion for any mass particle subject to forces other than electromagnetic ones. It can serve as a relativistic generalization of the newtonian formulation: $d\mathbf{p}/dt = \mathbf{F}$ for a particle in circumstances that can be described with the help of a 3-vector force **F**. Much of the power of the newtonian prescription comes from such representations of the interactions of a particle with "the rest of the universe." The relativistic generalization requires finding *4-vectors* K_μ that can serve the same purposes.† In the general context, K_μ is referred to as a *Minkowski force*.

Relativistic Motion in a Uniform Electric Field

The equation of motion (12.65) can now be used to find how such motions of a point charge in an external electromagnetic field as were reviewed in Chap. 5 are modified when the particle attains relativistic velocities, comparable to (but not exceeding) the speed of light.

The simple parabolic motion of a low-velocity particle in a purely electric (and hence, static) uniform field $\mathbf{E}_o$ was alluded to in connection with (2.34). Clearly, if the acceleration by the electric field is allowed to continue long enough, an otherwise unhindered charge will attain relativistic velocities (the sooner the stronger the field), and such phases of its motion will be considered now.

It will be sufficient to consider the motion from an initial phase at which the velocity is perpendicular to the field $\mathbf{E}_o$; that is, $\mathbf{u}_o = \mathbf{u}_\perp(0)$ and $\mathbf{u}_\parallel(0) = 0$. The subsequent acceleration can be expected to produce phases with all possible values for $\mathbf{u}_\parallel(t)$ parallel to $q\mathbf{E}_o$; motion from phases in which there is a velocity component antiparallel to $q\mathbf{E}_o$ can be found from evaluations for $t < 0$ of the trajectory to be obtained in the $\mathbf{u}_\parallel(0) = 0$ case.

With the force in (12.65) given by the constant $q\mathbf{E}_o$, the uniform growth of the momentum

$$\mathbf{p} \equiv \gamma m\mathbf{u} = q\mathbf{E}_o t + \gamma_o m\mathbf{u}_o \tag{12.69a}$$

† Since the spatial components of (12.68) reduce to $d\mathbf{p}/dt = \mathbf{K}/\gamma$, one mode of generalization might be to adopt $\mathbf{K} = \gamma\mathbf{F}$ and $K_4 = i\mathbf{u} \cdot \mathbf{K}/c$, with **F** the force that is found to serve in the nonrelativistic domain of velocities. Just such a procedure works in the Lorentz-force case, as indicated by (12.64). Situations are rarely so simple, however. Most of the power of the newtonian formulation resides in the possibility of adopting for **F** some force *function* of the particle positions and momenta. The transformation properties of such a function are then dictated by the behavior in frame changes of the variables it depends on, as the 4-vector character of K_μ is by the behavior of **u**, **E**, and **B** in the Lorentz-force case (12.64). In the end, about as complete a relativistic theory of force as the Maxwell-Lorentz one must be developed in each case before an explicit expression for the Minkowski force can be adopted with any confidence.

is the most immediate result. The concurrent mass dilatation follows most easily from using the fact that $\beta^2\gamma^2 = \gamma^2 - 1$ at every velocity and that $\mathbf{u}_o \cdot \mathbf{E}_o = 0$ in the terms being used here. As a result,

$$\gamma m = \gamma_o m \left[1 + \left(\frac{t}{\tau_o}\right)^2\right]^{1/2} \qquad \text{with } \tau_o \equiv \frac{\gamma_o mc}{q\mathsf{E}_o} \tag{12.69b}$$

Thus, as to be expected, the kinetic mass and its energy equivalent γmc^2 will grow indefinitely large as the acceleration continues. From the results (12.69), expressions for the velocity components parallel and perpendicular to $q\mathbf{E}_o$ are easy to construct:

$$\mathbf{u}_{\parallel}(t) = \frac{(q\mathbf{E}_o/\gamma_o m)t}{[1 + (t/\tau_o)^2]^{1/2}} \qquad \mathbf{u}_{\perp}(t) = \frac{\mathbf{u}_o}{[1 + (t/\tau_o)^2]^{1/2}} \tag{12.70}$$

It can be seen that as $t \to \infty$, $u_{\parallel} \to (q\mathsf{E}_o/\gamma_o m)\tau_o = c$ while $u_{\perp} \to 0$. Despite this diminution of the transverse velocity to a near-vanishing point, the transverse momentum $\mathbf{p}_{\perp} = \gamma m\mathbf{u}_{\perp}$ stays properly conserved at its initial value $\mathbf{p}_{\perp} = \gamma_o m\mathbf{u}_o$ because of the concurrent growth of the kinetic mass.

A parametric description of the trajectory can be obtained from simple integrations over time, after setting $u_{\perp}(t) = \dot{x}(t)$ and $u_{\parallel}(t) = \dot{y}(t)$. The results are

$$x(t) = u_o\tau_o \ln \left\{\frac{t}{\tau_o} + \left[1 + \left(\frac{t}{\tau_o}\right)^2\right]^{1/2}\right\} \qquad y(t) = c\tau_o\left\{\left[1 + \left(\frac{t}{\tau_o}\right)^2\right]^{1/2} - 1\right\} \tag{12.71}$$

Eliminating the time parameter from these two equations readily produces the purely geometric description

$$y(x) = 2c\tau_o \sinh^2 \frac{x}{2u_o\tau_o} \tag{12.72}$$

A sample of such a trajectory, including phases at $x(t < 0) < 0$, is shown in Fig. 12.1. It is compared with a simple parabola, the approximation $y \approx cx^2/2u_o^2\tau_o = (q\mathsf{E}_o/2mu_o^2)x^2$ of (12.72), that would be followed if the field were so weak that $u \ll c$ over the entire range of the displacements shown. The same parabola is derivable from the considerations in (2.34).

Notice that the relation of the kinetic energy $T = (\gamma - 1)mc^2$, as given by (12.69), to its minimum value $T_o = (\gamma_o - 1)mc^2$ can be put in terms of the coordinate y of (12.71) to produce

$$T_o = T - q\mathsf{E}_o y = T + q\phi \tag{12.73}$$

where $\phi = -\mathsf{E}_o y$ is just the electrostatic potential of the uniform field, gauged from the position at which the kinetic energy has its minimum. Thus the same energy as that given by (5.4), generalized to a relativistic evaluation of the kinetic energy, remains conserved.

For the evaluations here, only the force arising from the given, externally applied, field was taken into account. Thus, any recoils from, and energy losses to, the *radiation* that will inevitably accompany the accelerations of the particle

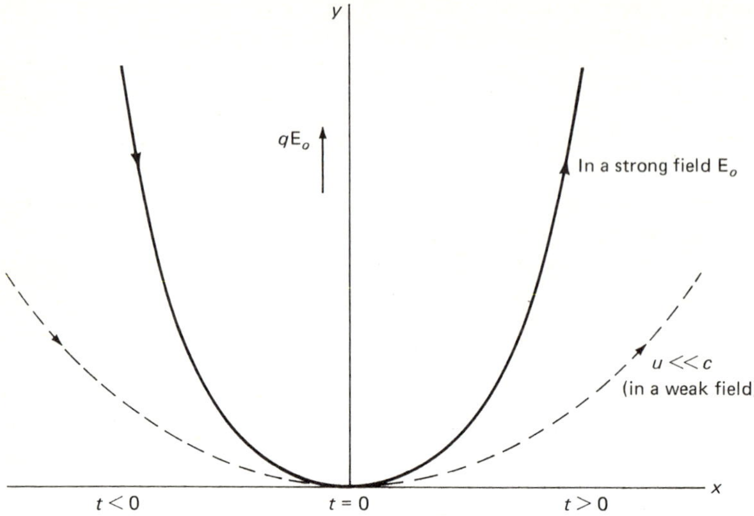

Figure 12.1

have been ignored. Such effects are usually considered separately, as radiative corrections to the motion, since they turn out to be so small as to be unobservable in practice for situations in which classical analysis is valid. Some attention will be paid to the rather complex consequences in Sec. 14.3.

Relativistic Motions in Purely Magnetic Fields

In any purely magnetic, and hence magneto*static*, field, the equation of motion (12.65) becomes $d\mathbf{p}/dt = q\mathbf{u} \times \mathbf{B}/c$, and scalar multiplication of it by $\mathbf{p} = \gamma m\mathbf{u}$ shows that

$$\gamma m\mathbf{u} \cdot \frac{d}{dt}\gamma m\mathbf{u} = 0 \rightarrow p^2 = \gamma^2 m^2 u^2 = \text{const} \tag{12.74}$$

Then the magnitudes u, γ, and p are each constants, and the equation of motion can be written

$$\dot{\mathbf{u}} = \mathbf{u} \times \frac{q\mathbf{B}}{Mc} \tag{12.75}$$

exactly as in the nonrelativistic treatments (5.8) or (5.42), except that the mass involved is to be interpreted as the constant kinetic mass $M = \gamma m$ rather than as the rest mass of the particle. A consequence of this is an increase by a factor γ of the radii (5.11) of the spirals about a magnetic field line to be expected for a given velocity and at a given field strength, for now

$$r_\perp = \frac{\gamma mcu_\perp}{q\mathrm{B}} = \frac{cp_\perp}{q\mathrm{B}} \tag{12.76}$$

Thus, it is the *relativistic* momentum in the spiraling motion $p_\perp = \gamma m u_\perp$ to which the radii are proportional. The radii can then be indefinitely large despite the limitation $u_\perp < c$. The angular velocity $\omega_o = q\mathsf{B}/\gamma mc$ can become very large in very strong fields, but the linear velocity $u_\perp = \omega_o r_\perp$ is independent of the field strength and remains within the limiting value c.

In the special case of a uniform field, any velocity component $u_\parallel$ parallel to $\mathbf{B} = \mathbf{B}_o$ stays separately constant. The spiral followed by the particle is then a uniform helix, and its radius (12.76) ought to be invariant to translations parallel to the field direction. The expression (12.76) is consistent with this, since neither $\mathbf{p}_\perp$ nor $\mathbf{B} \equiv \mathbf{B}_\parallel$ is changed by Lorentz boosts parallel to the field.

Because the conclusions from (12.74) are also valid in nonuniform fields, the discussion of the magnetic mirror effect in Sec. 5.4 is not changed by the relativistic considerations, except for the reinterpretation of the mass constant. This does not enter into the kinetic-energy-conserving relation (5.47) that is essential for the mirror effect. This relation makes it impossible for the particle to penetrate to a region in which the field strength is multiplied so greatly that $u_\perp^2(z) = u_\perp^2(0)\mathsf{B}_o(z)/\mathsf{B}_o(0) \to c^2$ in (5.45).

Relativistic E × B Drifts

When both electric and magnetic fields are present, interest turns to the relativistic effects on such cycloidal drifts as were discussed in connection with Fig. 5.1. Accordingly, the concern here will be with the motions of a point charge in crossed uniform fields, $\mathbf{E}_o \perp \mathbf{B}_o$.

As in the nonrelativistic considerations, the description will be simplest with respect to a frame having the drift velocity $\mathbf{v} \equiv \bar{\mathbf{u}} = c\mathbf{E}_o \times \mathbf{B}_o/\mathsf{B}_o^2$ of (5.17). Now, however, the Lorentz transformation must be used to find the connections to what will be observed in the initial (laboratory) frame of $\mathbf{E}_o$ and $\mathbf{B}_o$; since that is applicable only to $v < c$, and since $v = \bar{u} = c\mathsf{E}_o/\mathsf{B}_o$ in magnitude, the drifting frame being introduced here can serve only when $\mathsf{E}_o < \mathsf{B}_o$. The situations in which $\mathsf{E}_o > \mathsf{B}_o$ are left to Exercise 12.16.

Since the Lorentz field transformation (12.25) does not change the zero values of the field components parallel to the frame velocity, the entire field in the drifting frame is given by

$$\mathbf{E}' = \bar{\gamma}\left(\mathbf{E}_o + \bar{\mathbf{u}} \times \frac{\mathbf{B}_o}{c}\right) \qquad \mathbf{B}' = \bar{\gamma}\left(\mathbf{B}_o - \bar{\mathbf{u}} \times \frac{\mathbf{E}_o}{c}\right) \tag{12.77}$$

with $\bar{\gamma} = (1 - \mathsf{E}_o^2/\mathsf{B}_o^2)^{-1/2}$. Substitution of $\bar{\mathbf{u}} = c\mathbf{E}_o \times \mathbf{B}_o/\mathsf{B}_o^2$ leads to

$$\mathbf{E}' = 0 \qquad \mathbf{B}' = \frac{\mathbf{B}_o}{\bar{\gamma}} = \mathbf{B}_o\left(1 - \frac{\mathsf{E}_o^2}{\mathsf{B}_o^2}\right)^{1/2} \tag{12.78}$$

As in the nonrelativistic case, the electric force is reduced to zero† relative to the drifting frame; this is just the advantage provided by reference to it. The mag-

† See page 127*n*.

netic field is not changed in direction but is weakened in magnitude, an effect not found in the nonrelativistic approximation because the galilean transformation does not provide correctly for effects of order $\bar{u}^2/c^2$ $(= \mathsf{E}_o^2/\mathsf{B}_o^2$ here). This shows that the nonrelativistic treatment in Sec. 5.2 can remain valid only for $\mathsf{E}_o \ll \mathsf{B}_o$, even from a start with $u_o \ll c$. Stronger electric fields accelerate the particle sufficiently before it is turned back by the magnetic field (with the subsequent deceleration by $\mathbf{E}_o$) to make the relativistic effects important in each cycle.

With only a uniform magnetic field $\mathbf{B}' = \mathbf{B}_o/\bar{\gamma}$ existing in the drifting frame, the equation of motion with respect to it is

$$\frac{d\mathbf{p}'}{dt'} = q\mathbf{u}' \times \frac{\mathbf{B}_o}{\bar{\gamma}c} \qquad \text{with } \mathbf{p}' = \gamma' m\mathbf{u}' \tag{12.79}$$

and the transverse motion of the particle describes a circle of radius

$$R' = \frac{\gamma'\bar{\gamma}mcu'_\perp}{q\mathsf{B}_o} \tag{12.80}$$

according to the adaptation of (12.76) to $M' = \gamma' m$ and $\mathsf{B}' = \mathsf{B}_o/\bar{\gamma}$. Each factor in this expression is individually constant, and $u'_\perp$ is the component perpendicular to the magnetic field of the particle's velocity relative to the drifting frame. The angular velocity of the circling is

$$\omega'_o = \frac{q\mathsf{B}'}{M'c} = \frac{q\mathsf{B}_o}{\bar{\gamma}\gamma' mc} = \frac{u'_\perp}{R'} \tag{12.81}$$

and its phases are determined by $\omega'_o t'$, with t' measured on the time scale in the drifting frame. It is already evident that the motion as observed in the laboratory will not differ in quality from the cycloidal orbiting found in the nonrelativistic case, but the cycloids will be distorted by Lorentz contractions in space and by dilatations of the time scale.

The nature of the relativistic distortions is simplest to formulate and is well exemplified in the special case of a particle that starts from rest in the laboratory $(\mathbf{u}_o = 0)$. It is the distortions of the cusped cycloid of Fig. 5.1*b* that are thus found, in what then amounts to a generalization of it to a comparatively stronger electric field $(\mathsf{E}_o \to \mathsf{B}_o)$. When $\mathbf{u}_o = 0$, the velocity transformation (11.23) yields $\mathbf{u}'_o = -\bar{\mathbf{u}}$ for the initial velocity in the drifting frame, and then its magnitude stays constant at the value $u' \equiv u'_\perp = \bar{u}$ thereafter. With $\gamma' = \bar{\gamma}$ in this case, the radius (12.80) becomes

$$R' = \frac{(mc^2/q)\mathsf{E}_o}{\mathsf{B}_o^2 - \mathsf{E}_o^2} \tag{12.82}$$

which is larger than the value $R = mc^2\mathsf{E}_o/q\mathsf{B}_o^2$ found in the nonrelativistic approximation because of the weakening (12.78) of the magnetic field relative to the drifting frame. The transverse (to the drift) dimensions of the orbit are not altered by reference to the laboratory frame, and so the depth of penetration of the particle into the direction $q\mathbf{E}_o$ is increased from the $2R$ found in the nonrelativistic treatment to $2R' = 2\bar{\gamma}^2 R$ here, as indicated in Fig. 12.2. The distance of

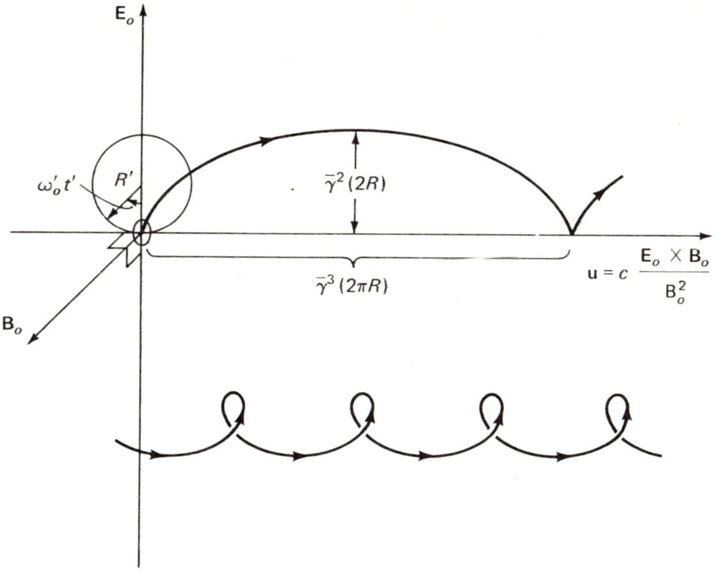

Figure 12.2 The cusped curve represents the orbit of a $q > 0$ particle started with $u_o = 0$, as in Fig. 5.1*b*. Its reflection in the $\bar{u}$ axis, having the same drift direction $\bar{\mathbf{u}}$, would be followed by a charge $-q$. The lower curve is a sample of an orbit followed by a negative charge with some $\mathbf{u}_o \neq 0$ in the orbital plane, with the result that never thereafter is $\mathbf{u} = 0$.

drift during each cycle of the motion is to be found as $\Delta z = \bar{u}T$, where T is a period dilated from $T' = 2\pi/\omega_o' = 2\pi R'/\bar{u}$, for one of the revolutions in the drifting frame, to $T = \bar{\gamma}T'$. Then $\Delta z = 2\pi\bar{\gamma}R'$, greater than the distance $2\pi R$ between cusps of the cycloid in Fig. 5.1 by a factor $\bar{\gamma}^3 = (1 - \mathsf{E}_o^2/\mathsf{B}_o^2)^{-3/2}$. Thus each of the cycles in space is enlarged by the relativistic corrections, by a factor $\bar{\gamma}^2$ in the transverse dimension and by $\bar{\gamma}^3$ in the drift direction. There is a relative distention, rather than contraction, of the cycle lengths in the drift direction, primarily because the kinetically massive high-velocity particle is more difficult for the electric field to bring back to the starting point of a cycle.

EXERCISES

12.1 Obtain the velocity field (10.79) by boosting the Coulomb field, as suggested on page 382.

12.2 Show that transformations of 4-vector potentials can be expressed analogously to (12.25) as

$$\mathbf{A}'_\perp = \mathbf{A}_\perp \qquad \mathbf{A}'_\parallel = \gamma(\mathbf{A}_\parallel - \boldsymbol{\beta}\phi) \qquad \phi' = \gamma(\phi - \boldsymbol{\beta} \cdot \mathbf{A})$$

12.3 Boost the magnetostatic dipole field (4.28) to show that an ideal *magnetic* dipole moment $\boldsymbol{\mu}$ moving with a uniform velocity $c\boldsymbol{\beta}$ generates a scalar potential

$$\phi = \frac{(\boldsymbol{\beta} \times \boldsymbol{\mu}) \cdot \mathbf{r}}{r^3} \frac{1 - \beta^2}{(1 - \beta^2 \sin^2 \vartheta)^{3/2}}$$

which, for $\beta^2 \ll 1$, describes a simple *electric* dipole field, of moment $\mathbf{D} = \boldsymbol{\beta} \times \boldsymbol{\mu}$, superposed on the magnetic one. The electric field here is only a curlless *part*, $\mathbf{E}_D = -\nabla\phi$, of the total electric field that must be expected. (Compare Exercise 6.11.)

12.4 (*a*) Show that the electric dipole field of the preceding exercise is supplemented by a divergenceless part, $\mathbf{E} - \mathbf{E}_D = -\mathring{\mathbf{A}}/c$, which, for $\beta^2 \ll 1$, can be expressed as $\boldsymbol{\epsilon}_\beta \times \boldsymbol{\mu}$, where $\boldsymbol{\epsilon}_\beta$ has a vector-dipole-field form of moment $\boldsymbol{\beta}$.

(*b*) Show that the $\beta^2 \ll 1$ results for the *total* electric field can also be obtained by giving the magnetostatic dipole field $\mathbf{B}_\mu$ of (4.11) a galilean boost that is the reciprocal of (4.69).

(*c*) Find the magnetic-flux change through a disk normal to $\boldsymbol{\beta}$ during a displacement by $c\boldsymbol{\beta}\,dt$ of a $\boldsymbol{\mu}$ parallel to $\boldsymbol{\beta}$ in order to check that the electric field supplementation in (*a*) is to be expected from Faraday induction.

12.5 An energy like (5.54) or (5.67) of an electron orbiting about an atomic nucleus is actually modified by the fact that each electron is also endowed with a magnetic moment $\boldsymbol{\mu} = -e\mathbf{s}/mc$, arising from an intrinsic spin of angular momentum $\mathbf{s}$, according to findings by Uhlenbeck and Goudsmit. The spin is supposed to have a constant magnitude $\frac{1}{2}\hbar$ relative to any instantaneous rest frame of the electron and must be expected to perform a simple Larmor precession like (5.52)

$$\frac{d\mathbf{s}}{d\tau} = \boldsymbol{\mu} \times \mathbf{B}^o = -e\mathbf{s} \times \frac{\mathbf{B}^o}{mc}$$

whenever some magnetic field $\mathbf{B}^o$ exists in such a frame. The corresponding supplementation of the electron's energy is $U^o = -\boldsymbol{\mu} \cdot \mathbf{B}^o$, as in (5.40).

Even when no external magnetic field is imposed on the atom, there will still be a $\mathbf{B}^o$ because the nuclear charge $+Ze$ has a velocity $-\mathbf{u}(t)$ relative to the electron when this has the orbital velocity $\mathbf{u}(t)$. Show that for $u \ll c$, as is usual for atomic electrons,

$$\mathbf{B}^o = -\frac{\mathbf{L}}{mc}\frac{d\phi}{r\,dr}$$

in terms of the electron's orbital angular momentum $\mathbf{L} = \mathbf{r} \times m\mathbf{u}$ and the electrostatic potential ϕ at the electron (this may differ from $\phi = Ze/r$ if there are companion electrons).

12.6 The instantaneous rest frame of the orbiting electron, considered in Exercise 12.5, rotates with an angular velocity $\boldsymbol{\omega}_T = (\gamma - 1)\dot{\mathbf{u}} \times \mathbf{u}/u^2$ relative to the frame of the nucleus at rest, according to findings in Exercise 11.4 (the Thomas precession).

(*a*) Show how it follows from (5.62) that the equation of motion given in Exercise 12.5 becomes

$$\frac{d\mathbf{s}}{dt} = -e\mathbf{s} \times \frac{\mathbf{B}^o + mc\boldsymbol{\omega}_T/e}{mc}$$

relative to the frame of observation and hence the interaction energy in this frame is transformed from U^o to $U = -\boldsymbol{\mu} \cdot \mathbf{B}^o + \mathbf{s} \cdot \boldsymbol{\omega}_T$.

(*b*) Show that when the acceleration $\dot{\mathbf{u}}$ is provided by the electrostatic force on the atomic electron, the Thomas-precession frequency can be expressed as

$$\boldsymbol{\omega}_T \approx \dot{\mathbf{u}} \times \frac{\mathbf{u}}{2c^2} \approx +\mathbf{L}\frac{e}{2m^2c^2}\frac{d\phi}{r\,dr}$$

for comparison with $\mathbf{B}^o$ of the Exercise 12.5.

(*c*) Show that a consequence is

$$U = \frac{1}{2m^2c^2}\left[\frac{d}{r\,dr}(-e\phi)\right]\mathbf{s} \cdot \mathbf{L}$$

called a *spin-orbit coupling energy*. (The reduction from U^o by the factor $\frac{1}{2}$, due to the Thomas precession, is needed for agreement with experimental observations.)

12.7 Hyperbolic trajectories describing deflections of electrons, as from beams scattered by passage through targets containing atomic nuclei, were discussed in connection with Fig. 10.11.

(*a*) Compare the directions of orbital angular momenta **L**, relative to a nucleus, of electrons passing the nucleus on opposite sides of it within some given orbital (scattering) plane.

(*b*) Only a coulombic interaction, $U_c = -e\phi$, between electron and atom was considered in the earlier discussion. The absolute magnitude of the interaction, and hence the intensity of scattering, can be enhanced or diminished by the spin-orbit coupling found in the Exercise 12.6. Compare these effects on beams in which the electron spins are polarized in some one of the two directions normal to the orbital plane with those on beams of the opposite polarization. Give arguments that one of the two polarizations will lead to more scattering to the right than to the left and the other will suffer more leftward deflections. (This provides a basis for an experimental discrimination between opposite transverse polarizations.)

12.8 Spin-orbit couplings of the type discussed in Exercise 12.6 have also been identified for orbitings of spinning nucleons inside atomic nuclei. Here they arise almost entirely from Thomas precession, since the nonelectromagnetic nuclear binding forces, representable by potentials $V(r) < 0$ that weaken with r in place of the atomic potentials $(-e\phi)$, are far stronger than any electromagnetic effects, including ones on the tiny nucleonic magnetic moments (approximately 10^{-3} times the electronic ones). Show that, as a consequence, the *signs* of the nuclear spin-orbit couplings are reversed relative to the net atomic ones. (They are said to be *inverted*.)

12.9 In expression (12.67) of the equation of motion for a point charge in an electromagnetic field, the 4-velocity $w_\mu \equiv p_\mu/m$ defined in (12.66) is used.

(*a*) Evaluate the invariants w_μ^2 and $w_\mu F_{\mu\nu} w_\nu$, which show that the energy $\frac{1}{2}mw_\mu^2$ will be conserved.

(*b*) Show how the 4-vector transformation $w'_\mu = a_{\mu\nu} w_\nu$ can yield the results in Exercise 11.1(*c*).

12.10 Equation (12.67) can also be applied to translational motions of a *spinning* electron, inasmuch as this can be treated as a point charge with $q = -e$, but must be supplemented by an equation guiding precessions of the spin (compare page 140*n*.). The latter equation ought to be one that reduces to the equation given in Exercise 12.5 in any instantaneous rest frame of the electron. The simplest generalization† to an arbitrary frame introduces a 4-vector S_μ to represent the spin, one that has spatial components $\mathbf{S}^o \equiv \mathbf{s}$ in any rest frame of the electron.

(*a*) Show that

$$\frac{dS_\mu}{d\tau} = \frac{e}{mc} S_\nu F_{\nu\mu}$$

has the proper spatial ($\mu = 1, 2, 3$) components in a rest frame but only if S_μ is defined to have $S_4^o \equiv 0$ in rest frames (otherwise the electron also has attributed to it an *intrinsic electric* moment, sensitive to $\mathbf{E}^o$).

(*b*) Boost $S_\mu^o(\mathbf{s}, 0)$ by some velocity $c\boldsymbol{\beta}$ to find the general connection between **S** and **s**, $\boldsymbol{\beta}$ and to show that $S_4 = i\boldsymbol{\beta} \cdot \mathbf{S} = i\gamma\boldsymbol{\beta} \cdot \mathbf{s}$ in an arbitrary frame.

(*c*) Check that the invariant condition $w_\alpha S_\alpha = 0$ (a 4-space orthogonality) can be used to represent the connection of S_4 to **S** that must exist if S_4 is to vanish properly in every rest frame, throughout arbitrary motions $w_\alpha(t)$ of the electron.

12.11 Consider accelerations of a spinning electron by some *nonelectromagnetic* force, representable by $dw_\mu/dt = K_\mu/m$ of (12.6), in place of (12.67). Show that:

(*a*) The spin property $w_\alpha S_\alpha = 0$, found in Exercise 12.10, requires a generation of a spin precession dS_μ/dt such that

$$w_\alpha \frac{dS_\alpha}{d\tau} = -S_\alpha \frac{dw_\alpha}{d\tau} \equiv -\gamma \mathbf{S} \cdot \frac{d\mathbf{u}}{dt}$$

even when there is no electromagnetic torque on the magnetic moment.

† Found by Bargmann et al. (Ref. 3), S_μ must actually have axial 3-vector components, so that it behaves like an ordinary 4-vector only as long as transformations between right- and left-handed frames are excluded.

(*b*) The equation of motion

$$\frac{dS_\mu}{d\tau} = \frac{w_\mu S_\alpha}{c^2}\frac{dw_\alpha}{d\tau}$$

is sufficient to represent the generation of the spin precession found in (*a*).

(*c*) The spatial components of (*b*) can be expressed as $\dot{\mathbf{S}} = \gamma^2 \boldsymbol{\beta}(\dot{\boldsymbol{\beta}} \cdot \mathbf{S})$ and the fourth component equation *follows* from this.

(*d*) Use the Lorentz connections of **S** to **s** found in Exercise 12.10(*b*) to show that (*c*) leads to $\dot{\mathbf{s}} = \boldsymbol{\omega}_T \times \mathbf{s}$, exactly the Thomas-precession effect incorporated in Exercise 12.6(*a*) and existing even when $\mathbf{B}^o = 0$. [This demonstrates that the torque in (*b*) is an inertial effect, arising whenever the rest-frame spin **s** is referred to a frame in which the electron is accelerated by *any* means.]

12.12 (*a*) Evaluate the purely inertial (Thomas) torque, found in Exercise 12.11(*b*) for the special case of acceleration by electromagnetic fields, as in Exercise 12.10(*a*).

(*b*) It must be supposed that the Thomas effect is not separately evident in Exercise 12.10(*a*), as it is in Exercise 12.6(*a*), only because part of the full electromagnetic torque on the magnetic moment has canceled it out. Show that this implies that the full electromagnetic torque must be given by the right side of

$$\frac{dS_\mu}{d\tau} - \frac{w_\mu}{c^2}\left(S_\alpha \frac{dw_\alpha}{d\tau}\right) = \frac{e}{mc}\left[S_\nu F_{\nu\mu} + \frac{w_\mu}{c^2}(S_\alpha F_{\alpha\nu} w_\nu)\right]$$

having the same rest-frame spatial components as the net torque in Exercise 12.10(*a*).

(*c*) The exact cancellation of the Thomas effect by the last part of the electromagnetic torque in (*b*) will occur only insofar as the magnetic moment $\boldsymbol{\mu} = -(ge/2mc)\mathbf{s}$ has exactly the $g = 2$ value used by Uhlenbeck and Goudsmit. (Exercise 12.5; the factor g in the more general expression, called the Landé *g factor*, is the gyromagnetic ratio in units of $e/2mc$.) Show that for the more general magnetic moment, (*b*) reduces to

$$\frac{dS_\mu}{d\tau} = \frac{ge}{2mc}\left[S_\nu F_{\nu\mu} + \left(1 - \frac{2}{g}\right)\frac{w_\mu}{c^2}(S_\alpha F_{\alpha\nu} w_\nu)\right]$$

as in Ref. 3, in place of the equation in Exercise 12.10(*a*).

12.13 Specialize the equation in Exercise 12.12(*c*) to the case of an otherwise free electron traveling in a purely magnetic field **B** when its translation is guided by (12.75) with $q = -e$ and $M = \gamma m$. Show that the fourth component equation leads to the conclusion that $\boldsymbol{\beta} \cdot \mathbf{S}$ and hence $\boldsymbol{\beta} \cdot \mathbf{s}$, the component of the rest spin parallel to the velocity, stays constant during the motion *if* $g = 2$ exactly. (The quantum electrodynamics predicts a deviation to $g = 2.00232\ldots$. Observation on the consequent *changes* in the longitudinal polarization $\boldsymbol{\beta} \cdot \mathbf{s}$ have made it possible to check the prediction to about 1 part in 3×10^7.)

12.14 (*a*) Form a description of the $\mathbf{u}_o = 0$ trajectory in Fig. 12.2 by rectangular coordinates as functions of the time t' in the drifting frame and find the connection of t' to t. Compare your results with (5.19).

(*b*) Find a particular starting velocity $\mathbf{u}_o \neq 0$ with which the particle will simply be transmitted along a straight line and retain its velocity unchanged. (This provides the basis for a device called a *velocity selector*, which deflects and discards, out of an initially nonmonoenergetic beam, particles of essentially all velocities except the particular one.)

12.15 Electrons that are products of particle disintegrations frequently have longitudinal polarizations (helicities), with the spins parallel or antiparallel to the directions of motion. They cannot easily be detected by means like those suggested in Exercise 12.7, which can discriminate only between transverse polarizations. That makes it desirable first to twist the spins relative to the electron trajectories. Use the equations of Exercise 12.5 or 12.10 to show that the spin can be twisted by sending the electrons through a properly adjusted velocity selector (Exercise 12.14) and that a twist through 90° requires transmission over the length

$$l = \frac{\pi mc^2 \mathsf{E}_o}{4e(\mathsf{B}_o^2 - \mathsf{E}_o^2)}$$

with the speed $c\mathsf{E}_o/\mathsf{B}_o$.

12.16 Consider motions of a point charge in uniform fields $E_o > B_o$ such that $\mathbf{E}_o \cdot \mathbf{B}_o = 0$.

(*a*) Show that no drifting frame exists in which $\mathbf{E}' = 0$, as in (12.78).

(*b*) Find a drift velocity for a frame in which $\mathbf{B}' = 0$, instead.

(*c*) One of the trajectories in Fig. 12.2 is followed by a point charge started with $\mathbf{u}_o = 0$ when $E_o < B_o$. Show how the results (12.71) can be used to determine a description of the trajectory followed when $E_o > B_o$, in terms of rectangular coordinates as functions of $\theta \equiv qE_o t'/\bar{\gamma}^2 mc$, where t' is the time on the scale of the drifting frame (*b*).

(*d*) Show that the trajectory follows a monotonic curve asymptotic (for t, $t' \to \infty$) to a line having the slope $[(E_o^2/B_o^2) - 1]^{1/2}$ to the direction of drift and with a speed approaching c.

12.17 Particle trajectories must also exist in crossed fields ($\mathbf{E}_o \cdot \mathbf{B}_o = 0$) when $E_o = B_o$, despite the indefinability of a drifting frame in which one of these fields vanishes, as in the $B_o \gtrless E_o$ cases of (12.78) and Exercise 12.16.

(*a*) Form laboratory-frame equations of motion for rectangular components of the particle position, taking the $\mathbf{E}_o$, $\mathbf{B}_o$ directions as x and y axes, and show that the product $\gamma(1 - \beta_z)$ stays constant during the motion.

(*b*) On the trajectory started with $\mathbf{u}_o = 0$ show that $p_x = mc\omega_o\tau$, $p_z = \frac{1}{2}mc(\omega_o\tau)^2$, and $E = mc^2[1 + \frac{1}{2}(\omega_o\tau)^2]$, where $\tau = t/\gamma$ is the particle's proper time and $\omega_o = qB_o/mc$.

(*c*) Find the constant of proportionality in the relation $x^3 \sim z^2$ which holds on the trajectory and show that both the $B_o \gtrless E_o$ trajectories found in Exercises 12.14 and 12.16 start (at t near zero) with the same proportionality and only later deviate to and from the $\mathbf{E}_o \times \mathbf{B}_o$ direction, respectively.

CHAPTER

THIRTEEN

FIELD DYNAMICS AND CONSERVATION LAWS

The lagrangian and the hamiltonian of a charged particle in an electromagnetic field · Its canonical energy-momentum 4-vector · Electromagnetic fields as examples of mechanical systems, with motions following from the variational principle · How the symmetries to be expected of any description in space-time lead to conservation laws · That invariance to translations in space-time leads to the existence of a conserved field energy-momentum · That invariance to Lorentz rotations leads to a conserved field angular momentum · Gauge invariance and charge conservation

Long before the photonic constitution of radiation fields was discovered, it was recognized that any electromagnetic field could be treated as a mechanical system evolving in time according to equations of motion identifiable with the field equations. The motions then consist of field redistributions in space and time. Like the motions of matter analyzed into discretely localizable particles, field redistributions transport energy-momentum, which can be exchanged with the particles and which endows the fields with kinetic mass. A difference between a field in motion and a classically conceived particle stems from the superposability of fields; those arising from distinct sources coalesce into a resultant field that can be treated as a single entity. However, like properties were also found for matter upon the development of quantum mechanics. The now appropriately conceived particles have become identifiable with rest-massive quanta of various superposable matter fields, just as the rest-mass-less photons have become quanta of electromagnetic field energy-momentum. The electromagnetic fields are not only superposable but also decomposable, and into alternative types of component modes, e.g., plane-wave vs. spherical-wave photons. An even greater variety of possibilities exists for the rest-massive particles, as seen when particle intertransformations were discussed in Chap. 11. All these ramifications cannot be developed without introducing the full quantum-mechanical formalism. However, a high degree of unification in the treatment of matter and fields had already been achieved before the quantum developments, especially in the form of a variation principle that can encompass not only the classical descriptions but also the quantum-mechanical ones. Its exposition will be approached here through a review of it as applied to systems of discrete particles.

13.1 ALTERNATIVE FORMULATIONS OF MECHANICS

Describing circumstances in which motion is to take place in terms of forces is characteristic of the newtonian formulation. Alternative formulations, in which the role of forces is replaced by potentials, or by a lagrangian, or a hamiltonian have also been advanced, each with certain advantages in the solution of, and lending new insight into, the problems of motion.

A typical advantage of replacing an electro*static* force on a particle $q\mathbf{E}$ with a potential energy $U_q = q\phi(\mathbf{r})$, from which the force is derivable, was seen in the discussion of the nonrelativistic equation of motion (5.1). That made it possible to define a conserved energy (5.4), despite the presence of an independently maintained external field that could provide *any* kinetic energy to the particle. The definability of such a conserved energy, in static fields, persists for relativistic velocities of the particle. It follows from forming the work rate $\mathbf{u} \cdot \mathbf{F}_q$ given by the relativistic equation of motion (12.65*a*), as (5.4) followed from (5.1), that

$$E = T + mc^2 + q\phi(\mathbf{r}) \tag{13.1}$$

is constant during the particle's motion in static electromagnetic fields if $T \equiv (\gamma - 1)mc^2$ is the relativistic kinetic energy of the particle. The same result follows even more directly from substituting $\mathbf{E} = -\nabla\phi$ into the fourth component (12.65*b*) of the relativistic equation. An example of $E - mc^2$ was shown in (12.73). Consideration of how the energy (13.1) might be fitted into a manifestly Lorentz-covariant description suggests that it can only be a component of an appropriate energy-momentum 4-vector. The latter will be found in (13.21), upon exploring the alternative formulations in the following subsections.

The Variational Principle

All the alternative formulations of mechanics are encompassed† in Hamilton's variation principle

$$\delta I = 0 \qquad \text{with } I = \int_{t_1}^{t_2} L\, dt \tag{13.2}$$

The integral I is thought of as some kind of *generalized action*, or *expenditure*, during a system's motion. The expenditure rate L, called a lagrangian, is to characterize the system and the circumstances in which its motions are to take place. The principle is applicable to the integrated expenditure over any time interval $t_2 - t_1$ that may be chosen for consideration. The symbol δI stands for the variation of I from the value it takes during the actual motions that will occur when it is evaluated instead for any alternatives to (variations from) the actual motions that could be contemplated. Then $\delta I = 0$ is a condition for the

† See, for example, Ref. 17, sec. 7.4. There the word "expenditure" was introduced because "action" has a more special meaning in many contexts.

actual expenditure to be a comparative extremum, asserted to be true of all motions if their characterizations by L have been properly constructed.

An analytic expression for the lagrangian can be constructed only after some decision is made concerning the variables during the motion that it is desirable to use for describing this. Whenever an analysis of a system into discrete point particles is adopted, the components of the particle position vectors $\mathbf{r}_i(t)$ can be used to describe their motions. However, substitutions can always be made, replacing $\mathbf{r}_1(t)$, $\mathbf{r}_2(t)$, ... with some set of new coordinates $q_1(t)$, $q_2(t)$, ..., $q_f(t)$. The number f of these generalized coordinates is said to be the number of degrees of freedom attributed to the system. The q's need not even be interpretable as ordinary geometrical entities like distances or angles but may be any kind of quantities that might be observed to vary during a motion. As will be seen in the next section, the generalized coordinates need not even be discretely enumerable but may consist of a continuous infinity in number (thus forgoing the analysis into discrete particles). For the present, however, only discrete degrees of freedom will be considered.

Classical mechanics can predict an explicit, unique motion for given circumstances only if additional information is also given: about the start given to the motion, as specified by the initial positions *and* initial velocities, or some equivalent of these data. On this account, a set of $2f$ values, for $q(t) \equiv q_1(t), \ldots, q_f(t)$ *and* for $\dot{q}(t) \equiv \dot{q}_1(t), \ldots, \dot{q}_f(t)$, is said to characterize a *phase*, or momentary *state*, of the motion being passed through at the time t. As indicated, the single symbol $q(t)$ is here being used to stand for the whole set of f degrees of freedom and $\dot{q}(t)$ for the corresponding generalized velocities. Motion is thus viewed as a series of transitions: from earlier phases $q(t_1)$, $\dot{q}(t_1)$ to later instantaneous states of motion $q(t_2)$, $\dot{q}(t_2)$ that become predictable from the earlier phases. The expenditure rate at a given moment is expected to depend on the phase the system is passing through at the moment and so is constructed as some function $L(q, \dot{q}, t)$ of the $2f$ variables. A further, explicit dependence on time is also generally allowed for since in some circumstances the rate might depend on the moment in time, relative to some external event, at which a given phase is being passed through.

A continuously varying set of coordinates $q(t)$ describes a *path* followed by the system as the motion goes on, and the integrated expenditure I of (13.2) is determined by this path. Variations $q_s(t) \to q_s(t) + \delta q_s(t)$ from the actual path can always be considered, and corresponding variations δI in the function of path I. A finding deducible (see page 418) for any lagrangian $L(q, \dot{q}, t)$ is that the extremum condition $\delta I = 0$ is satisfied if only the well-known Lagrange equations of motion

$$\frac{d}{dt}\left(\frac{\partial L}{\partial \dot{q}_s}\right) = \frac{\partial L}{\partial q_s} \qquad \text{for each } s = 1, 2, \ldots, f \tag{13.3}$$

are satisfied. Every sufficiently complete formulation of a motion measured by the time evolution of a given, discrete set of variables can be construed as a special case of this with a suitably constructed lagrangian. The correspondence with the nonrelativistic newtonian equations $m_i \ddot{\mathbf{r}}_i = \mathbf{F}_i$ is particularly simple whenever the forces are derivable from some potential energy $U(\mathbf{r}_1, \mathbf{r}_2, \ldots)$. Then the choice

$$L = T - U = \sum_i \tfrac{1}{2} m_i(\dot{x}_i^2 + \dot{y}_i^2 + \dot{z}_i)^2 - U \tag{13.4}$$

leads to $\partial L/\partial \dot{x}_i = m_i \dot{x}_i$, and (13.3) is transcribed to $m_i \ddot{x}_i = -\partial U/\partial x_i = (\mathbf{F}_i)_x$.

The simple illustration indicates why

$$p_s \equiv \frac{\partial L}{\partial \dot{q}_s} \tag{13.5}$$

is called the generalized momentum, *conjugate* to the degree of freedom q_s. It is not always merely the familiar linear momentum of a particle, even in a system of particles, but depends on the significance of the generalized coordinate q_s that is conjugate to it. For instance, the conjugate momentum of an angular degree of freedom becomes an angular, rather than a linear, momentum. An example in which the conjugate momentum consists not only of particle momentum but also of field momentum available to the particle was encountered in (5.7). One or more of the conjugate momenta may remain conserved during a motion. The equations of motion (13.3) show directly that $\dot{p}_s = 0$ whenever $\partial L/\partial q_s = 0$. This occurs when the description of a system has a symmetry that makes it independent of the particular values taken on by $q_s(t)$ as it changes during motions in which $\dot{q}_s(t) \neq 0$. Such variables $q_s(t)$, not appearing in the lagrangian explicitly, are customarily called *ignorable;* only the appearance of the corresponding generalized velocity $\dot{q}_s(t)$ is left to represent investments in an ignorable degree of freedom.

Relativistic Lagrangians of Point-Charge Motions

A lagrangian that makes the Lagrange equations (13.3) equivalent to the equation of motion (12.65) becomes simple to construct after a potential description of the field has been introduced, as done in the steps from (5.1) to (5.6) in the nonrelativistic case. For a relativistically described particle, (5.6) must be generalized to

$$\frac{d}{dt}\left(\gamma m \mathbf{u} + \frac{q}{c}\mathbf{A}\right) = -q\,\nabla\left(\phi - \frac{\mathbf{u}\cdot\mathbf{A}}{c}\right) \tag{13.6}$$

This is equivalent to (12.65) even when arbitrary, nonstatic fields are imposed, since

$$\frac{d}{dt}\mathbf{A}(\mathbf{r}, t) = \frac{\partial \mathbf{A}}{\partial t} + \sum_i \frac{\partial \mathbf{A}}{\partial x_i}\frac{\partial x_i}{\partial t} = \mathring{\mathbf{A}} + (\mathbf{u}\cdot\nabla)\mathbf{A}$$

and substitution into (13.6) yields

$$\frac{d}{dt}\gamma m \mathbf{u} = q\left\{\left(-\nabla\phi - \frac{\mathring{\mathbf{A}}}{c}\right) + \frac{1}{c}[\nabla(\mathbf{u}\cdot\mathbf{A}) - (\mathbf{u}\cdot\nabla)\mathbf{A}]\right\}$$

The square bracket here is just $\mathbf{u} \times (\nabla \times \mathbf{A}) = \mathbf{u} \times \mathbf{B}$, while $-\nabla\phi - \dot{\mathbf{A}}/c = \mathbf{E}(\mathbf{r}, t)$, so that the equivalence to (12.65) is evident.

The lagrangian for the case here, $L(\mathbf{r}, \mathbf{u}, t)$, must be so constructed that each of the rectangular components of the vector equation (13.6) takes on the form (13.3). Plainly, results like

$$\frac{\partial L}{\partial u_i} = \gamma m u_i + \frac{q}{c} A_i \qquad \text{and} \qquad \frac{\partial L}{\partial x_i} = -q \frac{\partial}{\partial x_i}\left(\phi - \frac{\mathbf{u} \cdot \mathbf{A}}{c}\right)$$

are needed. Since

$$u_i\, du_i = \tfrac{1}{2}\, du_i \frac{\partial u^2}{\partial u_i} = \tfrac{1}{2}\, du_i \frac{\partial}{\partial u_i} c^2\left(1 - \frac{1}{\gamma^2}\right) = -\frac{c^2}{\gamma} \frac{\partial}{\partial u_i} \frac{1}{\gamma}\, du_i$$

an appropriate construct is

$$L(\mathbf{r}, \mathbf{u}, t) = -\frac{mc^2}{\gamma} - q\left[\phi(\mathbf{r}, t) - \mathbf{u} \cdot \frac{\mathbf{A}}{c}\right] \tag{13.7}$$

This is a relativistic lagrangian insofar as it makes the Lagrange equations (13.3) equivalent to the relativistic equations of motion (13.6) or (12.65). The above expression for $\partial L/\partial u_i$ shows that the conjugate momentum (13.5) of a point charge in an electromagnetic field is given by

$$\mathbf{p}(\mathbf{r}, t) = \gamma m \mathbf{u}(t) + \frac{q\mathbf{A}(\mathbf{r}, t)}{c} \tag{13.8}$$

This has the nonrelativistic approximation already seen in (5.7) for the static field case.

The result (13.7) also shows that a neutral point particle or a charged one in a space free of fields is described by

$$L_o(\mathbf{u}) = -\frac{mc^2}{\gamma} = -mc^2\left(1 - \frac{u^2}{c^2}\right)^{1/2} \tag{13.9}$$

In this case, Hamilton's principle (13.2) reduces to

$$\delta \int_{t_1}^{t_2} \frac{dt}{\gamma(\mathbf{u})} = \delta \int_{\tau(t_1)}^{\tau(t_2)} d\tau = 0 \tag{13.10}$$

It becomes a principle of *least proper time* for the motions in free space. In this respect, it is comparable to Fermat's famous *principle of least time* for the paths of light rays.†

† See Ref. 17, eq. (8.119) and its discussion; also exercise 15.7 there.

The Hamiltonian Energy

Energies conserved in particle motion, like (13.1) or (5.4), are identifiable from the rate of kinetic energy change $\mathbf{u} \cdot d\mathbf{p}/dt$ that follows from the equation of motion $d\mathbf{p}/dt = \mathbf{F}$. More generalized expressions of energy conservation can follow from the Lagrange equations $\dot{p}_s = \partial L/\partial q_s$ of (13.3) in a similar way, by forming

$$\sum_s \dot{q}_s \frac{dp_s}{dt} = \sum_s \dot{q}_s \frac{\partial L}{\partial q_s}$$

The last expression can be recognized as a contribution, from just the dependence of $L(q, \dot{q}, t)$ on the q's, to the total rate of change with time

$$\frac{d}{dt} L(q, \dot{q}, t) = \sum_s \left(\frac{\partial L}{\partial q_s} \dot{q}_s + \frac{\partial L}{\partial \dot{q}_s} \ddot{q}_s \right) + \frac{\partial L}{\partial t}$$

$$= \sum_s (\dot{p}_s \dot{q}_s + p_s \ddot{q}_s) + \frac{\partial L}{\partial t}$$

It is then evident that

$$\frac{d}{dt}\left(\sum_s p_s \dot{q}_s - L \right) = -\frac{\partial L}{\partial t} \tag{13.11}$$

and that

$$H \equiv \sum_s p_s \dot{q}_s - L \tag{13.12}$$

is a conserved quantity whenever $L(q, \dot{q})$ depends only on the phase of the motion and not explicitly on the time at which this is passed through, so that $\partial L/\partial t = 0$. In the simple example (13.4), $\mathbf{p}_i = m_i \dot{\mathbf{r}}_i = m_i \mathbf{u}_i$ and

$$H = \sum_i \mathbf{p}_i \cdot \mathbf{u}_i - L = \sum_i \tfrac{1}{2} m_i u_i^2 + U(\mathbf{r}_1, \mathbf{r}_2, \ldots) \tag{13.13}$$

the familiar nonrelativistic energy that is conserved with the help of a potential energy, an expression of energy conservation that is possible when all the forces are derivable from a scalar potential. A much greater variety of expressions of energy conservation is made possible by identifying the general expression (13.12) for H as a generalized energy. This is conserved if only $\partial L/\partial t = 0$, a condition that must be expected, since admitting an explicit dependence on time corresponds to making changes independent of the motion going on, as in representing effects of external influences that can change any energy the system might have, however defined.

For the relativistic particle interacting with an electromagnetic field, as described by the lagrangian (13.7) and the conjugate momentum (13.8), $H \equiv \mathbf{p} \cdot \mathbf{u} - L$ becomes

$$H = \left(\gamma m \mathbf{u} + \frac{q}{c} \mathbf{A} \right) \cdot \mathbf{u} + \frac{mc^2}{\gamma} + q\left(\phi - \frac{\mathbf{u} \cdot \mathbf{A}}{c} \right)$$

$$= \gamma mc^2 \left(\frac{u^2}{c^2} + \frac{1}{\gamma^2} \right) + q\phi = \gamma mc^2 + q\phi(\mathbf{r}, t) \tag{13.14}$$

This is conserved if $\partial L/\partial t = -\partial H/\partial t = -q\,\partial\phi/\partial t = 0$, that is, in a static field. In the latter case, $H = E$ of (13.1), since $\gamma mc^2 = T + mc^2$. If the external field ϕ in (13.14) were varied arbitrarily in time, in addition to its variations with the positions $\mathbf{r}(t)$ that the particle takes during its motion, the energy could not be expected to be conserved.

Hamilton's Canonical Equations of Motion

A significant property of the generalized energy expression (13.12) is that despite the explicit appearance of the generalized velocities $\dot{q}_s$ in its connection to the lagrangian, it varies with any $\dot{q}_s$ only insofar as it varies with the conjugate momenta p; for if its rate of variation $\partial H/\partial \dot{q}_s$ is evaluated while keeping the p's fixed and also the q's in $L(q, \dot{q}, t)$, the result is $p_s - \partial L/\partial \dot{q}_s \equiv 0$ by the definition of p_s. Thus H should be expressible as a function of q, p, t only, not containing any $\dot{q}_s$. How this works out in the nonrelativistic example of (13.13) is easy to see from substituting $\mathbf{u}_i = \mathbf{p}_i/m_i$ into it:

$$H = \sum_i \frac{p_i^2}{2m_i} + U(\mathbf{r}_1, \mathbf{r}_2, \ldots) \tag{13.15}$$

(A salient importance of this type of expression for general physical theory stems from its use in representing the connection between the energy and momentum operators of quantum mechanics.)

When it is expressed as a function of (q, p, t), H is called the hamiltonian of the system. Its rates of variation $\partial H/\partial q_s$ and $\partial H/\partial p_s$ with each of the $2f$ variables q_s and p_s can be found from its connection (13.12) to the lagrangian, the definition (13.5) of the conjugate momenta, and the Lagrange equations of motion (13.3). The well-known results† are

$$\dot{q}_s(t) = \frac{\partial H}{\partial p_s} \qquad \dot{p}_s(t) = -\frac{\partial H}{\partial q_s} \qquad s = 1, 2, \ldots, f \tag{13.16}$$

These can now serve as the equations of motion in place of the Lagrange equations or newtonian ones. They are called *Hamilton's canonical equations* and the q, p are called *canonical variables.* In the special case of the nonrelativistic hamiltonian (13.15), they reduce to

$$\mathbf{u}_i = \frac{\mathbf{p}_i}{m_i} \qquad \dot{\mathbf{p}}_i = -\nabla_i U(\mathbf{r}_1, \mathbf{r}_2, \ldots) \tag{13.17}$$

a mere reiteration of the connections between momenta and velocities, together with newtonian equations in force fields $\mathbf{F}_i = -\nabla_i U$, derivable from a scalar-potential energy. The explicit statements of the velocity-momentum connections in the canonical equations make it possible to treat the p's as independent phase variables, on an equal footing with the q's and yet to maintain the correct relations of the $\dot{q}$'s to the q's, during the motion.

† See Ref. 17, eq. (8.5), for example.

When $H(q, p)$ is a *conserved* energy, constant during the motions $q(t)$, $p(t)$, its variations with the individual variables q_s and p_s must cancel each other. In general,

$$\frac{d}{dt}H(q, p, t) = \sum_s \left(\frac{\partial H}{\partial q_s}\dot{q}_s + \frac{\partial H}{\partial p_s}\dot{p}_s\right) + \frac{\partial H}{\partial t} = \frac{\partial H}{\partial t} \tag{13.18}$$

follows from the canonical equations (13.16), and so H is conserved in value whenever it is not made explicitly dependent on the time. This is the same criterion for energy conservation as the $\partial L/\partial t = 0$ of (13.11), now expressed in the terms of a hamiltonian rather than a lagrangian description. The conservation of an individual canonical momentum $\dot{p}_s = 0$ follows from $\partial H/\partial q_s = 0$ in (13.16), an invariance of the hamiltonian to changes in q_s, just as it followed from $\partial L/\partial q_s = 0$ in the discussion of (13.5).

The special hamiltonian for a point charge in an electromagnetic field can be found from the energy expression (13.14) after putting the isolated particle energy γmc^2 in terms of the kinetic momentum $\gamma m\mathbf{u}$, as in (11.46), with the result

$$H = c\sqrt{(\gamma m\mathbf{u})^2 + m^2c^2} + q\phi$$

Now, introduction of the canonical momentum (13.8) yields

$$H(\mathbf{r}, \mathbf{p}, t) = c\sqrt{\left(\mathbf{p} - \frac{q}{c}\mathbf{A}\right)^2 + m^2c^2} + q\phi(\mathbf{r}, t) \tag{13.19}$$

as the hamiltonian function of the conjugate variables $\mathbf{r}$, $\mathbf{p}$. It can be checked that with (13.19) the canonical equations (13.16) respectively become a restatement of the connection (13.8) between velocity and canonical momentum and the correct equation of motion (13.6).

Expression (13.19) can be put into the form

$$\left(\mathbf{p} - \frac{q}{c}\mathbf{A}\right)^2 - \frac{(H - q\phi)^2}{c^2} = -m^2c^2 \tag{13.20}$$

recognizable as just the invariant square (11.45) of the energy-momentum 4-vector when the connections (13.8) and (13.14) between $\mathbf{p}$, H, and the mass-particle momentum and energy $\gamma m\mathbf{u}$ and γmc^2 are recalled. Plainly, a *conjugate* momentum 4-vector

$$p_\mu\left(\mathbf{p}, \frac{iH}{c}\right) \tag{13.21}$$

can now be defined, since its components differ from those of the free-particle 4-vector $(\gamma m\mathbf{u}, i\gamma mc)$ by just the 4-vector components of $qA_\mu/c(q\mathbf{A}/c, iq\phi/c)$. The definition (13.21) constitutes the particle-in-field energy-momentum 4-vector referred to in the discussion of (13.1).

Manifestly Invariant Forms

In the special case of a point particle, its velocity with respect to any frame defines proper-time intervals that are the same for all observers, and that fact simplifies casting the Lagrange and Hamilton formulations of single-particle motions into manifestly invariant forms. Lagrange's generalized velocities should now be taken to be the components of the 4-vector velocity (12.66)

$$w_\mu = \frac{dx_\mu}{d\tau}(\gamma\mathbf{u}, ic\gamma) \qquad \text{with } w_\mu^2 = -c^2 \tag{13.22}$$

For a free particle (no fields present) it is identical with p_μ/m, but more generally the symbol p_μ has been preempted† for the *canonical* momentum 4-vector (13.21). This has the connection

$$p_\mu(x) = mw_\mu + \frac{qA_\mu(x)}{c} \tag{13.23}$$

to the 4-velocity of a point charge in an electromagnetic field.

If a lagrangian $\mathscr{L}(x, w)$ is now constructed by integrating $\partial\mathscr{L}/\partial w_\mu = p_\mu$ of (13.23), the frame-invariant form

$$\mathscr{L} = \tfrac{1}{2}mw_\rho^2 + \frac{qw_\rho A_\rho}{c} - \tfrac{1}{2}mc^2 \tag{13.24}$$

(with summations over ρ to be understood) has the requisite dependence on the 4-velocity components. That integration, by itself, permits adding an arbitrary function of $x_\mu(\mathbf{r}, ict)$ to (13.24), rather than just the absolute constant, $-\frac{1}{2}mc^2$, but the gauging of $\mathscr{L}$ in (13.24) is already complete enough to make the Lagrange equation

$$\frac{dp_\mu}{d\tau} \equiv \frac{d}{d\tau}\frac{\partial\mathscr{L}}{\partial w_\mu} = \frac{\partial\mathscr{L}}{\partial x_\mu} \tag{13.25}$$

have the correct equations of motion, (13.6) or (12.65), as its 3-space components. This 4-vector Lagrange equation possesses manifest Lorentz invariance and has a fourth component equivalent to the result $dH/dt = -\partial L/\partial t$ of (13.11), one that also follows from the other three component equations. The value $-\frac{1}{2}mc^2$ was chosen‡ for the additive integration constant in (13.24) because the fact that $w_\mu^2 = -c^2$ [Eq. (13.22)] during the actual motion§ makes the $\mathscr{L}$ so gauged take

† This has become customary in part because it is the canonical momentum (13.8), rather than $\gamma m\mathbf{u}$ or $m\mathbf{u}$, that must be given the Schrödinger operator representation $\mathbf{p} \to -i\hbar\nabla$ in transcriptions to quantum mechanics. (More generally, $p_\mu \to -i\hbar\,\partial_\mu$, with $H \to +i\hbar\,\partial/\partial t$.)

‡ Lagrangians can be gauged with certain degrees of arbitrariness, like the potentials they contain in such forms as (13.4). More will be made of this in the next section; see also Ref. 17, pages 173ff.

§ The expression $\mathscr{L}(x, w)$ in (13.24), with $\frac{1}{2}mw_\mu^2$ *not* replaced by $-\frac{1}{2}mc^2$, must be used for the Lagrange equations (13.25) because the expression of all such equations is predicated on defining the derivatives of $L(q, \dot{q}, t)$ with respect to $\dot{q}$'s to be carried out holding the q's constant, irrespective of their interrelationships during a specific motion.

on the value

$$\mathscr{L} = -mc^2 + \frac{qw_\rho A_\rho}{c} = -mc^2 + \frac{\gamma q(\mathbf{u} \cdot \mathbf{A} - c\phi)}{c} = \gamma L \tag{13.26}$$

simply proportional to the lagrangian L already used for gauging action in (13.7). This proportionality is consistent with the equivalence of (13.25) to the variational condition (13.2) on the invariant action

$$\delta \int_{t_1}^{t_2} L\, dt = \delta \int_{\tau(t_1)}^{\tau(t_2)} \mathscr{L}\, d\tau = 0 \tag{13.27}$$

Whereas L was an expenditure rate per unit of the time scale in the frame of observation, $\mathscr{L}$ gauges the frame-invariant rate per unit of the particle's proper time.

The invariant lagrangian $\mathscr{L}(x, w)$ of (13.24) leads to an invariant hamiltonian $\mathscr{H} = p_\rho w_\rho - \mathscr{L}$ that can be expressed as a function of the 4-vectors x_μ and p_μ:

$$\mathscr{H} = w_\rho\left(mw_\rho + \frac{qA_\rho}{c}\right) - \tfrac{1}{2}m(w_\rho^2 - c^2) - \frac{qw_\rho A_\rho}{c} = \tfrac{1}{2}m(w_\rho^2 + c^2) \tag{13.28}$$

$$\mathscr{H} = \frac{1}{2m}\left(p_\rho - \frac{q}{c}A_\rho\right)^2 + \tfrac{1}{2}mc^2 \tag{13.29}$$

Then

$$\frac{\partial}{\partial p_\nu}\mathscr{H}(x, p) = w_\nu \qquad \frac{\partial}{\partial x_\nu}\mathscr{H}(x, p) = -\frac{dp_\nu}{d\tau} \tag{13.30}$$

the last following from the 4-vector Lagrange equation (13.25) because $\partial\mathscr{H}/\partial x_\nu = -\partial\mathscr{L}/\partial x_\nu$. These constitute manifestly covariant, 4-vector canonical equations of motion. Because $w_\rho^2 = -c^2$ during the motion, $\mathscr{H} \equiv 0$ in value at every stage of it, exactly as required for the consistency of (13.29) with (13.20). It is $p_4 = iH/c$ of (13.21), occurring in (13.29), that describes the energy, on an equal footing with the momentum components, as to be expected of a manifestly invariant expression.

An invariant lagrangian for the variation principle in the form (13.27), the consequent Lagrange equations (13.25), and the 4-vector canonical equations (13.30) can also be constructed for the motion of a particle under nonelectromagnetic forces. Comparison of the Minkowski force equation $d(mw_\mu)/d\tau = K_\mu$ of (12.68) to the Lagrange form (13.25) shows that (12.68) can be applied in any circumstances representable by a Minkowski force that can be derived from a 4-scalar work function $\mathscr{W}(x, w)$ as†

$$K_\mu = \left(\frac{\partial}{\partial x_\mu} - \frac{d}{d\tau}\frac{\partial}{\partial w_\mu}\right)\mathscr{W}(x, w) \tag{13.31}$$

† A generalization of the derivability of general newtonian forces from 3-scalar work functions [Ref. 17, eq. (7.35)].

In this broad class of cases,

$$\mathscr{L} = \tfrac{1}{2}mw_\mu^2 + \mathscr{W}(x, w) \tag{13.32}$$

$$p_\mu = mw_\mu + \partial\mathscr{W}/\partial w_\mu \tag{13.33}$$

Expressions (13.24) and (13.23) are special cases of these equations, with $\mathscr{W} = qw_\rho A_\rho/c$ (plus a constant).

13.2 THE FIELD LAGRANGIAN

Whereas at least the classical particles are each conceived as having a configuration in 3-space that is completely specified by giving just the position $\mathbf{r}(t)$ of one point, fields are continuously extended over 3-space, each at best conceivable as a spread-out particle. There are also quite ordinary, classical-mechanical matter systems that are described by such continuous distributions, material media in which elastic waves can propagate or continuously extended deformable bodies that can house continua of internal vibrations. Treating them has led to experience with lagrangian and variational formulations of variable continua.† In such cases, fields of motion are described by some generalized coordinates $q_i(\mathbf{r}, t)$, a set for each of the continuous infinity of volume elements $dV(\mathbf{r})$ of the matter in motion. Then $\mathbf{r}$ serves as an independent variable, additional to t, labeling the various elements of the matter. The dependent variables $q_i(\mathbf{r}, t)$ at each $\mathbf{r}$ might, for example, be component displacements of the matter initially at $\mathbf{r}$. [In hydrodynamics, the generalized coordinates may be taken to be the components of a *velocity* field $\mathbf{u}(\mathbf{r}, t)$ distributed over the various points $\mathbf{r}$ in the space of the fluid moving through them.]

What goes on in an electromagnetic field is described by giving functions of the space-time 4-vector, $x_\mu(\mathbf{r}, ict)$, the field components $A_\nu(x)$ *or* $F_{\mu\nu}(x)$, which are distributions over the four dimensions that at once comprise variations in time, and also spatial variations in the 3-space distribution at each moment. The field components can be adopted as the generalized coordinates in terms of which the lagrangian of the system can be constructed. Moreover, it should be possible to continue treating all four components of $x_\mu(\mathbf{r}, ict)$ as independent variables on an equal footing, thus maintaining a Lorentz invariance; therefore the variation principle (13.2) should be formulated for a definite span $V_4 \equiv \int d^4x$ of the four-dimensional space-time, rather than just for a time span $t_2 - t_1$ between definite configurations. A restriction of the system treated to some volume of it, as well as to a definite time span, also turns out to be necessary for the elastic displacement and velocity fields of motion in matter, even when relativistic invariance is not imposed. It amounts to settling for equations of motion that need sup-

† See, for example, Ref. 17, chap. 15. A noteworthy feature is the identification of momentum associated with the transport of excitation energies during elastic wave propagation, quite apart from the oscillating momenta of vibrating rest-mass elements of the medium—all this even in a nonrelativistic description!

plementation not only by initial conditions but also by spatial boundary conditions. This is, of course, characteristic of any field equations.†

In accord with these considerations, a variation principle of the form

$$\delta I = 0 \qquad \text{with } I = \int_{V_4} d^4x \mathscr{L}[A_\nu(x), \partial_\mu A_\nu, x] \tag{13.34a}$$

is proposed. For it, a lagrangian *density* $c\mathscr{L}$ is defined, with the connection

$$\int_{t_1}^{t_2} c\, dt \int dV \mathscr{L} \equiv \int_{t_1}^{t_2} dt L \tag{13.34b}$$

to I of (13.2), as written for any one frame of observation. Whereas $\dot{q}$'s are included with the $q(t)$'s and t, as specifications of the phases in motions of discrete degrees of freedom when forming $L(q, \dot{q}, t)$, here the lagrangian must generally contain all 16 component derivatives $\partial_\mu A_\nu$ of the generalized coordinates $A_\nu(x)$ to complete specifications of the phases in field redistributions over space, as well as time. It is the Lorentz-potential components that are being adopted as the generalized coordinates here, although the field-tensor components $F_{\mu\nu}(x)$ might have been introduced in their place. A basis for field equations of motion in the form (12.11) is thus being sought, rather than an immediate emergence of the Maxwell equations (12.28). In neither approach can the field variables be treated as being unconstrained.‡ An electromagnetic field must be described by a special type of field tensor in the four-dimensional geometry, an antisymmetric one with the purely space-time geometric property (12.29), the homogeneous Maxwell requirements. They are automatically satisfied by $F_{\mu\nu} = \partial_\mu A_\nu - \partial_\nu A_\mu$ with *any* $A_\nu(x)$. However, $A_\nu(x)$ must also be constrained, by the purely geometric Lorentz condition $\partial_\nu A_\nu = 0$ of (12.13); otherwise the equations of motion $\partial_\mu^2 A_\nu = -4\pi j_\nu/c$ [Eq. (12.11)] cannot be expected for $A_\nu(x)$. The scalar Lorentz condition on $A_\nu(x)$ thus remaining is far simpler to enforce than the 64-component Maxwell requirement (12.29) on $F_{\mu\nu}(x)$, and, indeed, it is through the introduction of potentials that satisfaction of the condition (12.29) is quite generally approached.

Next, the variational procedure that makes equations of motion follow from the extremum condition $\delta I = 0$ should be examined. In the Lagrange equations (13.3), applying to discretely enumerable coordinates, the procedure is familiar.

† It is thus that the formulations gain adaptability to various specific physical situations, just as equations for particle motion gain such adaptability through the possibility of fitting various initial conditions. Boundary conditions on field equations of motion can represent energy-momentum transfers through the boundary, of the field system to be treated, from any fields that might exist outside the boundary of separation.

‡ Geometrical constraint conditions on *discrete* variables $q_s(t)$ are also used for some systems whenever the number of coordinates introduced exceeds the degrees of freedom actually available to a system. For example, there are problems in which variables x, $y(t)$ are convenient to use even though they are constrained to a circle $x^2 + y^2 = a^2$ or some other geometrically specified surface of constraint.

Each coordinate $q_s(t)$ is varied to some $q_s(t) + \delta q_s(t)$, so that variations $\delta \dot{q}_s = d[\delta q_s(t)]/dt$ follow. The consequent variation of the lagrangian $L(q, \dot{q}, t)$ is

$$\delta L = \sum_s \left(\frac{\partial L}{\partial q_s} \delta q_s + \frac{\partial L}{\partial \dot{q}_s} \frac{d}{dt} \delta q_s \right)$$
$$= \frac{d}{dt}\left(\sum_s \frac{\partial L}{\partial \dot{q}_s} \delta q_s \right) + \sum_s \delta q_s \left(\frac{\partial L}{\partial q_s} - \frac{d}{dt}\frac{\partial L}{\partial \dot{q}_s} \right)$$

Then the time integration in (13.2), needed to form δI, reduces the first summation to just the difference of the limit values indicated by

$$\left[\sum_s \frac{\partial L}{\partial \dot{q}_s} \delta q_s(t) \right]_{t_1}^{t_2}$$

This vanishes at each limit for any motion between definite configurations† $q(t_{1,2})$, with $\delta q(t_{1,2}) = 0$, and leaves

$$\delta I = \int_{t_1}^{t_2} dt \sum_s \delta q_s \left(\frac{\partial L}{\partial q_s} - \frac{d}{dt}\frac{\partial L}{\partial \dot{q}_s} \right) = 0$$

which can be satisfied at all times $t_1 < t < t_2$ and for arbitrary variations δq_s in each degree of freedom only if the (Euler-) Lagrange conditions (13.3) are satisfied.

The corresponding steps for the continua of component variables $A_\nu(x)$ require varying each function to some $A_\nu(x) + \delta A_\nu(x)$ at each space-time point within $V_4 \equiv \int d^4x$. Then $\delta(\partial_\mu A_\nu) = \partial_\mu[\delta A_\nu(x)]$, and an $\mathscr{L}$ containing A_ν's and $\partial_\mu A_\nu$'s is varied by

$$\delta \mathscr{L} = \sum_\nu \frac{\partial \mathscr{L}}{\partial A_\nu} \delta A_\nu + \sum_{\mu,\nu} \frac{\partial \mathscr{L}}{\partial(\partial_\mu A_\nu)} \partial_\mu(\delta A_\nu)$$
$$= \sum_{\mu,\nu} \partial_\mu \left[\frac{\partial \mathscr{L}}{\partial(\partial_\mu A_\nu)} \delta A_\nu \right] + \sum_\nu \delta A_\nu \left[\frac{\partial \mathscr{L}}{\partial A_\nu} - \sum_\mu \partial_\mu \frac{\partial \mathscr{L}}{\partial(\partial_\mu A_\nu)} \right] \tag{13.35}$$

Each of the perfect differential terms, coming first in the last expression, is reduced by one dimension‡ of the four-dimensional integration over $V_4 = \int d^4x$, to a result to be evaluated at boundary points of V_4, and every term of these results is proportional to some $\delta A_\nu(x)$ to be evaluated on the boundary. Whatever boundary conditions may be specified, $\delta A_\nu(x) = 0$ must be chosen to maintain consistency with them, leaving an outcome for the integral of (13.35)

† Thus, in formulations of the variation principle, a uniqueness of the motion is enforced by consistency with given boundary conditions in time $q(t_1)$ and $q(t_2)$ in place of the initial conditions $q(t_1)$ and $\dot{q}(t_1)$. This suggests why the variation principle, with appropriate choices for the generalized coordinates, is also applicable to quantum-mechanical descriptions, in which values for $q(t_1)$ and $\dot{q}(t_1)$ cannot be simultaneously known and only questions about transitions from any initial configuration $q(t_1)$ to any final one $q(t_2)$ can be answered.

‡ A generalization to space-time of the Gauss theorem (A.14) or the more general (A.15).

that can yield $\delta I = 0$ for arbitrary variations *within* V_4 only if the Euler-Lagrange conditions

$$\sum_\mu \partial_\mu \frac{\partial \mathscr{L}}{\partial(\partial_\mu A_\nu)} = \frac{\partial \mathscr{L}}{\partial A_\nu} \qquad \nu = 1, 2, 3, 4 \tag{13.36}$$

are satisfied. These constitute field equations of motion for whatever 4-vector field variables $A_\nu(x)$ have been introduced into $\mathscr{L}$.

For $A_\nu(x)$ to be the Lorentz potential field arising from a given source distribution $j_\nu(x)$, the lagrangian $\mathscr{L}(A_\nu, \partial_\mu A_\nu, x)$ that is put into (13.36) must be so constructed that the equation takes the form (12.11), which can be written

$$-\sum_\mu \partial_\mu \frac{\partial_\mu A_\nu}{4\pi} = \frac{j_\nu(x)}{c} \tag{13.37}$$

Thus $\partial_\mu A_\nu$ and A_ν must occur in $\mathscr{L}$ in such a way that

$$\frac{\partial \mathscr{L}}{\partial(\partial_\mu A_\nu)} = -\frac{\partial_\mu A_\nu}{4\pi} \qquad \frac{\partial \mathscr{L}}{\partial A_\nu} = \frac{j_\nu(x)}{c}$$

A sufficient construct is

$$\mathscr{L} = -(8\pi)^{-1}(\partial_\mu A_\nu)^2 + \frac{j_\nu A_\nu}{c} \tag{13.38}$$

for which the summation convention has been reintroduced, so that the result is a 4-scalar invariant.

Plainly, the variation principle permits $\mathscr{L}$ to be defined with any constant multiplier. The choice here was designed to make the lagrangian density for the field in interaction with a given charged-matter distribution comparable to the lagrangian (13.24) for a charged particle in interaction with a given field distribution. The interaction term in the field lagrangian

$$\frac{j_\nu A_\nu}{c} = \mathbf{j} \cdot \frac{\mathbf{A}}{c} - \rho\phi \tag{13.39a}$$

is just the negative of the interaction *energy* density, in a form which has so far been introduced only in connection with static fields, the electric interaction-energy density $\rho\phi$ in (2.46) and the magnetic one $-\mathbf{j} \cdot \mathbf{A}/c$ in (5.34). The corresponding term in the particle lagrangian (13.24) has been volume-integrated† to an expression representing the role of the interaction energy (5.53),

$$\frac{q w_\rho A_\rho}{c} = \gamma\left(q\mathbf{u} \cdot \frac{\mathbf{A}}{c} - q\phi\right) \tag{13.39b}$$

where the factor γ comes from the gauging relative to the particle's *proper* time intervals $d\tau = dt/\gamma$ that replace the units of dt used for the particle lagrangian

† The occurrence of the volume integration is clear from comparing the particle $\mathscr{L}$ in (13.27) to the field $\mathscr{L}$ of (13.34*b*).

$L \equiv \mathscr{L}/\gamma$ of (13.7) and for the field lagrangian density in (13.34*b*). Notice also that just as the particle lagrangian provides for the particle motions outside ranges of interaction through a term proportional to w_ρ^2, so the field lagrangian provides for the field redistributions, irrespective of interactions, through terms proportional to $(\partial_\mu A_\nu)^2$.

It should also be plain, in view of the vanishing of any perfect differential terms upon integration over V_4 like those made explicit in (13.35), that any space-time divergence $\partial_\mu \Omega_\mu(x)$ with any $\Omega_\mu(x)$ can be added to $\mathscr{L}$ without effect on results for the equations of motion. One thus regauged lagrangian can be given the form

$$\mathscr{L}' = -(16\pi)^{-1}F_{\mu\nu}^2 + \frac{j_\nu A_\nu}{c} \tag{13.40}$$

where $F_{\mu\nu} \equiv \partial_\mu A_\nu - \partial_\nu A_\mu$ must be taken to *define*† $F_{\mu\nu}$ in order to maintain the Maxwell constraint (12.29). The difference between $\mathscr{L}'$ and $\mathscr{L}$ can be seen from considering the double summation (over μ and ν)

$$F_{\mu\nu}^2 = (\partial_\mu A_\nu - \partial_\nu A_\mu)^2 = 2(\partial_\mu A_\nu)^2 - 2(\partial_\mu A_\nu)(\partial_\nu A_\mu)$$

where the factors 2 come from interchanging the dummy summation indices μ and ν in half the terms. Further,

$$F_{\mu\nu}^2 = 2(\partial_\mu A_\nu)^2 - 2\,\partial_\mu(A_\nu\,\partial_\nu A_\mu) + 2A_\nu\,\partial_\nu(\partial_\mu A_\mu)$$

The last term vanishes because of the Lorentz condition $\partial_\mu A_\mu = 0$, and the middle term represents a mere regauging of $\mathscr{L}$ by the gauge function $\Omega_u(x) = A_\nu \partial_\nu A_\mu/8\pi$. What remains permits replacing $(\partial_\mu A_\nu)^2$ in $\mathscr{L}$ of (13.38) with $\frac{1}{2}F_{\mu\nu}^2$ as in $\mathscr{L}'$ of (13.40). This invariance of the results for the equations of motion to the regauging $\mathscr{L} \to \mathscr{L}'$ can be checked directly, by starting from the regauged canonical conjugates to the field variables $A_\nu(x)$:

$$\begin{aligned}\frac{\partial \mathscr{L}'}{\partial(\partial_\mu A_\nu)} &= -\frac{\partial}{\partial(\partial_\mu A_\nu)}\sum_{\mu'\nu'}\frac{(\partial_{\mu'}A_{\nu'} - \partial_{\nu'}A_{\mu'})^2}{16\pi} \\ &= -\sum_{\mu'\nu'}\frac{\partial_{\mu'}A_{\nu'} - \partial_{\nu'}A_{\mu'}}{8\pi}(\delta_{\mu'\mu}\delta_{\nu'\nu} - \delta_{\mu'\nu}\delta_{\nu'\mu}) \\ &= -\frac{F_{\mu\nu} - F_{\nu\mu}}{8\pi} = \frac{F_{\nu\mu}}{4\pi}\end{aligned} \tag{13.41}$$

Substitution of this result into the Lagrange equations (13.36) with $\mathscr{L} \to \mathscr{L}'$ immediately yields the field equations of motion in the Maxwell form (12.28). Then the steps

$$\partial_\mu F_{\mu\nu} = \partial_\mu^2 A_\nu - \partial_\nu(\partial_\mu A_\mu) = \partial_\mu^2 A_\nu = -\frac{4\pi j_\nu}{c}$$

† This must be taken into account when the operations giving the Lagrange field equation (13.36) are applied to $\mathscr{L}'$ in place of $\mathscr{L}$.

complete demonstrating that the equations in the form (13.37) follow from $\mathscr{L}'$ as well as from $\mathscr{L}$.

The Lorentz-invariant double sum $F_{\mu\nu}^2$ was shown to be equivalent to $2(\mathsf{B}^2 - \mathsf{E}^2)$ in (12.27). Thus, the noninteracting, or free-field lagrangian (when $j_\nu \equiv 0$) becomes simply

$$\mathscr{L}_o = -\frac{F_{\mu\nu}^2}{16\pi} = \frac{\mathsf{E}^2 - \mathsf{B}^2}{8\pi} \tag{13.42}$$

This might be compared with the familiar $L = T - U$ of (13.4). The latter indicates that a particle's motion arises from exchanges of its kinetic and potential energies. The result (13.42) similarly indicates that redistributions of a field isolated from sources arise from exchanges between electric and magnetic energy distributions. The *integrated* energies balance to a sum that is constant in time once the field reaches an isolated equilibrium, as was shown in Sec. 7.1.

13.3 INVARIANCES AND CONSERVATION LAWS

To be considered now are profound consequences, for descriptions in space-time, of the frame invariances that such descriptions must be expected to have.

Invariance to Translations

The variation principle provides a highly unified standpoint from which to view and exhibit a fundamental connection between the identifiability of conserved energy-momenta and symmetries to be expected of any descriptions in space-time. This type of demonstration is called a *Noether theorem.*

One of the simplest expectations is that a system should be able to go through the same time development regardless of *when* it is started, given the same initial and boundary conditions in each case. A customary statement of this expectation is that the description should be invariant to translation in time. It should also be possible to displace the entire system bodily in space and have the same behavior ensue if only the same conditions (as before the displacement) are applied to the translated boundary. Such a spatial translation has already been considered in Sec. 9.1 but only as a basis for classifying plane waves. The time and space translations together have a natural expression as shifts by a constant 4-vector Δ_ρ in the four dimensions of space-time.

The translations can be equivalently discussed as leaving the system itself undisturbed but shifting the origin of the frame to which the space-time coordinates are referred

$$x_\rho \rightarrow x'_\rho = x_\rho + \Delta_\rho \tag{13.43}$$

for shifts of the origin by $-\Delta_\rho$. This is the simplest type of frame transformation, one that leaves each axis parallel to its original direction in space-time. (It belongs to the 4-parameter translation group of transformations, which has members distinguished by all possible choices for each of the four constants $\Delta_{1,2,3,4}$.) Since the system itself is being left undisturbed, it is a passive interpretation of the transformation that will be discussed here, as against the active one of the preceding paragraph.†

A *general* frame transformation $x \to x'$ (including the Lorentz rotation to be discussed in the next subsection) replaces the description of any field $A_\nu(x)$ by a correspondingly transformed description $A'_{1,2,3,4}(x')$ relative to the new frame. There will be an invariance, such as is expected here, if substituting the $A'_{1,2,3,4}(x')$ into the equations of motion (13.36) satisfies them as well as $A_\nu(x)$ does and if the same boundary conditions are satisfied, now reexpressed relative to the new frame. The same functional *form* is thus presumed for $\mathscr{L}$ in the two cases (at least within added space-time divergences, found above to have no effect on the equations of motion). The invariance can be entirely enforced by gauging the expenditures of each case in the same way and letting $I' = I$ of (13.34) when

$$I' = \int_{V'_4} d^4x' \mathscr{L}[A'_\nu(x'), \partial'_\mu A'_\nu, x'_\rho] \tag{13.44}$$

with V'_4 a symbol for the same space-time volume as V_4 but redescribed relative to the transformed frame. The requirement $I' - I = \Delta I = 0$ can then be treated like the vanishing of the variation δI in the preceding section if the transformation is taken to be an infinitesimal one, by a $\Delta_\rho \to 0$; this is sufficient since any continuous transformation can be built up of infinitesimal ones. Treating the shifts as infinitesimal means neglecting any effects proportional to Δ_ρ^2, compared with ones proportional to Δ_ρ only.

The effect $\Delta I = I' - I$ of a transformation is most readily found by changing from the integration variables in (13.44) to the ones in I of (13.34), through the substitution $x'_\rho = x_\rho + \Delta_\rho$. Since mere translation does not distort volume elements, $d^4x' = d^4x$ and the integration range V_4 can now be described in the same way for both I' and I. Moreover, with the system to be left undisturbed and with x' and x referring to the same space-time point, the invariance demands that $A'_\nu(x') = A_\nu(x)$ *for each component* on space-time axes left parallel by the translation.‡ It also requires that $\partial'_\mu A'_\nu = \partial_\mu A_\nu$, since $\partial'_\mu \equiv \partial_\mu$ for variables differing only by a constant. Thus,

$$\Delta I = \int d^4x\{\mathscr{L}[A_\nu(x), \partial_\mu A_\nu, x_\rho + \Delta_\rho] - \mathscr{L}[A_\nu(x), \partial_\mu A_\nu, x_\rho]\}$$

is the result of the translation. For an infinitesimal shift, the integrand here can be represented as

$$\left(\frac{\partial \mathscr{L}}{\partial x_\rho}\right)_A \Delta_\rho \tag{13.45}$$

† Active interpretations were employed in Secs. 9.1 and 9.2.

‡ The conventional procedure is to introduce a $\Delta A_\nu = A'_\nu(x') - A_\nu(x)$ and only after integration to concede that actually $\Delta A_\nu \equiv 0$ in translations. The procedure with a $\Delta A_\nu \neq 0$ will be found essential for the next section.

where the subscript A denotes differentiation with respect to the explicit dependence of $\mathscr{L}$ on x only and summation over ρ is to be understood. In the course of the *integration*, $A_\nu(x)$ and $\partial_\mu A_\nu(x)$ also vary, and to prepare for this

$$\partial_\rho \mathscr{L} \equiv \frac{\partial \mathscr{L}}{\partial x_\rho} = \left(\frac{\partial \mathscr{L}}{\partial x_\rho}\right)_A + \frac{\partial \mathscr{L}}{\partial A_\nu} \partial_\rho A_\nu + \frac{\partial \mathscr{L}}{\partial(\partial_\mu A_\nu)} \partial_\rho(\partial_\mu A_\nu)$$

is introduced. Because $A_\nu(x)$ is supposed to satisfy the equations of motion (13.36), the $\partial\mathscr{L}/\partial A_\nu$ here can be replaced by $\partial_\mu[\partial\mathscr{L}/\partial(\partial_\mu A_\nu)]$, with the result that

$$\frac{\partial \mathscr{L}}{\partial x_\rho} = \left(\frac{\partial \mathscr{L}}{\partial x_\rho}\right)_A + \partial_\mu \left[\frac{\partial \mathscr{L}}{\partial(\partial_\mu A_\nu)} \partial_\rho A_\nu\right] \tag{13.46}$$

Thus,
$$\Delta I = \Delta_\rho \int d^4x\, \partial_\mu \left[\mathscr{L}\delta_{\mu\rho} - \frac{\partial \mathscr{L}}{\partial(\partial_\mu A_\nu)} \partial_\rho A_\nu\right] = 0 \tag{13.47}$$

is required for the translational invariance. It is required for arbitrarily different component shifts $\Delta_{1,2,3,4}$ and however V_4 (and the system demarked by the choice for V_4) might have been chosen. Consequently,†

$$\partial_\mu \mathscr{H}_{\mu\rho} = 0 \tag{13.48a}$$

if $\mathscr{H}_{\mu\rho}$ is defined‡ by

$$\mathscr{H}_{\mu\rho} \equiv \frac{\partial \mathscr{L}}{\partial(\partial_\mu A_\nu)} \partial_\rho A_\nu - \mathscr{L}\,\delta_{\mu\rho} \tag{13.48b}$$

(with summation over ν understood), an entity sometimes called the *canonical tensor*. The vanishing of the space-time divergence in (13.48*a*) constitutes a continuity equation describing the *conservation* of something representable by $\mathscr{H}_{\mu\rho}$. It is a universal conservation law insofar as it can be expected to hold for *any* field $A_\nu(x)$ as a consequence of its translatability in homogeneous space-time whenever the field is isolated from any external effects that might have been left undescribed.

Consider the interpretation of the conserved quantity $\mathscr{H}_{\mu\rho}$ in the case of an *electromagnetic* field that is isolated from matter, derivable from the lagrangian $\mathscr{L}_o$ of (13.42). Reference to (13.41) shows that this leads to

$$\mathscr{H}_{\mu\rho} = \frac{F_{\nu\mu}\,\partial_\rho A_\nu + \frac{1}{4}\delta_{\rho\mu}F^2_{\alpha\beta}}{4\pi}$$

† As in the conclusions (6.7), (6.26), and (6.37). The type of argument employed in all these cases was first introduced in the step from the integral conservation law (A.18) to the continuity equation (A.19).

‡ The symbol $\mathscr{H}_{\mu\rho}$ is adopted here because $\mathscr{H}_{\mu 4}$ is a *hamiltonian density* [see Ref. 17, sec. 15.1, for an example applying to a one-component (scalar) field]. The other components are vector-field counterparts of the conjugate particle momentum in $p_\rho(\mathbf{p}, iH/c)$ of (13.21). The distinction between $\mathscr{H}_{\mu\rho}$ and $\partial\mathscr{L}/\partial(\partial_\mu A_\nu)$ is comparable to that between the wave and mass momenta alluded to on page 416*n*.

and the definition $F_{\rho\nu} = \partial_\rho A_\nu - \partial_\nu A_\rho = -F_{\nu\rho}$ allows a replacement of $\partial_\rho A_\nu$ that produces

$$\mathscr{H}_{\mu\rho} = \frac{1}{4\pi}(F_{\mu\nu}F_{\nu\rho} + \tfrac{1}{4}\delta_{\rho\mu}F_{\alpha\beta}^2) + \frac{F_{\nu\mu}\,\partial_\nu A_\rho}{4\pi} = T_{\mu\rho} + \frac{F_{\nu\mu}\,\partial_\nu A_\rho}{4\pi} \tag{13.49}$$

where $T_{\mu\rho}$ is just the symmetric stress-energy-momentum tensor of (12.44) and (12.46). Also consider what happens to the last term of (13.49) when the space-time divergence operation of (13.48*a*) is applied to it:

$$\partial_\mu(F_{\nu\mu}\,\partial_\nu A_\rho) = \partial_\mu\,\partial_\nu(F_{\nu\mu}A_\rho) - \partial_\mu(A_\rho\,\partial_\nu F_{\nu\mu}) \tag{13.50}$$

The last term here vanishes because of the Maxwell equation (12.28) for a field isolated from its sources ($j_\mu \equiv 0$) as here. The first term of the result (13.50) also vanishes because the antisymmetry of $F_{\mu\nu}$ makes it equivalent to

$$\partial_\mu\,\partial_\nu(F_{\nu\mu}A_\rho) \equiv \partial_\nu\,\partial_\mu(F_{\mu\nu}A_\rho) = -\partial_\nu\,\partial_\mu(F_{\nu\mu}A_\rho)$$

its own negative, and hence zero. Thus the vanishing of the space-time divergence (13.48*a*) reduces to

$$\partial_\mu T_{\mu\rho} = 0 \tag{13.51}$$

just the continuity equation (12.45) for the case of a free field (the force density f_ρ vanishes where no matter is present). The unsymmetrical tensor (13.49) is merely a regauged version of the field's energy-momenta and their currents; the space-time divergence of the difference $\mathscr{H}_{\mu\rho} - T_{\mu\rho}$ vanishes because regaugings of $\mathscr{L}_o$ by adding space-time divergences to it makes no difference to the motions for which $\mathscr{H}_{\mu\rho} - T_{\mu\rho}$ was evaluated in the step to (13.46).

It has thus been shown that conserved energy-momenta become identifiable because of the translational invariance that space-time descriptions must be expected to have. Notice that if only a shift in time ($\Delta_{1,2,3} = 0$) had been tried in producing the translating effect (13.47), only the components $T_{\mu 4}$ would have been identified. They describe the field energy and its flux, as found in (12.47*b*). Thus, there exist conserved energies because valid laws should be applicable at arbitrary times. Similarly, linear momenta are conserved because the laws should be equally applicable at different locations in a space that is homogeneous for systems isolated from external effects. Angular momenta are conserved because of an isotropy of space, as will be seen in the next subsection.†

The vanishing of the integrand in (13.47), equivalent to the conservation law (13.48*a*), implies the vanishing of the differential (13.45)

$$\left(\frac{\partial \mathscr{L}}{\partial x_\rho}\right)_A \Delta_\rho = 0 \tag{13.52}$$

for which it was substituted through the use of the relation (13.46). Thus the existence of the conservation law requires that the lagrangian contain no explicit

† See Ref. 17, sec. 7.5, for similar conclusions about particle systems.

dependence on x, that it be a functional of the field and its gradients only, as in the $\mathscr{L}_o$ just discussed. This is comparable to the need for the absence of an explicit time dependence in the $L(q, \dot{q}, t)$ of (13.11) for the hamiltonian energy (13.12) to be conserved. A conservation principle cannot be expected for the lagrangians (13.38) or (13.40), since they contain explicit x dependences due to the presence of source distributions,† $j_\nu(x)$, aside from the fields. The terms containing the $j_\nu(x)$ describe interaction energies (13.39) between matter and field, so that the field energy-momenta are no longer conserved to themselves, as described by the vanishing space-time divergence (13.51) in the absence of charged matter. The rate at which the field energy-momenta are changed by the interactions is now described by the *nonvanishing* space-time divergence found in (12.45), with forces $f_\nu \neq 0$.

The circumstance that the conservation principle (13.48*a*) applies only to lagrangians without explicit x dependence, like $\mathscr{L}_o$, calls attention to the fact that the expression for ΔI leading to (13.45) would have reduced to an uninformative identity if an explicit x dependence had not at least been tried out. The relation most essential to the translational invariance is the result (13.52). This amounts to a generalization of‡ d'Alembert's principle of virtual work, which is implemented by trying out virtual displacements like Δ_ρ. Indeed, the Noether theorems amount to elaborations of that venerable principle, rendered desirable by being based on the highly unifying variation principle.

Invariance to Rotations

Any physical system ought exhibit the same behavior no matter how it is oriented in space so long as it is screened from any external influences. Spatial reorientations are special cases of the rotations in space-time described by the general homogeneous Lorentz transformations§ $x'_\mu = a_{\mu\nu} x_\nu$ of (11.8). That there should be invariance to those was a chief burden of Chap. 12. It will now be seen how this invariance leads to the identification of conserved quantities.

Like the translations, the rotations can also be fitted into an infinitesimal transformation procedure via

$$x'_\rho = x_\rho + (a_{\rho\sigma} - \delta_{\rho\sigma})x_\sigma \rightarrow x'_\rho \equiv x_\rho + \Delta_\rho(x) \tag{13.53}$$

† Translational invariance might still be expected if the sources are translated together with the fields, so that not only $A'_\nu(x') = A_\nu(x)$ but also $j'_\nu(x') = j_\nu(x)$; but then the derivation of a conservation law would require the introduction of equations of motion for the $j_\nu(x)$, as was done for $A_\nu(x)$ in producing (13.46). The lagrangians (13.38) and (13.40) make no provision for such. They contain only the interaction energy $-j_\nu A_\nu/c$ between matter and field but no provision for energy-momenta of the matter itself, as they have for the field $A_\nu(x)$ by itself.

‡ See Ref. 17, sec. 7.1. An application is represented by (B.23) in this book.

§ Each instance is a member of the six-parameter (proper, orthochronous) rotation group in four-dimensional space-time, with the three Euler angles and the components of the relative frame velocity as the parameters, a fact already mentioned following (11.4). The joint group of space-time rotations and translations (13.43) forms the ten-parameter inhomogeneous Lorentz group of transformations $x'_\mu = a_{\mu\nu} x_\nu + \Delta_\mu$, also known as a *Poincaré group*.

with shifts now depending on the space-time point. That effects of spatial rotations depend on the distance of the transformed point from the rotation axis has already been pointed out in leading to (9.43). For a rotation in the xy plane (about the z axis and leaving the t axis untouched)

$$x' = x \cos \varphi + y \sin \varphi \qquad y' = -x \sin \varphi + y \cos \varphi \qquad z' = z \qquad t' = t$$

the only nonvanishing components of $\Delta_\rho(x)$ for $\varphi \to \delta\varphi \to 0$ follow from

$$x' \approx x + y\,\delta\varphi \qquad y' = y - x\,\delta\varphi \tag{13.54}$$

as $\Delta_1 = y\,d\varphi$ and $\Delta_2 = -x\,\delta\varphi$. For a general space-time rotation, $\Delta_\rho(x)$ can be represented with the help of a suitably defined antisymmetrical† tensor $\omega_{\rho\sigma} = -\omega_{\sigma\rho} \to 0$, as

$$\Delta_\rho(x) = \omega_{\rho\sigma} x_\sigma \tag{13.55}$$

Plainly $\omega_{12} = -\omega_{21} = \delta\varphi \to 0$ are the only nonvanishing components of the tensor $\omega_{\rho\sigma}$ in the special case of rotation in the xy plane. The special Lorentz rotation (11.6) in the (z, ict) plane becomes infinitesimal for $\beta \to 0$, $\gamma \to 1$ when $x_3' \approx x_3 + i\beta x_4$ and $x_4' \approx x_4 - i\beta x_3$. Here $\omega_{34} = -\omega_{43} = i\beta$ are the only nonvanishing components of the infinitesimal tensor. There is still a difference from a galilean transformation, despite $\beta \to 0$, because an infinitesimal time transformation is being allowed.

All 4-vectors undergo transformations like $A_\mu' = a_{\mu\nu} A_\nu$ in the space-time rotations. Whereas the invariance of the field to frame translations led to the conclusion $A_\nu'(x') = A_\nu(x)$, used in constructing the translational effect ΔI just above (13.45), what is now labeled as the ν component $A_\nu'(x')$ differs from $A_\nu(x)$ because the ν axis has a new direction. There is a nonvanishing first-order effect‡ as in (13.55)

$$\Delta A_\nu(x) \approx A_\nu'(x') - A_\nu(x) \approx \omega_{\nu\rho} A_\rho(x) \tag{13.56}$$

a consequence of an infinitesimal rotation for any 4-vector. Similarly

$$\Delta(\partial_\mu A_\nu) \approx \partial_\mu(\Delta A_\nu) + \omega_{\mu\rho}\,\partial_\rho A_\nu \tag{13.57a}$$

This follows from the transformation $\partial_\mu' = a_{\mu\rho}\,\partial_\rho$ (12.2) of the space-time gradients and the fact that $a_{\mu\rho} \approx \delta_{\mu\rho} + \omega_{\mu\rho}$ for the infinitesimal rotations [compare (13.53) and (13.55)]. Then

$$\partial_\mu' A_\nu' \approx (\delta_{\mu\rho} + \omega_{\mu\rho})\,\partial_\rho(A_\nu + \Delta A_\nu) \approx \partial_\mu A_\nu + \partial_\mu(\Delta A_\nu) + \omega_{\mu\rho}\,\partial_\rho A_\nu$$

† That this is characteristic of rotations follows from the necessity for $(x_\rho')^2$ to equal x_ρ^2:

$$(x_\rho')^2 \approx (x_\rho + \omega_{\rho\sigma} x_\sigma)^2 \approx x_\rho^2 + x_\rho(\omega_{\rho\sigma} + \omega_{\sigma\rho})x_\sigma$$

and hence $\omega_{\rho\sigma} + \omega_{\sigma\rho} = 0$.

‡ Remember that x' and x are still the same points in the undisturbed field, unlike the situation for the active rotation in (9.56). Only the *decompositions* of the field undergo change in the passive interpretation here.

giving just the result (13.57*a*) for the difference

$$\Delta(\partial_\mu A_\nu) \equiv \partial'_\mu A'_\nu(x') - \partial_\mu A_\nu(x) \tag{13.57b}$$

Notice the contrast of the result (13.57*a*) to the conclusion $\delta(\partial_\mu A_\nu) = \partial_\mu(\delta A_\nu)$ for the variations leading to (13.35); there was no frame change $x \to x'$ to consider in the earlier case.

The consequent difference $\Delta I = I' - I$ between the action integrals of (13.44) and (13.34) is best found, as in the case of the translations, by changing from the integration variables x' in I' to the same ones x as are used in I, now through the relation (13.53). The space-time volume elements $d^4x' = d^4x$ are invariant to *any* rotations, as was recognized after (12.10), and so

$$\Delta I = \int d^4x\, \Delta\mathscr{L} = 0 \tag{13.58}$$

if

$$\Delta\mathscr{L} \equiv \mathscr{L}[A_\nu + \Delta A_\nu, \partial_\mu A_\nu + \Delta(\delta_\mu A_\nu), x_\rho + \Delta_\rho(x)] - \mathscr{L}(A_\nu, \partial_\mu A_\nu, x_\rho) \tag{13.59a}$$

For infinitesimal changes, this can be expressed as

$$\Delta\mathscr{L} = \left(\frac{\partial\mathscr{L}}{\partial x_\rho}\right)_A \Delta_\rho + \frac{\partial\mathscr{L}}{\partial A_\nu}\Delta A_\nu + \frac{\partial\mathscr{L}}{\partial(\partial_\mu A_\nu)}\Delta(\partial_\mu A_\nu) \tag{13.59b}$$

replacing (13.45), which applied to translations for which $\Delta A_\nu = 0$ and $\Delta(\partial_\mu A_\nu) = 0$. As in that case, the first term here can be replaced from (13.46). As for the latter expression, the equations of motion (13.36) can be used to replace $\partial\mathscr{L}/\partial A_\nu$ of the second term with $\partial_\mu[\partial\mathscr{L}/\partial(\partial_\mu A_\nu)]$. Finally, the last factor in (13.59*b*) can be replaced by (13.57*a*), in connection with which it is profitable to notice that $\partial_\mu \Delta_\rho = \partial_\mu \omega_{\rho\sigma} x_\sigma = \omega_{\rho\sigma}\delta_{\mu\sigma} = \omega_{\rho\mu} = -\omega_{\mu\rho}$, according to (13.55). The result will be a sum of differentials which can be collected into the expression

$$\Delta\mathscr{L} = \partial_\mu\left\{\delta_{\mu\rho}\mathscr{L}\,\Delta_\rho + \frac{\partial\mathscr{L}}{\partial[\partial_\mu A_\nu]}(\Delta A_\nu - \Delta_\rho\,\partial_\rho A_\nu)\right\} \tag{13.60}$$

a space-time divergence as in the translational effect (13.47). Indeed, (13.60) properly reduces to the translation case for Δ_ρ a constant and $\Delta A_\nu \equiv 0$. As written in (13.60), the expression is actually valid for any continuous infinitesimal transformation $\Delta_\rho(x) = x'_\rho - x_\rho$, and not only for the translations and rotations.

To have the invariance $\Delta I = 0$ (13.58) to a transformation, the space-time divergence (13.60) must vanish, and it then constitutes a conservation law. In the case of the rotations $\Delta_\rho = \omega_{\rho\sigma}x_\sigma$, when $\Delta A_\nu = \omega_{\nu\sigma}A_\sigma$ (13.56), the conserved, divergenceless quantity bracketed in (13.60) can, with the help of the definition (13.48), be written as

$$\omega_{\rho\sigma}\left[-x_\sigma\mathscr{H}_{\mu\rho} + \frac{A_\sigma\,\partial\mathscr{L}}{\partial(\partial_\mu A_\rho)}\right]$$

Thus, for each of the six independently choosable constants† $\omega_{\rho\sigma} = -\omega_{\sigma\rho}$, (13.60) becomes a sum of six terms which must vanish independently. Then the separately vanishing coefficients of $\omega_{\rho\sigma} = -\omega_{\sigma\rho}$ are

$$\partial_\mu \left[-x_\sigma \mathscr{H}_{\mu\rho} + x_\rho \mathscr{H}_{\mu\sigma} + A_\sigma \frac{\partial \mathscr{L}}{\partial(\partial_\mu A_\rho)} - A_\rho \frac{\partial \mathscr{L}}{\partial(\partial_\mu A_\sigma)} \right] = 0 \tag{13.61}$$

This is a 16-component conservation law applicable to any system described by a 4-vector field $A_\nu(x)$ in a Lorentz-invariant theory.

Consider the interpretation of the conservation law (13.61) in the case of the free-field lagrangian $\mathscr{L}_o$ of (13.42) and (13.49). Plainly, the $T_{\mu\rho}$ part of $\mathscr{H}_{\mu\rho}$ in (13.49) leads to the introduction of the third-rank tensor $\Lambda_{\mu\rho\sigma}$ of (12.52) into (13.61). The remaining terms from $\mathscr{H}_{\mu\rho}$ can be combined with those following from the remainder of (13.61), after substitutions from (13.41) for the derivatives of the lagrangian, to produce

$$\partial_\mu \left\{ \Lambda_{\mu\rho\sigma} + \frac{\partial_\nu [F_{\nu\mu}(x_\rho A_\sigma - x_\sigma A_\rho)]}{4\pi} \right\} = 0$$

Interchange of the dummy summation indices μ, ν in the final, derivative terms proportional to $\partial_\mu \partial_\nu$ shows that they are their own negatives, and hence vanish, because $F_{\mu\nu} = -F_{\nu\mu}$. The conclusion is that

$$\partial_\mu \Lambda_{\mu\rho\sigma} = 0 \tag{13.62}$$

for an isolated field, just the expression (12.53) that describes the conservation of field angular momenta, together with a uniformity of the CM motion of the field.

Gauge Invariance and Charge Conservation

There is another type of transformation, not a frame change in space-time, to which physically meaningful results should be invariant. That is the gauge transformation $\bar{A}_\mu = A_\mu + \partial_\mu \chi$ (12.14) to which potential descriptions ought be invariant. Notice that the lagrangian $\mathscr{L}'$ of (13.40) for an electromagnetic field interacting with given sources is not invariant to the gauge change

$$\mathscr{L}' \to \bar{\mathscr{L}} = \mathscr{L}' + \frac{j_\nu \partial_\nu \chi}{c}$$

However, this change in the lagrangian can be expressed as

$$\Delta \mathscr{L} \equiv \bar{\mathscr{L}} - \mathscr{L}' = \frac{\partial_\nu (j_\nu \chi)}{c} - \frac{\chi \partial_\nu j_\nu}{c}$$

Thus, for conserved charges, with $\partial_\nu j_\nu = 0$, there is merely a regauging, by $\partial_\nu \Omega_\nu$ with $\Omega_\nu = j_\nu \chi / c$, of the lagrangian, like that found to have no effect on the consequent equations of motion [as in the discussion connected with $\mathscr{L}'$ (13.40)]. On this account, it is said that the charge-conservation principle is equivalent to the potential gauge invariance, in something like the same sense that the invariances to translations and rotations were found equivalent to energy- and momentum-conservation principles.

† Here serving as the six group parameters mentioned on page 425*n*.

CHAPTER

FOURTEEN

RADIATIVE MOTIONS OF A POINT CHARGE

The relativistic mass renormalization of a charged particle · Covariant reformulation of energy-momentum radiation by a charged particle · The supplementation of the Lorentz force by radiative reactions in the Lorentz-Dirac equation · An example of self-field distortions and their transfer to radiation · Preacceleration · The radiative collapse of the classical atom · The correspondence principle · Matching of energy and angular-momentum transfers in a radiation process · The origin of selection rules in angular-momentum conservation · Classical broadenings of radiated frequencies

The charged-particle equations of motion in Secs. 12.3 and 13.1, like those of Chap. 5, take into account only externally applied force fields. The fact that the particle itself is a source of field was ignored. At least the radiation field generated by the accelerated particle must be expected to modify its motion, through recoils from the radiation and energy losses to it. The self-field that continues to be attached to the particle must also be considered, and not only because of the mass renormalization made evident in (4.101), as evaluated in a $u \ll c$ approximation. There will also be temporary *distortions* of the attached field because of the retardations of its responses to accelerations, and it is just the recoveries from such distortions that generate the radiation, as will be seen. The formulation of all these effects will here be approached by first considering a joint *uniform* motion of a charged particle and its attached self-field.

14.1 THE CHARGE PLUS SELF-FIELD SYSTEM

When a point charge moves, it must carry along its Coulomb field, thus forming a charge plus self-field system in joint motion. The Coulomb field contains field energy, equivalent to mass, and is also expected to acquire field momentum when it moves with its source. These must form parts of the joint system's total energy-momentum, participating in the behavior of the entity treated as a charged particle. There has even been speculation that the *entire* rest mass of an electron, which has the least value among the known charged particles, is constituted of its electromagnetic self-field mass.

The Rest Mass of a Point Charge

In the proper frame of an unaccelerated simple point charge, its field is static, and the energy in a static Coulomb field has already been discussed on page 48, with a conclusion that the part outside any radius a has just the field energy

$W^o = q^2/2a$. Evaluating this for all the space down to a zero radius, as seems appropriate for a point charge, would result in $W^o \to \infty$, already noted earlier. To avoid such an infinite contribution to the rest mass, it seems necessary to interrupt the ideal Coulomb-field distribution at some small but finite radius a and to assume finite charge densities within it. Then $W^o = q^2/2a$ is the total electromagnetic energy only if the charge is presumed to be confined to the surface of the small sphere, which seems reasonable in view of the mutual repulsions between like charge elements. Even if the charge were homogeneously distributed within the radius a, there would only be a supplementation of the field energy by 20 percent, according to the discussion on page 48.

It is considerations like the ones just reviewed that were responsible for early expectations that a purely electromagnetic electron ($q = -e$), in which a field energy of order e^2/a_c is to account for its observed rest energy via $e^2/a_c = mc^2$, would have a radius of order

$$a_c = \frac{e^2}{mc^2} \approx 2.82 \times 10^{-13} \text{ cm} \tag{14.1}$$

ever since called the *classical radius of the electron.* Note the near coincidence of this magnitude with the radius suggested in connection with the Thomson cross section (10.29), on the basis of equating the electromagnetic reflections from a point charge to an elastic wave scattering from an impenetrable sphere. However, any observation of Thomson scattering cannot be said to *detect* such a radius, since the expectations for such observations were based on evaluations for a *pointlike* charge, carried out in a way valid only for incident waves much longer than the Compton wavelength: $\hbar/mc = (\hbar c/e^2)(e^2/mc^2) \approx 137e^2/mc^2$. No such process can have sufficient resolving power to detect any object of radius approximately equal to e^2/mc^2 only.

Plainly, observations with the greater resolution afforded by much shorter wavelengths are called for. Pertinent measurements† on very high energy electron processes, with resolving powers corresponding to wavelengths down to the order of 5×10^{-15} cm, have been performed, and without detecting the slightest deviation from a *point*-charge field behavior of the electron. This apparently indicates that any radius a at which departures from a point-charge field could

† Among the most recent were experiments in $e^{\pm}$ storage rings (see page 361*n*.) at Stanford, reported in Ref. 2, by a group of 30 investigators having surname initials ranging from A to Z! Observations on the processes $e^+ + e^- \to e^+ + e^-$ and $\mu^+ + \mu^-$, with CM energies up to 4.8 Gev, were made. The results were accurately consistent with expectations from a renormalized quantum electrodynamics based on a *point* charge. This applies to the muons as well as the electrons, both being types of particles that have no interactions stronger than the electromagnetic ones.

Inner structures within a finite radius of some 10^{-13} cm *have* been detected in the strongly interacting nucleons. Even this relatively large radius must be considered beyond the purview of classical analysis, such as is being attempted here, because quantum effects had to be taken into account in its detection.

begin has a magnitude $a < 5 \times 10^{-15}$ cm and hence a Coulomb-field energy exterior to a of $e^2/2a > 28\ mc^2$! Thus the problem of accounting for the electron's observed rest mass *appears* to become one of finding a large *negative* contribution, of a magnitude nearly equal to a large ($\gg mc^2$) positive contribution by the exterior Coulomb field. This is unlikely to be a purely electromagnetic problem, especially in view of the fact that the muon, differing from the electron in having a rest mass 207 times as great, emulates the point-charge behavior quite as far as the electron does.

Negative contributions to a rest mass can be envisaged, even on a classical basis. Even if the charge on a particle were dispersed to the surface of some small sphere, it still seems necessary to invoke the existence of some kind of nonelectromagnetic attractions† (sometimes formulated as Poincaré stresses, of unknown origin) in order to prevent a complete dispersal of the charge. Any interior field energy that might exist, besides the exterior Coulomb energy $W^o = e^2/2a$, might well be expected to have superposed on it negative potential energies generated by the nonelectromagnetic attractions. The net mass-energy that results from an equilibrium of the interior electromagnetic and nonelectromagnetic forces, together with any material rest mass that might still be held to exist inside a, could be treated as some practically imperturbable quantity $m_o c^2$ that could be negative. Then the total rest mass of the charged particle would be

$$m = m_o + \frac{W^o}{c^2} = m_o + \frac{e^2}{2ac^2} \tag{14.2}$$

A quantity like m_o is sometimes referred to as a bare mass of the particle, and adding to it the Coulomb energy contribution is said to renormalize the bare mass to the observed resultant m.

The separation (14.2) of a charged particle's rest mass, into putatively constant parts interior and exterior to some unknown radius a, is considered primarily because electromagnetic theory can yield expectations at least about the effects of the exterior part. It can then be hoped that any conclusions which may turn out to be independent of what the exact magnitude of a might be will have some range of validity, just as renormalization procedures have enabled *quantum* electrodynamics to yield new expectations in quite impressively accurate agreement with measurements, without having served to detect any deviations from point-charge behavior in the electron or muon.

† Already alluded to (page 166) in the discussions of the explosive forces in charged spheres. One known source of attraction is provided by gravitation, which exists even between the mass equivalents of purely electromagnetic energy elements (see page 368). However, it is at present not even clear that this could have a significant role. A need for finding large negative contributions to rest masses has become widespread in elementary-particle physics. The strongly interacting nucleons, mentioned in the preceding footnote, are each supposed to be constituted of quarks much heavier in rest mass than the nucleon they constitute!

The Self-Field in Uniform Motion

If the self-field energy $W^o = e^2/2a = (m - m_o)c^2$ is to be a separately constant part of a particle's constant rest mass (whatever a may be), it must contribute the energy $W = \gamma W^o$ and a momentum $(W/c^2)\mathbf{u} = \gamma W^o \boldsymbol{\beta}/c$ to the particle when it has a uniform velocity $\mathbf{u} = c\boldsymbol{\beta}$. These results of course follow from the transformation properties of any quantity of rest energy when it is viewed from a frame in which it has a constant velocity. However, a direct calculation from the self-field as it is described for a uniformly moving particle, the velocity field $\mathbf{E}_v$ and $\mathbf{B}_v = \boldsymbol{\beta} \times \mathbf{E}_v$ of (10.9) and (10.79), will throw some light on the responses of the self-field to accelerations, considered in following sections.

The evaluations required are the appropriate volume integrations that will yield the total energy-momentum in the velocity field outside a uniformly moving point charge. The earliest such evaluations (including ones by Abraham and by Lorentz!) were erroneous, showing that the problem is not a trivial one. They have been corrected several times, in papers† using a variety of approaches.

Earlier calculations of the velocity-field energy amounted to integrations of the energy density $w = (\mathsf{E}_v^2 + \mathsf{B}_v^2)/8\pi$ in the space outside the surface corresponding to a Lorentz-contracted sphere having a radius a in the proper frame of the point charge. Thus, the integration is carried out over the volume

$$\int dV = \oint d\Omega(\vartheta, \varphi) \int_{r(\vartheta)}^{\infty} dr r^2 \tag{14.3}$$

where $r(\vartheta) = a/\gamma(1 - \beta^2 \sin^2 \vartheta)^{1/2}$ describes an ellipsoidal surface like (10.86), and is indicated by the inner ellipse in Fig. 10.13b. In such terms (which require a minimum of formal presumptions), the expression (10.79) for the field distribution $\mathbf{E}_v$ can be used in the evaluation of

$$w = \frac{\mathsf{E}_v^2 + (\boldsymbol{\beta} \times \mathbf{E}_v)^2}{8\pi} = \frac{\mathsf{E}_v^2}{8\pi}(1 + \beta^2 \sin^2 \vartheta) \tag{14.4}$$

The integration of this quantity over the volume (14.3) is quite straightforward and yields an expression that has been standard as the final result in some textbooks,

$$\int dV w = \gamma \frac{e^2}{2a}(1 + \tfrac{1}{3}\beta^2) \tag{14.5}$$

instead of the expected $W = \gamma W^o = \gamma(e^2/2a)$. There also is a deviation from expectation when the momentum density, evaluated with the help of $\mathbf{E}_v = \hat{\mathbf{r}}\mathsf{E}_v$ of (10.79),

$$\mathbf{g} = \frac{\mathbf{E}_v \times \mathbf{B}_v}{4\pi c} = \frac{\boldsymbol{\beta}\mathsf{E}_v^2 - \mathbf{E}_v(\boldsymbol{\beta} \cdot \mathbf{E}_v)}{4\pi c} = \frac{\mathsf{E}_v^2}{4\pi c}(\boldsymbol{\beta} - \hat{\mathbf{r}}\beta \cos \vartheta) \tag{14.6}$$

† About the earliest is Ref. 11. A summary of approaches, including his own, is provided by Rohrlich (Ref. 26).

is integrated over the volume (14.3). The result is

$$\int dV\mathbf{g} = \frac{4}{3}\frac{\gamma(e^2/2a)\boldsymbol{\beta}}{c} \tag{14.7}$$

which is larger than the expected $\gamma(e^2/2ac^2)\mathbf{u}$ by just the factor $\frac{4}{3}$. This discrepancy remains even in the nonrelativistic limit $\beta^2 \ll 1$, when the mass-energy (14.5) does attain its expected value. [Compare (4.101).]

According to relativistic principles, *any* quantity of rest energy W^o will be attributed the value γW^o by an observer who sees it being transported with a velocity $c(1-\gamma^{-2})^{1/2}$, and so the above calculations are simply not referring to the same energy as the rest energy $W^o = e^2/2a$ in the proper volume outside a, which is in *motion* with respect to the second observer. What has been overlooked is that the integrations (14.5) and (14.7) extend over field points that are simultaneous to the second observer, whereas $W^o = e^2/2a$ refers to contributions from field points that are instead simultaneous to a first observer in the rest frame of the field volume. It is contributions of the latter type, from (field) points moving with the particle, that should properly† be taken to make up the *integrated* energy of the particle's self-field.

The corresponding revision of the calculation (14.5) is perhaps most elementary to find by taking advantage of the direct correspondences that exist between each factor used in the rest-frame evaluation

$$W^o = \int dV^o \frac{(\mathbf{E}^o)^2}{8\pi} = 4\pi \int_a^\infty dr_o r_o^2 \frac{e^2}{8\pi r_o^4} = \frac{e^2}{2a} \tag{14.8}$$

and such observables as $\mathbf{E}_v$ and $\mathbf{B}_v = \boldsymbol{\beta} \times \mathbf{E}_v$ (since $\mathbf{B}^o = 0$) in the second frame, provided by the field-transformation relations (12.25), which yield

$$(\mathbf{E}^o)^2 = \mathbf{E}_v^2 - \mathbf{B}_v^2 = \mathbf{E}_v^2 - (\boldsymbol{\beta} \times \mathbf{E}_v)^2$$

[as might have been expected from the invariance of $-\frac{1}{2}F_{\alpha\beta}^2$ in (12.27)]. When the rest-frame volume elements are now replaced by the corresponding contracted ones, $dV = dV^o/\gamma$, and the resulting expression for W^o is put into

$$W = \gamma W^o = \gamma^2 \int dV \frac{\mathbf{E}_v^2 - (\boldsymbol{\beta} \times \mathbf{E}_v)^2}{8\pi} \tag{14.9}$$

a means for calculating the $W = \gamma W^o$ in the second frame is found. The integrand here differs from (14.4) in the replacement of $1 + \beta^2 \sin^2 \vartheta$ by $\gamma^2(1 - \beta^2 \sin^2 \vartheta)$, and this is easily found to be just sufficient to replace the result (14.5) by $\gamma(e^2/2a)$.

More physical insight into the new way of evaluating integrated field energies—more general in that it also applies to a uniformly *moving* volume of

† The variational basis of the field equations, in Sec. 13.2, affirms that a field system is defined by the volume of points it occupies, and these points move with respect to relatively moving frames.

field points—comes from examining the relation of the factor multiplying each proper volume element $dV^o \equiv \gamma\, dV$ in (14.9)

$$\frac{\gamma[\mathbf{E}_v^2 - (\boldsymbol{\beta} \times \mathbf{E}_v)^2]}{8\pi} = \frac{\gamma(\mathbf{E}_v^2 - \mathbf{B}_v^2)}{8\pi}$$

to the energy density $w = (\mathbf{E}_v^2 + \mathbf{B}_v^2)/8\pi$ used in (14.4). The last expression is equivalent to

$$\gamma\left(w - \frac{\mathbf{B}_v^2}{4\pi}\right) = \gamma\left[w - \frac{(\boldsymbol{\beta} \times \mathbf{E}_v) \cdot \mathbf{B}_v}{4\pi}\right] = \gamma\left(w - \frac{\mathbf{u} \cdot \mathbf{N}}{c^2}\right) \tag{14.10}$$

Thus, quite unsurprisingly, the *flux* density $\mathbf{N} = c^2\mathbf{g}$ of the energy contributions simultaneous to the second observer play a role in evaluating an integrated sum of energies contributed by field points that are instead at rest with respect to the moving (proper) frame of the field volume.

Expression (14.10) should be recognized as a scalar product of the moving volume's uniform 4-velocities, $w_\rho(\gamma\mathbf{u}, i\gamma c)$, with components of the field's energy-momentum tensor (12.46), so that (14.9), now expressed as the fourth component of the field energy-momentum, becomes

$$P_4 = \frac{iW}{c} = \gamma \int dV \frac{w_\rho T_{\rho 4}}{c^2}$$

This can validly be treated as the fourth component of the 4-vector

$$P_\mu\left(\mathbf{P}, \frac{iW}{c}\right) = \gamma \int dV \frac{w_\rho T_{\rho\mu}}{c^2} \tag{14.11}$$

because the product $w_\rho T_{\rho\mu}$ is plainly a 4-vector, and $\gamma\, dV = dV^o$ is an *invariant* proper-volume element, given the same value by all relatively moving observers, just as such agree on the invariance (12.62) of the proper-time intervals $d\tau = dx_\rho/w_\rho$, factors in the invariant space-time volume elements $dV^o\, d\tau = (\gamma\, dV)(dt/\gamma) = d^4x/c$ (12.10). The 3-momentum components of (14.11) are to be calculated as

$$\mathbf{P} = \gamma^2 \int dV\left(\mathbf{g} + \mathbf{u} \cdot \frac{\mathbf{T}}{c^2}\right) \tag{14.12}$$

in place of the $\int dV\mathbf{g}$ in (14.7). Here, flux densities $-\mathbf{T}$ of field momentum replace the energy fluxes $\mathbf{N} = c^2\mathbf{g}$ of (14.10).

When the integrated field momentum (14.12) is evaluated for the particular case of a velocity field outside a proper radius a centered on a moving point charge, the Maxwell stress tensor (6.24) yields

$$\frac{\mathbf{u} \cdot \mathbf{T}}{c^2} = \frac{(\boldsymbol{\beta} \cdot \mathbf{E}_v)\mathbf{E}_v}{4\pi c} - \boldsymbol{\beta}\frac{w}{c}$$

since $\boldsymbol{\beta} \cdot \mathbf{B}_v = 0$ when $\mathbf{B}_v = \boldsymbol{\beta} \times \mathbf{E}_v$. Expressed in the variables used for the evaluation (14.7), the first term here just cancels the last term of the expression (14.6) for $\mathbf{g}$, and then the expression (14.4) for w leaves

$$\gamma^2\left(\mathbf{g} + \frac{\mathbf{u} \cdot \mathbf{T}}{c^2}\right) = \gamma^2 \frac{\mathbf{E}_v^2}{8\pi c} \boldsymbol{\beta}(1 - \beta^2 \sin^2 \vartheta)$$

to be integrated over the volume (14.3). This replacement for $\mathbf{g}$ of (14.6) is just sufficient to reduce to unity the factor of discrepancy, $\frac{4}{3}$, found in (14.7). The proper result $\mathbf{P} = \gamma(e^2/2ac^2)\mathbf{u}$ is thus obtained.

Notice that expression (14.11) reduces to the usual calculations in the *rest* frame of the particle, where $w_\rho(0, ic)$ leads to

$$P_\mu^o = \int dV^o \frac{icT_{4\mu}}{c^2} = \delta_{\mu 4} \frac{i}{c} \int dV^o w = \delta_{\mu 4} \frac{iW^o}{c}$$

since $T_{4k} = -icg_k = 0$, because $\mathbf{B}^o \equiv 0$ in this case and $\mathbf{P}^o = 0$.

Actually, expression (14.11) provides the proper manifestly covariant definition of the field energy-momentum 4-vector P_μ applicable to *any* field (including ones with $\mathbf{B}^o \neq 0$ in the rest frame of the field volume of interest). The expressions used heretofore

$$W(t) = \int_V dV(\mathbf{r})w(\mathbf{r}, t) \qquad \text{and} \qquad \mathbf{P} = \int_V dV\mathbf{g} \tag{14.13}$$

of (6.4) and (6.15), were derived for volumes V of field points that were presumed to be simultaneously *at rest* relative to the observer of the instantaneous (to him) distributions w and $\mathbf{g}$. In such cases $\mathbf{w} = 0$ and $w_4 = ic$ must be put into the more general definition (14.11), and then it reduces to

$$P_\mu = \frac{i}{c} \int dVT_{4\mu}$$

having the $\mathbf{P}$ and $P_4 = iW/c$ of (14.13) as its four components. A velocity $\mathbf{w} = \gamma c\boldsymbol{\beta} \neq 0$ need only be used by observers who wish to evaluate the energy-momenta (14.13) as transforms of observations in a frame relative to which the volume demarking the field has a velocity $c\boldsymbol{\beta}$ (also attributable to the field itself or to the field points to be considered simultaneous).

Expressions (14.13) were originally obtained (Chap. 6) essentially from volume integrations of the continuity equations that describe the conservation of field energies and momenta in their flows through a point. The same procedure can be modified to yield the manifestly covariant form (14.11) of P_μ, by working† in four-dimensional space-time and with the manifestly covariant form (12.45) of the continuity equations. This geometrical approach does not seem to

† As does Rohrlich (Ref. 26).

add to the clarity of the physical description. The above development of (14.11) was restricted to the special case of $\mathbf{B}^o = 0$ because the problem considered seems to be about the only application of the generalization from (14.13) to (14.11) that has ever been found necessary. Integrated field energy-momenta are almost always best calculated in the rest frame of the volume that demarks the field of interest, as in (14.13), and for the Coulomb contribution to the point-charge rest mass.

The developments here have implications for the distributions of the self-field energy-momenta of a particle that are instantaneous in a frame relative to which the particle moves. Expressions (14.10) and (14.12) most explicitly show that they are in flux about the moving particle's center—quite understandably, since the self-field must redistribute itself as it keeps adjusting itself to centrality about a point that keeps moving. During a *uniform* motion, the integrated energy and momentum are each constant, at the values $W = \gamma W^o$ and $\mathbf{P} = (W/c^2)\mathbf{u}$ calculated above, but their distributions continue in flux. These fluxes also exist at the surface of any imperturbable interior the particle may have, within the radius a discussed above, and may be regarded as net reflections off that surface.

Particularly the term proportional to $\mathbf{u} \cdot \mathbf{T}$ in (14.12) shows that stresses parallel to the particle's velocity (see Sec. 6.2) are being propagated into the self-field, adjusting it to new positions of the particle. They can be considered to be originating at the Lorentz-contracted sphere of proper radius a, showing that interactions between the charge and its attached self-field continue even during uniform motion; these are elastic-reflection interactions since the total energy-momentum of the self-field remains constant.

All these phenomena disappear when viewed in the proper frame of the particle, where $\mathbf{u} = 0$. Over field points that are simultaneous in the proper frame the energy distribution is static; the Coulomb field there is in static rather than dynamic equilibrium with the charge on the proper-time scale. However, for the accelerated motions to be considered in the next sections, attention cannot be restricted more than instantaneously to a proper frame, and no time is afforded to establish static equilibria. Then it will be essential to take into account the interactions between a charge and its instantaneous self-field in flux.

14.2 COVARIANT FORMULATIONS OF POINT-CHARGE RADIATIONS

The discussion of the charge plus self-field in the preceding section was appropriate to a uniformly moving system. Finding its responses to acceleration will require formulating the energy-momenta that are then imparted to radiation. For maintaining a consistency with the relativity principle, it will help to recast the findings about radiation rates, in Chap. 10, into manifestly invariant forms, the objective of this section.

It is simplest to start from the description of the radiation rate in an instantaneous rest frame of the particle, where its velocity is zero and so the simple

Larmor formula (10.18) becomes exact

$$\mathscr{P} = \frac{2e^2}{3c^3}\left(\frac{d\mathbf{u}^o}{d\tau}\right)^2 \tag{14.14}$$

Symbols for the velocity and time appropriate to the proper frame of the particle are used here. Note that the derivation of (10.18) was not based on the more general expression (10.15), which has not yet been derived as part of the text (but see Exercises 10.5 and 11.9); (10.18) was instead seen to follow from the expressions (10.16) and (10.17) for the acceleration fields and the angular distribution with $\mathbf{u} = 0$ but $\dot{\mathbf{u}} \neq 0$.

In considering the radiation rate relative to an arbitrary frame, it should be recognized that rates of energy loss are *invariants*

$$\mathscr{P} = -\left(\frac{dE^o}{d\tau}\right)_{\text{rad}} = -\left(\frac{dE}{dt}\right)_{\text{rad}} \tag{14.15}$$

because both dE^o and $d\tau$ are dilated by the same factor when observed with respect to a frame in which the radiating particle has velocity, making $dE = \gamma\, dE^o$ and $dt = \gamma\, dt^o \equiv \gamma\, d\tau$. Thus all that is needed to generalize expression (14.14) to an arbitrary frame is to reexpress the rest-frame accelerations in terms of corresponding accelerations as observed in the new frame (see Exercise 11.9). It is quite obvious that these acceleration transforms must be implicit in the relations between the proper-time derivatives of the 4-vector world-velocity components $w_\rho(\gamma\mathbf{u}, ic\gamma)$

$$\frac{dw_\rho^o}{d\tau}\left(\frac{d\mathbf{u}^o}{d\tau}, ic\frac{d\gamma^o}{d\tau}\right) = a_{\rho\sigma}\frac{dw_\sigma}{d\tau} = a_{\rho\sigma}\left(\gamma\frac{dw_\sigma}{dt}\right) \tag{14.16}$$

having the 4-vector transformation properties because $d\tau = dt/\gamma$ is an invariant. In any frame

$$\frac{d\gamma}{d\tau} = \frac{d}{d\tau}(1-\beta^2)^{-1/2} = \gamma^3\boldsymbol{\beta}\cdot\frac{d\boldsymbol{\beta}}{d\tau} \tag{14.17}$$

and, since $\boldsymbol{\beta}^o = \mathbf{u}^o/c = 0$ in the proper frame, $d\gamma^o/d\tau = 0$ there [already seen, for $\gamma^o mc^2$, in (12.60)]. Thus $dw_\rho^o/d\tau$ has no fourth component, and so $(d\mathbf{u}^o/d\tau)^2 = (dw_\rho^o/d\tau)^2$. Since the square of any 4-vector is invariant, the Larmor expression (14.14) becomes simply

$$\mathscr{P} = \frac{2e^2}{3c^3}\left(\frac{dw_\rho}{d\tau}\right)^2 \tag{14.18}$$

as the radiation rate in any frame.

The generalized Larmor form (14.18) becomes the more explicit Lienard one (10.15) when it is expressed in terms of the 3-vector acceleration $d\mathbf{u}/dt = c\dot{\boldsymbol{\beta}} = c\, d\boldsymbol{\beta}/(\gamma\, d\tau)$. Decomposing the 4-vector velocity in (14.18) yields

$$\mathscr{P} = \frac{2}{3}\frac{e^2}{c}\gamma^2\left[\left(\frac{d}{dt}\gamma\boldsymbol{\beta}\right)^2 - \left(\frac{d\gamma}{dt}\right)^2\right] \tag{14.19}$$

Since

$$\frac{d}{dt}\gamma\boldsymbol{\beta} = \gamma\dot{\boldsymbol{\beta}} + \boldsymbol{\beta}\gamma^3(\boldsymbol{\beta}\cdot\dot{\boldsymbol{\beta}}) \tag{14.20}$$

by (14.17), squaring it can be found to lead to

$$\mathscr{P} = \frac{2}{3}\frac{e^2}{c}\gamma^6\{\dot{\beta}^2 - \boldsymbol{\beta}\cdot[\boldsymbol{\beta}\dot{\beta}^2 - \dot{\boldsymbol{\beta}}(\boldsymbol{\beta}\cdot\dot{\boldsymbol{\beta}})]\}$$

The square bracket here is equivalent to the triple vector product $\dot{\boldsymbol{\beta}} \times (\boldsymbol{\beta} \times \dot{\boldsymbol{\beta}})$ and multiplying this by $\boldsymbol{\beta}$ gives $(\boldsymbol{\beta} \times \dot{\boldsymbol{\beta}})^2$, exactly as in (10.15). This is regarded as a more elegant way to obtain the result (10.15) than the brute-force integration of the angular distribution (10.14), as in Exercise 10.5.

The radiation rate (14.18) can also be expressed in terms of the particle's 4-momentum, $p_\rho = mw_\rho(\gamma m\mathbf{u}, i\gamma mc)$. Then the decomposition (14.19) is replaceable by

$$\mathscr{P} = \frac{2}{3}\frac{e^2}{m^2c^3}\gamma^2\left[\left(\frac{d\mathbf{p}}{dt}\right)^2 - \frac{1}{c^2}\left(\frac{dE}{dt}\right)^2\right] \tag{14.21}$$

Now $E^2 = c^2p^2 + m^2c^4$ [Eq. (11.46)], and $E\,dE = c^2\mathbf{p}\cdot d\mathbf{p} = cE\boldsymbol{\beta}\cdot d\mathbf{p}$, since $\boldsymbol{\beta} = c\mathbf{p}/E$ by (11.43). Then introducing the angle χ between the instantaneous velocity and acceleration (parallel to $d\mathbf{p}$) directions, as in the rate expression (10.31), yields

$$\mathscr{P} = \frac{2}{3}\frac{e^2}{m^2c^3}\gamma^2\left[\left(\frac{d\mathbf{p}}{dt}\right)^2 - \beta^2\cos^2\chi\left(\frac{dp}{dt}\right)^2\right]$$

For accelerations parallel to the velocity, $\chi = 0$ and

$$\mathscr{P}_\parallel = \frac{2}{3}\frac{e^2}{m^2c^3}\left(\frac{dp}{dt}\right)^2 = \frac{2}{3}\frac{e^2}{m^2c^5}\left(\frac{dE}{\beta\,dt}\right)^2 \tag{14.22a}$$

whereas for an instant at which the acceleration is perpendicular to the velocity $(\chi = \frac{1}{2}\pi)$,

$$\mathscr{P}_\perp = \frac{2}{3}\frac{e^2}{m^2c^3}\gamma^2\left(\frac{d\mathbf{p}}{dt}\right)^2 \tag{14.22b}$$

larger by a factor γ^2 ($\to \infty$ as $u \to c$) when equal *forces* proportional to $d\mathbf{p}/dt$ are applied in each case. This contrasts with the relative rates for equal *accelerations* (velocity, rather than momentum, changes), pointed out in connection with (10.31), where it was the parallel accelerations that provided the more intense radiation by the factor γ^2. A force transverse to a high velocity can produce much more acceleration than the same force directed parallel to that velocity because the transverse acceleration need not change a velocity *magnitude* that is already close to the limit c.

The contrasting radiation rates (14.22) from transverse and parallel momentum changes account for the importance of radiation losses in the circular-orbit devices for accelerating charged particles, compared with their complete negligibility in the linear accelerators. In both types, periodic energy boosts ΔE are provided by momentary potential differences applied to sections of particle trajectory.

For the linear accelerators (14.22*a*) gives

$$\frac{\mathscr{P}_{\parallel}}{\Delta E/\Delta t} = \frac{2}{3\beta}\frac{\Delta(E/mc^2)}{\Delta[x/(e^2/mc^2)]} \tag{14.23}$$

where $\Delta x = c\beta\,\Delta t$ is a length across which an energy boost ΔE is supplied. Before the radiation loss rate $\mathscr{P}_{\parallel}$ could approach the boost rate $\Delta E/\Delta t$, the latter would have to supply $\Delta E \approx mc^2 \approx \frac{1}{2}$ Mev in a distance $\Delta x \approx e^2/mc^2 \approx 2.8 \times 10^{-13}$ cm, equal to the unobservably tiny classical radius (14.1) of one electron! Actually, potential differences providing as much as $\Delta E \approx mc^2$ over a whole centimeter are difficult to achieve in practice, and at this practical extreme the fractional loss rate (14.23) is only about 2×10^{-13}, completely negligible.

In a synchrotron like the one considered on page 309 the momentum is given some revolution rate $|d\mathbf{p}/dt| = \omega_o(\gamma mc\beta) = \gamma mc^2\beta^2/R$ on some radius R, and (14.22*b*) then yields a radiation loss rate per revolution

$$-(\Delta E)_{\text{rad}} = \frac{2\pi}{\omega_o}\mathscr{P}_{\perp} = \frac{4\pi}{3}\frac{e^2}{R}\beta^3\gamma^4 \tag{14.24a}$$

In terms of the particle energy $E = \gamma mc^2$, this becomes

$$-\frac{(\Delta E)_{\text{rad}}}{E} = \frac{4\pi}{3}\frac{e^2/mc^2}{R}\left[\left(\frac{E}{mc^2}\right)^2 - 1\right]^{3/2}$$

$$\rightarrow \frac{4\pi}{3}\frac{e^2/mc^2}{R}\left(\frac{E}{mc^2}\right)^3 \qquad \text{for } E \gg mc^2 \tag{14.24b}$$

a result found by Schwinger (Ref. 28; see also Exercise 10.10). For the 100-Mev machine ($R \approx 1$ m) cited on page 309, thc loss amounts to about 10eV per revolution but increases as a *fourth* power of the energy $E = \gamma mc^2$. This fact seems to have made it impossible, with the energy boosts that are technically available, to use circular machines for accelerating electrons to more than 5 Gev or so when $\gamma \rightarrow 10^4$ and the radiation-loss rates approach 6 Mev per revolution on a 10-m radius. On the other hand, the relatively great intensity of the synchrotron radiations has made them useful laboratory sources of well-polarized radiations.

Expression in an arbitrary frame of the *momentum* being carried off by the radiation will also be important to formulating effects of radiative reactions on the motions of a charge. In the proper frame of the particle, *no* net momentum†

† This does not mean that a photon with a nonzero momentum in the proper frame cannot be radiated. In the quantum-mechanical interpretation, the angular distribution of the radiation yields the relative *probabilities* for a plane-wave photon to obtain various of the momentum directions insofar as the Maxwell equations constitute a relativistic wave mechanics of photons (see page 276*n*.).

is radiated, as is evident from the symmetry of the Larmor radiation pattern (10.17). Thus, the 4-vector energy-momentum in the proper frame has the radiation-rate components

$$\left(\frac{d\mathbf{p}^o}{d\tau}\right)_{\text{rad}} = 0 \qquad \left(\frac{dp_4^o}{d\tau}\right)_{\text{rad}} \equiv \frac{i}{c}\left(\frac{dE^o}{d\tau}\right)_{\text{rad}} = -\frac{i}{c}\mathscr{P} \tag{14.25a}$$

according to (14.15). Since proper-time intervals are invariants, the general rate $dp_\mu/d\tau$ per unit of proper time will merely be a 4-vector transform (boost) of (14.25*a*). The transformation is simple because $\mathscr{P}$ is also invariant, and then the Lorentz-transformation coefficients yield

$$\left(\frac{d\mathbf{p}}{d\tau}\right)_{\text{rad}} = -\frac{\gamma\boldsymbol{\beta}\mathscr{P}}{c} \qquad \left(\frac{dE}{d\tau}\right)_{\text{rad}} = \gamma\left(\frac{dE}{dt}\right)_{\text{rad}} = -\gamma\mathscr{P} \tag{14.25b}$$

[compare (12.8), for example] for a frame in which the particle has the instantaneous velocity $\mathbf{u}(t_q) = c\boldsymbol{\beta}(t_q)$. Note the consistency of the transformed energy rate with the invariance of $\mathscr{P}$ in (14.15). The expression†

$$-\left(\frac{d\mathbf{p}}{dt}\right)_{\text{rad}} = \frac{\mathscr{P}(t_q)}{c^2}\mathbf{u}(t_q) \tag{14.26}$$

for the instantaneous rate of field-momentum emission conforms with the notion that each *increment* of radiation field is immediately dissociated (freed) from the charge, i.e., isolated from further interaction with it, yet has the position of the particle as its mass center (12.56) and the particle velocity $\mathbf{u}(t_q)$ as its CM velocity.

A manifestly covariant expression of the energy-momentum being radiated, with the components (14.25*b*), can be obtained by introducing the 4-velocity $w_\mu(\gamma\mathbf{u}, ic\gamma)$ into it

$$-\left(\frac{dp_\mu}{d\tau}\right)_{\text{rad}} = \frac{\mathscr{P}}{c^2}w_\mu = \frac{2}{3}\frac{e^2}{c^5}\left(\frac{dw_\rho}{d\tau}\right)^2 w_\mu \tag{14.27}$$

the last after insertion of (14.18) for the invariant $\mathscr{P}$.

14.3 THE NONRELATIVISTIC LORENTZ-ABRAHAM EQUATION

The equations for accelerated motions of point charges must be modified if the reactions on the motions from their radiations are to be taken into account. The effects should be most conspicuous for electrons, which have far the largest charge-to-mass ratios among the known particles, and the radiation rates (14.22) are proportional to the squares $(e/m)^2$ of those ratios. Yet even for electrons,

† The same result is obtained in Exercise 10.6, from an integration over all directions of the field-momentum flux densities that are being radiated.

direct observation of the expected gradual deviations of their trajectories seems never to have been managed, the effects being too small in practical situations to be unambiguously identifiable. More indirect consequences can be said to have been verified, in broadenings of radiated spectra like those mentioned on page 310. Situations in which the radiative corrections are expected to be really appreciable tend to involve high energies and the intervention of quantum phenomena (like the replacement of some radiated photons by electron-positron pairs). Even for these, a proper classical treatment can have at least a conceptual importance.

A natural way to approach a satisfactorily complete equation of motion for a point charge is to start with a finite-density distribution of charge confined to a small radius a, as was done for the evaluation of finite self-energies in Sec. 2.4, before going to the limit $a \to 0$. For the extended-charge distribution, a sufficiently complete consideration of the electromagnetic forces requires taking into account not only the Lorentz forces contributed by externally applied fields, as already done in the development from (12.59), but also those arising from the retarded fields generated by every charge element at the site of every other one. Then the requisite integration over the Lorentz-force densities can be made tractable, at least for nonrelativistic velocities, by taking advantage of the fact that the interest is only in separation distances within a vanishingly small charge radius, $a \to 0$, to use a Taylor expansion into powers of the small retardation distances $c|t_s - t_s'| < a$ between elements. The result obtained both by Lorentz and by Abraham is

$$m_o \dot{\mathbf{u}} = \mathbf{F} - \frac{4}{3}\frac{e^2}{2ac^2}\dot{\mathbf{u}} + \frac{2}{3}\frac{e^2}{c^3}\ddot{\mathbf{u}} \tag{14.28}$$

where $\mathbf{F}$ is the applied force, and after terms proportional to positive powers of a are omitted because they vanish as $a \to 0$. More details of the derivation (see, for example, Ref. 13) will not be reviewed here because it has produced an ostensibly erroneous result and, moreover, the outcome can also be understood on more instructive grounds.

The middle term on the right of (14.28) has already been obtained for (4.101) as a result of nonrelativistic ($u \ll c$) considerations, and discussed in terms of a renormalization of the bare mass m_o by the addition of a self-field mass $4W^o/3c^2$. The latter supplementation is not the correct one of (14.2), being wrong by just the factor $\frac{4}{3}$ encountered in the erroneous evaluation (14.7) of the self-field momentum. According to the discussion on page 436, confining attention to self-field effects *inside* the charge, as done by Abraham and Lorentz, fails to take into proper account reactions on it from reflections of the outer self-field off the charge surface. However, it may still be held that the result (14.28) is faulty only in its representation of the mass renormalization from m_o to the observed mass m and that the version

$$m\dot{\mathbf{u}} = \mathbf{F} + \frac{2}{3}\frac{e^2}{c^3}\ddot{\mathbf{u}} \tag{14.29}$$

of it is a correct nonrelativistic relation, particularly since the term proportional to $\ddot{\mathbf{u}}$ was found to be independent of the charge radius $a \to 0$ responsible for the separation into inner and outer self-fields and the need for renormalization.

Supporting the conclusion (14.29) is the identifiability of the force term $(2e^2/3c^3)\ddot{\mathbf{u}}$ as just the expected reaction on the charge from radiation it must emit upon being accelerated, evaluated nonrelativistically (for $u \ll c$). A finite amount of radiation, emitted at the nonrelativistic Larmor rate $(2e^2/3c^3)\dot{u}^2$, must be considered and hence some finite time $t_2 - t_1$ during which the applied force $\mathbf{F}$ is producing accelerations $\dot{\mathbf{u}}(t_1 < t < t_2) \neq 0$ that keep the particle within a nonrelativisitic range of velocities $u \ll c$. Low-velocity radiation imparts no appreciable recoil momentum to the particle, as already noted in leading to (14.25a) and as in the Thomson vs. Compton scattering, yet the particle is losing energy to the radiation. The loss rate can be simulated as being due to a resistive force $\mathbf{F}_r$ that works on the particle at a rate $\mathbf{F}_r \cdot \mathbf{u}$ such that

$$\int_{t_1}^{t_2} dt \mathbf{u} \cdot \mathbf{F}_r = -\frac{2}{3}\frac{e^2}{c^3}\int_{t_1}^{t_2} \dot{\mathbf{u}} \cdot \dot{\mathbf{u}}\, dt$$

This can be partially integrated to

$$-\frac{2}{3}\frac{e^2}{c^3}\dot{\mathbf{u}} \cdot \mathbf{u}\Big|_{t_1}^{t_2} + \int_{t_1}^{t_2} dt\mathbf{u} \cdot \left(\frac{2}{3}\frac{e^2}{c^3}\ddot{\mathbf{u}}\right) \tag{14.30}$$

and so the radiative effects can be represented as in (14.29) if only interest is confined to periods of radiation (each forming one photon, say) with end times at which $\dot{\mathbf{u}}(t_{1,2}) = 0$, or just such a transition between initial and final stationary states of motion, characterized by *definite* momenta $\mathbf{p}_{1,2} = m\mathbf{u}(t_{1,2})$, as were considered for the photon emissions of Chap. 11. This feature of the representability of radiative effects as in (14.29) will have an important bearing on the admissibility of various possible explicit solutions of it [see (14.46)].

Dividing the expression (14.29) by the observable rest mass m results in the formation of a parameter

$$\tau_c = \frac{2}{3}\frac{e^2}{mc^3} \approx 6.3 \times 10^{-24} \text{ s} \tag{14.31}$$

with the dimensions of a time and makes it possible to reexpress (14.29) as

$$m(\dot{\mathbf{u}} - \tau_c \ddot{\mathbf{u}}) = \mathbf{F} \tag{14.32}$$

The constant τ_c can be recognized as just the very small time required for any signal of velocity c to cross a distance having the order of magnitude of the tiny classical electron radius a_c of (14.1). This makes it easy to see that the radiative reaction $m\tau_c \ddot{\mathbf{u}}$ is almost always negligible, as asserted in connection with (4.101), during accelerations by a force $\mathbf{F}$ that is at all amenable to classical analysis. A force producing acceleration variations violent enough to make $\tau_c \ddot{\mathbf{u}}$ comparable to $\dot{\mathbf{u}}$ would have to contain Fourier components with frequencies of the order $1/\tau_c$. Such are provided by γ-rays of energy $\hbar/\tau_c = \frac{3}{2}(\hbar c/e^2)mc^2 \approx 100$ Mev, and a quantum-mechanical pair production, in which some photons are replaced by $e^+ + e^-$ pairs, sets in at the much less violent energy $\gtrsim 2mc^2 \approx 1$ Mev.

Radiative effects should be expected to be most appreciable on particles with relativistic speeds $u \to c$ in view of their greatly enhanced radiation rates (Sec. 10.4). It therefore becomes important to generalize the nonrelativistic result (14.29) for applicability to high speeds. Moreover, a Lorentz-covariant formulation of the equation of motion is needed to enforce consistency with einsteinian relativity.

14.4 THE RELATIVISTIC LORENTZ-DIRAC EQUATION

If, as supposed in the discussion of (14.28), the faulty mass renormalization it contains is due to the neglect of reflections of outer self-field from the moving-charge surface, attention should be turned to the field-momentum fluxes outside the charge. Indeed, *all* the force on the field and matter within the charge surface should be obtainable from the transmission of field momenta to it, as in the much simpler static-field examples of Sec. 6.2. This approach is essentially one used by Dirac (Ref. 9) in deriving a properly Lorentz-covariant result [to be presented in (14.42)] now known as the *Lorentz-Dirac equation* because it reduces to (14.32) for nonrelativistic velocities. It is widely accepted to be as correct a classical formulation of radiative effects on point-charge motions as is feasible.

The approach is actually beset with ambiguities. Any attempt to restrict the description of the accelerated self-field to the *point*-charge solution, (10.9) plus (10.10), leads to infinite results unless a charge radius a is used to exclude the singularities in that solution. This introduction of a surface for an imperturbable interior of *accelerated* charge requires admitting some kind of boundary conditions that inevitably alter the field with reflections, and it is not at all clear what those conditions should be. Dirac summarily adopted an altered field description in which the singularities are *canceled*, by symmetrizing the point-charge solution; he replaced half the *retarded* field, (10.9) plus (10.10), with a corresponding *advanced* solution, of the type discussed and discarded on page 211. The consequent, nominally acausal effects are difficult to accept literally, but their results may still be acceptable, at least as a renormalization artifice.

Here it will be seen that the Lorentz-Dirac equation can also be derived without introducing advanced fields, although in a much more superficial way, forgoing any attempt to follow the redistributions of the self-field contributions to the particle's rest mass that are caused by the retarded responses to accelerations. The procedure deals directly with the rest mass presumed already renormalized (to its observed value m) then finds the reactions from the attached self-field that are *needed* in order to maintain that value against a mismatch that exists between the energy-to-momentum relation in the losses by a particle having rest mass and the corresponding instantaneous gains by radiation without rest mass.

Rest-Mass-Preserving Interactions

The electromagnetic forces on a charged particle arising only from an externally applied field are represented by the Minkowski 4-vector K_μ of (12.67). This has the readily discerned property (Exercise 12.9)

$$\sum_\mu K_\mu w_\mu = \frac{q w_\mu F_{\mu\nu} w_\nu}{c} = \frac{q w_\mu (-F_{\nu\mu}) w_\nu}{c} = 0 \tag{14.33}$$

following from the antisymmetry of the field tensor. The expression $K_\mu w_\mu \, d\tau = K_\mu \, dx_\mu$ is plainly associated with the work elements on the space-time trajectory (world line) of a particle, being traversed during proper-time intervals $d\tau$. When K_μ is decomposed as in (12.64), into terms making the Lorentz force $\mathbf{F}_q$ explicit,

$$K_\mu \, dx_\mu = \gamma[\mathbf{F}_q \cdot d\mathbf{r} - (\mathbf{u} \cdot \mathbf{F}_q) \, dt] = 0 \tag{14.34}$$

can be seen to be a consequence of the fact that $d\mathbf{r} = \mathbf{u} \, dt$ characterizes a moving point. The property of K_μ emphasized here was already implicit in (12.41).

Now consider any $K_\mu = dp_\mu/d\tau$ [Eq. (12.68)] as it imparts energy-momentum to a particle having a given rest mass m, so that $p_\mu = m w_\mu$. With the invariant $w_\mu^2 = (p_\mu/m)^2 = -c^2$, following from the decomposition (12.66) of w_μ or the energy-momentum invariance (11.37), it becomes evident that

$$w_\mu K_\mu = w_\mu \frac{dp_\mu}{d\tau} = \frac{d}{d\tau}(\tfrac{1}{2} m w_\mu^2) = -\tfrac{1}{2} c^2 \frac{dm}{d\tau} \tag{14.35}$$

Thus $w_\mu K_\mu = 0$ must characterize an interaction that does not change† the rest mass of the particle. Such interactions can only change the particle's momentum $\mathbf{p} = \gamma m \mathbf{u}$ and its instantaneous kinetic energy T in proportions consistent with the invariance of $m^2 c^2 = -p_\mu^2 = (T + mc^2)^2/c^2 - \mathbf{p}^2$.

During Losses to Radiation

While externally applied forces K_μ, accelerating a charged particle, may impart energy and momentum to it or deplete them in the right proportions to leave its rest mass unchanged, as represented by $w_\mu \, dp_\mu/d\tau = 0$ in (14.35), the depletions by radiation alone do not. This is apparent from the fact that the radiation rates (14.27) lead to

$$w_\mu \left(\frac{dp_\mu}{d\tau}\right)_{\text{rad}} = -\frac{\mathscr{P}}{c^2} w_\mu^2 = +\mathscr{P} \neq 0 \tag{14.36}$$

for the reaction $(dp_\mu/d\tau)_{\text{rad}}$ on the particle plus self-field system as it frees the permanently radiated energy-momenta—obviously a consequence of the fact

† Interactions that do change rest masses were implicit in the inelastic particle-intertransforming processes treated in Chap. 11.

that an electron can only *give up* energy and momentum in proportions consistent with $p_\mu^2 = -m^2c^2 \neq 0$, whereas radiation *takes up* proportions characteristic of a zero rest mass. Since electrons do emerge from radiation processes with their total rest masses intact, there must be some extra reactions, from the temporary distortions of the attached self-field, that cancel out the rest-mass-changing effects (14.36). If the extra reactions are represented by a Minkowski 4-vector κ_μ, it is $w_\mu \kappa_\mu = -\mathscr{P}$ that is needed. The instantaneous balance of energy-momentum changes (per unit of proper time)

$$\frac{dp_\mu}{d\tau} = K_\mu - \frac{\mathscr{P}}{c^2} w_\mu + \kappa_\mu \tag{14.37}$$

will have the rest-mass-preserving property $w_\mu \, dp_\mu/d\tau = 0$, when $w_\mu K_\mu = 0$ if only κ_μ is such that

$$w_\mu \kappa_\mu = -\mathscr{P} = -\frac{2}{3}\frac{e^2}{c^3}\dot{w}_\mu^2 \tag{14.38}$$

The last equality expresses the radiation rate (14.18) in a now convenient notation with $\dot{w}_\mu \equiv dw_\mu/d\tau$, dots on the 4-vector velocity standing for rate of change per unit of *proper* time.

The Minkowski 4-vector κ_μ needed to represent the extra force of reaction can be identified from considering a work element† by the latter, as this is implicit [compare (14.34)] in

$$\kappa_\mu \, dx_\mu = \kappa_\mu w_\mu \, d\tau = -\frac{2e^2}{3c^3}\dot{w}_\mu^2 \, d\tau \tag{14.39}$$

Now the consequences of the invariance $w_\mu^2 = -c^2$

$$w_\mu \dot{w}_\mu = 0 \qquad \text{and} \qquad w_\mu \ddot{w}_\mu = -\dot{w}_\mu \dot{w}_\mu \tag{14.40}$$

can be used to transform (14.39) into

$$\kappa_\mu \, dx_\mu = +\frac{2e^2}{3c^3}\ddot{w}_\mu w_\mu \, d\tau = \frac{2e^2}{3c^3}\ddot{w}_\mu \, dx_\mu$$

Thus

$$\kappa_\mu = \frac{2e^2}{3c^3}\ddot{w}_\mu \tag{14.41}$$

and the balance (14.37) becomes

$$m\dot{w}_\mu = K_\mu + \frac{2}{3}\frac{e^2}{c^3}\left[\ddot{w}_\mu - w_\mu\left(\frac{\dot{w}_\rho}{c}\right)^2\right] \tag{14.42}$$

This manifestly Lorentz-covariant result is just the Lorentz-Dirac equation.

† A procedure familiar as an application of d'Alembert's principle. A simpler example of its use is shown in (B.23).

The Lorentz-Dirac equation as expressed in (14.42) owes the relative simplicity of that form to the parametrization by the proper-time variable τ. The form most directly yields descriptions of particle motion by $x_\mu(\tau)$ with $w_\mu \equiv dx_\mu/d\tau$, trajectories in the four dimensions of space-time. The directly observable trajectories $\mathbf{r}(t)$ in 3-space are then obtainable by eliminating τ from the parametrized description $\mathbf{r}(\tau)$ and $t = x_4(\tau)/ic$. Equation (14.42) can itself be reexpressed in terms of the more directly observable velocities $c\boldsymbol{\beta}(t) = d\mathbf{r}/dt$, through the use of such relations as (14.17) and (14.20), but the results become inconveniently lengthy. The reexpressions do provide the most secure way of showing that the 3-space components of the Lorentz-Dirac equation reduce to (14.29) for low velocities, but this may already be sufficiently evident from inspecting (14.42).

The Instantaneous Energy of the Charge Plus Self-Field System

Especially interesting is the balance of energy changes described by the fourth component of the Lorentz-Dirac equation (14.42), easily shown to *follow* from its other three components, just as (12.65*b*) follows from (12.65*a*). Using $K_4 = i\gamma\mathbf{u} \cdot \mathbf{F}/c = i\mathbf{w} \cdot \mathbf{F}/c$ of (12.64) and $p_4 = mw_4 = i\gamma mc$ in (14.42) leads to

$$\frac{d}{d\tau}(\gamma - \tau_c\dot{\gamma})mc^2 = \mathbf{F} \cdot \mathbf{w} - \gamma\mathscr{P} \tag{14.43}$$

This indicates that it is an energy

$$E(\tau) = (\gamma - \tau_c\dot{\gamma})mc^2 \tag{14.44}$$

with $\dot{\gamma} \equiv d\gamma/d\tau$ a rate per unit of proper time, that should be considered the instantaneous energy of the charge plus self-field system, i.e., the quantity which is augmented when the applied force does positive work on the charge at the rate $\mathbf{F} \cdot \mathbf{w} = \mathbf{F} \cdot d\mathbf{r}/d\tau$ and which is depleted by losses to radiation, freed at the rate $\gamma\mathscr{P}$ per unit of proper time (14.25*b*). The instantaneous energy $E(\tau)$ reduces to γmc^2, the quantity ascribed to any system when *all* its rest mass is moving with the same velocity, only during phases of steady motion ($\dot{\gamma} = 0$). At instants of positive acceleration, parallel to the velocity, $\dot{\gamma} > 0$, and a *lesser* mass energy is instantaneously involved. This can be understood as a result of the fact that parts of a rest mass spread out over a self-field cannot acquire the accelerated velocity of the charge until all of it is reached by signals, i.e., disturbances in the instantaneous self-field, with the finite velocity c. Only after accelerations cease ($\dot{\gamma} = 0$) and the self-field regains equilibrium with the charge, can the energy $E(\tau)$ again be described as just γmc^2.

14.5 THE INTEGRODIFFERENTIAL EQUATION OF MOTION

Some of the motions† permitted by both the Lorentz-Dirac equation (14.42) and its nonrelativistic approximation (14.32) are evident from considering phases in which no external forces are being applied. Free motion of constant velocity is to

† A compendium of solutions is provided in Ref. 23.

be expected during such phases and, indeed, both the equations are satisfied by solutions with $\dot{\mathbf{u}} = 0$ and $\ddot{\mathbf{u}} = 0$ while $\mathbf{F} = 0$. However, putting $\mathbf{F} = 0$ into (14.32) makes it most directly evident that motions with

$$\frac{\ddot{\mathbf{u}}}{\dot{\mathbf{u}}} = \frac{1}{\tau_c} \rightarrow \dot{\mathbf{u}}(t) = \dot{\mathbf{u}}(t_0)e^{(t-t_0)/c} \tag{14.45}$$

are also permitted. This would indicate that, once a charge has been accelerated, it is consistent with those equations by themsclves for the acceleration to increase without limit, even after applied forces have ceased to accelerate the charge. Such runaway solutions have also been found in the continued presence of applied forces and for the relativistic equation (14.42).

The occurrence of the unphysical runaway solutions, corresponding to no actual physical situation there is to describe, can be traced to the circumstance that an instantaneous balancing of energy-momentum changes, like the Lorentz-Dirac equation, is not sufficient to yield a proper *equation of motion* when a part of the reactions is represented by $m\tau_c \ddot{w}_\mu$, as in (14.42), or by $m\tau_c \ddot{\mathbf{u}}$, as in its nonrelativistic approximation (14.32). An equation of motion must determine a unique trajectory $\mathbf{r}(t)$ from any given starting point $\mathbf{r}(0)$ and initial velocity $\mathbf{u}(0) = \dot{\mathbf{r}}(0)$, just two (vector) pieces of information, which should be sufficient to determine a phase of the motion according to relativistic principles.† The forces present are then to provide a *definite* value $\dot{\mathbf{u}}(t)$ for each phase, a rate at which the starting velocity and each succeeding one will change. The Lorentz-Dirac relation is insufficiently definite to do this; it can at best provide a definite rate at which the acceleration $\dot{\mathbf{u}}(t)$ will *change*. It is a third-order differential equation, in $\ddot{\mathbf{u}} \equiv \dddot{\mathbf{r}}(t)$, and so needs a third piece of information, like the $\dot{\mathbf{u}}(t_0)$ of (14.45), before it can determine a unique trajectory. The Lorentz-Dirac balance should be satisfied, but it must be supplemented by an additional condition, one that applies to all possible motions, before a proper equation of motion can be developed from it.

The supplementary condition needed is suggested by the discussion of the nonrelativistic development (14.30). It was found that reactions from the self-field can be represented as simply as by $m\tau_c \ddot{\mathbf{u}}$ only for processes that begin and end with zero acceleration. Corresponding restrictions can apply to *any* charged-particle trajectory, followed from as early to as late as desired, and are given a properly Lorentz-covariant form if expressed as‡

$$\dot{w}_\mu(\tau \rightarrow \mp\infty) \rightarrow 0 \tag{14.46}$$

This condition immediately eliminates the unphysical runaway solutions, with $\dot{\mathbf{u}}(t \rightarrow \infty) \rightarrow \infty$, like (14.45). It also prevents considering applied forces so idealized that they keep changing the energy-momentum $p_\mu = mw_\mu$ no matter

† See Ref. 17, page 38, and the discussion of equations of motion on page 408.

‡ Suggested by Rohrlich (Ref. 26). Such asymptotic boundary conditions (formulated as *in* and *out* stationary states of motion) characterize quantum-mechanical descriptions that deal with distributions spread over infinite space and time. See page 418*n*.

how far the trajectory may extend; such force formulations should be considered overidealized in this respect.

A standard way to incorporate boundary conditions with a differential equation like the Lorentz-Dirac relation is to integrate the equation partially and adopt limits on the resultant integral that embody the boundary conditions. That is most easily done for (14.42) after reexpressing it as

$$m\frac{d}{d\tau}(w_\mu - \tau_c \dot{w}_\mu) = \Gamma_\mu(\tau) \tag{14.47}$$

where

$$\Gamma_\mu(\tau) \equiv K_\mu(\tau) - m\tau_c w_\mu \left(\frac{\dot{w}_\rho}{c}\right)^2 \tag{14.48}$$

is a Minkowski 4-vector comprised of the external forces and the recoil effects of radiation. Multiplying both sides of (14.47) by the integrating factor $-\tau_c^{-1}\exp(-\tau/\tau_c)$ makes its left side equivalent to $md(e^{-\tau/\tau_c}\dot{w}_\mu)/d\tau$, and then one integration and a multiplication of both sides of the result by $\exp(+\tau/\tau_c)$ produces

$$m\dot{w}_\mu(\tau) = e^{\tau/\tau_c}\int_\tau^\infty \frac{d\tau'}{\tau_c} e^{-\tau'/\tau_c}\Gamma_\mu(\tau') \tag{14.49}$$

The limits on the integral must be chosen as indicated to ensure that the boundary condition $\dot{w}_\mu(\tau \to \infty) \to 0$ is met, whatever the forces, through the vanishing of the integration range as $\tau \to \infty$. The exponential outside the integral takes care that $\dot{w}_\mu(\tau \to -\infty) \to 0$ also if the forces are properly formulated. The result is an integrodifferential equation containing the space-time acceleration $\dot{w}_\mu$ not only as specified on its left but also within the integral, as reference to (14.48) shows. Solving such an equation requires finding a *function* $\dot{w}_\mu(\tau < \tau' < \infty)$ that will satisfy it and thus an entire range of accelerations at once. Since a definite value of $\dot{w}_\mu(\tau)$ is thus found for any moment τ, Eq. (14.49) can serve as a proper equation of motion. In practice, it is generally simpler to obtain a sufficiently general solution of the less restrictive Lorentz-Dirac differential equation first and only afterward impose the boundary conditions (14.46), thus eliminating any runaway terms.

Another form of the integrodifferential equation, more convenient for some purposes, is obtained by changing from the integration variable $\tau \le \tau' \le \infty$ to

$$0 \le s = \frac{\tau' - \tau}{\tau_c} = \frac{c(\tau' - \tau)}{3a_c/2} \le \infty \tag{14.50}$$

with $ds = c\,d\tau'/(3a_c/2)$ an interval on the space-time trajectory measured in units of three-halves the tiny classical electron radius (14.1). Then (14.49) becomes

$$m\dot{w}_\mu(\tau) = \int_0^\infty ds e^{-s}\Gamma_\mu(\tau + \tau_c s) \tag{14.51}$$

This form makes it easy to see how, when applied to a chargeless particle ($\tau_c \to 0$) subject to nonelectromagnetic forces, so that $\Gamma_\mu \to K_\mu$ in (14.48), it reduces to the equation of motion (12.68)

$$m\dot{w}_\mu(\tau) = K_\mu(\tau) \int_0^\infty ds e^{-s} = K_\mu(\tau) \tag{14.52}$$

as to be expected when there is no radiation to take into account.

An implication of the present formulations, made evident in both (14.49) and (14.51), that has been found disturbing is that the acceleration $\dot{w}_\mu(\tau)$ at any given instant τ is determined not only by the forces present at the instant but by all the forces that might be applied at future times $\tau \leq \tau' \leq \infty$, in contradiction to the principle of causality. Actually, this violation of causality is a very slight one, in a sense most immediately evident in (14.49), where the exponential inside the integral practically eliminates any necessity of knowing the forces farther into the "future" than τ_c [Eq. (14.31)], about the time it takes a signal of velocity c to cross the charge on an electron. Applied forces that can change appreciably in a time so short are far outside the purview of classical considerations, as pointed out in the discussion of (14.32).

The characteristic of the present formulations made most evident by (14.51), that the acceleration at any one point of a trajectory depends on the external forces and radiation recoils to which all subsequent parts of the trajectory may be subjected, is a natural outcome of having represented reactions from the self-field by κ_μ of (14.41). This treats the entire moving mass as being concentrated at one point of the trajectory at every moment, whereas the self-field contributions to the mass are actually spread out. Thus whenever acceleration is applied to the charge, it is only after forces still in the charge's "future" have come into action that the inertia of all the mass is overcome. The expression for κ_μ was so determined that it allows only an *ultimate* return of the mass contributions by the accelerated self-field to their equilibrium values, *after* radiations at *finite* rates have been completed. A determination of the acceleration at an instant only by effects present at the *same* instant presents the so far intractable problem of finding a simultaneous solution of *coupled* equations, one for the motion of the bare mass at a point, subject to a force like $\oint d\mathbf{S} \cdot \mathbf{T}$ of (6.27), and another, a field equation, for the redistributions of the self-field arising from the motion of the reacting charge. In any case, the formulation found is quite understandable as a suitable way to represent radiative effects and should be regarded as only *pseudo*-acausal.

14.6 LINEAR MOTIONS AND PREACCELERATION

Important features of the radiative effects become evident in cases of *linear* motion, like those expected when the charge is accelerated by some external force $\mathbf{F}$ having a constant direction and any initial velocity $\mathbf{u}_o$ the particle is given starts it into the same direction. The 4-velocity will then have only the two components indicated in $w_\mu(w, 0, 0, ic\gamma)$ with $w \equiv \gamma u (\gtrless 0)$. The fourth component, $w_4 = ic\gamma$, can also be put in terms of the magnitude $|w|$, since the invariance $p_\mu^2 = m^2 w_\mu^2 = -m^2c^2$ leads to

$$\gamma \equiv \left(1 - \frac{u^2}{c^2}\right)^{-1/2} = \left(1 + \frac{w^2}{c^2}\right)^{+1/2} \geq 1 \tag{14.53}$$

for the dilation factor.

As already stated, it is simplest to begin with a sufficiently general solution of the Lorentz-Dirac equation (14.42), for which the decompositions

$$\dot{w}_\mu \left[\dot{w},\ ic\dot{\gamma} = \frac{iw\dot{w}}{(w^2 + c^2)^{1/2}}\right] \tag{14.54}$$

$$\ddot{w}_\mu \left\{\ddot{w},\ i\left[\frac{w\ddot{w}}{(w^2 + c^2)^{1/2}} + \frac{c^2\dot{w}^2}{(w^2 + c^2)^{3/2}}\right]\right\} \tag{14.55}$$

are needed. Moreover, the radiation rate occurring in (14.42) is now given by

$$\mathscr{P} = \frac{2}{3}\frac{e^2}{c^3}\dot{w}_\mu^2 = m\tau_c \frac{c^2\dot{w}^2}{w^2 + c^2} \tag{14.56}$$

The first component of the Lorentz-Dirac equation becomes

$$\dot{w} = \frac{F}{m}\left(1 + \frac{w^2}{c^2}\right)^{1/2} + \tau_c\left(\ddot{w} - \frac{w\dot{w}^2}{w^2 + c^2}\right) \tag{14.57}$$

The square root multiplying F here is just the dilation factor (14.53) needed to form the Minkowski force K_1. The corresponding *fourth* component of the Lorentz-Dirac equation *follows* from (14.57) and so needs no attention while solving (14.57).

The equation of linear motion (14.57) can be greatly simplified by expressing it for a widely used variable symbolized by ξ and called the *rapidity* of the particle. It essentially replaces the momentum magnitude $p = mw$ and is defined by the equation

$$w\left(\equiv \gamma u = \frac{p}{m}\right) = c \sinh \xi \tag{14.58}$$

For nonrelativistic velocities $u \approx w \ll c$, $\sinh \xi \approx u/c \ll 1$ and so $\xi \approx u/c \ll 1$ becomes just the velocity ($\gtrless 0$) in units of c. However, whereas $|u| < c$ always, the rapidity must be allowed the full range of values $-\infty < \xi < +\infty$ in representing the momentum, through (14.58). It has a simple connection to the dilation factor (14.53),

$$\gamma = \left(1 + \frac{w^2}{c^2}\right)^{1/2} = [1 + (\sinh \xi)^2]^{1/2} = \cosh \xi > 1 \tag{14.59}$$

and so relativistic velocities are given by

$$u = \frac{w}{\gamma} = c \tanh \xi \qquad (\to c\xi \text{ for } u \ll c) \tag{14.60}$$

The proper-time derivatives needed for (14.57) become

$$\dot{w} = c\dot{\xi} \cosh \xi \qquad \ddot{w} = c\ddot{\xi} \cosh \xi + c\dot{\xi}^2 \sinh \xi \tag{14.61}$$

and then the equation is reduced to

$$\dot{\xi} - \tau_c \ddot{\xi} = \frac{F}{mc} \tag{14.62}$$

This is just like the nonrelativistic equation (14.32) except that it replaces the velocity (in units of c) with the rapidity and so properly has (14.32) as its nonrelativistic approximation, when $\xi \approx u/c \ll 1$.

The result (14.62) also has the same form as the version (14.47) of the Lorentz-Dirac equation and so can be integrated in a similar way, to

$$\dot{\xi}(\tau) = e^{\tau/\tau_c} \int_\tau^\infty \frac{d\tau'}{\tau_c} \frac{F}{mc} e^{-\tau'/\tau_c} \tag{14.63}$$

after enforcing the boundary condition $\dot{w}(\tau \to \infty) = 0$ of (14.46), which requires $\dot{\xi}(\tau \to \infty) = 0$ according to (14.61). This equation of motion, for the rapidity, has the advantage over (14.49) that it does not contain unprescribable effects of radiation recoils in its integrand. Whenever the given external force on the trajectory can be expressed as a function $F(\tau')$ of the trajectory parameter τ', (14.63) directly yields the rapidity itself as

$$\xi(\tau) = \xi(-\infty) + \int_{-\infty}^{\tau} d\tau' e^{\tau'/\tau_c} \int_{\tau'}^{\infty} \frac{d\tau''}{\tau_c} \frac{F(\tau'')}{mc} e^{-\tau''/\tau_c} \tag{14.64}$$

where $\xi(-\infty) = \tanh^{-1}(u_o/c)$ is the initial rapidity, as determined by the starting velocity. The subsequent velocities and momenta are then found as (monotonic) hyperbolic functions of this $\xi(\tau)$, in accordance with (14.60) and (14.58).

The important features of the radiative effects on linear motions become most evident in the simple special case of a *constant* force, of the kind provided by imposing a uniform electric field $\mathbf{E}_o$ on an electron when $F = +e\mathsf{E}_o$ if the $w > 0$ direction is taken to be that of $q\mathbf{E}_o = -e\mathbf{E}_o$. However, this force cannot be idealized to the extent of supposing it to have an unlimited range. Finite radiative effects can be obtained only when at least the boundary conditions (14.46) are satisfied, and those require that no accelerating force exist at the $\tau \to \pm\infty$ limits of the trajectory. Consequently, it is the case

$$F(\tau < 0) = 0 \qquad F(0 \le \tau \le \tau_1) = e\mathsf{E}_o \qquad F(\tau > \tau_1) = 0 \tag{14.65}$$

that will be explored, corresponding to the physically *attainable* situation in which the uniform field is provided by maintaining a constant potential difference between two plates at some distance apart (and this distance eventually determines the value of τ_1). The special case (14.65) can throw light on the general case insofar as various external forces the particle may meet along its trajectory can be regarded as a succession of such impulses as (14.65), with various signs and strengths and various durations of near constancy τ_1.

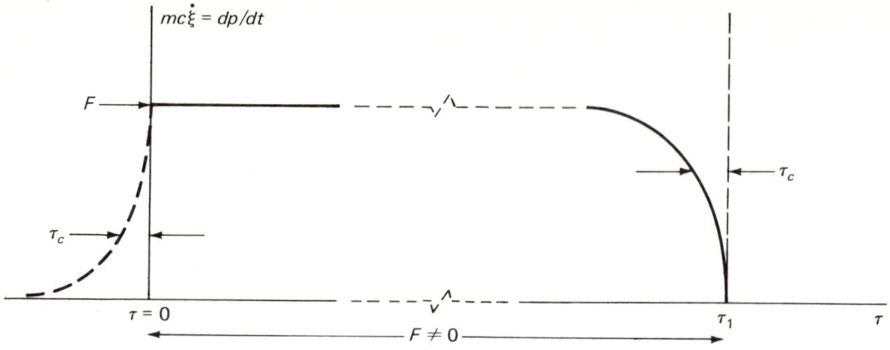

Figure 14.1

In (14.65), the integration for ξ in (14.63) extends up to $\tau' = \tau_1$ only, since $F = 0$ beyond that, and thus produces

$$mc\dot{\xi}(0 < \tau < \tau_1) = e\mathbf{E}_o(1 - e^{-(\tau_1 - \tau)/\tau_c}) \tag{14.66}$$

This is just the ordinary time rate of change of the particle momentum, since (14.61) and (14.59) together show that

$$\frac{dp}{dt} = \frac{d(mw)}{\gamma\, d\tau} = \frac{m\dot{w}}{\gamma} = mc\dot{\xi} \tag{14.67}$$

The behavior of this quantity along the part of the trajectory $0 \le \tau \le \tau_1$, during which the force is being applied to the particle center, is indicated in Fig. 14.1, as given by (14.66) for cases of $\tau_1 \gg \tau_c$ ($< 10^{-23}$ s!). Over almost its entire range, the constant force then imparts momentum to the particle as if there were *no* radiative effects [$\tau_c \to 0$ in (14.62)] up to a tiny interval, of order $\Delta\tau = \tau_1 - \tau \approx \tau_c$, at the very *end* of the force range. Of course, this interval can be much dilated, for very high velocities $u \to c$, on a laboratory time scale, to $\Delta t = \gamma\, \Delta\tau \approx \gamma\tau_c$, but even for 100-Mev electrons ($\gamma \approx 200$), the distance traversed during that final interval is only $\Delta x \lesssim c\, \Delta t \approx \gamma c\tau_c = \frac{2}{3}\gamma a_c \approx 4 \times 10^{-11}$ cm, about a thousandth of an atomic diameter! Even incident forces with variations of γ-ray frequency will be sensibly constant over ranges $\tau_1 \gg \Delta\tau \approx \tau_c$.

The conclusion is that, over almost its entire range, the constant force increases the particle's momentum at the full rate $dp/dt = e\mathbf{E}_o$. All its work goes into increasing the particle's *kinetic energy* during times when the particle is known to be losing *radiation* energy at the rate $\mathscr{P}_{\parallel} = (\tau_c/m)(dp/dt)^2$ of (14.22*a*). Only *after* that period of maximum radiation is there some allowance for a brief interval $\Delta\tau \ll \tau_1$ of work by the force that is not devoted solely to producing particle kinetic energy. The outcome presents the puzzle of accounting for how energy has been made available for the main part of the radiation.

The resolution of the puzzle lies in the difference between the equation of motion (14.62) that allows for radiative effects and one that does not include them ($\tau_c = 0$). That indicates that the force can work on the particle's field only at points where the rate of change of the particle's momentum is changing, where $(d/dt)(dp/dt) = (mc/\gamma)\ddot{\xi} \neq 0$. In (14.65) that happens at the *beginning* of the force range as well as at its end; the force provides impulse and energy to the particle and its field *before* its charge is launched on the constant force range. In the pseudo-acausal formulation on which the equation of motion (14.62) is based, that process is simulated as work by the force at times $\tau < 0$ when $\tau = 0$ marks the beginning of the force range. It must be recognized that the integral (14.63) for $\dot{\xi}$ has nonvanishing values for $\tau < 0$ as well as in the later phases $0 < \tau < \tau_1$ of (14.66). For $\tau < 0$ the integral is

$$\dot{\xi}(\tau < 0) = e^{\tau/\tau_c} \frac{e\mathsf{E}_o}{mc} \int_0^{\tau_1} \frac{d\tau'}{\tau_c} e^{-\tau'/\tau_c}$$

since $F \neq 0$ for $0 \leq \tau' \leq \tau_1$, and has the result

$$\dot{\xi}(\tau < 0) = \frac{e\mathsf{E}_o}{mc}(1 - e^{-\tau_1/\tau_c})e^{\tau/\tau_c} \tag{14.68}$$

already indicated in Fig. 14.1. Because $\xi \sim \dot{u}$ in nonrelativistic approximation, $\dot{\xi}(\tau < 0)$ is called a *preacceleration* of the charge plus self-field system by the force. Effects through which the force imparts energy-momentum to the particle's field are thus mapped out on the proper time scale preceding displacements of the particle's center within the external force field. For $\tau_1 \gg \tau_c$, the preacceleration (14.68) furnishes just the amount $\dot{\xi}(0) = e\mathsf{E}_o/mc$ needed for a proper initial response to the force of the *entire* mass m, which includes the part, $m - m_o$, of the rest mass that is spread out over the self-field. That the force is represented as affecting the particle before its charge enters the force range is not surprising when the findings discussed on page 160 are considered. Its self-field *precedes* the charge into the force field, and interactions of external and self-fields can be expected to begin immediately, altering the self-field momenta of those elements of the rest mass m that are to be contributed by the overlapping part of the self-field. All this comes from the fact that the inertial mass associated with a charge is not confined to the site of the charge but is spread out wherever its self-field reaches.

The example here has been most instructive because, in it, an initial phase in which the self-field is being distorted by the applied force is entirely separated from a phase in which radiation draws only on the energy-momenta of the distorted self-field and does *not* come directly from work by the applied force. This is the basis of the statement on page 429 that the radiation is generated by recoveries from self-field distortions.

The outcome of the above example is most interesting when the particle is started from *rest*, as when a heated plate releases a thermoelectron of negligible velocity which is accelerated toward a second plate, at some distance l, by a potential difference $\Delta\phi = \mathsf{E}_o l$. Then the preacceleration is mapped on a $\tau < 0$ part of a space-time trajectory corresponding to no 3-space displacements and hence parallel to the time axis of the frame in which the plates are at rest. All the preacceleration, overcoming the initial inertia of the entire mass m, takes place while the charge center is still at rest and over a mean time of the unobservably small magnitude τ_c.

Space will not be taken for further details (like the determination of the τ_1 in the solutions in terms of the given force range l) because the results are practically unobservable. It is already known from the discussion of the radiation rates (14.23) during linear motions that the energy lost to radiation is a fraction having the order of magnitude 10^{-13} of the largest work-energy boosts $e\,\Delta\phi = e\mathsf{E}_o l$ that are technically feasible.

All the above discussion was couched in a fully relativistic formulation, but all the phenomena also follow from the nonrelativistic approximation (14.32), as is evident from the fact the latter merely replaces the rapidity with the ordinary velocity $u(t)$.

The radiative effects should become relatively more conspicuous when the external force has components perpendicular to the directions of motion, in view of the greater radiation rates at such instants. The force is always perpendicular to the velocity of electrons traversing a uniform magnetic field, as in the synchrotrons mentioned in connection with (14.24*b*). However, even in a 5-Gev orbit the radiation loss per revolution was found to be only a fraction approximating 6/5000 times the electron's energy. It is not the consequent, practically unobservable deviations from circularity of orbit that limit the use of the devices to lower energies but the difficulties of making up small fractions of very large energies, as required of accelerators.

14.7 THE COLLAPSE OF THE CLASSICAL ATOM

A case of a force that is perpendicular to the velocity, at least insofar as the motion remains unperturbed by radiative effects, is presented by the classical circular orbits of electrons in the Coulomb field of a practically immovable atomic nucleus, like those considered in Sec. 8.3. Because the velocity of electrons in atoms can almost always be treated as small compared with the velocity of light, it will be sufficient to continue such a nonrelativistic treatment as that of Sec. 5.5. The instantaneous radiative effects on each orbit will therefore be very small, but their gradual *accumulation* over many orbitings gives them great importance to the basic concepts of physics. They led to the problem of accounting for the stability of atoms. On a classical basis, orbiting electrons must be expected to continue radiating, and should eventually fall into the oppositely charged nucleus as they lose the energy that can keep them in flight.

The Radiating Classical Atom

The dynamics of the unperturbed motion, as reviewed in Sec. 5.5, gives the energy in an orbit of radius r_n as

$$\epsilon_n = \tfrac{1}{2}mu_n^2 - \frac{Ze^2}{r_n} = -\tfrac{1}{2}mu_n^2 = -\frac{Ze^2}{2r_n} \tag{14.69}$$

The last equalities introduced here follow from the equilibrium between the centrifugal and attractive forces in circular orbits

$$\frac{mu_n^2}{r_n} = \frac{Ze^2}{r_n^2} \tag{14.70}$$

One revolution is performed during each (Kepler) period

$$T_n = \frac{2\pi r_n}{u_n} = 2\pi\left(\frac{m}{Ze^2}\right)^{1/2} r_n^{3/2} \tag{14.71}$$

and the angular momentum is conserved at a value $l_n = mu_n r_n$ with $l_n^2 = m(Ze^2)r_n$. To be noticed is that lower (more negative) energy orbits are characterized by smaller radii r_n and *larger* velocities u_n. The angular momentum in the smaller circular orbits is likewise smaller.

During each revolution of the electron in one of the circular orbits, it is given the steady centripetal acceleration u_n^2/r_n, and hence, insofar as the Coulomb force can maintain the steady motion, there is a steady radiation at a Larmor rate. This just equals the dipole rate (see the discussion of that equality in Sec. 10.3) already found in (8.50) for $\omega_o = u_n/r_n$

$$\mathscr{P}_n = \frac{2}{3}\frac{e^2}{c^3}\left(\frac{u_n^2}{r_n}\right)^2 = \frac{64}{9Z^2}\left(\frac{-\epsilon_n}{mc^2}\right)^4 \frac{mc^2}{\tau_c} \tag{14.72}$$

where $\tau_c = 2e^2/3mc^3$, as in (14.31). The connection to the (negative) energy in orbit ϵ_n follows from the expressions (14.69) for it. The loss rate to radiation is the greater the *lower* (more negative) the energy already is, as to be expected when the lower-energy motions have smaller radii, larger velocities, and therefore greater resultant accelerations u_n^2/r_n.

The energy loss in one revolution is given by

$$\mathscr{P}_n T_n = \frac{4\pi}{3}\frac{e^2}{r_n}\left(\frac{u_n}{c}\right)^3 = |\epsilon_n| \frac{8\pi}{3Z}\left(\frac{Ze^2/r_n}{mc^2}\right)^{3/2} \tag{14.73}$$

The first expression here coincides with the nonrelativistic approximation of the result already obtained in (14.24*a*). The equivalent final expression makes it simple to evaluate the fraction of the orbit energy that is lost while the electron makes one revolution at a typical atomic radius, of order $r_n \approx 10^{-8}$ cm

$$\frac{\mathscr{P}_n T_n}{|\epsilon_n|} = \frac{8\pi Z^{1/2}}{3}\left(\frac{a_c}{r_n}\right)^{3/2} \approx 10^{-6} Z^{1/2} \tag{14.74}$$

where $a_c = e^2/mc^2 \approx 2.8 \times 10^{-13}$ cm of (14.1). Such small (10^{-4} to 10^{-3} percent) losses in each revolution makes the approximation used (evaluating them as if the motion remained circular for the duration of each cycle) a very good one. As the orbitings continue, the losses will accumulate, and the energy remaining, as well as the radius of the motion, will gradually decrease. The result should be a very gradual inward spiraling, and (14.72) should remain an excellent approximation for the loss rate at each instant, an adiabatic approximation since it treats each momentary phase of the motion as part of an equilibrium orbit. There will then be some finite collapse time T_c from any initial radius r_n during which the instantaneous radius $r(t)$ is decreased to zero and $\epsilon \to -\infty$ in (14.69). That should be calculable from the rate (14.72) as

$$T_c = -\int_{\epsilon_n}^{-\infty} \frac{d\epsilon}{\mathscr{P}(\epsilon)} = -\frac{9Z^2}{64}\tau_c(mc^2)^3 \int_{\epsilon_n}^{-\infty} \frac{d\epsilon}{\epsilon^4} \tag{14.75}$$

The integration here can be made to describe more explicitly a falling into the nucleus ($-\int_{r_n}^{0} dr$) by using the adiabatic connection between instantaneous energy and radius, so that $-d\epsilon = +d(Ze^2/2r) = -(Ze^2/2r^2)\,dr$ and

$$T_c = -\frac{9\tau_c}{8Z}\left(\frac{mc^2}{e^2}\right)^3 \int_{r_n}^{0} dr r^2 = \frac{3\tau_c}{8Z}\left(\frac{r_n}{a_c}\right)^3 \tag{14.76}$$

The proportionality to r^2 of each element is important to notice since it shows that the final orbitings close to the nucleus ($r \to 0$) contribute very little to the total collapse time. The innermost orbitings attain more highly radiative relativistic velocities, invalidating the use of the Larmor rates for them, but the relatively briefer times spent in the inner orbits make the effect negligible. For a start from $r_n \approx 10^{-8}$ cm, since $\tau_c = 2a_c/3c \approx 6.3 \times 10^{-24}$ s, (14.76) gives

$$T_c \approx 10^{-10} Z^{-1} \text{ s} \tag{14.77}$$

for the life of an atom against radiative collapse. This is very long on the atomic scale, of periods $T = 2\pi r/u$, permitting the electron to complete many orbitings, but, of course, very short in comparison to the multibillion-year-long life of the universe.

The Correspondence Limit

Explaining the actual stability of atoms is a principal triumph of the quantum theory. Again, there is an intervention of quantum-theoretical restrictions on radiative effects, reducing the importance of their purely classical evaluation.

The quantum theory makes a lowest energy of finite amount stable against radiation through its description of the electron as a quantum of matter field, already mentioned in the opening paragraph of Chap. 13. That description, in effect, spreads the electron's charge over a stationary, *nonradiating* distribution as the least energetic electron-field state. Radiation does occur from higher-energy (excited) distributions that can also exist, but only with any one of a *discrete* set of energy values, as already considered for the treatment of recoils from radiation in Sec. 11.4. For the electron influenced only by a nuclear Coulomb field, the discrete energies

$$\epsilon_n = -\frac{mZ^2e^4}{2n^2\hbar^2} \qquad \text{with } n = 1, 2, 3, \ldots, \infty \tag{14.78}$$

are the quantum-mechanical counterparts of the classical energies† (14.69). The discreteness arises because of a restriction of the electron-field distributions to component wavelengths‡ that fit within the confines of the atom (as maintained by the nuclear Coulomb attraction), in much the same way that radiation-field distributions confined to cavities were found to be restricted to discrete frequencies and hence discrete energies $\hbar\omega$ in Sec. 7.3. Expression (14.78) is famous as the *Balmer formula*.

Equating the classical energy expressions in (14.69) to the quantum-mechanical ones (14.78) suggests a correspondence of the quantum-mechanical electron-field states of motion to a set of discrete classical orbits, having only the radii

$$r_n = \frac{n^2\hbar^2}{mZe^2} \approx \frac{n^2}{Z}(0.53 \times 10^{-8}) \qquad \text{cm} \tag{14.79}$$

Relying on such correspondences has made it possible to use classical radiation theory to derive many expectations that have proved to be consistent§ with the quantum theory. There must be a correspondence limit at which quantum-mechanical and classical expectations actually merge, since the successful classical theories do retain a range of validity.

The correspondences suggest that the radiation will consist of energy elements released to it upon the transition from one orbit to the next smaller one [$n \to n - 1$ in (14.79) and (14.78)], so that

$$-(\Delta\epsilon)_{\text{rad}} = \epsilon_{n-1} - \epsilon_n = \frac{mZ^2e^4}{2\hbar^2}\,\frac{2n-1}{n^2(n-1)^2} \to \frac{mZ^2e^4}{n^3\hbar^2} \qquad \text{for } n \gg 1 \tag{14.80}$$

If, as in Sec. 11.4, the transition is held to supply the energy for the formation of a single photon¶ it will have the frequency

$$\omega = \frac{(-\Delta\epsilon)_{\text{rad}}}{\hbar} \to \frac{mZ^2e^4}{(n\hbar)^3} = \left(\frac{Ze^2}{m}\right)^{1/2} r_n^{-3/2} = \frac{2\pi}{T_n} = \omega_o \tag{14.81}$$

for $n \gg 1$, just the period (14.71) and frequency in orbit $\omega_o = u_n/r_n$. This was also the classical expectation for the dipole radiation (8.50) and shows that the correspondence principle is most efficacious for the higher-energy ($n \gg 1$) orbits when the neighboring orbits connected by the transition differ comparatively little in radius ($\Delta r_n/r_n \sim 2\,\Delta n/n \to 0$ for $\Delta n = 1 \ll n$). For the large quantum numbers $n \gg 1$, the permitted radii and energies are both nearly continuous with each other in value, approaching the expectations of classical mechanics.

† Equations (8.113) and (8.114) of Ref. 17 exhibit a more explicit correspondence.

‡ Always definable, through Fourier analysis of any field distribution. The components here are known as *de Broglie waves*.

§ The main reason has been suggested on page 276*n*.

¶ As noted on page 369, splitting an available energy between the fewest possible photons is always favored, because of the rapid increase of radiation rate with frequency.

The above results also indicate that $\Delta n = 1$ is a selection rule for dipole photons, each of energy $\hbar\omega_o$ in the classical ($n \gg 1$) limit. There can also be radiations supplied by single-photon transitions with $\Delta n > 1$ but of higher multipole orders, called forbidden because they violate the dipole selection rule and are weak compared with the dominant dipole radiation. This has already been mentioned in connection with the small quadrupole rate (8.79), producing photons with $n \gg 1$ energies $\hbar\omega_Q \to 2\hbar\omega_o$ and so satisfying the quadrupole selection rule $\Delta n = 2$.

Angular-Momentum Conservation and Selection Rules

Another classical finding about the dipole radiation from the orbiting charge of Sec. 8.3 and for higher-order multipoles in Chap. 9 was that a field angular momentum is also radiated, in an amount that can now be expressed as $\Delta n/\omega$ for each unit of energy radiated.† When the latter is quantized into photons of energy $\hbar\omega$, an angular momentum $\Delta n\hbar$ is carried off by each photon. Because angular momentum must be conserved, the radiating system must concurrently lose an amount $\Delta l_n = \Delta n\hbar$ of mechanical angular momentum. Using the expression for l_n^2 cited just below Eq. (14.71) and the permitted values (14.79) for the radii yields‡

$$l_n = (mZe^2r_n)^{1/2} = n\hbar \tag{14.82}$$

and, indeed, there is a loss $\Delta l_n = \Delta n\hbar$ during a transition $n \to n - \Delta n$. On this account, such selection rules as the $\Delta n = 1$ for dipole radiation can be regarded as following from the necessity for angular-momentum conservation.

The necessity for conserving angular momentum jointly with the energy, just pointed out, presents a problem that can be treated classically at least in the correspondence limit. The forces responsible for the mechanical transitions between orbits must be able to match the angular momentum to the energy changes suitably. The connection required becomes clearest when the energy (14.69) is expressed in still another way, in terms of the l_n^2 quoted just below (14.71), which makes

$$\Delta\epsilon_n = -\Delta\left(\frac{mZ^2e^4}{2l_n^2}\right) = |\epsilon_n| \frac{2\Delta l_n}{l_n} \tag{14.83}$$

both decreasing together, and with the ratio 2 of the fractional energy change $\Delta\epsilon_n/|\epsilon_n|$ to the fractional angular-momentum change $\Delta l_n/l_n$. That this can be a problem was illustrated by the Larmor precession of Fig. 5.3, which is a result of the *inability* of a uniform magnetic field to align a dipole with itself—to cause a

† See page 227*n*.

‡ Essentially the *space-quantization principle* $l_n = n\hbar$ used by Bohr to transform the classical energy expression (14.69), in an equivalent form, $\epsilon_n = -mZ^2e^4/2l_n^2$, into the quantum-mechanical result (14.78). See Ref. 17, eqs. (8.113) and (8.114), for modifications that recognize the vector character of angular momentum, inessential for the considerations here.

transition to a state of the lower energy—because the magnetic force cannot change the angular momentum suitably.

The external Coulomb force by itself would merely keep the electron following one of the periodic orbits of constant energy. It is the force of radiative reaction, represented by $m\tau_c\ddot{\mathbf{u}}$ in the nonrelativistic equation of motion (14.32), that is responsible for the mechanical energy and angular-momentum *changes* in (14.83) accompanying the transitions from one orbital radius to another and concurrently for transferring work on the charge by the external (Coulomb) force to the electron's field (including its radiation). Equation (14.32) must have the solutions that provide a detailed description of the gradual spirals, connecting initial and final radii of each transition, anticipated above. Forming this explicit description is unnecessary for deriving what can be observed† (including the radiations already treated) or for the problem being considered now.

That it is work by the radiative reaction $m\tau_c\ddot{\mathbf{u}}$ that transforms energy supplied by the work of the external force into the radiated field energy has already been demonstrated in the discussion of (14.32). There the extra work temporarily stored in the electron's nonradiated self-field was not recognized, as in the discussion of the relativistic energy expression $E_\tau = (\gamma - \dot{\gamma}\tau_c)mc^2$ of (14.44). For a low-velocity electron, the self-field mass never departs appreciably from its rest value (14.2). The instantaneous particle energy reduces to

$$E_\tau \approx mc^2 + \tfrac{1}{2}mu^2 - m\dot{u}(u\tau_c) \tag{14.84}$$

where the magnitude of the last term is equal to the small fraction u/c of the already very small quantity $m\dot{u}(c\tau_c) \approx Fa_c$ of work that can be done by the external force over the tiny classical electron radius. In the discussion of the nonrelativistic equation of motion, the particle energy $E_\tau - mc^2$ was therefore treated as being indistinguishable from the kinetic energy $\frac{1}{2}mu^2$ of the full rest mass at every instant. The consequence for the case here is that

$$\frac{d\epsilon}{dt} \equiv \frac{d}{dt}\left(\tfrac{1}{2}mu^2 - \frac{Ze^2}{r}\right) \approx m\tau_c\ddot{\mathbf{u}}\cdot\mathbf{u} \tag{14.85}$$

the work by the nuclear Coulomb force being treated as an accumulation of potential energy; and, indeed, $\Delta\epsilon_n$ of (14.83) equals a work by the radiative reaction.

† As in quantum-theoretical procedures. The correspondence principle does require that there exist quantum-mechanical descriptions corresponding to the spirals, despite the fact that energy measurements on an atom can only yield the eigenvalues (14.78), as if there were a *discontinuous* jump from one orbit to another. The energy eigenstates of motion are only the stationary electron-field distributions, ideally monochromatic fields corresponding to the classical *equilibrium* orbits. When radiative effects are taken into account, distributions with nonstationary (nonmonochromatic) time evolutions also exist, describing average flows of the distributions between the stationary orbits and providing the radiating currents of charge. The nonmonochromaticities appear as observable line breadths of the radiated frequencies, as indicated in the next section.

The same radiative reaction must also produce the matching angular-momentum change in (14.83). How it can do so follows from crossing the radius vector into the equation of motion (14.32)

$$\mathbf{r} \times m\dot{\mathbf{u}} \equiv \frac{d\mathbf{l}}{dt} = m\tau_o \mathbf{r} \times \ddot{\mathbf{u}} \equiv \tau_c \frac{d}{dt}(\mathbf{r} \times m\dot{\mathbf{u}}) - m\tau_c \mathbf{u} \times \dot{\mathbf{u}}$$

after putting the torque $\mathbf{r} \times \mathbf{F} = 0$ for the central Coulomb force. Each of the factors $m\dot{\mathbf{u}}$ on the right can be replaced by $\mathbf{F}$ since each differs from $\mathbf{F}$ only by $m\tau_c\ddot{\mathbf{u}}$, and the correction would be proportional to $\tau_c^2 \sim 4 \times 10^{-47}\ \mathrm{s}^2$, making it at least as negligible as already neglected relativistic effects. Thus, up to terms linear in τ_c

$$\frac{d\mathbf{l}}{dt} \approx -\tau_c \mathbf{u} \times \mathbf{F} = +\tau_c \mathbf{u} \times \hat{\mathbf{r}}|\mathbf{F}| \tag{14.86}$$

directed oppositely to $\mathbf{l} = \mathbf{r} \times m\mathbf{u}$ because the Coulomb force $\mathbf{F} = -\hat{\mathbf{r}}Ze^2/r^2$ is radially inward. The radiative reaction does make the angular momentum diminish in magnitude and, in one of the circular orbits of energy ϵ_n (14.69), where the orbital velocity u_n is perpendicular to the Coulomb force, by the amount

$$-dl_n = \tau_c u_n \frac{Ze^2}{r_n^2}\, dt = \tau_c u_n \frac{2|\epsilon_n|}{r_n}\, dt$$

during a time interval of radiation $dt = -d\epsilon_n/\mathscr{P}_n$. The radiation rate is given by (14.72) and can be expressed as

$$\mathscr{P}_n = \frac{m\tau_c u_n^4}{r_n^2} = \frac{\tau_c}{m}\frac{(2\epsilon_n)^2}{r_n^2}$$

after use of (14.70) for a replacement of mu_n^2. This leads to

$$\frac{dl_n}{d\epsilon_n} = \frac{mu_n r_n}{2|\epsilon_n|} = \frac{1}{2}\frac{l_n}{|\epsilon_n|}$$

producing just the matching (14.83) that was required. The radiative reaction $m\tau_c\ddot{\mathbf{u}}$ does properly account for the mechanical changes during radiation.

Cascade Times

After the development of the quantum theory, calculations like (14.76) of the atomic collapse times became academic. There is no radiative collapse beyond the innermost permitted radius r_1 of (14.79), because this corresponds to a nonradiating electron-field distribution that remains stationary even when radiative reactions are taken into account. However, there is still radiation from any larger orbit $r_{n>1}$, and a cascade of radiative transitions can be initiated when the electron starts from any $n \gg 1$ orbit. Calculating a collapse time from larger r_n's to r_1 still has value; it is useful to know in several circumstances, if only roughly.

Cases in point are provided by the formation of what are sometimes called *exotic atoms*, like those formed by the capture of negatively charged muons or pions, for example, as such particles are slowed down by passage through a material. The capture into an orbit around an atomic nucleus generally takes place when a particle of suitable velocity penetrates sufficiently into the cloud of electrons around a neutral atom. The electrons screen most of the nuclear charge from the particle at the capture point, an effect representable by using a $Z_{\text{eff}} \ll Z$ to replace the charge number Z of the nucleus. From capture in an initial orbit of radius r_n, the cascade time is calculable from an obvious modification of (14.76)

$$T(n \to 1) \approx \frac{3\tau_c}{8Z_{\text{eff}}} \frac{r_n^3 - r_1^3}{a_c^3} \approx \frac{3\tau_c}{8Z_{\text{eff}}^4} \left(\frac{\hbar c}{e^2}\right)^6 (n^6 - 1) \tag{14.87}$$

Because even r_2 encloses a spherical volume proportional to r_2^3 larger by a factor 64 than r_1 does, (14.87) would differ very little from the total collapse time of (14.76), except that now the interest is in a $\tau_c = \frac{2}{3}e^2/mc^3$ evaluated with a muon or pion mass, about 207 and 285 times as great, respectively, as the electron mass. Thus, the cascade times of these particles are less than 10^{-12} s. An important point is that such times, whatever should be found suitable for Z_{eff} (1 to $\lesssim 30$), are short in comparison to the lifetimes of the muon and pion against spontaneous disintegration, about 2.2×10^{-6} and 2.6×10^{-8} s respectively. Thus exotic atoms can have a substantial existence before their constituents disintegrate. Moreover, the muon and pion are subject to capture (absorption) by the constituents of the nucleus. The short cascade times give assurance that most of these nuclear captures† will take place from the innermost orbit r_1 around the atomic nucleus.

14.8 CLASSICAL RADIATION WIDTHS

As treated in Sec. 8.3, an orbiting charge yielded dipole, quadrupole, . . . radiations of ideally discrete frequencies ω_o, $2\omega_o$, ... because it was presumed that a periodic motion of orbital frequency ω_o was perfectly maintained throughout each radiation process. In atoms, such well-defined periodicities can be maintained by the internal forces like the energy-conserving nuclear Coulomb force (external to the electron, as treated above, but internal to the entire atomic system) only insofar as the mechanical system remains unperturbed by the radiative reactions (or by other conceivable dissipative effects) and as the constant energy needed to maintain a perfect periodicity is conserved. At least the losses to radiation will not permit this, having consequences like the continuous inward spiraling‡ of the electrons, discussed in the preceding section. The radiatively perturbed motion becomes aperiodic, and its Fourier decomposition will have a continuum of frequencies§ replacing the unperturbed discrete frequencies.

† See Ref. 16, for example.

‡ With the quantum-mechanical counterpart mentioned on page 459*n*.

§ As the aperiodic unbound orbits do in the processes of bremsstrahlung, treated in Sec. 10.5.

Since radiative effects on motion are very small ($\sim \tau_c$), the continuum of departures from each unperturbed frequency will have a small effective range $\Delta\omega$, with the result a small broadening of each spectral line. Direct measurements of such radiation widths have value for conclusions about the dynamics of radiating systems.

Evaluations of the effect are generally approached through the solution of a simple prototype problem, the case of a nonrelativistic linear harmonic oscillator, consisting of a point electron subject to a Hooke's-law force $F = -m\omega_o^2 x$ that is proportional to a suitably defined displacement x from an equilibrium value $x = 0$. The corresponding nonrelativistic equation of motion (14.32)

$$\ddot{x} = -\omega_o^2 x + \tau_c \dddot{x} \tag{14.88}$$

is simple to solve through the usual trial of a time-dependence proportional to $x \sim \exp(-i\omega t)$, so that $\ddot{x} = -\omega^2 x$ and $\dddot{x} = i\omega^3 x$. Then the general solution is a superposition of the trial forms containing, for ω, the roots of the cubic

$$\omega^2 = \omega_o^2 - i\omega^3\tau_c \tag{14.89}$$

The unperturbed ($\tau_c \to 0$) roots of this are $\omega \to \pm\omega_o$, and at these values the radiative correction adds only a fraction of magnitude $\omega_o\tau_c$ to (14.89). This is very small, $\omega_o\tau_c \ll 1$, for all ω_o's short of very hard γ-ray frequencies [see the discussion of (14.32)]. Moreover, an oscillation with $\omega_o \to 1/\tau_c$ would have to be restricted to an amplitude $\ll c\tau_c = 2a_c/3 \approx 2 \times 10^{-13}$ cm in order to remain nonrelativistic ($|\dot{x}| \ll c$)! In all relevant situations it is more than adequate to evaluate the radiative term of (14.89) as $\mp i\omega_o^3\tau_c$ for the respective unperturbed roots $\pm\omega_o$, and then the results† are

$$\omega \approx \pm\omega_o(1 \mp i\omega_o\tau_c)^{1/2} \approx \pm\omega_o - \frac{i\omega_o^2\tau_c}{2} \tag{14.90}$$

Thus the solutions will have oscillations proportional to $x \sim \exp(\pm i\omega_o t)$, modulated by an exponentially decreasing amplitude proportional to $\exp(-\Gamma t/2)$, with

$$\Gamma \equiv \omega_o^2\tau_c \tag{14.91}$$

A sufficiently general solution, one containing two arbitrary constants, here

† Improving these results by using them to reevaluate $-i\omega^3\tau_c$ in (14.89), a so-called *reiteration procedure*, would add a *real* term proportional to τ_c^2 to each of the roots (14.90), describing a frequency shift from $\pm\omega_o$. However, corrections so tiny are hardly relevant, in view of larger relativistic corrections that can be expected from using the relativistic equation of motion (14.42) in place of (14.32). Larger (though still small) shifts are known to arise from purely quantum-mechanical effects, namely, the famous *Lamb shifts* generated by vacuum fluctuations in the radiation field and predictable with high accuracy from the renormalized quantum electrodynamics.

taken to be an initial amplitude A and a starting phase α, follows from the two roots† (14.90):

$$x(t > 0) = Ae^{-\Gamma t/2} \sin (\omega_o t + \alpha) \tag{14.92}$$

Such a motion is characteristic of damped oscillators, and so Γ is said to be a measure of the *radiative damping* of the motion.

The oscillator's energy $E = \frac{1}{2}m\dot{x}^2 + \frac{1}{2}m\omega_o^2 x^2$, which is conserved at its initial value $E(0) = \frac{1}{2}m\omega_o^2 A^2$ in unperturbed oscillations (Γ, $\tau_c = 0$), now diminishes exponentially as

$$E(t > 0) \approx (\tfrac{1}{2}m\omega_o^2 A^2)e^{-\Gamma t} = E(0)e^{-\Gamma t} \tag{14.93}$$

aside from oscillations of zero average and neglecting contributions proportional to τ_c^2 to remain consistent with what was done to obtain the solution (14.92). The fractional rate of this energy dissipation is just

$$-\frac{dE}{E\,dt} = \Gamma \tag{14.94}$$

leading to the name *decay constant* for Γ [compare the discussion of π^0 decay in connection with (11.125)]. The decay of the charged oscillator's energy is into radiation (photons) at an instantaneous rate that is given by (8.46) for an instantaneous amplitude $a(t) = A \exp(-\Gamma t/2)$ as

$$\mathscr{P} = \frac{1}{3}\frac{e^2\omega_o^4}{c^3} A^2 e^{-\Gamma t} = \omega_o^2\left(\frac{2}{3}\frac{e^2}{mc^3}\right)\tfrac{1}{2}m\omega_o^2 A^2 e^{-\Gamma t}$$

equal to $\Gamma E(t) = -dE/dt$ properly. Thus

$$\frac{1}{\Gamma} = \frac{E}{-(dE/dt)} = \frac{E}{\mathscr{P}} \tag{14.95}$$

is a mean decay time, defined in the same way (7.54) as for the energy losses in a system consisting of a cavity field, and sometimes called the mean lifetime of the excited energy state of the radiating system.

The spectral distribution $I(\omega > 0)$ of the radiated frequencies can be calculated from (10.67), an appropriately nonrelativistic form, prepared there for radiation from aperiodic unbound motions. That gives

$$\frac{dI}{d\Omega} = \frac{e^2}{c^3}\,|\langle\ddot{x}\rangle_{\omega>0}|^2 \sin^2\vartheta \tag{14.96}$$

† The cubic (14.89) naturally has a third root ω_3, which would lead to introducing a third arbitrary constant in an addition to (14.92). It can be found that the third root is purely imaginary, with a magnitude exceeding that in $-i\omega_3 > +1/\tau_c$. The corresponding addition to the solution (14.92) is then proportional to

$$\exp\left[\frac{+\text{const}\ t}{\tau_c}\right] \qquad \text{with const} > 1$$

a runaway solution of the kind eliminated by the boundary condition (14.46).

if ϑ is the angle from the direction of the oscillation to an emission direction. The radiation moment here is just the Fourier transform of the acceleration,

$$\langle \ddot{x} \rangle_\omega = \int_0^\infty \frac{dt}{2\pi} \ddot{x}(t > 0) e^{+i\omega t} \tag{14.97}$$

for the dominant dipole radiation.† Notice that the solution (14.92) furnishes accelerations only for times after a starting time $t = 0$, at which the depletion (14.93) of a given initial excitation energy $E(0)$ begins. The moment (14.97) introduces no further directional dependence, and so (14.96) can be integrated immediately, to give

$$I(\omega) = \frac{8\pi}{3} \frac{e^2}{c^3} |\langle \ddot{x} \rangle_{\omega > 0}|^2 \tag{14.98}$$

as the total energy radiated per unit of frequency range $d\omega$.

In evaluating the acceleration of the displacements $x(t)$ of (14.92), the variations of the slowly decreasing exponential should be neglected for consistency with the derivation of $x(t)$, in which neglecting effects proportional to τ_c^2 was justified. The radiation (14.98) is already proportional to $2e^2/3c^3 = m\tau_c$, and each time derivative of the exponential introduces a factor $\Gamma/2 = \omega_o^2 \tau_c/2$ that can only lead to adding a radiative correction proportional to τ_c^2, at most. The oscillation velocities are much greater, being proportional to $\omega_o \gg \Gamma = \omega_o^2 \tau_c$ for $\omega_o \tau_c \ll 1$. Thus $\ddot{x} \approx -\omega_o^2 x$ should be used in

$$\langle \ddot{x} \rangle_\omega = -\frac{\omega_o^2}{2\pi} \int_0^\infty dt x(t) e^{+i\omega t} = -\omega_o^2 \langle x \rangle_\omega \tag{14.99}$$

making it proportional to the dipole moment of the oscillation, as to be expected for the dominant dipole radiation. The integration is simple to perform after the trigonometric function in (14.92) has been analyzed into exponentials

$$\langle \ddot{x} \rangle_\omega = \frac{\omega_o^2 A}{4\pi i} \left(e^{-i\alpha} \int_0^\infty dt e^{i(\omega - \omega_o + i\Gamma/2)t} - e^{+i\alpha} \int_0^\infty dt e^{i(\omega + \omega_o + i\Gamma/2)t} \right)$$

The contributions to the first integral consist of oscillations of near-zero average except in the vicinity of $\omega \approx +\omega_o$, while the elements of the second integral can contribute only near $\omega \approx -\omega_o$. It must now be remembered that the definition of $I(\omega)$ in (10.37) requires that only the $\omega > 0$ Fourier components be used in (14.98); the periodicities represented by negative frequencies, $\omega < 0$, have already been allowed for, through the introduction of a factor 2 in the second line of (10.37). The last of the integrals above only has oscillations of near-zero average for all $\omega > 0$ and can be dropped. The remainder yields

$$\langle \ddot{x} \rangle_{\omega > 0} = \frac{\omega_o^2}{4\pi} \frac{A e^{-i\alpha}}{\omega - \omega_o + i\Gamma/2} \tag{14.100}$$

† See the discussion following (10.67). The much weaker higher multipole radiations cannot be treated adequately when using the nonrelativistic equation of motion (14.88) since they are relativistic corrections to the Larmor radiation, as discussed in connection with (10.20).

since the exponential damping factor, exp $(-\Gamma t/2)$, takes care that the contribution from the upper limit $t \to +\infty$ vanishes. The continuous spectrum (14.98) thus becomes

$$I(\omega) = \frac{e^2}{6\pi c^3} \frac{\omega_o^4 A^2}{(\omega - \omega_o)^2 + \Gamma^2/4} \tag{14.101}$$

having just the Lorentz line shape shown in Fig. 7.6, peaked at a resonance frequency $\omega = \omega_o$ and spread about it with a breadth at half maximum equal to Γ, which is consequently called the *radiation width*, just the inverse of the mean life against radiation of (14.95).

The validity of the result for $I(\omega)$ can be checked by integrating over the entire spectrum of frequencies. The result is

$$\int_0^\infty d\omega I(\omega) = \tfrac{1}{2} m\omega_o^2 A^2 = E(0) \tag{14.102}$$

just the initial excitation energy given the oscillator. The definition of $I(\omega)$ in (10.37) as being equal to the energy radiated per unit of frequency range over all time requires this outcome, since the classical oscillator must be expected to keep radiating until all its excitation energy has been dissipated and it comes to rest at its equilibrium point. That also checks with the integral

$$\int_0^\infty dt\mathscr{P} = \Gamma E(0) \int_0^\infty dt e^{-\Gamma t} = E(0) \tag{14.103}$$

of the instantaneous radiation rates in (14.95).

The treatment of the simple oscillator is usually regarded as an adequate demonstration of classical line broadening because small deviations from any equilibrium motion can be represented as generating a Hooke's-law restoring force. In the important example of the circular motions about an atomic nucleus, considered in the preceding subsection, the unperturbed radial equilibrium is the result of the balance (14.70) of Coulomb attraction and centrifugal repulsion. When considering perturbations by the radiative reaction, it is sufficient to consider only the radial components, perturbations of an orbit $r = r_n$ by an $x \equiv r - r_n \ll r_n$, since the developments from (14.86) demonstrated that the torques arising from nonradial force components will then take care of the angular-momentum changes. The radial perturbations have no torque about the nuclear center, and so the deviations from radial equilibrium should be considered for a given, constant value of the angular momentum $l_n = mu_n r_n$ rather than a given velocity in orbit u_n; the centrifugal force in (14.70) should now be expressed as $mu_n^2/r_n = l_n^2/mr_n^3$. Then the Hooke's-law force generated by a small departure from radial equilibrium is to be calculated as

$$\frac{l_n^2}{m(r_n + x)^3} - \frac{Ze^2}{(r_n + x)^2} \approx \frac{l_n^2}{mr_n^3}\left(1 - \frac{3x}{r_n}\right) - \frac{Ze^2}{r_n^2}\left(1 - \frac{2x}{r_n}\right)$$

The terms independent of x cancel at $r = r_n$, and the result is

$$-\frac{Ze^2}{r_n^2}\frac{x}{r_n} = \frac{-mu_n^2}{r_n^2}x \tag{14.104}$$

just the force used in the simple-oscillator equation of motion (14.88), with $\omega_o \equiv u_n/r_n$ properly!

The simple-oscillator approximation is not actually applicable to a *purely* classical atom, since it deals only with the broadening of a *discrete* frequency ω_o characteristic of the unperturbed motions whatever their amplitude of excitation as their energy is decreased by the radiation process. On the other hand, the classical atom has its electron spiraling inward continuously, passing through a continuum of possible unperturbed orbits, characterized by frequencies $u/r \sim r^{-3/2}$ [Eq. (14.71)] that increase *continuously* as the radii and the corresponding energies are decreased. The radiated spectrum is consequently expected to extend from whatever initial ω_o is excited to $\omega \to \infty$, and with a continuously increasing [see (14.72)] intensity. Such classical expectations—of a broad, continuous spectrum—were known to disagree with observations on actual atomic radiations long before the quantum theory was developed. Large ranges of the observed spectra exhibit well-separated nearly discrete frequencies, and the quantum theory not only accounted for the stability of atoms but also for their discrete spectra.

The simple-oscillator approximation became more pertinent to atoms, at least in a correspondence limit, after the advent of the quantum theory. As discussed in the preceding subsection, radiation with a discrete dipole frequency is completed while the orbital radius changes by a fraction of itself ($\Delta r/r = 2/n \ll 1$ was pointed out on page 457), as required for the approximation (14.104). Thus the simple-oscillator expectation of the line broadening $\Gamma \equiv \omega_o^2\tau_c$ (14.91), in (14.101), has some justification.

The calculation of radiation widths has been much refined by quantum-mechanical procedures, but the results are still weighed against the classical expression $\Gamma_{c1} = \omega_o^2\tau_c$ by defining *oscillator strengths* f such that

$$\Gamma_{qm}(\omega_{mn}) = \Gamma_{c1}\, f(\omega_{mn})$$

The oscillator strengths depend on the particular transition that yields a given frequency $\omega_{mn} = (\epsilon_m - \epsilon_n)/\hbar$.

Throughout this chapter it has been evident that quantum-mechanical procedures have superseded classical ones in almost all cases of substantially observable radiative effects. The review of the classical effects has nevertheless been valuable. Besides presenting a classical background against which quantum-mechanical results are better understood, it has indicated the limits of the applicability of classical radiation theory. In this chapter, the classical electromagnetic theory has been stretched to its limits, a fitting juncture at which to end a review of it.

SUPPLEMENTARY CHAPTERS

MATHEMATICAL DEVELOPMENTS AND MACROSCOPICALLY DESCRIBED MATTER

CHAPTER

A

THE CALCULUS OF FIELDS

Representations of gradient, divergence, and curl derivatives · Path integrals and scalar-potential differences · Fluxes and the Gauss theorem · Divergences as field sources · Continuity equations and conservation · Curl sources and the Stokes theorem · Potentials of divergenceless and curlless fields · Regaugings of potentials · Laplacians as measures of departures from smoothness · The Green theorems and solutions of Poisson equations

Since electromagnetic theory is expressed in the language of fields, it helps to be conversant with the mathematical description of fields in general. Familiarity with *scalar* potential-energy fields, often symbolized as $U(\mathbf{r})$, and with force-*vector* fields $\mathbf{F}(\mathbf{r})$ has already been gained in newtonian mechanics. Much of the vernacular associated with fields was adopted to describe hydrodynamics, which is concerned with the vector fields $\mathbf{u}(\mathbf{r})$ formed by the velocities at various points $\mathbf{r}$ of a fluid continuum. Such readily visualized examples will be used for illustrating the mathematical developments in this chapter.

The various ways of gauging *variations* of a field with position in space will receive primary attention. They gain their importance from the effort to represent the generation of a field at a point by the sources of it at the point, through the changes the sources make in it, from field values in the neighborhood of the point.

A.1 GRADIENT, DIVERGENCE, AND CURL DERIVATIVES

Continuous changes with position $\mathbf{r}$ in space of scalar fields like $U(\mathbf{r})$ or vector fields like $\mathbf{u}(\mathbf{r})$ can be represented with the help of what is called the *gradient operator*, symbolized by ∇ (read "del"). Operations by ∇ can be given various explicit expressions, depending on the particular set of coordinates used to specify positions $\mathbf{r}$ in the field. The most useful coordinates are the rectangular, cylindrical, and spherical ones, respectively implicit in the expressions for a displacement $d\mathbf{r}$ of $\mathbf{r}$

$$\begin{aligned} d\mathbf{r} &= \mathbf{i}\,dx + \mathbf{j}\,dy + \mathbf{k}\,dz \\ &= \mathbf{e}_s\,ds + \mathbf{e}_\varphi s\,d\varphi + \mathbf{k}\,dz \qquad \text{(A.1)} \\ &= \mathbf{e}_r\,dr + \mathbf{e}_\vartheta r\,d\vartheta + \mathbf{e}_\varphi r\sin\vartheta\,d\varphi \end{aligned}$$

(see Exercise A.1). Each line has three mutually perpendicular *unit* vectors pointing in a direction of displacement that increases a corresponding coordinate. The corresponding representations of the gradient derivatives are

$$\begin{aligned} \boldsymbol{\nabla} &= \mathbf{i}\frac{\partial}{\partial x} + \mathbf{j}\frac{\partial}{\partial y} + \mathbf{k}\frac{\partial}{\partial z} \\ &= \mathbf{e}_s\frac{\partial}{\partial s} + \mathbf{e}_\varphi\frac{\partial}{s\,\partial\varphi} + \mathbf{k}\frac{\partial}{\partial z} \\ &= \mathbf{e}_r\frac{\partial}{\partial r} + \mathbf{e}_\vartheta\frac{\partial}{r\,\partial\vartheta} + \mathbf{e}_\varphi\frac{\partial}{r\sin\vartheta\,\partial\varphi} \end{aligned} \tag{A.2}$$

Each component of this vector operator must have the dimensions of an inverse length.

Frame-independent results of the operations on fields are named and symbolized as follows:

$$\begin{aligned} \text{Gradient} &\equiv \boldsymbol{\nabla} U(\mathbf{r}) && \text{a vector} \\ \text{Divergence} &\equiv \boldsymbol{\nabla}\cdot\mathbf{u}(\mathbf{r}) && \text{a scalar} \\ \text{Curl} &\equiv \boldsymbol{\nabla}\times\mathbf{u}(\mathbf{r}) && \text{a vector} \end{aligned} \tag{A.3}$$

Divergences and curls of scalar fields like $U(\mathbf{r})$ are simply not definable, but the gradient of a vector field does sometimes occur; the result has nine components like $\nabla_x u_x$, $\nabla_x u_y$, ..., etc., and is called a *tensor of the second rank*.

The most useful explicit representations of the divergence and the curl are (see Exercise A.1), as applied to $\mathbf{u}(x, y, z)$,

$$\begin{aligned} \boldsymbol{\nabla}\cdot\mathbf{u} &= \frac{\partial u_x}{\partial x} + \frac{\partial u_y}{\partial y} + \frac{\partial u_z}{\partial z} \\ \boldsymbol{\nabla}\times\mathbf{u} &= \mathbf{i}\left(\frac{\partial u_z}{\partial y} - \frac{\partial u_y}{\partial z}\right) + \mathbf{j}\left(\frac{\partial u_x}{\partial z} - \frac{\partial u_z}{\partial x}\right) + \mathbf{k}\left(\frac{\partial u_y}{\partial x} - \frac{\partial u_x}{\partial y}\right) \end{aligned} \tag{A.4}$$

as applied to $\mathbf{u}(s, \varphi, z)$,

$$\begin{aligned} \boldsymbol{\nabla}\cdot\mathbf{u} &= \frac{\partial}{s\,\partial s}su_s + \frac{\partial u_\varphi}{s\,\partial\varphi} + \frac{\partial u_z}{\partial z} \\ \boldsymbol{\nabla}\times\mathbf{u} &= \mathbf{e}_s\left(\frac{\partial u_z}{s\,\partial\varphi} - \frac{\partial u_\varphi}{\partial z}\right) + \mathbf{e}_\varphi\left(\frac{\partial u_s}{\partial z} - \frac{\partial u_z}{\partial s}\right) + \mathbf{k}\left(\frac{\partial}{s\,\partial s}su_\varphi - \frac{\partial u_s}{s\,\partial\varphi}\right) \end{aligned} \tag{A.5}$$

and as applied to $\mathbf{u}(r, \vartheta, \varphi)$,

$$\begin{aligned}
\nabla \cdot \mathbf{u} &= \frac{\partial}{r^2\, \partial r} r^2 u_r + \frac{\partial}{r \sin \vartheta\, \partial \vartheta}(\sin \vartheta\, u_\vartheta) + \frac{\partial u_\varphi}{r \sin \vartheta\, \partial \varphi} \\
(\nabla \times \mathbf{u})_r &= \frac{1}{r \sin \vartheta}\left[\frac{\partial}{\partial \vartheta}(\sin \vartheta\, u_\varphi) - \frac{\partial u_\vartheta}{\partial \varphi}\right] \\
(\nabla \times \mathbf{u})_\vartheta &= \frac{\partial u_r}{r \sin \vartheta\, \partial \varphi} - \frac{\partial}{r\, \partial r} r u_\varphi \\
(\nabla \times \mathbf{u})_\varphi &= \frac{\partial}{r\, \partial r} r u_\vartheta - \frac{\partial u_r}{r\, \partial \vartheta}
\end{aligned} \tag{A.6}$$

The corresponding representations of ∇U merely require applying (A.2) to $U(\mathbf{r})$. The pre- and postmultiplications by functions of coordinates that occur with some of the derivatives in (A.5) and (A.6) are traceable to the fact that the directions $\mathbf{e}_s$, $\mathbf{e}_\varphi$, $\mathbf{e}_r$, and $\mathbf{e}_\vartheta$ themselves vary with position.

To be presented next are properties of gradients, divergences, and curls through which each gains significance.

A.2 GRADIENT FIELDS AND LINE INTEGRALS

A primary use of gradients like $\nabla U(\mathbf{r})$ is for the evaluation of field increments like the scalar

$$U(\mathbf{r} + d\mathbf{r}) - U(\mathbf{r}) \equiv dU(\mathbf{r}) = \nabla U \cdot d\mathbf{r} \tag{A.7}$$

The example

$$dU(r, \vartheta, \varphi) = \frac{\partial U}{\partial r}\, dr + \frac{\partial U}{\partial \vartheta}\, d\vartheta + \frac{\partial U}{\partial \varphi}\, d\varphi \tag{A.8}$$

representable as a scalar product of the last lines in (A.1) and (A.2), applied to U, illustrates just why denominators like those in (A.2) are needed.

Familiar examples of the resultant gradient vector fields are provided by newtonian forces, $\mathbf{F}(\mathbf{r}) = -\nabla U(\mathbf{r})$, that are (thus) derivable from potential energies $U(\mathbf{r})$. Then negatives of increments like (A.7), $-dU = -\nabla U \cdot d\mathbf{r} = \mathbf{F} \cdot d\mathbf{r}$, represent elements of work by the force during displacements like $d\mathbf{r}$.

Help in an overall visualization of a scalar field like $U(\mathbf{r})$ can come from diagrams of equipotential surfaces, each a continuum of points $\mathbf{r}_C$ at which $U(\mathbf{r}_C) = C$, a constant value that is determined as soon as some one point is chosen for an equipotential to pass through (Exercise A.5). The diagram may be supplemented with field lines, of force when U is a potential energy, i.e., continuous curves constructed so that at each point they cross they are parallel to the $\mathbf{F}(\mathbf{r}) = -\nabla U$ there. The field lines will always intersect each equipotential they cross at right angles. This is apparent from the fact that for any displacement

$d\mathbf{r} \equiv d\mathbf{r}_{\parallel}$ from a point of intersection and into any one of the directions along the equipotential surface (parallel to it)

$$F \cdot d\mathbf{r}_{\parallel} = -\nabla U \cdot d\mathbf{r}_{\parallel} = -dU = 0 \tag{A.9}$$

since U does not change from point to point of an equipotential. Work is done only in displacements having components $d\mathbf{r}_{\perp}$ (perpendicular to the surfaces) as the potential steps up or down from the value on one equipotential surface to that on another.

Next consider work done during a continuous succession of displacements $d\mathbf{r}$ that follow some curve C,

$$W(C) = \int_C d\mathbf{r} \cdot \mathbf{F}(\mathbf{r}) \tag{A.10}$$

Whatever the vector field here, the result is called a *line integral*; in general it can depend on just what path C is followed from a starting point $\mathbf{r}_o$ to an endpoint $\mathbf{r}$. However, in the special case that a scalar potential exists from which $\mathbf{F} = -\nabla U$ is derivable,

$$W = -\int_{\mathbf{r}_o}^{\mathbf{r}} d\mathbf{r} \cdot \nabla U = -\int_{U(\mathbf{r}_o)}^{U} dU = U(\mathbf{r}_o) - U(\mathbf{r}) \tag{A.11}$$

Now the result depends only on the starting and endpoints of the integration and *not* on just what path is followed between them. This independence of the path characterizes work by forces derivable from scalar potentials (Exercise A.4).

A related way of characterizing gradient vector fields like $\mathbf{F} = -\nabla U$ is by the property

$$\oint d\mathbf{r} \cdot \mathbf{F}(\mathbf{r}) = 0 \tag{A.12}$$

Here the integration path is to be any closed loop, as indicated by the symbol $\oint d\mathbf{r}$. The vanishing of loop integrals plainly follows from (A.11), since $\mathbf{r}$ and $\mathbf{r}_o$ are now identical points [and it is equally plain that (A.12) actually holds only if the potentials are defined with a single value at every point of the loop].

A.3 DIVERGENCES AND FIELD SOURCES

A divergence derivative like $\nabla \cdot \mathbf{u}(\mathbf{r})$, as evaluated at some point $\mathbf{r}$ in the field, can be defined in a way that represents its primary significance by considering an element of volume $\Delta V(\mathbf{r}) \to 0$ enclosing the point $\mathbf{r}$. Let the vectors $d\mathbf{S}$ be area elements on the infinitesimal surface enclosing ΔV and facing *outward* from it. Then

$$\nabla \cdot \mathbf{u} \equiv \lim_{\Delta V \to 0} \frac{\oint d\mathbf{S} \cdot \mathbf{u}}{\Delta V} \tag{A.13}$$

where the symbol $\oint d\mathbf{S}$ stands for integration over the entire surface ($\to 0$) enclosing the volume ΔV ($\to 0$). How the representations in (A.4), (A.5), and (A.6) follow from this can be found as in Exercise A.1.

Any *finite* volume V can be divided up into contiguous elements ΔV. Then, from the fact that the area integrations over surfaces separating contiguous elements ΔV have opposing signs and will cancel, a summation of contributions ΔV $(\nabla \cdot \mathbf{u})$ leads to the well known *Gauss theorem*,

$$\int_V dV(\nabla \cdot \mathbf{u}) = \oint d\mathbf{S} \cdot \mathbf{u} \tag{A.14}$$

where the area integration now extends only over the outer surface enclosing the entire volume V. The operator equation

$$\int_V dV\nabla = \oint d\mathbf{S} \tag{A.15}$$

implicit in (A.14) actually holds more generally (Exercise A.3).

Flux and Field-Line Representations

An element like $\mathbf{u} \cdot d\mathbf{S}$ is called a *flux* through the area $d\mathbf{S}$. The name seems to have been suggested by the hydrodynamical example, when $\mathbf{u}(\mathbf{r})$ is the velocity of fluid passing through $\mathbf{r}$, the field lines are formed of fluid streamlines, and the flux $\mathbf{u} \cdot d\mathbf{S}$ is the volume of fluid passing through $d\mathbf{S}$ per unit time. The last assertion can be understood in the terms of Fig. A.1, which indicates an arbitrarily oriented area element $d\mathbf{S}$ cutting through streamlines. A cylinder of side length $\mathbf{u}\ dt$ and right cross section $dS \cos \alpha$ is extended back from $d\mathbf{S}$, parallel to the streamlines there. It will contain just the volume of fluid, expressed with the help of the angle α in the diagram,

$$(u\ dt)(dS \cos \alpha) = (\mathbf{u} \cdot d\mathbf{S})\ dt$$

that will flow through $d\mathbf{S}$ during the time interval dt. The expression also shows that $\mathbf{u}(\mathbf{r})$ can be called a *flux density*, i.e., a flow per unit time *and* per unit cross-sectional area ($dS \cos \alpha$) transverse to $\mathbf{u}$.

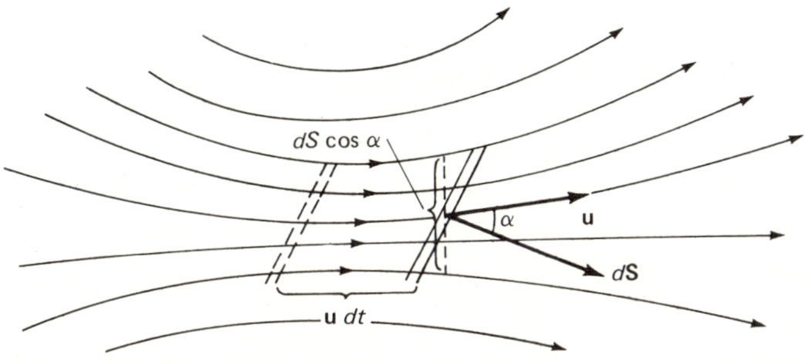

Figure A.1

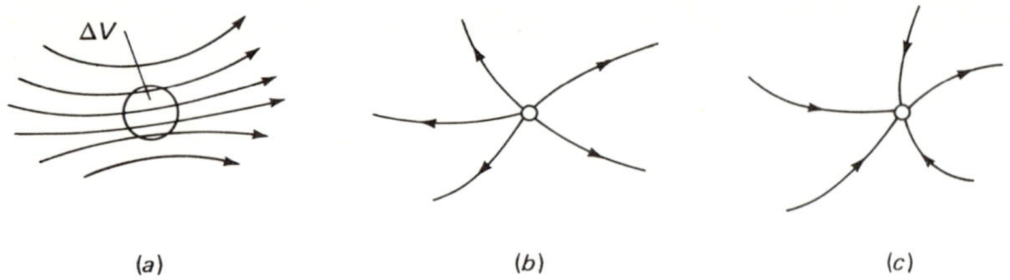

Figure A.2 (*a*) $\nabla \cdot \mathbf{u} = 0$; (*b*) $\nabla \cdot \mathbf{u} > 0$; (*c*) $\nabla \cdot \mathbf{u} < 0$.

A field-line representation can be adopted for any vector field, as done in the discussion of (A.7) for the force field $\mathbf{F}(\mathbf{r})$, now regardable as a flux density of impulse. The representation can also be made quantitative by using a number of lines per unit cross section in the neighborhood of each point proportional to the *magnitude* of the field at the point.† Then the flux through any given area is measured by the number of lines passing through that area.

The quantitative and directed field-line representation makes visualizations like those in Fig. A.2 possible. A region having $\nabla \cdot \mathbf{u} = 0$ must have a line leaving every point which a line enters so that the flux integral in the definition (A.13) will vanish at the point. Conversely, any space in which the field lines are continuous must have $\nabla \cdot \mathbf{u} = 0$ everywhere in it; a region with $\nabla \cdot \mathbf{u} \neq 0$ cannot properly be given a continuous field-line representation in a density proportional to the field strength (Exercise 2.4). Figure A.2*b*, *c* indicates *isolated* points at which $\nabla \cdot \mathbf{u} \neq 0$, surrounded by $\nabla \cdot \mathbf{u} = 0$ regions of field representable by continuous lines. If $\nabla \cdot \mathbf{u} > 0$ at an isolated point, the definition (A.13) requires a net emanation of field lines into the surrounding region; a finite divergence confined to a point thus introduces a discontinuity in field lines through the point, a singularity there.

Clearly, any point of $\nabla \cdot \mathbf{u} > 0$ serves as a source of field. There is a net emanation of lines out of it, and hence "new" lines are started there. In the hydrodynamical example, the point may represent an idealized end of an open standpipe out of which fresh fluid flows. A point at which $\nabla \cdot \mathbf{u} < 0$ has lines ending there; flux is disappearing and it is called a *sink* of the field. It is actually more convenient to think of sinks as negative sources. The net result is a measure of field sources by its positive and negative divergence derivatives.

The field sources need not be confined to singular points but may be continuously distributed in finite densities over extended regions. The definition (A.13) of the divergence at a point can be read as the efflux (or influx) per unit of continuous volume at the point.

The result (A.14) of the Gauss integration is spoken of as the *net* flux out of the enclosed volume, since the various elements $\mathbf{u} \cdot d\mathbf{S}$ may have opposite signs on different parts of the enclosing surface.

† The possibility of doing this with *continuous* lines has an implication discussed on page 10 and exemplified in Exercise 2.4.

Current Densities and the Continuity Equation

An example closer to one use of field divergences for describing electromagnetism is provided by the hydrodynamics of *compressible* fluid flow. Now a variable mass density $\rho_m(\mathbf{r})$ is introduced and a flux $\rho_m(\mathbf{u} \cdot d\mathbf{S})$ defined to measure the *mass* of fluid passing through $d\mathbf{S}$ per unit time. The flux density at a point is then representable by the vector field

$$\mathbf{j}_m(\mathbf{r}) \equiv \rho_m(\mathbf{r})\mathbf{u}(\mathbf{r}) \tag{A.16}$$

called a *current density*—of mass, in the hydrodynamical example.

Next consider a more general (than just steady) flow of the compressible fluid, one in which both the velocity and density vary in time as well as over space, so that $\rho_m \mathbf{u} = \mathbf{j}_m(\mathbf{r}, t)$. In general, the mass of fluid $m(t) = \int dV\rho_m(\mathbf{r}, t)$ in any chosen volume V will change during the motions, and since mass is expected to be conserved, there must be a compensating flow of mass $\mathbf{j}_m = \rho_m \mathbf{u}$ per unit time and per unit cross-sectional area into or out of V through its surface:

$$\frac{dm}{dt} = \frac{d}{dt}\int dV\rho_m = -\oint d\mathbf{S} \cdot \mathbf{j}_m \tag{A.17}$$

The area elements $d\mathbf{S}$ here are defined to point outward from V, and hence each contribution $\mathbf{j}_m \cdot d\mathbf{S} > 0$ leads to a *decrease* of the enclosed mass. The mass *increases* only if there are more or greater negative elements, $\mathbf{j}_m \cdot d\mathbf{S} < 0$, than positive ones, so that there is a *net* influx: $\oint d\mathbf{S} \cdot \mathbf{j}_m < 0$.

With the help of the Gauss theorem (A.14), now applied to the vector field $\mathbf{j}_m(\mathbf{r}, t)$ at a given moment t, the mass-conservation equation (A.17) can be expressed as the balance to zero:

$$\frac{d}{dt}\int dV\rho_m(\mathbf{r}, t) + \oint d\mathbf{S} \cdot \mathbf{j}_m = \int dV\left(\frac{\partial \rho_m}{\partial t} + \nabla \cdot \mathbf{j}_m\right) = 0 \tag{A.18}$$

This must hold whatever the V chosen (it may be a $V \to 0$ enclosing just one point), and hence

$$\frac{\partial \rho_m}{\partial t} + \nabla \cdot \mathbf{j}_m = 0 \tag{A.19}$$

must be valid at every point. Relations like this are called *continuity equations* and are used to express the conservation of all sorts of quantities that may be distributed over space (and time). Elsewhere in this treatise, similar equations are employed to represent the conservation of electric charge (1.14), and of energy, momentum, and angular momentum spread out over an electromagnetic field [(6.7), (6.26), and (6.37)].

A.4 FIELD CURL

It will next be found that the curl derivatives, like $\nabla \times \mathbf{u}$, of a vector field can serve as measures of other properties than those measured by its divergences $\nabla \cdot \mathbf{u}$.

Whereas the primary significance of $\nabla \cdot \mathbf{u}$ emerged after it was integrated over a volume, as in (A.14), for $\nabla \times \mathbf{u}$ it emerges after integration over an area, i.e., after a flux of curl is formed. The definition at a point, analogous to (A.13), can be expressed for a flux element $(\nabla \times \mathbf{u}) \cdot \Delta\mathbf{S}$ through an arbitrarily oriented area element at the point. Let the direction faced by $\Delta\mathbf{S}$ be indicated by a unit vector $\mathbf{n}$, so that $\Delta\mathbf{S} = \mathbf{n}\,\Delta S$ and the flux element becomes $(\nabla \times \mathbf{u}) \cdot \mathbf{n}\,\Delta S$. Now the component of the vector curl in the arbitrary direction $\mathbf{n}$ can be defined as

$$(\nabla \times \mathbf{u}) \cdot \mathbf{n} \equiv \lim_{\Delta S \to 0} \frac{\oint d\mathbf{r} \cdot \mathbf{u}}{\Delta S} \tag{A.20}$$

where the loop integral is to follow the succession of elements $d\mathbf{r}$ forming the *perimeter* of $\Delta\mathbf{S}$ and in the sense a right-handed screw must be rotated in order to progress in the direction $\mathbf{n}$ of $\Delta\mathbf{S}$. How the representations in (A.4) to (A.6), with the signs indicated, follow from the definition (A.20) can be found as in Exercise A.1.

The evaluation of the flux of curl can be extended over any finite area S by dividing it up into contiguous area elements ΔS. Then, from the fact that line integrals on contiguous sides of the ΔS will cancel when they are all carried out in the senses required for (A.20), a summation of the flux elements leads to the well-known *Stokes theorem*,

$$\int_S d\mathbf{S} \cdot (\nabla \times \mathbf{u}) = \oint d\mathbf{r} \cdot \mathbf{u} \tag{A.21}$$

in which the loop integral extends only over the outside perimeter of the entire area S. The operator equation

$$\int_S d\mathbf{S} \cdot (\nabla \times \cdots) = \oint d\mathbf{r} \cdot (\cdots) \tag{A.22}$$

implicit in (A.21) holds quite as generally as (A.15).

In hydrodynamical cases, circulations of the fluid can be started that have velocity field lines forming complete loops. On each such loop, the elements $\mathbf{u} \cdot d\mathbf{r}$ have the same sign all around the loop, and the so-called circulation integral

$$\Gamma \equiv \oint d\mathbf{r} \cdot \mathbf{u} \tag{A.23}$$

cannot vanish. This is a situation in which the Stokes theorem (A.21) requires that nonvanishing curls of the velocity field exist in the region. The theorem may now be spoken of as an equality of the flux of curl, through any surface, to the circulation† around the edge of that surface. The curl derivative *at a point* (A.20) amounts to an *area density* of circulation about the point.

† In electromagnetism, there occur such loop integrals as (A.23) called emf (1.32) and mmf (1.19).

The Divergencelessness of Curls

Any given loop to which the Stokes theorem (A.21) can be applied encloses any number of different surfaces, all having the same loop as perimeter and each continuously extended from it (think of variously shaped caps worn on the same perimeter of your head). Yet the theorem has it that the same loop integral equals the flux of curl evaluated on different examples of that continuous variety of surfaces, a surface independence analogous to the path independence (A.11) found for line integrals of gradients. Here, if S_1 and S_2 are any pair of surfaces based on the same perimeter, their flux *difference* vanishes,

$$\int_{S_1} d\mathbf{S} \cdot (\nabla \times \mathbf{u}) - \int_{S_2} d\mathbf{S}' \cdot (\nabla \times \mathbf{u}) = 0 \equiv \oint d\mathbf{S} \cdot (\nabla \times \mathbf{u}) \qquad \text{(A.24)}$$

just as loop integrals of gradient fields vanish (A.12), if $d\mathbf{S}$ and $-d\mathbf{S}'$ are elements facing outward from the volume V enclosed between S_1 and S_2. Now the vanishing total flux integral is equivalent to a volume integral over V of $\nabla \cdot (\nabla \times \mathbf{u})$, according to the Gauss theorem (A.14). Since the perimeter, S_1, S_2 (and hence V) could be chosen quite arbitrarily, the divergence of any curl field must vanish everywhere, and

$$\nabla \cdot (\nabla \times \cdots) \equiv 0 \qquad \text{(A.25)}$$

must be a mathematical identity. This might have been expected from the isomorphism to the identity $\mathbf{A} \cdot (\mathbf{A} \times \mathbf{B}) \equiv 0$ for ordinary vectors, valid because $\mathbf{A} \times \mathbf{B}$ is by definition perpendicular to each of its vector factors. It is easy to check that the identity (A.25) is satisfied by each of the explicit representations (A.4), (A.5), and (A.6).

The findings here have an important consequence: any vector field which is divergenceless ("sourceless" or "solenoidal") in a region can be expressed as the curl of another field definable in that region. A *general* solution of the equation

$$\nabla \cdot \boldsymbol{\omega}(\mathbf{r}) = 0 \qquad \text{is} \qquad \boldsymbol{\omega} = \nabla \times \mathbf{A}(\mathbf{r}) \qquad \text{(A.26)}$$

for which $\mathbf{A}(\mathbf{r})$ can be chosen quite arbitrarily, insofar as $\boldsymbol{\omega}(\mathbf{r})$ need only be divergenceless. Whatever the $\mathbf{A}(\mathbf{r})$ chosen, it is called a *vector potential* from which the divergenceless field in derivable, just as the $\mathbf{F} = -\nabla U(\mathbf{r})$ discussed above is derivable from a scalar potential (via a gradient, rather than curl, operation). Of course, the identity (A.25) by itself only shows that representability by a curl is *sufficient* for divergencelessness. However, the necessity for some vector potential to exist, for any given divergenceless field, can also be demonstrated.†

† Readers might construct such a demonstration for themselves by starting with the fact that a divergenceless $\boldsymbol{\omega}(\mathbf{r})$ has the same flux through any of the surfaces that can be defined with the same perimeter. Then some continuum of values $d\mathbf{r} \cdot \mathbf{A}(\mathbf{r})$ must be assignable along that perimeter, such that $\oint d\mathbf{r} \cdot \mathbf{A}(\mathbf{r}) = \int d\mathbf{S} \cdot \boldsymbol{\omega}(\mathbf{r})$. However, the demonstration can safely be left to mathematicians.

The Curllessness of Gradients

Another mathematical identity relating space derivatives of a field is one comparable to the ordinary vector identity $\mathbf{A} \times \mathbf{A} \equiv 0$, valid because, by definition, vector products vanish when their vector factors are parallel to each other. Thus, the isomorphism

$$\nabla \times \nabla \equiv 0 \tag{A.27}$$

is to be expected and is easily verified for any one of the explicit representations of ∇ in (A.2) by substituting it for the operand $\mathbf{u}$ in the corresponding curl expression of (A.4), (A.5), or (A.6). The identity is necessary in order that the Stokes theorem (A.21), as applied to any gradient field like $\mathbf{F} = -\nabla U$, be consistent with the vanishing of the loop integral (A.12), much as (A.25) is necessary to (A.24).

Because of the automatic curllessness of any gradient, the general solution of the equation

$$\nabla \times \mathbf{u}(\mathbf{r}) = 0 \qquad \text{is} \qquad \mathbf{u} = -\nabla\phi(\mathbf{r}) \tag{A.28}$$

with a $\phi(\mathbf{r})$ that can be chosen quite arbitrarily insofar as $\mathbf{u}(\mathbf{r})$ need only be curlless. Moreover, a scalar potential like $\phi(\mathbf{r})$ can be found for any given curlless field $\mathbf{u}(\mathbf{r})$ simply by constructing an integral like (A.11), which is written for a scalar potential $U(\mathbf{r})$.

The integral in (A.11) actually measures only a potential *difference*. This implies that the definition of any scalar potential can be changed by adding to it any constant without altering the field it represents if the field is characterized only by giving a gradient, like the $\mathbf{u}(\mathbf{r})$ in (A.28). This *gauge-invariance property* is characteristic of all descriptions by scalar potentials and is familiar from the example of the scalar potential energy $U(\mathbf{r})$. Such an energy can be gauged by work done from any chosen starting point $\mathbf{r}_o$ of (A.11); changing the choice of starting point merely adds a constant to the definition of the potential energy without altering any energy *differences*—all that is ever observable.

The finding (A.27) that all gradient fields are curlless also has implications for gauging vector potentials, like $\mathbf{A}(\mathbf{r})$ of (A.26), representing a field characterized by giving a curl distribution $\boldsymbol{\omega}(\mathbf{r})$. Plainly, any gradient field like $\nabla\chi(\mathbf{r})$ can be added to $\mathbf{A}(\mathbf{r})$ without altering its curl. Thus

$$\mathbf{A}(\mathbf{r}) \to \mathbf{A}'(\mathbf{r}) = \mathbf{A}(\mathbf{r}) + \nabla\chi(\mathbf{r}) \tag{A.29}$$

with $\chi(\mathbf{r})$ arbitrary, characterizes the gauge-invariance property characteristic of descriptions by vector potentials.† $\nabla \times \mathbf{A} \equiv \nabla \times \mathbf{A}'$ because $\nabla \times (\nabla\chi) \equiv 0$. The point here will be enlarged upon in the development from (A.51) to (A.54).

† Similarly wide choices of gauge characterize the definitions of conjugate momenta, hamiltonian energies, and lagrangians in ordinary classical mechanics (Ref. 17, pages 173–175; see also Chap. 13 in this book).

A.5 THE LAPLACIAN

Consider the forms in which the lines of a vector field $\mathbf{u}(\mathbf{r})$ may arise. There may be source points from which they emanate, as in Fig. A.2*b*, and this form of their generation can be represented by giving

$$\nabla \cdot \mathbf{u}(\mathbf{r}) = s(\mathbf{r}) \tag{A.30}$$

The points at which these divergences have values need not be isolated ones but may form some continuous distribution of source points. Moreover, the source strengths $s(\mathbf{r})$ may be negative at some points, so that they actually form sinks of the field at which field lines terminate, as in Fig. A.2*c*.

It is also possible that, superposed, are some field lines which arise at once in complete loops, neither starting up nor terminating anywhere, as indicated in Fig. A.3. Such everywhere-continuous field lines are divergenceless (solenoidal). On the other hand, they plainly lead to nonvanishing loop integrals; certainly, loop integration along any one of the complete field loops, itself, will yield a nonzero result since every segment of such makes a contribution of the same sign. To get these nonvanishing circulations, the field loops must be threaded by flux of curl, according to the Stokes theorem. There must be points at which the curl of the vector field has value

$$\nabla \times \mathbf{u}(\mathbf{r}) = \mathbf{c}(\mathbf{r}) \tag{A.31}$$

In hydrodynamics, such vectors are used to measure vortices, or vortex flows, which can generate a circulating distribution of fluid flows in the regions surrounding them.

It is of general interest to see how far *giving* all the scalar sources $s(\mathbf{r})$ of a vector field and its circulation densities $\mathbf{c}(\mathbf{r})$ everywhere can determine the field via Eqs. (A.30) and (A.31). The emanations and terminations of field lines at points of positive and negative divergence, respectively, and their generation as closed loops, without the discontinuities, are the only way their generation can be visualized, and hence the importance of the problem.

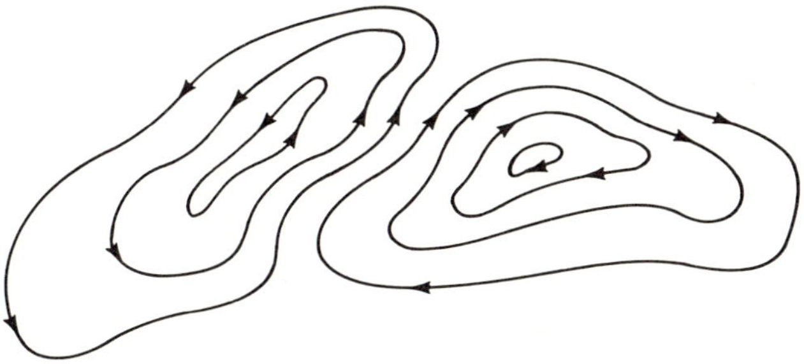

Figure A.3

A division into two simpler problems can be obtained by splitting the field $\mathbf{u}(\mathbf{r})$ into a curlless part $\mathbf{u}_s(\mathbf{r})$ and a divergenceless part $\mathbf{u}_c(\mathbf{r})$:

$$\mathbf{u}(\mathbf{r}) = \mathbf{u}_s(\mathbf{r}) + \mathbf{u}_c(\mathbf{r}) \tag{A.32}$$

This still permits $\mathbf{u}$ to have any divergence and any curl, and it will have the given scalar sources $s(\mathbf{r})$ and given circulation densities $\mathbf{c}(\mathbf{r})$ if

$$\nabla \cdot \mathbf{u}_s = s(\mathbf{r}) \qquad \nabla \times \mathbf{u}_s = 0 \tag{A.33}$$

and

$$\nabla \cdot \mathbf{u}_c = 0 \qquad \nabla \times \mathbf{u}_c = \mathbf{c}(\mathbf{r}) \tag{A.34}$$

are satisfied by the component fields. The splitting will always be permissible since the resultant $\mathbf{u}$ will prove to be a *unique* solution of (A.30) and (A.31).

The Laplacian Derivative

Any curlless field can be represented as a gradient of some scalar potential (A.28), and hence the solution of Eqs. (A.33) for the curlless field $\mathbf{u}_s(\mathbf{r})$ can be reduced to finding a scalar field $\phi(\mathbf{r})$ such that

$$\mathbf{u}_s = -\nabla\phi(\mathbf{r}) \qquad \text{and} \qquad \nabla^2\phi = -s(\mathbf{r}) \tag{A.35}$$

where $\nabla^2 \equiv \nabla \cdot \nabla (\equiv \text{div grad})$. The second-order differential equation is of a type called a *Poisson equation.*

Encountered here is an extension of the field calculus to *second* derivatives with respect to a vector. The type ∇^2, known as a *Laplace operator* or *laplacian*, has the cartesian representation

$$\nabla^2 = \frac{\partial^2}{\partial x^2} + \frac{\partial^2}{\partial y^2} + \frac{\partial^2}{\partial z^2} \tag{A.36}$$

as follows from forming $\nabla \cdot \nabla$ in accordance with (A.4) and (A.2). The laplacian occurs with great frequency in the formulation of physical descriptions and should receive special attention.

The frequency of occurrence can be traced to the fact that $\nabla^2\phi(\mathbf{r})$ amounts to a measure of departures from smoothness in variations of the field over a space. When it vanishes everywhere in some region, i.e., the field satisfies the so-called *Laplace equation*

$$\nabla^2\phi(\mathbf{r}) = 0 \tag{A.37}$$

there, then $\phi(\mathbf{r})$ varies from any *given* values on the borders of the region as smoothly as is consistent with matching up with these border values. The variations allowed have minimum curvature, as is especially evident for one-dimensional variations when, for example, $\phi(x)$ need be a function of x only, so that

$$\frac{d^2\phi}{dx^2} = 0 \rightarrow \phi = ax + b \tag{A.38}$$

with a, b the arbitrary constants of integration which make the solution a general one. Only straight-line variation, with no curvature at all, is being permitted, in joining any pair of values $\phi_0 = ax_0 + b$ and $\phi_1 = ax_1 + b$ which might be given at the edges of a region $x_0 \leq x \leq x_1$. In more general cases, with more complicated boundaries or value distributions on them, the variations joining these will naturally have to be more complex but may be regarded as smooth, consistent with the boundary conditions, if they satisfy the Laplace equation.†

Besides the cartesian representation (A.36), it will be important to have available the spherical and cylindrical coordinate representations of the laplacian (and see Exercise A.1).

The spherical representation follows readily from applying the divergence operation of (A.6) to the gradient as represented in (A.2):

$$\nabla^2 = \frac{\partial}{r^2\,\partial r}\left(r^2\frac{\partial}{\partial r}\right) + \frac{\partial}{r^2\sin\vartheta\,\partial\vartheta}\left(\sin\vartheta\frac{\partial}{\partial\vartheta}\right) + \frac{\partial^2}{r^2\sin^2\vartheta\,\partial\varphi^2} \qquad \text{(A.39}a\text{)}$$

The radial derivative operation here can be given alternative forms,

$$\frac{\partial}{r^2\,\partial r}\left(r^2\frac{\partial}{\partial r}\right) \equiv \frac{\partial^2}{\partial r^2} + \frac{2}{r}\frac{\partial}{\partial r} \equiv \frac{\partial^2}{r\,\partial r^2}r \equiv \left(\frac{\partial}{r\,\partial r}r\right)^2 \equiv \left(\frac{\partial}{\partial r} + \frac{1}{r}\right)^2 \qquad \text{(A.39}b\text{)}$$

one or the other of which may be most convenient in various contexts.

The cylindrical representation is equally easy to form:

$$\nabla^2 = \frac{\partial}{s\,\partial s}\left(s\frac{\partial}{\partial s}\right) + \frac{\partial^2}{s^2\,\partial\varphi^2} + \frac{\partial^2}{\partial z^2} \qquad \text{(A.40)}$$

The radial derivative term here can be expanded into $\partial^2/\partial s^2 + \partial/(s\,\partial s)$.

The Green Theorems

Just as the Gauss theorem enables conclusions to be drawn about divergence derivatives and the Stokes theorem about curls, so *Green theorems* reveal properties of fields following from their Laplace derivatives, e.g., the ones given by the Poisson equation (A.35). Initially, the Green theorems may be regarded as being concerned with some *pair* of scalar fields, $\phi(\mathbf{r})$ and $\chi(\mathbf{r})$, say. In application, it will turn out that when $\phi(\mathbf{r})$ is a potential arising from sources according to the Poisson equation, $\chi(\mathbf{r})$ can serve as a propagator of the source influence to various field points. It is sometimes given the more neutral name Green's function, instead of propagator. The aptness of the physical name will become more fully evident in the applications to electromagnetic theory (Sec. 3.1). For the present, a more purely mathematical viewpoint will be maintained.

† For more detailed justifications of this description, consult the first paragraphs of chap. 13 and exercise 14.1 in Ref. 17. An additional reason for the frequent occurrence of the laplacian can be found there; it helps represent first departures from *equilibria*, in a Hooke's-law approximation of restoring forces aroused by departures from smoothness.

The first Green theorem amounts to an application of the Gauss theorem to the divergences of a vector field formed as $\chi\nabla\phi$ (Exercise A.13):

$$\nabla \cdot (\chi\nabla\phi) = (\nabla\chi) \cdot (\nabla\phi) + \chi\nabla^2\phi$$

Integration over any volume V yields

$$\int_V dV[\chi\nabla^2\phi + (\nabla\chi) \cdot (\nabla\phi)] = \oint d\mathbf{S} \cdot (\chi\nabla\phi) \tag{A.41}$$

with the right side a flux through the surface enclosing V. This result will prove useful for investigating conditions for the *uniqueness* of solutions of the Poisson equation.

The second Green theorem gives symmetrical roles to the pair of arbitrary scalar fields $\phi(\mathbf{r})$ and $\chi(\mathbf{r})$. It can be obtained by forming an expression like (A.41) with ϕ and χ interchanged, then subtracting it from (A.41) itself

$$\int_V dV(\chi\nabla^2\phi - \phi\nabla^2\chi) = \oint d\mathbf{S} \cdot (\chi\nabla\phi - \phi\nabla\chi) \tag{A.42}$$

This result will be put to use immediately.

Solution of the Poisson Equation

The second Green theorem can be made to yield the desired solution of the Poisson equation (A.35) through suitable choices for the scalar fields, ϕ and χ, and the integration volume V.

The scalar ϕ will be chosen to be the solution desired, and then the term $\int dV\phi\nabla^2\chi$ of (A.42) can be made to yield the value of ϕ at an arbitrarily chosen point $\mathbf{r}_f$, inside the volume of integration, by choosing the second scalar χ to be such that $\nabla^2\chi = 0$ everywhere except at that point, called the *field point*.

A sufficient choice of χ is easy to find. It should suffice for it to be a function of just the distance R from the chosen field point; the spherical coordinate representation (A.39), with the distance R as its radial variable, can be used,

$$\nabla^2\chi(R) = \frac{d}{R^2\,dR} R^2 \frac{\partial\chi}{\partial R} = \frac{d^2}{R\,dR^2} R\chi = 0 \qquad \text{except at } R = 0 \tag{A.43a}$$

$R\chi = \text{const} + (\text{const}')R$ is the general solution, and $\chi = 1/R$ will be a sufficient one. It has the desired property

$$\nabla^2 \frac{1}{R} = 0 \qquad \text{except at } R = 0 \tag{A.43b}$$

because $\chi = 1/R \to \infty$ as $R \to 0$, and also its derivatives do not vanish there. The property leads to the conclusion that, whatever the volume V,

$$\int_V dV\phi\nabla^2\chi \to \phi(\mathbf{r}_f) \int_V dV\nabla^2 \frac{1}{R} \tag{A.44}$$

since only the value $\phi(\mathbf{r}_f)$ of ϕ is not multiplied by zero when $R = 0$ at the field point $\mathbf{r}_f$.

The integral remaining in (A.44) can be evaluated by applying the Gauss theorem to a sphere of any radius, $R = \epsilon$ say, centered on $R = 0$, since only the point $R = 0$ in it can give a nonvanishing contribution. Thus

$$\int_V dV \boldsymbol{\nabla} \cdot \boldsymbol{\nabla} \frac{1}{R} = \oint d\mathbf{S}(\epsilon) \cdot \left(\boldsymbol{\nabla} \frac{1}{R} \right)_{R=\epsilon} = 4\pi\epsilon^2 \left(-\frac{1}{\epsilon^2} \right) = -4\pi \tag{A.45}$$

Only the radial component, $\partial(R^{-1})/\partial R = -1/R^2$, of the gradient, which is constant on the spherical surface here, contributes to the flux through it. The conclusion is that

$$\int_V dV \phi \nabla^2 \chi = -4\pi\phi(\mathbf{r}_f) \tag{A.46}$$

can be put into the Green-theorem expression (A.42), if $\chi = 1/R$ is taken, with R a distance from $\mathbf{r}_f$.

The remaining volume integral in (A.42) can have $\nabla^2\phi = -s(\mathbf{r})$ put into it, and hence the theorem yields the expression

$$\phi(\mathbf{r}_f) = \int_V dV(\mathbf{r}_s) \frac{s(\mathbf{r}_s)}{4\pi R} + \oint \frac{d\mathbf{S}}{4\pi} \cdot \left(\frac{1}{R} \boldsymbol{\nabla}\phi - \phi \boldsymbol{\nabla} \frac{1}{R} \right) \tag{A.47}$$

The notation $\mathbf{r}_s$ has been introduced here for the points to be ranged over in the volume integration, in order to distinguish the source points, at which $s(\mathbf{r}_s) \neq 0$ exists, from the field points $\mathbf{r}_f$, at which values of the solution are desired. The values of $\chi(\mathbf{r}_s) = 1/R$ in that integral have $R = |\mathbf{r}_f - \mathbf{r}_s|$, the distance of each source point from the field point. In the area integration, $R = |\mathbf{r}_f - \mathbf{r}_S|$ when the $\mathbf{r}_S$ are positions on the outer surface of the volume of integration V.

Expression (A.47) may not yet be considered an explicit solution for ϕ since ϕ is also involved in the surface integral. However, the interest here is in the solution determined when the sources of it are given everywhere, and $V \to \infty$ is taken in order to include them all. Then the surface is a very remote one, with $\chi = 1/|\mathbf{r}_f - \mathbf{r}_s| \to 0$ on it, and it can be presumed that the surface integral vanishes. This still entails the assumption that sources in any finite portion of space will have no field effects, $\mathbf{u}_s = -\boldsymbol{\nabla}\phi$, infinitely far away, but that is not difficult to accept.† The conclusion is that the curlless field $\mathbf{u}_s = -\boldsymbol{\nabla}\phi$, with

$$\phi(\mathbf{r}) = \oint \frac{dV(\mathbf{r}_s)s(\mathbf{r}_s)}{4\pi|\mathbf{r} - \mathbf{r}_s|} \tag{A.48}$$

is determined when all the sources $s(\mathbf{r}_s)$ are given, everywhere. The symbol $\oint$ here denotes integration over all space, but in practice it only need extend far enough to include sources with detectable effect at the field points of interest.

† This point is discussed in greater detail in connection with electromagnetism, in Sec. 3.1, but the conclusion is the same.

Because it may have seemed a bit arbitrary to have chosen the particular Green's function $\chi = 1/R$ in extracting the expression for $\phi(\mathbf{r}_f)$ above from Green's theorem, a check of the uniqueness of this solution is desirable. A standard procedure is to assume, for the sake of the argument, that two solutions do exist, calling the one already obtained ϕ_1 and letting ϕ_2 be a second one. Then choose $\phi = \chi = \phi_1 - \phi_2$ for *both* the scalar fields in expression (A.41) of the first Green theorem, which then yields

$$\int_V dV |\nabla(\phi_1 - \phi_2)|^2 = \frac{1}{2} \oint d\mathbf{S} \cdot \nabla(\phi_1 - \phi_2)^2 \tag{A.49}$$

since $\nabla^2(\phi_1 - \phi_2) = 0$ when both ϕ_1 and ϕ_2 satisfy $\nabla^2 \phi_{1,2} = -s$. In the case of interest here, the right side is taken to vanish because of the remoteness of the surface, and then the contributions of every element to the integral on the left vanish, since none can be negative. Thus $\mathbf{u}_1 = -\nabla\phi_1 = \mathbf{u}_2 = -\nabla\phi_2$ at every point, and the solution found is the unique one under the boundary conditions which have been assumed here. Of course, the potentials ϕ_1 and ϕ_2 may still differ by a constant, but scalar potentials are defined only within an arbitrary additive constant in the first place.

The Divergenceless Part of the Field

The preceding discussion has shown how giving all the scalar sources $s(\mathbf{r})$ of a field $\mathbf{u}(\mathbf{r})$, determines its curlless part $\mathbf{u}_s(\mathbf{r})$ of (A.32). The remaining, divergenceless, part $\mathbf{u}_c(\mathbf{r})$ is to be determined from the given circulation densities $\mathbf{c}(\mathbf{r})$ by Eqs. (A.34).

The problem can be reduced from solving the simultaneous equations in (A.34) to a matter of solving a single vector equation for a vector potential, denoted $\mathbf{A}(\mathbf{r})$ here, by using the fact (A.26) that any divergenceless field is representable as the curl of some vector potential

$$\mathbf{u}_c = \nabla \times \mathbf{A}(\mathbf{r}) \tag{A.50}$$

Substituting this solution of the first of Eqs. (A.34) into the second one yields

$$\nabla \times (\nabla \times \mathbf{A}) = \mathbf{c}(\mathbf{r}) \tag{A.51}$$

as the equation from which a suitable vector potential is to be found.

Encountered here is another variety of second derivative with respect to a position vector, additional to the laplacian ∇^2. The two types are related as in

$$\nabla \times (\nabla \times \mathbf{A}) = \nabla(\nabla \cdot \mathbf{A}) - \nabla^2 \mathbf{A} \tag{A.52}$$

to be expected from the well-known vector identity (Exercise A.15)

$$\mathbf{A} \times (\mathbf{B} \times \mathbf{C}) = \mathbf{B}(\mathbf{A} \cdot \mathbf{C}) - (\mathbf{A} \cdot \mathbf{B})\mathbf{C} \tag{A.53}$$

When such isomorphisms are used for conclusions about vector *operators*, like ∇, care must be taken to maintain a proper order of the factors. For (A.52), it must be recognized that both derivative operations apply to the field $\mathbf{A}(\mathbf{r})$ and must continue to appear to its left. The result (A.52) can also be verified directly, e.g., by using cartesian representations for all the operations. When curvilinear coordinates like the spherical or cylindrical ones are used, it should be observed that, for example, $(\nabla^2 \mathbf{A})_r \neq \nabla^2 A_r$ for the radial components (or curvilinear ones, see Exercise A.16). However, for decompositions on fixed directions in space, any cartesian ones, $(\nabla^2 \mathbf{A})_i = \nabla^2 A_i$. Quite generally, it is only resolutions like the latter that are really essential.

It is known from (A.29) that a variety of vector potentials, differing from each other by an arbitrary gradient field, are equally suitable for representing a given divergenceless field. Advantage can be taken of this freedom to simplify the determination of a suitable vector potential $\mathbf{A}(\mathbf{r})$ from Eq. (A.51). Only the circulation densities $\nabla \times \mathbf{A} = \mathbf{u}_c(\mathbf{r})$ of $\mathbf{A}$ have been specified so far, in (A.50), and what divergences $\mathbf{A}$ may have has been left open. None will be introduced, and $\nabla \cdot \mathbf{A} = 0$ will be chosen. This involves no real loss of generality for after the divergenceless $\mathbf{A}$ has been found, it can always be altered to have any divergence desired without any change of its curl from $\mathbf{u}_c(\mathbf{r})$. All that is necessary to have a vector potential $\mathbf{A}'$ with an arbitrarily chosen divergence $\nabla \cdot \mathbf{A}' = d \neq 0$ is to add a $\nabla\chi$ to the divergenceless $\mathbf{A}$ so that $\mathbf{A}' = \mathbf{A} + \nabla\chi$, with χ a solution of $\nabla^2\chi = d$. All this is irrelevant to the solution really wanted, $\mathbf{u}_c(\mathbf{r}) = \nabla \times \mathbf{A} = \nabla \times \mathbf{A}'$, since $\nabla \times \nabla\chi \equiv 0$ anyway.

With the problem narrowed to a matter of finding some *divergenceless* vector potential, the operation (A.52) is simplified to $-\nabla^2\mathbf{A}$, and Eq. (A.51) to

$$\nabla^2 \mathbf{A} = -\mathbf{c}(\mathbf{r}) \tag{A.54}$$

Thus, each of the components $A_{x,y,z}$ on any set of steady, mutually orthogonal directions in space, is to satisfy a Poisson type of equation. Each of the component equations must have a solution $A_i(\mathbf{r})$ like the one (A.48) formed for the scalar potential ϕ of the equation in (A.35), except that $s(\mathbf{r})$ is to be replaced by $c_i(\mathbf{r})$:

$$A_i(\mathbf{r}_f) = \oint \frac{dV(\mathbf{r}_s)c_i(\mathbf{r}_s)}{4\pi|\mathbf{r}_f - \mathbf{r}_s|}$$

The complete result can be written more economically in vector form.

$$\mathbf{A}(r_f) = \oint \frac{dV(\mathbf{r}_s)\mathbf{c}(\mathbf{r}_s)}{4\pi|\mathbf{r}_f - \mathbf{r}_s|} \tag{A.55}$$

if it is understood that the component of $\mathbf{A}$ on any given direction is to be obtained from the integration of elements projected on the same steady direction.

The solution $\mathbf{u}_c = \nabla \times \mathbf{A}$, with $\mathbf{A}$ given by (A.55), is a unique one, on the same basis that $\mathbf{u}_s = -\nabla\phi$, with ϕ as given by (A.48), was shown to be unique. Using the same procedure first shows only that any derivative $\nabla_i A_j$ is unique, but $\nabla \times \mathbf{A}$ is a definite linear combination of such derivatives and hence is unique also.

It has been shown that the solution (A.55) satisfies the Poisson equations (A.54), but this is not enough, for a Poisson equation is relevant only to a *divergenceless* vector potential. It must also be seen to that $\nabla \cdot \mathbf{A} = 0$ is satisfied. Therefore form (Exercise A.12)

$$\begin{aligned}\nabla_f \cdot \mathbf{A}(\mathbf{r}_f) &= \oint \frac{dV(\mathbf{r}_s)\mathbf{c}(\mathbf{r}_s)}{4\pi} \cdot \nabla_f \frac{1}{|\mathbf{r}_f - \mathbf{r}_s|} \\ &= \oint \frac{dV(\mathbf{r}_s)}{4\pi} \mathbf{c}(\mathbf{r}_s) \cdot \left(-\nabla_s \frac{1}{|\mathbf{r}_f - \mathbf{r}_s|}\right) \\ &= -\oint \frac{dV(\mathbf{r}_s)}{4\pi} \left\{\left[\nabla_s \cdot \frac{\mathbf{c}(\mathbf{r}_s)}{|\mathbf{r}_f - \mathbf{r}_s|}\right] - \frac{1}{|\mathbf{r}_f - \mathbf{r}_s|} \nabla_s \cdot \mathbf{c}(\mathbf{r}_s)\right\} \end{aligned} \tag{A.56}$$

Now $\nabla_s \cdot \mathbf{c} = 0$ by definition of any circulation density ($\equiv$ a curl), or as follows formally from $\mathbf{c} = \nabla \times \mathbf{u}$.(A.31) or (A.34). Then the Gauss theorem permits expressing the result of (A.56) as

$$\nabla_f \cdot \mathbf{A} = -\oint \frac{d\mathbf{S} \cdot \mathbf{c}(\mathbf{r}_S)}{4\pi |\mathbf{r}_f - \mathbf{r}_S|}$$

a flux to be evaluated on the remote surface "enclosing" all space, $\oint dV$. Now *all* the circulation densities are supposed to have been given and restricted to some finite portion of space, so that they are sure to vanish ($\mathbf{c} = 0$) on a surface which is sufficiently remote. In this way, the solution (A.55) for $\mathbf{A}(\mathbf{r})$ does indeed become divergenceless.

The final conclusion of this section is that giving all the sources $s(\mathbf{r})$ and $\mathbf{c}(\mathbf{r})$ of a vector field is sufficient to determine it, as

$$\mathbf{u}(\mathbf{r}) = -\nabla\phi + \nabla \times \mathbf{A} \tag{A.57}$$

in terms of the potentials calculable from (A.48) and (A.55), respectively. The conclusion is an important one for electromagnetic theory, since this has been based on equations for the divergences and curls of electromagnetic fields, the famous Maxwell equations, introduced in Chap. 1.

EXERCISES

A.1 Cylindrical and spherical coordinates are special cases of curvilinear coordinates q_1, q_2, q_3 which respectively increase during displacements into orthogonal $\mathbf{e}_1$, $\mathbf{e}_2$, $\mathbf{e}_3$ directions ($e_i^2 = 1$, $\mathbf{e}_i \cdot \mathbf{e}_j = 0$ for $j \neq i$ and, for example, $\mathbf{e}_1 \times \mathbf{e}_2 = \mathbf{e}_3$). Then each line in (A.1) is a special case of

$$d\mathbf{r} = \mathbf{e}_1 h_1\, dq_1 + \mathbf{e}_2 h_2\, dq_2 + \mathbf{e}_3 h_3\, dq_3$$

where each h_i may be a function of $q_{1,2,3}$.

(*a*) Show how the definition (A.7) leads to $\nabla_i = \partial/h_i\, \partial q_i$.

(*b*) By applying the definition (A.13) to a $dV = h_1 h_2 h_3\, dq_1\, dq_2\, dq_3$, having faces like $d\mathbf{S}_1 = \mathbf{e}_2 h_2\, dq_2 \times \mathbf{e}_3 h_3\, dq_3$, show that

$$\nabla \cdot \mathbf{u} = \frac{1}{h_1 h_2 h_3}\left(\frac{\partial}{\partial q_1} u_1 h_2 h_3 + \frac{\partial}{\partial q_2} u_2 h_3 h_1 + \frac{\partial}{\partial q_3} u_3 h_1 h_2\right)$$

(*c*) By applying (A.20) to $d\mathbf{S}_1$ show that

$$(\nabla \times \mathbf{u})_1 = \frac{1}{h_2 h_3}\left(\frac{\partial}{\partial q_2} h_3 u_3 - \frac{\partial}{\partial q_3} h_2 u_2\right)$$

(*d*) Show that

$$\nabla^2 = \frac{1}{h_1 h_2 h_3}\left[\frac{\partial}{\partial q_1}\left(\frac{h_2 h_3}{h_1}\frac{\partial}{\partial q_1}\right) + \frac{\partial}{\partial q_2}\left(\frac{h_3 h_1}{h_2}\frac{\partial}{\partial q_2}\right) + \frac{\partial}{\partial q_3}\left(\frac{h_1 h_2}{h_3}\frac{\partial}{\partial q_3}\right)\right]$$

(*e*) Check the consistency of the results here with the special cases in the text.

A.2 In view of the remarks following (A.6), show that in a *radial* displacement $\mathbf{e}_r\,\Delta r$, the unit vectors $\mathbf{e}_{r,\,\vartheta,\,\varphi}$ remain steady but in rotations by $\Delta\vartheta$ and $\Delta\varphi$

$$\Delta\mathbf{e}_r = \mathbf{e}_\vartheta\,\Delta\vartheta + \mathbf{e}_\varphi\,\Delta\varphi\sin\vartheta \qquad \Delta\mathbf{e}_\vartheta = -\mathbf{e}_r\,\Delta\varphi + \mathbf{e}_\varphi\,\Delta\varphi\cos\vartheta \qquad \Delta\mathbf{e}_\varphi = -\Delta\varphi(\mathbf{e}_r\sin\vartheta + \mathbf{e}_\vartheta\cos\vartheta)$$

[These results could be used in deriving the spherical coordinate forms of ∇, $\nabla\cdot$ and $\nabla\times$ from the simpler cartesian ones, as a laborious alternative to using the definitions (A.7), (A.13), and (A.20).]

A.3 A *tensor of second rank* is a quantity resolvable into *nine* components T_{kl} with $k, l = x, y, z$, on any cartesian frame. Show that the Gauss and Stokes theorems carry through for fields with tensor components as well as they do for vector fields, in such forms as

$$\sum_k \int dV \nabla_k T_{kl} = \sum_k \oint dS_k T_{kl} \qquad \text{for each } l$$

$$\sum \int dS_i\,(\nabla_j T_{kl} - \nabla_k T_{jl}) = \sum_k \oint dx_k T_{kl}$$

where the summation on the left of the last expression is restricted to $ijk = 123, 321, 312$.

A.4 A force field E has the spherical components

$$\mathsf{E}_r = \frac{2D\cos\vartheta}{r^3} \qquad \mathsf{E}_\vartheta = \frac{D\sin\vartheta}{r^3} \qquad \mathsf{E}_\varphi = 0$$

(*a*) Evaluate *by line integration* the work it does in taking a point (particle) from point A in the diagram to point B via the quarter circle $r = a$, $0 < \vartheta < \frac{1}{2}\pi$.

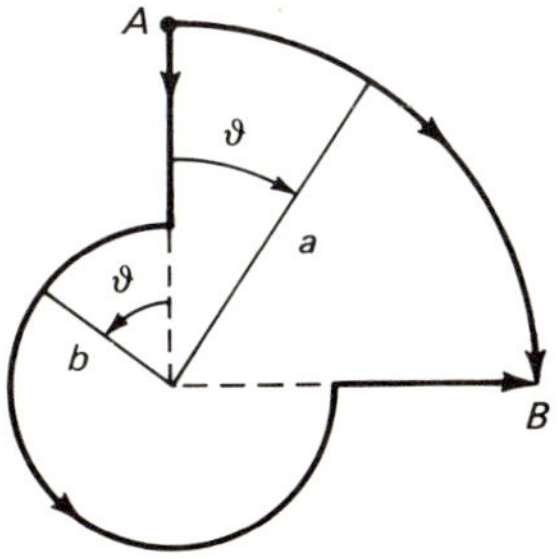

Exercise A.4

(*b*) Do the same for the path that includes the sector $r = b < a$, $0 < \vartheta < 3\pi/2$.

(*c*) You should conclude that E is derivable from a scalar potential via $\mathsf{E} = -\nabla\phi$. Find ϕ in a gauge such that $\phi(r \to \infty) = 0$.

(*d*) Show that each piece of work you evaluated in (*a*) and (*b*) equals the potential difference between the endpoints of the piece.

A.5 (*a*) Show that the lines of force in the field of Exercise A.4 are described geometrically by $r(\vartheta) = a\sin^2\vartheta$, where a specifies a field line that passes through the equatorial plane at a distance $r = a$.

(*b*) Similarly, find $r(\vartheta)$ for lines which intersect every field line perpendicularly to it.

(*c*) Show that the lines in (*b*) are curves formed by the intersection of equipotential surfaces with meridional planes.

A.6 (*a*) Find the $\nabla \cdot \mathbf{E}$ distribution everywhere in the field of Exercise A.4.

(*b*) Imagine a sphere of any radius $a > 0$ centered on $r = 0$ of the field. Evaluate the fluxes through each of the hemispherical parts $0 < \vartheta < \frac{1}{2}\pi$ and $\frac{1}{2}\pi < \vartheta < \pi$ of the spherical surface. Account for your results in view of (*a*).

(*c*) Evaluate the flux through an annular ring, $0 < b < r < a$, in the equatorial plane ($\vartheta = \frac{1}{2}\pi$) and account for its magnitude by comparisons with fluxes through hemispheres of radii $r = a$ and b.

A.7 (*a*) Find ∇r^n, $\nabla \cdot \mathbf{r}$, $\nabla \times \mathbf{r}$, $\nabla \cdot \hat{\mathbf{r}}$, and $\nabla \times \hat{\mathbf{r}}$, where $\hat{\mathbf{r}} \equiv \mathbf{r}/r$.

(*b*) Find the corresponding results with $\mathbf{r}$ replaced by $\mathbf{R} \equiv \hat{\mathbf{R}}R \equiv \mathbf{r} - \mathbf{r}_s$.

(*c*) Repeat (*b*) with ∇ replaced by ∇_s, the gradient derivative with respect to $\mathbf{r}_s$ (instead of $\mathbf{r}$).

A.8 Find the gradient and curl fields respectively derivable from the potentials

$$\phi = -\mathbf{E}_o \cdot \mathbf{r} \qquad \text{and} \qquad \mathbf{A} = \tfrac{1}{2}\mathbf{B}_o \times \mathbf{r}$$

where $\mathbf{E}_o$ and $\mathbf{B}_o$ are constant vectors. Show that $\nabla \cdot \mathbf{A} = 0$ here.

A.9 Let $\mathbf{u}(\mathbf{r})$ be the velocity field generated inside a homogeneous spherical volume of *rigid* material by rotating it about one of its diameters with the uniform angular velocity $\boldsymbol{\omega}_o$.

(*a*) Find the circulation (A.23) on a circle of radius a inside the sphere and coaxial with the rotation.

(*b*) Find $\nabla \cdot \mathbf{u}$ and $\nabla \times \mathbf{u}$. What do you expect for the flux of this curl through a spherical sector based on the circle in (*a*) as perimeter and in view of your result in (*a*)?

A.10 (*a*) Show that the velocity field in Exercise A.9 is derivable from the vector potential $\mathbf{A} = -\frac{1}{2}\boldsymbol{\omega}_o r^2$.

(*b*) Show that $\mathbf{A}$ of (*a*) can be made divergenceless by augmenting it with the gradient of the scalar gauge field $\chi(\mathbf{r}) = r^2(\boldsymbol{\omega}_o \cdot \mathbf{r})/10$ and that this does not change the field $\mathbf{u}(\mathbf{r})$ derived from it.

A.11 (*a*) Show that the field $\mathbf{u} = \nabla \times \mathbf{A}$ derived from $\mathbf{A} = \frac{1}{2}(\mathbf{i}xz^2 + \mathbf{j}yx^2 + \mathbf{k}zy^2)$ is *also* derivable from a scalar potential and find the potential.

(*b*) Is the vector potential $\mathbf{A}$ divergenceless? If not, find a scalar gauge field $\chi(\mathbf{r})$ such that $\mathbf{A}' = \mathbf{A} + \nabla\chi$ *is* divergenceless.

A.12 If $f(R)$ is any function of just the *magnitude* of $\mathbf{R} = \hat{\mathbf{R}}R = \mathbf{r} - \mathbf{r}_s$, show that

$$\nabla f = \hat{\mathbf{R}}\frac{df}{dR} \qquad (\mathbf{R} \cdot \nabla)f = R\frac{df}{dR}$$

$$\nabla \cdot (\hat{\mathbf{R}}f) = \frac{df}{dR} + \frac{2f}{R} \qquad \nabla^2 f = \frac{d^2 f}{dR^2} + \frac{2}{R}\frac{df}{dR}$$

$$\nabla \times (\hat{\mathbf{R}}f) = 0$$

A.13 If $f(\mathbf{r})$ is a scalar field and $\mathbf{g}(\mathbf{r})$ a vector field, show that

$$\nabla \cdot (f\mathbf{g}) = f\nabla \cdot \mathbf{g} + (\mathbf{g} \cdot \nabla)f \qquad \text{and} \qquad \nabla \times (f\mathbf{g}) = f\nabla \times \mathbf{g} - (\mathbf{g} \times \nabla)f$$

A.14 (*a*) For any three vectors $\mathbf{A}$, $\mathbf{B}$, $\mathbf{C}$, show that

$$\mathbf{A} \cdot (\mathbf{B} \times \mathbf{C}) = (\mathbf{A} \times \mathbf{B}) \cdot \mathbf{C}$$

(*b*) Show that, for any pair of vector fields $\mathbf{f}$ and $\mathbf{g}$

$$\nabla \cdot (\mathbf{f} \times \mathbf{g}) = \mathbf{g} \cdot (\nabla \times \mathbf{f}) - \mathbf{f} \cdot (\nabla \times \mathbf{g})$$

A.15 Show that:

(*a*) $\mathbf{A} \times (\mathbf{B} \times \mathbf{C}) = (\mathbf{A} \cdot \mathbf{C})\mathbf{B} - (\mathbf{A} \cdot \mathbf{B})\mathbf{C}$

(*b*) $\mathbf{f} \times (\nabla \times \mathbf{g}) = \sum_{i=x,y,z} f_i \nabla g_i - (\mathbf{f} \cdot \nabla)\mathbf{g}$

(*c*) $\nabla \times (\mathbf{f} \times \mathbf{g}) = \mathbf{f}(\nabla \cdot \mathbf{g}) - \mathbf{g}(\nabla \cdot \mathbf{f}) + (\mathbf{g} \cdot \nabla)\mathbf{f} - (\mathbf{f} \cdot \nabla)\mathbf{g}$

(*d*) $\nabla \times (\mathbf{r} \times \mathbf{g}) = \mathbf{r}(\nabla \cdot \mathbf{g}) - r\left(\frac{\partial}{\partial r} + \frac{2}{r}\right)\mathbf{g}$

A.16 In view of the remark on page 485, show that even if **A** has *only* a radial component $A_r = A$,

$$(\nabla^2 \mathbf{A})_r = \nabla^2 A_r - \frac{2A_r}{r^2}$$

If $A_{\vartheta,\varphi} \neq 0$, the A_r in the last term must be replaced by

$$A_r + \frac{\partial}{\sin\vartheta\,\partial\vartheta}(\sin\vartheta A_\vartheta) + \frac{\partial A_\varphi}{\sin^2\vartheta\,\partial\varphi}$$

A.17 Starting from the definition (9.44), derive the results (9.47) and (9.48).

A.18 Operation on a field by $\mathbf{L} \equiv -i\mathbf{r} \times \nabla$ is called *rotational displacement* of it [to be encountered in (9.44)]. Show that:

(*a*) The gradient operation is resolvable into radial and transverse components, as

$$\nabla = \hat{\mathbf{r}}\frac{\partial}{\partial r} - r^{-1}(i\hat{\mathbf{r}} \times \mathbf{L})$$

(*b*) $\nabla^2 = \left(\frac{\partial}{\partial r} + \frac{1}{r}\right)^2 - \frac{L^2}{r^2}$

(*c*) $\nabla \times \mathbf{L} = \frac{i}{r}\hat{\mathbf{r}}L^2 + (\hat{\mathbf{r}} \times \mathbf{L})\left(\frac{\partial}{\partial r} + \frac{1}{r}\right)$

(All these resolutions are useful in both the quantum-mechanical and electromagnetic theory of spherical waves.)

CHAPTER

B

THE ELECTROSTATICS OF CONDUCTORS†

The equipotentials presented by conductors in static equilibrium with fields · Charges induced on conductors · The simulation of induced charges by images · Dipole moments induced in conductors · Shielding by conductors · The electrostatic forces and stresses on conductors

Circumstances in which the sources of a field cannot all be given initially but must be codetermined with the field arise from the presence of materials. Even initially neutral volumes of material can strongly affect the fields that are observed. This can be understood as a consequence of neutral matter's having indefinitely large, equal quantities of positive and negative charge which are normally well superposed but may be shifted relative to each other by the force fields.

The types of behavior observed have led to a classification of materials into two broad kinds, the *conductors* and the *insulators* (or, only a little more specially, the *dielectrics*). Attention here will be confined to idealized "perfect" conductors, and ideal dielectrics will be considered in Chap. D. The descriptions to be adopted for these materials permit deriving the main electrostatic effects exhibited by most actual materials.

B.1 CONDUCTING SPACES

The materials classed as conductors behave as if charges within them shift about quite freely. As soon as a conductor is placed in a preexistent electric field or whenever charges are placed near or on conductors, motions of charge, detectable as transient currents, ensue. The charge distributions inside the conductors and the resultant electric field both keep changing until an electrostatic equilibrium is attained. It is the resultant equilibrium situation that is the sole concern of electro*statics*, and in it the field must be such that

$$\mathbf{E} = 0 \qquad \text{and} \qquad \phi = \phi_C = \text{const inside conductors} \tag{B.1}$$

No $\mathbf{E} = -\nabla\phi$ can exist under the equilibrium conditions anywhere inside a perfect conductor because a net electric force would immediately restart motions of its charges. It should also be recognized that, during the establishment of the

† Chapters B to D are designed to be read after Chap. 3.

equilibrium, the charges could not normally† have been sent beyond the confines of the conducting medium, and hence accumulations of charge on the surfaces of the conductors may well be expected. However, within the surfaces, no point can bear net charge, for an $\mathbf{E} \neq 0$ would originate from it and a nonstatic dispersal of the charge ensues [as will eventually be described in (G.4)].

The two equations (B.1) can be regarded as the complete vector-field and scalar-potential descriptions, respectively, of the parts of an electrostatic field within conducting spaces. They respectively replace the general equations (2.1) and (2.7), being the only solutions of them appropriate to such media.

B.2 CONDUCTOR BOUNDARIES

In problems with some parts of space occupied by conducting media while others are empty except for charges fixed in them, there is need to join solutions (B.1) to solutions of (2.1), or (2.7), at the interfaces of the regions, so that the two types of solutions *together* describe a unique, overall electrostatic field. Boundary conditions on the conductor surfaces, describing the transitions of field values from conducting media to free space, must be considered.

Perhaps the most noteworthy characteristic of any conductor boundary is that it forms an equipotential surface (in electro*static* situations). Indeed, the entire conductor forms an equipotential *volume*, according to (B.1). This means that when the Poisson equation of (2.7) is solved for the potential in free spaces between conductors, the solution $\phi(\mathbf{r})$ must be subjected to the boundary condition that $\phi(\mathbf{r}_C) \to \phi_C$, a constant, for points $\mathbf{r}_C$ on the surface of a conductor. Of course, unconnected conductors may have different constant potentials ϕ_C.

The electric field $\mathbf{E} = -\nabla\phi$ at the surface of a conductor must be normal to it, $\mathsf{E}(\mathbf{r}_C) = \mathsf{E}_n = -\nabla_n \phi$ if $\mathbf{n} \equiv d\mathbf{S}/dS$, since field gradients are always perpendicular to equipotentials (A.9). Considered more physically, no field components $\mathbf{E}_{\parallel}$ *tangential* (parallel) to the surface can be admitted because they would immediately start currents along the surface. Of course, any $\mathsf{E}_n \neq 0$ must be considered to be "just outside" the conductor surface, since $\mathbf{E} = 0$ inside. There is, in general, a discontinuity $\Delta\mathsf{E}_n = \mathsf{E}_n$ at the surface, originating at surface charges of area density $\sigma(\mathbf{r}_C) = \mathsf{E}_n(\mathbf{r}_C)/4\pi$ [Eq. (2.31)]. They must be regarded as results of the redistribution of conductor charges, sent as far as they could be conducted out of the volume (to the surface) during the establishment of the equilibrium with the field. The total $\oint dS\sigma = q_C$ on a whole, insulated conductor can have a value $q_C \neq 0$ only if a net charge had been placed on the conductor in the first place, since charge is conserved. If the insulated conductor is initially neutral, the

† Very strong fields can cause *cold emission* of electrons into free space, a phenomenon which does not ordinarily influence the final electrostatic equilibria appreciably. Another occurs when some insulating medium borders on a conductor; strong enough fields can then cause a breakdown of the insulation, marked by a sparking which discharges the conductor. However, this involves a nonideal behavior of insulators that is to be ignored here.

ultimate charge distribution on it must consist of equal amounts of positive and negative charges, integrating to $q_C = 0$.

The boundary conditions on the electric vector field can be regarded as reductions of the field equations (2.1) to a two-dimensional space making up the surface continuum of a conductor. Thus

$$\mathsf{E}_n = 4\pi\sigma \qquad \mathbf{E}_{\parallel} = 0 \qquad \text{on conductors} \tag{B.2}$$

where $\mathbf{E}_{\parallel}$ stands for field components along (parallel to) the surface, can be regarded as the respective equations into which $\nabla \cdot \mathbf{E} = 4\pi\rho$ and $\nabla \times \mathbf{E} = 0$ degenerate.

The reduction of $\nabla \cdot \mathbf{E} = 4\pi\rho$ to $\Delta\mathsf{E}_n = 4\pi\sigma$ was demonstrated in (2.31), and this accounts for the discontinuity $\Delta\mathsf{E}_n \equiv \mathbf{n} \cdot \Delta\mathbf{E}$ being called a surface divergence at $d\mathbf{S} = \mathbf{n}\, dS$. On similar grounds, any discontinuity $\Delta\mathbf{E}_{\parallel} \equiv \mathbf{n} \times \Delta\mathbf{E}$ of field components tangential to a surface is called a *surface curl*. It measures a flux of curl confined to a surface layer, as can be seen from applying the Stokes theorem to any vanishingly narrow strip of area which penetrates the surface normally and encloses a *line* element (the intersection) on the surface within its vanishing breadth. This can be visualized with the help of a diagram which is indistinguishable from Fig. 2.1 if the rectangle indicated in it is now taken to represent a rectangular *area* (rather than volume) having a breadth perpendicular to the surface, vanishing with the surface thickness, and having a length Δl parallel to the surface. The Stokes theorem applied to this area gives for the flux of $\nabla \times \mathbf{E}$ through it

$$\int d\mathbf{S} \cdot (\nabla \times \mathbf{E}) = \oint d\mathbf{r} \cdot \mathbf{E} = \mathsf{E}_l^o\, \Delta l - \mathsf{E}_l^i\, \Delta l \tag{B.3}$$

or $\Delta\mathsf{E}_l = \mathsf{E}_l^o - \mathsf{E}_l^i$ per unit width of surface, where the subscript l denotes field components along the line element on the surface chosen for enclosure. Thus, in any electrostatic situation, when $\nabla \times \mathbf{E} = 0$ everywhere, all surface curls also vanish, and this can be represented by

$$\Delta\mathbf{E}_{\parallel} \equiv \mathbf{n} \times \Delta\mathbf{E} = \mathbf{E}_{\parallel}^o - \mathbf{E}_{\parallel}^i = 0 \tag{B.4}$$

if $\mathbf{n} = d\mathbf{S}/dS$. There is a continuity $\mathbf{E}_{\parallel}^o = \mathbf{E}_{\parallel}^i$, at any surface, of any field components parallel to the surface, and since $\mathbf{E}^i = 0$ in conductors, the result $\mathbf{E}_{\parallel}^o = \mathbf{E}_{\parallel} = 0$ [Eq. (B.2)] can be understood as a reduction of $\nabla \times \mathbf{E} = 0$ to a conductor surface.

It is their equivalence to the Maxwell equations in boundary media that makes such boundary conditions as (B.2) clearly sufficient for representing the entire physical role of surfaces in making solutions properly unique (see page 64).

The condition $\mathbf{E}_{\parallel} = 0$ of (B.2) is equivalent to $-\nabla_{\parallel}\phi = 0$ for gradients along the conductor surface, and so the condition is entirely equivalent to $\phi = \phi_C$, a constant, on conductors. This is part of a potential description of the boundary conditions (B.2)

$$\phi = \phi_C \qquad \nabla_n\phi = -4\pi\sigma \qquad \text{on conductors} \tag{B.5}$$

appropriate for use together with the potential description (2.7) of the outside field. The condition $\phi = \phi_C$ is the only stricture on the electrostatic field that can exist outside a conductor; $\sigma = -\nabla_n \phi/4\pi$ represents an adjustment of the conductor to the incident field.

Whenever a definite value for just ϕ_C can be prescribed, this is sufficient, according to (3.14), to represent the part played by the conductor boundary in determining a unique electrostatic field. Problems in which conductor potentials ϕ_C can be prescribed form one of two main types encountered.

When specific values of the conductor potentials are not given, it is still known that they must have some *constant* value on each conductor. As a consequence, the relation (3.13) between any two possible solutions becomes

$$\int_V dV\,|\mathbf{E}_1 - \mathbf{E}_2|^2 = \sum_C (\phi_1^C - \phi_2^C) \oint dS_C(\mathrm{E}_{1n} - \mathrm{E}_{2n}) \tag{B.6}$$

when only conducting spaces are excluded from the volume of definition V. ϕ_1^C and ϕ_2^C are the possibly different constant values assigned to the conductor labeled C by the two solutions. In general, two different solutions may also lead to different surface charge distributions $\sigma_{1,2} = \mathrm{E}_{1,2n}/4\pi$, respectively. However, suppose that the solution wanted is to correspond to a given *total* charge $q_C = \oint dS_C \sigma_1 = \oint dS_C \sigma_2$ on each conductor. This requirement makes the right side of (B.6) vanish, and so a unique solution, with $\mathbf{E}_1 = \mathbf{E}_2$ everywhere, is actually determined. Thus, the prescription of the total charge on a conductor in place of the potential on it constitutes a second way in which information sufficient to determine a unique solution may be given.

B.3 IMAGE CHARGES IN CONDUCTORS

The simplest demonstration of the effect of a conductor and an electric field on each other is perhaps provided by putting a point charge q in front of a conductor. The result of an arbitrary charge distribution outside the same conductor is then obtainable by superposing the fields arising from point-charge elements $dq = \rho(\mathbf{r}_s)\, dV(\mathbf{r}_s)$.

The simplest conductor configuration to try is one with a plane boundary of extent large enough compared with the distance of the point charge from it to permit edge effects to be neglected and the plane to be treated as infinite. The conductor is thus connected to infinity, and it is appropriate to specify $\phi_C = 0$ on it (so large a conductor has a capacity for holding charge without an appreciable rise of its potential, comparable to that of the whole earth, and the conductor is in effect grounded to $\phi_C = 0$). The interest is in the equilibrium field formed in the free space containing the point charge, and then the situation can be formulated by taking the half-infinite space $z < 0$ to be occupied by a conducting medium while the half-infinite space $z > 0$ is empty except for the presence of a point charge q at some distance $z = l$ from the conductor boundary $z = 0$, as indicated in Fig. B.1.

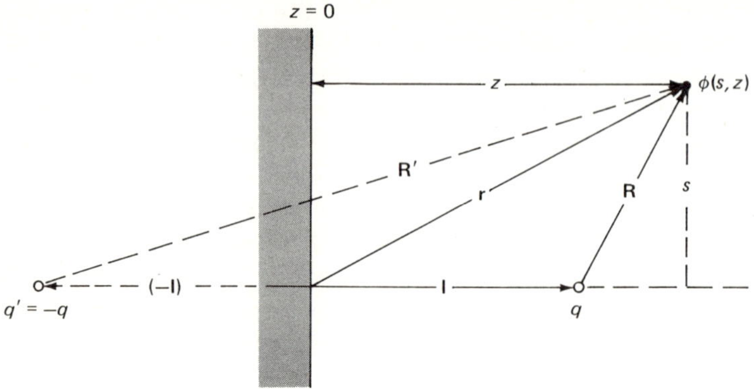

Figure B.1

It will be convenient to describe the field as $\phi(s, z)$ in terms of the cylindrical coordinates s, (φ), z around the axis of symmetry formed by the perpendicular to the conducting plane dropped from the point charge. Then $\phi(s, z \le 0) = 0$, and $\phi(s, z \ge 0)$ is a solution of

$$\nabla^2\phi = -4\pi q\delta(\mathbf{r} - \mathbf{l}) \tag{B.7}$$

in terms of the vectors $\mathbf{r}$ and $\mathbf{l}$ indicated in Fig. B.1. The general solution $\phi = \phi_\rho + \phi_o$, (3.1) with (3.10), has $\phi_\rho(s, z \ge 0) = q/R$ with $R = [s^2 + (z - l)^2]^{1/2}$. The supplement ϕ_o must be a solution of the Laplace equation (3.2) such that $\phi_\rho + \phi_o = \phi(s, 0) = 0$ on the boundary, and hence $\phi_o(s, 0) = -q/(s^2 + l^2)^{1/2}$. The latter is just a potential characteristic of a point charge $q' = -q$ at a position $\mathbf{r}' = -\mathbf{l}$. If now the total potential which arises solely from two point charges $\pm q$ at $\mathbf{r} = \pm\mathbf{l}$, respectively, is considered, it can be seen that it must satisfy

$$\nabla^2\phi = -4\pi q\delta(\mathbf{r} - \mathbf{l}) + 4\pi q\delta(\mathbf{r} + \mathbf{l})$$

This equation is identical with (B.7) everywhere in the $z > 0$ region, since $\delta(\mathbf{r} + \mathbf{l}) = 0$ everywhere in it, and hence the desired solution can be expressed in the two pieces:

$$\phi(s, z \le 0) = 0 \qquad \text{and} \qquad \phi(s, z \ge 0) = \frac{q}{R} - \frac{q}{R'} \tag{B.8}$$

where $R' = [s^2 + (z + l)^2]^{1/2}$. It properly satisfies both (B.7) and the boundary condition $\phi(s, 0) = 0$. Thus, the effect of the conductor is to supplement $\phi_\rho = q/R$ with $\phi_o = -q/R'$, properly satisfying $\nabla^2\phi_o = 0$ for $z > 0$. The effects outside the conductor are being simulated by what is called an *image* $q' = -q$ at $\mathbf{r}' = -\mathbf{l}$ of the point charge q at $\mathbf{r} = \mathbf{l}$ in the conducting plane.

Of course, the *actual* sources of field presented by the conductor are charges induced on its surface, now calculable from the electric field at the conductor $\mathsf{E} = -(\partial\phi/\partial z)_{z=0}$. The resultant distribution

$$\sigma(s) = \frac{\mathsf{E}}{4\pi} = -\frac{ql}{2\pi(s^2 + l^2)^{3/2}} \tag{B.9}$$

is centrally symmetric about the point $s = 0$, opposite the point charge, and has a maximum value at this point. Simple integration shows that the total induced charge is $q_C = 2\pi \int_0^\infty ds\ s\sigma(s) = -q$, just equal to the image charge, in conformity with the Gauss law. An infinite, or grounded, conductor has infinite reservoirs of charge of both signs available for such inducement without any appreciable change in its potential.

It is possible to check directly that $\phi_o(s, z > 0) = -q/R'$ arises from the surface charge distributions $\sigma(s)$, as in Exercise 2.11. Moreover, the separate potentials due to σ and q cancel exactly at all points *inside* the conductor.

The special device used in the problem here, to obtain the proper solution ϕ_o of the Laplace equation, was a choice of a point-charge configuration that by itself possesses an equipotential surface of the same shape as the conductor boundary. Plainly, this approach could be extended to the use of other configurations of charge for which the equipotentials are known. Any of the equipotentials can be replaced by a conductor boundary at the corresponding potential, and then the field between equipotentials correctly represents the field there will be between the conductors in the configurations thus formed. However, the problem is usually the more difficult one of starting with *prescribed* conductor shapes and finding some image-charge configuration which has equipotentials with just those shapes.

More phenomena arise in connection with *finite* conductors, e.g., a *sphere*. Let the sphere radius be a, and place a point charge q at a distance r_o from the center of the sphere, as indicated in Fig. B.2. It will be appropriate to express the field as $\phi(r, \vartheta)$, in terms of spherical coordinates with $r = 0$ at the sphere center and the vector $\mathbf{r}_o$ as polar axis.

Suppose first that the conductor is grounded to $\phi_C = 0$ (as through a wire connection to the earth). Then $\phi(r \le a, \vartheta) = 0$ and $\phi(r \ge a, \vartheta) = q/R + \phi_o$, where $R = (r^2 + r_o^2 - 2rr_o \cos \vartheta)^{1/2}$. ϕ_o must be a solution of the Laplace equation such that $\phi_o(a, \vartheta) = -q/R_C$, with $R_C = (a^2 + r_o^2 - 2ar_o \cos \vartheta)^{1/2}$. Again, a representation of it as arising from a point charge image can be attempted if a magnitude q' and a position $\mathbf{r}'_o$ can be found which will make the sphere a $\phi_C = 0$ equipotential between the real point charge and the image. The result can have

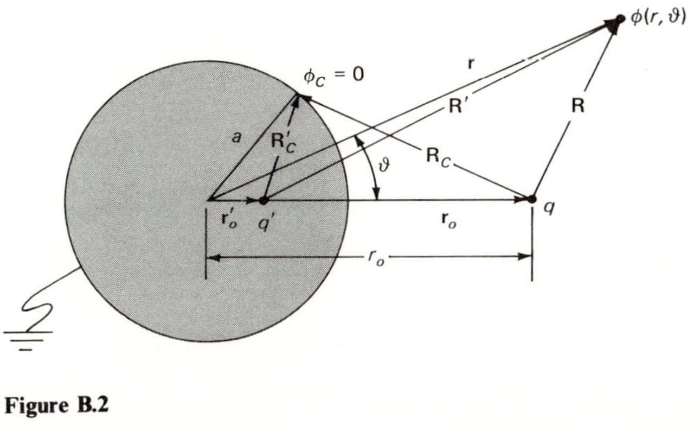

Figure B.2

the proper azimuthal symmetry only with the help of an image at a position $\mathbf{r}'_o$ somewhere on the line from sphere center to point charge, so that

$$\phi_o(a, \vartheta) = -\frac{q}{R_C} = \frac{q'}{R'_C} \qquad \text{with } R'_C = [a^2 + (r'_o)^2 - 2ar'_o \cos \vartheta]^{1/2}$$

This will yield a total potential $\phi(a, \vartheta) = 0$ all around the sphere only if the equality

$$(R'_C)^2 = a^2 + (r'_o)^2 - 2ar'_o \cos \vartheta = \left(\frac{q'}{q}\right)^2 (a^2 + r_o^2 - 2ar_o \cos \vartheta)$$

holds independently of ϑ, and that will be so if

$$a^2 + (r'_o)^2 = \left(\frac{q'}{q}\right)^2 (a^2 + r_o^2) \qquad \text{and} \qquad r'_o = \left(\frac{q'}{q}\right)^2 r_o$$

independently. Thus provided are two equations for determining q' and r'_o, with the solution†

$$r'_o = \frac{a^2}{r_o} \qquad \text{and} \qquad q' = -\frac{qR'_C}{R_C} = -\frac{qa}{r_o} \tag{B.10}$$

Since a suitable point charge image has thus been found, the field which is sought has

$$\phi(r \leq a, \vartheta) = 0 \qquad \phi(r \geq a, \vartheta) = q\left(\frac{1}{R} - \frac{a}{r_o R'}\right) \tag{B.11}$$

where $R' = [r^2 + (a^4/r_o^2) - 2(a^2/r_o)r \cos \vartheta]^{1/2}$.

The charge distribution induced on the sphere can be calculated as $\sigma = -(\partial\phi/\partial r)_{r=a}/4\pi$, with the result

$$\sigma(\vartheta) = \frac{\mathsf{E}}{4\pi} = -\frac{q(r_o^2 - a^2)}{4\pi a R_C^3} \tag{B.12}$$

varying inversely as the cube of the distance R_C from surface point to point charge. The total induced charge $q_C = 2\pi a^2 \int_{-1}^{1} d(\cos \vartheta)\sigma(\vartheta) = -qa/r_o$ is equal to the image charge again, as required by the Gauss law.

Next suppose that, instead of being grounded, the sphere is insulated from the earth and was initially neutral. Being insulated, it must also be neutral as a whole after electrostatic equilibrium has been attained, and so the net flux of electric field at the surface must continue to vanish.

Although the sphere is not being kept at $\phi_C = 0$, it is still an equipotential surface of a shape obtainable from the joint effects of the point charge q and the same image, $q' = -qa/r_o$ at $r'_o = a^2/r_o$, as before. The shape remains undisturbed if now a *second* image charge is imagined placed at the *center* of the sphere. To

† See also Exercise 3.15.

get the vanishing total flux characteristic of the neutral sphere, this second image must be given the magnitude $q'' = -q' = +qa/r_o$, equal and opposite in sign to that of the first image. Now the potential outside the sphere can be represented as

$$\phi(r \geq a,\, \vartheta) = \frac{q}{R} + \frac{q'}{R'} + \frac{q''}{r} = q\left(\frac{1}{R} - \frac{a}{r_o R'} + \frac{a}{r_o r}\right) \tag{B.13a}$$

at the radial distance r from the sphere center, where the second central image has been located. Meanwhile, the potential of the conductor has been changed to

$$\phi(r \leq a,\, \vartheta) \equiv \phi_C = \frac{q}{r_o} \tag{B.13b}$$

just the value of the potential at the sphere center in the absence of conducting medium (as when $a = 0$).

The flux from the central image $q'' = qa/r_o$ alone is uniform over the sphere, and so it merely adds a constant, $q''/4\pi a^2 = q/4\pi a r_o$, to the charge distribution (B.12). As a result,

$$\sigma(\vartheta) = -\frac{q(r_o^2 - a^2)}{4\pi a R_C^3} + \frac{q}{4\pi a r_o} \tag{B.14}$$

for an insulated neutral sphere in the presence of the point charge q. Of course, this integrates to $q_C = 0$.

The results (B.13) and (B.14) are easy to modify to the case in which the insulated sphere is initially charged to some total $q_C \neq 0$. It is only necessary to increase the central image charge by q_C, to a total $q'' = qa/r_o + q_C$. Then the potential outside the sphere is changed by q_C/r and on it by q_C/a. The charge density (B.14) is changed by $q_C/4\pi a^2$. These changes obviously correspond to the superposition of a field such as arises from a uniformly charged sphere in the absence of any exterior charge (compare page 33).

B.4 THE CONDUCTING SPHERE IN A UNIFORM FIELD

The effect of putting the sphere into a uniform electric field $\mathbf{E}_o$, in place of the point charge field of q, can be obtained from the latter case by letting the distance r_o to q approach infinity and choosing q equal to $\mathsf{E}_o r_o^2 \to \infty$. The field lines from the infinite point charge at infinity will become parallel to each other in the vicinity of the finite sphere, and representation of a uniform field $\mathbf{E}_o$ there will be approached.

The most interesting case is that of the neutral insulated sphere, and so it is the limits of the formulas (B.13) and (B.14) that will be investigated. Infinite potentials can be avoided by changing their gauge by the constant $\mathsf{E}_o r_o$ ($\to \infty$). Then the potential on the sphere (B.13) is reduced to $\phi_C = \mathsf{E}_o r_o^2/r_o - \mathsf{E}_o r_o = 0$, and so the potentials in the new gauge will be measured by work per unit charge done upon starting from the conductor itself. In the expression (B.13a), for the potential outside the sphere, the limiting process gives

$$\frac{q}{R} = \frac{\mathsf{E}_o r_o^2}{r_o}\left[1 - 2\frac{r}{r_o}\cos\vartheta + \left(\frac{r}{r_o}\right)^2\right]^{-1/2} \to \mathsf{E}_o r_o\left(1 + \frac{r}{r_o}\cos\vartheta\right)$$

$$-\frac{q'}{R'} = \mathsf{E}_o r_o \frac{a}{r}\left(1 - 2\frac{a^2}{r_o r}\cos\vartheta + \frac{a^4}{r_o^2 r^2}\right)^{-1/2} \to \mathsf{E}_o r_o \frac{a}{r}\left(1 + \frac{a^2\cos\vartheta}{r_o r}\right)$$

so that

$$\phi(r \geq a, \vartheta) \to \mathsf{E}_o r\cos\vartheta - \frac{\mathsf{E}_o a^3\cos\vartheta}{r^2} \tag{B.15}$$

relative to $\phi(a, \vartheta) = \phi_C = 0$. The first term here represents the potential $\phi' = \mathsf{E}_o z$ of the uniform field by itself, existing as $\mathsf{E}_z' = -\mathsf{E}_o$ even in the absence of the sphere (as when $a = 0$). The last term represents the change due to the sphere, and can be recognized as an ideal dipole potential (2.26), such as arises from a dipole moment of magnitude $D = \mathsf{E}_o a^3$ parallel to the incident field ($D_z = -D$ just as $\mathsf{E}_z' = -\mathsf{E}_o$) and proportional to it.

The effect of a uniform field on a conducting sphere is thus to induce a dipole moment in it. The magnitude of the induced moment can be understood in terms of the two image charges, $q' = -aq/r_o \to -a\mathsf{E}_o r_o$ at $r_o' = a^2/r_o \to 0$, and the central image $q'' = -q'$, which together yield the dipole moment $D = |q'|r_o' \to \mathsf{E}_o a^3$. Of course, the dipole field actually arises from charges induced on the surface of the sphere, distributed according to

$$\sigma(\vartheta) = -(4\pi)^{-1}\left(\frac{\partial\phi}{\partial r}\right)_a = -\frac{3}{4\pi}\mathsf{E}_o\cos\vartheta \tag{B.16}$$

when evaluated from (B.15). The same result follows from the limit of (B.14). A direct evaluation of the moments of the distribution (B.16) would show that the dipole moment (2.20) has the magnitude

$$D = \left|2\pi a^2\int_{-1}^{1} d(\cos\vartheta)\sigma(a\cos\vartheta)\right| = \mathsf{E}_o a^3 \tag{B.17}$$

and that no other moments exist relative to the sphere center (3.101).

B.5 SHIELDING BY CONDUCTORS

The preceding illustrations of the effect of perfect conductors on fields outside them suggest that it is only the shapes of, and conditions on, the *outer boundaries* of the conductors that play any role. Conditions behind those boundaries can presumably be changed without affecting the exterior field, and regions which are completely separated by conducting surfaces are electrostatically independent. The validity of such presumptions will now be examined.

It can first be seen that arbitrary hollows within the conductors can be made without affecting the electrostatic field anywhere if no charges are placed in the free spaces created by the removal of conducting material. The potential within such charge-free hollow spaces must satisfy $\nabla^2\phi = 0$, subject to the boundary condition $\phi = \phi_C$, the conductor potential on the surface enclosing the hollow. A solution is $\phi = \phi_C = \text{const}$ everywhere in the hollow, and it is known from (3.14) that this is the unique solution. Thus, no change has been made anywhere from the case in which the hollow was filled with conducting medium.

With $\mathbf{E} = 0$ continuing everywhere in the hollows and also in the conductor bordering each hollow, there are no inner-surface divergences $\Delta \mathsf{E}_n = 4\pi\sigma = 0$, and none of the charges on the conductor are distributed along the *inner* boundaries. All charge on the conductors accumulates only on outer surfaces, facing into regions with field sources, in all electrostatic situations.

A very practical aspect of the same conclusions is that any space enclosed by conductors is field-free if it itself does not contain sources whatever the electrostatic conditions outside the conductors may be. The enclosed region is said to be *shielded* from the exterior fields by the enclosing conductor. In laboratories such spaces are called *Faraday cages*.

It can next be seen that if the conductor is grounded to $\phi_C = 0$, any charges can be placed in hollows within it without affecting the exterior field, simply because the boundary condition $\phi_C = 0$ is not changed thereby. Moreover, the field that will now exist in the hollow will be determined solely by the charges in it and the condition $\phi_C = 0$ on its boundary; it is thus quite independent of the exterior electrostatic situation (Exercise B.8). A *grounded* conductor can indeed separate regions into electrostatically independent ones.

Finally, consider an *insulated* conductor. If charge is placed in a hollow inside it, an equal and opposite charge will be induced on the inner boundary, for there can be no net flux through surfaces lying in the fieldless conducting medium immediately outside the boundary of the hollow. The induced charge must come from somewhere in the insulated conductor, and hence a charge must appear on the *exterior* boundary of the conductor, equal and opposite to the charge induced on the inner surface. There will be a charge on the outer boundary equal to the charge placed in the hollow, and so the existence of the charge in the hollow can be detected as part of the exterior field. On the other hand, the *configuration* of the exterior field will be completely independent of the way the charge inside the hollow is distributed there. Only the shape of the outer boundary and the value of the *total* charge on it determine the exterior field and the distribution of the charge on the outer surface. To this extent, the exterior field is still independent of the electrostatic situation on the other side of its conductor boundaries. Placing charge in a hollow of an insulated conductor is equivalent to placing it on the conductor itself, insofar as the field detected outside the conductor is concerned. Quantitative evaluations corresponding to these various assertions are left to Exercise B.21.

B.6 FORCES ON CONDUCTORS

Valuable insights into the force balances in an electrostatic system can be gained from the simple examples of interaction between a point charge and a conductor reviewed in connection with Figs. B.1 and B.2.

In each case charge is induced on the conductor and has a sign opposite that of the point charge when the conductor is grounded. Consequently there must be a mutual force of attraction between point charge and conductor which should be calculable equally well by considering the force $\mathbf{F}_C$ acting on the conductor or by evaluating the reaction on the point charge $\mathbf{F}_q = -\mathbf{F}_C$. Actually, considering the $\mathbf{F}_q$ will be found the less ambiguous, since it is quite clear that the force on a point charge q must be calculated as $\mathbf{F}_q = q\mathbf{E}'$, where $\mathbf{E}'$ is the field arising from all sources present *except q itself*, exactly as in evaluating the Coulomb law (when $\mathbf{E}' = q'/r^2$ arises from a second *point* charge). This emphasizes that a distinction must be maintained† between the *entire* field $\mathbf{E}$ characterizing a given electrostatic system and various partial fields, like $\mathbf{E}'$, representing forces on charges that are parts of the entire system. Whereas the $\mathbf{E} = -\nabla\phi$, with ϕ as calculated in (B.8) or (B.11) for example, represents force that would be detected by an extraneous *test* charge, introduced in such a way that it does not alter the system being tested, the force $\mathbf{F}_q = q\mathbf{E}'$ to be evaluated now is part of the balance of forces existing within the system, upon charge that *has* altered it, and producing stresses and strains between the constituents.

Each of the specific systems being considered here consists of a conductor in addition to the point charge q. Then the field $\mathbf{E}'$ giving the force $\mathbf{F}_q = q\mathbf{E}'$ on the point charge must arise solely from charges on the conductor. It is particularly simple to evaluate for a *grounded* conductor, for then it is simulated by the field of a single image in the conductor. In the case of the conducting *plane*, the magnitude of $\mathbf{E}'$ at q is simply $q/(2l)^2$, where l is the distance from point charge to plane; thus, $F_q = q^2/4l^2$ is the mutual force of attraction between the point charge and the plane. For the grounded *sphere*

$$F_q = q\mathsf{E}' = q\,\frac{qa/r_o}{(r_o - a^2/r_o)^2} = \frac{q^2 a r_o}{(r_o^2 - a^2)^2} \tag{B.18}$$

according to the findings in connection with Fig. B.2.

The results just obtained give the *total* force on the conductor as a reaction from the force on the point charge: $\mathbf{F}_C = -\mathbf{F}_q$. However, there is interest also in knowing how the total force $\mathbf{F}_C$ is *distributed* over the conductor, and this requires considering the forces acting at various points of the conductor itself. Since a conductor is sensitive to electrostatic forces only because of surface charge distributions it may bear, the forces on a conductor are distributed over its *surface*, having a resultant

$$\mathbf{F}_C = \oint dS(\mathbf{r}_C)\mathbf{T}(\mathbf{r}_C) \tag{B.19}$$

† See page 501*n*.

of forces $\mathbf{T}(\mathbf{r}_C)$ *per unit area* at various points $\mathbf{r}_C$ of the surface. Forces per unit area, like **T**, are called *stresses* and have interest not only to engineers (who might, for example, be concerned with possible spallations of material from conductor surfaces by strong fields) but also because of what can be learned from them about the propagation of force within an electrostatic system.

The calculation of the force at a given point $\mathbf{r}_C$ of a conductor surface can proceed from considering the element of force

$$d\mathbf{F}_C \equiv \mathbf{T}(\mathbf{r}_C)\, dS(\mathbf{r}_C) = dS\sigma(\mathbf{r}_C)\mathbf{E}''(\mathbf{r}_C)$$

acting on the charge element $\sigma\, dS$ at $\mathbf{r}_C$. Here $\mathbf{E}''(r_C)$ must be the field at $\mathbf{r}_C$ arising from all sources present in the system *other than*† the charge element $\sigma\, dS(\mathbf{r}_C)$ itself. It is this omission of just one element of the continuously distributed conductor charge $q_C = \oint dS\sigma$ in evaluating $\mathbf{E}''$ that makes the problem ambiguous; it is not immediately clear just how this is to be done.

First it should be made clear that the force at a point on the conductor arises not only from charges outside the conductor but also from neighboring charges on the conductor itself. It is just these neighboring charges that must be held responsible for canceling out any components of field incident from outside which are parallel to the conductor surface. (The cancellations are necessary if no surface currents are to be started and the electrostatic equilibrium maintained.) An immediate conclusion is that whatever the force per unit charge $\mathbf{E}''(r_C)$ at $\mathbf{r}_C$ may be, it must be *normal* to the conductor surface and the stress $\mathbf{T}(\mathbf{r}_C)$ must be a normally directed pull on the surface.

The last conclusion settles completely the question of calculating $\mathbf{T} = \sigma\mathbf{E}''$ on a *plane* conductor boundary, for the neighboring charges in this case can only contribute force components parallel to the surface and have *solely* the function of reducing to zero the resultant of all parallel components of force. Then $\mathbf{E}''(r_C)$ is just the *normal* component of the field incident from exterior sources. At the distance s from the point $s = 0$ opposite a point charge, $\mathsf{E}''(s) = (q/R_C^2)(l/R_C)$, where $R_C = (s^2 + l^2)^{1/2}$ in Fig. B.1. Comparison with the charge density (B.9) induced on the conducting plane shows that $\mathsf{E}''(s) = 2\pi|\sigma(s)| = \frac{1}{2}\mathsf{E}(s)$, where $\mathsf{E}(s)$ is the *entire* (normal) field at the surface point. Thus

$$T = |\sigma\mathsf{E}''| = 2\pi\sigma^2 = \frac{\mathsf{E}^2}{8\pi} \tag{B.20}$$

is expressible in terms of the *entire* field $\mathbf{E}$ characterizing the system. It is easy to check that this result integrates properly to the total force $F_C = 2\pi \int_0^\infty ds\, sT(s) = q^2/4l^2$, already found. A significant point to notice is that $T \neq |\sigma\mathsf{E}|$, despite the fact that E is the total field at σ; instead $T = \frac{1}{2}|\sigma\mathsf{E}|$. The reduction from E to $\mathsf{E}'' = \frac{1}{2}\mathsf{E}$, for the force per unit charge on σ, is the consequence of leaving out the force of the charge element on itself.

† Such distinctions are really necessary only when the charges on which the force acts are overidealized by putting finite charges into zero volumes, as is done in the representation of distributions on surfaces of zero thickness and of test charges as occupying points. This is confirmed in Sec. 6.2.

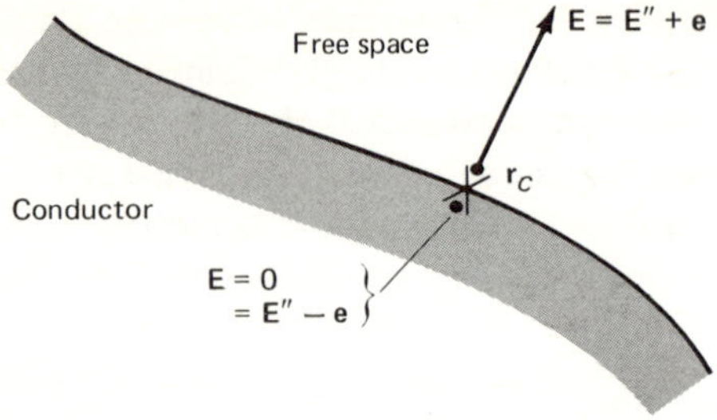

Figure B.3

The situation in the case of a *curved* conductor surface, such as the sphere, is less clear because now the charge elements neighboring a point *can* contribute to the normal force $\mathbf{E}''(\mathbf{r}_C)$ at the point. However, it might well be expected that the result implicit in (B.20), $\mathbf{E}'' = \frac{1}{2}\mathbf{E}$, holds as well at a point on a smoothly curved surface as it does on the plane conductor; in each case, a field point right at a surface element is concerned, and even a single element approaches equivalence to an infinite plane for a point close enough to it. The same conclusion can be argued in several ways (see Exercise 2.5), and perhaps the cleverest is an argument attributed to Laplace (see page 166). Consider a point just inside and another just outside a given point $\mathbf{r}_C$ of a conductor boundary, as indicated in Fig. B.3. The entire (normal) field $\mathbf{E}$ just outside must be a resultant, $\mathbf{E} = \mathbf{E}'' + \mathbf{e}$, of the field $\mathbf{E}''(r_C)$ due to all sources except the charge element $\sigma\ dS(\mathbf{r}_C)$ itself and of $\mathbf{e}(\mathbf{r}_C)$ arising from this element. The field at the point just inside the conductor boundary, being only infinitesimally removed from the point just outside, must have exactly the same contribution $\mathbf{E}''(r_C)$. However, it is on the opposite side of the source element which contributes $\mathbf{e}(\mathbf{r}_C)$, and hence the resultant is $\mathbf{E}'' - \mathbf{e}$ at the point just inside the boundary; moreover this resultant must vanish and $\mathbf{e} = \mathbf{E}''$ since the point is inside a conductor. The conclusion is that $\mathbf{E}'' = \mathbf{E} - \mathbf{e} = \mathbf{E} - \mathbf{E}'' = \frac{1}{2}\mathbf{E}$, exactly as is implicit in (B.20), for any conductor boundary, plane or curved.

Now the stress at any point on the conducting sphere of Fig. B.2 can be calculated as in (B.20); with σ as given in (B.12),

$$T(\vartheta) = \frac{q^2(r_o^2 - a^2)^2}{8\pi a^2 R_C^6} \tag{B.21}$$

It can be checked that the integral

$$F_C = 2\pi a^2 \int_{-1}^{1} d(\cos\vartheta) T(\vartheta) \cos\vartheta$$

yields just the result (B.18), as it should.

The result of these studies which has general interest is that an equipotential surface of an electrostatic system has transmitted through each point of it a (pulling) stress,

$$T = \frac{\mathbf{E}^2}{8\pi} \tag{B.22}$$

where E is the entire (normal) field at that point of the equipotential. The same result is usually obtained in a more formal way by deriving it from the field (interaction) energy in a manner which is quite analogous to the derivation of the forces from interaction potential energies in Sec. 2.4. The work which would be done by the force T on a unit area of conductor surface is considered when the unit area is *imagined* to undergo a virtual displacement δx_n normal to the surface. The work done by T is then just $T\,\delta x_n$. At the same time, the displacement reduces the volume of field by δx_n, since the unit area of conductor boundary has encroached on the field-containing region by a distance δx_n. Thus, the work has been done at the cost of the field energy

$$T\,\delta x_n = \delta x_n \frac{\mathsf{E}^2}{8\pi} \tag{B.23}$$

since $\mathsf{E}^2/8\pi$ [Eq. (2.61)] is known to be the energy density in the field. The result (B.22) has again emerged. In such considerations as these there is always a judicious neglect of various effects, like the changes that would occur in an actual displacement; however, it should be clear that such effects are of second order or higher, always negligible for *small enough* virtual displacements.

EXERCISES

B.1 Two equal and opposite point charges $\pm q$ are fixed at a distance l from each other. Is the force on each diminished or increased and by what factor when:

(*a*) An infinite plane conducting slab of thickness $\frac{1}{2}l$ is placed (symmetrically) midway between them?

(*b*) A neutral conducting sphere of diameter $\frac{1}{2}l$ is centered at the midpoint? Does it make a difference if this sphere is grounded? Would it make a difference if the sphere were placed closer to one of the charges?

B.2 A simple pendulum is suspended above a horizontal conducting plane so that it can swing above it. Suppose negative charge is added to the pendular bob. Will this increase or decrease the frequency of small oscillations of the pendulum? What if the charge were positive?

B.3 A point charge q is equidistant from each of two half-infinite grounded conducting planes that intersect at right angles to each other.

(*a*) Find the force on the charge. Should it be *expected* to be less than $\sqrt{2}(q^2/4l^2)$ when l is the distance of q from each plane?

(*b*) Find the charge distributions on each plane. Show that they vanish at the intersection and have maxima at some distance from the intersection that is greater than l.

(*c*) Suppose that the point charge is replaced by an infinite line with charge λ per unit length and parallel to the intersection of the planes. Find the force on each unit length of the charge.

B.4 A point charge q lies within a wedge of free space bordered by grounded conducting planes making an opening angle $\alpha = \pi/3$ with each other. Find the positions of all the images needed to represent the field in the wedge when s and $0 < \varphi < \alpha$ are the position coordinates of the point charge. (A student so inclined might find that a finite number of images suffices for any of the opening angles $\alpha = \pi/n$, where $n = 1, 2, 3, \ldots$. If α were some integer or noninteger multiple of a π/n, satisfying the boundary conditions would require putting images into the free space, thus introducing singularities into the field there that invalidate the image method for such cases.)

B.5 A point charge q has a position $s = 0$, $z = l > 0$ between parallel conducting planes at $z = 0$ and $z = L > l$.

(*a*) Construct simple arguments that the charge induced on the $z = 0$ plane is just $-q(1 - l/L)$, partially based on the fact that any $f(\zeta)$ such that $f(0) = 1$ and $f(\zeta) + f(1 - \zeta) = 1$ at every $0 < \zeta < 1$ can only be $f(\zeta) \equiv 1 - \zeta$ [as can be concluded by considering a Taylor expansion of $f(\zeta)$, for example]. What should be expected for the charge on the $z = L$ plane?

(*b*) Show that the charge *distribution* on the $z = 0$ plane following from the method of images can be written as the convergent series

$$\sigma_o = -\frac{q}{2\pi}\left(\frac{l}{(s^2 + l^2)^{3/2}} - \sum_{n=1}^{\infty}\left\{\frac{2nL - l}{[s^2 + (2nL - l)^2]^{3/2}} - \frac{2nL + l}{[s^2 + (2nL + l)^2]^{3/2}}\right\}\right)$$

and that the distribution on the $z = L$ plane differs from this only by the replacement $l \to L - l$, as expected (easiest to do after rewriting σ_o as a sum $\sum_{n=-\infty}^{\infty}$).

(*c*) Integrate σ_o to the result of (*a*). Caution is necessary when integrating the $n \to \infty$ terms, since integration over s and summation over n can be interchanged only for *any finite* number of terms, $n < N$. However, it is easy to show that the charge induced on the $z = 0$ plane within any very large radius $s < R$ can be expressed as

$$-q\left(1 - \int_0^R dss \sum_{n=N}^{\infty}\{\cdots\}\right)$$

where the contents of the braces are the same as those in the above expression for σ_o. This is valid only if $R \gg 2NL$, but in the limit $R \to \infty$, N can still be large enough ($N \gg \Delta n \equiv 1$) for the remaining summation to be replaceable by an integration $\int_N^\infty dn\{\cdots\}$ that *precedes* the integration over s properly.

B.6 Suppose that instead of being grounded the sphere of Fig. B.2 is maintained at a potential ϕ_o above ground. What is the total charge on the sphere?

B.7 (*a*) In the situation of Fig. B.2 but with the conductor insulated from ground, how much charge q_C should be placed on it to ensure that the point charge will remain at r_o even if it is freed to move along a radius? Will this be a *stable* equilibrium position after q_C is supplied?

(*b*) Can the same trick be used to maintain an off-center position of a point charge *inside* a hollow conducting sphere? Is even the center position here stable to radial displacements?

B.8 For a point charge q *inside* a grounded spherical conducting surface with radius a and fixed at a distance $r_o < a$ from the sphere center:

(*a*) Find the induced charge distributions on the inner and outer surfaces of the sphere.

(*b*) Find the force on the charge.

(*c*) Suppose now that the sphere is insulated and initially neutral. Any alteration of the inner field plainly cannot be represented by adding an image at the center, as was done in the text for the outer field, since that would introduce an unallowable singularity into the inner field. What is the solution of the problem?

B.9 Suppose that the point charge of Fig. B.1 is replaced by a conducting sphere of radius $a < l$ centered at a distance l from the plane and bearing the charge q.

(*a*) Show that equipotentialities of each conductor can be maintained by an infinite sequence of images generated from an initial one, $q_1 = a\phi_a$, at the sphere center. Here ϕ_a is the ultimate potential of the sphere relative to the $\phi = 0$ plane. (Reference to Exercise B.6 may help.)

(*b*) Show that ϕ_a can be expressed in terms of q as

$$\phi_a = \frac{q}{a}\left[1 + \sum_{n=2}^{\infty}\prod_{p=1}^{n-1}\frac{a}{l + z_p}\right]^{-1}$$

where the nth term of the sum consists of a product of $n - 1$ factors having the forms shown and each z_p is a distance of an image from the plane, to be generated from $z_1 \equiv l$ via the recurrence relation

$$z_{p+1} = l - \frac{a^2}{l + z_p}$$

(*c*) Show that the representation converges sufficiently well for even as large a sphere radius as $a = \frac{1}{2}l$ to yield the result $\phi_a \approx \frac{3}{4}(q/a)$ within a 1 percent accuracy in that case.

(*d*) Show that within an accuracy of 6 percent the force on the sphere is double the point charge value $q^2/4l^2$ when $a = \frac{1}{2}l$. (The approach here can be generalized to the case of the plane replaced by a second sphere but with results of poor convergence unless the sphere radii are much smaller than the distance apart.)

B.10 (*a*) Calculate the potential energy $U(z)$ of a point charge q at a distance z from a conducting plane by evaluating the minimum work needed to remove the point charge to infinity.

(*b*) The result of (*a*) is just *half* the result of calculating U from the interaction energy expression (2.46), which gives $q\phi_\sigma$, where ϕ_σ is the potential arising from the induced charge distribution σ (or from the image $-q$). This happens because the work evaluation (*a*) includes the self-energy $\frac{1}{2}\oint dS_C\sigma\phi_\sigma$ of inducing σ, whereas all self-energies were omitted in the definition of (2.46). By starting from the particularly unambiguous expression (2.60) of the total field energy and subtracting from it only the self-energies of exterior charges *in isolation*, show that for any charges ρ exterior to any conductor surfaces $\oint dS_C$, the potential energy is given by

$$U = \int dV\rho\phi_\sigma + \frac{1}{2}\oint dS_C\sigma\phi_\sigma = \frac{1}{2}\int dV\rho\phi_\sigma$$

(exclusive of the self-energy of the ρ distribution). The final expression is also obtainable by considering afresh the work of bringing charges into the field of induced charges, just as (2.39) was obtained.

B.11 Compare such a work evaluation as in Exercise B.10(*a*) with integrations like those in (*b*) for a point charge and an *insulated sphere*.

B.12 An ideal electric dipole of moment **D** is placed at a distance l from a conducting plane and oriented so that **D** deviates by an angle α from pointing directly at the plane.

(*a*) Show how expression (2.27*b*) for the fields of dipoles can be used to find the charge distribution induced on the plane and also the total induced.

Find the locations of the nodes and the maxima in the induced charge distributions for a dipole (*b*) normal to the plane and (*c*) parallel to it.

B.13 In the situation of Exercise B.12:

(*a*) Use (2.54) to find the translational force on the dipole in terms of **D**, l, and α.

(*b*) Use (2.55) to find the torque. What are the stable orientations?

(*c*) Show that both the force and the torque are derivable from a potential energy equal to just half a dipole-dipole interaction energy like (2.57). By considering total field energies develop a very simple argument for the factor $\frac{1}{2}$ here.

B.14 In the situation of Exercise B.12 find the work needed to twist the dipole from $\alpha = 0$ to $\alpha = \frac{1}{2}\pi$:

(*a*) By evaluating work against the torque

(*b*) By evaluating appropriate work against the translational force alone

(*c*) From the potential-energy difference

B.15 An ideal dipole of moment **D** is put at a position $\mathbf{r}_o$ relative to the center of a conducting sphere having a radius $a < r_o$.

(*a*) Do you expect it to make any difference whether the sphere is grounded or insulated?

(*b*) Show that a dipole image of moment

$$\mathbf{D}' = -[\mathbf{D} - 2\hat{\mathbf{r}}_o(\hat{\mathbf{r}}_o \cdot \mathbf{D})]\left(\frac{a}{r_o}\right)^3$$

relative to a center at $\mathbf{r}_o(a/r_o)^2$, is induced in the sphere. Check that this reduces to the plane-conductor result of Exercise B.12 as $a \to \infty$ but $r_o - a \to l$.

(*c*) Find the ratio of the attractions to the sphere of the dipole orientations normal to the nearest part of the sphere surface and parallel to it. How does this ratio compare with that of the plane-conductor case?

B.16 The stress on a conducting plane due to a point charge in front of it is the same whether the charge is positive or negative. Suppose that to the single point charge is added a second one, some distance away, equal in magnitude to the first charge but opposite in sign. Will the total force on the plane be augmented or diminished?

B.17 For the dipole in front of the conducting plane considered in Exercise B.12:

(*a*) Find the stress distribution on the plane for an arbitrary orientation of the dipole and check your result by integration.

(*b*) If the dipole lies in an xz plane, find the torque on the conductor about a y axis opposite the dipole's position.

B.18 Suppose that the three walls bordering the field described in (3.30) and (3.34) are conductors.

(*a*) Find the charge distribution on each wall.

(*b*) Show that the total charge, per unit length of channel, on the wall at the potential ϕ_o is just $\phi_o/2\pi$ and that the charges on the grounded walls have the opposite sign (and hence some field lines emanating from the ϕ_o wall terminate at the sides). (Finite results for the total charges on all sides are obtainable when the side walls are terminated at any finite length l by a fourth grounded wall opposite the ϕ_o side, as in Exercise 3.3. Then there are only negative charges on all the grounded walls, adding up to the same magnitude as that of the positive charge on the ϕ_o wall, as to be expected from the electric fluxes.)

B.19 Show how the result (B.15) can be more simply obtained from applying appropriate boundary conditions to the general solution (3.72).

B.20 A conducting plane has a hemispherical protrusion (*boss*) of radius a from its surface. Incident on this is a field that infinitely far away has the uniform magnitude E_o and a direction normal to the plane. Without the boss the plane would bear the uniform charge density $|\sigma| = \mathsf{E}_o/4\pi$. How does σ differ near the protrusion and on it, and what is the ratio of the stress at the boss tip to the stress on the plane without the boss?

B.21 An insulated and initially neutral spherical conductor of radius c has a spherical hollow of radius b in it, off center by a distance $d < c - b$. The hollow contains, centered in it, a spherical charge distribution

$$\rho(r_s \le a, \vartheta_s) = \frac{3q_o}{4\pi a^3}\left(1 + \frac{4D_o}{q_o a}\cos\vartheta_s\right)$$

maintained within the radius $a < b$. Find the field everywhere and the charge distributions induced on the conductor surfaces. How is the outermost field influenced by shifting the sphere of charge from the center of the hollow by a distance $s < b - a$?

B.22 Two conducting spheres, each of radius a, have a large distance $L \gg a$ between their centers.

(*a*) Just one of the spheres is charged, to a total q. Find the dipole moment induced on the second sphere, to the lowest order of approximation in $a/L \ll 1$.

(*b*) Give arguments that the data in (*a*) are sufficient for an evaluation of the total field energy present within a fraction $<2(a/L)^4$.

(*c*) Now the two spheres are connected by a conducting wire of negligible capacity and a new equilibrium awaited. Show that the energy

$$\frac{q^2}{4a}\left[1 - \left(\frac{a}{L}\right)^2 - 2\left(\frac{a}{L}\right)^4 + \cdots\right]$$

must be dissipated in I^2R losses before the new equilibrium is reached. What is going on in the interval?

CHAPTER

C

CYLINDRICAL LAPLACE FIELDS

A series representation of Laplace fields invariant along one direction and interpretations of its terms · Properties of the Bessel, Neumann, and Hankel functions · Fourier-Bessel series and their uses · Modified Bessel functions and their uses

Next in importance to the situations of rectangular and spherical symmetries, described in Chap. 3 with the help of Laplace fields $\phi(x, y, z)$ and $\phi(r, \vartheta, \varphi)$, are those of an intermediate cylindrical symmetry, to be described by functions $\phi(s, \varphi, z)$ of the cylindrical coordinates.

Bounded variations with the azimuthal coordinate φ can again be represented by the orthogonal set proportional to $e^{im\varphi}$ of (3.59). It is the accompanying variations of a Laplace field with distance s from the axis and coordinate z along that axis that will be the chief concern here.

C.1 POLAR COORDINATE REPRESENTATIONS

The description of Laplace fields $\phi(s, \varphi)$ in polar coordinates, with no variations along the z direction, will be considered first. The resolution

$$\phi(s, \varphi) = \sum_{m=-\infty}^{\infty} R_m(s) e^{im\varphi} \tag{C.1}$$

presents the problem of finding radial distributions $R_m(s)$ that will make this form satisfy the Laplace equation. It is easy to see that the same problem would be presented by a search for solutions which are separable in the polar coordinates. Applying the Laplace operation in the form (A.40), expressed in the appropriate coordinates, shows that the result for each of the linearly independent terms of the above decomposition will vanish properly if only

$$s \frac{d}{ds}\left(s \frac{d}{ds}\right) R_m = m^2 R_m \tag{C.2}$$

Trying a proportionality to a power, $R_m \sim s^\alpha$, as a solution shows that either $\alpha = +m$ or $\alpha = -m$ is satisfactory. A solution $R_m(s) = a_m s^m + b_m/s^m$ is thus obtained, having the two arbitrary constants needed for generality, except for the case of $m = 0$ (when $a_o + b_o \equiv A_o$ amounts to a single arbitrary constant). In that special case, $s\, dR_o/ds =$ some constant B_o, and so $R_o(s) = A_o + B_o \ln s$ in general. Moreover, *real* fields are wanted, and they can be expressed without any explicit appearances of the imaginary unit i after the substitution $e^{im\varphi} = \cos m\varphi + i \sin m\varphi$. Then

$$\phi(s, \varphi) = A_o + B_o \ln s + \sum_{m=1}^{\infty} (A_m s^m + B_m s^{-m})(C_m \cos m\varphi + D_m \sin m\varphi) \tag{C.3}$$

is a way to present the general† two-dimensional Laplace field expressed in polar coordinates. Terms formerly labeled by the negative integers are now enumerated by positive ones (after substitutions $m \leftrightarrow -m$) because this more clearly distinguishes terms that are *regular* as $s \to 0$ (with $m \geq 0$ and constants A_m) from *irregular* terms that are singular at $s = 0$ (those proportional to B_m). Detailed reduction of the initial expression (C.1) to the final one (C.3) is unnecessary, since it is clear that the latter contains all the $m > 0$ terms and that negative m values in (C.3) would produce no new types of terms linearly independent of the ones already present.

The result (C.3) for the potential can be explored by considering what each of its linearly independent terms can be used to describe. Various choices of A_o merely determine the gauge of the potential. The term $B_o \ln s$ can be recognized as arising from net charge $\lambda = -\frac{1}{2}B_o$ per unit of the z dimension. This charge need not be confined to the line through $s = 0$ but may be distributed in any way, uniform in z, that produces a net outward flux $4\pi\lambda = -2\pi B_o$ (per unit of z) at the radius s of the field evaluation. None of the other terms can contribute to this *net* flux $\int_0^{2\pi} \mathsf{E}_s s \, d\varphi$, since their angle dependence (nonisotropic) makes their contributions to $\mathsf{E}_s = -\partial\phi/\partial s$ average to zero. When the charge is not distributed uniformly in angle, the term plainly represents an analog of the monopole effect, discussed in Sec. 2.3 for more general charge distributions. The analogs of the multipole corrections are represented by the remaining singular terms, proportional to the various inverse powers $1/s^m$. For example, the contributions proportional to B_1 are characteristic of a potential arising from the equivalent of an ideal dipole line. This consists of ideal dipole elements distributed uniformly along the z direction and oriented normally to that direction. If α is the angle this orientation makes with the $\varphi = 0$ radius and δ is the dipole moment per unit length of the line, it produces a potential $2\delta \cos (\varphi - \alpha)/s$ (see Exercise 2.20). Thus, the correspondent constants in the general expression (C.3) have the interpretations $B_1 C_1 = 2\delta \cos \alpha$ and $B_1 D_1 = 2\delta \sin \alpha$.

The charge distribution just discussed as determining the strengths B_o, $B_m C_m$, and $B_m D_m$ of the singular contributions to the potential need not be given as such but can be induced‡ in materials also confined within (toward $s \to 0$ from) the sourceless region in which (C.3) applies. The inducing fields may

† When this is required to be single-valued.

‡ An example in spherical geometry will be treated in Sec. D.8.

be incident from the regions beyond the sourceless one (toward $s \to \infty$). For example, a uniform field $\mathbf{E}_o$ incident from some direction $\varphi = \alpha$ can be represented by the terms proportional to $A_1 s$. It is easy to see that values of the constants that do just this are $A_1 C_1 = \mathsf{E}_o \cos \alpha$ and $A_1 D_1 = \mathsf{E}_o \sin \alpha$. If there are no such outer sources of field present, so that the sourceless region extends to $s \to \infty$, all the constants $A_{m>0}$ must be set equal to zero.

Various terms with $A_m \neq 0$ are needed whenever the region of interest is enclosed by some boundary on which there are conditions to be satisfied. For example, suppose that the ideal dipole line discussed above is enclosed by a grounded conducting cylinder of radius a centered on the line. It is simple to conclude that it then produces the potential

$$\phi = 2\delta \cos (\varphi - \alpha)\left(\frac{1}{s} - \frac{s}{a^2}\right) \tag{C.4}$$

in $s < a$; the boundary condition $\phi(a, \varphi) = 0$, for all φ, has here required that a choice $A_1 = B_1/a^2$ be made. The resulting term proportional to $A_1 s$ here represents a uniform field that arises from charges induced on the conductor by the dipole line.

The sketchy exploration here has shown that the unbounded power-law behavior proportional to $s^{\pm m}$ along radii orthogonal to the arcs along which the bounded variations proportional to $e^{\pm im\varphi}$ take place has a role like that of the unbounded exponential variations (3.25) and (3.52) along orthogonal cartesian directions.

C.2 CYLINDRICAL HARMONICS

Now suppose that there are also variations along the z dimension to be described. Appropriate Laplace fields $\phi(s, \varphi, z)$ can be sought in forms that are separable in the three cylindrical coordinates

$$\phi = Z(s)\Phi(\varphi)F(z)$$

The choice of the symbol Z for the radial, rather than the axial, distribution here conforms to a standard notation used for the types of solutions that will develop (called *Zylinder-funktionen* in German works). Substitution of the form into the Laplace equation, with the operator ∇^2 expressed in the cylindrical coordinates (A.40), leads to

$$s^2 \frac{\nabla^2 \phi}{\phi} = \frac{s}{Z} \frac{d}{ds}\left(s \frac{dZ}{ds}\right) + \frac{\Phi''(\varphi)}{\Phi} + s^2 \frac{F''(z)}{F} = 0$$

The multiplication by s^2 here has made it more evident that Φ''/Φ must be some constant, following the same type of argument as that used for the separation of the cartesian variables in (3.16). If the constant is called $-m^2$, then $\Phi'' = -m^2\Phi$

has solutions proportional to $e^{\pm im\varphi}$. In them, the constant m must be restricted to real, integer values if the sourceless space continues uninterrupted in a circling of the central axis; the points with φ and $\varphi + 2\pi$ on each radius are identical, and then the continuous definition of the field demands that $e^{\pm im(\varphi + 2\pi)} = e^{\pm im\varphi}$. Thus, the azimuthal variations are to be described by the members of the complete set (3.59). Now the above equation can be reduced to

$$-\frac{1}{Z}\frac{d}{s\,ds}\left(s\frac{dZ}{ds}\right) + \frac{m^2}{s^2} = \frac{F''(z)}{F}$$

a form which makes it plain that F''/F must also be put equal to some constant. It will be called k^2 without any immediate decision whether it is to be taken positive or negative. There remain the two ordinary differential equations for $F(z)$ and $Z(s)$

$$F''(z) = k^2 F \tag{C.5}$$

$$\left[-\frac{d}{s\,ds}\left(s\frac{d}{ds}\right) + \frac{m^2}{s^2}\right]Z_m(ks) = k^2 Z_m \tag{C.6}$$

As indicated, the solution for Z will depend on the choices made for m and k; moreover, it should be expressible in terms of the product $\zeta \equiv ks$ as independent variable, since dividing the equation through by k^2 allows rewriting it as

$$\frac{d^2 Z_m}{d\zeta^2} + \frac{dZ_m}{\zeta\,d\zeta} + \left(1 - \frac{m^2}{\zeta^2}\right)Z_m = 0 \tag{C.7}$$

This is known as the *Bessel equation.* Its various solutions for real and complex argument and for integer and noninteger m have the generic name *cylindrical harmonics.*

The general solution of the Bessel equation for any given m can be written as an arbitrary linear combination of the regular Bessel function of order m, denoted $J_m(\zeta)$, and the linearly independent irregular Neumann function $N_m(\zeta)$

$$Z_m(\zeta) = a_m J_m(\zeta) + b_m N_m(\zeta) \tag{C.8}$$

containing the two arbitrary constants needed for generality. Extensive numerical tabulations of lower-order functions of both types exist,† just as they do for the low-order $(m = 1)$ trigonometric functions cos $m\varphi$ and sin $m\varphi$. There are recurrence relations, analogous to $\cos(m + 1)\varphi = \cos m\varphi \cos\varphi - \sin m\varphi \sin\varphi$, through which higher-order functions can be evaluated from the tabulated ones; the relations are recorded in many places,† Watson (Ref. 30) being one of the most exhaustive works.

† For example, in Ref. 14. Electronic computers have made great tomes of tables available. The most frequently used properties are collected in Ref. 17, sec. 13.2 and exercise 13.4.

The Bessel and Neumann functions are said to be regular and irregular, respectively, because of their contrasting behavior as $\zeta \to 0$, here expressed† for real $m > 0$

$$J_m(\zeta \to 0) \to \frac{\zeta^m}{2^m m!} \tag{C.9}$$

$$N_{m \geq 1}(\zeta \to 0) \to -\frac{2^m(m-1)!}{\pi\zeta^m} \tag{C.10}$$

These properly agree with the behavior proportional to $s^{\pm m}$ found for the linearly independent solutions in the $k = 0$ case of the Bessel equation (C.6), investigated as (C.2). Corresponding to the $m = 0$ solution proportional to ln s of the $k = 0$ case is

$$N_o(\zeta \to 0) \to \frac{2}{\pi} \ln \zeta \tag{C.11}$$

In the normalization adopted, the asymptotic behavior of the functions is

$$J_m(|\zeta| \to \infty) \to \left(\frac{2}{\pi\zeta}\right)^{1/2} \cos\left[\zeta - \frac{(m+\frac{1}{2})\pi}{2}\right] \tag{C.12}$$

and

$$N_m(|\zeta| \to \infty) \to \left(\frac{2}{\pi\zeta}\right)^{1/2} \sin\left[\zeta - \frac{(m+\frac{1}{2})\pi}{2}\right] \tag{C.13}$$

That these satisfy the Bessel equation where $|\zeta^2| \gg m^2 - \frac{1}{4}$ is simple to check.

The last forms suggest defining the linear combinations

$$H_m^{(1)}(\zeta) \equiv J_m + iN_m \to \left(\frac{2}{\pi\zeta}\right)^{1/2} e^{i[\zeta - (m+1/2)\pi/2]} \tag{C.14}$$

$$H_m^{(2)}(\zeta) \equiv J_m - iN_m \to \left(\frac{2}{\pi\zeta}\right)^{1/2} e^{-i[\zeta - (m+1/2)\pi/2]} \tag{C.15}$$

likewise linearly independent of each other. These are known as *Hankel functions of the first and second kinds*, respectively. It is frequently more convenient to put the general solution (C.8) into the form

$$Z_m(\zeta) = \tfrac{1}{2}(a_m - ib_m)H_m^{(1)} + \tfrac{1}{2}(a_m + ib_m)H_m^{(2)} \tag{C.16}$$

for such reasons as sometimes make linear combinations of $e^{\pm im\varphi}$ more convenient than expressions in cos $m\varphi$ and sin $m\varphi$.

For real values of m and the argument $\zeta = ks$, the two kinds of Hankel functions are simply each other's complex conjugates, and the linear combinations

$$J_m = \tfrac{1}{2}(H_m^{(1)} + H_m^{(2)}) \qquad N_m = \frac{1}{2i}(H_m^{(1)} - H_m^{(2)}) \tag{C.17}$$

† Readers unfamiliar with factorials of numbers that need not be integers may prefer the notation $m! \equiv \Gamma(m+1)$.

become *real* functions. They have an oscillatory character, as their asymptotic behavior (C.12) and (C.13) indicates, with the oscillations modulated by decreasing amplitudes proportional to $s^{-1/2}$ as $s \to \infty$. The wavelengths of the oscillations approach the uniform value $2\pi/k$ only as $s \to \infty$; for smaller values of ks, the nonuniform distances between nodes can be found by consulting† tables of roots, arguments $\zeta_{m1}, \zeta_{m2}, \ldots, \zeta_{mn}, \ldots$ at which the function of a given order m becomes zero. For example, tabulations show that $J_0(\zeta_{0n}) = 0$ for

$$\zeta_{01} = 2.405, \zeta_{02} = 5.520, \zeta_{03} = 8.654, \zeta_{04} = 11.792, \ldots \tag{C.18}$$

The half wavelength between the last two nodes listed here is $\Delta s = 3.138/k$, compared with the asymptotic approximation $(\pi = 3.142)/k$.

C.3 FOURIER-BESSEL SERIES

An *orthogonal* set, useful for representing radial dependences on finite intervals $0 \le s \le a$, can be developed from *each* of the real functional forms $J_m(ks)$ of given nonnegative m. The sets arise naturally when considering the electrostatic fields that can exist in a sourceless region enclosed by a grounded cylinder, much like the sine-wave set (3.19) developed for representing fields enclosed between grounded conducting planes. The restriction to the regular Bessel functions $J_m(ks)$ is necessary to avoid infinite field values on the axis $s = 0$ of the sourceless region. The outer boundary condition, $\phi(a, \varphi, z) = 0$ for all φ, z at some radius a, can be satisfied by restricting k to discrete values such that

$$J_m(k_{mn} a) = 0 \tag{C.19}$$

The discrete $k_{m1}, k_{m2}, \ldots, k_{mn}, \ldots$ can be found as $k_{mn} = \zeta_{mn}/a$, from the roots ζ_{mn} of the mth-order Bessel function discussed in the preceding paragraph. The tables of the roots need not be consulted when the asymptotic approximation (C.12) of $J_m(ks)$ is considered adequate. This has zeros for

$$\zeta_{mn} = k_{mn} a \approx [n + \tfrac{1}{2}(|m| - \tfrac{1}{2})]\pi \tag{C.20}$$

yielding, for example, $k_{01} \approx \frac{3}{4}\pi/a = 2.374/a$, $k_{02} \approx 5.515/a$, $k_{03} \approx 8.657/a$, $k_{04} \approx 11.799/a$, ... in place of the values $2.405/a$, $5.520/a$, $8.654/a$, $11.792/a$, ... from the table (C.18).

The imposition of the boundary conditions on the interval $0 \le s \le a$ produces a discrete orthogonal set

$$J_m(k_{mn} s) \equiv J_m\left(\frac{\zeta_{mn} s}{a}\right) \qquad n = 1, 2, 3, \ldots \tag{C.21}$$

† See Ref. 14 and Ref. 17, sec. 13.2 and exercise 13.4.

for every $m \geq 0$. The orthogonality property is expressed† in

$$\int_0^a dss J_m(k_{mn}s) J_m(k_{mn'}s) = \delta_{nn'} \tfrac{1}{2} a^2 [J_{m+1}(k_{mn}a)]^2 \tag{C.22}$$

Each set (C.21) for any one m value is by itself complete enough to represent any radial function $f(s)$, defined on the interval $0 \leq s \leq a$, that is continuous with some finite $f(0)$ value

$$f(s) = \sum_{n=1}^{\infty} c_n J_m(k_{mn}s) \qquad c_n = \frac{\int_0^a dss J_m(k_{mn}s) f(s)}{\frac{1}{2}a^2 [J_{m+1}(\zeta_{mn})]^2} \tag{C.23}$$

The order of the Bessel functions used here need not even be restricted to an integer m.

Bessel functions of *all integer* orders are needed for representing both the radial and the azimuthal dependences of functions $f(s, \varphi)$. If f is continuous and finite on $0 \leq s \leq a$ and $0 \leq \varphi < 2\pi$,

$$f(s, \varphi) = \sum_{m=-\infty}^{\infty} \sum_{n=1}^{\infty} c_{mn} J_m(k_{mn}s) e^{im\varphi} \tag{C.24}$$

with

$$c_{mn} = \frac{\int_0^a dss \int_0^{2\pi} d\varphi f(s, \varphi) J_m(k_{mn}s) e^{-im\varphi}}{\pi a^2 [J_{m+1}(\zeta_{mn})]^2} \tag{C.25}$$

Bessel functions of negative integer order are defined for this series; they are not linearly independent of those with $m > 0$, but

$$J_{-m} = (-)^m J_m \qquad \text{for } m = 0, \pm 1, \pm 2, \ldots \tag{C.26}$$

The terms of the representation (C.24) having opposite signs of m are nevertheless linearly independent of each other because each of the pair of functions $e^{\pm im\varphi}$ is. The property (C.26) is consistent‡ with (C.8)'s being a general solution for arbitrary m, since only m^2 appears in the Bessel equation (C.7).

The basis used for the representation (C.24) makes it most directly adaptable to finding what electrostatic fields can exist within grounded cylindrical walls. Any such field $\phi(s, \varphi, z)$ will be representable with coefficients $c_{mn} \equiv F_{mn}(z)$ that depend on z as solutions of the equation (C.5) do, since this was found as the restriction on the z dependence demanded by the Laplace equation, on such a basis as that of the separable terms in (C.24). The basis uses only real k values k_{mn}, and so $F_{mn}(z)$ must be some linear combination of the real exponentials $e^{\pm k_{mn}z}$ to accompany the bounded variations in the orthogonal directions of s and φ. The resulting series can be written in terms of positive integers m only,

† See Ref. 14 and Ref. 17, sec. 13.2.

‡ For details, see ibid.

because of (C.26), and can also be expressed in terms of real quantities after replacements of $e^{\pm im\varphi}$ by $\cos m\varphi \pm i \sin m\varphi$,

$$\phi(s, \varphi, z) = \sum_{m=0}^{\infty} \sum_{n=1}^{\infty} (A_{mn} e^{k_{mn}z} + B_{mn} e^{-k_{mn}z}) J_m(k_{mn} s) \times (C_{mn} \cos m\varphi + D_{mn} \sin m\varphi) \tag{C.27}$$

This reduces to $\phi(a, \varphi, z) = 0$ on the cylinder at $r = a$ and is adaptable to arbitrary boundary conditions at the ends of the cylindrical cavity.

Suppose that the interest is in the transmission down an indefinitely long cylindrical tube (with grounded walls) of a field that is applied as a potential distribution $\phi_o(s, \varphi)$ at one end, which may be taken to be $z = 0$. With the sourceless cavity extending to $z \to \infty$, all A_{mn} in (C.27) must be set equal to zero and $B_{mn} \equiv 1$ can be taken, for the constants C_{mn} and D_{mn} are now sufficient to represent the remaining arbitrariness. They are to be determined from the given $\phi_o(s, \varphi)$ as†

$$\left.\begin{matrix} C_{mn} \\ D_{mn} \end{matrix}\right\} = \frac{2}{\pi a^2 [J_{m+1}(k_{mn} a)]^2} \int_0^{2\pi} d\varphi \left\{\begin{matrix} \cos m\varphi \\ \sin m\varphi \end{matrix}\right\} \int_0^a ds s J_m(k_{mn} s) \phi_o(s, \varphi) \tag{C.28}$$

expressions that follow quite readily from the orthogonalities when the indicated operations are applied to the series (C.27) with $A_{mn} = 0$, $B_{mn} = 1$, and $z = 0$. [They are also obtainable as suitable linear combinations of the coefficients (C.25), with $f \to \phi_o(s, \varphi)$.] Perhaps the most interesting result is the simple one that the mode proportional to $e^{-k_{01}z} = e^{-2.4z/a}$ penetrates farthest [here $k_{01} a = \zeta_{01} = 2.405$ of (C.18) is the smallest root]. That can be compared to the transmission down a square tube with the same cross section, obtainable from (3.36) for $L_x = L_y = (\pi a^2)^{1/2}$, $n_x = n_y = 1$, and yielding a proportionality to $\exp(-\sqrt{2\pi}\, z/a)$. The mean penetration depth is only slightly greater down the circular tube, by the factor $\sqrt{2\pi}/2.40 \approx 1.04$.

The form (C.27) is adaptable to *two* independently arbitrary potential distributions applied to the two ends of a finite cylinder, say $\phi_L(s, \varphi)$ at $z = L$ in addition to $\phi_o(s, \varphi)$ at $z = 0$. In this situation, the two formulas (C.28) respectively give $(A_{mn} + B_{mn})C_{mn}$ and $(A_{mn} + B_{mn})D_{mn}$ rather than C_{mn} and D_{mn}. Two similar formulas, different only in having ϕ_L in place of ϕ_o, will give

$$A_{mn} e^{+k_{mn}L} + B_{mn} e^{-k_{mn}L}$$

multiplied by C_{mn} and D_{mn}, respectively. The resulting relations are just sufficient to determine all the unknown constants in the series representation (C.27) of the field. It is easy to check that exactly the same result can be obtained as a superposition of two separately determined fields, one arising from $\phi_o(s, \varphi)$ at $z = 0$ and satisfying the boundary condition $\phi(s, \varphi, L) = 0$ at $z = L$, the other from $\phi_L(s, \varphi)$ at $z = L$ with the $z = 0$ end grounded.

† $D_{0n} = 0$, and the factor 2 should be omitted for C_{0n}.

The last remark suggests an approach to the problem posed by applying arbitrary potentials to *all* sides of a finite cylinder, say $\phi_a(\varphi, z)$ to the curved surface at $s = a$, in addition to the $\phi_o(s, \varphi)$ and $\phi_L(s, \varphi)$ at the cylinder ends. The contributions of the latter to the total field can be taken to be those already dealt with in the preceding paragraph. The boundary conditions of the complete problem will clearly be satisfied if now there is added a field which reduces to $\phi_a(\varphi, z)$, instead of zero, at $s = a$ and satisfies the conditions $\phi = 0$ at both $z = 0$ and $z = L$. The solution of such a problem is considered next.

C.4 MODIFIED BESSEL FUNCTIONS

Now posed is a problem of finding a general Laplace field $\phi(s, \varphi, 0 \le z \le L)$ that vanishes at both ends, $z = 0$ and L, of a cylindrically symmetric system. These boundary conditions will be satisfied automatically if the z dependence of ϕ is analyzed into sine-wave modes, $\sin(n\pi z/L)$, like those of (3.19). In the cylindrical variables that corresponds to using solutions $F(z) \sim \sin(n\pi z/L)$ of the differential equation (C.5) with $k^2 = -(n\pi/L)^2$. The accompanying radial functions Z_m must now be taken to be solutions of the Bessel equations (C.7) with *imaginary* argument:

$$\zeta_n \equiv ks = i\frac{n\pi s}{L} \tag{C.29}$$

The consequent Laplace field will have the form

$$\phi(s, \varphi, z) = \sum_{m=-\infty}^{\infty} \sum_{n=1}^{\infty} Z_m(\zeta_n) e^{im\varphi} \sin\frac{n\pi z}{L} \tag{C.30}$$

bounded on the intervals $0 \le \varphi < 2\pi$, $0 \le z \le L$ of the azimuthal and paraxial variables.

The general solution of the Bessel equation for a given m and the imaginary argument $\zeta_n = i(n\pi s/L)$ can be given the same forms, (C.8) and (C.16), as before. Now, however, the behavior of all the solutions as functions of the radius s is quite different from the cases of real $\zeta = ks$. The new behavior can best be seen by inserting the argument (C.29) into the asymptotic forms (C.14) and (C.15)

$$\begin{aligned} H_m^{(1)}(|\zeta_n| \to \infty) &\to \frac{(-i)^{m+1}}{\pi}\left(\frac{2L}{ns}\right)^{1/2} e^{-n\pi s/L} \\ H_m^{(2)}(|\zeta_n| \to \infty) &\to \frac{(+i)^{m}}{\pi}\left(\frac{2L}{ns}\right)^{1/2} e^{+n\pi s/L} \end{aligned} \tag{C.31}$$

The bounded radial oscillations that characterized the functions of the real variable are replaced by decreasing and increasing exponentials of real argument when the variable ζ is imaginary. This means that neither $H_m^{(2)}(\zeta_n)$ nor $J_m(\zeta_n)$ nor $N_m(\zeta_n)$ of (C.17) can be used for Z_m of (C.30) if the sourceless space extends to $s \to \infty$; only $Z_m \sim H_m^{(1)}(\zeta_n)$ can then be permitted. On the other hand, if there are no sources on the axis $s = 0$, then, because of (C.14), (C.15), and the irregularity (C.10), neither $H^{(1)}(\zeta_n)$ nor $H_m^{(2)}(\zeta_n)$ nor $N_m(\zeta_n)$ can be used for Z_m if the sourceless space extends through the axis. Only the regular function $J_m(\zeta_n)$ is then available.

It is now clear how the problem proposed in the last paragraph of the preceding section can be solved. This deals with the interior of a finite cylinder having grounded ends, with a given potential distribution $\phi_a(\varphi, z)$ on the curved surface of radius $s = a$ serving as the only source of the field. The interior field will have the form (C.30) with Z_m replaced by $a_{mn} J_m[i(n\pi s/L)]$. The coefficients a_{mn} are to be obtained from a decomposition of the given $\phi_a(\varphi, z)$ into members of the complete orthogonal set $e^{im\varphi} \sin (n\pi z/L)$.

It is also simple to find the field exterior to the same cylinder if it extends from $s = a$ to $s = \infty$ between grounded planes at $z = 0$ and L [a modification of the problem of a sphere that led to the result (3.73)]. $\phi(s > a, \varphi, 0 \le z \le L)$ will be given by (C.30) with $Z_m \sim H_m^{(1)}(in\pi s/L)$.

It may still be of interest to look at a problem in which singularity on the axis is actually needed, to help describe the presence of a source there. The simplest situation of this kind is presented by a uniform line segment of charge λ per unit length of the $s = 0$ axis stretched between two grounded planes (at $z = 0$ and L). The sourceless region will be allowed to extend from $s = 0$ to $s = \infty$, so that the only admissible radial solutions are the Hankel functions, $H_m^{(1)}(in\pi s/L)$, which tend to vanish as $s \to \infty$ (C.31). Only the functions of order $m = 0$ are actually needed here since the situation is isotropic about the axis and there are no variations with azimuth φ to be represented. The field will therefore have the form

$$\phi(s, z) = \sum_{n=1}^{\infty} c_n H_o^{(1)}\left(\frac{in\pi s}{L}\right) \sin \frac{n\pi z}{L} \tag{C.32}$$

with the coefficients c_n to be determined from the given charge distribution. With this charge uniform along the line $s = 0$ and having strength λ per unit length, the potential must approach $\phi(s \to 0, z) \to -2\lambda \ln s$ on the axis, within additive constants. Each $H_o^{(1)}(in\pi s/L)$ also approaches proportionality to $\ln s$ as $s \to 0$, within relatively negligible additive constants. This can be seen from the expression $H_o^{(1)} \equiv J_o + iN_o$ [Eq. (C.14)], in which J_o stays finite and negligible as $s \to 0$, and from (C.11), which shows that $H_o^{(1)}(\zeta_n \to 0) \to (2i/\pi)[\ln s + \ln (in\pi/L)]$ with the final term negligible. Thus the coefficients in (C.32) are to be determined so that

$$\frac{2i}{\pi} \ln s \sum_{n=1}^{\infty} c_n \sin \frac{n\pi z}{L} = -2\lambda \ln s$$

The summation here is to add up to a constant, $i\pi\lambda$, independently of z. Just such a problem has already been solved in producing (3.29) and (3.30), from which it can be concluded that needed here are $c_n = 0$ for all even n and $c_n = 4i\lambda/(2N + 1)$ for $n = 2N + 1 = 1, 3, 5, \ldots$. The final result is

$$\phi(s, z) = 4\lambda \sum_{N=0}^{\infty} \frac{\sin [(2N + 1)\pi z/L]}{2N + 1} \left[iH_o^{(1)}\left(\frac{i(2N + 1)\pi s}{L}\right)\right] \tag{C.33}$$

Sometimes the real combination $iH_o^{(1)}$ occuring here is denoted $(2/\pi)K_o[(2N + 1)\pi s/L]$, and K_o is called a *modified Bessel function.*

It has been more customary to use special notations for the solutions of the Bessel equation (C.7) with imaginary $\zeta \equiv i\xi$. They are given as real functions of ξ called modified Bessel functions and are related to the forms used above by

$$I_m(\xi) \equiv i^{-m}J_m(i\xi) \qquad K_m(\xi) \equiv \tfrac{1}{2}\pi i^{m+1}H_m^{(1)}(i\xi) \tag{C.34}$$

which are linearly independent and have the asymptotic behavior

$$I_m(\xi \to \infty) \to \frac{e^{+\xi}}{(2\pi\xi)^{1/2}} \qquad K_m(\xi \to \infty) \to \left(\frac{\pi}{2\xi}\right)^{1/2} e^{-\xi} \tag{C.35}$$

that follows from (C.12) and (C.14), respectively. The function $I_m(\xi)$ is regular

$$I_m(\xi \to 0) \to \frac{1}{m!}\left(\frac{\xi}{2}\right)^m \tag{C.36}$$

as can be seen from (C.9), whereas $K_m(\xi)$ has the irregularity

$$K_{m\neq 0}(\xi \to 0) \to \frac{(m-1)!}{2}\left(\frac{2}{\xi}\right)^m \tag{C.37}$$

following from (C.14) and (C.10). $K_o \to -\ln \xi$.

The function $K_m(\xi)$ will prove useful for representing radiated spectra in Chap. 10. In that connection, the known† integral

$$\int_0^\infty d\xi \xi^2 K_m^2(\xi) = \frac{\pi^2}{8}\frac{\frac{1}{4} - m^2}{\cos m\pi} \tag{C.38}$$

will be needed.

EXERCISES

C.1 A uniform field $\mathbf{E}_o$ is incident on an infinitely long conducting cylinder, radius a, from a direction normal to the cylinder axis. Show that the effect of the conductor is to superpose on $\mathbf{E}_o$ the field arising from a dipole line of moment $\delta = \frac{1}{2}\mathbf{E}_o a^2$ per unit length of the cylinder's axis. Find the induced charge distribution $\sigma(\varphi)$ responsible for the moment.

C.2 Two halves, $0 < \varphi < \pi$ and $-\pi < \varphi < 0$, of an indefinitely long hollow circular cylinder, radius a, are kept at equal and opposite potentials $\pm\phi_o$.

(*a*) Find series (C.3) that represent $\phi(s \gtrless a, \varphi)$.

(*b*) Show how the series can be summed into the closed form

$$\phi(s \gtrless a, \varphi) = \frac{2\phi_o}{\pi}\tan^{-1}\left(\frac{2\,as\sin\varphi}{|a^2 - s^2|}\right)$$

(*c*) Find the dipole effect in terms of the field strength on the axis and compare it with the result of Exercise C.1.

† A recent review of its validity appears in Ref. 1.

C.3 A line of charge λ per unit length is parallel to a solid conducting cylinder of radius a and centered at a distance $s_o > a$ from the line.

(*a*) By paralleling the procedure used for (B.10) show that the effect of the conductor is to superpose on the charged-line field a Laplace field ϕ_o that is equivalent to the field of an image line at an $s < a$ and having a strength *independent* of the cylinder radius.

(*b*) Evaluate the force per unit length of the line and its limit in a filament approximation $(a \to 0)$ of the cylinder.

(*c*) Show that if $\lambda = -\frac{1}{2}\mathsf{E}_o s_o$ is put at a distance $s_o \to \infty$ from the cylinder axis, the result for $\phi(s > a, \varphi)$ reduces to that of Exercise C.1.

(*d*) Show how the results for a charged line in front of a conducting plane follow from those of (*a*) and (*b*) when $a \to \infty$ in such a way that $s_o - a \to l$, the finite distance of the line from the plane.

C.4 (*a*) For a line of charge at a distance s_o from an axis of cylindrical coordinates show how the analog

$$-\ln(s^2 - 2ss_o \cos\varphi + s_o^2) = -\ln s_>^2 + 2\sum_{m=1}^{\infty}\left(\frac{s_<}{s_>}\right)^m \frac{\cos m\varphi}{m}$$

of the point-charge multipole expansion (3.81) can be developed. (Reference to the results quoted in Exercise C.15 may help with the integrations.)

(*b*) Show how (*a*) can help in an alternative approach to obtaining the field $\varphi(s > a, \varphi)$ of the preceding exercise, starting with a series representation (C.3). (Compare Exercise 3.15.)

C.5 Show that $Z_{m-1} = \zeta^{-m}(d/d\zeta)(\zeta^m Z_m)$ satisfies an appropriate Bessel equation if Z_m does. [One of many useful properties that can be found in any of numerous references. Compare results quoted in Exercise 9.1(*a*).]

C.6 The flat end disks of a grounded circular cylinder of length L and radius a are kept at equal and opposite potentials, $\pm\phi_o$. Show that the field inside the cylinder can be represented by

$$\phi(s, z) = 2\phi_o \sum_{n=1}^{\infty} \frac{J_o(k_n s)}{(k_n a)J_1(k_n a)} \frac{\sinh k_n z}{\sinh (k_n L/2)}$$

where $J_o(k_n a) = 0$ for all the k_n. (Exercise C.5 can help with the integration.)

C.7 Suppose that the cylinder of Exercise C.6 instead has its flat ends grounded and that opposite semicircular halves of its curved surface are kept at equal and opposite potentials. Adapt (C.30) to describe the field inside the cylinder and check your result by showing that it reduces to the field found in Exercise C.2 for $L \to \infty$.

The following exercises continue the development of the complex representations begun in Exercises 3.17 to 3.22, now to include the *polar*, or *Argand*, *representations* of the field points, as in

$$\zeta = x + iy = r(\cos\varphi + i\sin\varphi) \equiv re^{i\varphi}$$

C.8 (*a*) Show that the two potential distributions considered in Exercise 3.18 can be expressed respectively in terms of r and φ as

$$\frac{\phi_o}{x_o^2} r^2 \cos 2\varphi \qquad \text{and} \qquad \frac{\phi_o}{x_o^2} r^2 \sin 2\varphi.$$

Note how obvious these make the conclusion of Exercise 3.18(*c*).

(*b*) Find how $\mathsf{E}_{r,\varphi}(r, \varphi)$, in either of the two fields, are related to the results of Exercise 3.20, including showing that multiplication by $e^{i\varphi}$ rotates any complex number through an angle φ.

C.9 With $f(\zeta) = -2\lambda \ln(\zeta/r_o)$, where λ and r_o are real constants:

(*a*) Describe the charge distribution that yields $\phi(r) \equiv \text{Re}\,[f(\zeta)]$.

(*b*) Show that $\phi(\varphi) = \text{Im}\,[f(\zeta)]$ can be generated by a uniform double layer (2.32), with $\tau = \lambda$, spread over just the half plane $(x > 0, z)$ and that $\phi(\varphi)$ has the expected discontinuity (2.33). [The resulting line of discontinuity in $f(\zeta)$, consisting of the positive real axis in the complex plane, is called a *cut* emanating from a *branch point* $x = y = 0$.]

(*c*) Show that $-f'(\zeta) = \mathsf{E}_x - i\mathsf{E}_y$ yields the correct $\mathbf{E} = -\nabla\phi(r)$ for the field in (*a*) (see Exercise 3.20).

(*d*) Since Im $[f(\zeta)] \equiv$ Re $[-if(\zeta)]$, $if'(\zeta) = \mathsf{E}_x - i\mathsf{E}_y$ should give the correct $\mathbf{E} = -\nabla\phi(\varphi)$ of the field in (*b*). Check this and note that the sources of (*a*) and (*b*), if only $\tau = \lambda$, yield equal magnitudes for $|\mathbf{E}|$ but mutually orthogonal field lines.

(*e*) Show that, quite generally, multiplication by i rotates any complex number through 90°. [Compare Exercise C.8(*b*).]

C.10 (*a*) Show that the points $\zeta = \zeta_o + Re^{i\varphi}$, where $\zeta_o = x_o + iy_o$ and R is a real constant, follow a circle of radius R centered on x_o, y_o, as φ is increased continuously from 0 to 2π (called the counterclockwise sense).

(*b*) Show that loop integrations (see Exercise 3.22) over circles can be carried out explicitly as in

$$\oint d\zeta f(\zeta) = iR\int_0^{2\pi} d\varphi e^{i\varphi} f(\zeta_o + Re^{i\varphi})$$

once an $f(\zeta)$ is given.

(*c*) Let $f(\zeta) \to \mathsf{E}(\zeta) = (2\phi_o/x_o^2)\zeta$, which helps describe the fields of Exercises 3.18 and C.8. By explicit integration over any circle show that the Cauchy theorem of Exercise 3.22 is satisfied, as it should be for an $\mathsf{E}(\zeta)$ analytic inside the integration loop.

C.11 (*a*) Repeat a loop integration like that of Exercise C.10 for $\mathsf{E}(\zeta) = 2\lambda/\zeta$, which helps represent the fields of Exercise C.9, over a circle centered at $x_o = y_o = 0$ ($\zeta_o = 0$), followed counterclockwise.

(*b*) What would be the result over a smaller circle, radius $\epsilon \ll R$, followed *clockwise* ($\varphi = 2\pi$ to 0 or $\varphi = 0$ to -2π)?

(*c*) The sum of the integrations (*a*) and (*b*) constitutes integration over a contour having two pieces, which are said to *enclose* between them the annular region between the two circles. Taking satisfaction of the Cauchy theorem as the criterion, argue that $\mathsf{E}(\zeta)$ is analytic only *outside* the point $\zeta = 0$, which is called a simple *pole* of $\mathsf{E}(\zeta)$. [In view of the interpretability of λ as the charge per unit length of a line passing through $\zeta = 0$, a simple pole in $\mathsf{E}(\zeta) = \mathsf{E}_x - i\mathsf{E}_y$ corresponds to a monopolar source of $\mathbf{E}$. Compare the interpretation of the cut in Exercise C.9.(*b*).]

C.12 (*a*) Show that

$$\oint \frac{d\zeta}{\zeta - \zeta_o} = 2\pi i$$

by integration over a circle centered on ζ_o.

(*b*) Since $\mathsf{E}(\zeta) = 2\lambda/(\zeta - \zeta_o) = \mathsf{E}_x - i\mathsf{E}_y$ can describe the field of a line of charge, show that (*a*) can be regarded as a consequence of the Gauss law (see Exercise 3.22).

(*c*) Represent the magnetostatic field (1.20) by an analytic function in such a way that it leads to the result in (*a*) as a consequence of the Ampère law (1.21).

(*d*) Show that if $f(\zeta)$ is analytic everywhere about the point ζ_o of the integral in

$$\oint d\zeta \frac{f(\zeta)}{\zeta - \zeta_0} = 2\pi i f(\zeta_o)$$

the right side follows because the integration loop can be taken to be a circle of vanishingly small radius centered on ζ_o.

C.13 There exist complex functions that are analytic only outside some finite central region about some point ζ_o and are representable by convergent *Maclaurin series*

$$\mathsf{E}(\zeta) \equiv \sum_n a_n(\zeta - \zeta_o)^n$$

with constant coefficients a_n, including negative powers $n < 0$, so that $\mathsf{E}(\zeta_o), \mathsf{E}'(\zeta_o), \ldots \to \infty$. Through term-by-term integration over a circle lying entirely in the region of validity for the series representation show that

$$\oint d\zeta \mathsf{E}(\zeta) = 2\pi i a_{-1}$$

only, even if some $a_{n<-1} \neq 0$. The coefficient a_{-1} is then called the *residue* of $\mathsf{E}(\zeta)$ at ζ_o.

C.14 Refer $\mathsf{E}(\zeta)$ of the preceding exercise to an origin $\zeta_o \equiv 0$. It is then known, from Exercise C.9, that the residue term a_{-1}/ζ of $\mathsf{E}(\zeta) = \mathsf{E}_x - i\mathsf{E}_y$ can be interpreted as arising from a charged line of $\lambda = \frac{1}{2}a_{-1}$. Now show that a_{-2}/ζ^2 is a similar representation of the field due to such a *dipole* line as was alluded to on page 508. [Thus, the singular terms of the Maclaurin series can be interpreted as constituting a multipole expansion of the field due to sources in the finite region of nonanalyticity. (The $m \geq 0$ terms may represent external fields that can be superposed.) It then becomes clear that the residue theorem describes the fact that only the monopolar effect of a source can yield net flux out of it (the Gauss law).]

C.15 (*a*) Integrate *even* powers of cos φ around the unit circle $\zeta = e^{i\varphi}$ to obtain the result quoted on page 92. What happens to the integral over *odd* powers?

(*b*) By integration around the unit circle, show for $\alpha < 1$ that

$$\int_0^{2\pi} d\varphi \cos m\varphi \ln(1 - 2\alpha\cos\varphi + \alpha^2) = \begin{cases} 0 & \text{for } m = 0 \\ -2\pi\alpha^m/m & \text{for } m > 0 \end{cases}$$

Note that the Taylor expansion

$$\ln(1 - \alpha\zeta) = -\sum_{k=0}^{\infty} \frac{(\alpha\zeta)^k}{k}$$

is analytic for $\alpha < 1$, like the $f(\zeta)$ of Exercise C.12(*d*).

CHAPTER

D

THE ELECTROSTATICS OF DIELECTRICS

Polarization densities and bound charges induced in dielectrics · The electric displacement field and the phenomenological Maxwell equations of electrostatics · Polarizabilities and the distinction between local and average fields in dielectrics · Electric susceptibilities, dielectric constants, and the Clausius-Mossotti equation · Potential fields and Coulomb's law in dielectric media · Capacitance · The refraction of fields at dielectric boundaries · Images in dielectrics · Field energies and forces in dielectrics · The renormalization of free charges embedded in dielectrics

So far, the materials interacting with electrostatic fields have been taken to be conductors. Imposing an electric field on an isolated conducting body separates normally canceling charges in it to the outermost limits of the whole body. The separated charges form *macroscopic* surface distributions with equal amounts of opposite sign on opposite sides of the body, relative to the dominant direction of the imposed field. The body then has a dipole effect as its largest one at sufficient distances from it. The case of the insulated sphere in a uniform field, treated in Sec. B.4, leads to a general expectation that a dipole moment of order $\mathbf{D} = \mathbf{E}_o a^3$ will be induced, where $\mathbf{E}_o$ is some mean of the field at the body and a some average radial dimension of it.

Those results help in understanding the almost perfect insulators that are classed as dielectrics. Absolutely perfect insulators, conceived as permitting no deviations from a perfect superposition of their positive and negative charges, would have no electrostatic effects at all. It is the small deviations which must be expected in any actual insulator that become important to evaluate. Small relative displacements of opposite charges proportional to the imposed electric force must be allowed for—inducements of charge as in conductors but only on a *microscopic* scale, with the resultant charge separations confined to tiny domains behaving as if they were insulated from each other. The electrostatic effects of actual macroscopic insulators become understandable when it is supposed that there are charge displacements in them so small (extending only over domains of molecular order $a \lesssim 10^{-6}$ cm) that each has only dipole effects on the macroscopic scale—across any classically distinguishable increments of distance. With conduction† thus permitted only within domains of some very small mean

† This mode of describing charge displacements induced in molecules is perhaps not as apt in the relatively rarer cases when each molecule already has a *permanent* electric dipole moment. This superposes (often larger) contributions to a resultant polarization, through mere net alignments of preexisting dipoles by the imposed field. See pages 548 and 570 and Sec. F.8 for more details.

radius a, each domain can be expected to contribute a very small dipole moment of order $\mathbf{E}_o a^3$ if $\mathbf{E}_o$ is the electric field that penetrates to the point of the domain. There will be some very large number N of such domains ("molecules") per unit volume of the material and so a resultant vector density of dipole moment $\mathbf{P} = Na^3\mathbf{E}_o$. This might become quite appreciable, despite the smallness of a, since N may be large enough to compensate.

On such grounds, an ideal dielectric is defined as a medium that can be characterized by giving some finite continuous dipole moment density distribution $\mathbf{P}(\mathbf{r})$, called the *polarization* of the medium. It may vary from point to point and is expected to adapt itself, in magnitude as well as distribution, to imposed electric fields, thereby producing the equilibria that are the concern of electro*statics*. The magnitude reduces to $P = 0$ in such "perfect" insulators as empty space.

D.1 BOUND CHARGES

The field $\phi_P(\mathbf{r}_f)$ which a given polarization $\mathbf{P}(\mathbf{r}_s)$ will by itself generate has already been presented in (2.28). The expression applies to all field points $\mathbf{r}_f$, inside dielectrics as well as outside them. This field can be analyzed into volume effects and surface effects through a relation implicit in (1.27)

$$\frac{\hat{\mathbf{R}}}{R^2} = -\nabla_f \frac{1}{R} = +\nabla_s \frac{1}{R} \tag{D.1}$$

where $\mathbf{R} \equiv \mathbf{r}_f - \mathbf{r}_s$. This allows rewriting (2.28) as

$$\begin{aligned}\phi(\mathbf{r}_f) &= \int_V dV(\mathbf{r}_s)\mathbf{P}(\mathbf{r}_s) \cdot \nabla_s \frac{1}{R} \\ &= \int_V dV\left(\nabla_s \cdot \frac{\mathbf{P}}{R} - \frac{1}{R}\nabla_s \cdot \mathbf{P}\right) \\ &= \oint d\mathbf{S} \cdot \frac{\mathbf{P}}{R} - \int_V dV \frac{\nabla_s \cdot \mathbf{P}}{R}\end{aligned} \tag{D.2}$$

The area integration here is to extend over the surface enclosing the volume V chosen for evaluation and is a result of applying the Gauss theorem.

The volume integral in (D.2) yields the entire effect whenever $\mathbf{P}(\mathbf{r}_s)$ decreases to zero at the surface, as when this is a very remote one ($V \to \infty$) or when it simply lies outside all dielectrics. The volume effect can be seen to be equivalent to a field arising from a charge distribution defined by

$$\rho_P(\mathbf{r}_s) = -\nabla_s \cdot \mathbf{P}(\mathbf{r}_s) \tag{D.3}$$

existing at points of divergence in the polarization. It vanishes in regions of uniform polarization, ostensibly because the positively charged head of each dipole element has its effect canceled by an equally negative tail of a contiguous element. When the polarization instead increases along its own direction, each next contiguous *negative* tail is larger, and this accounts for the sign in (D.3).

How nonuniformities in a dipole distribution give rise to real, net charge at points of the nonuniformities is perhaps more easily understood when the polarization is uniform up to the surface of a dielectric material, there is only empty space beyond that surface, yet $\mathbf{P}$ has a nonvanishing normal component P_n^i at the surface. If $P_n^i > 0$, the positively charged heads of elementary dipoles are exposed on the surface while their negatively charged tails are neutralized in the interior of the material. The resultant net charge can be represented by a surface density σ_P calculable from the charge on an element $d\mathbf{S} = \mathbf{n}\, dS$ of the surface: $\sigma_P\, dS = dV\rho_P$, where $dV = dS\delta$ and $\delta \to 0$ is a vanishingly small surface thickness. With ρ_P given by the divergence expression (D.3), the evaluation yields a surface divergence for σ_P, as in (2.31),

$$\sigma_P = -\mathbf{n} \cdot \Delta\mathbf{P} = -(P_n^o - P_n^i) \tag{D.4}$$

with P_n^o the normal component of the polarization outside the surface (the side faced by $d\mathbf{S} = \mathbf{n}\, dS$). When there is only free space outside, $P_n^o = 0$ and $dS\sigma_P = dSP_n^i = d\mathbf{S} \cdot \mathbf{P}$, just such an element as gives the surface contribution to the potential (D.2). The volume integral in (D.2) must be taken to extend only *up to* the surface, and the surface integral is an extension of it that includes the surface domain when nothing outside the volume of evaluation V is being taken into account. The surface integral in (D.2) can be dropped and the entire effect represented by ρ_P of (D.3) if the volume integration is taken to extend *through* the surface, since *any* discontinuity $\Delta\mathbf{P} \neq 0$ met in the passage through any surface would automatically have to be treated as in (D.4), with the help of an appropriate σ_P replacing ρ_P.

The charges represented by ρ_P, and by σ_P when they are confined to a surface, are called *bound charges* in (or on) dielectric material. They are presumed to arise from the microscopic charge separations discussed above, and hence each element is constrained to stay within a microscopic domain. (However, the polarizations, and therefore the net sizes of the bound charges, change upon imposition of changes in the incident fields. The consequent changes in the locations at which net bound charge exists have the effect of transfers of charge, producing effective currents. These will be given some attention in Sec. F.1.)

D.2 THE ELECTRIC DISPLACEMENT FIELD

A distinction can now be *maintained* between the bound-charge sources of field ρ_P and such charges as those heretofore considered, fixed either in free space or in dielectric surroundings or resident on conductors. All the latter are called the free charges, and will be represented by ρ' to emphasize that they are no longer all the sources of field. All the sources $\rho = \rho' + \rho_P$ are taken into account in the expression

$$\nabla \cdot \mathbf{E} = 4\pi(\rho' + \rho_P) = 4\pi(\rho' - \nabla \cdot \mathbf{P})$$

The last equality follows from (D.3) and can be rewritten as

$$\nabla \cdot (\mathbf{E} + 4\pi\mathbf{P}) = 4\pi\rho' \tag{D.5}$$

This suggests defining a new field

$$\mathbf{D}(\mathbf{r}) = \mathbf{E}(\mathbf{r}) + 4\pi\mathbf{P}(\mathbf{r}) \tag{D.6}$$

which has only the free charges as its divergences

$$\nabla \cdot \mathbf{D} = 4\pi\rho' \tag{D.7}$$

It is called the *electric displacement*, for not particularly illuminating reasons. It must be considered only a *part* of the actual electric force field $\mathbf{E} = \mathbf{D} - 4\pi\mathbf{P}$, despite the fact that $|\mathbf{D}| \geq |\mathbf{E}|$, as will be seen.

It may be mentioned at this point that the set of equations

$$\nabla \cdot \mathbf{D} = 4\pi\rho' \qquad \nabla \times \mathbf{E} = 0 \qquad \mathbf{D} = \mathbf{E} + 4\pi\mathbf{P} \tag{D.8}$$

is sometimes put at the basis of the theory, in place of (2.1), primarily because of its greater mathematical generality. However, mathematical generality does not always coincide with an apter physical basis. The way **P** and **D** were introduced above reveals a sense in which these should be regarded as *derived* concepts, in a theory with Eqs. (2.1) at its basis. Moreover, they are only macroscopic concepts, which ignore the fact that closer analysis, to an atomic level, would make them only representative of certain gross effects, averages derivable from more detailed atomic effects. The detailed analyses into atoms are not undertaken here because their really proper treatment requires inquiry into quantum phenomena (although progress was made, far beyond the crude analysis $\mathbf{P}/\mathbf{E}_o = Na^3$, long before quantum effects were recognized as such. See Sec. F.8, for example.)

D.3 THE DIELECTRIC CONSTANT

Practical application of the theory requires taking into account the fact that the polarization $\mathbf{P}(\mathbf{r})$ can rarely be given initially but must be *co*determined with the field. The preliminary discussion of the distinctively dielectric behavior indicated that a connection is to be expected between polarization and field which can be represented as

$$\mathbf{P}(\mathbf{r}) = \alpha\mathbf{E}_o(\mathbf{r}) \tag{D.9}$$

where $\mathbf{E}_o$ is the electric force, per unit charge, on the element of dielectric at **r**. This relation is in effect an example of the well-known† Hooke's law, according to which there is always a proportionality between small displacements from normal configurations and the restoring forces responsible for maintaining the normal equilibrium. The proportionality constant α is called the polarizability characteristic of the dielectric material. According to the preliminary discussion,

† See Ref. 17, secs. 3 and 11.

it can be analyzed as $\alpha = Na^3$, where N is a number density of molecules or atoms and a is an effective radius of conduction provided by each.

The force $\mathbf{E}_o(\mathbf{r})$ responsible for inducing the microscopic charge separations in the element of dielectric at $\mathbf{r}$ is *not* the same as the average field $\mathbf{E}(\mathbf{r})$ which passes through the point in the dielectric. $\mathbf{E}_o$ should not include force on the element generated by itself, whereas $\mathbf{E}$ includes the effect of every element (on an extra, test, charge). To find the connection, consider subtracting the effect of a *spherical* element having a vanishingly small radius $s \to 0$. Subtracting such an element from the entire dielectric should in the limit $s \to 0$ have a negligible effect on the field distribution $\mathbf{E}(\mathbf{r})$ as a whole but can have a decided effect on the force $\mathbf{E}_o(\mathbf{r})$ right at the point $\mathbf{r}$ of the subtraction, since it is dielectric sources of field closest to the point that are being removed. A spherical cavity of radius $s \to 0$ is being formed, and hence the difference of the local field $\mathbf{E}_o$ from the average $\mathbf{E}$ stems from bound surface charges σ_P on the cavity boundary. No appreciable volume effects ρ_P are involved since a vanishingly small cavity can be treated as having a *uniform* polarization $\mathbf{P}(\mathbf{r})$, in the effective vicinity of the point $\mathbf{r}$. A surface element $dS = 2\pi s^2\, d(\cos\vartheta)$ of the cavity wall, located at the angle ϑ from the direction of $\mathbf{P}$ at $\mathbf{r}$, will have

$$\sigma_P = -P_s^o = -P\cos\vartheta$$

according to (D.4), since $\mathbf{P}^i = 0$ inside the cavity. $P \equiv |\mathbf{P}(\mathbf{r})|$ is the magnitude of the undisturbed polarization at $\mathbf{r}$, to be taken as constant over the cavity wall. The element $d\mathbf{E}_s$ of field at the center of the cavity, due only to the surface element $\sigma_P\, dS$ of bound charge, will have the magnitude $\sigma_P\, dS/s^2$, and an inward radial direction from elements with $\sigma_P > 0$. Only the components parallel to the axis $\hat{\mathbf{P}} \equiv \mathbf{P}/P$ will survive the surface integration, so that

$$\mathbf{E}_o - \mathbf{E} = -\hat{\mathbf{P}} \oint \left(dS\,\frac{\sigma_P}{s^2}\right)\cos\vartheta = +2\pi\mathbf{P}\int_{-1}^{1} d(\cos\vartheta)\cos^2\vartheta = +\frac{4\pi\mathbf{P}}{3} \tag{D.10}$$

In terms of the polarizability α, $\mathbf{E}_o = \mathbf{P}/\alpha$ and hence

$$\mathbf{P} = \frac{\alpha}{1-(4\pi/3)\alpha}\mathbf{E} \equiv \chi_e \mathbf{E} \tag{D.11}$$

The result is a proportionality between the polarization at a point and the average field $\mathbf{E}(\mathbf{r})$ in the undisturbed dielectric at the point. The proportionality constant χ_e, defined as $\mathbf{P}/\mathbf{E}$ by (D.11), is called the *electric susceptibility* of the medium. It is the polarizability α that is the more directly *calculated* when deriving dielectric characteristics from the properties of constituent atoms and molecules (as in the rough evaluation $\alpha = Na^3$, above, but better calculated by using quantum statistics); the susceptibility χ_e is more often used for expressing the results of *measurements* performed on a macroscopic, classical, scale.

The constitutive property of a dielectric material is usually expressed by giving a value to a dielectric constant ϵ, defined through

$$\mathbf{D} = \mathbf{E} + 4\pi\mathbf{P} = (1 + 4\pi\chi_e)\mathbf{E} \equiv \epsilon\mathbf{E} \tag{D.12a}$$

Its connection to the polarizability is evidently

$$\epsilon = 1 + \frac{4\pi\alpha}{1 - 4\pi\alpha/3} \tag{D.12b}$$

known as a *Clausius-Mossotti equation.* In free space, $\epsilon = 1$ since $\alpha = 0$. In dielectrics, $\epsilon > 1$ always, as can be seen from noting that when the analysis $\alpha = Na^3$ is used, $(4\pi/3)\alpha = NV_a$, where $V_a = (4\pi/3)a^3$ is the effective conducting volume of each molecule. When the size of V_a approaches the entire volume alloted in the material to each molecule, that is, $V_a \to 1/N$, then $\epsilon \to \infty$; this can be taken to be characteristic of conductors,† in which the entire space consists of a continuous conducting volume.

A uniform and isotropic dielectric will generally have some uniform value ϵ throughout its extent, at a given temperature and pressure. Sometimes nonuniform dielectric compositions can be represented by a varying continuous distribution $\epsilon(\mathbf{r})$. A *nonisotropic* dielectric is generally representable by a tensor, with components ϵ_{ij}, such that $\mathsf{D}_i = \sum_j \epsilon_{ij} \mathsf{E}_j$, and then $\mathbf{D}$ and $\mathbf{E}$ are generally not parallel to each other. No further attention will be paid to such elaborations here.

The discussion of the connection between polarization and field has made it clear why $\mathsf{D} > \mathsf{E}$ within dielectrics, yet $\mathbf{D}$ constitutes only a part of the total electric force field $\mathbf{E} = \mathbf{D} - 4\pi\mathbf{P}$. The point is that $\mathbf{E}$ induces dipoles with moment densities $\mathbf{P} = \chi_e \mathbf{E}$ parallel to itself, a direction such as to provide sources tending to *cancel* the effects of the free sources. This is a moderated version of the effect of *conducting* media, which undergo charge adjustments reducing $\mathbf{E}$ to the *vanishing* point everywhere inside them.

D.4 THE POTENTIAL DESCRIPTION

A scalar potential $\phi(\mathbf{r})$ can be introduced as a step in solving the field equations (D.8), just as in Sec. 2.2 for Eqs. (2.1). It is the full force field $\mathbf{E}$, rather than $\mathbf{D}$, that is curlless, and hence $\mathbf{E} = -\nabla\phi$ still. When $\mathbf{D} = \epsilon\mathbf{E}$ rather than $\mathbf{D} = \mathbf{E} + 4\pi\mathbf{P}$ is used as the constitutive relation, substitution into the divergence equation yields $\nabla \cdot (\epsilon\nabla\phi) = -4\pi\rho'$ and a Poisson equation

$$\nabla^2\phi = -\frac{4\pi\rho'}{\epsilon} \tag{D.13}$$

to hold everywhere in a dielectric medium characterized by a *uniform* ϵ. It follows that a free-charge element $\rho'\,dV$ embedded in the dielectric will generate $d\phi = \rho'\,dV/\epsilon R$ at any field point well inside the same medium.

† Such a statement seems to be avoided in most textbooks, despite the fact that electrostatic effects of dielectrics do reduce to those of conductors for $\epsilon \to \infty$, as will be found throughout this chapter. A more detailed justification will become evident in connection with (G.68).

A point charge q embedded in the uniform dielectric will produce a field $\phi_q = q/\epsilon r$ within the dielectric. The force on a test charge q' in this field will then be

$$\mathbf{F} = \frac{\hat{\mathbf{r}}qq'}{\epsilon r^2} \tag{D.14}$$

This is the modification of Coulomb's law holding within a uniform medium characterized by ϵ. Since $\epsilon > 1$, the repulsion or attraction between the pair of charges q, q' is weaker than in free space (of $\epsilon = 1$). That is the consequence of the partial neutralizations by the microscopic charge adjustments constituting the polarization of the medium lying between and around the charges. If a conducting medium intervened instead, the charges would be completely shielded from each other, as for $\epsilon \to \infty$ in (D.14). Of course, charges *embedded* in conducting material would immediately neutralize themselves locally by drawing equal and opposite conductor charges upon themselves. (Compare the charge-containing conductor hollows of Exercises B.8 and B.21.)

D.5 DIELECTRIC EFFECTS ON CONDUCTOR CAPACITANCE

The useful concept of the capacitance C of a conductor will be given only passing attention here because its most frequent uses are already widely familiar in elementary physics. Some consideration of it is appropriate at this point because it has provided one of the simplest means for determining dielectric constants of various materials.

The capacitance of an entire isolated (and hence insulated) single conductor is defined by

$$C = \frac{q_C}{\Delta\phi_C} \tag{D.15}$$

if its potential is increased by $\Delta\phi_C$ whenever free charge q_C is added to it. The rise of potential per unit of charging is regarded as a practical measure of its capacity to hold charge because higher potential differences from its surroundings make it more liable to discharge by breakdown of its insulation (see page 491*n*.).

In the simple case of a conducting sphere in free space, the potential difference from a zero at infinity is just $\phi_C = q_C/a$ if a is the sphere radius. Thus the sphere in free space has the capacitance $C_1 = a$. Next, suppose that the free space is replaced by a dielectric medium of constant ϵ. The potential will now be $\phi_C = q_C/\epsilon a$ and $C_\epsilon = \epsilon a$. This demonstrates how, in principle, direct measurements of capacitances (of potentials and charges) can yield determinations of ϵ—as $\epsilon = C_\epsilon/C_1$, for example.

More practical measurements require only finite expanses of dielectric, enclosed between pairs of conductors bearing equal and opposite charge. Then $\Delta\phi_C$ of (D.15) refers to the resulting potential difference between the conductors. (See Exercises D.1 and D.2.)

D.6 DIELECTRIC BOUNDARY CONDITIONS

Solutions of the field equations (D.8) or the Poisson equation (D.13) are not uniquely determined until they have also been made to satisfy suitable boundary conditions, sufficient to delimit a specific physical situation. Sufficient boundary conditions can be obtained from Eqs. (D.8) themselves by adapting them to surfaces, as discussed for the conductor boundary conditions (B.2). The first two equations in (D.8) respectively reduce to the surface divergence and surface curl

$$\mathbf{n} \cdot \Delta \mathbf{D} = 4\pi\sigma' \qquad \text{and} \qquad \mathbf{n} \times \Delta \mathbf{E} = 0 \tag{D.16a}$$

or

$$\Delta D_n \equiv D_n^o - D_n^i = 4\pi\sigma' \qquad \text{and} \qquad \Delta \mathbf{E}_{\|} = \mathbf{E}_{\|}^o - \mathbf{E}_{\|}^i = 0 \tag{D.16b}$$

where σ' is the area density of any free charge that might be residing on the surface. The surface curl condition is just (B.4), already asserted to be valid in *any* electrostatic situation.

Values of $\sigma' \neq 0$ usually occur at interfaces between *conductors and dielectrics.* Since $\mathbf{E}^i = 0$ inside conductors, $\mathbf{E}_{\|}^o = 0$ on the dielectric side and the field is normal to the conductor surface exactly as when it is a boundary of a free space. The total charge at the interface must still have $\sigma = E_n^o/4\pi$, but only a part $\sigma' = D_n^o/4\pi$ represents *free* charge induced from the conducting material, according to (D.16). The remainder, $\sigma - \sigma' = (E_n^o - D_n^o)/4\pi = -P_n^o = \sigma_P$ [Eq. (D.4)], is induced from the dielectric material and is bound to it. If the dielectric is characterized by the constant ϵ^o, then $\mathbf{D}^o = \epsilon^o \mathbf{E}^o$ and that bound surface charge is $\sigma_P = (1 - \epsilon^o)E_n^o/4\pi = -(\epsilon^o - 1)\sigma$, showing that the charge induced from the dielectric has a sign opposite to that of the inducing free charge on the conductor, $\sigma' = \epsilon^o\sigma$. Notice that for $\epsilon^o = 1$ all these results reduce to the conditions on the interface of a conductor and free space.

On a surface separating *two different dielectrics*, it is exceptional to meet with circumstances in which free charges are placed precisely on the boundary, and so $\sigma' = 0$ will be assumed. The boundary conditions (D.16) then become

$$D_n^o = D_n^i \qquad \text{and} \qquad \mathbf{E}_{\|}^o = \mathbf{E}_{\|}^i \tag{D.17}$$

so that the normal component of $\mathbf{D}$ is *continuous* across the boundary and so are the transverse components of $\mathbf{E}$ (those parallel to the surface). There are then discontinuities in the *directions* of the field lines crossing the boundary, as indicated in Fig. D.1. The connection between the angles indicated as i and r, respectively called the *angles of incidence* and *refraction*, is

$$\frac{\tan i}{\tan r} = \frac{E_{\|}^i/E_n^i}{E_{\|}^o/E_n^o} = \frac{D_{\|}^i/D_n^i}{D_{\|}^o/D_n^o} = \frac{\epsilon^i}{\epsilon^o} \tag{D.18}$$

as follows from (D.17) when $\mathbf{D}^i = \epsilon^i\mathbf{E}^i$ and $\mathbf{D}^o = \epsilon^o\mathbf{E}^o$. The $\mathbf{D}$ lines must be refracted by the same angles as $\mathbf{E}$ lines since the two fields are parallel to each other in each medium.

The boundary conditions (D.16) also permit a demonstration that, within dielectric media, the field $\mathbf{D}$ is in principle as directly measurable as $\mathbf{E}$ is, in each

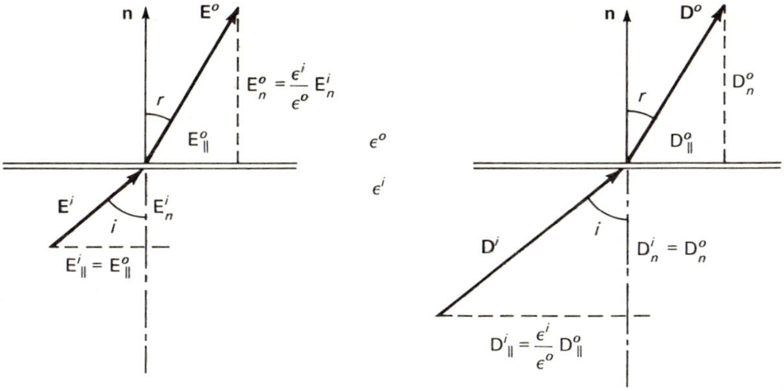

Figure D.1

case as force per unit test charge. For any such measurement, the test charge must displace dielectric material; a *cavity* is made in the dielectric, and a field $\mathbf{E}^i \equiv \mathbf{D}^i$ is then measured in the free space of the cavity. That will permit conclusions about the field in the undisturbed dielectric if the cavity is made vanishingly small. Any degree of accuracy is in principle achievable by extrapolating to a limit of measurements in progressively smaller cavities (some such limiting procedures are characteristic of all physical measurements on continuously variable quantities). Suppose first that the cavities are given long, thin needlelike shapes, parallel to the field $\mathbf{E}$, to be made small by contracting only their lateral dimensions, so that the end effects are always negligible. Then the field just outside each cavity, in the dielectric nearest its center, is $\mathbf{E}^o_\parallel \equiv \mathbf{E} = \mathbf{E}^i$. Conclusions about the field $\mathbf{E}(\mathbf{r})$ at that point of the dielectric are thus reached. Suppose next that the cavities are given flat cylindrical shapes with axes again parallel to $\mathbf{E}$. The radii of the cylinders are kept large enough for edge effects to remain negligible as the vanishing of the cavity is approached by making the axis lengths shorter and shorter. Now $\mathbf{D}^o_n \equiv \mathbf{D} = \mathbf{D}^i_n = \mathbf{E}^i$, and conclusions about the field $\mathbf{D}(\mathbf{r})$ are reached. (Making diagrams of these situations will be helpful in understanding those assertions.) Notice that, for $\mathbf{D} = \mathbf{E} + 4\pi\mathbf{P}$ in the undisturbed dielectric, the field at the center of a long, thin cylindrical cavity is $\mathbf{E}^i = \mathbf{E}$, in a flat cylindrical cavity it is $\mathbf{E}^i = \mathbf{E} + 4\pi\mathbf{P}$, and in an isotropic (spherical) cavity it was found in (D.10) to have the intermediate value $\mathbf{E}^i = \mathbf{E} + 4\pi\mathbf{P}/3$.

A separate statement of the boundary conditions on the potential ϕ, like that made for conductor boundaries in (B.5), is not really essential since they follow quite directly from (D.16) and (D.17) through the use of the relation $\mathbf{E} = -\nabla\phi$. However, the surface curl condition $\Delta\mathbf{E}_\parallel = 0$ is always somewhat more simply stated in terms of the potential. It always reduces to $\phi = \phi_C$, a constant, on conductor surfaces as in (B.5), whether they are bordered by free space or dielectric. For an interface between two different dielectrics, the second condition (D.17) becomes

$$\frac{\partial\phi^o}{\partial l} = \frac{\partial\phi^i}{\partial l} \qquad \text{or} \qquad \phi^o(\mathbf{r}_S) = \phi^i(\mathbf{r}_S) \tag{D.19}$$

if l is a length coordinate along the boundary, parallel to the direction of $\mathbf{E}_{\parallel}^{o} = \mathbf{E}_{\parallel}^{i}$. The potentials on the two sides of each surface point could ostensibly still differ by a constant, but that is merely a matter of choosing gauges. The gauges can always be so chosen that the variable potentials *match* on the two sides of any surface separating two dielectrics.

D.7 IMAGES IN DIELECTRICS

Suppose that, in the situation indicated by Fig. B.1, the conductor filling the half-infinite space $z < 0$ is replaced by dielectric with $\epsilon \geq 1$. The problem will not be appreciably more complicated if the other half of the space ($z > 0$), containing the free point charge q, is also filled with dielectric of different composition characterized by $\epsilon_o \neq \epsilon$. The case of free space in *either* half can then be obtained simply by reducing the corresponding dielectric constant to unity.

About the field $\phi(s, z > 0)$, it is known that it must approach $q/\epsilon_o R$ in the limit $\epsilon \to \epsilon_o$, since then all space is filled with a uniform dielectric. It can be hoped that the effect of an $\epsilon \neq \epsilon_o$ on $\phi(s, z > 0)$ can be represented by some image q' at $\mathbf{r} = \mathbf{l}'$, as in the case of the conductor, so that the solution is sought in the form

$$\phi(s, z > 0) = \frac{q}{\epsilon_o R} + \frac{q'}{R'} \tag{D.20a}$$

with $R' = [s^2 + (z + l')^2]^{1/2}$. Any dielectric constant factor that may belong in the last term can be incorporated into the definition of q', to be determined together with l'. The field $\mathbf{D} = -\epsilon_o \nabla\phi$ will properly have divergence only at the free charge q if $l' > 0$; the position of the image must be outside the space $z > 0$ to which (D.20a) applies.

Since no free charges have been given in the $z < 0$ region, the field $\phi(s, z < 0)$ must satisfy the *Laplace* equation, $\nabla^2\phi = 0$, there. Any constant multiple of q/R is a solution of the Laplace equation (everywhere except right at the point charge itself), and it can be supposed that the field $\phi(s, z < 0)$ will be some fraction f of q/R because of the expected partial neutralization by the polarizations of the intervening dielectrics. It is thus hopeful to try as the remainder of the solution which is partly (for $z > 0$) given by (D.20a)

$$\phi(s, z < 0) = \frac{fq}{R} \tag{D.20b}$$

It has already been pointed out that $1/f = \epsilon_o$ is to be expected when $\epsilon \to \epsilon_o$, and some mean of ϵ_o and ϵ may well be expected for $1/f$ when $\epsilon \neq \epsilon_o$. The approach here will be successful if values for q', l', and f can be found which will satisfy the boundary conditions on the interface between the two dielectrics.

According to the boundary condition (D.19), the two potential expressions (D.20a) and (D.20b) must match at $z = 0$ and all s:

$$\frac{q}{\epsilon_o(s^2 + l^2)^{1/2}} + \frac{q'}{[s^2 + (l')^2]^{1/2}} = f\,\frac{q}{(s^2 + l^2)^{1/2}} \tag{D.21a}$$

At $s = 0$ this gives $q'/l' = (f - \epsilon_o^{-1})q/l$, and at $s = \infty$, $q' = (f - \epsilon_o^{-1})q$. Thus $l' = l$, just as for the image in a conductor plane; this could have been anticipated, since no new length dimension is introduced by the change from conductor to dielectric. With $l' = l$, (D.21*a*) is satisfied everywhere on the boundary (that is for all s) by

$$\frac{q'}{q} = f - \frac{1}{\epsilon_o} \tag{D.21b}$$

The surface curl boundary condition of (D.17) is hereby satisfied.

The surface divergence boundary condition in (D.17) requires a matching at $z = 0$ of

$$\mathsf{D}_z(s, z > 0) = -\epsilon_o \frac{\partial \phi}{\partial z} = \epsilon_o \frac{q(z - l)}{\epsilon_o R^3} + \epsilon_o \frac{q'(z + l)}{(R')^3} \tag{D.22a}$$

and

$$\mathsf{D}_z(s, z < 0) = -\epsilon \frac{\partial \phi}{\partial z} = \epsilon f \frac{q(z - l)}{R^3} \tag{D.22b}$$

Since $R' = R$ at $z = 0$ (now that $l' = l$ is known), the matching yields

$$\epsilon_o q' = (1 - \epsilon f)q$$

This and (D.21*b*) constitute two equations for f and q' having the solution

$$\frac{q'}{q} = -\frac{\epsilon - \epsilon_o}{\epsilon_o(\epsilon + \epsilon_o)} \qquad f = \frac{2}{\epsilon + \epsilon_o} \tag{D.23}$$

It thus turns out that $1/f$ is just the *arithmetical mean*, $\frac{1}{2}(\epsilon + \epsilon_o)$, of ϵ and ϵ_o. The magnitude q' of the image charge approaches the value $q' \to -q$, found before for the conducting plane, in the limit $\epsilon \to \infty$ and $\epsilon_o = 1$ (free space outside the conductor). It is $q = -q/\epsilon_o$ for dielectric with $\epsilon_o > 1$ outside the conductor (in any case, $f \to 0$, so that $\phi \to 0$ *inside* the conductor).

The conclusion about the field in the two dielectrics is represented by

$$\begin{aligned} \phi(s, z > 0) &= \frac{q}{\epsilon_o R} - \frac{\epsilon - \epsilon_o}{\epsilon + \epsilon_o} \frac{q}{\epsilon_o R'} \\ \phi(s, z < 0) &= \frac{2}{\epsilon + \epsilon_o} \frac{q}{R} \end{aligned} \tag{D.24}$$

Notice that both expressions properly reduce to $q/\epsilon_o R$ for $\epsilon = \epsilon_o$. The consequent field (**E**) lines for a case of $\epsilon > \epsilon_o$ are compared with (dotted) lines of the conductor case in Fig. D.2*a*. It can be seen that approach toward a conductor concentrates them the more, but then the lines are dispersed much more widely (to zero density) *inside* the conducting medium. A case of $\epsilon < \epsilon_o$ is shown in Fig. D.2*b*, where approach toward the boundary tends to disperse the lines; the different signs of curvature in the cases $\epsilon \gtrless \epsilon_o$ allow for the proper refractions (D.18) at the boundary in each case.

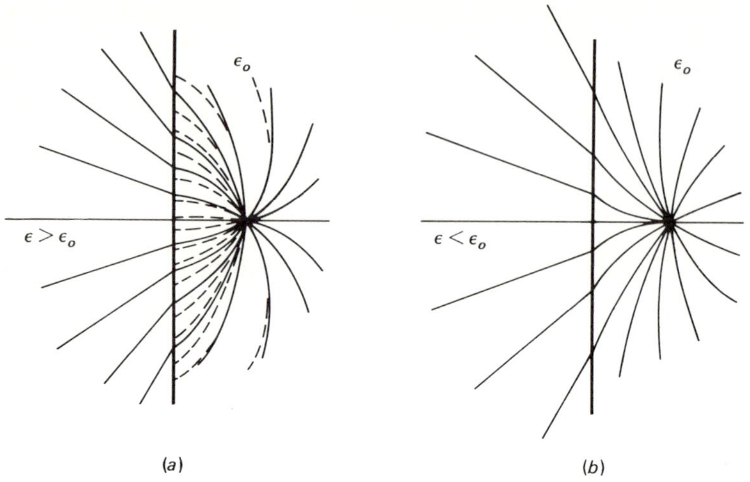

Figure D.2

D.8 A DIELECTRIC SPHERE IN A UNIFORM FIELD

Suppose that dielectric material of constant ϵ forms a sphere in free space and that there is incident on it an external field that was uniformly $\mathbf{E}_o$ before the introduction of the sphere. The corresponding problem with conducting rather than dielectric material was solved in Sec. B.4 as the limit of an image problem. It will be found in the next section that a dielectric sphere does not merely produce simple images even for the case of a point charge. The methods to be applied here are more general ones, applicable also to the conductor cases (Exercise B.19).

The fields inside and outside the dielectric must each satisfy a Laplace equation according to (D.13), applied to uniform sourceless regions. With a uniform field $\mathbf{E}_o$ incident from infinity, the situation that arises will be azimuthally symmetric about the field line directed through the sphere center. It becomes appropriate to use the form (3.72) for the Laplace field $\phi(r, \vartheta)$. In the form for the field $\phi(r \geq a, \vartheta)$ in the free space outside the sphere of radius a, the coefficients of all the positive powers r^l must be taken to vanish except for the linear term, which must be $\mathsf{E}_o z = \mathsf{E}_o r \cos \vartheta$ to serve as the potential at infinity of an incident field $\mathsf{E} = -\mathsf{E}_z = \mathsf{E}_o$. Thus

$$\phi(r \geq a, \vartheta) = \mathsf{E}_o r P_1(\cos \vartheta) + \sum_{l=0}^{\infty} \frac{B_l}{r^{l+1}} P_l \tag{D.25a}$$

while the interior field must have the form

$$\phi(r \leq a, \vartheta) = \sum_{l=0}^{\infty} A_l r^l P_l(\cos \vartheta) \tag{D.25b}$$

having no terms that become singular as $r \to 0$ if there are no free sources inside the dielectric.

To satisfy the boundary condition (D.19) at the dielectric surface $r = a$, the two potential forms must have matching values there for all angles ϑ. The linearly independent terms must match independently, so that

$$A_1 a = \mathsf{E}_o a + \frac{B_1}{a^2} \qquad A_l a^l = \frac{B_l}{a^{l+1}} \qquad \text{for } l \neq 1$$

The imposition of the remaining, surface divergence boundary condition in (D.16) requires expressions for the radial-displacement field components D_r, and since there is no distinction between D_r and $\mathsf{E}_r = -\partial\phi/\partial r$ in the free space outside the dielectric sphere,

$$\mathsf{D}_r(r \geq a, \vartheta) = -\mathsf{E}_o P_1 + \sum_{l=0}^{\infty} \frac{(l+1)B_l}{r^{l+2}} P_l$$

Inside the sphere of dielectric constant ϵ

$$\mathsf{D}_r(r \leq a, \vartheta) = -\epsilon \frac{\partial\phi}{\partial r} = -\epsilon \sum_{l=1}^{\infty} l A_l r^{l-1} P_l$$

These match at $r = a$ only if

$$-\epsilon A_1 = -\mathsf{E}_o + \frac{2B_1}{a^3} \qquad -\epsilon l A_l = \frac{(l+1)B_l}{a^{2l+1}} \qquad \text{for } l \neq 1$$

The two relations found between A_l and B_l for each $l \neq 1$ can both be consistent with an arbitrary $\epsilon > 1$ only if $A_l = B_l = 0$ $(l \neq 1)$. This might have been anticipated from the fact that an exciting agency [the incident field $\phi(\infty) = -\mathsf{E}_o r \cos\vartheta$ here] proportional to $P_1(\cos\vartheta)$ cannot have excited linearly independent effects proportional to $P_{l\neq 1}(\cos\vartheta)$. The two relations between A_1 and B_1 require that

$$A_1 = \frac{3\mathsf{E}_o}{\epsilon + 2} \qquad B_1 = -\frac{\mathsf{E}_o a^3(\epsilon - 1)}{\epsilon + 2}$$

The resultant field is therefore to be described by

$$\begin{aligned} \phi(r \geq a, \vartheta) &= \mathsf{E}_o r \cos\vartheta \left[1 - \frac{\epsilon - 1}{\epsilon + 2}\left(\frac{a}{r}\right)^3\right] \\ \phi(r \leq a, \vartheta) &= \frac{3}{2 + \epsilon} \mathsf{E}_o r \cos\vartheta \end{aligned} \tag{D.26}$$

Both parts properly reduce to the expression $\phi(r \gtrless a, \vartheta) = \mathsf{E}_o z$, describing an undisturbed incident field, when $\epsilon = 1$, which represents a continuation of the free space through the sphere.

As in the case of the conducting sphere (B.15), the exterior field in (D.26) has superposed on it the effect of a dipole moment induced in the sphere. The moment $\mathsf{E}_o a^3$ found for the conductor is here multiplied by the fraction

$(\epsilon - 1)/(\epsilon + 2)$, which properly approaches unity for $\epsilon \to \infty$. Whereas the macroscopic charge separation to the surface of the conducting sphere reduced the field inside it to zero, the uniform field $\mathbf{E}_o$ penetrates into the dielectric sphere. It is only weakened, by the factor $3/(2 + \epsilon)$ because of the polarizations induced in the medium.

D.9 THE INTERACTION OF A POINT CHARGE WITH A DIELECTRIC SPHERE

To be considered now is the effect on a dielectric sphere of a point charge q set in front of it, replacing the uniform field treated in the preceding section. Figure B.2 can be adapted to representing the situation here, and there will be no extra difficulty in taking the space outside the sphere to be filled with dielectric, of constant ϵ_o, in addition to the dielectric within the sphere, to be characterized by ϵ. With q replacing $\mathbf{E}_o$ as the exciting agency, the field (D.25*a*) outside the sphere should be replaced by

$$\phi(r \geq a, \vartheta) = \frac{q}{\epsilon_o|\mathbf{r} - \mathbf{r}_o|} + \sum_{l=0}^{\infty} \frac{B_l}{r^{l+1}} P_l(\cos\vartheta) \tag{D.27}$$

This incorporates the singularity at the position $\mathbf{r}_o$ of the point charge and may contain in addition only a solution of the Laplace equation that does not become infinite as $r \to \infty$. The field inside the sphere should again be represented as in (D.25*b*), and its matching to (D.27) can be eased by expanding the part arising from the point charge as in (3.81):

$$\phi(a \leq r < r_o, \vartheta) = \sum_{l=0}^{\infty} \left(\frac{qr^l}{\epsilon_o r_o^{l+1}} + \frac{B_l}{r^{l+1}} \right) P_l(\cos\vartheta)$$

Matching the potentials at $r = a$ then demands that

$$A_l = \frac{B_l}{a^{2l+1}} + \frac{q}{\epsilon_o r_o^{l+1}}$$

and the normal displacement fields $\mathbf{D}_r$ also match if

$$-\epsilon l A_l = \epsilon_o(l+1)\frac{B_l}{a^{2l+1}} - \frac{lq}{r_o^{l+1}}$$

It is now straightforward to conclude that

$$\begin{aligned} \phi(r \leq a, \vartheta) &= \frac{q}{r_o} \sum_{l=0}^{\infty} \frac{2l+1}{l\epsilon + (l+1)\epsilon_o} \left(\frac{r}{r_o}\right)^l P_l(\cos\vartheta) \\ \phi(r \geq a, \vartheta) &= \frac{q}{\epsilon_o|\mathbf{r} - \mathbf{r}_o|} - \frac{qa}{\epsilon_o r_o} \sum_{l=1}^{\infty} \frac{l(\epsilon - \epsilon_o)}{l\epsilon + (l+1)\epsilon_o} \frac{(a^2/r_o)^l}{r^{l+1}} P_l \end{aligned} \tag{D.28}$$

Both expressions properly reduce to just the point charge field for $\epsilon = \epsilon_o$ and to the results (B.11) found for the conducting sphere if $\epsilon \to \infty$, $\epsilon_o = 1$. Moreover, the results (D.26) for the case of the uniform field incident from infinity follow from (D.28) with $\epsilon_o = 1$, $r_o \to \infty$, and $q = \mathsf{E}_o r_o^2 \to \infty$; as in the similar considerations leading to (B.15) of the conductor case, here also the gauge of the potential is shifted by the infinite constant $\mathsf{E}_o r_o$.

The final summation in (D.28) indicates that the images of a point charge in a dielectric sphere consist of a sequence of multipoles. The lowest-order one is a dipole of moment

$$D = -D_z = \frac{qa^3}{\epsilon_o r_o^2} \frac{\epsilon - \epsilon_o}{\epsilon + 2\epsilon_o}$$

This becomes the same as the dipole moment in (D.26), induced by a uniform field, for $q = \mathsf{E}_o r_o^2$ and $\epsilon_o = 1$.

D.10 FIELD ENERGIES AND FORCES IN DIELECTRICS

To complete a review of the electrostatics of dielectrics, the energies and forces in systems containing dielectrics are still to be considered.

The expression for the energy density already obtained, $w = \mathsf{E}^2/8\pi$ of (2.61), is supposed to be valid for any electrostatic system, including cases in which some of the field sources consist of bound charges in dielectric materials. However, the expression was derived on the basis that any electrostatic system consists of some *definite* distribution of charges. This can be presumed ultimately true of dielectrics but corresponds to microscopic (atomic or molecular) descriptions of them. Such descriptions are unwieldy and unnecessarily detailed for treating *macroscopically* observable phenomena. For most of such phenomena it has been found both adequate and useful to treat dielectrics as being completely defined once polarization distributions $\mathbf{P}(\mathbf{r}) = (\mathsf{D} - \mathsf{E})/4\pi$ in them have been obtained. The treatment puts no restrictions on the divergences that $P(\mathbf{r})$ may have and so presumes that arbitrarily different amounts of bound charge (D.3) may be present in different states (configurations) of the same system. As a consequence, it becomes unclear in that description just how much of the energy $w = \mathsf{E}^2/8\pi$ is tied up in invariant self-energies of charge and how much is available for work by the field during changes in the configurations of the free charges and dielectric bodies.

An energy expression having well-defined and correct variations with changes described by polarizations can be derived by reconsidering such evaluations as (2.39) and (2.60), of the work needed to construct an electrostatic system. What is now wanted is the energy needed to establish a definite field which arises from some definite, final configuration described in terms of *free* charges and *polarizations*. The energy can be gauged from a zero value in an initial state having none of the free charges in place as yet and therefore none of the polarizations yet induced. The polarizable dielectrics can be assumed to be already in place, since they produce no field before they are polarized. Then the desired final configuration, including its polarizations, can be established by doing just the work of bringing the free charges ρ' into position

$$W = \frac{1}{2}\int dV \rho' \phi \tag{D.29}$$

Here ϕ must be the actual potential arising from both the final configuration of free charges and the polarizations they have automatically induced. This is the appropriate modification of (2.39) when dielectric effects are described by polarizations rather than through microscopic bound charges.

The expression (D.29) can be transformed into a field energy in the same way that (2.39) was reexpressed as (2.60), but now $\rho' = \nabla \cdot \mathbf{D}/4\pi$ [Eq. (D.8)], and so

$$W = \oint dV \phi \frac{\nabla \cdot \mathbf{D}}{8\pi} = -\oint dV \frac{(\mathbf{D} \cdot \nabla)\phi}{8\pi}$$

Thus, the expression (2.61) for the energy density is replaced by

$$w(\mathbf{r}) = \frac{\mathbf{E} \cdot \mathbf{D}}{8\pi} \tag{D.30}$$

when dielectrics described by $\mathbf{P} = (\mathbf{D} - \mathbf{E})/4\pi \neq 0$ are present. It is valid to use even when the *inequivalent* expression $\mathbf{E}^2/8\pi$ agrees with what would have been obtained from detailed, *microscopic* descriptions of the dielectrics. Different energies are being counted in the two expressions, and they are differently gauged. It is only *changes* of energy that are observable and useful, and (D.30) can be expected to change correctly as long as only alterations in free charge and polarization distributions are in question.

Questions can also be raised about the really proper evaluation of forces on free charges embedded in dielectric materials. The standard practice is to consider

$$\mathbf{F} = q\mathbf{E}(\mathbf{r}_q) \tag{D.31}$$

to be *the* force on any point of free charge q at a position $\mathbf{r}_q$ when $\mathbf{E}(\mathbf{r}) = \mathbf{D} - 4\pi\mathbf{P}$ is the field in the dielectric. That was taken for granted in the Coulomb-field expression (D.14), an example in which $\mathbf{E} = \hat{\mathbf{r}}q'/\epsilon r^2$ arises from a second point charge in the medium.

It is considerations like those discussed on page 529 that raise questions about (D.31). Embedding free charge in a dielectric necessarily displaces some of it, and so it is the field $\mathbf{E}^i$ in a dielectric *cavity* that determines the force on the free charge by itself. The discussion showed that $\mathbf{E}^i$ is generally greater than $\mathbf{E}$, ranging up to $\mathbf{E} + 4\pi\mathbf{P}$ (depending on the shape of the cavity) even in the limit of a vanishingly small cavity, such as is sufficient to accommodate a point charge. Thus the force $q\mathbf{E}^i$ on q itself must generally be expected to depend on the mode of embedment and to be greater than $q\mathbf{E}$.

What $\mathbf{F} = q\mathbf{E}$ of (D.31) does give unambiguously and correctly is *not* the force on the bare charge q but the force on the actual charge at the point of embedment $q + q_P < q$, where q_P is bound charge, of sign opposite to that of q, with which q automatically coats itself upon being embedded in dielectric material. It may be said that q is renormalized to $<q$ by charges induced from the dielectric. In the special case of a medium characterizable by some ϵ, results of Exercise D.8 show that $q + q_P = q/\epsilon$. Then $\mathbf{F} = q\mathbf{E} = (q + q_P)\mathbf{D}$. Demonstrating that it is $(q + q_P)\mathbf{D} = q\mathbf{E}$, and *not* $(q + q_P)\mathbf{E}$, that is the net force on $q + q_P$ requires inquiry into how forces are transmitted by fields, to be taken up in Chap. 6. The application to transmissions through dielectrics is left to Exercises D.14 to D.17.

EXERCISES

D.1 A parallel-plate capacitor is a system of two conducting planes facing each other across a space (between $z = 0$ and $z = a$, here). Suppose that the part of the space between $z = 0$ and $z = l < a$ is filled with dielectric of constant ϵ, the remainder ($l < z < a$) being left empty. Neglecting edge effects:

(*a*) Find the capacitance per unit area $\sigma'/\Delta\phi$ of the system, where σ' is the free-charge density on each conductor whenever a potential difference $\Delta\phi$ between the plates is applied. Compare the expectations for $l = 0$ and $l = a$.

(*b*) Find the ratios to σ' of bound-charge densities σ_P at each of the interfaces $z = 0$, l, and a.

(*c*) Compare the free charges on the plates in the two cases $l = 0$ and $l = a$ when the same $\Delta\phi$ is maintained for each.

(*d*) Compare the potential differences between the plates in the two cases $l = 0$ and $l = a$ when the same free charges are maintained in each (as by disconnecting the empty capacitor from a source of emf, and then filling it with the dielectric).

D.2 A spherical capacitor consists of two thin conducting concentric spherical shells enclosing between them a space with inner and outer radii a and b. Suppose that the inner shell bears a charge $q > 0$ and the outer shell is charged up to a total $Q > q$.

(*a*) How will the charges be distributed at equilibrium?

(*b*) Now suppose that the outer shell only is connected to ground. Find the capacitance $C = q/\Delta\phi$ of the system, both when the space $a < r < b$ is left empty and when it is half-filled, in the hemispherical space between $\vartheta = \frac{1}{2}\pi$ and $\vartheta = \pi$, with dielectric fluid of constant ϵ.

(*c*) For the half-filled case in (*b*), find the free and bound charge *distributions* at the conductor faces and also on the annular interface between the dielectric and the empty remaining space.

D.3 A point charge q is placed at $z = l$ in the free space outside the $z = 0$ plane boundary of dielectric ϵ that extends to $z = -\infty$. A second point charge $-q$ is embedded in the dielectric at $z = -l$. Find the field everywhere and the force on $+q$.

D.4 (*a*) From the field (D.26) find the bound-charge distribution $\sigma_P(\vartheta)$ induced on the dielectric sphere by the uniform field $\mathbf{E}_0$.

(*b*) Since dielectrics fundamentally have effect only through the charges induced in them, the field $\phi_P(r < a)$ due to the σ_P of (*a*) should be responsible for the difference between $\phi(r < a)$ of (D.26) and the incident field $E_0 z$. Show that this is so for an arbitrary point inside the sphere.

(*c*) Show that the inner electric field due to σ_P is just $\mathbf{E}_P(r < a) = -4\pi\mathbf{P}/3$, where $\mathbf{P} = (\epsilon - 1)\mathbf{E}(r < a)/4\pi$. [Compare (D.10).]

(*d*) Evaluate the total dipole moment $\mathbf{D}$ of the σ_P distribution. Show that this is responsible for the contribution of the sphere to the outer field and that $\mathbf{D} = \mathbf{P}V$ with V equal to the sphere volume.

D.5 A point charge q is embedded at the center of a homogeneous dielectric sphere having constant ϵ and radius a.

(*a*) Find ϕ, $\mathbf{E}$, and $\mathbf{D}$ everywhere.

(*b*) Show that the *same* ϕ, $\mathbf{E}$ can arise in an otherwise empty space if a suitable point charge is placed at $r = 0$ and an appropriate uniform distribution of free charge on a surface $r = a$. How do the appropriate free charges here compare with the free and bound charges in (*a*)?

(c) Now suppose that a second point charge, *equal to* the q in (a), is placed at $r_o = 2a$ outside the dielectric sphere. Evaluate the force on it in terms of q and r_o for an $\epsilon = 2$, to an accuracy of 1 percent or so. (You may also want to consider the field of a point charge at the center of a spherical hollow in a dielectric sphere.)

D.6 Suppose that the point charge inside the dielectric sphere of Exercise D.5 is instead placed at some distance $r_o < a$ from the center.

(a) Find the field everywhere, checking your results against expectations for $r_o = 0$ (see Exercise D.5) and for $r_o = a$ [see (D.28)].

(b) Show that if the sphere is coated with a grounded conductor shell, the field becomes representable with the help of a single point image; find the image charge.

D.7 A large expanse of dielectric ϵ has in it a small spherical hollow of radius a. A field is imposed, approaching uniformity, $\mathbf{E}_o = \mathbf{D}_o/\epsilon$, far from the hollow.

(a) Show that the resultant field distribution differs from (D.26), which has dielectric and free space interchanged, by the substitution $\epsilon \leftrightarrow \epsilon^{-1}$.

(b) Compare the direction (relative to $\mathbf{E}_o$) of the dipole moment that is induced in the hollow with that induced in the dielectric sphere of (D.26). Correlate your finding with the signs of the bound charge distributions you must expect in the two cases.

(c) Evaluate the charge distributions $\sigma_P(\vartheta)$ in the two cases.

D.8 Suppose that a point charge q is placed at the center of the hollow in the situation of Exercise D.7.

(a) Show that the force on q itself is independent of the radius a of the hollow, persisting unchanged as $a \to 0$. Note that this force is greater than $q\mathbf{E}_o$ of (D.31), usually taken as the force on a point charge embedded in dielectric. (The discrepancy is investigated in Exercise D.16.)

(b) Does the fact that q also polarizes the dielectric make any difference to the conclusion in (a)? Find the net bound charge q_P that is distributed on the surface of the hollow in the presence of q. (For $a \to 0$, this bound charge may be said to coat the point charge q, reducing the total charge at the point to $|q + q_P| < |q|$.)

D.9 A uniform electric field is imposed on a dielectric sphere, of constant ϵ, containing a concentric spherical hollow.

(a) Show that the outer (free) space and the dielectric shell contain dipole effects of opposite sign superposed on uniform field components.

(b) Find the force on a point charge placed at the center of the hollow and check your result against expectations for $\epsilon \to \infty$ (dielectric replaced by conductor).

(c) The field found in connection with (a) is uniform inside the hollow. Does this mean that the force on the point charge of (b) would be the same if the charge were placed off center?

D.10 A uniform field $\mathbf{E}_o$ is incident normally on a line of charge, λ per unit length, that is sheathed by a hollow dielectric cylinder. The cylinder has the charged line as axis and has an outer radius a; the concentric hollow has radius $b < a$. Find the force per unit length on the charge if $b \ll a$ ($\approx 8\lambda\mathbf{E}_o/9$ for $\epsilon = 2$).

D.11 The usual capacitor consists of a pair of conductors having some potential difference $\Delta\phi$, constructed so that practically all the field lines originating at one conductor surface terminate at the other. Show how expression (D.29) of the field energy leads to

$$W = \tfrac{1}{2}q\,\Delta\phi = \tfrac{1}{2}C\,\Delta\phi^2 = \frac{q^2}{2C}$$

where q is all the charge on the conductor at the higher potential and C is a capacitance defined as in (D.15).

D.12 Two like conducting plates, each with a large rectangular area of sides a and b, face each other a distance $l \ll a, b$ apart. Inserted partway between them and filling the space to a distance $x < b$ is a rectangular slab of solid dielectric ϵ having the same area ab and thickness l.

(a) Suppose that at the stage just described, a potential difference $\Delta\phi$ is established and maintained between the plates. Find the minimum pressure on the outer slab face al that is consequently needed to increase the penetration distance x. [See the connection (B.23) between a stress and a field energy density. Neglect edge effects.]

(*b*) Now suppose that the supply of emf is cut off, free charges $\pm q$ remaining on the plates. Show that there will now be a suction tending to draw the dielectric slab to increased x and amounting to a tension

$$T = 2\pi\left(\frac{q}{ab}\right)^2 \frac{\epsilon - 1}{[1 + (\epsilon - 1)x/b]^2}$$

per unit area al of the entering slab face as long as edge effects can be neglected.

D.13 Dielectric bodies are brought into a preexistent field $\mathbf{E}_o$ arising from *fixed* sources ρ_o, thereby changing the field to some $\mathbf{D} = \mathbf{E} + 4\pi\mathbf{P}$.

(*a*) Show that the change in the field energy density can be expressed as

$$(8\pi)^{-1}[(\mathbf{E} + \mathbf{E}_o) \cdot (\mathbf{D} - \mathbf{E}_o) - 4\pi\mathbf{P} \cdot \mathbf{E}_o]$$

(*b*) Show that the part of this proportional to $\mathbf{E} + \mathbf{E}_o = -\nabla(\phi + \phi_o)$ vanishes upon integration over all space because $\mathbf{D}$ and $\mathbf{E}_o$ have the same sources ρ_o.

(*c*) Show that the consequent interaction energy of the dielectric with the initial field can be evaluated as

$$U = -\frac{1}{2}\oint dV\mathbf{P} \cdot \mathbf{E}_o$$

compared with $U_D = -\mathbf{D} \cdot \mathbf{E}_o$ [Eq. (2.53)], without the factor $\frac{1}{2}$, for a fixed dipole moment magnitude D. (There is a similar factor $\frac{1}{2}$ in the comparison of $U = \frac{1}{2}\oint dV\rho_o\phi_\sigma$, as found in Exercise B.12, with $q\phi_\sigma$ when ϕ_σ arises from induced charges on a conductor.)

(*d*) A particle is subject to a Hooke's-law force proportional to its displacement x from an equilibrium point $x = 0$. Suppose that the particle bears a charge q and that when a uniform field $\mathbf{E}_o$ is imposed, the equilibrium point is shifted to $x = x_o$. Show that the work done by $q\mathbf{E}_o$ during this displacement to x_o is just $\frac{1}{2}\mathbf{D} \cdot \mathbf{E}_o$, where $|\mathbf{D}| = qx_o$. [This should help clarify the factor $\frac{1}{2}$ in (*c*).]

The following exercises are to be done after Chap. 6 has been read.

D.14 Specialized to electrostatic situations, the relations (6.26) and (6.25) become $\nabla \cdot \mathbf{T}(\mathbf{E}) = \rho\mathbf{E}$ and $\int dV\rho\mathbf{E} = \oint d\mathbf{S} \cdot \mathbf{T}(\mathbf{E})$, respectively. Use the procedures that led to them for the following.

(*a*) In the special case of dielectric media, with free charges $\rho' = \nabla \cdot \mathbf{D}/4\pi$ present, show that the stress tensor defined by $\int dV\rho'\mathbf{E} = \oint dS \cdot \mathbf{T}_D$ is

$$\mathbf{T}_D = \frac{\mathbf{ED}}{4\pi} - \mathbf{1}\frac{\mathbf{E} \cdot \mathbf{D}}{8\pi}$$

whenever linear relationships like $\mathbf{D} = \epsilon\mathbf{E}$ hold (and the medium is essentially incompressible). This so-called *Minkowski stress tensor* evidently measures force transfers to bound charges induced in the dielectric as reactions from the forces on the free charges that induce them.

(*b*) Since the effects of dielectrics were supposed due to the bound charges $\rho_P = -\nabla \cdot \mathbf{P}$ induced in them, and $\mathbf{D}$ need not have been defined at all, it should be possible to find $\mathbf{T}_D$, with $\mathbf{D}$ replaced by $\mathbf{E} + 4\pi\mathbf{P}$, by substituting $\rho = \rho' - \nabla \cdot \mathbf{P}$ into $\rho\mathbf{E} = \nabla \cdot \mathbf{T}(\mathbf{E})$ and demonstrating that

$$\rho'\mathbf{E} = \nabla \cdot \mathbf{T}(\mathbf{E}) + \mathbf{E}(\nabla \cdot \mathbf{P}) = \nabla \cdot \mathbf{T}_D$$

Show that this is so wherever $\mathbf{P}$ is simply proportional to $\mathbf{E}$, as in (D.11).

D.15 Apply $\mathbf{T}_D$ of Exercise D.14 in the case of a point charge q outside a plane boundary of dielectric ϵ generating the field (D.24) with $\epsilon_o \equiv 1$.

(*a*) Evaluate the vector stress $\mathbf{n} \cdot \mathbf{T}_D = \mathbf{n} \cdot \mathbf{T}(s, z = 0+)$ being transmitted from the free space into the dielectric surface; then check that its surface integral is just the reaction from the force on q, as expected.

(*b*) Show that the stresses $\mathbf{n} \cdot \mathbf{T}_D(s, z = 0-)$ being transmitted into the body of dielectric beyond its surface have a vanishing resultant (surface integral). [Thus the force (*a*) is confined to the bound charges induced on the surface. The nonvanishing *elements* (having the zero resultant) represent stresses between *field* lines in the dielectric.]

D.16 Apply $\mathbf{T}_D$ of Exercise D.14 to investigating the discrepancy noted in Exercise D.8.

(*a*) Show that the resultant force being transmitted from the dielectric to the surface of the cavity $\oint d\mathbf{S} \cdot \mathbf{T}_D(a+, \vartheta)$ is just $q\mathbf{E}_o$.

(*b*) Show that the net force, the integrated surface divergence

$$\oint d\mathbf{S} \cdot [\mathbf{T}_D(a+, \vartheta) - \mathbf{T}(a-, \vartheta)]$$

transmitted to the bound charge q_P induced on the cavity surface is independent of the cavity radius a.

(*c*) Show that adding the force on q_P to the force on the bare q found in Exercise D.8 has just the resultant $q\mathbf{E}_o$ of (D.31). This indicates that evaluations like (D.31), applicable when $a \to 0$ and hence $\mathbf{E}(r > a) \to \mathbf{E}_o$, give the force on the renormalized charge $q + q_P$ rather than the bare charge q itself.

D.17 Show that the conclusions of Exercise D.16 are unchanged when the free charge q is homogeneously distributed over the entire volume of a finite spherical cavity in the dielectric.

D.18 Classical expectations about the polarizability α of (D.9) can be derived by taking advantage of the fact that whenever some mass m is given a small displacement $\mathbf{r}$ from an equilibrium configuration, a Hooke's-law restoring force $-m\omega_o^2\mathbf{r}$ arises in first approximation, ω_o being a natural frequency with which m can oscillate about equilibrium. Applying a field $\mathbf{E}_o$ to a dielectric can then be expected to produce an equilibrium in which electrons (m, $q = -e$) are displaced and contribute dipole moments $-e\mathbf{r}_o$ related to their natural frequencies in molecules of dielectric. Show that:

(*a*) $\alpha = Ne^2/m\omega_o^2$ if each of N molecules per unit volume has just one electron that is appreciably displaced.

(*b*) $(\epsilon - 1)/(\epsilon + 2) = (4\pi Ne^2/3m) \sum_i \omega_i^{-2}$ if each molecule has several electrons, with natural frequencies $\omega_{1,2,3,\ldots}$.

(*c*) Hooke's-law forces are frequently described in terms of equivalent springs with various stiffnesses. What expectations does this picture yield for the relative contributions of different frequencies to the result in (*b*)?

CHAPTER

E

THE MAGNETOSTATICS OF MATERIALS†

Magnetization and its sources in magnetization (Ampère) currents · The phenomenological magnetic field and magnetic induction · The phenomenological Maxwell equations of magnetostatics · Diamagnetic and paramagnetic permeabilities · Ferromagnetism and hysteresis · The atomic origins of diamagnetism, paramagnetism, ferromagnetism, and ferroelectricity · Magnetic shielding and the fields of permanent magnets

Most materials affect magnetostatic fields less conspicuously than they do electrostatic fields, although there is a highly restricted class, the *ferromagnetic materials*, including the naturally occurring magnetic ores, that do have striking effects.

An ordinary material owes its influence to sources induced in it by the fields. Its relatively small magnetic influence can in part be ascribed to the fact that any magnetic source has at most a dipole effect and nothing quite comparable to the monopole effects electric sources can have; it was pointed out just below (4.48) that a dipole is the simplest element of a magnetic source. The magnetic effects are thus more nearly comparable to those of dielectrics (than to the more conspicuous effects of conductors) on electrostatic fields.

E.1 MAGNETIZATION

On grounds quite comparable to those on which induced dielectric sources are represented by the electric polarization densities introduced in (2.28) and (D.2), induced sources of magnetism can be represented by magnetic polarization, or magnetization, distributions $\mathbf{M}(\mathbf{r}_s)$. Each volume element $dV(\mathbf{r}_s)$ in which $\mathbf{M} \neq 0$ is supposed to behave like a magnetic dipole of moment $\mathbf{M}\, dV$, and this gives rise to a vector-potential field

$$\mathbf{A}_M(\mathbf{r}_f) = \int_V dV(\mathbf{r}_s)\mathbf{M} \times \frac{\hat{\mathbf{R}}}{\mathbf{R}^2} \qquad \text{(E.1)}$$

with $\mathbf{R} = R\hat{\mathbf{R}} \equiv \mathbf{r}_f - \mathbf{r}_s$, according to (4.28).

† Designed to be read after Chap. 4.

For any finite integration volume V, the field can be separated into a volume effect and a surface effect, much as this was done in (D.2) for the electric polarization, with the help of the equivalence (D.1) $\hat{\mathbf{R}}/R^2 = +\nabla_s(1/\mathrm{R})$. Then

$$\mathbf{A}_M = \int_V dV(\mathbf{M} \times \nabla_s)\frac{1}{R} = \int_V dV\left(-\nabla_s \times \frac{\mathbf{M}}{R} + \frac{1}{R}\nabla_s \times \mathbf{M}\right)$$

Just as the Gauss theorem gives $\int (dV\nabla_s) \cdot \mathbf{M} = \oint (d\mathbf{S}) \cdot \mathbf{M}$, so $\int (dV\nabla_s) \times \mathbf{M} = \oint (d\mathbf{S}) \times \mathbf{M}$ can be proved [see (A.15)], with the surface integration to be extended over the entire envelope of V. The result

$$\mathbf{A}_M = \frac{1}{c}\int_V dV \frac{c\nabla_s \times \mathbf{M}}{R} + \frac{1}{c}\oint \frac{c\mathbf{M} \times d\mathbf{S}}{R} \tag{E.2}$$

follows.

The volume effect may be the only appreciable one far in the interior of a material, at points remote from the surface. Comparing it to the vector-potential expression (4.4) shows that it is equivalent to a field arising from a current-density distribution

$$\mathbf{j}_M(\mathbf{r}) = c\nabla \times \mathbf{M}(\mathbf{r}) \tag{E.3}$$

This exists only where the magnetization is nonuniform; it can represent a resultant of microscopic circulations of charge in atomic domains (alluded to on page 17 and there presumed to account for the magnetism found in the magnetic ores).

The surface effect arises because $\mathbf{M}(\mathbf{r})$ is treated as discontinued (as if $\mathbf{M} = 0$ outside V) when the integration is restricted to any finite volume V. Such a nonuniformity in $\mathbf{M}$ also contributes to a current like (E.3); $c\mathbf{M} \times d\mathbf{S}$ is equivalent to $c\nabla \times \mathbf{M}$ after this has been partially integrated over a volume $dV = dS\delta$, of vanishing thickness $\delta \to 0$, that encloses the discontinuity in $\mathbf{M}$ at the surface element $d\mathbf{S}$. Indeed, $c\mathbf{M} \times \mathbf{n}$, where $\mathbf{n} = d\mathbf{S}/dS$, can be recognized as just a *surface* curl like that considered in connection with the electrostatic boundary condition (B.4).

All the results of this section have paralleled those concerning the bound charges in dielectrics, discussed in connection with (D.3) and (D.4).

E.2 MAGNETIC INDUCTION

In treating the electric polarizations, a distinction was maintained between free charges and the no less real bound charges that are induced in dielectrics. A similar distinction is customarily maintained between such, more directly controllable, currents as are sent through wires and the magnetization currents $\mathbf{j}_M = c\nabla \times \mathbf{M}$. Ampère's law is then written as

$$\nabla \times \mathbf{B} = \frac{4\pi}{c}(\mathbf{j}' + c\nabla \times \mathbf{M})$$

with $\mathbf{j}'$ standing for all currents other than $\mathbf{j}_M$. Rewriting this as

$$\nabla \times (\mathbf{B} - 4\pi\mathbf{M}) = \frac{4\pi\mathbf{j}'}{c} \tag{E.4}$$

invites defining a new field

$$\mathbf{H}(\mathbf{r}) \equiv \mathbf{B}(\mathbf{r}) - 4\pi\mathbf{M}(\mathbf{r}) \tag{E.5}$$

which has curl only at the usually macroscopically distinguishable currents $\mathbf{j}'$

$$\nabla \times \mathbf{H} = \frac{4\pi\mathbf{j}'}{c} \tag{E.6}$$

It has long been customary to refer to $\mathbf{H}$ as the *magnetic field* (it is indistinguishable from $\mathbf{B}$ in free space, where $\mathbf{M} \equiv 0$) and to call $\mathbf{B}$ the *magnetic induction.*

The field $\mathbf{H}$ can have points of divergence just as electric charge provides points of divergence in the electric field $\mathbf{E}$. Since $\nabla \cdot \mathbf{B} = 0$,

$$\nabla \cdot \mathbf{H} = -4\pi\nabla \cdot \mathbf{M} \equiv 4\pi\rho_M(\mathbf{r}) \tag{E.7}$$

if $\rho_M(\mathbf{r})$ is defined on the model of the bound charge density $\rho_P = -\nabla \cdot \mathbf{P}$ of (D.3). Then $\rho_M(\mathbf{r})$ acts as a density distribution of pole strength that can reduce to a $q_M\,\delta(\mathbf{r} - \mathbf{r}_o)$, like the distribution characteristic of a point charge, wherever a discrete point $\mathbf{r}_o$ of divergence in the magnetization distribution can be discriminated.

Because this provides explicitly for the description of materials as sources of magnetism additional to the macroscopic currents, the equations

$$\nabla \cdot \mathbf{B} = 0 \qquad \nabla \times \mathbf{H} = \frac{4\pi\mathbf{j}'}{c} \qquad \mathbf{B} = \mathbf{H} + 4\pi\mathbf{M} \tag{E.8}$$

are sometimes put at the basis of magnetostatic theory in place of Eqs. (4.1). However, it has become possible (most properly with the help of quantum statistics) to *derive* the magnetizations from descriptions of microscopic currents in atomic domains. On this basis, $\mathbf{M}(\mathbf{r})$ becomes representative only of gross, average effects to be evaluated from a theory that uses Eqs. (4.1) as the starting basis.

It is the field $\mathbf{B}(\mathbf{r})$, rather than $\mathbf{H}$, that provides the basic description of the magnetostatic field. This cannot be decided by seeing which of the two fields wherever they differ is detected by a test dipole, because such tests must be carried out in cavities where $\mathbf{B} \equiv \mathbf{H}$, much as in the attempts to discriminate between $\mathbf{E}$ and $\mathbf{D}$ described on pages 528 and 529. However, it is the field $\mathbf{B}$ that has the important property of divergencelessness, describing the basic lack of any such thing as "true" magnetic charge. Moreover, it is changes in the flux of $\mathbf{B} = \mathbf{H} + 4\pi\mathbf{M}$, and not of $\mathbf{H}$ alone, that induce the currents observed by Faraday.

The specialized forms (E.8) of the basic equations remain useful for macroscopic descriptions, of only classical detail, whenever the representation of materials by magnetization distributions in them is adequate. As usual, the differential equations can yield explicit descriptions only after boundary conditions sufficient to define a specific physical situation are imposed. As last illustrated for the dielectric boundary conditions (D.16), the ones needed here can be obtained by reducing the divergence and curl in (E.8) to a surface. Then

$$\Delta \mathsf{B}_n \equiv \mathsf{B}_n^o - \mathsf{B}_n^i = 0 \qquad \text{and} \qquad \Delta \mathbf{H}_{\|} \equiv \mathbf{H}_{\|}^o - \mathbf{H}_{\|}^i = 0 \tag{E.9}$$

with the latter written for boundary surfaces on which no macroscopic currents have been introduced. Thus it is the component of **B** normal to a surface separating two media that is continuous through the surface ($\mathsf{B}_n^o = \mathsf{B}_n^i$), whereas it is the components of **H** parallel to the surface that are continuous ($\mathbf{H}_{\|}^o = \mathbf{H}_{\|}^i$).

E.3 PERMEABILITY

A typical way to investigate the effect of a material on a magnetic field is to make use of the field that exists inside a long circularly cylindrical solenoid formed by windings of current-bearing wire, say n turns per unit length, each carrying the same current I, as indicated in Fig. E.1. Conclusions about the field **H** then follow from integrating the Ampère law in (E.8) and the symmetry of the situation. With any field lines outside the solenoid having to be continuations of ones emerging from a remote end and terminating at another,† and with the lines dispersed over an infinite space, the exterior field must be expected to have a vanishing intensity. Inside, the field can only be parallel to the axis if there are no variations either in any material that may be filling the space or in the current around it. Then such a rectangular geometric loop as the one indicated in the figure will have mmf along only the inmost of its four sides. The flux integral of $\nabla \times \mathbf{H} = 4\pi \mathbf{j}'/c$ through this loop will yield $\mathsf{H}l = 4\pi(nIl)/c$ if the contributing side has the length l. This is true independent of how far the contributing side has been put from the axis, and hence the field must be uniform everywhere inside the solenoid. The result (see also Exercise 4.5) $\mathsf{H} = 4\pi nI/c$ is determined solely by the current I sent through the solenoid and is independent of any material with which the inner space might have been filled, provided that this is uniformly distributed and isotropic. Any magnetization induced in it can only be uniform and parallel to **H** and can be evaluated as $M = (\mathsf{B} - \mathsf{H})/4\pi$ from measurements on the field **B**. These measurements might, in principle, be carried out in such flat, cylindrical cavities as were described on page 529 for the measurement of **D** (since $\Delta\mathsf{B}_n = 0$ just as $\Delta\mathsf{D}_n = 0$, on the flat, near surfaces of those types of cavities). However, conclusions about **B** are also obtainable, e.g., by wrapping secondary windings about the solenoid and measuring Faraday currents induced in them when the current I in the primary windings is varied. Of course, ingenious experimenters have developed many other, and more accurate, procedures for measuring the magnetization.

† In practice, ending a solenoid is avoided by shaping the cylinder into a large closed ring solenoid of relatively negligible local curvature (see Exercise 4.22).

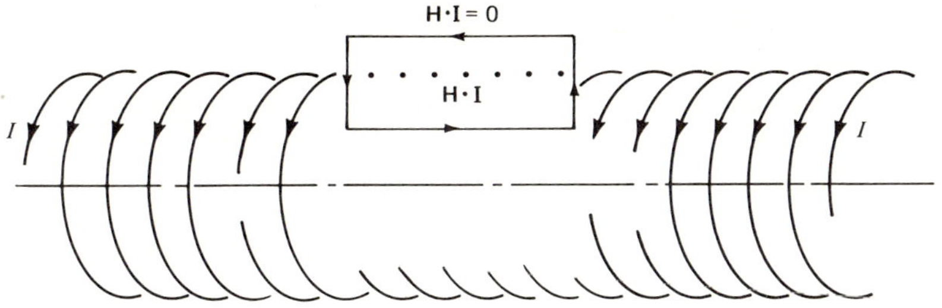

Figure E.1

The procedure sketched here suggests why **H** is treated as an independent variable, controlled through choices of current for $\mathbf{H} = 4\pi nI/c$, and the resultant $\mathbf{B}(\mathbf{H})$ as a dependent variable representing the effect of the material. The functional dependence found in most materials is the linear one

$$\mathbf{B} = \mu \mathbf{H} \tag{E.10}$$

where μ is a constant characteristic of the substance called its *permeability* (not to be confused with any dipole moment, for which the symbol μ is also customarily used). The linearity might have been anticipated as a first term in the Taylor expansion of the function $\mathbf{B}(\mathbf{H})$. The codirectionality of **B** and **H** [and of $\mathbf{M} = (\mathbf{B} - \mathbf{H})/4\pi$] might have been expected on the basis that there is no other direction of **B** or **M**, relative to **H**, that can be singled out in a uniform isotropic medium.

The representation of a material by a $\mu = \mathbf{B}/\mathbf{H}$ is comparable to the representation of dielectrics by constant values for $\epsilon = \mathbf{D}/\mathbf{E}$. Actually, since **B** is the complete magnetic field in the same sense that **E** is the electric one, it is $\mu^{-1} = \mathbf{H}/\mathbf{B}$ that is the more comparable to $\epsilon = \mathbf{D}/\mathbf{E}$. Whereas ϵ is always greater than unity, it is found that $\mu < 1$ in some materials, called *diamagnetic*, but $\mu > 1$ in others, the *paramagnetic* substances. In both cases, the deviation from unity is usually found to be very small, i.e., only a few parts in 10^5, as against the quite common occurrence of substantial deviations from unity among dielectric constants. (The exceptional ferromagnetic substances, in which μ values as large as 10 to 10^6 have been found, will be discussed in the next section.)

The diamagnetic behavior, $\mu < 1$, corresponds to magnetizations

$$\mathbf{M} = \frac{\mathbf{B} - \mathbf{H}}{4\pi} = \frac{(\mu - 1)\mathbf{H}}{4\pi} \tag{E.11}$$

that are *antiparallel* to the inducing field. They tend to reduce the field $(\mathbf{B} < \mathbf{H})$, just as dielectric behavior tends to reduce the electrostatic field in a medium $(\mathbf{E} < \mathbf{D})$. A distribution of microscopic dipole moments antiparallel to the inducing field is exactly what is to be expected, on the basis of the Lenz law, for the dipole effect of microscopic currents induced by the Faraday effect. Presumably the imposition of a magnetic field starts such currents, confined within atoms where they meet no macroscopic resistance and so can continue as long as the external field is not decreased again.

The paramagnetic effects, $\mu > 1$, are understood as arising from permanent microscopic current distributions (within atomic domains) that have permanently nonvanishing dipole moments, behaving like tiny elementary magnets dispersed throughout the material. Presumably, before an external field is imposed, incessant thermal motions keep the elementary dipoles in random relative orientations, without any macroscopic dipole effect. The imposition of the external field tends to align the dipoles *parallel* to the field. The thermal jostling continues, but the field establishes a definite direction that more of the dipoles take, on the average, than any other direction. Such an interpretation is supported by the fact that paramagnetic effects are found to decrease as the temperature of the material is increased. This finding is usually expressed for what is called the *magnetic susceptibility*

$$\kappa \equiv \frac{\mathbf{M}}{\mathbf{H}} = \frac{\mathbf{B} - \mathbf{H}}{4\pi \mathbf{H}} = \frac{\mu - 1}{4\pi} \tag{E.12}$$

and is representable by

$$\kappa = \frac{C}{T + \theta} \tag{E.13}$$

where T is the absolute temperature and C and θ are constants characteristic of the material ($\theta \approx 0$ for most gases). The empirical relation is known as *Curie's law*, and there have been various theoretical derivations of it, based on successively more satisfactory statistical treatments, since its first discovery.

Diamagnetic effects are actually superposed on any paramagnetic ones, so that $\mu > 1$ is the result only when the paramagnetism is the stronger. Diamagnetism is not as sensitive to temperature changes because the induced Faraday currents responsible for it are average drifts superposed in a linearly independent way on the temperature motion. All the effects of magnetic fields on the motions of charge alluded to here can perhaps be better understood after familiarity with some of the most basic effects, as treated in Chap. 5.

E.4 FERROMAGNETISM AND PERMANENT MAGNETS

A highly restricted class of materials, iron and its alloys being its most prominent members, can exhibit permeabilities of the order of thousands. These extraordinarily high permeabilities are field-dependent, in that $\mu = \mathbf{B}/\mathbf{H}$ itself varies with the strength of the imposed field. A plot of $\mathbf{B}$ vs. $\mathbf{H}$ typical of these cases is shown in Fig. E.2. As the field $\mathbf{H}$ imposed on an unmagnetized specimen is increased, there is initially a linear growth of $\mathbf{B}$ with a slope that may be anywhere from $\mu = 10$ to 10^4, depending on the constitution of the ferromagnet. Eventually, a saturation value of $\mathbf{B}$ is reached, presumably because all available elementary dipoles are completely aligned. The interesting point is that when, at

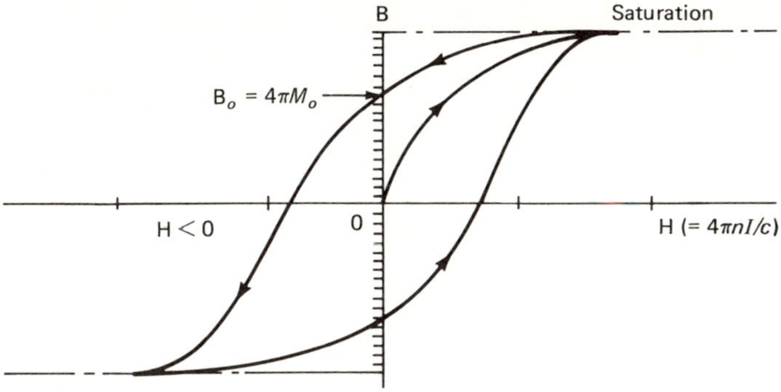

Figure E.2

any stage, the imposed field is removed again, some magnetization remains, as at the value $M_o = \mathsf{B}_o/4\pi$ indicated in the diagram. Demagnetization to $\mathsf{B} = 0$ requires the imposition of a reverse field, as indicated ($\mathsf{H} < 0$), and a continued increase in the reverse field will eventually yield saturation in the reversed direction. The complete loop drawn in the figure is called a *hysteresis cycle* through which the specimen may be taken. Complete demagnetization, so that $\mathsf{B} = 0$ when $\mathsf{H} = 0$ again, requires putting the specimen through a whole series of successively smaller cycles (obtained by reducing the amplitude of H for each reversal).

The specimen can be dissociated from the apparatus for imposing the field H at a stage when it has a substantial magnetization M_o and then it serves as a quasi-permanent magnet like those found in nature. The relative permanence of this state is understood as a result of elementary dipoles strong enough to *help each other* maintain any initial alignment (spoken of as a *cooperative phenomenon*).

The discussions of the magnetostatic properties of materials here and of the electrostatic properties in preceding chapters have kept reference to details about the atomic structure of the materials to a minimum. That has the advantages of greater generality and of demonstrating how much can be understood without going into details. However, it makes the explanations relatively superficial, in view of how much has by now been learned about atoms, individually and in aggregates.

It is now quite generally known that atoms contain orbiting electrons and that each electron also has an intrinsic-spin† angular momentum. Charges with angular momentum give rise to magnetic moments [as in (4.40), for example], and considering these makes it possible to understand more deeply the contrasts between diamagnetism, paramagnetism, and ferromagnetism. Most of the atomic electrons are paired into a *cancellation* of their contributions to resultant magnetic moments. It is the superposition of extra, Faraday-induced, motions that is

† Some quantitative detail about this will be found in Exercise 12.5.

responsible for the diamagnetic effects produced by imposing magnetic fields [the fundamental basis of this will be considered in connection with (5.13)]. Paramagnetic effects arise in materials which normally have some unpaired electrons in their atoms and molecules, acting like elementary dipole magnets that can be aligned by the imposition of magnetic fields, and the sensitivity of such alignments to temperature has already been mentioned. A much smaller class of materials has atoms which, in their (normal) states of lowest energy, permit some electrons to orbit and spin in *parallel*, producing extra large magnetic moments. These can be large enough to influence neighborhood atoms appreciably, and such interactions can lead to a *spontaneous* magnetization of domains containing many atoms. This characterizes the ferromagnetic substances. Before bulk magnetization, the (competitively formed) domains are each only a small part of the matter in bulk and are randomly oriented. An imposed magnetic field now has relatively large domain magnetic moments to work on and can align them throughout the whole bulk of material. The resulting bulk magnetization can become as self-sustaining as the spontaneous domain magnetizations and so tends to persist after the field has been removed. That is what accounts for the large permeabilities and hysteresis effects in the ferromagnets.

Analogous phenomena can be expected to appear during *electro*static effects on substances consisting of anisotropic polar molecules that sustain *electric* dipole moments in their normal states. Indeed, materials called ferroelectrics (even when they contain no iron) exhibit hysteresis and residual electric polarizations analogous to those of ferromagnetism.

Of course, most of the assertions in discussions like these are statements about the outcomes of much more detailed investigations. They cannot be undertaken here because they require calling on bodies of principles (including those of quantum mechanics) that lie far outside electromagnetic theory itself. They have become the concern of whole disciplines still in development, like the burgeoning field of solid-state physics (see, for example, Ref. 15).

E.5 MAGNETIC SHIELDING

The effect of an induced magnetization on a field can be evaluated from Eqs. (E.8). This is simplest to do when the relation $\mathbf{B} = \mu\mathbf{H}$ can be used with a constant μ. Cases of substantial permeability naturally have the greatest interest, and for these μ is often field-dependent. However, the relation can still be used with some suitably averaged *constant* $\bar{\mu}$ if the variations of the field within the permeable material turn out to be moderate.

The specific example now to be treated will demonstrate the shielding effect of a highly permeable material ($\bar{\mu} \gg 1$) on a space it encloses that is devoid of sources. Let this space be the hollow inside an indefinitely long cylindrical shell, having an inner radius a and an outer radius $b > a$, as indicated in Fig. E.3. The imposed field will be taken to be a uniform one $\mathbf{B}_o$ incident from infinity.

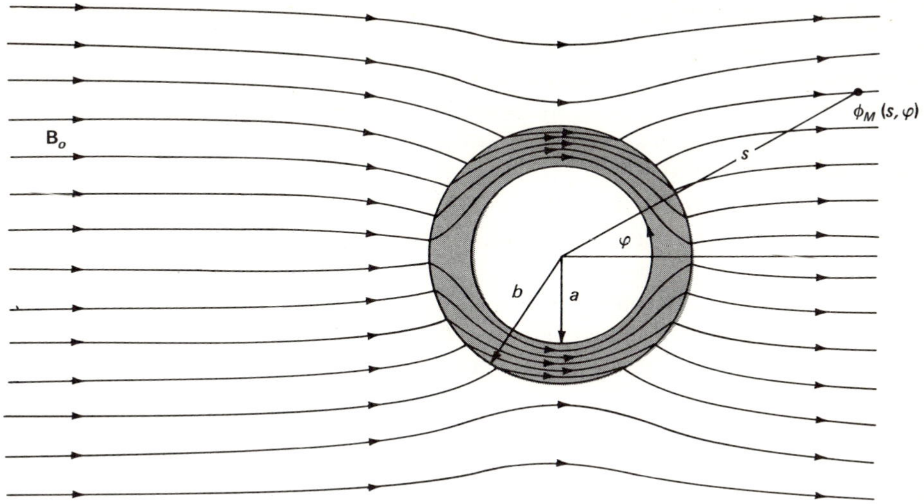

Figure E.3

No macroscopic currents $\mathbf{j}'$ have been introduced, so that the equations (E.8) reduce to $\nabla \times \mathbf{H} = 0$ and $\nabla \cdot \mu\mathbf{H} = 0$. The curl equation has a general solution

$$\mathbf{H} = -\nabla\phi_M(\mathbf{r}) \tag{E.14}$$

according to (A.28); like any field that is curlless in a region, $\mathbf{H}$ is derivable from a scalar potential there. The divergence equation can also be satisfied by making ϕ_M obey a Laplace equation, $\nabla^2\phi_M = 0$, *in each region of uniform permeability* ($\mu = 1$ in the free spaces inside and outside the material shell). Consequently ϕ_M will have to be presented in a piecewise manner, as $\phi_M(s > b, \varphi)$, $\phi_M(a < s < b, \varphi)$, and $\phi_M(s < a, \varphi)$ in terms of the polar coordinates indicated in the figure.

The three expressions will all be parts of the unique solution if they are properly matched up at the boundaries $s = a$ and $s = b$, in accordance with the conditions (E.9), here expressible as $\Delta\mathbf{B}_s = 0$ and $\Delta\mathbf{H}_\varphi = 0$. The latter type of condition requires the azimuthal gradients, $\partial\phi_M/(s\,\partial\varphi) = -\mathbf{H}_\varphi$, to match all around (at all φ on) the surfaces $s = a$ and $s = b$, and this is equivalent to demanding a continuity $\Delta\phi_M(s = a, b; \varphi) = 0$ of the potentials themselves, just as in the case of the dielectric boundary conditions (D.19). There is also a boundary condition at infinity, describing the incidence of the uniform field $\mathbf{B}_o$, to be met. If the $\varphi = 0$ radius (the $+x$ axis) is chosen parallel to the incident field, as is indicated in the figure,

$$\phi_M(s \to \infty, \varphi) = -\mathbf{B}_o x = -\mathbf{B}_o s \cos\varphi \tag{E.15}$$

is the requirement, since $\mathbf{B} = \mathbf{H} = -\nabla\phi_M(s > b, \varphi)$ in the free ($\mu = 1$) outer space.

The boundary conditions on the cylindrical surfaces will be most simply fitted by the form (C.3), the general Laplace field in polar coordinates. A systematic imposition of all the boundary conditions (including the requirement of finiteness at the axis $s = 0$) on forms as general as (C.3) would be found to determine unique values for all the arbitrary coefficients in them. However, most of the coefficients will simply turn out to be zero because of the obvious symmetries the result must have. Thus, $\phi_M(s, -\varphi) = \phi_M(s, \varphi)$ must be expected, and so all the linearly independent odd-parity terms proportional to $\sin m\varphi$ will inevitably vanish. Moreover, the only directional dependence introduced into the problem is the proportionality to $\cos \varphi$ in (E.15), and so the coefficients of all the *other*, linearly independent, $\cos m\varphi$ (including $m = 0$) can be expected to vanish. This means that the forms proportional to $\cos \varphi$,

$$\phi_M(s > b, \varphi) = -\mathsf{B}_o s \cos \varphi + C_1 \frac{\cos \varphi}{s}$$

$$\phi_M(a < s < b, \varphi) = \left(A_1 s + \frac{B_1}{s}\right) \cos \varphi \tag{E.16}$$

$$\phi_M(s < a, \varphi) = c_1 s \cos \varphi$$

should be general enough.† Now the requirement that ϕ_M be continuous through each of the surfaces $s = a$ and $s = b$ yields

$$c_1 = A_1 + \frac{B_1}{a^2} \qquad C_1 = B_1 + A_1 b^2 + \mathsf{B}_o b^2$$

The continuity of $\mathsf{B}_s = -\mu\, \partial\phi_M/\partial s$ demands that

$$c_1 = \bar{\mu}\left(A_1 - \frac{B_1}{a^2}\right) \qquad \text{and} \qquad C_1 = \bar{\mu}(B_1 - A_1 b^2) - \mathsf{B}_o b^2$$

with $\bar{\mu}$ here characterizing the shell of material. Thus obtained are four equations just sufficient to determine all the arbitrary coefficients c_1, C_1, A_1, and B_1 in the forms (E.16).

The results will be discussed in terms of the total field $\mathbf{B} = -\mu\nabla\phi_M$ that arises in each of the three regions. In each, $\mathbf{B}$ is derivable as a negative gradient of a form

$$\mu\phi_M = -\mathsf{B}'x + 2D_M \frac{\cos \varphi}{s} \tag{E.17}$$

where the μ on the left differs from unity only within the material of the shell, $\mathsf{B}' \equiv \mathsf{B}_o$ in the outer region, and $D_M \equiv 0$ inside the hollow.

† Much as in the problem yielding (D.26).

Plainly, the value of B' in each region represents a *uniform* component, parallel to the incident $\mathbf{B}_o$, of the total field in the region. A uniform component is the only one inside the hollow, and its magnitude will have been found to be

$$\mathsf{B}(s < a) = -c_1 = \frac{4\bar{\mu}b^2}{(\bar{\mu}+1)^2b^2 - (\bar{\mu}-1)^2a^2}\,\mathsf{B}_o \tag{E.18}$$

which properly reduces to B_o for $\bar{\mu} = 1$ or $a = b$ (the material replaced by free space). The interesting observation is that for large $\bar{\mu}$ values

$$\mathbf{B}(s < a) \to \frac{4b^2}{b^2 - a^2}\frac{\mathbf{B}_o}{\bar{\mu}} \qquad \bar{\mu} \gg 1 \tag{E.19}$$

so that the inner field tends to vanish in proportion to $1/\bar{\mu}$, especially if the shell is not too thin. A highly permeable material *shields* the interior from the incident field $\mathbf{B}_o$.

The nonuniform field components in the two outer regions each have the form of a field due to a line of dipoles with a moment D_M per unit length, oriented parallel to the incident field (when $D_M > 0$). That can be confirmed by reference to the discussion on page 508 of an electric dipole line. In the outermost space

$$D_M(s > b) = \tfrac{1}{2}C_1 = \frac{1}{2}\frac{\mathsf{B}_o b^2(b^2 - a^2)(\bar{\mu}^2 - 1)}{(\bar{\mu}+1)^2b^2 - (\bar{\mu}-1)^2a^2} \tag{E.20}$$

negative for a diamagnetic material ($\bar{\mu} < 1$) but positive for paramagnetic materials. In highly permeable materials ($\bar{\mu} \gg 1$), the incident field induces moments with $D_M \to \frac{1}{2}\mathsf{B}_o b^2$, independent of any hollow in the cylinder. The superposition of the paramagnetism on the incident field causes the field lines to converge upon the cylinder in the way indicated in Fig. E.3; as many as possible of the always continuous lines of $\mathbf{B}$ take a path through the permeable material.

The field within the material of the shell has the uniform component

$$\mathsf{B}'(a < s < b) = -\bar{\mu}A_1 = \frac{2\bar{\mu}(\bar{\mu}+1)b^2}{(\bar{\mu}+1)^2b^2 - (\bar{\mu}-1)^2a^2}\,\mathsf{B}_o \tag{E.21a}$$

and superposed on this a nonuniform component such as is due to an axial line of dipoles with the moment per unit length

$$D_M(a < s < b) = \tfrac{1}{2}\bar{\mu}B_1 = -\frac{2\bar{\mu}(\bar{\mu}-1)a^2}{(\bar{\mu}+1)^2b^2 - (\bar{\mu}-1)^2a^2}\,\tfrac{1}{2}\mathsf{B}_o b^2 \tag{E.21b}$$

The nonuniformity properly disappears for $\bar{\mu} = 1$, and $\mathbf{B}' \to \mathbf{B}_o$ at the same time. The nonuniformity also vanishes if there is no hollow ($a = 0$); the field within a *solid* cylinder is entirely uniform with the magnitude $\mathsf{B}' = 2\bar{\mu}\mathsf{B}_o/(\bar{\mu}+1)$, which becomes twice B_o for very large permeabilities. The negative sign of the moments (for $\bar{\mu} > 1$), that produce the nonuniformity when a hollow does exist, serves to make the resultant field lines, including the uniform component, continue through the shell of permeable material instead of entering the hollow, as indicated in Fig. E.3 for a case of high permeability. Thus, the shielding by the *para*magnetic effects comes from giving the field lines more permeable paths to follow, and so arises in a way that contrasts with the shielding by *di*electric effects in an electrostatic field. (The contrasting behavior of dielectricity makes the field inside dielectric with very large ϵ approach zero, as in a conductor, instead of twice the incident field strength, as in the highly permeable ferromagnetic body.)

E.6 THE FIELD OF A PERMANENT MAGNET

Whereas dielectric polarizations must almost without exception be codetermined with the fields that induce them, like the magnetization in the example just above, the existence of (quasi-) permanent magnetism sometimes makes it possible to give beforehand at least part of the magnetization distribution that will exist in a given situation. To be considered now are cases in which the entire field present arises from a *given* magnetization $\mathbf{M}(\mathbf{r})$.

Either of two primary approaches can be applied to such problems. One is addressed to finding directly the entire field $\mathbf{B}$ everywhere, essentially by solving the basic equations $\nabla \cdot \mathbf{B} = 0$, $\nabla \times \mathbf{B} = 4\pi\mathbf{j}_M/c$ with $\mathbf{j}_M = c\nabla \times \mathbf{M}$ given. Their solution can be presented as $\mathbf{B} = \nabla \times \mathbf{A}_M$, where $\mathbf{A}_M$ is to be obtained by integrating expression (E.2).

A second approach first obtains the field $\mathbf{H}$ as a solution of the equations $\nabla \times \mathbf{H} = 0$ [to which (E.6) reduces when an $\mathbf{M}(\mathbf{r})$ provides the only sources] and $\nabla \cdot \mathbf{H} = 4\pi\rho_M$, with $\rho_M = -\nabla \cdot \mathbf{M}$, as in (E.7). Then, outside the given magnets, $\mathbf{B} \equiv \mathbf{H}$, whereas inside them $\mathbf{B}$ is found merely by adding the given $4\pi\mathbf{M}$ to the $\mathbf{H}$ there. This approach reduces the problem to one of an electrostatic type. The curlless $\mathbf{H}$ is everywhere derivable from a magnetic *scalar* potential, as $\mathbf{H} = -\nabla\phi_M$. The scalar potential can be obtained from integrating an expression like (D.2), with $\mathbf{P}$ replaced by the given $\mathbf{M}$.

Applied to a simple situation like that presented by a circularly cylindrical magnet that has been prepared with a uniform magnetization $\mathbf{M}_o$ throughout its volume and parallel to its axis, the two approaches allow coming to fairly detailed conclusions about the field without complete, formal calculation (an example of the latter is left to Exercise E.7). The uniformly magnetized body has $\mathbf{j}_M = c\nabla \times \mathbf{M}_o = 0$ and $\rho_M = -\nabla \cdot \mathbf{M}_o = 0$ throughout its volume but has *surface* curls and divergences where $\mathbf{M}$ changes discontinuously from $\mathbf{M}_o$ to 0, these then being the only sources of $\mathbf{B}$ and $\mathbf{H}$.

Consider first the lines of $\mathbf{H}$ that arise everywhere, inside as well as outside the cylindrical magnet. Their sources are solely the surface divergences $-\mathbf{n} \cdot \Delta\mathbf{M} \equiv \sigma_M$ [compare the surface charges on dielectrics, (D.4)] that exist at the flat top and flat bottom of the cylinder, as indicated in Fig. E.4. At the top, $\sigma_M = +M_o$, and $\sigma_M = -M_o$ at the bottom. These sources of $\mathbf{H}$ produce a field of the same form as the electrostatic field arising from two coaxial uniformly

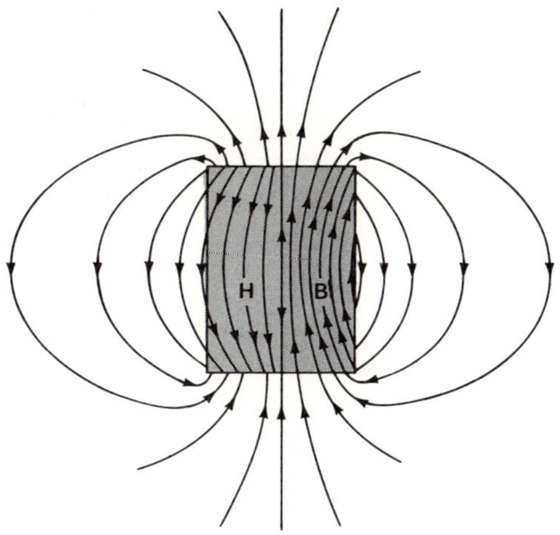

Figure E.4

charged disks (see Exercise 2.14), the cylinder length L apart and respectively having total charges $q_M = \pm\pi a^2 M_o$ for a cylinder of radius a. If the effect of the bottom disk at the top one could be neglected ($L \gg a$), the flux $4\pi\sigma_M$ out of each unit of the top area would, as in (2.5a), be shared equally by upward emanations in density $\mathsf{H} = 2\pi\sigma_M = 2\pi M_o$ and downward ones corresponding to $\mathsf{H}_z = -2\pi M_o$ into the body of the cylinder. The effect of the bottom disk near the center of the top one can be evaluated from the electrostatic result for a disk (2.4). It diminishes the flux density out of the top by a factor $L/(L^2 + a^2)^{1/2}$ and strengthens the downward flux density to correspond to

$$\mathsf{H}_z(s = 0,\ z = \tfrac{1}{2}L-) = -2\pi M_o\left[2 - \frac{L}{(L^2 + a^2)^{1/2}}\right] \tag{E.22}$$

Both effects will be somewhat moderated toward the edges of the disk. The result (2.4) can also be used to evaluate the effect of both disks near the center of the entire cylinder:

$$\mathsf{H}_z(s = 0,\ z = 0) = -4\pi M_o\left[1 - \frac{L}{(L^2 + 4a^2)^{1/2}}\right] \tag{E.23}$$

This has a smaller magnitude than the flux density nearer the top or bottom of the cylinder, corresponding to an outward (from the axis) dispersal of the field lines into the convex shapes indicated in the figure. The lines pay no attention to the side boundary of the cylinder as they cross it, since $\mathbf{M}_o$ has no component normal to that boundary and so there is no surface divergence there. The convexity of the lines near the central plane continues indefinitely outward from the axis as the form of the field merges into one characteristic of an ideal dipole. It is plain that indefinitely far away the field will be characteristic of a dipole with the moment $q_M L = M_o(\pi a^2 L) = M_o V_o$, where V_o is the volume of the cylinder. The total moment is thus equal to the volume integral of the moment per unit volume $\mathbf{M}_o$.

The lines of **H** have been indicated in the left half of Fig. E.4, and the corresponding **B** lines are shown in the right half. Outside the cylinder, the lines of **B** and **H** are identical. Inside, $\mathbf{B} = \mathbf{H} + 4\pi\mathbf{M}_o$ and since $-\mathrm{H}_z$ is everywhere less than $4\pi M_o$, the lines have their z components reversed in sign by the addition of $4\pi M_o$. This provides the most striking contrast between the **H** and **B** lines—their essentially opposite directions inside the cylinder. A switch in curvature from convexity to concavity from the axis is also engendered by the addition of $4\pi\mathbf{M}_o$. At the top or bottom of the cylinder, the addition of $4\pi\mathbf{M}_o$ to the results for **H** there yields

$$\mathrm{B}(s = 0, z = \pm\tfrac{1}{2}L) = \frac{2\pi M_o L}{(L^2 + a^2)^{1/2}} \tag{E.24}$$

continuous through each of the flat surfaces, while the central density obtained from (E.23) is

$$\mathrm{B}(s = 0, z = 0) = \frac{4\pi M_o L}{(L^2 + 4a^2)^{1/2}} \tag{E.25}$$

This is the larger, corresponding to a concentration of the **B** lines toward the axis as the central plane is approached, and represented by the concavity of the lines. This curvature also permits the **B** lines to cross the top and bottom surfaces continuously, without even a change of slope.

The **B** lines should pay no attention to the top and bottom surfaces. The sources of **B** are the magnetization currents $\mathbf{j}_M = c\nabla \times \mathbf{M}$, and these exist, degenerated to surface curls $c\mathbf{M} \times \mathbf{n}$ (as discussed on page 542), on the curved sides of the cylinder only. The complete **B** line picture could have been built up from considering these sources of it alone. The surface curl $|c\mathbf{M} \times \mathbf{n}| = cM_o$ is equivalent to a current $I/L = cM_o$ per unit of the cylinder's height confined to the convex surface of the cylinder and having the azimuthal direction (of φ). The **B** lines everywhere must form continuous loops enclosing this current or sections of it. The convex lines just outside the cylinder sides, found earlier as $\mathbf{B} \equiv \mathbf{H}$ lines, intersect those sides and must be continuous with **B** lines inside, to form the closed loops. For a very long cylinder ($L \gg a$), a central portion of it must have the field known to be characteristic of a very long solenoid $4\pi(I/L)/c = 4\pi M_o$, agreeing with (E.25) for $L \gg a$. In the opposite limit, $L \ll a$, the finding (4.23) for a circular filament leads to an expectation of $\mathrm{B} = 2\pi I/ca = 2\pi M_o L/a$ at the center; this agrees with both (E.24) and (E.25) for $L \ll a$, as it should. The magnetic dipole moment of the current should be $\mu = \pi a^2 I/c = \pi a^2 M_o L$, and this is exactly the result already pointed out above.

The overall picture must also be consistent with the generally applicable boundary conditions (E.9). At the top and bottom of the cylinder, the conditions require $\Delta\mathrm{B}_z = 0$ and $\Delta\mathrm{H}_s = 0$. Since $M_s = 0$ everywhere, the last condition leads to $\Delta\mathrm{B}_s = 0$ also, so that the lines should not even suffer a slope change in crossing those surfaces. On the sides, $\Delta \mathsf{B}_s = 0$ and $\Delta \mathsf{H}_z = 0$. The latter leads to $\Delta \mathsf{B}_z = -4\pi M_o$, and this accounts for the slope discontinuities in the **B** loops as they cross the cylinder sides.

EXERCISES

E.1 Steady-current densities $\mathbf{j}(x, y, z)$ are distributed in the free ($z > 0$) space outside the plane ($z = 0$) boundary of material occupying all the $z < 0$ space and having a uniform permeability μ. Show that:

(*a*) $\mathbf{B}(x, y, z > 0) = \mathbf{B}_1 + \mathbf{B}'$, where $\mathbf{B}_1$ is the field that would arise from $\mathbf{j}$ alone (as if $\mu = 1$) and $\mathbf{B}'$ is calculable from image currents distributed in $z < 0$, each component of which is representable by $j'_i = C_i(\mu)j_i(x, y, -z)$ with some value for each of the constants $C_{x,y,z}$ (to be found).

(*b*) $\mathbf{B}(x, y, z < 0)$ is calculable from $[2\mu/(\mu + 1)]\mathbf{j}(x, y, z)$.

E.2 Suppose that the current distribution of Exercise E.1 forms a filamentary circular loop having radius a and carrying a current I.

(*a*) Do you expect the loop to be attracted to, or repelled from, the permeable material when the plane of the circle is parallel to the material boundary? Perpendicular to it?

(*b*) Suppose that the center of the loop is at a large distance $l \gg a$ from the material boundary and that the plane of the loop makes some angle α with it. Find the resultant force on the loop and also the torque on it. Compare your results to those of Exercises B.13 and 4.15.

E.3 An indefinitely long wedge of free space, $0 \le \varphi \le \alpha < 2\pi$, is bordered by half-infinite planes which serve as boundaries of material occupying all the $\alpha < \varphi < 2\pi$ space and having permeability μ. A filament of steady current I runs along the vertex of the wedge.

(*a*) Use the symmetry of the situation and Ampère's integral law to find the field **B** everywhere.

(*b*) Show how the same result follows from the sufficiently general (and multivalued) solution $\phi_M = (\text{const})\varphi$ of the Laplace equation.

(*c*) Show how the findings in Exercise E.1 reduce to the result for the case $\alpha = \pi$.

(*d*) In the case $\alpha = \pi$, modify the result to the situation in which the free space is replaced by a medium of permeability $\mu_o \neq \mu$.

E.4 Ferromagnetic materials can have permeabilities so large ($> 10^3$) that $\mu \to \infty$ limits of results in various situations can provide adequate approximations.

(*a*) Show that the surface of an infinitely permeable material becomes an equipotential having a constant value for the ϕ_M from which $\mathbf{H} = -\nabla\phi_M$ is derivable, the magnetic flux lines leaving or entering the surface normal to it.

(*b*) With ϕ_M constant everywhere in a $\mu \to \infty$ medium, $\mathbf{H} \to 0$ in it. How about **B**?

(*c*) Show that the results of Exercise E.1 approach conformity with the findings (*a*) and (*b*) and that the magnitudes of the image currents become equal to those of the currents which induce them.

E.5 A field that is uniformly $\mathbf{B}_o$ at infinity is incident on an homogeneously permeable sphere of radius a. Find the dipole moment induced in the sphere, in terms of $\mathbf{B}_o$, a, and the permeability μ, and also the field strength inside the sphere for $\mu \gg 1$.

E.6 Suppose that the sphere of Exercise E.5 has a concentric empty hollow of some radius $b < a$ made in it. Find the field that penetrates into the hollow when $\mu \gg 1$.

E.7 A sphere of radius a contains a uniform magnetization $\mathbf{M}_o$.

(*a*) Find the magnetic-charge distribution $\sigma_M(\vartheta)$ that can be taken to serve as the source of **H**.

(*b*) Show how (3.81), modified by (3.99), leads to simple results for $\phi_M(r, \vartheta)$, **H**, and **B** here, both inside and outside the sphere.

E.8 For the spherical magnet of Exercise E.7:

(*a*) Find the magnetization-current distribution $\mathbf{j}_M$ that can serve as the source of **B**.

(*b*) Show how the results for the spinning shell of electric charge found in Exercise 4.1 can be adapted to give the same results as those of Exercise E.7(*b*).

(*c*) Show that the boundary conditions on **H** and **B** are satisfied by the solution obtained here.

E.9 Apply the results of Exercise E.1 to the following situation. An indefinitely long cylindrical magnet with radius a and uniformly magnetized by $\mathbf{M}_o$ parallel to its axis has its south face parallel to a plane at distance l that has material of permeability μ filling the space beyond it. Find the field everywhere along the axis.

E.10 Suppose that $\mathbf{B} = \nabla \times \mathbf{A}$ is the result of establishing free currents $\mathbf{j}' = c(\nabla \times \mathbf{H})/4\pi$ in the presence of magnetizable materials. By paralleling the arguments that led from (D.29) to (D.30) show that a free magnetostatic field-energy density

$$\frac{\mathbf{H} \cdot \mathbf{B}}{8\pi}$$

follows from the suitable modification of (4.96).

E.11 Consider the situation of Exercise E.9. By evaluating the field-energy change that would take place if an area element dS of the permeable surface at just the central point (on the axis of the system) were displaced (virtually) by δz, as for (B.23), show that the force per unit area at the central point, attracting the surface to the magnet, is just

$$T = 2\pi M_o^2 \frac{\mu(\mu - 1)}{(\mu + 1)^2}\left(1 - \frac{l}{\sqrt{l^2 + a^2}}\right)^2$$

This indicates that the maximum force that would be needed to pull a magnet of pole face area $\pi a^2 \to S$ from contact with a highly permeable $(\mu \to \infty)$ material is $2\pi M_o^2 S$. [Compare (B.20).]

E.12 Suppose $\mathbf{B}_{\text{ext}}$ has steady sources that lie entirely outside a body of material containing a magnetization distribution $\mathbf{M}(\mathbf{r})$. Then the net translational force on the body can be calculated from the following adaptation of (4.53):

$$\mathbf{F}_M = \int_V dV \mathbf{j}_M \times \frac{\mathbf{B}_{\text{ext}}}{c} \qquad (a)$$

with $\mathbf{j}_M = c\nabla \times \mathbf{M}$. This integral can be held to include contributions $\oint dS(\mathbf{n} \times \mathbf{M}) \times \mathbf{B}_{\text{ext}}$ by the body's surface itself if the integration volume V extends beyond that surface, to where $\mathbf{M} \equiv 0$. By using this to eliminate several integrals $\int dV\nabla \cdots \equiv \oint d\mathbf{S} \cdots$ over the surface of V show that (a) can be transformed into

$$\mathbf{F}_M = \int_V dV \rho_M \mathbf{B}_{\text{ext}} \qquad (b)$$

which can be held to include contributions $\oint dS\sigma_M \mathbf{B}_{\text{ext}}$ from the body's surface itself. [Thus elements in (a) that are each *perpendicular* to $\mathbf{B}_{\text{ext}}$ are found to yield the same resultant as elements in (b) each *parallel* to $\mathbf{B}_{\text{ext}}$!]

E.13 Test the equivalence of the two integrals in Exercise E.12 for a bar magnet lying along the z axis of a circular loop of current (radius a). Take the magnet to be a circular cylinder (radius b), and let its magnetization be uniform and parallel to the axis. Let the cylinder radius be small enough $(b \ll a)$ for the component B_z of the loop's field to be well approximated throughout the magnet by just the axial values found in Exercise 4.3. Moreover, use correspondingly approximated values for the radial component B_s, as deduced from $\partial(s\mathrm{B}_s)/\partial s = -s\, \partial\mathrm{B}_z/\partial z$ or from (4.22).

CHAPTER

F

WAVES IN TRANSPARENT MATERIALS†

The phenomenological Maxwell equations for nonstatic fields · Propagation velocities and refractive indexes in dielectrics · Field energies and their flows in material media · Reflection and refraction · The Fresnel formulas · The Brewster angle and polarization by reflection · Reflection and transmission coefficients · Total reflection and its consistency with nonvanishing penetrations into the reflecting medium · Normal dispersions of frequencies · Group velocities of signals

Among the most conspicuous phenomena that must be accounted for by any theory of light are its reflections and refractions at interfaces between different media. For this, the effect of material media on the electromagnetic waves should be examined.

F.1 THE GENERAL MAXWELL EQUATIONS IN DIELECTRICS

Transparent media are generally nonconducting and may be classed as dielectrics. As such, their effects can be represented by induced polarization distributions $\mathbf{P}(\mathbf{r}, t)$, which need not be static as in Sec. D.1. Since the electromagnetic waves are also constituted of magnetic fields, the possible induction of magnetization distributions $\mathbf{M}(\mathbf{r}, t)$ in the material should also be taken into account.

As found in (D.3), the first effect of dielectric polarizations is to present bound charges $\rho_P = -\nabla \cdot \mathbf{P}(\mathbf{r}, t)$ to supplement any free charges ρ' that might be present. The effective redistributions of the bound charges that take place during their inductions by nonstatic fields will give rise to effective currents, already mentioned on page 523, which can be identified from the bound charge conservation $\nabla \cdot \mathbf{j}_P = -\partial\rho_P/\partial t = +\nabla \cdot \mathring{\mathbf{P}}$. The conclusion $\mathbf{j}_P \equiv \mathring{\mathbf{P}}$ follows,‡ since $\mathbf{j}_P = 0$

† Designed to be read after Sec. 7.2.

‡ An additional assumption is actually implicit here. The result $\mathbf{j}_P = \mathring{\mathbf{P}}$ has the correct divergence, but a $\mathbf{P}$ with curl, resulting in an additional *divergenceless* current contribution $c\nabla \times \mathbf{P}$, analogous to $c\nabla \times \mathbf{M}$ of (E.3), might be conceived. The extra assumption is that a linear proportionality of polarization to $\mathbf{E}$ (D.11) is to be expected and that $c\nabla \times \mathbf{P} = c\chi_e\nabla \times \mathbf{E} = -\chi_e\mathring{\mathbf{B}}$, a result of Faraday induction already taken into account (as diamagnetism) in the definition of $\mathbf{M}$. Actually, $\mathbf{j}_P = \mathring{\mathbf{P}}$ formed part of Maxwell's initial hypothesis in (1.37), through expressing his displacement current as $\mathring{\mathbf{D}}/4\pi = \mathring{\mathbf{E}}/4\pi + \mathring{\mathbf{P}}$, since he treated empty and material-filled spaces on the same footing, as different media merely. See Exercise F.10.

unless $\mathring{\mathbf{P}} \neq 0$ according to the meaning given to $\mathbf{j}_P$. In view of these findings, the two Maxwell equations (1.40) that depend on charges present can be written:

$$\nabla \cdot \mathbf{E} = 4\pi(\rho' - \nabla \cdot \mathbf{P})$$

$$\nabla \times \mathbf{B} = \frac{4\pi}{c}(\mathbf{j}' + c\nabla \times \mathbf{M} + \mathring{\mathbf{P}}) + \frac{\mathring{\mathbf{E}}}{c}$$

where $c\nabla \times \mathbf{M} \equiv \mathbf{j}_M$ is the divergenceless magnetization current first introduced in (E.3). The equations are usually written in terms of the partial fields

$$\mathbf{D} = \mathbf{E} + 4\pi\mathbf{P} \qquad \mathbf{H} = \mathbf{B} - 4\pi\mathbf{M} \tag{F.1a}$$

defined in (D.6) and (E.5), and serve as part of a full panoply of Maxwell equations

$$\begin{aligned} \nabla \cdot \mathbf{D} &= 4\pi\rho' \qquad & \nabla \times \mathbf{E} &= -\frac{\mathring{\mathbf{B}}}{c} \\ \nabla \cdot \mathbf{B} &= 0 \qquad & \nabla \times \mathbf{H} &= \frac{4\pi\mathbf{j}'}{c} + \frac{\mathring{\mathbf{D}}}{c} \end{aligned} \tag{F.1b}$$

The interest here is not in the effect of the medium on the fields arising from given sources ρ', $\mathbf{j}'$ but its effect on propagations away from such sources, where $\rho' \equiv 0$ and $\mathbf{j}' \equiv 0$, to be assumed henceforth.

The effect of a medium on the transmission of electromagnetic waves is simplest to formulate over expanses of medium that can be represented by a uniform dielectric constant $\epsilon = \mathbf{D}/\mathbf{E}$ and a uniform permeability $\mu = \mathbf{B}/\mathbf{H}$. Then Eqs. (F.1*b*), as written for $\mathbf{E}$ and $\mathbf{B}$, come to differ from those in free space (7.1) only in that $\mathring{\mathbf{E}}/c$ in the last equation is multiplied by the product $\mu\epsilon$. The consequence for the wave equation (7.2) is to change it into

$$\left(\nabla^2 - \frac{\mu\epsilon}{c^2}\frac{\partial^2}{\partial t^2}\right)\begin{Bmatrix}\mathbf{E}\\ \mathbf{B}\end{Bmatrix} = 0 \tag{F.2}$$

Thus the primary effect on the electromagnetic waves is to change their propagation velocity from c to

$$c_1 = \frac{c}{\sqrt{\mu\epsilon}} \tag{F.3}$$

a velocity characteristic of the medium.

The light velocity c_1 in any medium is always less than it is in free space because the dielectric constant ϵ is always greater than unity by a much wider margin than μ is ever less than unity (as when arising from diamagnetic effects, see page 545). Of course, deeper reasons lie in the atomic pictures of matter. Fundamentally, materials present charges that can be influenced by incident fields and, in turn, serve as sources of field. Thus, the passage of electromagnetic waves is expected to be attended by absorptions and reemissions of field energy. On the atomic scale, matter can be regarded as mostly empty, with light having the same velocity as in free space from atom to atom. It is the delays attending the absorptions and coherent reemissions of the light energy by the many atoms that are supposed to be responsible for the lower effective velocities c_1 found in direct measurements on the resultant propagations.

F.2 PLANE WAVES IN DIELECTRICS

The wave equation (F.2) has plane-wave solutions exactly like those in free space except that the relation between wave number and frequency is $k_1 = \omega/c_1$ rather than $k = \omega/c$. The wave number is convenient to write as

$$k_1 \equiv \frac{\omega}{c_1} \equiv nk = \frac{n\omega}{c} \tag{F.4}$$

in terms of the velocity ratio

$$n \equiv \frac{c}{c_1} = \sqrt{\mu\epsilon} \tag{F.5}$$

called the *index of refraction* of the medium. It is also convenient to work with such complex Fourier amplitudes $\mathbf{E}(\mathbf{r})$, $\mathbf{B}(\mathbf{r})$ as are defined by (6.63). Then a plane monochromatic wave field in the medium can be described by

$$\mathbf{E}(\mathbf{r}) = \mathbf{a}e^{in\mathbf{k}\cdot\mathbf{r}} \qquad \mathbf{B}(\mathbf{r}) = n\hat{\mathbf{k}} \times \mathbf{E}(\mathbf{r}) \tag{F.6}$$

in which $\mathbf{a}$ must be chosen transverse to $\mathbf{k}$ (that is, $\mathbf{k} \cdot \mathbf{a} = 0$) for the field to be properly divergenceless, as required by the Maxwell equation for $\mathbf{D} = \epsilon\mathbf{E}$ in (F.1*b*), over spans of uniform medium without free charges: $\rho' \equiv 0$. The result for $\mathbf{B}(\mathbf{r})$ comes from the Maxwell equation for $\nabla \times \mathbf{E}(\mathbf{r})$ appropriate to the Fourier (frequency-eigenfunction) component, as in (6.57). The field $\mathbf{B}(\mathbf{r})$ is properly divergenceless and is orthogonal not only to $\mathbf{k}$ but also to $\mathbf{E}$, as in free space. However, its magnitude is not equal to $|\mathbf{E}|$ but $|\mathbf{B}| = n|\mathbf{E}|$, instead. These results satisfy the final Maxwell equation, $\nabla \times \mathbf{B}(\mathbf{r}, t) = \mu\mathring{\mathbf{D}}/c$ in the absence of free currents, because of the preliminary satisfaction of the wave equation. It should also be obvious that the plane wave in the dielectric, being transverse, can have polarizations exactly as in free space.

There will be occasion to evaluate the energies of wave fields in the medium. It was found in the final pages of Chap. D that the free electric-energy density available in a dielectric, after the automatic polarizations of the medium have been taken into account, is given by $w(\mathbf{E}) = \mathbf{E} \cdot \mathbf{D}/8\pi$, at least in the limit $\omega \to 0$, the static situation. This can be confirmed for nonstatic circumstances, together with a corresponding conclusion, $w(\mathbf{B}) = \mathbf{B} \cdot \mathbf{H}/8\pi$, about the magnetic energy (see Exercise E.10), by showing that the integral of these densities over an entire field is indeed a quantity that is conserved to itself. This will be done here only for fields lying entirely within a medium of uniform dielectric constant and permeability, when the Maxwell equations can be written exactly as in (7.1), for free space, except that $\nabla \times \mathbf{B} = \mu\epsilon\mathring{\mathbf{E}}/c$. Consider therefore

$$\frac{d}{dt}\oint \frac{dV(\mathbf{r})}{8\pi}\left(\epsilon \mathsf{E}^2 + \frac{\mathsf{B}^2}{\mu}\right) = \oint \frac{dV}{4\pi}\left[\epsilon\mathbf{E}\cdot\frac{c\nabla\times\mathbf{B}}{\mu\epsilon} + \frac{\mathbf{B}}{\mu}\cdot(-c\nabla\times\mathbf{E})\right]$$

The mathematical result (6.1) shows that this is just the volume integral of a divergence

$$\frac{c}{4\pi\mu}\nabla\cdot(\mathbf{E}\times\mathbf{B}) = \nabla\cdot\frac{c}{4\pi}(\mathbf{E}\times\mathbf{H})$$

Extended over the entire field, this becomes equivalent to a surface integral that vanishes, and so

$$W = \oint dV\frac{\mathbf{E}\cdot\mathbf{D} + \mathbf{B}\cdot\mathbf{H}}{8\pi} = \oint\frac{dV}{8\pi}(\epsilon\mathsf{E}^2 + \mu\mathsf{H}^2) \tag{F.7}$$

is indeed a conserved energy. It has also been learned that

$$\mathbf{N} = \frac{c}{4\pi\mu}\mathbf{E}\times\mathbf{B} = \frac{c}{4\pi}\mathbf{E}\times\mathbf{H} \tag{F.8}$$

is the suitable modification of the Poynting vector (6.6) for describing the energy flux in the medium.

For the plane-wave field (F.6), the time-averaged energy density is given by

$$\langle w\rangle = \frac{1}{2}\frac{\epsilon\mathbf{E}^*\cdot\mathbf{E} + \mu^{-1}\mathbf{B}^*\cdot\mathbf{B}}{8\pi} = \frac{\epsilon\mathbf{E}^*\cdot\mathbf{E}}{8\pi} = \frac{\mathbf{B}^*\cdot\mathbf{B}}{8\pi\mu} = \frac{\epsilon|a|^2}{8\pi} \tag{F.9}$$

as the suitable modification of (6.65) and (7.15). The corresponding modification of the energy-flux density (6.66) is

$$\begin{aligned}\langle\mathbf{N}\rangle &= \tfrac{1}{2}\,\mathrm{Re}\left(\frac{c}{4\pi\mu}\mathbf{E}^*\times\mathbf{B}\right) = \hat{\mathbf{k}}\frac{cn}{8\pi\mu}\mathbf{E}^*\cdot\mathbf{E}\\ &= \frac{\hat{\mathbf{k}}\langle w\rangle cn}{\mu\epsilon} = \hat{\mathbf{k}}\langle w\rangle c_1\end{aligned} \tag{F.10}$$

as to be expected of an energy density propagated with the velocity c_1 characteristic of the medium.

F.3 REFLECTION AND REFRACTION ANGLES

The effects on electromagnetic waves of variations in the medium can be studied by considering two contiguous half-infinite media separated by a plane interface $z = 0$, as indicated in Fig. F.1. Each of the two media by itself will be taken to be uniform, with constants ϵ_1, μ_1 characterizing the space $z > 0$ and ϵ_2, μ_2 the constants in $z < 0$. Since the variations are confined to a discontinuity at the surface $z = 0$, their effect will be represented by boundary conditions resulting

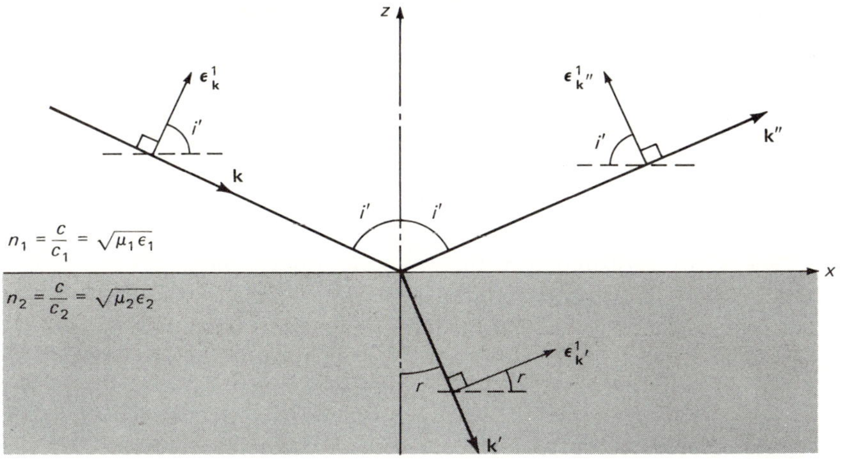

Figure F.1

from the reduction of the Maxwell equations (F.1) to surface divergences and surface curls, as in the static cases (D.16) and (E.9). No free charges or currents are to be introduced, so that not only **B** but also **D** is divergenceless and the components normal to the interface must be continuous:

$$\Delta \mathsf{D}_z = 0 \qquad \text{and} \qquad \Delta \mathsf{B}_z = 0 \tag{F.11a}$$

These entail *dis*continuities in E_z and H_z if $\epsilon_1 \neq \epsilon_2$ and $\mu_1 \neq \mu_2$. The curl equations require continuity of the x and y components (parallel to the interface) of **E** and **H**:

$$\Delta \mathsf{E}_{x,\,y} = 0 \qquad \text{and} \qquad \Delta \mathsf{H}_{x,\,y} = 0 \tag{F.11b}$$

It is true that the nonstatic **E** and **H** are not curlless in space, so that the surface curls (F.11*b*) should respectively be equated to fluxes of $\mathring{\mathsf{B}}_{y,\,x}$ and $\mathring{\mathsf{D}}_{y,\,x}$ along the surface. However, a reconsideration of the surface curls as discussed in connection with Fig. 7.3 shows that the fluxes along the surface vanish in the limit of an ideally vanishing surface thickness across which the discontinuities take place.

The effects on electromagnetic wave fields can be investigated adequately by examining the results of sending in a given incident plane wave from $z = +\infty$, as represented by

$$\mathsf{E}_i(z > 0) = \mathbf{a}_i e^{i n_1 \mathbf{k} \cdot \mathbf{r}} \qquad \mathsf{B}_i(z > 0) = n_1 \hat{\mathbf{k}} \times \mathsf{E}_i \tag{F.12}$$

A wave vector with $k_z < 0$ will have to be prescribed if this is to be directed from $z = +\infty$. Besides **k**, and hence the angle of incidence marked i' in Fig. F.1, the transverse amplitude vector $\mathbf{a}_i$ is prescribed, and all the results should finally be expressed in terms of these given quantities.

The continuation into the second ($z < 0$) medium of a solution having (F.12) as its only incident component must be represented by a field of the *same frequency* ω, since components with different frequencies are linearly independent at all times and will remain unconnected by the linear relationships (F.11). Thus, a transmitted plane wave

$$\mathbf{E}_t(z < 0) = \mathbf{a}_t e^{in_2\mathbf{k}' \cdot \mathbf{r}} \qquad \mathbf{B}_t(z < 0) = n_2 \hat{\mathbf{k}}' \times \mathbf{E}_t \tag{F.13}$$

with $|\mathbf{k}'| = \omega/c$ will be tried as the continuation.

A trial of the incident wave (F.12) as the *entire* solution in the first $(z > 0)$ medium, with any such transmitted wave as (F.13) as its continuation to $z < 0$, immediately results† in a mismatch in the boundary conditions (F.11). One way to satisfy them is to introduce a second, linearly independent wave, with $k_z > 0$ if $k_z < 0$ for the first one, into the $z < 0$ medium, but this corresponds to sending in a second incident wave, this one from $z = -\infty$, instead of confining attention to results of just the incidence (F.12). For the latter, it is necessary to admit the possibility of an additional, reflected wave, with $|\mathbf{k}''| = \omega/c$ and $k_z'' > 0$, into the *first* $(z > 0)$ medium

$$\mathbf{E}_r(z > 0) = \mathbf{a}_r e^{in_1\mathbf{k}'' \cdot \mathbf{r}} \qquad \mathbf{B}_r(z > 0) = n_1 \hat{\mathbf{k}}'' \times \mathbf{E}_r \tag{F.14}$$

helping form the complete field

$$\mathbf{E}(z > 0) = \mathbf{E}_i + \mathbf{E}_r \qquad \mathbf{B}(z > 0) = \mathbf{B}_i + \mathbf{B}_r$$

$$\mathbf{E}(z < 0) = \mathbf{E}_t \qquad \mathbf{B}(z < 0) = \mathbf{B}_t$$

The necessity of having a reflected wave because of the above mismatch will be demonstrated by an outcome that generally $a_r \neq 0$ whenever $c_1 \neq c_2$.

Each one of the boundary conditions (F.11) will now require a matching like

$$(\text{ampl})_i\, e^{in_1(k_x x + k_y y) - i\omega t} + (\text{ampl})_r\, e^{in_1(k_x'' x + k_y'' y) - i\omega t} = (\text{ampl})_t\, e^{in_2(k_x' x + k_y' y) - i\omega t} \tag{F.15}$$

at $z = 0$, to hold for all x, y, t. The time dependence cancels out properly because of the equality of the frequencies; this confirms the independence of different frequencies in the satisfaction of the linear relations. The x and y dependences are canceled out similarly if

$$n_1 k_{x,y} = n_1 k_{x,y}'' = n_2 k_{x,y}' \tag{F.16}$$

Now $|\mathbf{k}| = |\mathbf{k}'| = |\mathbf{k}''| = \omega/c$ by the definitions of $n_{1,2}$, and so a consequence of $k_{x,y}'' = k_{x,y}$ is $(k_z'')^2 = k_z^2$. For the reflected wave to be other than just a supplementation of the incident-wave intensity, it is necessary that

$$k_z'' = -k_z \tag{F.17}$$

This amounts to the *law of reflection*, more usually expressed: *the angle of reflection equals the angle of incidence* (i' of Fig. F.1).

† Such a mismatch is explicitly demonstrated in Ref. 17, sec. 14.1, for the simpler case of *scalar* elastic waves. Light also was treated as consisting of scalar disturbances in the earliest investigations, following Young's discovery of its interference effects. The discovery of its polarizability, and hence the *vector* character of the waves, came later.

There will be no real loss of generality if, for presenting the remaining consequence of (F.16), an x axis that lies in the *plane of incidence*, i.e., the plane defined by **k** and the direction normal to the interface, is chosen. Then $k_y = k'_y = k''_y = 0$, and

$$\frac{k'_x/k}{k_x/k} = \frac{\sin r}{\sin i'} = \frac{n_1}{n_2} = \frac{c_2}{c_1} \tag{F.18}$$

in terms of the angles indicated in Fig. F.1. This can be recognized as the elementary Snell's law of refraction.

Notice that the equal frequencies in the two media correspond to unequal wavelengths

$$\lambda_2 = \frac{2\pi c_2}{\omega} = \frac{c_2}{c_1}\lambda_1 = \frac{n_1}{n_2}\lambda_1$$

The wavelength is the longer in the medium with the larger propagation velocity because a given phase in the disturbance can travel farther in it over any one period $2\pi/\omega$.

F.4 THE FRESNEL FORMULAS

The six component boundary conditions (F.11) should be just sufficient to determine the six components of the reflected and transmitted amplitude vectors $\mathbf{a}_r$ and $\mathbf{a}_t$ of (F.14) and (F.13) in terms of the incident amplitude $\mathbf{a}_i$ of (F.12). If forms are used that *presume* the necessary transversalities, $\mathbf{k}'' \cdot \mathbf{a}_r = 0$ and $\mathbf{k}' \cdot \mathbf{a}_t = 0$, two of the conditions will turn out to be satisfied automatically, since they include conditions that impose divergencelessnesses.

If the decompositions are made on the cartesian frame indicated in Fig. F.1, it will be found that the y components a_{ry}, a_{ty}, a_{iy} are related by equations entirely independent of the equations determining the x and z components. This is to be expected from the fact that the incident-wave component polarized normally to the plane of incidence xz is linearly independent of the part polarized parallel to that plane, after an appropriate resolution of the type (7.24). A significant conclusion is that the polarizations resolved in this particular way are not mixed by the reflection and transmission processes and that each can be treated separately.

Consider first the components polarized normally to the plane of incidence, as represented by the y components of the electric vectors. About these, the condition $\Delta \mathsf{E}_y = 0$ gives

$$a_{iy} + a_{ry} = a_{ty}$$

The corresponding magnetic vectors, like the incident one proportional to $n_1 \mathbf{k} \times \mathbf{a}_i$, have only x and z components. Since $k_y = k'_y = k''_y = 0$, the condition $\Delta \mathsf{B}_z = 0$ gives

$$n_1(k_x a_{iy} + k''_x a_{ry}) = n_2 k'_x a_{ty}$$

which merely repeats the above result in view of (F.16), and $\Delta \mathsf{H}_x = \Delta(\mu^{-1}\mathsf{B}_x) = 0$ gives

$$\mu_1^{-1} n_1 (k_z a_{iy} + k_z'' a_{ry}) = \mu_2^{-1} n_2 k_z' a_{ty}$$

Since $k_z = -k_z'' = -k \cos i'$ and $k_z' = -k \cos r$, this is equivalent to

$$(a_{iy} - a_{ry}) \cos i' = \mu^{-1} n (\cos r) a_{ty}$$

where $\mu \equiv \mu_2/\mu_1$ is a relative permeability and

$$n \equiv \frac{n_2}{n_1} = \frac{c_1}{c_2} \tag{F.19}$$

is a relative index of refraction of the two media. The conditions applied so far, together with the law of refraction (F.18), are sufficient to determine

$$\begin{aligned} \left(\frac{a_r}{a_i}\right)_\perp &= \frac{\mu \cos i' - (n^2 - \sin^2 i')^{1/2}}{\mu \cos i' + (n^2 - \sin^2 i')^{1/2}} \\ \left(\frac{a_t}{a_i}\right)_\perp &= \frac{2\mu \cos i'}{\mu \cos i' + (n^2 - \sin^2 i')^{1/2}} \end{aligned} \tag{F.20}$$

for the components polarized perpendicularly to the plane of incidence, as indicated by the symbol $\perp$ (in place of y).

The subscript $\|$ will indicate component amplitudes parallel to the plane of incidence, each transverse to the corresponding propagation direction. Then waves polarized in the plane of incidence ($\sim \boldsymbol{\epsilon}^1$ in Fig. F.1) have electric amplitudes with the cartesian components

$$\begin{aligned} a_{ix} &= (a_i)_\| \cos i' & a_{iz} &= (a_i)_\| \sin i' \\ a_{rx} &= -(a_r)_\| \cos i' & a_{rz} &= (a_r)_\| \sin i' \\ a_{tx} &= (a_t)_\| \cos r & a_{tz} &= (a_t)_\| \sin r \end{aligned}$$

About these, the condition $\Delta \mathsf{E}_x = 0$ yields

$$(a_i - a_r)_\| \cos i' = (a_t)_\| \cos r$$

and it follows from $\Delta \mathsf{D}_z = 0$ that

$$\epsilon_1 (a_i + a_r)_\| \sin i' = \epsilon_2 (a_t)_\| \sin r$$

which at once reduces to

$$(a_i + a_r)_\| = \mu^{-1} n (a_t)_\|$$

when $\epsilon_2/\epsilon_1 = (\mu_1/\mu_2)(n_2/n_1)^2 = \mu^{-1} n^2$ is substituted and the law of refraction is invoked. The one remaining condition, $\Delta \mathsf{H}_y = \Delta(\mu^{-1}\mathsf{B}_y) = 0$, merely reiterates the last finding. The consequent ratios are

$$\begin{aligned} \left(\frac{a_r}{a_i}\right)_\| &= \frac{n^2 \cos i' - \mu (n^2 - \sin^2 i')^{1/2}}{n^2 \cos i' + \mu (n^2 - \sin^2 i')^{1/2}} \\ \left(\frac{a_t}{a_i}\right)_\| &= \frac{2n\mu \cos i'}{n^2 \cos i' + \mu (n^2 - \sin^2 i')^{1/2}} \end{aligned} \tag{F.21}$$

for waves polarized parallel to the plane of incidence. These, together with (F.20), are known as the *Fresnel formulas.*

In the special case of normal incidence ($i' = 0$), the two sets of results (F.20) and (F.21) ought to reduce to the same one, since there is then no plane of incidence defined relative to which the two types of polarization could be distinguished. It is easy to see that both sets do yield

$$\left|\frac{a_r}{a_i}\right| = \left|\frac{n-\mu}{n+\mu}\right| \qquad \frac{a_t}{a_i} = \frac{2\mu}{n+\mu} \qquad \text{for } i' = 0 \tag{F.22}$$

There is actually a trivial difference of sign in the formulas for a_r/a_i; it stems from the conventions adopted in the above formulations [$(a_r/a_i)_\perp$ was taken to be positive when the electric vectors are parallel, whereas $(a_r/a_i)_\parallel$ was so defined that it is positive when the electric vectors are *anti*parallel at normal incidence ($\epsilon^1_{\mathbf{k}''} \to -\epsilon^1_{\mathbf{k}}$ as $i' \to 0$ in Fig. F.1)].

For the remainder of this chapter, $\mu = 1$ will be assumed, since the deviations from unity of the permeabilities in almost all dielectrics are too small to warrant more space for their discussion.

F.5 POLARIZATION BY REFLECTION

The above results confirm that a mismatch $c_2 \neq c_1 \equiv nc_2$ of the velocities in the two media generally leads to reflections $\mathbf{a}_r \neq 0$. The formula in (F.20), for the reflected amplitude of light polarized normally to the plane of incidence, vanishes *only* for $n = 1$ (with $\mu = 1$), when there is no change of medium after all. The formulas in (F.21), for light polarized in the plane of incidence, also reduce to $a_r = 0$, $a_t = a_i$ for $n = 1$, as they should, but in addition yield a vanishing reflected amplitude for $n \neq 1$ and a particular angle of incidence that depends on n. This special angle is readily seen to be representable (for $\mu = 1$) as

$$i_B(n) = \tan^{-1} n \tag{F.23}$$

and is called *Brewster's angle.*

The occurrence of the transmission without reflection at the Brewster angle is associated with the fact that the reflection direction would have to make a right angle with the direction of the transmission. The law of refraction leads to

$$\sin r_B = \frac{\sin i_B}{n} = \frac{1}{\sqrt{1+n^2}} = \cos i_B$$

so that $r_B + i_B = \frac{1}{2}\pi$. As a consequence, the transverse electrical oscillations in the transmitted wave, when confined to the plane of incidence as indicated in Fig. F.2, have no projections on the transverse to the reflection direction. Such projections, corresponding to a visibility of the oscillations from the possible reflection direction, will later be found to be essential for radiation into that direction to be generated (by the electrical oscillations that can be expected in the reflecting medium). Thus the nonoccurrence of reflected radiation at the

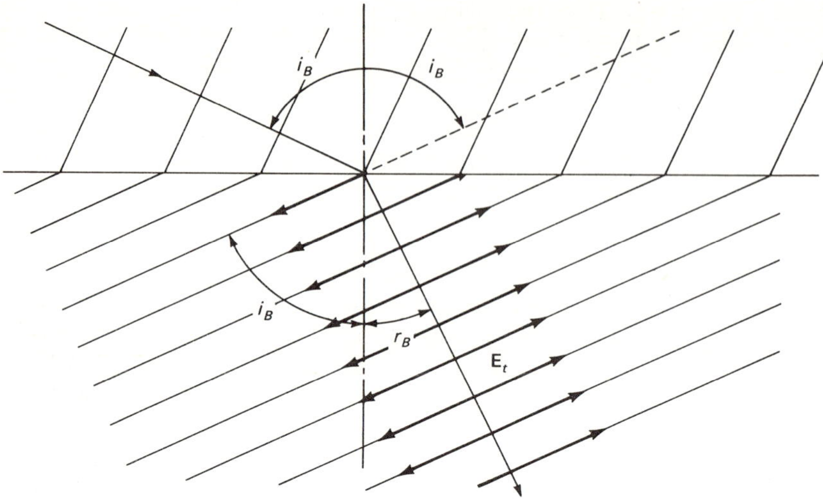

Figure F.2

Brewster angle can be understood on the basis that the media have effect through reemissions following upon absorptions of, and excitations by, incident electromagnetic waves, a type of picture already found, on page 559, to make the slowing down of propagations by a medium more plausible.

It is only light polarized in the plane of incidence that disappears from the reflections at Brewster's angle. The net effect on initially unpolarized light is a *polarization by reflection*, an emergence of light polarized normally to the plane of incidence and hence parallel to the reflecting surface. The suppression of the polarizations in the plane of incidence is actually complete only at the discrete Brewster angle itself,† but the variation of $(a_r/a_i)_{\parallel}$ with incidence angles around $i \approx i_B$ is quite moderate, and so there is a substantial range of angles for which $(a_r/a_i)_{\parallel} \ll (a_r/a_i)_{\perp}$. This accounts for the efficacy of Polaroid eyeglasses in eliminating glare from horizontal reflecting surfaces; the polarized film on the glasses has its optic axis (see the discussion on page 191) so oriented that it blocks most of the reflected light, which has the horizontal polarization (as long as the wearer stays erect!).

It may be noticed that as $(a_r/a_i)_{\parallel} \to 0$ at the Brewster angle, $(a_t/a_i)_{\parallel} \to 1/n \equiv c_2/c_1$ rather than unity. Such a ratio is exactly what is needed for energy conservation when $c_2 \neq c_1$, as will be seen in the following considerations.

F.6 REFLECTED AND TRANSMITTED INTENSITIES

The intensity of each wave is measured by the flux of energy it carries, as evaluated in (F.10). The component of each energy flux *parallel* to the interface is simply continued independently and uniformly (from $x = -\infty$ to $x = +\infty$ in

† For reflections of light in air from water, $n \approx \frac{4}{3}$ and $i_B = \tan^{-1} \frac{4}{3} \approx 53°$.

the idealized formulation) because there is no change of medium along its course. However, the *normal* (z) components are interrupted by a change of medium. The resultant normal flux is carried by incident and reflected waves in the $z > 0$ region and by a transmitted wave in the $z < 0$ space. The shares carried by each part must then be expected to be related by energy conservation.

The incident energy flux per unit area of interface is

$$\langle N_{iz} \rangle = \hat{k}_z \langle w_i \rangle c_1 = (-\cos i') \frac{\epsilon_1 |a_i|^2}{8\pi} c_1 \tag{F.24}$$

according to (F.10) and (F.9). The reflected† flux has the same expression except in sign and in the replacement of $|a_i|^2$ by $|a_r|^2$. The ratio of the reflected to incident intensities

$$R \equiv \frac{|a_r|^2}{|a_i|^2} \tag{F.25}$$

called the *reflection coefficient*, is given by

$$R_\perp = \left(\frac{\cos i' - n \cos r}{\cos i' + n \cos r} \right)^2 \qquad R_\parallel = \left(\frac{n \cos i' - \cos r}{n \cos i' + \cos r} \right)^2 \tag{F.26}$$

where

$$n \cos r = +(n^2 - \sin^2 i')^{1/2} \tag{F.27}$$

for the independent components (F.20) and (F.21), respectively (with $\mu = 1$). The normal component of the *transmitted* flux density is

$$\langle N_{tz} \rangle = \hat{k}_z \langle w_t \rangle c_2 = (-\cos r) \frac{\epsilon_2 |a_t|^2}{8\pi} c_2 \tag{F.28}$$

and so the *transmission coefficient* is

$$T \equiv \frac{\langle N_{tz} \rangle}{\langle N_{iz} \rangle} = \frac{\cos r}{\cos i'} \frac{\epsilon_2}{\epsilon_1} \left| \frac{a_t}{a_i} \right|^2 \frac{c_2}{c_1} \tag{F.29}$$

Since $\epsilon_2/\epsilon_1 = n^2 = (c_1/c_2)^2$, (F.20) and (F.21) respectively yield

$$T_\perp = \frac{4n \cos r \cos i'}{(\cos i' + n \cos r)^2} \qquad T_\parallel = \frac{4n \cos r \cos i'}{(n \cos i' + \cos r)^2} \tag{F.30}$$

In each of the cases $R + T = 1$, and this represents the requisite consistency with energy conservation. At the Brewster angle $i_B(n)$, for which $\cos r_B = n \cos i_B$ ($R_\parallel = 0$), it has already been asserted that $a_t/a_i = 1/n$, and it can now be seen that this does yield $T_\parallel = 1$ properly.

† There are no interference ($\sim a_i^* a_r$) terms in the resultant $z > 0$ normal flux, as the results here will bear out through their consistency with energy conservation. Formal confirmation is not difficult and is explicitly demonstrated for scalar disturbances in Ref. 17, sec. 14.1.

F.7 TOTAL INTERNAL REFLECTION

In such cases as the incidence of light in air upon the surface of water ($n_2 > n_1$ and $c_2 < c_1$), the refractive index $n = c_1/c_2$ is greater than unity, and the law of refraction (F.18) yields a real value for the refraction angle r in the entire range of incidences $0 \leq i' \leq \frac{1}{2}\pi$:

$$0 \leq r \leq \sin^{-1} n^{-1} \qquad (< \tfrac{1}{2}\pi \text{ for } n^{-1} < 1) \tag{F.31}$$

However, for the reverse transmission of light from water to air n is less than unity, and $\sin r = (\sin i')/n > 1$ for incidence angles

$$i' > i_c \equiv \sin^{-1} n \qquad (< \tfrac{1}{2}\pi \text{ for } n < 1) \tag{F.32}$$

that are closer to glancing incidence ($i' \to \frac{1}{2}\pi$) than a critical angle $i_c < \frac{1}{2}\pi$. It becomes meaningless to set $k'_x = k \sin r$, as in (F.18), for $i' > i_c$, since no real r exists in this range, and the treatment of $\mathbf{k}'$ as a real vector comes into question.

All conditions on the solution are still satisfied by the *forms* (F.13) for $\mathbf{E}$, $\mathbf{B}(z < 0)$, with

$$n_2 \mathbf{k}' \cdot \mathbf{r} \equiv n_2(k'_x x + k'_z z)$$

The matchings at $z = 0$ still require that $n_2 k'_x = n_1 k_x$, as in (F.16), so that k'_x is still a real wave number like k_x. However,

$$k'_z = -[k^2 - (k'_x)^2]^{1/2} = -\left[k^2 - \left(\frac{n_1}{n_2}\right)^2 k_x^2\right]^{1/2}$$

$$= -\frac{k}{n}(n^2 - \sin^2 i')^{1/2} = -\frac{k}{n}(\sin^2 i_c - \sin^2 i')^{1/2}$$

obviously becomes imaginary for $i' > i_c$. Therefore let

$$n_2 k'_z \equiv -i\kappa \qquad \text{with } \kappa(i') = n_1 k(\sin^2 i' - \sin^2 i_c)^{1/2} \tag{F.33}$$

a real inverse length. Now the real transmitted field can be written with the help of the forms (F.13) for the complex $\mathbf{E}_t(z < 0)$, as in

$$\mathbf{E}(\mathbf{r}, t) = \text{Re}\,(\mathbf{E}_t e^{-i\omega t}) = e^{\kappa z}\,\text{Re}\,(\mathbf{a}_t e^{i(n_1 k_x x - \omega t)})$$

$$\mathbf{B}(\mathbf{r}, t) = \text{Re}\,(n_2 \hat{\mathbf{k}}' \times \mathbf{E}_t e^{-i\omega t}) \tag{F.34}$$

The components of the cross product in the last line are to be formed as for real vectors, but then $\hat{k}'_z = -i\kappa/n_2 k$ is to be substituted. The result describes a wave propagated parallel to the interface, with a wavelength $\lambda_x = 2\pi/n_1 k_x = \lambda_1/(\sin i')$ and with an amplitude that decreases exponentially with depth $z \to -\infty$. The length κ^{-1} serves as a *mean penetration depth*, a distance in which the amplitude is reduced by a factor $1/e$.

The interesting outcome for incidences closer to grazing than a critical angle $i_c < \frac{1}{2}\pi$ can be found by using (F.33) to substitute $+i\kappa/n_1 k$ for each radical in the expressions for the reflected amplitudes of (F.20) and (F.21) with $\mu = 1$:

$$\left(\frac{a_r}{a_i}\right)_\perp = \frac{n_1 k \cos i' - i\kappa}{n_1 k \cos i' + i\kappa} \qquad \left(\frac{a_r}{a_i}\right)_\parallel = \frac{n^2 n_1 k \cos i' - i\kappa}{n^2 n_1 k \cos i' + i\kappa} \tag{F.35}$$

In each case the reflection coefficient $R = |a_r/a_i|^2$ is unity, and there is *total reflection*. This happens beyond the same critical angle for any polarization, and, of course, there is no polarization of unpolarized incident light by total reflection.

Energy conservation naturally demands that the transmission coefficients vanish when there is total reflection. Showing that the *non*vanishing expressions for a_t/a_i in (F.20) and (F.21) are consistent with this requires a reconsideration of the final form given the transmission coefficient in (F.29) since that becomes imaginary when $\cos r \equiv -\hat{k}'_z \rightarrow i|\kappa|/n_2 k$. For a $\mathbf{k}'$ with an imaginary component, a reversion to the starting formula in (F.10) is called for, and the form $\mathbf{B}_t = n_2 \hat{\mathbf{k}}' \times \mathbf{E}_t(z < 0)$ in (F.34) leads to

$$\langle \mathbf{N}_t \rangle = \frac{cn_2}{8\pi\mu} \operatorname{Re} [\hat{\mathbf{k}}'(\mathbf{E}_t^* \cdot \mathbf{E}_t) - \mathbf{E}_t(\hat{\mathbf{k}}' \cdot \mathbf{E}_t^*)] \tag{F.36}$$

where $\hat{\mathbf{k}}' = \mathbf{k}'/k$. The consequence of divergencelessness, $\mathbf{k}' \cdot \mathbf{E}_t = 0$, led to the vanishing of the last term for the evaluation in (F.29) because $\mathbf{k}'$ was there taken to be real. Now, in the frame of Fig. F.1, k'_x and $k'_y = 0$ remain real, but $k'_z = -i\kappa/n_2$ has become purely imaginary. Thus, a present conclusion from the divergencelessness is that

$$(\mathbf{k}' \cdot \mathbf{E}_t)^* = k'_x \mathsf{E}_{tx}^* - k'_z \mathsf{E}_{tz}^* = 0$$

This permits eliminating $k'_x \mathsf{E}_{tx}^*$ from

$$\mathbf{k}' \cdot \mathbf{E}_t^* = k'_x \mathsf{E}_{tx}^* + k'_z \mathsf{E}_{tz}^* = 2k'_z \mathsf{E}_{tz}^*$$

Consequently, the normal component of the flux density (F.36) becomes

$$\langle N_{tz} \rangle = \frac{cn_2}{8\pi\mu k} [\mathbf{E}_t^* \cdot \mathbf{E}_t - 2\mathsf{E}_{tz} \mathsf{E}_{tz}^*] \operatorname{Re}(k'_z)$$

since the terms in the square bracket are real. The expression then vanishes because k'_z has no real part. The transmission coefficient $T = \langle N_{tz} \rangle / \langle N_{iz} \rangle$ does indeed properly vanish when there is total reflection, despite the penetration of the fields into the reflector.

The penetration consists of the shallow propagation parallel to the interface to which attention was called in the discussion of the transmitted fields (F.34). It accounts for a *non*vanishing of the corresponding component $\langle N_{tx} \rangle$ of the flux density (F.36). This is an energy flux, uninterrupted from $x = -\infty$ to $x = \infty$ like parallel fluxes on the opposite side of the interface, that must be established at $x = -\infty$ before the steady and infinitely extended situation here being described by the ideally monochromatic formulation can be approximated. As discussed in connection with the complementarity (7.23), an ideally monochromatic formulation presumes an indefinitely long time of existence, a steady situation in which the incident wave and its products are spread all over their spaces. [Those can nevertheless be components of narrow rays, as mentioned in connection with (7.22).]

F.8 DISPERSION IN TRANSPARENT DIELECTRICS

So far, the dielectric constant ϵ and the refractive index $n = (\mu\epsilon)^{1/2}$, of (F.5), have been treated as numbers with some *constant* value characteristic of a material in a given situation. It was presumed that each value is to be obtained from appropriate measurements. Much progress in theoretical evaluations has also been made, on the basis of the atomic constitutions of the materials, but mostly by accounting for the *observed* values, in some approximation. The results of direct measurements are still far more reliable and are extensively recorded in many handbooks.† Consulting them reveals that the ϵ and n of a substance generally depend on its state, its temperature, its density [as already suggested by the analysis $\alpha \sim N$ discussed in connection with the Clausius-Mossoti equation (D.12*b*)], and also on the *frequency* ω of the field that is interacting with the material.

How variable ϵ and n may be is well illustrated by the naturally important case of water under atmospheric pressure, even while it remains in its liquid state and so is confined to negligible variations of density. Electro*static* ($\omega \to 0$) measurements yield ϵ's ranging from 88 down to 56 between the freezing and boiling points, and $\epsilon \approx 81$ at room temperature. Since $\mu \approx 1$ for water, the $\epsilon \approx 81$ corresponds to $n \approx 9$, but this large refractive index is found to apply only to frequencies lower than about $\omega \approx 10^9\ \text{s}^{-1}$. As the frequency is raised beyond this value, the index decreases, settling around and close to the much smaller value $n \approx \frac{4}{3}$ ($\epsilon \approx 1.8$), quoted on page 566*n*., over a range of high frequencies that includes the visible ones: $\omega \approx 2.7 \times 10^{15}$ to $4.7 \times 10^{15}\ \text{s}^{-1}$.

The unusually large static dielectric constant of water is due to the fact that the anisotropic water molecules have substantial permanent electric dipole moments, which need only be aligned by an imposed field to produce contributions to a bulk polarization much larger than those from freshly induced charge displacements; the alignments must be maintained against thermal jostlings and so a decrease of ϵ with temperature, as it is raised from 0 to 100°C, is to be expected. The permanent moments involve the whole, heavy molecules; realignments of them cannot keep in phase with high-frequency reversals of the field and so tend to be randomized. Thus, at the highest frequencies, only the induced displacements of the light individual electrons are left to yield the much smaller $\epsilon \approx 1.8$, a value more nearly that of the static $\epsilon \approx 2$ to 6 characteristic of most condensed materials having no appreciable permanent moments.

It is also pertinent to mention that water, like all condensed materials, absorbs some of the radiations passing through it, being warmed in the process. Absorption is greatest for incident frequencies near resonance with natural oscillations of the material charges around their equilibrium configurations. The oscillations are coupled to lower-frequency motions, including thermal ones, and

† A conveniently brief and usually sufficient one is Ref. 25.

so the incident energy is passed on, at least partially as an irreversible process that constitutes absorption. It happens that water has wide absorption bands with frequencies just below those of visible light and also just above them, in the far ultraviolet. The lower-frequency absorptions reduce the mean penetration depth of conventional radar to as little as 10^{-3} cm; that is why sonar is used instead for the underwater detection of schools of fish and submarines. Water is much more transparent, allowing mean penetrations of 10 m or so, to the narrow band of frequencies between the infrared and the ultraviolet. It has been speculated that this is why natural selection found it advantageous to develop retinas sensitive to what has thus become the visible band, in our remote, water-dwelling progenitors.

Such greater detailing of dielectric properties indicates that the considerations in the preceding sections remain valid only for the numerous processes in which the condition of the materials stays more or less constant and when they can be classed as transparent (absorption negligible throughout each process). Since the treatments dealt only with monochromatic radiations, variations with frequency were irrelevant. However, it does become important to formulate the refractive index as a function of frequency when superposing fields of various frequencies, as undertaken below.

Refraction as a Function of Frequency

Variations of polarization with frequency are most important to formulate for situations in which the principal contributions come from inducing displacements of electrons (mass m, charge $-e$), the most numerous constituents of any material (except pure hydrogen) and more readily displaced than the much heavier and relatively immovable ions. Whenever an electron suffers some displacement† $\mathbf{r}_i$ from an equilibrium configuration, the forces responsible for the normal equilibrium become unbalanced and so a net restoring force arises, proportional to $-\mathbf{r}_i$ in a first, Hooke's-law, approximation.‡ This is representable as in the simple oscillator equation of motion $m\ddot{\mathbf{r}}_i = -m\omega_i^2\mathbf{r}_i$, which expresses the force strength in terms of the frequency§ ω_i of the natural oscillations that will ensue unless some additional force intervenes. When a *static* field $\mathbf{E}_o$ is applied (Exercise D.18), a new equilibrium ($\ddot{\mathbf{r}}_i = 0$) is established, at a static displacement $\mathbf{r}_i$ such that $-e\mathbf{E}_o$ balances the restoring force. The consequence is

† Spoken of as conduction within an atomic domain in the undetailed discussion, leading to $\alpha = Na^3$, at the beginning of Chap. D.

‡ Valid for small enough displacements. Beyond this elastic limit, there arise nonlinear effects that necessitate replacing (D.9) with further parametrizations, as in $P_k = \alpha_k \mathsf{E}_k + \sum_l \alpha_{kl} \mathsf{E}_k \mathsf{E}_l + \cdots$. They can become important in extremely intense fields, approaching those of material-destructive laser beams, and are the concern of nonlinear optics.

§ Coincidences of such with orbital frequencies of the electrons in atoms is made evident in connection with the Hooke's-law approximation of (14.104).

an induced dipole moment $-e\mathbf{r}_i = (e^2/m\omega_i^2)\mathbf{E}_o$ which contributes to a resultant moment density

$$\mathbf{P} = \sum_i N_i(-e\mathbf{r}_i) = \sum_i \frac{N_i e^2}{m\omega_i^2}\mathbf{E}_o = \alpha_o \mathbf{E}_o \tag{F.37}$$

if there are N_i electrons with the natural frequency ω_i per unit volume of the dielectric. The $\mathbf{E}_o$ here is clearly *not* the averaged field $\mathbf{E}$ of (D.11) but the local field that penetrates to electrons in a cavity of dielectric, as in (D.9), and hence the coefficient of $\mathbf{E}_o$ in (F.37) is just a static polarizability α_o, now analyzed in terms of natural frequencies characterizing the substance. As to be expected, the largest contributions come from the lowest natural frequencies, since they correspond to the most weakly bound, and hence most easily displaceable, electrons.

The interest here is in results of applying a field like $\mathbf{E}_o \cos \omega t$ of some frequency ω. As this persists, it will force steady oscillations, $\mathbf{r}_i \sim \cos \omega t$, of its own frequency. Now the $\ddot{\mathbf{r}}_i = 0$ equilibrium of the static case above is replaced by

$$\ddot{\mathbf{r}}_i = -\omega^2\mathbf{r}_i = -\omega_i^2\mathbf{r}_i - \frac{e\mathbf{E}_o}{m}\cos \omega t \tag{F.38}$$

which in effect replaces the ω_i^2 in the static expressions with $\omega_i^2 - \omega^2$. The polarizability thus becomes

$$\alpha(\omega) = \frac{\mathbf{P}}{\mathbf{E}_o \cos \omega t} = \sum_i \frac{N_i e^2}{m(\omega_i^2 - \omega^2)} \tag{F.39}$$

a function of the applied frequency ω, which rises with increasing ω as this approaches resonances with the natural frequencies ω_i. It can fall off again as ω is increased just beyond any sufficiently resolved resonance peak.

The falloff may even allow $\alpha(\omega)$ to reach negative values if the next higher frequency resonance is sufficiently removed, but negative values are found to be accompanied by a substantial absorption. As discussed in connection with Fig. F.2, the propagation of a given frequency can be continued without energy loss only if the energy imparted to the electron oscillations is immediately reradiated (as in an *elastic* scattering process). When the oscillations are in a phase opposite to that of the propagation, as is described by $\alpha(\omega) < 0$, they diminish rather than sustain the propagation.

Absorption cannot be avoided near the resonances (compare the discussion of water above). The wide oscillations generated near resonance with any one electron's degree of freedom $\mathbf{r}_i$ quite understandably tend to excite neighboring electrons, a dispersal of energy that starts at least a partially irreversible passage out of the incident field. The energy losses to this resonance absorption tend to lower and broaden the resonance peaks in $\alpha(\omega)$, much as for the electromagnetic cavity resonances investigated in connection with Fig. 7.6. Such effects were not taken into account for the result (F.39), and that is responsible for the infinities it yields at each resonance. The more proper formulation, yielding finite and broadened peaks, is usually carried out by introducing still another phenomenological parameter, to help represent the magnitudes of the absorptions. That is left to Chap. G, which deals with the general effects of absorption on electromagnetic waves. This chapter continues to be concerned with transparency, with the propagation of frequencies far from regions of any substantial absorption by the materials transmitting them.

The frequency dependence of the dielectric constant ϵ and of the refractive index $n \sim \epsilon^{1/2}$ is obtainable from the polarizability $\alpha(\omega)$ through the Clausius-Mossoti equation (D.12*b*). This requires evaluating the susceptibility in $\epsilon = 1 + 4\pi\chi_e$ [Eq. (D.12*a*)] as $\chi_e = \alpha/(1 - 4\pi\alpha/3)$, but for the purposes here it is a frequent practice to use $\chi_e \equiv \alpha(\omega)$ instead. That can be justified on the grounds that the division by $1 - 4\pi\alpha/3$ is meant to correct for the difference between the local field $\mathbf{E}_o = \mathbf{E} + 4\pi\mathbf{P}/3$ and the average field $\mathbf{E}$, yet each of the electrons of different natural frequency ω_i should actually be expected subject to a somewhat different local field. It then seems preferable to redefine the restoring force $-m\omega_i^2 r_i$ to include perturbation by the difference between $\mathbf{E}$ and each local field. This will make the $\chi_e = \alpha_{\text{eff}}$ so redefined somewhat field-dependent but only negligibly so except near resonance, in view of how much stronger the atomic binding forces, which determine natural frequencies, are than the applied fields in the range of strengths to which parametrization by just an α is applicable (see page 571*n*.). It should be recognized that the values of ω_i must generally be found from observations in fields comparable to ones in which the results are used, and additional parametrizations (for the absorptions) must be determined by observation for the fields near resonance. An expression

$$\epsilon = 1 + 4\pi\chi_e = 1 + \frac{4\pi e^2}{m} \sum_i \frac{N_i}{\omega_i^2 - \omega^2} \tag{F.40}$$

so justified cannot be seriously misleading about the frequency dependence of ϵ away from resonance, the point of interest here. It is at least sufficient to show that the variations with frequency of $\epsilon - 1$, and hence $n - 1 \approx \sqrt{\epsilon} - 1$, must be expected to have the same features as those already pointed out for $\alpha(\omega)$ itself.

It is a matter of common observation that transparent glasses, the most used optical materials, have refractive indexes that rise in value as the frequency increases from the red to the blue-violet end of the visible spectrum. This happens because there is thus an approach toward known resonances in the ultraviolet. (The accompanying absorption makes glass a shield against the sun's ultraviolet rays.)

What is directly observed follows from the fact that beams of blue light, in passing from air ($n \approx 1$) into glass ($n \approx 1.5$ to 1.7) are bent through a larger angle [$i' - r$ of Snell's law (F.18)] than the lower-frequency red beams are. The consequence is that the mixture of frequencies constituting a beam of white light is *dispersed* into colors upon refraction, ranging from red to violet in the order expected from the greater bending of the violet components. Of course, observations through flat panes of glass do not show the dispersion, since the refracted passage out of the glass again restores the original white-light beam (fundamentally because of the symmetry described by refraction expressions like $n_1 \sin \alpha_1 = n_2 \sin \alpha_2$, used in Exercise F.4). Dispersion into colors is usually demonstrated by using a wedge of glass (part of a prism), as Newton himself did, in a way familiar from elementary physics (see Exercises F.8 and F.9).

The order in which different frequencies are dispersed when the refractive index rises with increasing frequency is called *normal dispersion*. The order of the dispersed frequencies is reversed when the refractive index falls off beyond its resonance values, and that is called *anomalous dispersion*. However, as discussed above, it is always accompanied by substantial absorption.

Dispersion in Propagations

Fields consisting of superpositions of various frequencies, as in (7.7) and (7.8), have both theoretical and practical importance. Their theoretical importance is due primarily to the fact, discussed in Chap. 7, that no actual field can be purely monochromatic. Among the practically important superpositions are those needed to represent finite pulses, which, for example, may constitute electromagnetic signals in various situations.

The concern here will be with describing a pulse being propagated through a dispersive medium that can be assumed essentially transparent to it. Since absorption can be negligible only during normal dispersion, this will also be assumed, thus limiting the *main* frequencies constituting the pulse to ones lying below resonances. With these assumptions, the pulse can be analyzed into a superposition of such plane waves as have been constructed for transparent media in Sec. F.3.

A pulse characteristically has, at any given moment, some finite extent Δz along its propagation direction, here being taken to be the z direction. It can always be Fourier-analyzed into the plane-wave components, each defined by a wave number k_1, and those must have a minimum range of k_1 values: $\Delta k_1 > 2\pi/\Delta z$ according to the complementarities (7.22). In free space, the various components would all travel with the same velocity c and would thus stay together to maintain the shape and extent of the pulse (Exercise 7.2). However, in a dispersive medium, the various components have different velocities, $c_1(\omega) = \omega/k_1 = c/n(\omega)$ of (F.4) and (F.5), when the refractive index varies with frequency. Every pair of components in the continuum must then be expected to come apart in the course of the propagation, and a broadening of the resultant pulse should ensue. It is measures of the way the space distribution of a pulse changes with time that are to be formulated.

Since only spatial distribution along the propagation direction is in question, it will be sufficient to consider a pulse component that is formed of the collinearly polarized plane-wave constituents in

$$\mathsf{E}(z, t) = \mathrm{Re}\left[\int_{-\infty}^{\infty} dk_1\, a(k_1) e^{i(k_1 z - \omega t)}\right] \tag{F.41}$$

where $\omega = c_1(\omega)k_1$. Adopt such coefficients $a(k_1)$ as will describe concentration in some range $\Delta z = 2\pi/\Delta k_1$ and let $k_o = \omega_o/c_1(\omega_o)$ be the wave number in the middle of the range Δk_1. Then the result of the integration (F.41) can always be represented as

$$\mathsf{E}(z, t) = \text{Re}\,[f(z, t)e^{i(k_o z - \omega_o t)}] \tag{F.42}$$

if the function $f(z, t)$ is suitably chosen. The pulse is thus formulated as a propagation in the z direction, modulated by an amplitude that is to vary with position and time.

The aim is to describe a pulse of some mean spread Δz, and for that purpose it is convenient to adopt such coefficients $a(k_1)$ for (F.41) that at some moment taken to be $t = 0$ the modulation factor in (F.42) will turn out to have the gaussian shape

$$f(z, 0) = e^{-z^2/(2\Delta z)^2} \tag{F.43}$$

describing a concentration symmetrical about some point, taken to be the origin of z, and falling off from the peak value at $z = 0$ to the fraction $1/e$ of that value at $z = \pm 2\,\Delta z$. The factor 2 is inserted into the definition of the constant Δz so that it will be equal to the rms value of a mean spread of the *energy* (proportional to E^2)—obtainable as the square root of the average $\langle z^2 \rangle$ over the energy distribution (see Exercise 7.2). The total energy in the pulse will not be in question, and so, for simplicity of expression, a unit amplitude at $z = 0$ has been chosen for (F.43). Investigation of gaussian shapes is of particular interest because, according to statistical theory, unavoidable fluctuations during the generation† of a pulse generally lead to those shapes, often called normal distributions.

The equation of (F.41) to (F.42) at $t = 0$ shows that the Fourier-integral theorem of (3.44) can be used to obtain the desired coefficients $a(k_1)$ as

$$a(k_1) = (2\pi)^{-1} \int_{-\infty}^{\infty} dz e^{-z^2/(2\,\Delta z)^2} e^{-i(k_1 - k_o)z} \tag{F.44a}$$

The total exponent here is proportional to

$$z^2 - 4i(k_1 - k_o)(\Delta z)^2 z = [z - 2i(k_1 - k_o)(\Delta z)^2]^2 + 4(k_1 - k_o)^2(\Delta z)^4 \tag{F.44b}$$

The result is an integral over a gaussian in the variable $\zeta = z - 2i\,(k_1 - k_o)(\Delta z)^2$, as given in all integral tables, and leading to

$$a(k_1) = \pi^{-1/2}(\Delta z)e^{-(k_1 - k_o)^2(\Delta z)^2} \tag{F.44c}$$

Inserting this into (F.41) can now show how the pulse $\mathsf{E}(z, t > 0)$ evolves with time after $t = 0$.

Finding the eventual modulation $f(z, t > 0)$ in (F.42) requires integration over the wave number k_1 in (F.41), and since the integrand contains exp $(-i\omega t)$,

† Giving (F.43) is not by itself sufficient as an initial condition for calculating the *generation* of a pulse, but the interest here is only in analyzing a pulse already in being. Electromagnetic signals are generated in a great variety of ways, mostly in some hardware constituting detection equipment, before their propagation in some transparent medium begins.

it will be necessary to express ω as a function of k_1, one that follows from $\omega = ck_1/n(\omega)$. It is possible to carry out the integration analytically in an approximation that best shows the important physical characteristics of the result. For this, it is supposed that at the moment $t = 0$ the pulse width Δz is many wavelengths, $2\pi/k_o \ll \Delta z$, broad. Then the expected range $\Delta k_1 = 2\pi/\Delta z$ for the effective wave numbers (and hence the range of main frequencies) can be sufficiently small for a validity of

$$\omega(k_1) \approx \omega_o + \omega_o'(k_1 - k_o) + \tfrac{1}{2}\omega_o''(k_1 - k_o)^2 \tag{F.45}$$

which amounts to a truncated Taylor expansion having as its constant coefficients the values of $\omega' \equiv d\omega/dk_1$ and

$$\omega'' = \frac{d^2\omega}{dk_1^2} \qquad \text{at } k_1 = k_o = \frac{n(\omega_o)\omega_o}{c}$$

The integral over k_1 is now reducible to one containing exponentials only, and with a total exponent that has only terms linear and quadratic in the integration variable, as in (F.44*a*). Consequently a procedure like that from (F.44*a*) to (F.44*c*) need only be repeated to find the modulating factor

$$f(z, t) = \frac{\Delta z}{[(\Delta z)^2 + \frac{1}{2}i\omega_o''t]^{1/2}}\, e^{-(z - \omega_o' t)^2/[(\Delta z)^2 + (1/2)i\omega_o'' t]} \tag{F.46}$$

for the form (F.42). After the real part has been obtained, the final result for the propagated pulse can be expressed as

$$\mathsf{E}(z, t) = \left[\frac{\Delta z}{(\Delta z)_t}\right]^{1/2} e^{-(z - \omega_o' t)2/[2(\Delta z)_t]2} \cos\,[k_o z - \omega_o t + \varphi(z, t)] \tag{F.47}$$

where

$$(\Delta z)_t = \left[(\Delta z)^2 + \left(\frac{\omega_o'' t}{2\Delta z}\right)^2\right]^{1/2} \tag{F.48}$$

and

$$\varphi(z, t) = \frac{1}{8}\frac{\omega_o'' t}{(\Delta z)^2}\frac{(z - \omega_o' t)^2}{(\Delta z)_t^2} - \frac{1}{2}\tan^{-1}\frac{\omega_o'' t}{2(\Delta z)^2}$$

describing a shifting phase in the oscillations, without influence on their enveloping shape. Notice that for $t = 0$ expression (F.47) properly reduces to the result of putting $f(z, 0)$ of (F.43) into (F.42) with $t = 0$.

The peak of the pulse amplitude in (F.47) travels in the course of the propagation, to $z = \omega_o' t$ by the time t. It thus has what is called a *group velocity*

$$\bar{c} = \omega' \equiv \frac{d\omega}{dk_1} \tag{F.49a}$$

evaluated at the frequency ω_o in the wave packet. From $\omega = ck_1/n(\omega)$ it follows that

$$c\, dk_1 = d\omega\left(n(\omega) + \omega\frac{dn}{d\omega}\right)$$

and hence that

$$\bar{c} = \frac{c}{n(\omega) + \omega\, dn/d\omega} \tag{F.49b}$$

This shows that the group velocity is always less than the phase velocity $c_1 = c/n(\omega)$ in the medium during normal dispersion, when $dn/d\omega > 0$. Notice that for a pulse in free space, when $\omega = ck$, the group velocity $d\omega/dk$ equals the phase velocity $c = \omega/k$ (Exercise 7.2).

As it travels, the pulse also broadens, from the mean spread Δz of (F.43) to the $(\Delta z)_t$ given by (F.48). The fractional increase of the broadening can be measured by

$$\left[\frac{(\Delta z)_t^2 - (\Delta z)^2}{(\Delta z)^2}\right]^{1/2} = \frac{|\omega_o''|\,t}{2(\Delta z)^2}$$

The inverse proportionality to $(\Delta z)^2$ shows that an initially sharp pulse broadens more rapidly than one that is wider at the outset.

EXERCISES

F.1 Unpolarized light is incident on water ($n = \frac{4}{3}$) at just the Brewster angle $i' = i_B$. Find the numerical percentage of the total incident intensity that is reflected.

F.2 By forming the flux-density component $\langle N_z \rangle$ for the entire field $\mathbf{E}(z > 0) = \mathbf{E}_i + \mathbf{E}_r$ of (F.12) and (F.14) verify the assertions in the footnote to page 567.

F.3 Light is incident normally on a uniform layer of water ($n = \frac{4}{3}$, $\mu \approx 1$) of depth d, covering a perfectly reflecting metal plane (see page 193 and/or Exercise 7.6). Find the fraction of the incident intensity that reaches the metal, showing that it can be *increased* by as much as a factor $\frac{16}{9}$ through *increasing* the depth of the water a quarter wavelength.

F.4 Consider the situation indicated by the diagram.

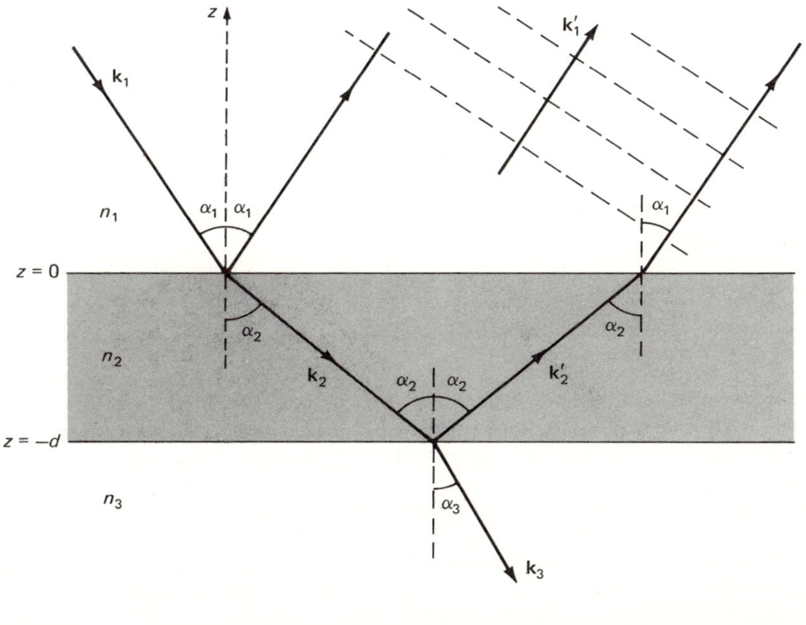

(*a*) Show that Snell's law $(\sin \alpha_1)/(\sin \alpha_3) = n_1/n_3$ is satisfied despite the intervention of the n_2 layer.

(*b*) Show that the reflection coefficient can be given the form

$$R = \frac{(D_r - N_r)^2 + (D_i - N_i)^2}{(D_r + N_r)^2 + (D_i + N_i)^2}$$

where, for an incident polarization *parallel* to the plane of incidence,

$$D_r \mp N_r = n_2 \cos \alpha_2 \, (n_3 \cos \alpha_1 \mp n_1 \cos \alpha_3) \cos \phi$$

$$D_i \mp N_i = (n_2^2 \cos \alpha_1 \cos \alpha_3 \mp n_1 n_3 \cos^2 \alpha_2) \sin \phi$$

and $\phi \equiv n_2(kd) \cos \alpha_2$.

F.5 In the situation of Exercise F.4 show that incidence at the Brewster angle $\alpha_1 = \tan^{-1} (n_2/n_1)$ leads to $R_{\|} = 0$ for *any* thickness $(0 < d < \infty)$, of the n_2 layer only if the media on both sides of it are the same $(n_3 \equiv n_1)$.

F.6 In the situation of Exercise F.4, let the incidence be normal $(\alpha_1 = 0)$ and then find a special choice of $n_2(n_1, n_3)$ for which $R = 0$ when $d = \lambda_2/4$. (The n_2 layer is then called a *quarter-wave plate*. The same phenomenon is also used to reduce reflections from lenses by coating them suitably.)

F.7 Suppose that two *like* half-infinite media, $z > 0$ and $z < -d$, are separated by a gap of free space. A monodirectional wave in the $z > 0$ medium is incident on the gap at an angle i' exceeding the critical angle $i_c = \sin^{-1} (1/n)$.

(*a*) Show that the propagations within the gap are then directed parallel to its boundaries (none toward the $z = -d$ surface) but a wave codirectional with the incident one is transmitted into the $z < -d$ medium.

(*b*) For a gap of sufficient width d show that, the transmitted intensity becomes proportional to $\exp(-2\kappa d)$, where $\kappa = k(n^2 \sin^2 i' - 1)^{1/2}$. (The exponential is called a *barrier penetrability* in the context of quantum mechanics.)

F.8 A monochromatic beam of light passes through a glass prism, traveling along a triangular cross section of it. The beam has an angle of incidence α_1 on a front face and emerges from a back face that makes an acute angle A with the front face.

(*a*) Find the angle δ by which the emerging beam is deflected from the incidence direction as a function of α_1, A, and the refractive index n of the glass.

(*b*) Suppose that the prism is rotated about an axis in the prism perpendicular to the beam plane only insofar as it keeps presenting the same front face to the incident beam. Show that the consequently varied deflections will pass through a *minimum* value δ_m, and describe the simple symmetry of the situation at this minimum.

F.9 Finding the minimum deflection δ_m of each color in an incident beam of white light as it is dispersed by the prism of Exercise F.8 provides a means for measuring the refractive index of each color separately.

(*a*) Derive an expression for n, in terms of δ_m and A, that can be used to evaluate the n's from observed values of the δ_m's.

(*b*) For a very thin prism $(A \ll \frac{1}{2}\pi)$ the minimum deflection can readily be found to be very small also, at least for the glasses with $n \lesssim 2$. Show that the approximate relation $\delta_m \approx (n - 1)A$ is a result.

F.10 An elementary example of displacement currents $\dot{\mathbf{D}}/4\pi = \dot{\mathbf{E}}/4\pi + \dot{\mathbf{P}}$ [see page 557*n*. and (1.37)] is provided by a line that is transmitting slowly varying (alternating) current $I(t)$ despite its interruption by a parallel-plate capacitor (like one of Exercise D.1).

(*a*) Show that the rate $dq/dt = I(t)$ at which one plate is being charged and the other discharged leads to a total flux of displacement current across the space between the plates; evaluate its connection to $I(t)$.

(*b*) The mmf in any one circuit enclosing the line can be evaluated from the current flux through a surface intersected by $I(t)$ itself or, equivalently, over a surface (with the same perimeter) that passes through the space between the plates instead. Why? How does this demonstrate the necessity for such displacement currents to exist, both when the space between the plates is filled with dielectric and when it is empty?

CHAPTER

G†

CONDUCTORS AND WAVE FIELDS

Finite conductivity and the consequent energy losses by fields · The complex index of refraction and the attenuation of propagations into conductors · Reflection and transmission by imperfect conductors · Good-conductor behavior and the quasi-static approximation of it · Current distributions in conductors and the skin effect · Energy transfers to currents as mediated by fields · Losses in wave guides · The reflection of optical frequencies · Resonance absorption · The merging of conduction and polarization currents · Cutoff frequencies for the reflections from metals and the ionosphere

The transparency exhibited by the materials as described in Chap. F follows from the omission from that description of any mechanism by which the electromagnetic waves could lose field energy permanently and so be absorbed inelastically. It can be held that the description used does represent absorptions, resulting in the excitation of oscillating polarizations in the medium, but only in such a way that a dynamical equilibrium is maintained with reemissions by the excited medium, as discussed in connection with Fig. F.2. The overall effect is comparable to an *elastic* scattering of the electromagnetic waves, since their field energy was represented as conserved throughout.

One way to formulate irreversible energy losses by a field to a material medium is suggested by the example of Fig. 6.1, which illustrates the transfer of field energy to thermal motions in matter by way of its ohmic resistance. This can be taken to characterize *imperfect* conductors, and wave fields that can exist in such media will be considered now.

G.1 CONDUCTIVITY

For the electrostatics of conductors as treated in Chap. B, it was sufficient to characterize them simply as media in which charge is free to redistribute itself until a fieldless equilibrium within them is established. If a conductor is not quite perfect because it presents an ohmic resistance to the requisite motions of the charge, this can be expected to protract the time needed to reach an equilibrium distribution but not to affect appreciably the final outcome for the static configuration (after things have cooled down; extra energy will have had to be expended in a Joule heating of the material).

† Designed to be read after Chap. 7 (and Chap. F).

A quantitative measure of the resistance will have to be introduced in describing *nonstatic* fields within a conductor. They will start *varying* currents of charge, and the results at any moment will depend on just how rapidly the resistance will allow the charges to *keep* redistributing themselves in response to the changing fields. Point-to-point variations within the conducting space will come into question, and so the Ohm law (1.30) that defines the resistance will have to be reduced to a point form.

Consider a volume element $\Delta V \equiv \Delta\mathbf{S} \cdot \Delta\mathbf{r} \to 0$ within the conductor, choosing $\Delta\mathbf{r}$ parallel to the current density $\mathbf{j}(\mathbf{r}, t)$ at the element so that $\mathbf{j}\,\Delta V = (\mathbf{j} \cdot \Delta\mathbf{S})\,\Delta\mathbf{r} = (\Delta I)\,\Delta\mathbf{r}$, as in (4.36). Ohm's finding was that the current has the direction of the electric force field $\mathbf{E}$ (the negative potential gradient) and that the resistance encountered in a length $\Delta l \equiv |\Delta\mathbf{r}|$ of material, having the potential difference $\mathsf{E}\,\Delta l$ across it, is measured by the ratio

$$R = \mathsf{E}\frac{\Delta l}{\Delta I} = \frac{\mathsf{E}\,\Delta l}{j\,\Delta S}$$

Thus

$$\mathbf{j} = \mathbf{E} \lim \frac{\Delta l}{R\,\Delta S} \equiv \gamma\mathbf{E} \tag{G.1}$$

is an expression of Ohm's law at a point. The quantity γ defined thus is called the (specific) *conductivity* of the medium. Its relation to the resistance in a volume $\Delta V = \Delta S\,\Delta l$

$$R = \frac{1}{\gamma}\frac{\Delta l}{\Delta S} \tag{G.2}$$

exhibits a direct proportionality to the length Δl of intervening material and an inverse proportionality to the cross section ΔS of conductor,† which are empirically confirmed over wide ranges of circumstance. The existence of a current $\mathbf{j}' = \gamma\mathbf{E}$ wherever $\mathbf{E}$ exists, in material with $\gamma \neq 0$, must be allowed for even if no free currents are deliberately sent into it.

It is possible to deduce an immediate connection between the conductivity γ and such times as are needed for establishing electrostatic equilibria within uniform conducting material of indefinite extent. The processes require redistributions of charge, and so a measure is obtainable by putting extra, free charge with some initial density distribution ρ_o into the otherwise fieldless conductor and seeing how long it takes to be dispersed to negligible density. The dispersal will occur in the form of some current distribution $\mathbf{j}' = \gamma\mathbf{E}'$, where $\mathbf{E}'$ is the field arising at each moment from any yet undispersed free charge ρ' according to $\nabla \cdot \epsilon\mathbf{E}' = 4\pi\rho'$. The current is a result of the charge conservation represented by

$$-\frac{\partial\rho'}{\partial t} = \nabla \cdot \mathbf{j}' = \gamma\nabla \cdot \mathbf{E}' = \frac{4\pi\gamma}{\epsilon}\rho' \tag{G.3}$$

† The connection of these proportionalities to the elementary series and parallel resistances in wire circuits should be quite evident (see Exercise G.1).

in a medium of uniform γ and ϵ. This has the solution

$$\rho' = \rho_o e^{-(4\pi\gamma/\epsilon)t} \equiv \rho_o e^{-t/\tau} \tag{G.4}$$

where $\tau \equiv \epsilon/4\pi\gamma$ serves as a mean decay time in which any initial density is reduced by a factor $1/e$. Notice that γ^{-1} has the dimensions of a time, as follows from the relation $\gamma = j/\mathbf{E}$ because a current density has the dimensions of $(q/L^3) \times (L/T)$ and an electric field of q/L^2.

Since γ represents the effects of the freely movable charges (the conduction electrons), the ϵ introduced here is meant to represent the polarizabilities of only such charges as remain bound to atoms. This ϵ will have such magnitudes† as are characteristic of the nonconducting dielectrics. The magnitudes of γ that follow from measurements on the resistance of the good metallic conductors (like copper or silver) are found to be of the order of $\gamma \gtrsim 10^{17}\ \mathrm{s}^{-1}$. Thus relaxation times of an order as small as $\tau \lesssim 10^{-18}$ s can be expected for the transient currents attending the imposition of an electrostatic field on a good conductor and the eventual electrostatic equilibrium.

The standard result (G.4) yields a valid connection of decay time τ to conductivity γ insofar as γ remains essentially constant throughout the charge-dispersal process. It does do so as long as the decaying distribution does not develop any substantial Fourier components of frequency higher than $10^{13}\ \mathrm{s}^{-1}$ or so, since γ may diminish for such high frequencies (Sec. G.5). This can lead to a longer-persisting residue of very high frequency oscillations, but they average to zero over classically distinguishable increments of distance.

G.2 WAVE FIELDS INSIDE CONDUCTING MEDIA

When dealing with fields inside conducting media, the possibility should be allowed for that the media may be magnetizable (particularly conductors like iron!) and electrically polarizable [as already done in (G.3), through the introduction of ϵ]. Accordingly, the field equations (F.1) characteristic of a dielectric are again to be used, with the additional proviso that a free conduction current $\mathbf{j}' = \gamma\mathbf{E}$ will exist wherever $\mathbf{E}$ exists. The equations also imply the existence of the additional effective currents $\mathbf{j}_M = c\nabla \times \mathbf{M}$ and $\mathbf{j}_P = \mathring{\mathbf{P}}$, as before, but these currents will continue to be independent of the resistance proportional to γ^{-1}. They are to represent bound charge adjustments within atomic domains, whereas resistance affects only the freely movable charges as they make their way among atoms (a picture to be examined more closely in Sec. G.5).

Fields that can be sustained in an expanse characterized by uniform values of γ, ϵ, and μ will now be discussed. *Sustained* fields must be expected to be

† Taking $\epsilon \to \infty$ was found in Chap. D to yield correct results for electrostatic effects in conductors. However, a limit $\epsilon \to \infty$ can be reached only if effects of *all* charges are included in ϵ, as is plain from the discussion of the point on page 526.

divergenceless, with $\rho' = \epsilon\nabla \cdot \mathbf{E}/4\pi = 0$, since any $\rho' \neq 0$ that might be introduced will immediately be dispersed in accordance with $\partial\rho'/\partial t = -\gamma(\nabla \cdot \mathbf{E}) = -\rho'/\tau$ of (G.3). A $\rho' \neq 0$ would thus have only such a transitory existence as is described by the exponential decay (G.4) and would play no role in any steadily sustained field as it persists after all transient effects have died out. It is therefore solutions of the equations

$$\begin{aligned} \nabla \cdot \mathbf{E} &= 0 \qquad & \nabla \times \mathbf{E} &= -\frac{\mathring{\mathbf{B}}}{c} \\ \nabla \cdot \mathbf{B} &= 0 \qquad & \nabla \times \mathbf{B} &= \frac{4\pi\mu}{c}\gamma\mathbf{E} + \frac{\mu\epsilon}{c}\mathring{\mathbf{E}} \end{aligned} \tag{G.5}$$

that are to be sought. The conduction and displacement currents in the last equation will be individually divergenceless in the divergenceless field, and that is what makes a nonstatic $\mathbf{j}' = \gamma\mathbf{E} \neq 0$ consistent with $\rho' = 0$.

A first conclusion from Eqs. (G.5) is that within a uniform conducting medium the wave-propagation Eqs. (F.2) are to be replaced by

$$\left(\nabla^2 - \frac{\mu\epsilon}{c^2}\frac{\partial^2}{\partial t^2} - \frac{4\pi\gamma\mu}{c^2}\frac{\partial}{\partial t}\right)\begin{Bmatrix}\mathbf{E}\\ \mathbf{B}\end{Bmatrix} = 0 \tag{G.6}$$

They are obtainable through such a procedure as led to the wave equation in free space (7.2). A striking feature of the outcome here is that both **E** and **B** are propagated according to the same equation, despite the modification that arises from a current distribution $\gamma\mathbf{E}$ proportional to **E** alone.

A second conclusion is that the field energy (F.7) is no longer conserved when resistance proportional to γ^{-1} exists, but any initially given quantity of it diminishes according to

$$\frac{dW}{dt} = -\oint \gamma\mathbf{E}^2\, dV \equiv -\oint \mathbf{j}' \cdot \mathbf{E}\, dV < 0 \tag{G.7}$$

This can be confirmed simply by evaluating the consequences of the specialized Maxwell equations (G.5) for the time rate of change of the expression (F.7) for W. The result (G.7) is expressed for an entire field, enclosable by a surface through which there is no energy flux. Wherever there is a flow of energy, it is again representable by the Poynting vector (F.8). The right side of (G.7) describes a rate of energy loss to the material exactly as in (6.5).

The expression may seem paradoxical in that the larger the resistance proportional to γ^{-1} the lower the loss rate by a given **E** it yields. This of course happens because a smaller resistance permits a given field **E** to generate a greater conduction current $\gamma\mathbf{E}$ subject to the resistance. The connection of (G.7) to the familiar I^2R losses, e.g., those represented in (6.9), can be seen from reexpressing it for a volume $\Delta V = \Delta S\, \Delta l$ as

$$-\frac{\partial w}{\partial t}\Delta V = \gamma^{-1}(j')^2\, \Delta S\, \Delta l = I^2R \tag{G.8}$$

using the relation (G.2) to the resistance. There were no such losses in the dielectrics as idealized in the preceding chapter because they were attributed an infinite resistance ($\gamma = 0$), so that they permit no conduction currents at all.

Transverse-Plane-Wave Solutions

Monochromatic solutions of the form (F.6), each varying along only one direction in space as specified by a wave vector $\mathbf{k} = \hat{\mathbf{k}}\omega/c$, can again be sought. Then substitution of the form for **E** or for **B** into the wave equation (G.6) yields the condition

$$-\frac{n^2\omega^2}{c^2} - \frac{\mu\epsilon}{c^2}(-\omega^2) - \frac{4\pi\gamma\mu}{c^2}(-i\omega) = 0$$

which can generally be satisfied only by a *complex* index of refraction having the square

$$n^2 = \mu\epsilon + i\frac{4\pi\gamma\mu}{\omega} \tag{G.9}$$

The form for **E** in (F.6) will be divergenceless if only the arbitrary amplitude **a** is chosen transverse to $\hat{\mathbf{k}}$. The remaining Maxwell equations (G.5) will also be satisfied if $\mathbf{B} = n\hat{\mathbf{k}} \times \mathbf{E}$ as in (F.6) but now with the complex-valued n. Thus, the electric and magnetic vectors are again perpendicular to each other, as well as to the direction $\hat{\mathbf{k}}$, in these solutions.

The modifications of the waves that have arisen from the finite conductivity are implicit in the complexity of the refractive index in (G.9). This can be separated into real and imaginary parts as represented by

$$n = n_r + in_i \tag{G.10}$$

and then equating the square of this to (G.9) yields

$$n_r^2 - n_i^2 = \mu\epsilon \qquad \text{and} \qquad n_r n_i = \frac{2\pi\gamma\mu}{\omega} \tag{G.11}$$

A quadratic equation for n_r^2 and another for n_i^2 follow from these relations and will be found to have the positive real roots

$$\left.\begin{matrix} n_r \\ n_i \end{matrix}\right\} = \left[\sqrt{\left(\frac{\mu\epsilon}{2}\right)^2 + \left(\frac{2\pi\gamma\mu}{\omega}\right)^2} \pm \frac{\mu\epsilon}{2}\right]^{1/2} \tag{G.12}$$

with the upper sign giving n_r and the lower one n_i. Notice that these expressions properly reduce to the results found earlier for $\gamma = 0$: $n_r = \sqrt{\mu\epsilon}$ and $n_i = 0$. The absolute value of the complex index can be calculated as

$$|n| = \sqrt{n_r^2 + n_i^2} = \sqrt{\mu\epsilon}\left[1 + \left(\frac{4\pi\gamma}{\epsilon\omega}\right)^2\right]^{1/4} \tag{G.13}$$

and the phase of the complex number $n \equiv |n| e^{i\alpha}$ is given by

$$\alpha = \tan^{-1} \frac{n_i}{n_r} = \frac{1}{2} \tan^{-1} \frac{4\pi\gamma}{\epsilon\omega} \tag{G.14}$$

as follows from $\tan 2\alpha = (2 \tan \alpha)/(1 - \tan^2 \alpha)$.

Substitution of $n = n_r + in_i = |n| e^{i\alpha}$ into the field forms (F.6) yields the monochromatic representatives

$$\begin{aligned} \mathbf{E}(\mathbf{r}) e^{-i\omega t} &= \mathbf{a} e^{-n_i \mathbf{k} \cdot \mathbf{r}} e^{i(n_r \mathbf{k} \cdot \mathbf{r} - \omega t)} \\ \mathbf{B}(\mathbf{r}) e^{-i\omega t} &= |n| \hat{\mathbf{k}} \times \mathbf{a} e^{-n_i \mathbf{k} \cdot \mathbf{r}} e^{i(n_r \mathbf{k} \cdot \mathbf{r} - \omega t + \alpha)} \end{aligned} \tag{G.15}$$

These describe plane waves having the wavelength $\lambda_\gamma = 2\pi c/n_r \omega$ and amplitudes that decrease exponentially with the distance $\hat{\mathbf{k}} \cdot \mathbf{r}$. A *mean attenuation length*, in which the amplitudes decrease by a factor $1/e$, is given by

$$(n_i k)^{-1} = \frac{n_r}{n_i} \frac{\lambda_\gamma}{2\pi} \tag{G.16}$$

The magnetic field is out of phase with the electrical oscillations by the phase angle α, an effect of self-induction by the magnetic flux being generated by the currents, as can be confirmed most explicitly from considerations like those of Exercise G.10.

High-Frequency and High-Conductivity Limits

The degree to which the behavior found here will differ from that found for the nonconducting dielectrics, in which $n_i = 0$ and $n_r = n = \sqrt{\mu\epsilon}$, will be determined by the magnitude of the ratio

$$\frac{4\pi\gamma}{\epsilon\omega} = (\omega\tau)^{-1} = \frac{T}{2\pi\tau} \tag{G.17}$$

where $T = 2\pi/\omega$ is the period of the oscillations and $\tau = \epsilon/4\pi\gamma$ is the relaxation time associated with charge redistributions in (G.4). This role of $\omega\tau$ can be seen most directly in the ratios

$$\frac{n_{r,i}}{\sqrt{\mu\epsilon}} = 2^{-1/2}[\sqrt{1 + (\omega\tau)^{-2}} \pm 1]^{1/2} \tag{G.18}$$

of the expressions (G.12) for n_r and n_i to the nonconducting dielectric index $\sqrt{\mu\epsilon}$.

It is at once evident that the conductivity has negligible effects not only when the resistance proportional to γ^{-1} is as large as it is in the transparent dielectrics but also when dealing with frequencies so high that

$$(\omega\tau)^{-1} = \frac{4\pi\gamma}{\epsilon\omega} \ll 1 \tag{G.19}$$

even in a medium of substantial conductivity. Then $n_r \approx |n|$ differs from $\sqrt{\mu\epsilon}$ only by a fraction of order $(\omega\tau)^{-2} \ll 1$ and

$$n_i \approx \frac{|n|}{2\omega\tau} \approx \frac{2\pi\gamma}{\epsilon\omega}\sqrt{\mu\epsilon} \ll |n| \tag{G.20}$$

as follows from expanding (G.18) in the small quantity $(\omega\tau)^{-1} \ll 1$. The mean attenuation length (G.16) becomes

$$(n_i k)^{-1} \approx \frac{(|n|/n_i)\lambda_\gamma}{2\pi} \approx \frac{\epsilon c/2\pi\gamma}{\sqrt{\mu\epsilon}} \tag{G.21}$$

spanning many wavelengths and independent of the precise value of the high frequency. The small effect of the currents $\mathbf{j}' = \gamma\mathbf{E}$ in high-frequency fields can be understood from the fact that they average to zero in much smaller periods $T = 2\pi/\omega$ than the times $\tau = \epsilon/4\pi\gamma$ they would need to produce substantial charge redistributions. Actually, the results (G.20) and (G.21) are applicable only when the condition $(\omega\tau)^{-1} = 4\pi\gamma/\epsilon\omega \ll 1$ can be satisfied by frequencies lower than about $\omega \approx 10^{13}\ \mathrm{s}^{-1}$, and hence only in the relatively poor conductors with $\gamma \lesssim 10^{12}\ \mathrm{s}^{-1}$. At higher frequencies, γ itself becomes a function of frequency, with results to be investigated in Sec. G.5.

The deviations from the nonconducting dielectric behavior will be most pronounced in good conductors for which

$$(\omega\tau)^{-1} = \frac{4\pi\gamma}{\epsilon\omega} \gg 1 \tag{G.22}$$

Additions of unity to a number supposed to be so large can be neglected, and then (G.18) yields

$$n_r \approx n_i \approx (2\omega\tau)^{-1/2}\sqrt{\mu\epsilon} = \sqrt{2\pi\gamma\mu/\omega} \gg 1$$

and

$$n \approx (1+i)\sqrt{\frac{2\pi\gamma\mu}{\omega}} = \sqrt{\frac{4\pi\gamma\mu}{\omega}}\, e^{i\pi/4} \tag{G.23}$$

The attenuation distance (G.16) is now reduced to less than a wavelength λ_γ. The phase difference (G.14) between the magnetic and electric oscillations approaches the extreme $\alpha \approx \frac{1}{4}\pi = 45°$. Moreover, the ratio $|n|$ of the magnetic to electric amplitudes in (G.15) becomes very large: $|n| \approx \sqrt{4\pi\gamma\mu/\omega} \gg 1$. Most of the field energy is thus invested in the magnetic form $w(\mathbf{B}) = \mathbf{B}^2/8\pi\mu$. This can be understood as a result of the fact that whatever energy is being supplied to sustain the fields is passed on to thermal motions through work $\mathbf{j}\cdot\mathbf{E}$ by the *electric* field and in direct proportion to the energy fraction in the electric form $w(\mathbf{E}) = \epsilon\mathbf{E}^2/8\pi$, since the loss rate per unit volume in (G.7) can be expressed as

$$-\frac{\partial w}{\partial t} = \gamma\mathbf{E}^2 = 2\frac{4\pi\gamma}{\epsilon}w(\mathbf{E}) = \frac{2}{\tau}w(\mathbf{E}) \tag{G.24}$$

At the same time, the large current $\gamma\mathbf{E}$ permitted to exist by a good conductivity γ generates the large magnetic field.

The solutions (G.15) resemble the static Laplace fields of Chap. 3 in that they become unbounded if they are extended indefinitely far along one direction of space (where $n_i \mathbf{k} \cdot \mathbf{r} \to -\infty$). Again the representation can be used only if the extension of the field is interrupted by sources of it before the infinite values are reached. Such sources are necessary for sustaining the steady oscillations described in (G.15) because of the steady losses (G.24). Sources as represented by the incidence of electromagnetic waves on a conductor surface will be considered next.

G.3 REFLECTION AND TRANSMISSION BY CONDUCTORS

To be investigated now is the field transmitted into a conductor confined to the half-infinite space $z < 0$ when an exterior electromagnetic wave is incident on its surface $z = 0$. The situation is to be representable by such a diagram as that in Fig. F.1, and distraction from the most essential points will be avoided by taking the exterior space $z > 0$ to be free; that is, $n_1 = \epsilon_1 = \mu_1 = 1$ will be assumed. The conducting $z < 0$ space will be characterized by ϵ, μ, and a conductivity γ.

The exterior field will be taken to be a superposition of the incident and reflected waves (F.12) and (F.14) with $n_1 \equiv 1$. The plane of incidence will at once be chosen as the xz plane so that the incident wave-vector components are $k_x = k \sin i'$, $k_y = 0$, and $k_z = -k \cos i'$. The given incident wave must be taken to be transverse, with $\mathbf{k} \cdot \mathbf{a}_i = 0$, and a transverse reflected wave, with $\mathbf{k}'' \cdot \mathbf{a}_r = 0$, is also necessary for the Maxwell equations in free space to be satisfied.

The Transmitted Field

The attempt will be to continue the solution into the conductor, through appropriate boundary conditions, in the monodirectionally propagated form

$$\mathbf{E}_t(z < 0) = \mathbf{a}_t e^{i\mathbf{K} \cdot \mathbf{r}} \qquad \mathbf{B}_t = \frac{c}{\omega} \mathbf{K} \times \mathbf{E}_t \tag{G.25}$$

The amplitude vector $\mathbf{a}_t$ must be restricted by

$$\mathbf{K} \cdot \mathbf{a}_t = 0 \tag{G.26}$$

for the fields to be properly divergenceless solutions of the Maxwell equations (G.5). It will be more convenient to work in terms of the wave vector $\mathbf{K}$ than to replace it with $n\mathbf{k}$, as for the solution (G.15). However, satisfaction of the wave equation (G.6) that prevails in the conductor still requires

$$K^2 = \left(\frac{\omega}{c}\right)^2 \left(\mu\epsilon + i\frac{4\pi\gamma}{\omega}\right) \tag{G.27}$$

as for $n^2(\omega/c)^2$ of (G.9). The remaining stipulations in (G.25) and (G.26) guarantee the satisfaction of all the Maxwell equations (G.5) if (G.27) is taken for K^2.

It can be anticipated that the appropriate boundary conditions will again require matchings of the type (F.15), with $n_2\,\mathbf{k}'$ to be replaced by the wave vector $\mathbf{K}$ of (G.25). Consequently, the conclusions (F.16) are now to be replaced by

$$k_x = k_x'' = K_x \qquad \text{and} \qquad k_y = k_y'' = K_y = 0 \tag{G.28}$$

since $n_1 = 1$ and $k_y = 0$ here. The simple law of reflection (F.17) still holds.

A conclusion from (G.28) is that the wave-vector components $K_{x,y}$ must be real despite the complexity of the magnitude K in (G.27). This situation differs from that in the solution (G.15), where all components of $\mathbf{K} \to n\mathbf{k} = (n_r + in_i)\mathbf{k}$ are complex, and so that solution cannot by itself be matched to the conditions met here.† This should not be surprising, since a solution is not uniquely specified when just a monodirectionality of propagation is presumed; only the complexity of $K_x^2 + K_y^2 + K_z^2$ in (G.27) is then required, without specification of just how the complexity is to be apportioned among the three components until boundary conditions are met. The solution needed here must have

$$K_x = k_x = \frac{\omega}{c}\sin i' \qquad K_y = k_y = 0 \qquad K_z = -\sqrt{K^2 - K_x^2} = -\frac{\omega}{c}n(i') \tag{G.29a}$$

where $$n(i') \equiv +\sqrt{n^2(0) - \sin^2 i'} \qquad n^2(0) \equiv \mu\epsilon + i\left(\frac{4\pi\gamma\mu}{\omega}\right) \tag{G.29b}$$

The complex index of refraction defined in (G.9) is now denoted $n(0)$, since a generalization of it to $n(i' \neq 0)$, depending on the incidence angle, is needed here. Inspection of (G.29*b*) shows that the generalization amounts to a replacement

$$\epsilon \to \epsilon - \mu^{-1}\sin^2 i' \tag{G.30}$$

Thus the formulas (G.12) will give the real and imaginary parts of $n(i') = n_r(i') + in_i(i')$ if the combinations $\mu\epsilon$ in them are replaced by $\mu\epsilon - \sin^2 i'$.

The transmitted wave forms (G.25) become proportional to

$$\mathbf{E}_t(\mathbf{r}) = \mathbf{a}_t e^{n_i(i')kz} e^{ik[x\sin i' - n_r(i')z]} \tag{G.31}$$

where $k \equiv \omega/c$, as usual. This describes an inhomogeneous plane wave in the sense that its planes of constant phase generally do not coincide with the planes of constant amplitude. The phase planes are propagated in a direction making a refraction angle r with the normal to the interface as given by

$$r = \tan^{-1}\frac{k\sin i'}{kn_r(i')} = \sin^{-1}\frac{\sin i'}{[n_r^2(i') + \sin^2 i']^{1/2}}$$

† Except when $k_{x,y} = 0$ (normal incidence), a case *included* by the more general solution now being developed.

Restated in terms of $n_i^2(i') = n_r^2(i') - (\mu\epsilon - \sin^2 i')$, which is less sensitive to i' than $n_r^2(i')$ is, this leads to the modification

$$\frac{\sin i'}{\sin r} = [\mu\epsilon + n_i^2(i')]^{1/2} \tag{G.32}$$

of Snell's law of refraction. It properly reduces to the unmodified law when there is no conductivity and so $n_i(i') = 0$. Whereas the phase planes have the generally oblique refraction direction depending on the incidence angle i', the planes of constant amplitude always remain parallel to the interface for any angle of incidence. The size of the amplitude decreases exponentially with depth $-z$ into the conductor, giving the field a mean penetration depth $[n_i(i')k]^{-1}$.

Imperfect-Conductor Boundary Conditions

To find the reflected and transmitted amplitudes, appropriate boundary conditions will have to be applied. As usual, they are equivalent to Maxwell equations reduced to a surface layer, treated as a space of vanishing thickness $\delta \to 0$. Equations (F.1*b*) with $\mathbf{j}' = \gamma \mathbf{E}$ are the ones appropriate here, since they allow for changes in γ, $\epsilon = \mathbf{D}/\mathbf{E}$ and $\mu = \mathbf{B}/\mathbf{H}$ across the boundary. The material-independent equations again lead to

$$\Delta \mathsf{B}_z = 0 \qquad \text{and} \qquad \Delta \mathsf{E}_{x,\,y} = 0 \tag{G.33}$$

as in (F.11). The surface flux of $\nabla \times \mathbf{H}$ can also be expected to vanish again, so that

$$\Delta \mathsf{H}_{x,\,y} = \Delta(\mu^{-1}\mathsf{B}_{x,\,y}) = 0 \tag{G.34}$$

if the conductivity is finite and $(\gamma\delta)\mathbf{E} \to 0$ as $\delta \to 0$. However, the experience with the electrostatics of conductors suggests that accumulations of surface charge should be allowed for. Accordingly, the reduction of $\nabla \cdot \mathbf{D} = 4\pi\rho'$ by integration over a surface volume $\delta(\to 0)$ per unit area is represented as

$$\Delta(\epsilon \mathsf{E}_z) = 4\pi(\rho'\delta) \to 4\pi\sigma' \tag{G.35}$$

where σ' is to stand for any free-surface charge density that might accumulate. This will be determined by the necessity of the charge conservation (G.3), which leads to the surface divergence

$$\Delta(\gamma \mathsf{E}_z) = -\frac{\partial}{\partial t}\rho'\delta \to -\frac{\partial \sigma'}{\partial t} \tag{G.36}$$

The right side here can be replaced by an $i\omega\sigma'$, for any one monochromatic representative, and then σ' can be eliminated with the help of (G.35) to give

$$\Delta\left[\left(\epsilon + i\frac{4\pi\gamma}{\omega}\right)\mathsf{E}_z\right] = \Delta[\mu^{-1}n^2(0)\mathsf{E}_z] = 0 \tag{G.37}$$

in terms of the complex n^2 defined in (G.9) or (G.29). The conditions (G.33), (G.34), and (G.37) should be as sufficient to yield a unique solution as (F.11) was for $\gamma = 0$, and then (G.35) can be used to evaluate the surface charge accumulations.

The boundary conditions (G.33), (G.34), and (G.37) differ from the ones (F.11) used for the nonconducting dielectrics only by the addition of the imaginary part in (G.9) to the $n^2 \to \mu\epsilon$ used when $\gamma = 0$. It is not surprising therefore that the Fresnel formulas, (F.20) and (F.21), for the amplitude ratios (a_r/a_i) emerge again, except that $n^2 = \mu\epsilon$ is to be replaced by the complex-valued expression $n^2(0)$ of (G.29) and (G.9). The similarly transformed $(\mathbf{a}_t)_{\parallel}$ of (F.21) is no longer simply a transverse amplitude (unless $i' = 0$) because of the exponential modulation along a direction different from the phase-propagation direction. However, it is not difficult to show (Exercise G.2) that

$$a_{tx} = \frac{n(i')}{n(0)} (\mathbf{a}_t)_{\parallel} \qquad a_{ty} \equiv (\mathbf{a}_t)_{\perp} \qquad a_{tz} = \frac{\sin i'}{n(0)} (\mathbf{a}_t)_{\parallel} \tag{G.38}$$

when $(\mathbf{a}_t)_{\perp}$ and $(\mathbf{a}_t)_{\parallel}$ are obtained from (F.20) and (F.21) as modified by using the complex-values $n(0)$. For $i' = 0$, when there is no distinction between $(\mathbf{a}_i)_{\parallel}$ and $(\mathbf{a}_i)_{\perp}$, so also for a_{tx} and a_{ty} (with $a_{tz} \to 0$).

Reflection and Transmission Coefficients

The coefficients of reflection from the conductor $R = |a_r/a_i|^2$ can be evaluated from the Fresnel formulas in (F.20) and (F.21), with due respect for the fact that the symbol n in them now stands for the *complex* expression $n = n_r + in_i = n(0)$ of (G.10), (G.9), and (G.29*b*). The results will be simplest to present after a replacement of the radicals occurring in the Fresnel formulas with their equivalents $n(i') = n_r(i') + in_i(i')$ of (G.29). This makes it possible to separate each numerator and denominator into real and imaginary parts that are to be individually squared and added in forming the absolute squares needed for the reflection coefficients:

$$\begin{aligned} R_{\perp} &= \frac{[\mu \cos i' - n_r(i')]^2 + n_i^2(i')}{[\mu \cos i' + n_r(i')]^2 + n_i^2(i')} \\ R_{\parallel} &= \frac{[\epsilon \cos i' - n_r(i')]^2 + [(4\pi\gamma/\omega) \cos i' - n_i(i')]^2}{[\epsilon \cos i' + n_r(i')]^2 + [(4\pi\gamma/\omega) \cos i' + n_i(i')]^2} \end{aligned} \tag{G.39}$$

For the second of these expressions, the results (G.11) have been used to eliminate $n_r^2(0) - n_i^2(0)$ and $n_r(0)n_i(0)$. A similar elimination of $n_{r,i}(i')$ in favor of the more immediately given μ, ϵ, γ, ω would require substitution of the expressions (G.12), with $\epsilon \to \epsilon - \mu^{-1} \sin^2 i'$ in them. Both expressions (G.39) properly reduce to unity for $i' \to \frac{1}{2}\pi$. They also reduce to the results (F.26) for $\gamma = 0$ (and $\mu = 1$).

Expressions for the transmission coefficients need not be separately presented since the energy conservation that must be expected in just the surface layer, because of its vanishing thickness, guarantees that $T = 1 - R$. A check that the transmitted amplitudes (G.38) are in fact consistent with this is left to Exercise G.3, from which it can also be learned that the important normal component of the energy flux into the conductor (the $-z$ direction here) is given by

$$-\langle N_{tz}\rangle = \frac{cn_r(i')}{8\pi\mu}\,|\mathbf{a}_t|^2 e^{-2n_i(i')k|z|} \tag{G.40}$$

Only the values at the surface $z = 0$ are actually needed for the transmission coefficients $T = \langle N_{tz}\rangle/\langle N_{iz}\rangle$.

At Good Conductors

The results of this section will now be discussed for the most important case of the good conductors, characterized by $(\omega\tau)^{-1} \equiv 4\pi\gamma/\epsilon\omega \gg 1$ [Eq. (G.22)] and consequently differing appreciably in their behavior from that of the $\gamma = 0$ dielectrics.

The index $n(i') = n_r(i') + in_i(i')$ used in the descriptions is to be obtained from the expressions (G.12) for $n_{r,i}(0)$ by replacing the ϵ in them with $\epsilon - \mu^{-1}\sin^2 i'$ according to (G.30), and this makes the expressions (G.23) even better approximations for $n_{r,i}(i' > 0)$ than they are for $n_{r,i}(0)$. Thus all these numbers approach equality to each other,

$$n_r(i') \approx n_r(0) \approx n_i(i') \approx n_i(0) \approx \sqrt{\frac{2\pi\gamma\mu}{\omega}} \gg 1 \tag{G.41}$$

They become large in comparison to the $n = \sqrt{\mu\epsilon}$ of the $\gamma = 0$ dielectrics and independent of the incidence angle i'. An immediate consequence is that the refraction angle defined in (G.32) becomes

$$r \approx \sin^{-1}\left(\sqrt{\frac{\omega}{2\pi\gamma\mu}}\sin i'\right) \to 0 \tag{G.42}$$

tending to vanish for $4\pi\gamma\mu/\omega \gg 1$. The waves transmitted into a good conductor thus appear to be refracted inward into a direction normal to the interface, regardless of the incidence angle! The planes of constant phase in the field expression (G.31) approach coincidence with the planes of constant amplitude parallel to the conductor surface. Corresponding to this, the transmitted field components normal to the interface (the z components) tend to vanish relative to the parallel (x and y) components (Exercise G.2).

The transmitted waves can now be regarded as simple transverse waves, propagated normally inward from the surface, except for the relatively much more gradual variations parallel to the surface that are described by the modulating-wave factor $\exp i(k \sin i')x$ in (G.31). The wavelength of this modulation $\lambda_x = 2\pi/k \sin i' \geq 2\pi/k$ is extremely long compared with the variations normal to the interface, described by $\lambda_\gamma = 2\pi/n_r k \ll 2\pi/k$ and a penetration depth $(n_i k)^{-1} \approx \lambda_\gamma/2\pi$. Thus the modulating wave merely furnishes a nearly constant phase factor over lengthwise spans much larger than the penetration depth transverse to these spans but does remain important to consider in dealing with conductor lengths $\gtrsim 2\pi/k$, as occur in transmission-line and in antenna-excitation problems.

Finally, the results for the reflection coefficients R and the transmission coefficients T should be evaluated for the good conductors being discussed here. When the approximations (G.41) for the refractive indexes are substituted into the exact expressions (G.39) for the reflection coefficients and, for consistency of approximation, the results are expanded as power series in the small quantity $(\omega/2\pi\gamma\mu)^{1/2} \approx n_r^{-1} \approx n_i^{-1}$, the outcome is

$$\begin{aligned} R_\perp &\approx 1 - 2\mu \cos i' \left(\frac{\omega}{2\pi\gamma\mu}\right)^{1/2} \approx 1 - T_\perp \\ R_\parallel &\approx 1 - \frac{2\mu}{\cos i'} \left(\frac{\omega}{2\pi\gamma\mu}\right)^{1/2} \approx 1 - T_\parallel \end{aligned} \tag{G.43}$$

The latter is correct to the order shown if the incidence angle is not too close to glancing, i.e., to $i' = \frac{1}{2}\pi$, so that $\cos i' \gg (\mu|n|^{-1} \ll 1)$. The transmission coefficients as given by (G.43) tend to vanish for very good conductors, and so the reflections from them are nearly total.

The approximations used for (G.43) are valid only for frequencies up to about 10^{13} s^{-1} because γ itself decreases rapidly for higher frequencies. Reflectivities of the visible frequencies $\omega = 2.7$ to 4.7×10^{15} s^{-1} must be evaluated from the more precise formula (G.39) in a way that will be discussed in Sec. G.5. The conclusion will again be that most of the radiation is reflected, accounting for the shiny appearance of well-defined (smooth or specular) metal surfaces, like those of silver or copper. Their directly measured reflectivities are 0.97 and 0.63, respectively.

The notion that reflections arise as reemissions by oscillations that are excited in the reflecting medium can be extended to the conductors. Added to the oscillating microscopic polarizations characteristic of the nonconducting dielectrics, the conductors have sources of reemission provided by the macroscopic oscillating conduction currents γ Re $[\mathbf{E}_t(\mathbf{r})e^{-i\omega t}]$ that are started when $\gamma \neq 0$. It appears that the currents excited in good conductors are sufficient to turn back most of the incident energy, to account for the nearly total reflections (G.43).

It is plain that highly conducting metals are opaque mostly because they *reflect* nearly all the flux incident on them. Materials with more resistance proportional to γ^{-1} admit more field energy but then attenuate it through the resistive losses (G.8). This leads to a fractional *spatial* attenuation of the energy flux, per unit penetration distance, that can be calculated directly from the flux expression (G.40):

$$-\langle N_{tz}\rangle^{-1} \frac{d}{d|z|} \langle N_{tz}\rangle = 2n_i k \tag{G.44}$$

a quantity called an *absorption coefficient* in this connection. It should be consistent with (G.7) and (G.24) in giving the energy-loss rate per volume of thickness dz and unit cross section, and that is demonstrated explicitly by the equalities

$$\frac{d}{dz}\langle N_{tz}\rangle = 2n_i k\langle N_{tz}\rangle = \frac{4\pi\gamma\mu}{cn_r}\langle N_{tz}\rangle = \tfrac{1}{2}\gamma\langle \mathbf{E}_t^* \cdot \mathbf{E}_t(\mathbf{r})\rangle = \gamma\langle \mathbf{E}_t^2(\mathbf{r}, t)\rangle = -\frac{\partial w}{\partial t} \tag{G.45}$$

which follow from the relations $n_i n_r = 2\pi\gamma\mu/\omega$ of (G.11) and (G.41) and the connection (G.40) between flux and field.

A really proper treatment of the absorption by a *finite* thickness of material requires considering the additional reflections that must be expected to attend the reemergence of the transmitted wave through the far side of the conducting layer. The investigation of this is left to Exercise G.7.

G.4 CURRENT DISTRIBUTIONS IN CONDUCTORS

There are large domains of practical problems in which the approximations valid for $(\omega\tau)^{-1} = 4\pi\gamma/\epsilon\omega \gg 1$ of (G.22), also given attention after (G.41), are quite adequate. The much used alternating currents mentioned in connection with (4.73) are typically given $\omega/2\pi = 60$ Hz (cycles per second) and are usually sent through copper, which has $\gamma \approx 5 \times 10^{17}\ \mathrm{s}^{-1}$. In this case, $4\pi\gamma/\epsilon\omega \approx 1.7 \times 10^{16}$, many orders of magnitude greater than unity. Even the very much higher frequencies of radar ($\omega \approx 10^{11}\ \mathrm{s}^{-1}$, corresponding to $\lambda = 2\pi c/\omega$ of a few centimeters) allow $(\omega\tau)^{-1}$ to attain values greater than 10^7. Intermediate values obtain for radio waves (λ in the kilometer range).

The Quasi-Static Field Equations

The approximations in question correspond to neglecting the displacement current $\mathring{\mathbf{D}}/4\pi$ in the last of the Maxwell equations (G.5), which can be written

$$\nabla \times \mathbf{H} = \frac{4\pi}{c}\left(\gamma - i\frac{\epsilon\omega}{4\pi}\right)\mathbf{E} \tag{G.46}$$

for representatives of frequency ω. This makes it evident that the displacement current has a magnitude with just the ratio $\omega\epsilon/4\pi\gamma$ to the conduction current (besides being out of phase by 90°), making its effects practically undetectable in the situations reviewed in the preceding paragraph. It is consequently sufficient to take

$$\nabla \times \mathbf{B} = \frac{4\pi\mu}{c}\mathbf{j}(\mathbf{r}, t) = \frac{4\pi\mu}{c}\gamma\mathbf{E} \tag{G.47}$$

as the Maxwell equation, inside conductors, appropriate to those situations. This amounts to using the Ampère law of magnetostatics for the magnetic fields generated by the oscillating divergenceless currents; with $\omega \ll 4\pi\gamma/\epsilon$, the currents can be considered to be varying slowly enough for what amounts to an adiabatic approximation and are said to be quasi-static. A simple application of the Stokes theorem to (G.47) shows that Ampère's *integral* law (1.21) becomes valid over regions within the conductor. This means that up to the conductor boundaries the magnetic field that exists at any given moment is determined by the currents flowing at the same moment. There is an effectively *instantaneous* response of the field to all the current elements, regardless of their distances from the field point, as a consequence of the displacement-current contribution to (G.47) having become negligible.

Another consequence is that the propagation equation is modified from (G.6) to what is called a *diffusion-propagation* equation (see Exercise G.4)

$$\nabla^2 \mathbf{E} = \frac{4\pi\gamma\mu}{c^2}\frac{\partial \mathbf{E}}{\partial t} \qquad \left(= \frac{\mathring{\mathbf{E}}}{\tau c_1^2}\right) \tag{G.48}$$

for **E**, and so also for **B**. The parenthesis calls attention to connections with the charge-redistribution time $\tau = \epsilon/4\pi\gamma$ of (G.4) and the propagation velocity $c_1 = c/\sqrt{\mu\epsilon}$ [Eq. (F.3)] as modified by the polarizabilities of the medium alone. The equations still have the attenuated plane-wave solutions (G.31), but now the restriction to $n(0) = (1 + i)(2\pi\gamma\mu/\omega)^{1/2}$ is immediate.

In many problems, the interest is in the divergenceless current distributions $\mathbf{j} = \gamma\mathbf{E}$ of (G.47), for which

$$\nabla^2 \mathbf{j} = \frac{4\pi\gamma\mu}{c^2}\frac{\partial \mathbf{j}}{\partial t} \tag{G.49}$$

follows from (G.48). Using this to find current distributions inside conductor boundaries still requires satisfying the boundary conditions (G.33) to (G.36). They involve the fields **E**, **B**, both inside the conductor and outside it, that support the currents and/or are generated by them, as **B** is inside the conductor through the Ampère law (G.47). Even when the currents are initiated by applying potential differences (static or nonstatic) directly only to the conductor itself, there are generated fields outside it that support the currents through carrying energy fluxes $\mathbf{N} = c\mathbf{E} \times \mathbf{B}/4\pi$ from the sources of the applied potentials to the sites of the ohmic losses, as already seen for steady currents in connection with Fig. 6.1.

Direct Current Distributions

A first point made evident by (G.49) is that steady, direct current distributions over conducting spaces must be solutions of the Laplace equation $\nabla^2\mathbf{j}(\mathbf{r}) = 0$. Since static currents must be driven by electrostatic fields $\mathbf{E}(\mathbf{r}) = \mathbf{j}/\gamma$ and such are curlless, they are derivable from scalar potentials, as in $\mathbf{j} = -\gamma\, \nabla\phi$. Thus the problem of finding the divergenceless steady-current distributions is reduced to solving the Laplace equation $\nabla^2\phi = 0$, just as in the electrostatic problems of Chap. 3. However, the boundary conditions to be applied now are different, since they are to admit nonvanishing fields $\mathbf{E}^i(\mathbf{r}) = \mathbf{j}(\mathbf{r})/\gamma \neq 0$ into the interiors of conductors. Only interior components $\mathbf{E}_n^i$ normal to the interface must still be excluded; such would generate currents $j_n = \gamma \mathsf{E}_n^i = d\sigma'/dt$ (G.36) leading to *non-static* charge accumulations that cannot be maintained after the situation becomes steady. However, a steady $\mathbf{j} = \gamma \mathbf{E}_{\parallel}^i = \mathbf{j}_{\parallel} = -\gamma \nabla_{\parallel} \phi$ need not vanish at the surface, and so this is not generally an equipotential, as the conductors of electrostatics are.

The example of Fig. 6.1 is a case in which the Laplace equation $\nabla^2 \mathbf{j} = 0$ is satisfied merely by a uniformity of $\mathbf{j}$ throughout the conductor, the solution that fits the boundary conditions (G.33) and (G.34), together with the $\mathsf{E}_n^i = 0$ following from (G.36). Nonuniform steady current distributions are most often of concern when the conductor has a nonuniform cross section.

Alternating Currents and the Skin Effect

The full diffusion equation (G.49) must be satisfied by any time-dependent (but divergenceless) current distribution $\mathbf{j}(\mathbf{r}, t)$ that might be sustained in a good conductor. As usual, it is sufficient to pay attention to a representative of definite frequency, as in $\mathbf{j}(\mathbf{r}, t) = \text{Re}\,[\mathbf{j}(\mathbf{r})e^{-i\omega t}]$. The conditions for such a simple harmonic oscillation by itself are of interest to alternating-current theory, dealing with currents like (4.73), i.e., products of rotatory generators. The typical interest is in currents directed parallel to the long side of an elongated conductor like a wire, and such situations are most properly approached by considering a cylindrical conductor (Exercise G.8), as in the steady-current situation of Fig. 6.1. However, direct use of the findings in the preceding section can be made by considering a current $j = j_x = \gamma \mathsf{E}_x$ parallel to the $z = 0$ surface of a half-infinite conducting space $z < 0$. This can at once be expected to yield conclusions valid in the immediate vicinity of a surface *element* on any conductor, and their character will turn out to be such as to permit corresponding conclusions about alternating currents sent through lengths of conductor having a finite cross section.

Alternating current $j = j_x$, parallel to a conductor's surface, is typically generated by applying a potential difference of definite frequency to a stretch of it, thus providing such a field inside the conductor as was denoted $\mathsf{E}_{tx} = j_x/\gamma$ in the preceding section. The resulting current $j = \gamma \mathsf{E}_{tx}$ will generate a magnetic field B_{ty} via the Ampère law (G.47) and together with the sources of the applied potentials themselves will also produce fields $\mathbf{E}$, $\mathbf{B}$ outside the conductor. Any exterior fields can be analyzed into linearly independent plane-wave modes, making various incidence and reflection angles with the $z = 0$ conductor surface, just such modes as were denoted $\mathbf{E} = \mathbf{E}_i + \mathbf{E}_r$, $\mathbf{B} = \mathbf{B}_i + \mathbf{B}_r$ in the preceding section. They must be connected to the interior field $\mathbf{E}_t$, $\mathbf{B}_t$ in agreement with the boundary conditions (G.33) to (G.37) again, and so findings in the preceding section need merely be suitably adapted to the situation of interest now.

One finding discussed for good conductors, in connection with (G.42), was that whatever the exterior incidence angles might be, the interior fields $\mathbf{E}_t$, $\mathbf{B}_t$ become parallel to the $z = 0$ conductor surface, just as required for a current $\mathbf{j} = \gamma \mathbf{E}_t$ parallel to it. For this current to have the chosen direction, of $j = j_x$, the fields should be polarized in the xz plane, and only components so polarized have any connection with $j = j_x$. Such components can be expected to be the principal ones excited by an efficiently applied emf. The outcome for the real current distribution that now follows from (G.31), under the good-conductor conditions that will make it satisfy the diffusion equation (G.49), can be written

$$j = j_x = e^{-|z|/\delta}\ \mathrm{Re}\ [\gamma a_t e^{-iz/\delta}(e^{i(k_x x - \omega t)})] \qquad \text{(G.50)}$$

Here δ is a standard symbol for the mean penetration depth (G.16)

$$\delta \equiv (n_i k)^{-1} \approx \frac{c}{\sqrt{2\pi\gamma\mu\omega}} \qquad \text{(G.51)}$$

under the good-conductor conditions (G.41). It also equals $(n_r k)^{-1} = \lambda_\gamma/2\pi$, where λ_γ is the wavelength of the inward propagation. The last exponential in (G.50) represents a comparatively long wavelength ($\lambda_x \approx 2\pi/k_x \gg \lambda_\gamma = 2\pi\delta$) modulation, as discussed on page 590, practically constant over stretches long compared with δ.

The distribution (G.50) indicates that, at any conductor boundary sufficiently remote from far sides, oscillating current distributions within the boundary are concentrated near the surface, being exponentially attenuated with distance inward and having a mean penetration depth $\delta(\omega)$ given by (G.51). This is known as the *skin depth* of the conductor for currents of frequency ω, and the behavior it helps describe is called the *skin effect*. For 60-cycle current in copper, which has $\gamma \approx 5.2 \times 10^{17}\ \mathrm{s}^{-1}$ and $\mu \approx 1$, the skin depth is $\delta = 0.8$ cm. For the much higher frequency microwaves mentioned on page 592, it is reduced to about 10^{-4} cm!

With the current essentially confined to a skin of thickness δ, it makes use of only a cross-sectional area ($\approx 2\pi a\delta$) of a circularly cylindrical conductor with radius $a \gg \delta$. Since the resistance to the current is inversely proportional to the transverse area of its flux (G.2), the resistance is multiplied by about a factor $\pi a^2/2\pi a\delta = a/2\delta(\omega)$ from its value for a uniformly distributed (direct) current and is frequency-dependent. Whereas direct-current carriers are given as large a cross section as feasible (to reduce resistance and hence I^2R losses), there is no such advantage to be gained from increasing the radius to more than twice the skin depth for alternating currents. This is why cables designed for high frequencies are often fractionated into thin wire strands.

Alternating currents in wires of diameter less than or approximately equal to 4δ are distributed quite uniformly over the wire cross sections, especially since there are also small interior reflections of the fields (see page 592); after transmission through one side, they are partially reflected from other sides, back into the interior. The distribution approaches the uniformity expected of direct currents, and that is why such calculations as (4.95) of mutual and self-inductances are usually carried out for constant current distributions despite their relevance only to the inductive effects of varying currents. When the same conductors are used for frequencies high enough to make the skin effect important, the current distributions in the integral (4.95) involve $\delta(\omega)$ and the inductances become frequency-dependent.

Energy Transfers to the Currents

With $\mathbf{E}_t = \mathbf{j}/\gamma$ parallel to the current, there is *no* energy flux $\mathbf{N}_t = c\mathbf{E}_t \times \mathbf{B}_t/4\pi$ along the flow direction of the current and inside the conductor. It is instead directed transversely inward from the conductor surface, as a continuation of normally incident flux $\mathbf{N}_\perp$ carried by field *outside* the conductor. Thus it is by transmission across spaces outside the conductor that sources provide the energy needed to maintain current.

As pointed out in the preceding subsection, the outside field is always analyzable into incidences on, and reflections from, the surface. Then the finding (G.43) indicates that most of the energy identified as incident is reflected away again; only a small fraction is transmitted to the current. This should *not* be taken to mean that oscillating currents can be maintained only at the cost of reflecting most of the applied energy out into free space. Some of this can occur, and the discussion on page 591 of reflections as reemissions by the induced currents suggests that it represents radiation losses (like those treated in the antenna-theory subsection of Chap. 8). However, the *total* free-space field here must satisfy boundary conditions at the primary source of the applied energy as well as at the current-carrying conductor. It is thereby largely concentrated between source and current, in a limited intervening space that can be of much smaller dimensions than even a single free-space wavelength (like the $\lambda = 2\pi c/\omega \approx 5000$ km of 60-cycle current!). The analyses into incidences and reflections remain valid, since any field is so resolvable, but their mutually interfering superpositions produce near-standing waves joining source and currents. In effect, the field-energy fluxes are passed back and forth, with a net transmission to the currents of most the energy finally expended.

Radiation losses can become serious from transmission lines much longer than a free-space wavelength $2\pi c/\omega$ and carrying fluxes $I = I_o \cos(k_x x - \omega t)$ with the longitudinal modulation noted in (G.50). It is to avoid such losses that use is made of devices like coaxial cables (Exercise 7.15), which prevent the escape of radiated field energy by confining the field between inner and outer cylindrical conductors.

Imperfectly Conducting Wave Guides

When the transmission of power at frequencies as high as those of microwaves is contemplated, skin depths approach the infinitesimal ($\delta \approx 10^{-4}$ cm was quoted on page 595) and fields are practically excluded from conductors. In any case it is outside the conductors that the energy is primarily carried, by fields that keep undergoing almost perfect reflections from the conductor surfaces. Attention then turns to power transmission directly by the fields themselves but without allowing them to escape as radiation losses. For this, conductor boundaries remain useful, to help guide the field-energy flux along desired channels, and hence a practical interest in the wave guides of Sec. 7.3.

To confine the fields within a channel, its conductor walls must respond to incidences on them with induced currents that help to reflect the fields back into the channel interior. Such currents were found in Sec. 7.3 and were there represented as fluxes $\mathbf{J} = c\mathbf{n} \times \mathbf{B}/4\pi$ [Eq. (7.33)] per unit breadth of wall surface that faces in a direction given by the unit vector **n**.

The representation $\mathbf{J} = c\mathbf{n} \times \mathbf{B}/4\pi$ is based on treating the channel walls as perfect ($\gamma \to \infty$) conductors having skin depths $\delta \to 0$, and hence excluding all fields from their interiors. Since even the best conductors have some $\gamma < \infty$, the fields will actually penetrate into the walls to a depth $\delta(\omega) \ll \lambda = 2\pi c/\omega$, the free-space wavelength, and so the effective channel cross section is enlarged, but only by a practically undetectable amount on the scale of the field wavelengths. The channel field configurations found from the perfect-conductor treatment should be practically unaltered, like the resultant current fluxes $\mathbf{J} = c\mathbf{n} \times \mathbf{B}/4\pi$ needed to contain them.

The small field penetrations into the channel walls, with $\mathbf{E}_t = \mathbf{j}/\gamma \neq 0$ inside them, need be taken into account only when considering ohmic losses from the power $\mathscr{P}$ being transported by the channel field. It must then be recognized that **J** is actually distributed over a cross-sectional area of unit width and mean depth δ in densities $\mathbf{j} = \gamma\mathbf{E}_t$ just under, and parallel to, elements of wall surface. Thus **J** encounters a resistance $R_1 = 1/\gamma\delta$ [Eq. (G.2)] per unit of length along its direction, and the corresponding I^2R loss rate under each unit of wall area should be expected to amount to $\langle J^2 \rangle R_1 = \frac{1}{2}(\mathbf{J}^* \cdot \mathbf{J})/\gamma\delta$ when $\mathbf{J} = c\mathbf{n} \times \mathbf{B}/4\pi$ is a monochromatic representive. This evaluation may seem to presume a uniform distribution **j**, confined to exactly the depth δ, but it is also the outcome of a proper treatment of **j** as an exponentially attenuated distribution like (G.50), a point left to Exercise G.11. Finally, the loss rate can be put in terms of the channel field components $\mathbf{B}_\parallel$, evaluated at the wall elements and parallel to their surfaces, the only ones involved in $\mathbf{J} = c\mathbf{n} \times \mathbf{B}/4\pi$ and in

$$\mathbf{J}^* \cdot \mathbf{J} = \left(\frac{c}{4\pi}\right)^2 \mathbf{n} \cdot [\mathbf{B}_\parallel^* \times (\mathbf{n} \times \mathbf{B}_\parallel)] = \left(\frac{c}{4\pi}\right)^2 |\mathbf{B}_\parallel|^2$$

The result for the power loss per unit length of channel is

$$-\frac{d\mathscr{P}}{dz} = \frac{c^2}{32\pi^2\gamma\delta} \oint dl\,|\mathbf{B}_\parallel|^2 = \frac{\mu\omega\delta}{16\pi} \oint dl\,|\mathbf{B}_\parallel|^2 \qquad \text{(G.52)}$$

with the integration to extend over the perimeter of a channel cross section, consisting of scalar elements dl such that $dl\,dz$ is an area element of wall. The last of the equalities is the outcome of using the definition (G.51) to eliminate γ in favor of the skin depth δ. More explicit evaluations of (G.52) depend on the particular channel shape employed and the modes of propagation (TM or TE) excited, together with their characteristic cutoff frequencies. An example is left to Exercise G.14. It must also be acknowledged that special measures become necessary for propagation frequencies ω that are close to cutoff frequencies. The latter are no longer quite discrete when the wall conductivity is finite, as can be understood from the fact that they are determined by the lengths of transverse half waves that fit into the channel cross section and these fits are loosened when the waves can penetrate into the walls to a frequency-dependent mean depth

$\delta(\omega)$. That a nondiscreteness of frequencies is associated with energy losses has been illustrated for cavity-resonance frequencies in Sec. 7.3.

The energy for the expenditures (G.52) is supposed to be provided by net field-energy fluxes, remainders after almost perfect reflections, representable as $\mathbf{N}_\perp = c\mathbf{E}_\parallel \times \mathbf{B}_\parallel/4\pi$, directed normally into the wall surfaces and hence determined solely by channel field components $\mathbf{E}_\parallel$, $\mathbf{B}_\parallel$ that are parallel to the surface at each wall element. For evaluating them the solutions obtained in the perfect-conductor ($\gamma \to \infty$) approximation of Sec. 7.3 are inadequate, since they were made to satisfy the perfect-conductor boundary condition $\mathbf{E}_\parallel = 0$ of (7.32), thus yielding $\mathbf{N}_\perp = 0$, as expected for the total reflections characteristic of perfect conductors. When $\gamma < \infty$, the field is allowed to penetrate into a skin depth of conductor, thus providing the $\mathbf{E}_t = \mathbf{j}/\gamma$ needed to form the distribution of each wall flux $\mathbf{J}$, now spread over a mean skin depth. Moreover, the corrected exterior field component at the wall is $\mathbf{E}_\parallel = \mathbf{E}_t \neq 0$, according to the imperfect-conductor boundary condition (G.33). The next one of these, (G.34), adds the requirement $\mathbf{B}_\parallel = \mathbf{B}_t/\mu$ at the wall, and it is known that $\mathbf{B}_t$ is orthogonal to $\mathbf{E}_t$, with $\mathsf{B}_t = n(0)\mathsf{E}_t$. Thus $\mathsf{B}_\parallel = n(0)\mathsf{E}_t/\mu = n(0)\mathsf{E}_\parallel/\mu$, with $n(0) = n_i + in_r$ the complex index of refraction, and the result is

$$N_\perp = \frac{c}{8\pi}\,\mathrm{Re}\,(\mathsf{E}_\parallel^* \mathsf{B}_\parallel) = \frac{c\mu}{8\pi}\,|\mathsf{B}_\parallel|^2\,\mathrm{Re}\left(\frac{1}{n(0)}\right) = \frac{\mu\omega\delta}{16\pi}\,|\mathsf{B}_\parallel|^2 \tag{G.53}$$

The last of these equalities follows from the expressions (G.41) for $n_{i,r}$, which yield

$$n(0) = \frac{1+i}{k\delta} \qquad \text{for } k\delta \ll 1 \tag{G.54}$$

and $1/n(0) = (1-i)\omega\delta/2c$ when γ is eliminated in favor of δ(G.51). The last of the equalities (G.53) also shows that the field-energy flux $N_\perp$ is exactly sufficient to provide for the ohmic loss under each unit of wall area in (G.52).

It should still be understood why the *only* significant correction of the channel field from the perfect-conductor approximations of Sec. 7.3 consists of a superposition on the latter of a small component making $\mathbf{E}_\parallel \neq 0$ on the walls, as long as these are good conductors. For (G.52) it was presumed that $\mathbf{B}_\parallel$ needed no such correction. The point can be understood by referring to findings in Sec. G.3, where fields exterior to the conductor were analyzed into incident and reflected plane-wave components. For $\mathbf{E}_\parallel = \mathbf{E}_i + \mathbf{E}_r$ and $\mathbf{B}_\parallel = \mathbf{B}_i + \mathbf{B}_r$, it is only a normally incident ($i' = 0$) plane wave that need be considered. Then the magnitudes at the surface are representable as $\mathsf{E}_\parallel = a_i + a_r$ and $\mathsf{B}_\parallel = -a_i + a_r$, the negative sign in the latter ensuring the correct energy flux directions in the incidences and reflections. For the $i' = 0$ incidence on a good conductor, (F.20) gives

$$a_r = -\frac{1-\mu/n(0)}{1+\mu/n(0)}\,a_i \approx -\left[1 - \frac{2\mu}{n(0)}\right]a_i \approx -a_i + (1-i)\mu k\delta a_i \tag{G.55}$$

This yields the expected perfect-conductor limit $\mathsf{E}_{\parallel} = 0$ for $\delta = 0$ and the small result $\mathsf{E}_{\parallel} \approx (1-i)\mu k\delta a_i$ for $\gamma < \infty$. On the other hand, the perfect-conductor limit $\mathsf{B}_{\parallel} = -2a_i$ already has a much larger magnitude, and correcting it would lead to only a second-order ($\sim \delta^2$) correction to the flux (G.53). Thus, retaining the perfect-conductor values for the $\mathbf{B}_{\parallel}$ in the loss expression (G.52) is more than adequate.

G.5 CONDUCTIVITIES AT HIGH FREQUENCIES

The preceding sections have given primary attention to what is called good conductor behavior, characterized by $(\omega\tau)^{-1} \equiv 4\pi\gamma/\epsilon\omega \gg 1$ of (G.22) and page 592. This encompasses by far the widest areas of application in engineering and in most laboratory devices. However, there also is interest in high frequencies at which the above condition is not pertinent even to metals, especially because the conductivity itself begins to decrease with frequency before wavelengths as short as the important optical ones of light are reached.

An expectation of that follows from looking into the microscopic origins of conductivity and resistance. As when the frequency sensitivity of dielectrics was examined in Sec. F.8, it is responses of electrons in the material that are important to consider. A relatively free motion of at least some of the electrons in conductors is supposed to be responsible for the ready generation of macroscopic currents $\mathbf{j} = \gamma\mathbf{E}$ in them. If there are N_o conduction electrons ($q = -e$) per unit volume, then $\mathbf{j} \equiv -N_o e\mathbf{v}$, where $\mathbf{v}$ must be an average drift velocity superposed on the random thermal motions† whenever a directed current is generated by some field.

Consider first the result of imposing a *static* field $\mathbf{E}_o$ that generates a *steady* current $\mathbf{j} = \gamma_o \mathbf{E}_o$. The steady force $-e\mathbf{E}_o$ on an otherwise free electron would keep accelerating it to indefinitely high velocities unless that motion itself gives rise to some sort of effective resistive force $\mathbf{F}_r$, one that is absent until the drift motion begins and is just sufficient to cancel the imposed force at some *steady* velocity $\mathbf{v}$. Thus, an effective force equal to $+e\mathbf{E}_o = e(\mathbf{j}/\gamma_o) = -(N_o e^2/\gamma_o)\mathbf{v}$ is required, one that is proportional to $-\mathbf{v}$, like‡

$$\mathbf{F}_r = -\nu_o(m\mathbf{v}) \tag{G.56}$$

† A consequent proportionality of the electric and thermal conductivities is known as the *Wiedemann-Franz law*.

‡ The representation of resistive or frictional effects in this way has been found to have wide applicability. The consequent damping effects on various types of motion, observed in many situations, are treated in Ref. 17, chap. 3 and exercises 11.5 and 11.6. A more thoroughgoing theory of electrical resistance might be expected to yield a derivation of ν_o in terms of the collision frequency of electrons trying to make their way among the atoms of the material. However, classical treatments of this point are practically valueless because quantum-mechanical effects intervene seriously. Typically quantum-mechanical penetrations of interionic potential barriers by the de Broglie waves of the electrons and the exclusion effects of the Pauli principle are particularly essential to an understanding of what really goes on. See Ref. 15.

where ν_o is a phenomenological parameter to be determined by comparison of results with observations on each material and to serve as a measure of the effective force of resistance per unit of the electron momentum $m\mathbf{v}$ that gives rise to it. An advantage of parametrizing the resistance by ν_o, in place of γ_o, is that it can be expected to be independent of just how that momentum is generated and hence of the field that is generating it. For a *static* field, the connection

$$\gamma_o = \frac{N_o e^2}{m\nu_o} \tag{G.57}$$

is expected from the steps that led to (G.56). It correctly gives the rate of energy loss, $\gamma_o \mathsf{E}_o^2$ per unit volume according to (G.7), as the work rate against the effective resistive force: $-\mathbf{F}_r \cdot \mathbf{v} = \nu_o m v^2 = \nu_o m(-e\mathsf{E}_o/m\nu_o)^2$ per electron and $N_o(e^2\mathsf{E}_o^2/m\nu_o) = \gamma_o \mathsf{E}_o^2$ per unit volume. The directly observed static conductivity of copper, $\gamma_o \approx 5 \times 10^{17}\ \text{s}^{-1}$ as cited on page 595, and the known number $N_o \approx 8 \times 10^{22}$ per cubic centimeter of conduction electrons (about one per atom) lead to $\nu_o \approx 4 \times 10^{13}\ \text{s}^{-1}$ for this case.

The concern now is with the conductivity to be expected in a field of some definite frequency, like $\mathbf{E} = \text{Re}\,(\mathbf{E}_\omega e^{-i\omega t})$. This would eventually impart a momentum of the same frequency, $m\mathbf{v} = \text{Re}\,(m\mathbf{v}_\omega e^{-i\omega t})$ to each conduction electron, in accordance with the equation of motion

$$m\dot{\mathbf{v}} \to -i\omega m\mathbf{v}_\omega = -e\mathbf{E}_\omega - \nu_o(m\mathbf{v}_\omega) \tag{G.58}$$

Because of the phase difference between the velocity and acceleration thus made evident, it becomes advantageous to work with a *complex* conductivity defined by the ratio $\mathbf{j}_\omega/\mathbf{E}_\omega$ of monochromatic representatives

$$\gamma_c(\omega) = -\frac{N_o e\mathbf{v}_\omega}{\mathbf{E}_\omega} = \frac{N_o e^2}{m(\nu_o - i\omega)} = \frac{\gamma_o}{1 - i\omega/\nu_o} \tag{G.59}$$

The complexity helps represent a phase difference that develops between the applied field and the real current responding to it

$$\begin{aligned} \mathbf{j}(\mathbf{r}, t) &= \text{Re}\,[\gamma_c(\omega)\mathbf{E}_\omega(\mathbf{r})e^{-i\omega t}] \\ &= |\gamma_c(\omega)|\ \text{Re}\,(\mathbf{E}_\omega e^{-i(\omega t - \eta)}) \end{aligned} \tag{G.60a}$$

where $$|\gamma_c(\omega)| = \frac{\gamma_o}{(1 + \omega^2/\nu_o^2)^{1/2}} \qquad \eta = \tan^{-1}\frac{\omega}{\nu_o} \tag{G.60b}$$

The outcome is a frequency dependence of the conductivity, but for the good conductors this remains negligible for an enormous range of frequencies that are not too high. In copper, $\nu_o \approx 4 \times 10^{13}\ \text{s}^{-1}$ was found above; then $(\omega/\nu_o)^2$ is a negligible addition to unity [and $\gamma_c(\omega) \approx \gamma_o$, as supposed for the evaluations on page 591] up through the microwave frequencies ($\omega \approx 10^{11}\ \text{s}^{-1}$) and beyond them up to the far infrared. However, for the yellowish-red light reflected by copper, $\omega \approx 3 \times 10^{15}\ \text{s}^{-1}$ and

$$|\gamma_c(\omega)| \approx \frac{\nu_o}{\omega}\gamma_o \approx 10^{-2}\gamma_o \tag{G.61}$$

only about 1 percent of the static conductivity. This, for example, makes the approximation (G.43) for the reflection coefficient, which presumed the lower frequencies and larger conductivities, invalid for light reflected from copper. A reversion to the more precise formulas (G.39) is quite generally required for calculating the reflectivity of the optical frequencies from metals. That does not change the conclusion that most of the light is reflected.

All the results of this chapter which, like (G.39), were *not* predicated on $(\omega\tau)^{-1} \gg 1$ [as is (G.43) and *all* of Sec. G.4] can be retained for the optical frequencies in the pertinent situations. They are as precise as Maxwell's phenomenological equations can make them, despite their having been derived for a γ treated as real. The γ occurs only in the Maxwell equation expressing the Ampère-Maxwell law, as given by (G.46) for monochromatic representatives. Putting $\gamma \to \gamma_c = \gamma_r + i\gamma_i$ into this, with

$$\gamma_r = \frac{\gamma_o}{1 + (\omega/v_o)^2} \qquad \gamma_i = \frac{\omega}{v_o}\gamma_r \tag{G.62}$$

according to (G.59), has the outcome

$$\nabla \times \mathbf{H}_\omega = \frac{4\pi}{c}\left[\gamma_r - i\frac{\omega}{4\pi}\left(\epsilon - \frac{4\pi\gamma_r}{v_o}\right)\right]\mathbf{E}_\omega \tag{G.63}$$

It thus turns out that the imaginary part of the conductivity is equivalent merely to a supplementation, by the conduction electrons, of the polarizability that is measured by ϵ, when this is taken to represent contributions from only the bound electrons.† That the relatively free conduction electrons should also contribute significantly at the high frequencies is quite understandable. The field-ordered motions of otherwise free electrons that consist of high-frequency oscillations are limited to such finite amplitudes as are characteristic of the bound-electron oscillations and so can produce like effects.

The result (G.63) shows that the *formal* results derived from monochromatic Maxwell equations remain unchanged. Quite naturally, it is the values of the material constants in them, those treated as real numbers γ and ϵ, that must be chosen appropriately for each application. For the high frequencies at which it makes a difference, the real γ should not be taken to be the static conductivity γ_o but the real part γ_r of $\gamma_c(\omega)$. Moreover, if ϵ_o represents only contributions from bound electrons, it should be replaced by the constant that is actually effective

$$\epsilon = \epsilon_o - \frac{4\pi\gamma_r}{v_o} \tag{G.64a}$$

The results (G.15) for the wave fields inside conductors still stand. Now, however, the indices n_r, n_i helping determine the complex wave vectors $\mathbf{K} = (n_r + in_i)\mathbf{k}$ can be made more explicit, through replacing (G.11) with

$$n_r^2 - n_i^2 = \mu\left(\epsilon_o - \frac{4\pi\gamma_r}{v_o}\right) \qquad n_r n_i = \frac{2\pi\mu\gamma_r}{\omega} \tag{G.64b}$$

† As done from the beginning, on page 581.

It is these forms that have been used (Ref. 22) in the reflection-coefficient formulas (G.39) for comparisons with direct observations. The conductivity was treated as an adjustable parameter in fitting the measured reflectivities as functions of frequency, and this provided the first experimental evidence that the resistance ($\sim \gamma^{-1}$) to optical frequencies is indeed much increased from the static values applying to the lower-frequency fields treated on page 591.

It should still be noticed that with such values as were used for the estimate (G.61), the contribution to the effective dielectric constant (G.64) by the conduction electrons, $4\pi\gamma_r/v_o \approx 4\pi\gamma_o v_o/\omega^2$, becomes much larger ($\approx 15$) than the ϵ_o that can normally be contributed by bound electrons, making the effective ϵ large and negative. How this can be understood will be better evident after the following.

The discussions after (G.63) suggest that for high frequencies the effects of the conduction electrons approach indistinguishability from those of bound electrons. Moreover, attention was called in Sec. F.8 to the necessity of taking the energy losses constituting absorption into account in dielectrics also, at least for frequencies near resonances. Starting with Lorentz himself, it has been customary to introduce absorption through a phenomenological resistive force again, parametrized as for conductors in (G.56). Thus, the equation of motion for a bound electron in (F.38), after its reexpression in terms of monochromatic representatives, is modified to

$$-\omega^2\mathbf{r}_j = -\omega_j^2\mathbf{r}_j - \frac{e\mathbf{E}_\omega}{m} - v_j(-i\omega\mathbf{r}_j) \tag{G.65}$$

when allowance is made for the possibility that bound electrons with different natural frequencies ω_j may also be subject to different absorptive effects as measured by the v_j. Notice that this equation also applies to the conduction electrons of (G.58) if it is supposed that $\omega_j = 0$ and $v_j = v_o$ for them.

With the complex expression (G.65) now replacing (F.38), it becomes as expedient to define a complex polarizability $\alpha_c(\omega)$ [a ratio of complex representatives, here $\sum N_j(-e\mathbf{r}_j)/\mathbf{E}_\omega$] as in the case of the conductivity; it is just what is needed for the Maxwell equation (G.63), relating monochromatic representatives. Then the result (F.39) is corrected to

$$\begin{aligned}\alpha_c(\omega) &= \sum_j \frac{N_j e^2}{m(\omega_j^2 - \omega^2 - i\omega v_j)} \\ &= \sum_j \frac{N_j e^2}{m[(\omega_j^2 - \omega^2)^2 + \omega^2 v_j^2]}(\omega_j^2 - \omega^2 + i\omega v_j)\end{aligned} \tag{G.66}$$

The last line makes it especially evident that the absorption makes the resonance peaks finite, as anticipated on page 572. It must be expected that $\omega \gg v_j$, as was found for the conduction electrons. Then for the resonances in the range of the optical frequencies or greater, $\omega_j \gg v_j$ also. This makes it simple to estimate the breadth of the corresponding resonances through seeing what departure $\Delta\omega = |\omega - \omega_j|$ from $\omega = \omega_j$ doubles the real denominator in (G.66) from its value $\omega_j^2 v_j$ at $\omega = \omega_j$. Since $\omega \approx \omega_j$ anywhere in a range expected to be narrow, the doubled denominator can be approximated as $(\Delta\omega)^2(2\omega_j)^2 + \omega_j^2 v_j^2 = 2\omega_j^2 v_j^2$, yielding a breadth $2\Delta\omega \approx v_j$. Thus observations on resonance breadths can help determine appropriate values for the parameters v_j.

The point of interest in connection with conductivities can be made by examining the resultant complex ϵ_c needed for expressing the displacement current in the Maxwell equation (G.46). The needed ϵ_c is given by the replacement of (F.40) with

$$\epsilon_c = 1 + \frac{4\pi e^2}{m} \sum_j \frac{N_j}{\omega_j^2 - \omega^2 - i\omega v_j} \tag{G.67}$$

according to (G.66). Next suppose that not only the bound electrons of various resonance frequencies ω_i but also the N_o conduction electrons (per unit volume) are counted in (G.67). For these $\omega_i = 0$ and $v_i = v_o$ should be taken, as noted in connection with (G.65), yielding

$$\epsilon_c = \epsilon_c' - \frac{4\pi N_o e^2}{m\omega(\omega + iv_o)} = \epsilon_c' + i\frac{4\pi}{\omega}\gamma_c(\omega) \tag{G.68}$$

if ϵ_c' is the complex contribution of the bound electrons only and $\gamma_c(\omega)$ is just the complex conductivity† (G.59). In this way, the displacement current in (G.46) by itself,

$$-\frac{i\omega}{4\pi}\epsilon_c \mathbf{E} = -\frac{i\omega}{4\pi}\epsilon_c' \mathbf{E} + \gamma_c(\omega)\mathbf{E} \tag{G.69}$$

already includes the conduction current, and it need not have been put in separately, as done in (G.46) and (G.63), if ϵ had been allowed complex values ϵ_c and had from the beginning been understood to include effects of both the bound and conduction electrons.‡

The results based on using both a real γ and a real ϵ can still be retained, simply by equating the coefficient of $\mathbf{E}$ in (G.69) to the one in (G.46) as in

$$-i\frac{\omega}{4\pi}\epsilon_c = \gamma - i\left(\frac{\omega}{4\pi}\right)\epsilon \tag{G.70}$$

Then ϵ is real, as supposed for (G.64*a*), if the complex ϵ_c here and in (G.68) is analyzed into its real and imaginary parts as $\epsilon_c = \epsilon + i\epsilon_i$. It now becomes evident that not only do conduction electrons contribute to the real ϵ but also the bound electrons can contribute to what has been called the conduction current $\gamma\mathbf{E}_\omega$,

† The last term in (G.68) helps account for the fact that electro*static* ($\omega \to 0$) effects in conductors are obtainable as $\epsilon \to \infty$ limits of dielectric results, as found in Chap. D.

‡ These electrons can remain separately identifiable, at least in the good conductors. The conduction electrons are nowadays understood to be those with a valence band of energies, one that is only partly filled. The bound electrons occupy filled inner-shell bands instead, the only kind supposed to exist in dielectrics that are good insulators. The separation into energy bands is a purely quantum-mechanical conception, and filling is an effect of the exclusion principle.

through the imaginary part of ϵ_c'. Whereas it was found that the conduction-electron contribution to ϵ was its largest part at optical frequencies, the bound-electron contribution to $\gamma\mathbf{E}_\omega$ is usually negligible in that range because $\nu_j \ll |\omega_j^2 - \omega^2|/\omega$ in (G.66), except close to resonances that, in metals, usually lie outside the visible range.

The development of (G.67) and (G.68) has also made it evident just why the contribution to ϵ by the conduction electrons at optical ($\omega \gg \nu_o$) frequencies becomes its largest part. It was already pointed out in connection with (F.39) that it is the loosest, most weakly bound (smallest ω_i) electrons that make the largest contributions. Nothing is looser than an unbound ($\omega_i = 0$) conduction electron!

Observations on reflectivities like those mentioned in connection with (G.64) led to a striking discovery upon being carried to higher frequencies than the visible ones. The reflections remain little less than total up to a certain fairly well defined cutoff frequency in the ultraviolet and then come close to disappearing just beyond it. Moreover, it was earlier discovered† that metals become practically transparent to the ultraviolet radiations beyond the same cutoff frequency. An explanation follows from examining the expression (G.68) for frequencies so high that $\nu_o \ll \omega$ can be neglected in it and in a region far from resonances (none were observed in the above investigations) so that the bound-electron contribution ϵ_c' is practically indistinguishable from the real one ϵ_o. Then the complex ϵ_c, representing all the effects of both the bound and conduction electrons, becomes the real one

$$\epsilon = \epsilon_o - \frac{4\pi N_o e^2}{m\omega^2} \equiv \epsilon_o - \frac{\omega_p^2}{\omega^2} \tag{G.71}$$

in which a quantity independent of any phenomenological parameter like ν_o and called the *plasma frequency*

$$\omega_p = \left(\frac{4\pi N_o e^2}{m}\right)^{1/2} \tag{G.72}$$

has been defined. Many metals have about $N_o \approx 10^{22}$ cm^{-3} conduction electrons (very closely one per atom), and $\omega_p \approx 5.7 \times 10^{15}$ s^{-1}, in the ultraviolet, for them.

The important point about the result (G.71) for ϵ is that this changes sign as ω is increased through the value $\omega = \omega_p/\sqrt{\epsilon_o}$ and that this frequency is found to agree very closely with the observed cutoff value, at which the reflections disappear and transparency begins. Just why these phenomena should attend the sign change can be better understood after the following considerations.

Taking the real $\epsilon_c \to \epsilon$ of (G.71) to represent *all* the effects of the electrons corresponds to evaluations with $\nu_o \to 0$, thus neglecting any resistive forces (G.56) and energy losses ($\sim \gamma\mathbf{E}^2$) to them. There is no imaginary part of ϵ to be

† By Wood (Ref. 31). The simple explanation was offered by Zener (Ref. 32).

represented by a real† $\gamma \neq 0$ as in (G.70), and hence the wave equation (G.6) in the medium reduces to just the one (F.2) characteristic of the $\gamma = 0$ dielectrics. This means that the wave vectors $\mathbf{K} = n\mathbf{k}$ of plane-wave field components in the medium will have to satisfy the real relation

$$K^2 \equiv \frac{n^2\omega^2}{c^2} = \frac{\epsilon\omega^2}{c^2} = \frac{\epsilon_o\omega^2 - \omega_p^2}{c^2} \tag{G.73}$$

(with $\mu \approx 1$, as in most metals) rather than a complex one like (G.27). On the other hand, the wave number $K = n\omega/c$ will be real (and‡ $n \equiv n_r$) only for positive values of ϵ; for the frequencies lower than $\omega_p/\sqrt{\epsilon_o}$, when ϵ is negative, the index $n = i\sqrt{|\epsilon|} \equiv in_i$ becomes *purely* imaginary! The notations n_r and n_i are reintroduced at this point in order to prepare for use of the formula (G.39) giving the reflection coefficient R. For the purpose here it will be sufficient to consider the reflections and transmissions ($T = 1 - R$) only at normal ($i' = 0$) incidence, when

$$R = R_\perp = R_\parallel = 1 - T = \frac{(n_r - 1)^2 + n_i^2}{(n_r + 1)^2 + n_i^2} \tag{G.74}$$

It is for frequencies below the cutoff at $\omega_p/\sqrt{\epsilon_o}$ that the full ϵ is negative and the index $n = in_i$ is purely imaginary (insofar as energy losses to a resistive force proportional to $v_o \ll \omega$ can be neglected). Then $n_r = 0$, and the reflection coefficient (G.74) approaches $R = 1$, accounting for the good reflectivities observed for those frequencies. The reflection becomes perfect for a *purely* imaginary wave vector because the penetration into the medium of the waves incident on it is then proportional to $\exp(-n_i\mathbf{k}\cdot\mathbf{r})$, an exponential attenuation of waves that does not permit the formation, and hence propagation, of a single wavelength within the medium. When this becomes the equilibrium situation, all the incident field energy is being sent back in reflected waves. That is strictly so only insofar as resistive losses are negligible but does demonstrate that the attenuation and consequent reflection are *not* primarily due to energy losses. Instead, it is akin to the attenuation in wave guides of frequencies below cutoff, as discussed in connection with (7.45) and existing even when no losses in the wave-guide walls are taken into account.

† In the terms of (G.63), $\gamma_r \to 0$ but $\gamma_r/v_o(\to 0/0) = \omega_p^2/4\pi\omega^2$ (with $\epsilon \to \epsilon_o$ there). The expressions (G.62) have put γ_r into terms of the *static* conductivity $\gamma_o \equiv \omega_p^2/4\pi v_o$ of (G.57) only in order to exhibit the connection to the *low*-frequency limit $\gamma_r \to \gamma_o$, inappropriate for the high frequencies $\omega \gg v_o$ to which (G.71) applies. It is now more appropriate to replace (G.62) with $\gamma_r \approx (v_o/\omega)^2\gamma_o \to v_o\omega_p^2/4\pi\omega^2$, which vanishes as $v_o \to 0$ while $\gamma_i = (\omega/v_o)\gamma_r \to \omega_p^2/4\pi\omega$ remains finite, to help represent the high-frequency contributions to polarizability by the conduction electrons.

‡ This refers to the $n = n_r + in_i$ notation last employed in (G.64), which now reduces to

$$n_r^2 - n_i^2 = \epsilon_o - \frac{\omega_p^2}{\omega^2} \equiv \epsilon \qquad n_r n_i = 0$$

as can be confirmed from the considerations in the preceding footnote. The proviso $n_r n_i = 0$ is satisfied by $n_i = 0$ for $\epsilon > 0$ and by $n_r = 0$ for $\epsilon < 0$.

The frequencies exceeding the cutoff at $\omega_p/\sqrt{\epsilon_o}$ have real wave vectors $\mathbf{K} = n_r \mathbf{k}$, insofar as the energy losses continue to be negligible, and allow transmission of unattenuated propagations proportional to $\exp(in_r \mathbf{k} \cdot \mathbf{r})$ through the medium. The $n_r^2 = \epsilon = \epsilon_o - \omega_p^2/\omega^2$ is reduced from the ϵ_o characteristics of a transparent dielectric (having bound electrons only) by an amount ω_p^2/ω^2 that is less than unity. The resultant $n_r^2 = \epsilon$ may even approximate the $n^2 = \epsilon = 1$ characteristic of free space ($\epsilon_o \approx 1.41$ is supposed to characterize solid potassium, for example). Thus $n_i = 0$ and n_r is in the neighborhood of unity, making the reflection coefficient R of (G.74) quite small, and the transmission $T = 1 - R$ approach unity, about as observed.

Phenomena like those just reviewed are also supposed to occur when radio- and video-frequency waves impinge on the ionospheric layer in the earth's upper atmosphere. The lower radio frequencies ($\omega \lesssim 10^5\ \mathrm{s}^{-1}$) can be received at very long range, beyond the horizons of their transmitting stations, because they are reflected from the ionosphere and can thus follow the earth's curvature. The higher-frequency video waves (VHF with $\omega \gtrsim 10^8\ \mathrm{s}^{-1}$) are not so reflected, and their reception is essentially limited to the transmitter's horizon. This can be explained through an occurrence of a cutoff frequency in a range given by $\omega_p \approx 10^6$ to $10^7\ \mathrm{s}^{-1}$ since ϵ_o is practically indistinguishable from unity in the highly rarefied upper atmosphere.

An ionospheric cutoff frequency ω_p is called a plasma frequency because the ionosphere resembles laboratory plasmas in that its electromagnetic interactions are due principally to ambient electrons freed by ionizing effects of the sun's higher-frequency radiations. This makes the ionospheric ω_p quite variable, depending on whether it is day or night and on phases of the sun-spot cycles. However, it is quite clear why it must be much lower than the ω_p's found for the solids. The number density N_o of the ionization electrons in an attenuated atmosphere can easily be as low as the $N_o = (\omega_p^2/4\pi)(m/e^2) \approx 3 \times 10^2$ to $3 \times 10^4\ \mathrm{cm}^{-3}$ implied by the values of the ionospheric ω_p assumed just above. The estimates of ω_p have actually been found somewhat sensitive to precessions induced by the earth's magnetic field, which make ϵ anisotropic. The highly specialized details are left to Exercise G.16.

EXERCISES

G.1 Find what follows from (G.2) alone for the resistance added to a circuit when *two* unequal resistances $R_{1,2}$ are inserted if (*a*) $R_{1,2}$ differ only in lengths $\Delta l_{1,2}$ and are connected in series; (*b*) $R_{1,2}$ differ only in cross sections $\Delta S_{1,2}$ and are connected in parallel.

G.2 (*a*) Derive expressions for the components $a_{tx,y,z}$ of the transmitted amplitude in (G.31) from the boundary conditions (G.33) to (G.37).

(*b*) Show that the same results follow from (F.20) and (F.21) with $n \to n(0)$, if $\tan r = a_{tz}/a_{tx}$ of the $\gamma = 0$ case is replaced by $-K_x/K_z$ of (G.26).

(*c*) For the good conductors to which (G.41) applies show that E_{tz} and B_{tz} are negligible compared with E_{tx} and B_{tx} even when $i' \neq 0$.

G.3 (*a*) Derive (G.40) from (F.8).

(*b*) Show that $T = \langle N_{tz} \rangle / \langle N_{iz} \rangle$, as evaluated from $\mathbf{a}_t$, is consistent with $R = 1 - T$ of (G.39).

G.4 The equation of a laplacian to a *first*-order time derivative, as in (G.48) and (G.49), is well known to govern diffusion processes of many kinds, leading to a type of propagation that differs markedly from the usual wave propagation.

(*a*) Show that the diffusion equation has the special solution

$$f(x, t) = (4\pi c_1^2 \tau t)^{-1/2} e^{-(x-x_o)^2/4c_1{}^2\tau t} \qquad \text{with} \int_{-\infty}^{\infty} dx f = 1$$

at all times $t > 0$.

(*b*) Find the mean spread Δx of the distribution (*a*) at any $t > 0$ from the definition $(\Delta x)^2 = \int dx(x - x_o)^2 f$, thus showing that $f(x, t)$ describes a growing spread of an initial pulse $f(x, 0) = \delta(x - x_o)$, $\int dx\, \delta = 1$, and reaches out to $x = \pm\infty$ *immediately* after $t = 0$ but with vanishingly small amplitudes. (Such wave solutions as those used in the text would be results of Fourier-analyzing the solution here.)

(*c*) Show that, from the start $f(x, 0) = \delta(x - x_o)$, $\mathring{f}(x, 0) = 0$, the wave equation $\partial^2 f/\partial x^2 = \partial^2 f/(c_1^2\, \partial t^2)$ yields two sharp pulses with the *finite* velocities $\pm c_1$. (See Exercise 3.16 for an approach using $ict \equiv y$.)

G.5 The equations for monochromatic representatives (Fourier-transform equations like those of Sec. 6.5) inside good conductors can be put in terms of the skin depth $\delta(\omega)$ in place of the conductivity γ. Show that:

(*a*) The Maxwell equations lead to

$$\mathbf{B} = \frac{\nabla \times \mathbf{E}}{ik} \qquad \text{and} \qquad \mathbf{E} = \tfrac{1}{2} k\delta^2 (\nabla \times \mathbf{B})$$

(*b*) The diffusion equations for $\mathbf{E}$, $\mathbf{B}$, and $\mathbf{j}$ become equivalent to requiring that operation by $-\nabla^2$ result in multiplication by $+2i/\delta^2$ (a purely imaginary eigenvalue of $-\nabla^2$, replacing the real k^2 of free space).

(*c*) The complex index of refraction is $n = (1 + i)/k\delta$, with $k\delta = 2\pi\delta/\lambda \ll 1$ characterizing the good-conductor conditions.

G.6 (*a*) Use equations in Exercise G.5 to derive results of a normally incident wave on the plane surface of a conductor occupying all the $z < 0$ space and compare them to the good-conductor limits of the $i' = 0$ results in Sec. G.3.

(*b*) Show that the $\mathbf{E} = \mathbf{E}_i + \mathbf{E}_r$ at the $z = 0$ surface (and continuous through it) can be represented as $\mathbf{E}(t) = \mathbf{E}_o \cos \omega t$ with a real $\mathbf{E}_o \approx (1 - i)\mu k\delta \mathbf{a}_i$ if the arbitrary $\mathbf{a}_i$ is suitably chosen.

(*c*) Show that *just outside* the conductor the corresponding $\mathbf{B}$ has the magnitude

$$\mathbf{B} \approx \frac{2^{1/2}\mathbf{E}_o}{\mu k\delta} \cos\left(\omega t - \tfrac{1}{4}\pi\right)$$

What is $\mathbf{B}$ *just inside*?

(*d*) Evaluate the energy flux into the surface in terms of $\mathbf{E}_o$ (and c, μ, k, δ).

G.7 A free-space plane wave is incident normally on a plane conductor of uniform thickness l and characterizable by a skin depth $\delta(\omega)$.

(*a*) Show that the absolute magnitude of the interior reflection amplitude is a fraction $\exp(-2l/\delta)$ of the wave amplitude for the incidence direction. (Compare remarks on page 592.)

(*b*) Find the fraction of the incident intensity that is transmitted beyond the conductor, showing that for thicknesses great enough to make $\exp(-2l/\delta) \ll 1$ it varies with l in conformity to the absorption coefficient (G.44).

G.8 A current flux $I = I_o \cos(kz - \omega t)$ carried by a transmission line consisting of an homogenous cylindrical conductor, radius a, can be expected to have the current distributed with circular symmetry, so that

$$j(s, z, t) = \text{Re}\left[j(s) e^{i(kz - \omega t)}\right]$$

(*a*) Show that under the good conductor conditions for which (G.49) is valid, $j(s) = AJ_o[(1+i)s/\delta]$, where J_o is the regular $m = 0$ solution of (C.7) and (see Exercise C.5)

$$A = \frac{(1+i)I_o}{2\pi a\delta J_1[(1+i)a/\delta]}$$

(*b*) Show that for $a \gg \delta$ (the skin depth) the distribution just under the surface $s = a$ becomes proportional to (G.50) suitably transformed ($x \to z$, $z \to s - a < 0$). (This helps bear out the assertions made on page 594.)

G.9 The field $\mathbf{E} = \mathbf{E}_o \cos \omega t$ along a conductor surface, considered in Exercise G.6, may be the result of applying a potential difference $\Delta\phi = \mathbf{E}_o \,\Delta l \cos \omega t$ to a stretch Δl of the conductor.

(*a*) Show that the integrated flux of current under a unit-wide strip of the conductor surface can be given the form $\mathbf{J} = \text{Re}\,[(\mathbf{E}_o \,\Delta l/Z_1)e^{-i\omega t}]$, where Z_1 is a complex impedance, and find the result for this in terms of Δl, γ, and the skin depth δ.

(*b*) The same form of J is deducible from the equation

$$\Delta\phi = IR + L\dot{I}$$

(a generalization of the equation in Exercise 4.24 to a case of $\Delta\phi \neq 0$) of elementary alternating-current theory. Thereby show that the real part of Z_1 is just the resistance $R_1 = \Delta l/\gamma\delta$ [see (G.2)] under a unit-wide strip of conductor surface and $L_1 = \Delta l/\omega\gamma\delta$ is a contribution to the self-inductance in the system. [The results here begin to suggest how the lumped constants (like R, L) of alternating-current theory are related to field distributions and their energy fluxes.]

G.10 (*a*) Show that the current flux found in Exercise G.9 can also be expressed as $J = J_o \cos(\omega t - \frac{1}{4}\pi)$, with the phase shift relative to that of the applied potential arising from the self-inductance, and find the peak current J_o in terms of the peak potential difference, $(\Delta\phi)_o \equiv \mathbf{E}_o \,\Delta l$, and R_1.

(*b*) Show that the corresponding magnetic field just inside the conductor surface, found in Exercise G.6(*c*), is just $\mathbf{B} = 4\pi\mu J/c$ [exactly as expected from a suitable application of the Ampère law (G.47)]. Thus, the phase shift $\alpha = \pi/4$ found in (G.23) is to be attributed to self-induction.

G.11 (*a*) Evaluate the ohmic losses from the current distribution of Exercise G.9 under each unit of conductor area through suitable integrations of (G.8) or (G.45).

(*b*) Show that the result can be expressed as $\frac{1}{2}J_o^2 R_1$, where J_o is the peak current found in the preceding exercise and R_1 is the resistance under a unit area ($\Delta l = 1$). ($J_o/\sqrt{2}$ is called the rms current because $\langle J^2 \rangle = \frac{1}{2}J_o^2$.)

(*c*) Check that the losses are exactly compensated by the exterior-field energy flux into the conductor of Exercise G.6(*c*).

G.12 (*a*) In connection with (G.55), surface values for the exterior fields $\mathbf{E}_\parallel$ and $\mathbf{B}_\parallel$ were found in terms of an incidence amplitude a_i. Use the imperfect-conductor boundary conditions to find the corresponding surface values of the *interior* fields $\mathbf{E}_t$ and $\mathbf{B}_t$.

(*b*) In the terms of Sec. G.3, the surface values are $\mathbf{E}_t = a_t$, $\mathbf{B}_t = -n(0)a_t$. Show that a_t/a_i ($i' = 0$) of (F.20) or (F.21), evaluated for a good conductor, give results agreeing with (*a*).

(*c*) Show that these results yield the *perfect*-conductor limit $\mathbf{E}_t = 0$ properly but $\mathbf{B}_t = \mu\mathbf{B}_\parallel \neq 0$ in that limit. How can the latter finding be consistent with the exclusion of all $\omega \neq 0$ fields from the interior of perfect conductors?

G.13 Continuing the investigation of the perfect-conductor ($\gamma \to \infty$) conditions as limits of the imperfect ones in Exercise G.12:

(*a*) Show that even though $\mathbf{E}_t \to 0$, $\mathbf{j} = \gamma\mathbf{E}_t \to \infty$ at the surface.

(*b*) Using the connection $\mathbf{J} = \mathbf{j}\delta/(1-i)$ between the distribution $\mathbf{j}$ and its integrated flux density representative $\mathbf{J}$, found in Exercise G.9(*a*), show that $\mathbf{J}$ reaches the proper finite limit $c\mathbf{n} \times \mathbf{B}_\parallel/4\pi$ for a perfect conductor.

G.14 For any one of the TM modes in a rectangular channel described by (7.42), with (7.40):

(*a*) Evaluate the power $\mathcal{P}$ being sent along the channel (z) direction and the loss rate (G.52), both for a given amplitude of excitation a of (7.42).

(*b*) Show that the fractional loss rate $-d\mathscr{P}/(\mathscr{P}\,dz)$ can be minimized by choosing a propagation frequency $\omega = \sqrt{3}\,\omega_c$, where ω_c is the cutoff frequency of the mode.

G.15 Most laboratory plasmas, like the ionospheric ones, are sufficiently tenuous for the polarizabilities of the electrons still bound in ions to be negligible ($\epsilon_o \approx 1$, as in air) and for encounters of the freed electrons with the ions to be rare enough for resistive forces also to be negligible ($v_o \approx 0$).

(*a*) Show how the $\epsilon = 1 - \omega_p^2/\omega^2$ characteristic of them follows directly from the equation of motion (G.65), suitably adapted, and from definition (G.72).

(*b*) Show how the same result for ϵ follows from the equation for velocities (G.58), suitably adapted, through a preliminary evaluation of the imaginary conductivity $\gamma_c = N_o(-e\mathbf{v}_\omega)/\mathbf{E}_\omega$.

(*c*) Show that the imaginary conductivity in (*b*) is just the ratio of a monochromatic representative of the polarization current $\mathbf{j}_P = \mathring{\mathbf{P}}$ to the field representative $\mathbf{E}_\omega$.

G.16 The equation for velocities used in Exercise G.15(*b*) is perhaps the most convenient one for incorporating effects of the earth's constant magnetic field $\mathbf{B}_o$ on electrons in the ionosphere.

(*a*) Find ϵ for plane waves incident on the ionosphere from a direction *normal* to the magnetic field lines there.

(*b*) Show that the left and right circularly polarized components of waves traveling *parallel* to the earth's field lines will be affected differently, in accordance with

$$\epsilon_\pm = 1 - \frac{\omega_p^2}{\omega(\omega \pm \omega_B)} \qquad \omega_B \equiv \frac{eB_o}{mc}$$

for the positive and negative helicities, respectively.

(*c*) Clearly, with any given ω_p, there will be a *range* of cutoff frequencies for arbitrarily directed incidences. Find the extremes of this range in terms of ω_p and ω_B.

(*d*) Evaluate ω_B for $B_o \approx 0.3$ gauss (Exercise 5.21), thus showing that ω_B can have the same order of magnitude as the ω_p's quoted for the ionosphere on page 606.

REFERENCES

1. Armstrong, B. H.: *Phys. Rev.*, **130**:2506 (1963).
2. Augustin, J.-E., et al.: *Phys. Rev. Lett.*, **34**:233 (1975).
3. Bargmann, V., L. Michel, and V. Telegdi: *Phys. Rev. Lett.*, **10**:435 (1959).
4. Becker, R., and F. Sauter: "Electromagnetic Fields and Interactions," Blaisdell, Waltham, Mass., 1964.
5. Bohm, D., and Y. Aharanov: *Phys. Rev.*, **115**:485 (1959).
6. Bronzan, J. B.: *Am. J. Phys.*, **39**:1357 (1971).

6*a*. Calkin, M. G.: *Am. J. Phys.*, **34**:921 (1966).

7. Crenshaw, T. E., J. P. Shiffer, and A. B. Whitehead: *Phys. Rev. Lett.*, **4**:163 (1960).
8. Dirac, P. A. M.: *Proc. R. Soc.*, **A133**:60 (1931).
9. Dirac, P. A. M.: *Proc. R. Soc.*, **A167**:148 (1938).
10. Edmonds, A. R.: "Angular Momentum," Princeton University Press, Princeton, N.J., 1957.
11. Fermi, E.: *Phys. Z.*, **23**:340 (1922).
12. Heitler, W.: "Quantum Theory of Radiation," Oxford University Press, London, 1944.
13. Jackson, D.: "Classical Electrodynamics," 2d ed., Wiley, New York, 1975.
14. Jahnke, E., and F. Emde: "Funktionen Tafeln," Teubner, Leipzig, 1933.
15. Kittel, C.: "Introduction to Solid-State Physics," Wiley, New York, 1971.
16. Konopinski, E. J.: "Theory of Radioactivity," Oxford University Press, London, 1966.
17. Konopinski, E. J.: "Classical Descriptions of Motion," Freeman, San Francisco, 1969.
18. Konopinski, E. J.: *Am. J. Phys.*, **46**(5):499 (1978).
19. Lewis, G. N., and R. C. Tolman: *Phil. Mag.*, **18**:510 (1909).
20. Moon, P. B., and W. G. Davey: *Proc. Phys. Soc.*, **A66**:956 (1953).
21. Mössbauer, R. L.: *Z. Phys.*, **151**:124 (1958).
22. Mott, N. F., and H. Jones: "Properties of Metals and Alloys," Clarendon Press, Oxford, 1936.
23. Plass, G. N.: *Rev. Mod. Phys.*, **33**:37 (1961).
24. Pound, R. V., and G. A. Rebka: *Phys. Rev. Lett.*, **3**:439 (1959); **4**:337 (1960).
25. Robson, J.: "Basic Tables of Physics," McGraw-Hill, New York, 1967.
26. Rohrlich, F.: "Classical Charged Particles," Addison-Wesley, Reading, Mass., 1965.
27. Rose, M. E.: "Theory of Angular Momentum." Wiley, New York, 1957.
28. Schwinger, J.: *Phys. Rev.*, **75**:1912 (1949).
29. Terrell, J.: *Phys. Rev.*, **116**:1041 (1959).
30. Watson, G. N.: "Bessel Functions," Cambridge University Press, London, 1922.
31. Wood, R. W.: *Phys. Rev.*, **44**: 353 (1933).
32. Zener, C. M.: *Nature*, **132**:968 (1933).

INDEX